Volcanic Plumes

Volcanic Plumes

R. S. J. SPARKS
University of Bristol, UK

M. I. BURSIK
State University of New York, USA

S. N. CAREY
University of Rhode Island, USA

J. S. GILBERT
Lancaster University, UK

L. S. GLAZE
NASA/Goddard Space Flight Center, USA

H. SIGURDSSON
University of Rhode Island, USA

A. W. WOODS
University of Bristol, UK

JOHN WILEY & SONS
Chichester · New York · Weinheim · Brisbane · Singapore · Toronto

Baffins Lane, Chichester,
West Sussex PO19 1UD, England

National 01243 779777
International (+44) 1243 779777
e-mail (for orders and customer service enquiries): cs-books@wiley.co.uk
Visit our Home Page on http://www.wiley.co.uk
or http://www.wiley.com

Other Wiley Editorial Offices

John Wiley & Sons, Inc., 605 Third Avenue,
New York, NY 10158-0012, USA

WILEY–VCH Verlag GmbH, Pappelallee 3,
D-69469 Weinheim, Germany

Jacaranda Wiley Ltd, 33 Park Road, Milton,
Queensland 4064, Australia

John Wiley & Sons (Asia) Pte Ltd, 2 Clementi Loop #02-01,
Jin Xing Distripark, Singapore 129809

John Wiley & Sons (Canada) Ltd, 22 Worcester Road,
Rexdale, Ontario M9W 1L1, Canada

Library of Congress Cataloging-in-Publication Data

Volcanic plumes / R.S.J. Sparks ... [et al.].
p. cm.
Includes bibliographical references (p. –) and index.
ISBN 0-471-93901-3
1. Volcanic plumes. I. Sparks, R. S. J. (Robert Stephen John), 1949– .
QE527.7.V65 1997
551.2′3—dc21 96-39034
CIP

British Library Cataloguing in Publication Data

A catalogue record for this book is available from the British Library

ISBN 0-471-93901-3

Typeset in 10/12pt Times from the authors' disks by Techset Composition Ltd, Salisbury, Wilts
Printed and bound in Great Britain by Bookcraft (Bath) Ltd, Midsomer Norton, Somerset
This book is printed on acid-free paper responsibly manufactured from sustainable forestation, for which at least two trees are planted for each one used for paper production.

Contents

Preface

Volcanic plumes are one of the most spectacular of natural phenomena. Fundamentally they represent convective transfer of substantial amounts of heat from the Earth's interior in a very short period of time. They also transfer particles, gases and aerosols to the surface environments of the Earth. The focused convective flows that constitute volcanic plumes involve a rich variety of dynamical processes that are both fascinating and challenging to study. The transient formation of volcanic plumes is important in the ocean and atmosphere and impacts on many geological, physical, environmental and biological processes. Volcanic plumes are relevant to problems as diverse as climate change, aircraft safety, volcanic hazards mitigation, global geochemical cycles and speciation in the deep oceans.

This book presents the major developments in understanding volcanic plumes and the application of this knowledge to a variety of scientific problems. The subject will be of interest to those working in many scientific fields including geology, volcanology, geochemistry, climatology, oceanography, remote sensing, fluid mechanics and life sciences. The authors' expertise covers many but not all of these disciplines. The emphasis is therefore on volcanic plumes as physical phenomena. The book has been written particularly for practising geologists and atmospheric scientists or advanced graduate students in those fields who need to be brought up to speed on the basic physics underlying volcanic plumes. We have assumed that the reader will have a mathematics and physics background that includes differential equations and basic mechanics, although much of the text could be mastered with much less preparation.

The plan of the book is to introduce the phenomena in the global context of volcanism, to describe the basic physical principles that govern plume behaviour and then to apply these principles to different kinds of volcanic plumes. The authors have assumed that the book will be read by scientists from many different disciplines who may be unfamiliar with all the relevant terminology and literature. Basic terms are defined and explained when they are first introduced. To some extent each chapter is self-contained and has been written partly with the objective of explaining a particular topic to non-specialists. There are of course areas of the subject which are poorly understood so that the book looks forward to areas where future progress is needed.

All the authors were motivated by the desire to understand one of the most dramatic and dynamic of geological phenomena. Plumes have a certain beauty and certainly inspire awe in those who witness them at first hand. Like other catastrophic geological phenomena they remind us that we live on a dynamic living planet which sustains life and variety by continuous cycles of change.

Acknowledgements

The writing of the book was foremost supported by a grant from NATO (CRG 910096, Tephra dispersal during explosive volcanic eruptions) which gave the authors the chance to meet as a writing team on several occasions. In particular two very enjoyable summer periods, each of about a month, were spent at the Graduate School of Oceanography, University of Rhode Island, when all seven authors worked together on the book. The work was greatly facilitated by a generous donation from Dr Tom Casadevall associated with the First International Symposium on Volcanic Ash and Aviation Safety (Seattle, Washington, 1993) sponsored by the US Federal Aviation Administration. A generous book grant from the Royal Society of London was also an important contribution. All the authors were generously supported both by their own institutions and various funding agencies which provided resources for some of the original research work which has been included in the book. The National Science Foundation (USA), the National Aeronautics and Space Administration (USA), the Natural Environment Research Council (UK), the Leverhulme Trust (UK) and the Royal Society of London have been prominent supporters of our research.

Many people helped us to complete the project. Steve Lane and Jon Venn were very thorough and skilled in the arts of Canvas graphics and played a major role in preparing diagrams. Reviews of some chapters by colleagues helped greatly and we particularly acknowledge very thorough reviews by Tom Casadevall, Russell Blong, Gerald Ernst and Colin Wilson. Many colleagues were generous in providing photographic material and their help is acknowledged in figure captions. Rick Hoblitt and Jim Moore are thanked for providing original photographs of the 1980 Mount St Helens eruption. Mike James is thanked for his help compiling the reference list. Perhaps the greatest burden was placed on the partners and family members of "Team Plume" so we are particularly grateful to Connie Carey, Nikki Cervantes, Terry Glaze, Steve Lane, Sharon Woods and Ann Sparks for putting up with us all. RSJS would also like to recognize the warm and friendly people of Rhode Island, in particular the Bentley, Reilly and Kelly families.

1 Explosive Volcanism and the Generation of Volcanic Plumes

"It rose to a great height on a sort of trunk and then split off into branches, I imagine because it was thrust upwards by the first blast and then left unsupported as the pressure subsided, or else it was borne down by its own weight so that it spread out and gradually dispersed. Sometimes it looked white, sometimes blotched and dirty, according to the amount of soil and ashes it carried with it. My uncle's scholarly acumen saw at once that it was important enough for closer inspection."

Description of the plume from the AD 79 eruption of Vesuvius volcano by Pliny the Younger.

1.1 INTRODUCTION

The generation of large-scale volcanic plumes is a key feature of explosive volcanism on the Earth and other planets. Plumes consist of turbulent buoyant mixtures of hot volcanic particles, gases exsolved from magma and water or atmospheric gases which are incorporated during the eruption process. To the people who live close to active volcanoes these plumes represent both a spectacle of nature and a significant hazard. Fallout of volcanic particles can cause roof collapse, agricultural disruptions, communication malfunctions and respiratory problems (Blong 1984; Tilling 1989). With the increase in commercial air traffic over volcanically active areas, such plumes also pose a threat to aircraft safety. Ingestion of volcanic particles by jet engines can lead to sudden shutdown with potentially catastrophic consequences (Casadevall 1993). Several close encounters have already occurred in Alaska, Indonesia and Colombia, although no loss of life or aircraft has yet been reported. A more dangerous phenomenon is the generation of hot, destructive flows of particles and gas called pyroclastic flows which can travel tens of kilometres from source at hurricane speeds. These can be generated by the gravitational collapse of volcanic plumes. In 1902 the city of St Pierre on the island of Martinique was completely destroyed by a pyroclastic flow from an explosive eruption of Mont Pelée volcano. All of the city's 30 000 inhabitants save two were killed in the span of only a few minutes. Similarly, the fossil casts of humans at Pompeii were discovered in the deposits of such flows, formed late in the AD 79 eruption of Vesuvius volcano.

Volcanic plumes also hold the potential for producing global environmental effects. Large-scale explosive eruptions can inject massive quantities of particles and gases to stratospheric elevations. The resulting veil of aerosols and particulates absorb incoming solar radiation resulting in stratospheric warming and tropospheric cooling (Lamb 1970;

Rampino and Self 1982; Sigurdsson 1990a). In 1815 the explosive eruption of Tambora volcano in Indonesia discharged 50 km^3 of magma. Volcanic plumes from this event lofted material above 40 km in the atmosphere and triggered anomalously cold weather during the summer of 1816 in the north-eastern United States, the Maritime Provinces of Canada, and Europe. The climatic disturbance caused widespread crop failures and has been implicated in accelerated immigration from New England and even the initiation of the great cholera pandemic (Stommel and Stommel 1979). Climatic perturbations and enhanced damage to the ozone layer have also been attributed to the recent 1991 eruption of Mount Pinatubo volcano in the Philippines.

In most cases plumes are generated when magma is disrupted explosively into small particles in a process called fragmentation. Many different types of explosive eruptions can produce volcanic plumes and their behaviour is dependent upon flow conditions, temperature, gas composition and the concentration of volcanic particles. In this book the term "volcanic plume" is used in a general sense to describe several aspects of the explosive ejection of material from volcanoes. Their rise and interaction with the atmosphere near the volcano and dispersal by advection, diffusion and flow downwind of the source are discussed. Much progress has been made in the last two decades towards understanding the origin, structure and behaviour of volcanic plumes. Sophisticated models of plumes have now been developed and are being evaluated based on observations of recent and ancient explosive activity.

The purpose of this first chapter is to provide readers with limited backgrounds in geology and volcanology, an introductory overview of explosive volcanism from which they can progress to the more comprehensive discussions within the book. Magma is the fundamental starting-point in the discussion of volcanic plumes and the chapter begins with its characterization, both on the basis of composition and physical properties. The remainder of the chapter focuses on the processes that cause explosive volcanism and generate a wide variety of volcanic plumes.

1.2 COMPOSITION AND VOLATILE CONTENT OF MAGMA

Volcanic activity results from the eruption of magma at the Earth's surface. Magmas are complex mixtures of melt, suspended crystals and gas bubbles. Melt is usually a major component and magma consequently behaves as a fluid. Most magmas, with the exception of some rare carbonate magmas, are complex silicate systems whose SiO_2 content spans the range of 45–77 wt.%. Other major elements include aluminium, calcium, iron, magnesium, potassium, sodium and titanium. Many types of classification schemes have been developed for the volcanic rocks that result from the cooling and crystallization of magma. The reader is referred to Wilson (1989) for a complete treatment of this topic. For the purposes of discussing explosive volcanism we shall use a simplified classification based on the chemical composition of the magma (Figure 1.1). SiO_2 content is used to subdivide magma types into four broad categories: silicic (>63 wt.% SiO_2), intermediate (52–63 wt.% SiO_2), basic (<52 to >45 wt.% SiO_2), and ultrabasic (<45 wt.% SiO_2). These are further subdivided on the basis of the amount of Na_2O and K_2O. Although specific names are assigned based on compositional parameters, a continuous gradation exists between different magma types.

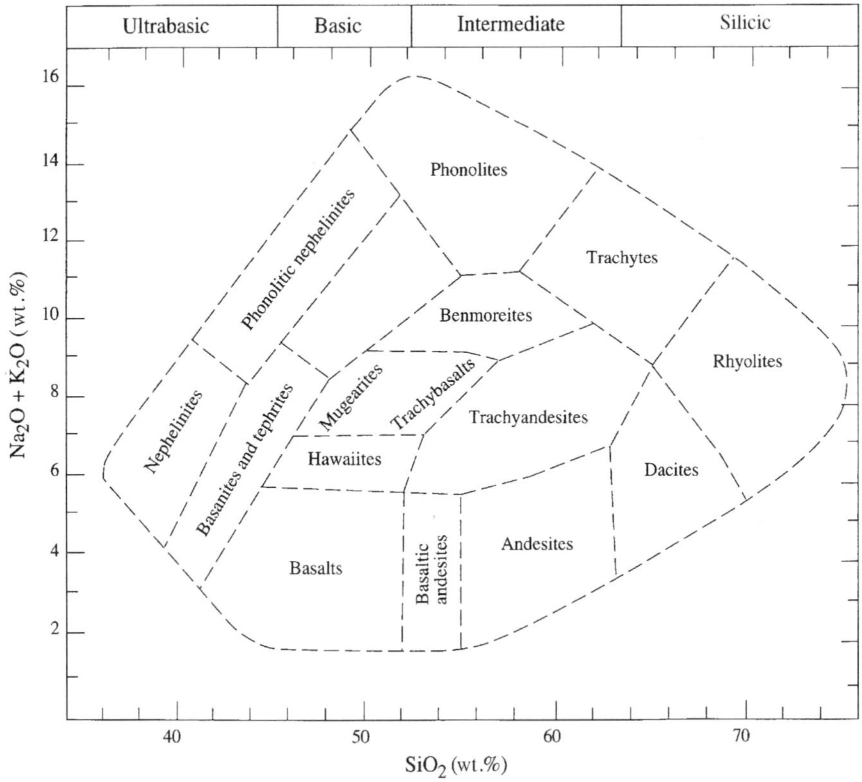

Figure 1.1 Classification of volcanic rocks based on the weight % $Na_2O + K_2O$ and weight % SiO_2 modified from Wilson (1989). Terms at the top of the figure are used as general descriptors for rocks or magmas within certain ranges of SiO_2 content

Magmas typically occur as distinct compositional suites that suggest a genetic relationship between the members. Different suites are associated with specific geological environments and styles of eruption. For example, the calcalkaline suite is commonly associated with explosive volcanism in island arcs and continental margins which rim the Pacific Ocean. Calcalkaline magmas span the compositional range basalt, andesite, dacite and rhyolite, or a SiO_2 content of about 50–77 wt.%. The compositional variations are the result of many interrelated processes. The most important processes are partial melting of source rocks, the chemical differentiation of magma by crystallization, melting and assimilation of crustal rocks by magmas and mixing of magmas. The study of the chemical variations among volcanic rocks and the interpretation of the causative processes forms the discipline of igneous petrology. Fuller discussion of the mechanisms of magma generation and evolution can be found in Sparks (1994).

Magmas typically contain small amounts of volatile components such as water, carbon dioxide, sulphur and halogens, primarily Cl and Fl. These volatiles are of great importance to explosive volcanism because they influence the rheological properties of magma and provide a driving force for eruption through their exsolution and expansion. All of the important volatiles in magmatic systems are more soluble at high pressure and thus as magma approaches the surface the volatiles form a separate gas phase as bubbles. Only a few weight per cent of dissolved volatiles is necessary to drive highly explosive eruptions because of the huge change in volume that occurs during decompression to atmospheric pressure. The volatiles are inherited from source rocks during magma generation processes, are incorporated from surrounding rocks as magma rises to the surface and are further concentrated by crystallization.

Water is the dominant volatile which drives explosive volcanism. It is dissolved in magma both as hydroxyl groups and molecular water (Stolper 1982). Figure 1.2 shows experimentally determined water solubilities for a range of magmatic compositions. The solubility increases with increasing pressure and SiO_2 content of magma. Pre-eruption water contents have been estimated for a number of explosive eruptions using a combination of melt inclusion analyses and experimental petrology. For example, the dacite magma erupted during the May 18, 1980 Mount St Helens eruption in Washington, USA, contained about 4.5% H_2O (Rutherford *et al.* 1985; Rutherford and Devine 1988). Similar melt inclusion analyses from rhyolitic inclusions in the Bishop Tuff deposit of California showed H_2O contents as high as 7.0% (Anderson *et al.* 1989). These values are consistent with magma storage at a pressure of at least 100 MPa ($\sim$9 km depth) prior to eruption. Basic magmas, on the other hand, usually contain relatively low water contents ($<$1.0%) although some basalts are inferred to contain up to 4% (Sisson and Grove 1993; Stolper and Newman 1994).

Carbon dioxide is typically the second most abundant volatile given off during volcanic eruptions. Its abundance as a dissolved species is considerably less than water in magmas of intermediate to silicic composition. In basic magmas, such as basalt, CO_2 dissolves in

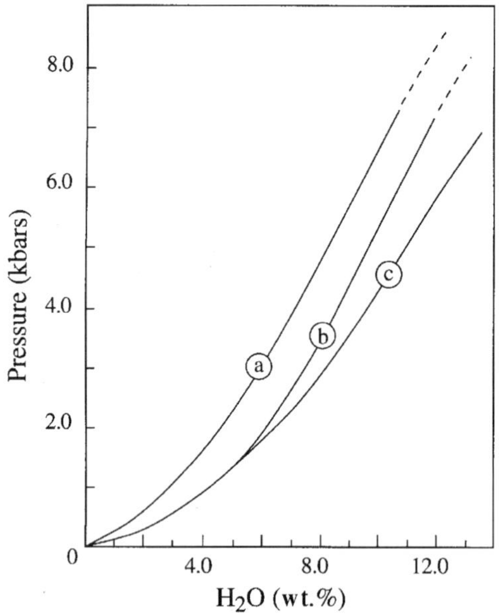

Figure 1.2 Solubility of H_2O as a function of pressure in (a) basalt, (b) andesite and (c) rhyolite melt compositions. Modified from Burnham (1979). All magmas are able to dissolve larger amounts of water at higher pressures

the form of carbonate ions (Stolper and Holloway 1988) whereas in silicic magmas, such as rhyolite, it is present solely in molecular form (Fogel and Rutherford 1990). Figure 1.3 shows the solubility of CO_2 in basalt and rhyolite magma at pressures up to 2 kbar. With increasing pressure the solubility of CO_2 also increases like that of water, but a comparison with Figure 1.2 shows that at similar pressures and magmatic composition the absolute solubility of CO_2 is substantially less. Because of its low solubility relative to water, the first gas phase formed by magma rising to the surface may consist of CO_2 instead of water, even though water is the more abundant volatile component. Estimates of pre-eruption CO_2 abundances are scarce for intermediate to silicic magmas.

Sulphur is a particularly important volatile component because its release can result in significant climatic effects (Rampino *et al.* 1988; Sigurdsson 1990a). Its solubility is a function of melt composition, especially FeO content and whether the conditions are highly oxidized or reducing (Carroll and Rutherford 1988). Under relatively reducing conditions sulphur is dissolved primarily as sulphide. As magmas become more oxidized, solubility goes through a minimum and then increases. Under relatively oxidizing conditions the solution of sulphur occurs as sulphate. Some magmas, such as those erupted during the 1982 eruption of El Chichon volcano in Mexico and the 1991 eruption of Mount Pinatubo, have sufficient sulphur and are oxidized enough to stabilize the mineral anhydrite ($CaSO_4$). Figure 1.4 shows the sulphur content of melt inclusions from magmas involved in explosive eruptions. There is a general inverse correlation between sulphur content and SiO_2 content with basalts containing up to 1000 ppm (parts per million) and rhyolites only about 20 ppm sulphur. These values represent the minimum amount of sulphur in magma prior to eruption. Separate sulphur gas species can be present at the time of eruption in addition to the amount dissolved in the melt.

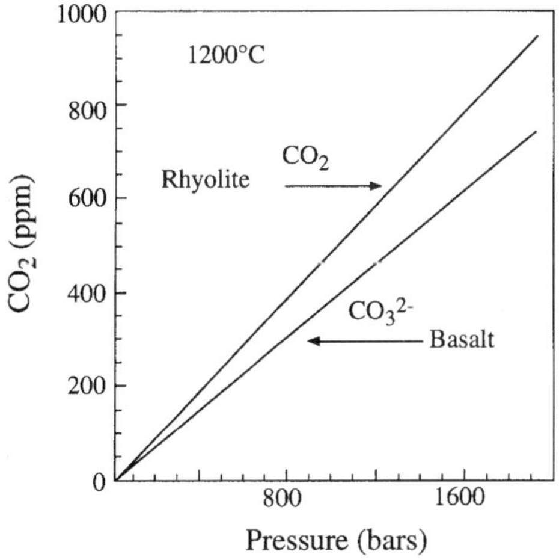

Figure 1.3 Solubility of CO_2 in rhyolitic and basaltic magmas as a function of pressure at 1200 °C. In rhyolite magmas CO_2 goes into solution dominantly as molecular CO_2, whereas in basaltic compositions it is dissolved as a carbonate species ($CO_3^{=}$). Modified from Fogel and Rutherford (1990)

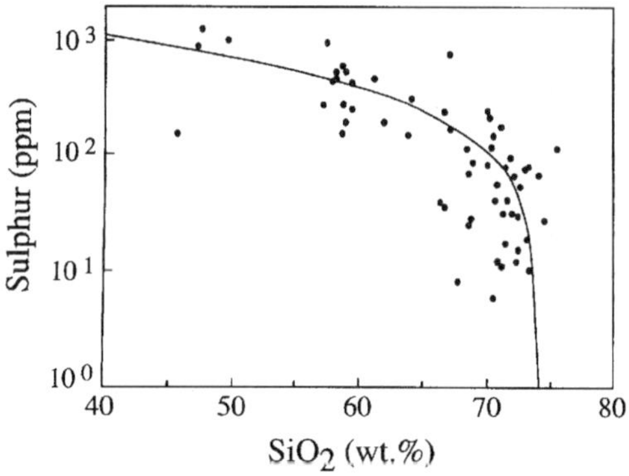

Figure 1.4 Concentration of sulphur in glass inclusions from minerals in volcanic rocks as a function of inclusion SiO_2 content. Modified from Sigurdsson (1990b). More evolved magmas (higher SiO_2) carry less sulphur in solution than less evolved magmas such as basalt

1.3 PHYSICAL PROPERTIES OF MAGMA

The style of eruption depends strongly on the physical properties of magma. These properties are principally related to temperature, melt composition, proportion of crystals, amount of dissolved volatiles and the abundance of gas bubbles. Temperature varies considerably among different magma types. In general, there is an inverse correlation between eruption temperature and SiO_2 content. Basic magmas, such as basalt (50% SiO_2), erupt at temperatures of about 1200 °C, whereas silicic magmas such as rhyolites (75 wt.% SiO_2) are considerably cooler and typically erupt in the temperature range 700–900 °C.

Viscosity is a key property that determines the behaviour of magma during an eruption. It is a measure of the amount of internal resistance to flow that fluid exerts when a force is applied to it. Figure 1.5 shows the relationship of shear stress to strain rate for a number of different types of fluid. Viscosity is the ratio of these two parameters and is thus the slope of the various lines on Figure 1.5. If viscosity is constant and an infinitesimally small amount of shear stress results in flow, the fluid is called Newtonian. Fluids which show a non-linear relationship between stress and strain rate are described as non-Newtonian. A Bingham fluid shows an intercept with the stress axis at zero strain rate and a linear variation of stress and strain rate. In a Bingham fluid a specific yield strength must be exceeded before flow begins (Figure 1.5). A line connecting the origin with a point on a rheological curve is called the apparent viscosity.

Pure, crystal-free silicate melts are always Newtonian (Ryan and Blevins 1987; Dingwell *et al.* 1993). The viscosity of silicate melts is a strong function of temperature and composition (Bottinga and Weill 1972; Shaw 1972; Ryan and Blevins 1987; Dingwell *et al.* 1993). Basaltic melts have relatively low viscosities (10^2–10^3 Pa s) compared to silicic magmas such as rhyolites (10^6–10^{12} Pa s). Silicate melts can show hyperexponential dependence of viscosity on temperature (Spera *et al.* 1988). Figure 1.6 shows the strong effect of temperature on melt viscosity for different compositions.

The presence of crystals and gas bubbles have important effects on magma rheology. Viscosity increases as crystal abundance increases according to a power law (Pinkerton and Stevenson 1992). Magmas with low to moderate suspended crystal contents still

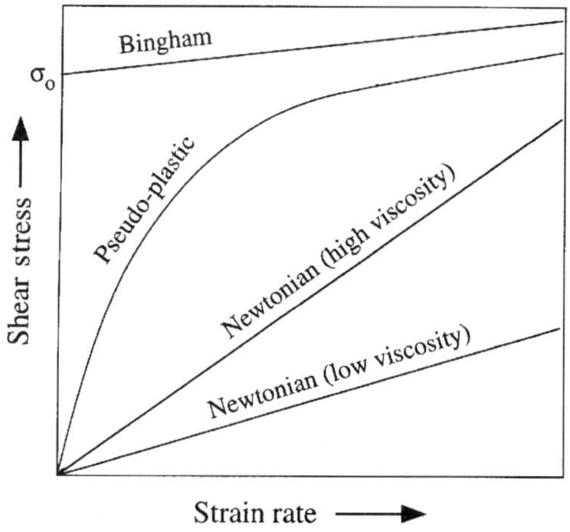

Figure 1.5 Relationship between shear stress and strain rate for different types of fluid behaviours. Modified from Cas and Wright (1987). Magmas without crystals behave as Newtonian fluids where the strain rate can be linearly related to the shear stress

behave as Newtonian fluids, but crystal-rich magmas can behave in a highly non-Newtonian manner. Empirically available data suggest that non-Newtonian properties can develop between 20 and 30% by volume of suspended crystals (Pinkerton and Stevenson 1992). Bubbles, on the other hand, can cause bulk viscosity either to decrease (Dingwell and Webb 1989) or increase depending on factors such as bubble size, surface tension of

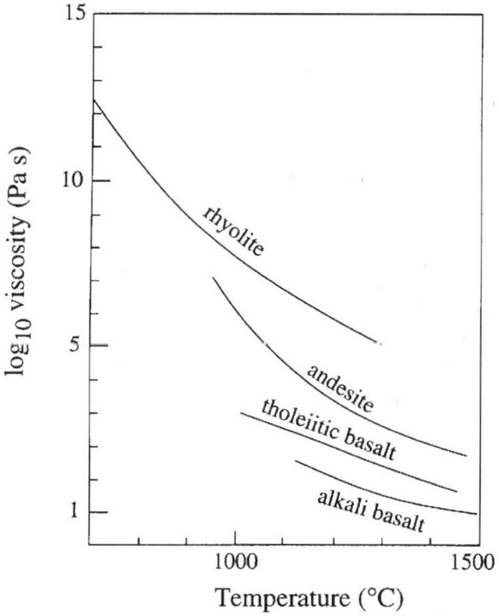

Figure 1.6 Variation in viscosity as a function of temperature for magmas of various composition. Modified from Williams and McBirney (1979). At a specific temperature the viscosity of magma increases with increasing SiO_2 content. All magma types experience a significant increase in viscosity as they cool

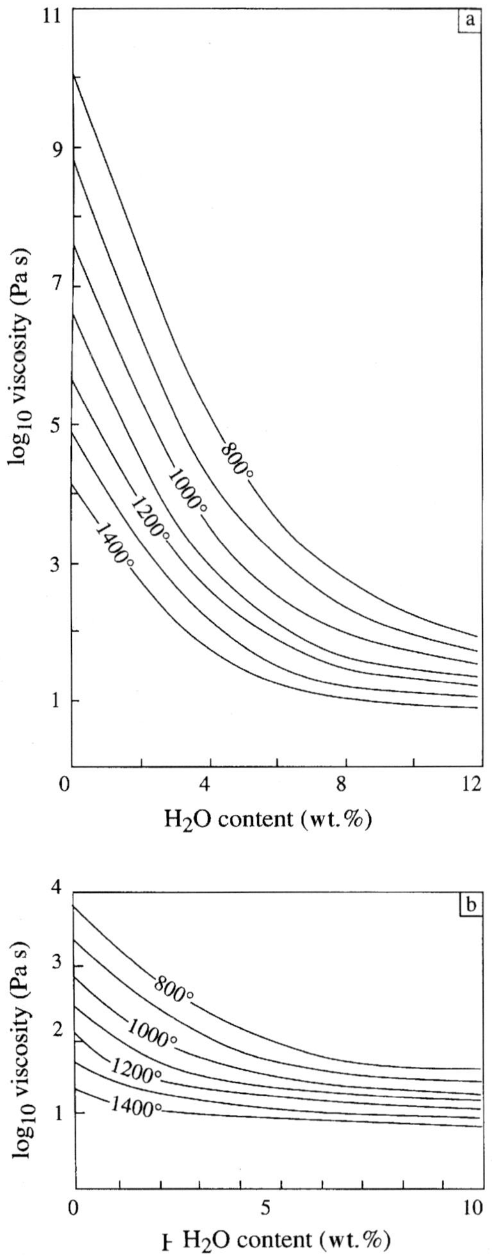

Figure 1.7 Effect of dissolved water content on the viscosity of (a) rhyolitic and (b) basaltic magmas at temperatures between 800 and 1400 °C. Modified from Williams and McBirney (1979). As magma loses water by degassing large increases in viscosity occur

the melt and strain rate. Presence of bubbles can cause non-linear rheology (Spera *et al.* 1988). More detailed discussions of magma rheology can be found in recent reviews (Pinkerton and Stevenson 1992; Dingwell *et al.* 1993).

Volatiles have a significant impact on the viscosity of magmas as a result of their modification of silicate melt structure. Solution of water, for example, results in a depolymerization of silicate melts and the breaking of strong Si–O bonds. This causes a reduction in the internal resistance to flow and thus a reduction in viscosity. Figure 1.7 shows the effect of H_2O on the viscosity of rhyolite and basalt magma. A rhyolite with 4 wt.%. H_2O will experience a five order of magnitude increase in viscosity if it loses all of its volatile component. Basalt also increases in viscosity as it loses H_2O but the magnitude of change is considerably less. The decrease of liquidus temperature with water content results in strong undercooling of magmas upon degassing and triggering of crystallization with associated increase in viscosity. Volatile loss is thus a very effective means of changing the rheological properties of magmas as they ascend towards the surface.

As magmas cool they develop mechanical strength. Many explosive eruptions involve the fragmentation of magma, and therefore the complex transition of magma from a Newtonian fluid to an elastic solid is of considerable importance. It is useful to recognize two end-member responses to cooling. First, low-viscosity magmas such as basalts crystallize as they cool, except under extreme conditions. If the magmas remain in thermodynamic equilibrium then the crystal content can be determined for known conditions of temperature, pressure, oxidation state and volatile content. When crystal content becomes so high that the crystals form a touching framework then the system becomes a partially molten rock rather than a magma with significant mechanical strength. Available evidence suggests that the transition from magma to partially molten rock typically occurs at around 60–65% crystals by volume. However, the transition is strongly dependent on crystal shape, size and flow conditions and may occur at less than 40% crystals and at over

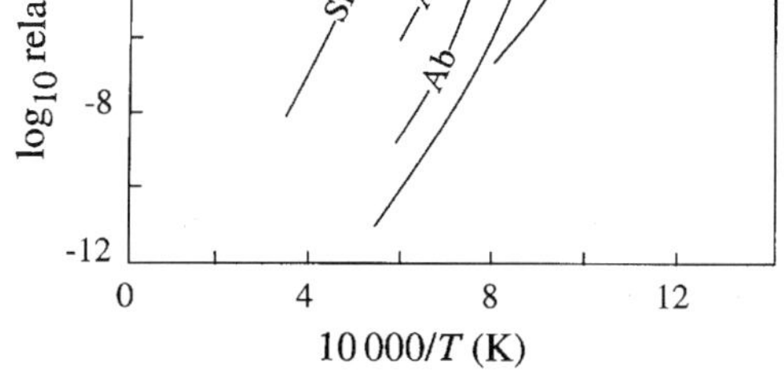

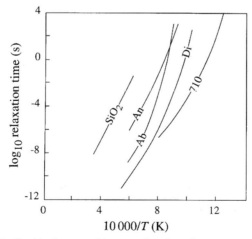

Figure 1.8 Plot of the liquid–glass transition as a function of temperature and relaxation times for silicate melts of different compositions. Modified from Dingwell and Webb (1990). Ab = albite, An = anorthite, Di = diopside, 710 = soda lime composition

80% crystals in different circumstances. Second, high-viscosity magmas commonly fail to crystallize and instead cool to form supercooled melts and glasses. In this case the glass transition temperature represents the boundary between viscous and brittle behaviour. The glass transition temperature is dependent on melt viscosity and flow conditions. Dingwell and Webb (1990) suggest that the transition occurs when the time scale associated with viscous strain (the reciprocal of the strain rate) equals the relaxation time scale of the melt (defined as the ratio of viscosity to elastic modulus). Figure 1.8 shows the glass transition using this criterion for different melts. The transition temperature and viscosity thus strongly depend on shear rate. Magmas in a slow lava flow can behave as viscous fluids at much lower temperatures than in an explosive eruption where strain rates are much higher.

1.4 CAUSES OF EXPLOSIVE VOLCANISM

Explosive volcanism is caused by (1) degassing of juvenile volcanic gases dissolved in magma at high pressure, (2) the interaction of magma with external water (ground, lake, or sea) or (3) some combination of both. Determination of the mechanism of explosive activity is often difficult and requires careful observations of an eruption and study of the characteristics of the resulting products.

1.4.1 Degassing of Juvenile Volatiles

The degassing of juvenile volatile components from magma is one of the most important causes of explosive volcanism and the generation of volcanic plumes. This is clearly evident from an examination of the products of this type of activity. The fragments typically resemble a stiff froth with abundant bubbles caused by gas coming out of solution as magma is depressurized. However, this process is extremely complex and there are many poorly understood details which are still under investigation. Degassing is controlled by the physical properties of the magma, the volatile content, and the rate of magma rise. Water and CO_2 in particular are the principal gases that control the dynamics of explosive eruptions. In the following discussion the dominant volatile component is assumed to be water. A major driving force of explosive volcanism is the tremendous change in volume that the volatile component undergoes from being dissolved in the magma at depth to its exsolution as a separate gas phase when it reaches the surface. As an example, 1 m^3 of rhyolite magma at 900 °C with 5 wt.% dissolved H_2O at depth will occupy a volume of 670 m^3 as a degassed mixture of water vapour and magma at 1 atm pressure (surface). A common analogy for this process is when the top is removed from a bottle of beer and dissolved CO_2 comes out of solution and forces a mixture of bubbles and liquid out of the bottle as a spray.

Figure 1.9 shows a cross-section of a volcano undergoing a sustained explosive eruption by degassing of volatiles. This generalized picture applies to the eruption of viscous, volatile-rich magma such as dacites and rhyolites or their alkaline counterparts, trachyte and phonolite. Prior to an eruption, magma rises buoyantly within the Earth's crust and commonly accumulates in magma chambers at depths between 3 and 30 km beneath volcanic areas (Figure 1.9). Within these chambers volatile components are dissolved in the magma, or if the magma is volatile-saturated, they may also exist as gas bubbles.

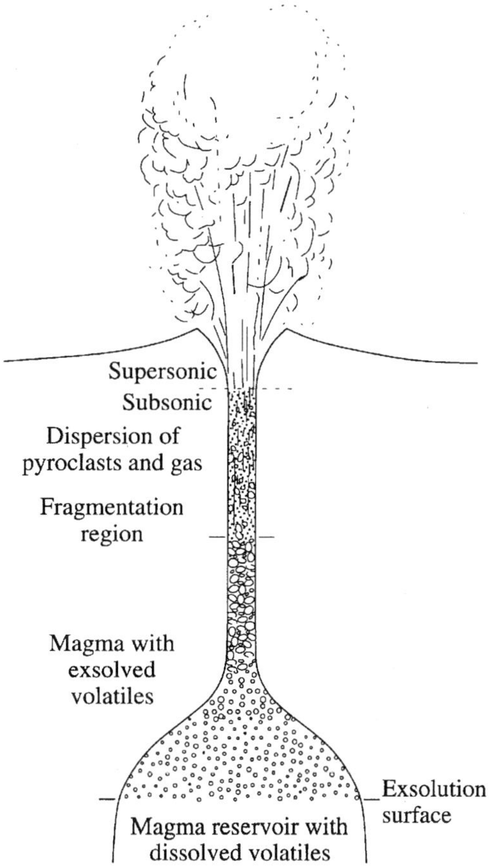

Figure 1.9 Schematic cross-section through a volcano undergoing an explosive eruption. As magma rises to the surface, exsolution of dissolved volatiles form gas bubbles. Close to the surface, bursting of bubbles generates a dispersion of pyroclasts and gas which is ejected from the vent at high velocity. Modified from Fisher and Schmincke (1984) and Sparks (1978a)

Three regions of gas exsolution can be identified in Figure 1.9. The saturation level is the depth where the magma is saturated with one of its volatile components (assumed to be water). It can occur in the conduit or in the magma chamber depending on the chamber depth and magma water content. Chamber pressure decreases with time during an eruption (Druitt and Sparks 1984; Stasiuk *et al.* 1993) and as a consequence the saturation level can move from the conduit into the chamber. Formation of gas bubbles depends critically on whether nucleation takes place homogeneously or heterogeneously (Sparks 1978a). Heterogeneous nucleation can take place on crystals at supersaturation pressures of only a few mega pascals, but homogeneous nucleation requires supersaturation pressures of the order of 100 MPa (Sparks 1978a; Hurwitz and Navon 1994). Bubble growth occurs by a combination of mass transfer of volatile components from magma to bubbles and by depressurization. Sparks (1978a) carried out numerical modelling of bubble growth in ascending basaltic and rhyolitic magmas using relevant physical properties and a range of ascent rates. A general result of the modelling is that mass transfer dominates the growth of bubbles at depth while decompression becomes important close to the surface.

In the lower region of slow degassing a mixture of melt, crystals and gas bubbles ascends slowly to the surface within a relatively narrow conduit (Figure 1.9). Conduit sizes are believed to vary from a few metres to tens of metres in diameter and typical rise velocities at depths in excess of 1 km are a few metres per second (Wilson *et al.* 1980). For example, during the May 18, 1980 eruption of Mount St Helens the conduit radius at depth has been estimated as 9.5 m and magma ascended at a speed of about 1 m s^{-1} from a chamber at 7 km depth (Scandone and Malone 1985; Carey and Sigurdsson 1985). Because of the viscous nature of the magma, slow ascent rates and gas exsolution rates, the flow regime is laminar and the magma and gas bubbles are assumed to be in thermal and mechanical equilibrium.

Closer to the surface, bubble volume begins to increase in a region of accelerated bubble growth (Figure 1.9). Magma still forms the continuous phase of the mixture but the gas volume may now exceed the melt volume. In the fragmentation region a transition occurs from magma with bubbles to an accelerated mixture of fragmented magma and gas, with the gas now forming the continuous phase. As yet the thicknesses of these regions are poorly known and are discussed more fully in a review by Sparks *et al.* (1994) and in Chapter 3.

Two different models can be considered for conditions in the upper part of a conduit at and below the fragmentation region. The conventional model has been developed in the work of Sparks (1978a) and Wilson *et al.* (1980). Most pumices contain bubble densities that approach close packing with volume void fractions typically in the range 0.7–0.8. This observation suggested to Sparks (1978a) that prior to fragmentation most bubbles have stopped growing because of overcrowding and the difficulty in driving highly viscous liquid that has been depleted in volatiles through narrow interbubble pathways. In this model a static non-expanding foam forms and gas continues to diffuse into the bubbles from the melt because the total pressure is still being reduced by magma ascent and eruption. Pressure builds up in the gas phase of the static foam and at the fragmentation level, the magmatic foam disintegrates. Wilson *et al.* (1980) assumed that the fragmentation level occurs when the gas fraction reaches a value of about 0.8, giving gas pressures of a few megapascals at the fragmentation level.

An alternative model is suggested by experimental work on bubble nucleation (Hurwitz and Navon 1994) and on explosive degassing in shock tubes (Mader *et al.* 1994, 1996) and new theoretical work (Sparks *et al.* 1994; Chapter 3). Homogeneous nucleation of bubbles at large supersaturation pressures leads to violent gas evolution, acceleration of the flow and explosive accelerations due to the positive feedback between bubble deformation, exsolution rates and foam expansion. The accelerating foam then disintegrates into the pyroclast-gas dispersion. Observations of this process in shock tube experiments (Mader *et al.* 1994) indicate a diffuse fragmentation region rather than a well-defined interface. This alternative concept implies that the region of explosive degassing and fragmentation occurs over a very short region in comparison to the length of the conduit.

The fragmentation region for most eruptions takes place at depths less than 1 km below the surface, although there is some evidence that this level can sometimes penetrate somewhat deeper and even into the magma chamber in the later stages of some eruptions. Above the fragmentation region the sudden expansion of the gas phase accelerates the mixture upwards through the remaining conduit and out of the vent at high velocities and under turbulent conditions (Figure 1.9). During some eruptions the geometry of the conduit and vent may be such that the erupting mixture can only achieve sonic velocities and exit at pressures greater than atmospheric (Woods and Bower 1995). As a consequence of the high exit pressure the vent can erode and expand to a flared configuration and approach the condition where the exit pressure is close to atmospheric. Numerical solutions for this case show that the exit velocity of material at the vent is strongly dependent upon the starting water content of the magma. Water contents in the range 1–5 wt.% will result in exit velocities of 100–500 m s^{-1} respectively for a magma discharge rate of 10^6 kg s^{-1} (Wilson *et al.* 1980; Dobran 1992). These source conditions for volcanic plumes are discussed in much greater detail in Chapter 3.

Degassing of low-viscosity magma such as basalt occurs in a very different style compared to viscous silicic magmas. In basic melts bubbles are able to move relative to liquid, grow to large sizes because of higher diffusion rates of magmatic volatiles and coalesce more efficiently. One model for degassing of basaltic magma involves accumulation of gas at the top of a shallow magma reservoir followed by a rise through a conduit as a bubbly flow (Vergniolle and Jaupart 1990). Near the surface, magma is disrupted into a spray of magma clots and gas, referred to as fire fountaining. Because of the fluidity of basaltic magma and the generally low volatile contents, the extent of fragmentation is generally less than the degassing associated with intermediate and silicic magmas. Explosive volcanism may also take place with relatively minor movement of basic magma from depth. Large bubbles, up to 10 m in diameter, are able to rise through an established conduit to a surface lava lake where they burst, breaking the magma up into small fragments which are accelerated by the expanding gas (Figure 1.10).

1.4.2 Interaction of Magma with External Water

Another mechanism for the generation of explosive volcanism is the interaction of magma with ground, lake or seawater. Magmas are a tremendous source of thermal energy because of their large heat capacity and high eruption temperatures (700–1250 °C). A kilogram of

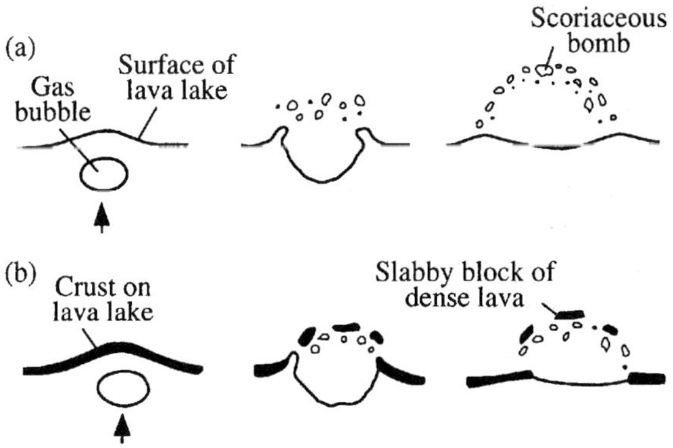

Figure 1.10 Strombolian activity is produced by the bursting of large gas bubbles as they reach the surface of a lava lake. If the surface is largely liquid (a) then bubble bursting produces scoriaceous bombs. In the case where a crust has developed on the lake surface by cooling (b), dense slabby blocks will also be produced during bubble bursting

magma contains 1.6×10^6 J of energy and water at 0 °C heated to 1000 °C at fixed volume generates a pressure of 500 MPa. If magma comes in contact with water at or near the Earth's surface then rapid conversion of water to steam can occur explosively. This type of activity is referred to as hydrovolcanism. Eruptions which involve the ejection of juvenile magma and explosively heated external water are called phreatomagmatic, whereas eruptions which eject only fragments of country rock and steam are called phreatic.

The conditions by which explosive activity is generated by magma–water interactions are complex and as yet not completely understood. Not all magma–water interactions are explosive. For example, some lava flows are able to flow into the sea without explosive disruption. Whether or not an explosion will take place depends on many factors such as the ratio of water mass to magma mass, the dynamic interaction between magma and water and the extent to which the interaction is confined by geometry.

Sustained explosive activity resulting from magma–water interaction has been compared to industrial explosions known as fuel–coolant reactions (Peckover *et al.* 1973). Fuel–coolant reactions involve mixing of two fluids with different temperatures in such a way that the cooler liquid vaporizes and triggers explosive eruptions. Wohletz (1986) has postulated a model for magma–water explosions in which generation of explosive activity takes place in a cyclic fashion on very short time scales (microseconds). The cycle begins

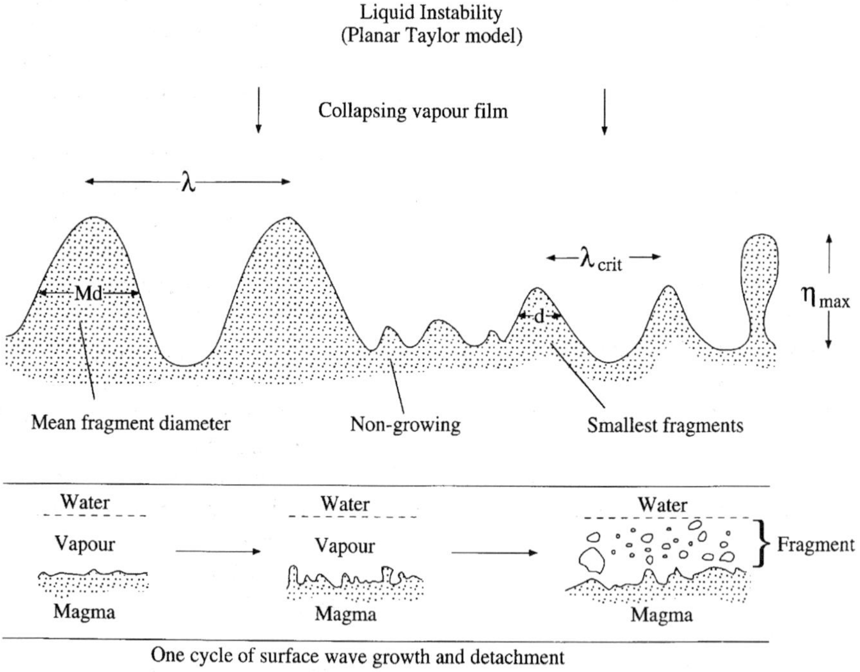

Figure 1.11 Dynamic behaviour of the interface between hot magma and water (upper diagram). Collapse of a thin vapour film results in the production of instabilities in the magma surface defined by their wavelength (λ) and amplitude (η). These instabilities break up into individual fragments. Instabilities smaller than some critical size (λ_{crit}) will not break up into discrete fragments. Vapour film production, collapse and fragmentation occur in a cyclic fashion as shown in the lower part of the figure

with the development of a thin vapour film at the boundary between magma and water (Figure 1.11). Heat transfer causes the film to expand rapidly outwards. Expansion ceases when condensation triggers local collapse of the vapour film back towards the magma. Energy from the collapsing film is utilized to distort and disrupt the magma interface resulting in fragmentation. As fragmentation occurs, heat transfer is enhanced because of the larger surface area of the magma fragments and a new vapour film is produced. The maximum efficiency of thermal energy transfer during this process takes place when the ratio of water mass to magma mass is about 0.3 (Wohletz 1986; Kokelaar 1986).

Interactions of magma with water often take place under confined conditions. For example, magma rising into a volcanic edifice might encounter a local aquifer. Transfer of heat from the magma can result in conversion of the water to superheated steam with a build-up of pressure. If the excess pressure exceeds the strength of the country rocks and the overburden pressure then the system can fail catastrophically to trigger an explosive eruption. Such eruptions probably differ from the fuel–coolant type in that they are an instantaneous release and are not sustained for any period of time.

Spectacular magma–water interactions occur during the growth of volcanic islands in the ocean. In deep water ($\sim$200 m) the growth is usually non-explosive because the water pressure inhibits the conversion of seawater to steam. As growth causes magma to approach the surface, however, water pressure becomes sufficiently low to allow the explosive expansion of seawater to steam. The emergence of Surtsey volcano south of Iceland in 1963 (Figure 1.12) is an example of this process. The exact mechanism of this Surtseyan-style activity remains controversial but may include a combination of fuel–coolant interactions, confined heating of seawater or the expulsion of a magma/water/steam slurry.

Interaction of magma with water also commonly occurs near the top of volcanoes. Many subaerial volcanoes contain craters or calderas formed by previous explosive activity. If drainage and rainfall conditions are appropriate these can act as basins to trap runoff and stabilize permanent or semipermanent lakes. Subsequent eruptions may then lead to the interaction of magma and lake water. However, it appears that certain conditions must be attained before explosive activity can take place. For example, in 1971 a volcanic dome grew up through a crater lake at the Soufrière volcano on St Vincent. The magma welled up slowly from beneath the lake until it breached the surface and continued to build a small island. Heat from the magma converted some of the lake water to steam but the activity was not explosive. Eight years later activity resumed and culminated with a series of explosive eruptions which led to the removal of the lake and the earlier-formed dome. Examination of the products of the explosive eruptions indicates that degassing of juvenile volatiles probably contributed to the explosive nature of the magma–water interaction.

Magma does not always reach the Earth's surface and instead intrudes at depth where it cools and crystallizes. The heat released from magma intrusions heats up ground water and releases gases. Large-scale geothermal systems form as a consequence of convective circulation and lead to a wide range of surface phenomena including steam plumes, hot springs, geysers, fumaroles and boiling mud pools. Volcanoes on the sea-floor heat sea-water which circulates through the hot rock. This leads to strong discharges of hot water containing many dissolved metals and chemicals. The resulting plumes are known as black smokers and can rise hundreds of metres in the ocean. Hydrothermal plumes at the world's ocean ridges are now recognized as one of the most important components of global

Figure 1.12 A Surtseyan-style eruption of Surtsey volcano in Iceland. (Photograph courtesy of H. Sigurdsson)

geochemical fluxes. Plumes formed by hydrovolcanism and geothermal activity are discussed in Chapter 8. Sea-floor hydrothermal plumes are discussed in Chapter 9.

1.5 VOLCANIC EJECTA

The fragmentation of magma and parts of the volcanic system through which the magma passes generates particles which are collectively known as pyroclasts, or by a synonymous term, tephra. Together with volcanic gases and entrained atmospheric air, they form the major constituents of volcanic plumes. Fragmentation can generate a substantial range of pyroclast sizes from several metres in diameter down to only a few micrometres. Specific terminology is applied to pyroclasts based on their size. All fragments greater than 64 mm are called blocks or bombs. Lapilli is used for those between 64 and 2 mm, while ash is applied to all fragments less than 2 mm. The composition of pyroclasts can be conveniently subdivided into three categories: juvenile fragments, crystals and lithics. Juvenile fragments consist of portions of magma that have fragmented into pieces, quenched and then deposited. Most magma that is erupted during explosive events consists of a liquid silicate melt that carries with it crystals and gas bubbles. Thus juvenile fragments will also contain varying amounts of these components.

The shape and morphology of juvenile pyroclasts reflects the processes of fragmentation, magma composition and transport. Fragments that are produced by volatile exsolution carry the remnants of the degassing process in the form of vesicles, or quenched bubble structures. Pumice is a highly vesicular rock produced by the fragmentation of relatively viscous intermediate and silicic magmas. Its density typically ranges from 300 to about 1100 kg m^{-3}, and thus many varieties can float on water. The vesicularity of pumice is measured by the volumetric fraction of voids. Pumices commonly have vesicularities in the rather narrow range of 0.7–0.8. The existence of pumice indicates that during the fragmentation process it is possible to exsolve gas on the scale of the fragment without destroying it. Studies of natural pumice samples have shown that individual fragments exhibit interconnectedness between vesicles (Whitham and Sparks 1986; Westrich and Eichelberger 1994). This results from the formation of small holes in the bubble walls which allows gas to escape locally. It is not clear whether interconnectedness develops before or after the fragmentation process.

Scoria is a term applied to vesicular juvenile clasts formed from the degassing of more basic magmas such as basalt or basaltic andesite. They are typically darker in colour, have larger vesicles and are more dense than pumice. Because of the more fluid nature of the magma, large fragments of scoria often develop surficial features, such as a ropy or stringy texture, that are caused by aerodynamic forces during transport. Eruptions of basic magma with low volatile content can also produce small melt droplets that take on smooth aerodynamic shapes, known as achneliths.

Pumice and scoria are used to describe pyroclasts from blocks to lapilli size. However, as grains are reduced to ash size the terms glass or pumice shard are generally used to describe vesicular juvenile fragments. Heiken and Wohletz (1985) have assembled an excellent collection of scanning electron microscope (SEM) photographs that illustrate the tremendous diversity in volcanic ash morphology. Shards from silicic magmas are delicate particles that consist of bubble wall membranes and the y-shaped junctions between adjacent bubbles (Figure 1.13). Pumice shards resemble miniature pumice in that they contain numerous small vesicles. Shards from more mafic magmas tend to be darker in colour with thicker bubble walls.

Eruptions that involve an interaction with external water produce juvenile fragments that have a distinctive morphology. The principal effects of water interaction are to promote quenching, inhibit vesiculation and cause thermal fracturing. These effects are best seen in ash-size juvenile particles. Studies of experimentally produced ash and natural samples have revealed complex morphologies depending on magma composition and the extent of water interaction. The dominant signature of phreatomagmatic fragmentation is the development of more blocky and less vesicular fragments compared to those produced solely by volatile exsolution (Wohletz 1983; Heiken and Wohletz 1985).

Crystals are the second major category of pyroclasts. These are usually derived from magma and are released during the fragmentation process. However, some crystals may represent pre-existing non-juvenile material (xenocrysts). Crystals typically serve as the nucleation sites of gas bubbles and can thus be separated from the melt phase by having gas bubbles grow around them. As the magma fragments by bubble bursting the crystals are then released to the eruptive mixture. Common crystals in the products of explosive volcanism are pure SiO_2 (quartz), aluminosilicates (e.g. feldspars), ferromagnesian silicates (e.g. orthopyroxene, clinopyroxene and hornblende), and oxides (titanomagnetite and ilmenite). Their size varies from a few millimetres to a few tens of micrometres. Often they contain a thin selvage of glass that is inherited from the melt phase.

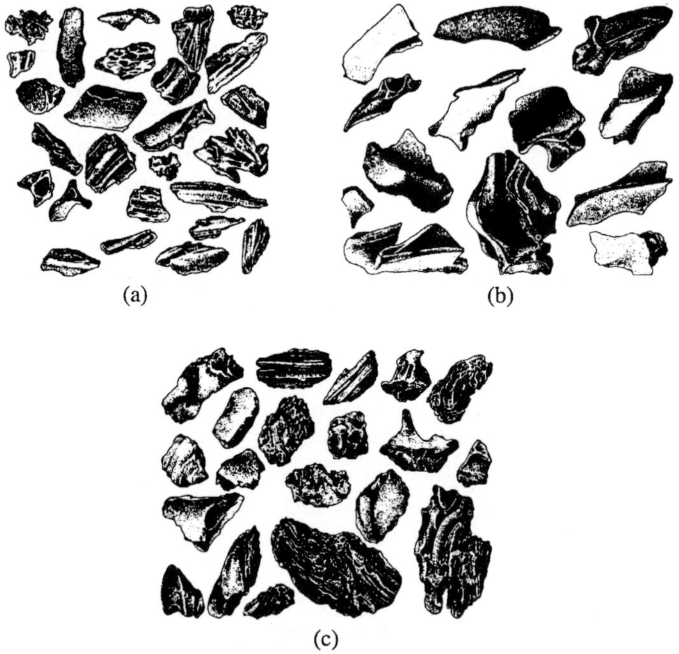

Figure 1.13 Examples of volcanic glass shard morphologies from rhyolitic eruptions of the western United States from Izett (1981). (a) Mesa Falls ash, Nebraska, (b) Lava Creek B ash, Utah, and (c) Bishop ash, Wyoming

Lithics are the third category of pyroclasts. These form by a variety of mechanisms. Most commonly they are pieces of the volcanic edifice that were torn off during an explosive eruption or pieces of country rock that were incorporated in the magma as it rose to the surface. Such accidental lithic clasts range in size from blocks to ash and can consist of igneous, sedimentary or metamorphic fragments. Fragmentation of lithic material may also contribute free crystals to the erupting mixture and it may be difficult to discriminate these from those contributed by juvenile magma. Dense lithic clasts can also have a juvenile origin such as when a lava dome is formed in the same eruption. Interpretation of glassy lithic clasts as juvenile or accidental can sometimes be problematic.

1.6 DISTRIBUTION OF EXPLOSIVE VOLCANISM

The majority of volcanic activity is not randomly distributed over the Earth's surface but is localized along linear belts that are the boundaries separating the surface into a series of major plates. The theory of plate tectonics provides a dynamic framework within which the distribution of different types of volcanism can be understood. Briefly, the theory states that the Earth's surface is divided up into large plates of cold rigid material known as the lithosphere above a warmer ductile interior, or asthenosphere (Figure 1.14). The lithosphere is defined by its thermomechanical properties and can be regarded as the outer layer of the Earth which is sufficiently rigid and cold to transmit stresses without significant permanent internal deformation. The lithosphere varies from about 80 km thickness in oceanic regions to 200–300 km thickness beneath continents. Lithospheric plates are in constant relative motion and interact along three different kinds of boundaries. At diverging boundaries the plates pull apart and new crust is created. At converging boundaries the plates push together and old crust is destroyed. At conservative boundaries the plates slide past one another with no net gain or loss. Earthquakes and volcanism are largely

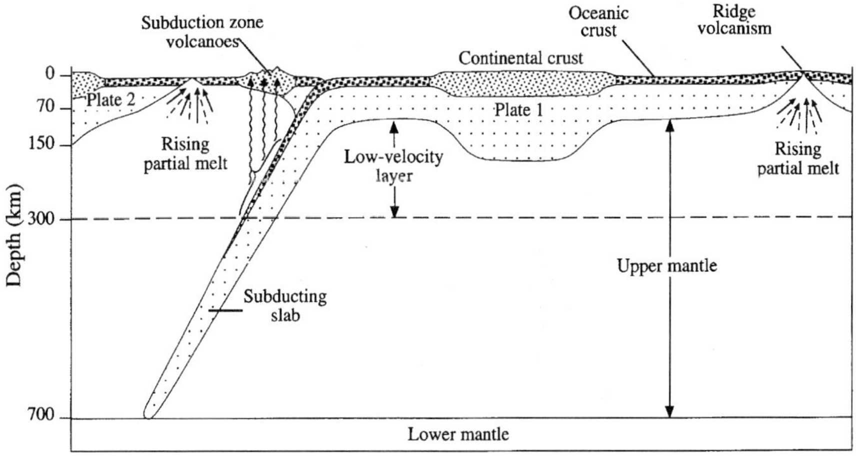

Figure 1.14 Cross-section through the crust and mantle illustrating the sites of mid-ocean ridge and subduction zone volcanism and their relationship to the movement of lithospheric plates. Modified from Wilson (1989)

confined to the narrow plate boundaries. Details on plate tectonics and the relationship to magmagenesis can be found in numerous geological textbooks (e.g. Press and Siever 1990; Sparks 1994).

At divergent plate boundaries, such as ocean ridge spreading centres, volcanism is dominated by the eruption of basalt. The majority of this activity occurs at great depth in the ocean and is non-explosive. It leads to the construction of new ocean floor which spreads slowly away from mid-ocean ridges (Figure 1.14). Basalt is, in fact, the most common magma type on earth and divergent boundary volcanism accounts for more than 80% of the total annual volcanic activity of the planet. Magma generation occurs mainly by uprise and adiabatic melting of mantle peridotite beneath the spreading ridge (McKenzie 1984). The basaltic magma generated at ridges typically has very low volatile contents (<0.5 wt.% H_2O and a few hundred parts per million CO_2). The low volatile content and the high water pressure at the depth of eruption severely inhibits the degassing of primary volatiles, thus eliminating an important cause of explosive activity. However, some explosive activity does occur along divergent boundaries where the ridge is unusually shallow or emerges above sea-level, as in the case of Iceland, situated along the Mid Atlantic Ridge.

A different variety of volcanic plume occurs at great water depth within oceanic spreading centres but is not directly related to eruption of magma at the surface. Hydrothermal plumes (Chapter 9) are generated by the discharge of hot seawater on the sea-floor which has been heated within the oceanic crust in areas close to axial magma chambers. These plumes play a key role in the removal of heat from newly formed oceanic crust, the exchange of major and minor elements between seawater and volcanic rocks and the formation of economically significant ore bodies. In addition, they have been found to host an exotic biological community based primarily on the chemosynthetic bacteria which thrive on the highly reduced and sulphur-rich plume fluids.

Explosive volcanism is largely concentrated at convergent boundaries where one lithospheric plate is being subducted beneath another (Figure 1.15). The subducted plate dips steeply as it descends into the mantle, triggering earthquakes as it fractures and decouples from the overlying plate. Island arcs are formed where the overriding and the subducting plate are both oceanic lithosphere. Examples include the Lesser Antilles, Aleutians and Marianas islands. They consist of arcuate chains of volcanic islands with a

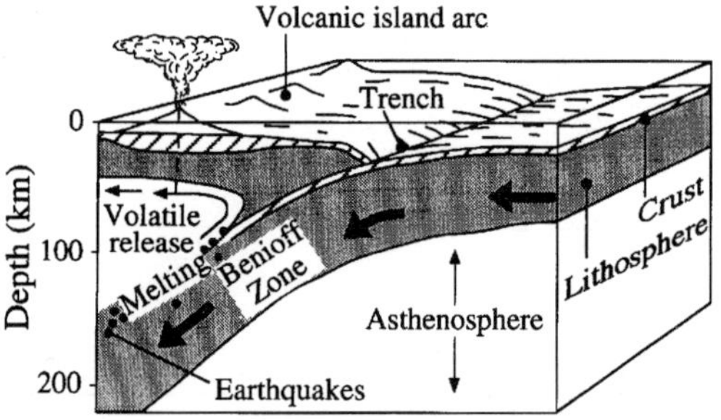

Figure 1.15 Block diagram illustrating the configuration and movement of lithospheric plates at a convergent boundary. Volcanism is triggered in a narrow band above the subducted slab where the release of volatiles triggers melting in the mantle wedge just above the slab surface. Modified from Wilson (1989)

deep oceanic trench on their convex side. Convergent boundaries are also created where oceanic lithosphere is thrust beneath continental margins such as along the west coast of South America and the north-west coast of the United States. In these environments explosive volcanism leads to the construction of majestic volcanic peaks such as Mount St Helens and Mount Rainier of the Cascade Range. The margin of the Pacific Ocean is dominated by convergent boundaries and has been referred to as the "Ring of Fire" because of the abundance of explosive volcanism (Figure 1.16).

Magmas erupted at convergent plate boundaries are typically higher in SiO_2 content, more viscous and generally contain a much higher content of dissolved gases than magmas erupted at divergent boundaries. At convergent margins magma generation appears to be a paradox because thrusting of a cold lithospheric plate into the mantle would not be expected to trigger melting. The important factor at convergent margins is the effect that relatively small amounts of water have in causing melting in the mantle above the subducted plate. At ocean ridges the crust becomes strongly hydrated by circulation of seawater through the newly formed hot volcanic rock. The uppermost 2–3 km of the ocean crust probably contains about 2 wt.% H_2O by the time it is subducted back into the mantle. The release of this water during subduction plays a crucial role in the explosivity of the associated volcanoes. Dehydration reactions take place as the descending plate enters a regime of higher temperature and pressure (Figure 1.15). Hydrous fluids that are given off migrate upwards into the mantle wedge just above the plate. These fluids reduce the melting temperature of the mantle wedge and generate magma. During the melting process

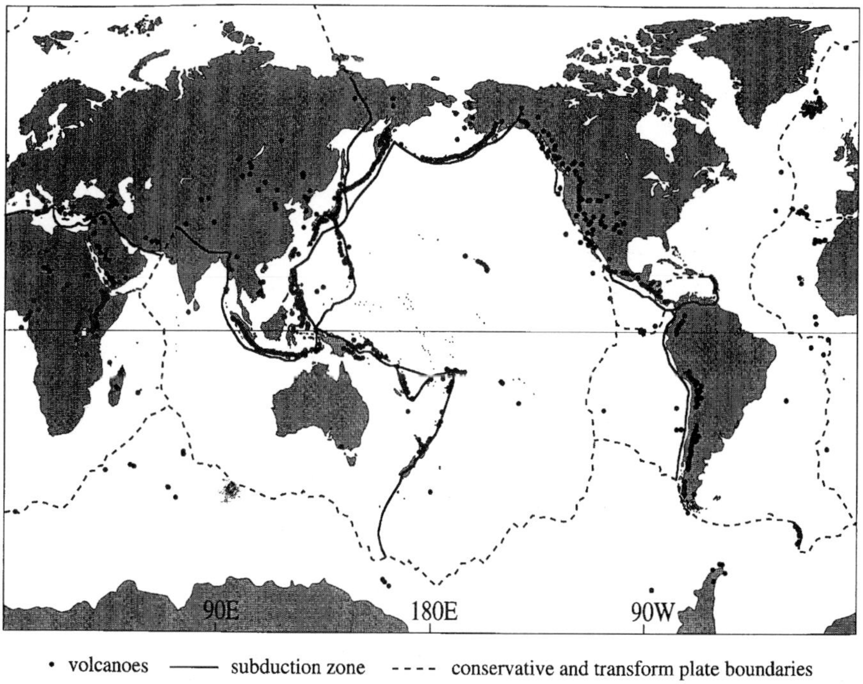

Figure 1.16 Map showing the global distribution of explosive volcanism. Modified from Wilson (1989)

the volatile components are preferentially partitioned into the magma which then rises to crustal levels. Basalts generated below island arcs can contain quite variable volatile contents in the range 1 to over 4 wt.% H_2O (Stolper and Newman 1994), in contrast to the relatively dry basalts of the spreading centres.

The compressional stress regime of convergent margins is conducive to the staging of magmas in crustal reservoirs beneath convergent margin volcanoes. Under conditions of elevated volatile components these magmas evolve by fractional crystallization, assimilation of country rock and mixing with other magmas to produce the complex suite of calcalkaline magmatic compositions associated with this tectonic environment. A common magma produced in such an environment has a SiO_2 content of about 60 wt.% and is called andesite, after the volcanically active Andes range of South America.

A certain amount of volcanism also occurs within the interior of the lithospheric plates away from both the divergent or convergent boundaries. This activity is fed by mantle plumes, which are localized areas of excess volcanism that are rooted deep in the Earth's mantle. They are generally believed to be caused by convective upwelling of hot mantle with anomalous composition, which melts copiously as it approaches the Earth's surface. A distinctive feature of mantle plumes is that they remain virtually fixed relative to the movement of lithospheric plates for long periods of geologic time and thus can be used to track absolute plate motions. As the plates move over these plumes, eruptive activity on the surface produces large-scale linear volcanic features that increase in age away from the plume source. In the ocean this takes the form of a linear chain of volcanic islands such as the Hawaiian group. Explosive volcanism on oceanic hot-spot volcanoes like Hawaii is usually of minor importance because most of the magmas are fluid basalts which release their volatiles easily. However, hot-spot magmas do appear to be enriched in volatiles compared to ridges and significant explosive eruptions can occur in more viscous chemically evolved magmas on ocean islands formed above mantle plumes. Examples include active volcanoes in the Azores and the Canary Islands which have had numerous large explosive eruptions of trachyte and phonolite magma in their history.

Highly explosive volcanism can also be triggered when continental crust moves over a deep-seated mantle plume. Large volumes of high-temperature basaltic magma are produced by convective uprise and melting of mantle peridotite. Some of this magma is directly erupted in a continental setting, but because of its high density relative to the crust some also comes to rest at a level of neutral buoyancy. Heat is then transferred to the surrounding crustal rocks leading to melting and the generation of high-silica viscous magmas. The plume-derived basaltic magmas appear to be volatile enriched and these volatiles can be transferred to and concentrated in the high SiO_2 melts. These magmas are often erupted explosively in some of the largest events that are known from the geologic record. Enormous explosive eruptions at Yellowstone National Park over the last 2 Ma are an example of the type of activity resulting from the passage of continental crust over a mantle plume (Christiansen 1984).

1.7 STYLES OF EXPLOSIVE VOLCANISM AND PLUME GENERATION

There is great diversity in the styles of explosive volcanism owing to the variety of magmatic compositions, the large range of magma volumes available for eruption, the wide variations in tectonic stress regime and magma chamber conduit geometry and the degree of interaction with external water. Much of the terminology in the classification of explosive eruptions reflects an early reliance on observations of eruptive phenomena at specific volcanoes. Many of the terms are neither rigidly defined nor entirely consistent and their usage among volcanologists follows in part from historical precedent rather than a completely rational classification. However, progress in the field of physical volcanology has now allowed various eruption styles to be understood based on the physical processes driving the eruption (Wilson *et al.* 1987).

Volcanic plumes are generated when explosive eruptions discharge a mixture of pyroclasts and gas into the atmosphere (Chapter 3). These mixtures can behave in two fundamentally different ways (Figure 1.17). They may either form convective plumes which rise to great heights in the atmosphere or they may collapse to form fountains at relatively low heights to produce flows of hot particles and gas that flow down the slopes of a volcano at high speed (Figure 1.17). These are referred to as pyroclastic flows or surges (Chapter 6) and can themselves generate large convective plumes or clouds (Chapter 7). The nature of a volcanic plume also depends on how the mixture of pyroclasts and gas is released during an eruption. Maintained plumes are generated if the duration of the sustained discharge is large relative to the ascent time of the plume (Chapter 4). Instantaneous plumes, or thermals, result from discharges that are short relative to the ascent time.

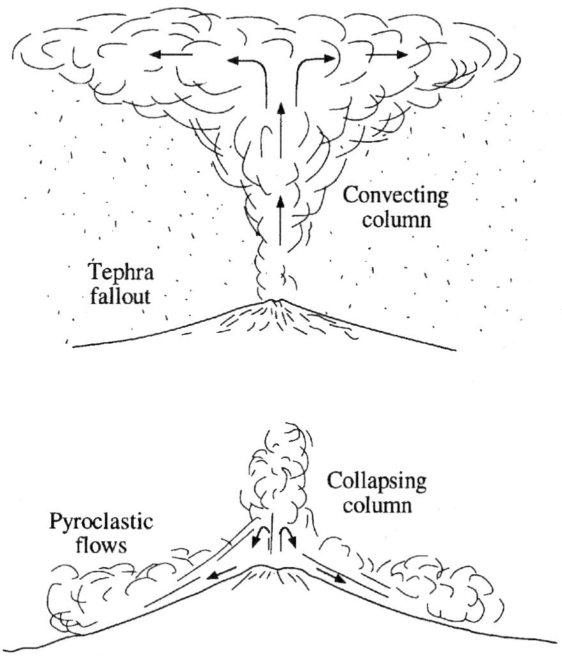

Figure 1.17 Two styles of eruption column behaviour during explosive eruptions. In the upper figure a buoyant convective column is produced by entrainment of atmospheric air as the mixture of pyroclasts exits the vent. In the lower figure the eruptive mixture is unable to rise buoyantly above the gas thrust region and consequently collapses as a low fountain to form pyroclastic flows

1.7.1 Plinian Eruptions

A sustained, explosive eruption which generates a high-altitude (stratospheric) plume is referred to as Plinian, in honour of Pliny the Younger who described the AD 79 Vesuvius eruption. Plinian eruptions typically involve high-viscosity silicic magma such as dacite or rhyolite, although some basaltic Plinian eruptions have been documented (Williams 1983). Discharge of magma occurs at rates of 10^6–10^9 kg s^{-1} for periods of a few hours to perhaps several days (Carey and Sigurdsson 1989). The fragmentation of magma occurs primarily as a result of juvenile volatile degassing (Figure 1.9). An important characteristic of Plinian activity is that gas bubbles travel to the surface at the same rate as the magma because of its highly viscous nature.

The basic structure of a Plinian plume can be subdivided into three parts based on the dominant forces that control plume motion (Figure 1.18). At the base is the gas thrust region where a mixture of pyroclasts and hot gas is ejected turbulently at velocities up to 600 m s^{-1} (Wilson 1976; Sparks and Wilson 1976; Sparks 1986). The vents that feed Plinian eruptions are of the order of tens of metres in diameter and are thus very small compared to the total height of the plume (tens of kilometres). Fragmentation is commonly high during such eruptions and the gas and particles are assumed to be at the same temperature. As the mixture first exits the vent its bulk density (particles and gas) is always greater than the surrounding atmosphere. Within the gas thrust region the momentum of the flow dominates over buoyancy and the flow behaves as a momentum jet. Rapid deceleration occurs and atmospheric air is entrained and heated. If the jet is able to entrain and heat enough air it can reduce its density below that of the surrounding atmosphere. Buoyancy forces will then begin to dominate the motion of the plume and strong vertical convection will allow it to continue to rise. The height of this transition is generally between a few hundred metres and a few kilometres for very large eruptions (Sparks 1986; Woods 1988). If the jet is unable to reduce its density below that of the atmosphere then it will collapse as a low fountain and generate pyroclastic flows and surges (Figure 1.17; Chapters 4 and 6).

The convective portion of the plume represents the trunk of the great umbrella pine that Pliny the Younger described over Vesuvius in AD 79. It typically makes up the majority of the rising plume and can extend tens of kilometres into the atmosphere (Figure 1.18). Within this part of the plume buoyant forces control its behaviour. Entrainment and heating of air are the most important processes that take place in this region and result in the generation of buoyancy. They also cause the plume to grow in width as it rises. Some pyroclasts fall out along the plume margin, but because of the high degree of fragmentation in many eruptions most of the particles are able to be transported to the upper portion of the plume. Above the gas thrust region vertical velocities in the convective region increase with height because of the large buoyancy force acting on the column. The velocity then goes through a maximum before decreasing towards the top of the column (Figure 1.18). The central velocities within the convective region can be of the order of a few tens to over 200 m s^{-1} depending on the magma temperature, exit velocity and vent radius.

In the density stratified atmosphere, a volcanic plume will reach an elevation where its bulk density is equal to that of the surrounding atmosphere. At this level, H_b, the plume will begin to spread out laterally. However, because the rising convective region possesses

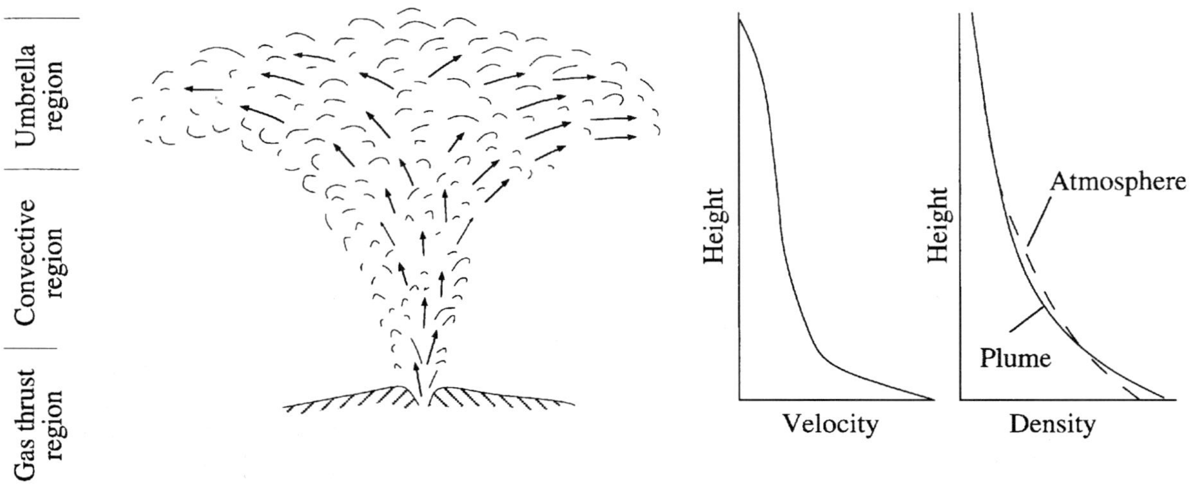

Figure 1.18 Subdivision of a sustained eruption column into the gas thrust, convective, and umbrella regions. The variation in column velocity and density as a function of height is shown schematically to the right. In the gas thrust region the column density is greater than the atmosphere. Throughout the convective region the column density is less than the atmosphere until it reaches the umbrella region where it is equal to the surrounding atmosphere

substantial momentum the plume will overshoot the neutral density level and attain a final height, H_t. The plume spreads out laterally between these two levels as a forced intrusion to form the umbrella region (Figure 1.18). This is part of the plume that Pliny equated with the top of the Mediterranean umbrella pines that are common in the Vesuvius area. The umbrella region is a visually spectacular feature of eruption plumes. Observations of recent eruptions and laboratory experiments have illustrated that the motion of the umbrella cloud can be usefully treated as an enormous expanding gravity current that is modified by its interaction with the local windfield. The umbrella region is of great importance to the deposition of pyroclasts because the majority of sedimentation occurs from the base of the laterally expanding cloud (Chapters 11 and 14).

Fallout of volcanic particles from Plinian eruption plumes can blanket large areas (several thousand km^2) with up to several metres of pyroclastic material. The AD 79 eruption of Vesuvius is considered to be the type example of a Plinian event. It lasted for a period of about 19 hours during which time an eruption plume was sustained at an elevation between 27 and 33 km above the volcano (Sigurdsson *et al.* 1985; Carey and Sigurdsson 1987). Fallout of pyroclasts occurred to the south-east resulting in the partial burial of the city of Pompeii. The May 18, 1980 eruption of Mount St Helens is an example of a small Plinian event. It lasted for about nine hours and generated a sustained eruption column at a height of roughly 16 km (Lipman and Mullineaux 1981, Figure 1.19). Fallout of tephra occurred over a large area of Washington, Idaho and Montana.

Phreato-Plinian eruptions are Plinian events where there is evidence for an interaction with external water in addition to the degassing of juvenile volatiles. The interaction results in enhanced fragmentation of pyroclasts and consequently a larger fraction of very fine ash in the eruption plume. This ash is flushed out of the plumes by condensing water and other aggregation processes to form a conspicuous fine ash component to the fall deposits (Self and Sparks 1978).

1.7.2 Ignimbrite-forming Eruptions

Another variety of highly explosive eruption involves the formation of large volume pyroclastic flows, or ignimbrites (Chapter 6). These ignimbrite-forming eruptions are similar to Plinian events in that a discharge of highly fragmented viscous magma occurs, but instead of forming a high-altitude eruption column the activity is characterized by the development of a low collapsing fountain (Figure 1.17). Some events have an initial Plinian phase although the majority of erupted material is in the form of pyroclastic flows. These events include the largest individual volcanic eruptions to occur on Earth. Single eruptions can discharge up to several thousand cubic kilometres of magma and commonly result in the formation of a large topographic depression known as a caldera, as the evacuated magma chamber collapses inwards. The explosive eruptions of Krakatoa (1883), Tambora (1815) and Pinatubo (1991) in Indonesia are examples of moderate-sized historic ignimbrite-forming eruptions.

During these eruptions a low fountain develops over the vent and the transition from the gas thrust to the convective region no longer occurs. A maintained collapsing fountain produces enormous volumes of pyroclastic flows that spread out to distances of up to 100 km from the vent. As the flows race outwards they entrain and heat atmospheric air. In addition, dense particles segregate to the base of the flow and are deposited. These two

Figure 1.19 Plinian eruption column developed over Mount St Helens during the May 18, 1980 eruption. (Courtesy of the US Geological Survey.)

processes cause the upper part of the flow to become buoyant and begin to lift off as a giant convective plume, referred to as a co-ignimbrite plume (Chapter 7). This transition can occur over large areas, of a few square kilometres to thousands of square kilometres, and thus the source area is markedly different compared to that for convective plumes produced by Plinian eruptions. A distinctive feature of co-ignimbrite plumes is their hourglass profile (Figure 1.20). When the plume becomes buoyant the mass flux per unit area increases faster than the addition of mass by entrainment and the column radius decreases to conserve mass (Woods and Wohletz 1991; Calder *et al.* 1997).

Co-ignimbrite plumes have no gas thrust region in their basal area because they become buoyant only through the heating of entrained air and the loss of particles in a laterally moving flow. A typical vertical velocity profile thus begins at low velocities, increases to a maximum as a result of the increasing buoyancy flux and then decreases as the plume approaches the density of the surrounding atmosphere. Co-ignimbrite plumes do develop an umbrella cloud in a fashion similar to Plinian plumes.

Figure 1.20 Photograph of co-ignimbrite plume from June 15, 1991 eruption of Mount Pinatubo in the Philippines. (Courtesy of National Oceanic and Atmospheric Administration.)

1.7.3 Strombolian Eruptions

Strombolian eruptions are explosive eruptions that are driven primarily by the exsolution of juvenile volatiles from low-viscosity magmas. In Strombolian eruptions the magmas are typically basic (basalt, basaltic andesite) and fluid. Gas exsolution creates bubbles which can rise through the magma and coalesce to sizes up to 10 m (Wilson 1980). When these large bubbles burst at the surface, magma is ejected as a spray of poorly fragmented incandescent particles. The activity is thus characterized by a series of discrete explosions, representing the bursting of individual large bubbles. Thermal exchange between ejected particles and the atmosphere is poor because of their large size. Consequently, strong convective plumes are not well developed during such eruptions.

Particles follow ballistic trajectories and are distributed within only a few kilometres of the vent (Figure 1.21), constructing a local cinder cone. Observed ejecta velocities range up to 260 m s^{-1} (Chouet *et al.* 1974; Blackburn *et al.* 1976). This suggests pressures of about 0.3 MPa in bubbles of 10 m diameter (Wilson 1980). The 1973 eruption of Heimaey volcano in Iceland is a typical example of Strombolian activity (Self *et al.* 1974). The eruption lasted over six months and in its early stages involved closely spaced explosions with intervals of a few seconds. As the gas content waned the intervals between explosions increased and much of the magma erupted as lava. The type volcano of Stromboli in the Mediterranean is in a state of continuous activity. Since Roman times, Strombolian explosions have occurred at intervals of minutes to hours from an open column of lava.

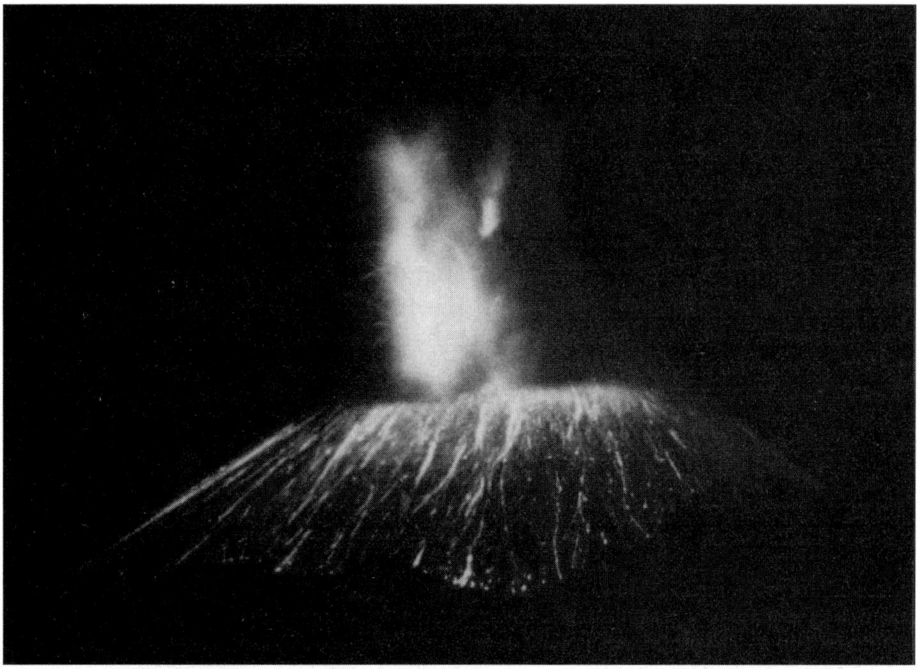

Figure 1.21 Photograph of Strombolian-style activity at Paricutin volcano, Mexico. (Courtesy of R.E. Wilcox, US Geological Survey.)

Violent Strombolian eruptions can be distinguished as a style in which a greater proportion of fine-grained ejecta is produced and the activity varies from discrete explosions to short periods of continuous discharge. These eruptions are characterized by somewhat more viscous magma (basaltic andesite and andesite) and can generate quite high plumes as well as abundant coarse ejecta which falls out to form a local cinder cone. The eruption of Paricutin in Mexico in 1943 displayed violent Strombolian activity. In terms of their behaviour violent Strombolian eruptions can be regarded as intermediate in character between Plinian and Strombolian.

1.7.4 Vulcanian Eruptions

Vulcanian eruptions are explosive events that occur as a series of discrete, energetic explosions at intervals of a few minutes to several hours. The explosions eject a mixture of fragmented magma, lithic debris, volcanic gases and steam at velocities of up to 400 m s^{-1} (Self *et al.* 1979). These eruptions are common at volcanoes that produce intermediate composition magmas such as basaltic andesite or andesite. The term "Vulcanian" is, however, very much a "dustbin" term, as the actual causes and character of discrete volcanic explosions vary greatly from one eruption to another. Some Vulcanian explosions occur in settings where explosive interaction of magma and groundwater is suspected (Fagents and Wilson 1993). The abundance of lithic material can be quite high in Vulcanian eruptions of hydrovolcanic origin and can even exceed the amount of juvenile fragments. On the other hand, some Vulcanian explosions appear to be related to explosive disruption of growing lava domes by exsolution of magmatic gas and thus discharge predominantly juvenile ejecta.

Individual plumes from these events have been observed to rise to elevations as high as 20 km (Nairn and Self 1978) although most are usually less than 10 km (Figure 1.22). Fallout of pyroclasts results in relatively thin, widespread deposits with volumes much less than 1.0 km^3. Vulcanian explosions are often notable for ejecting large bombs or blocks substantial distances. For example, the 1991 Vulcanian explosions at Lascar volcano in Chile ejected 2 m diameter blocks to 5 km from the volcano. Some Vulcanian plumes undergo collapse to form pyroclastic flows. The 1975 eruptions of Ngauruhoe volcano in New Zealand produced dense plumes which partially collapsed at a height of 500 m above the vent to produce small-volume pyroclastic flows (Nairn and Self 1978).

Figure 1.22 A Vulcanian eruption plume generated during the 1975 activity of Ngauruhoe volcano in New Zealand. The event produced both a vertically rising plume over the vent and a pyroclastic flow which can be seen descending the slopes of the volcano. (Courtesy of University of Colorado.)

The mechanism of Vulcanian eruptions is thought to be either hydrovolcanic interactions or sudden release of pressure within a conduit or lava dome that is blocked by a carapace of cooled lava or a pre-existing dome (Cas and Wright 1987). Quantitative evaluations of the dynamics of Vulcanian eruptions have provided insights into the process of pressure build-up prior to eruption. Wilson (1980) has shown that the ejection of material at velocities up to 200 m s^{-1} can be explained by the confined degassing of up to a few weight % water of juvenile volatiles from magma (Wilson 1980), but higher velocities (~ 1400 m s^{-1}), as observed in some eruptions, seems to require an additional source of volatiles such as the heating of external water by magma (Fagents and Wilson 1993).

Plumes generated by vulcanian-type activity are considered thermals if the release of the pyroclast–gas mixture occurs on a time scale much less than the ascent time of the plume. The interval between explosions must also be appropriately long. Closely spaced explosions can merge to form a discrete sustained plume. These plumes can undergo an evolution through each of the major subdivisions shown by maintained plumes (Figure 1.18). At the vent, the motion of the initial mixture may be dominated by momentum as a high-speed explosion. As the mixture ascends over the vent it will entrain and heat atmospheric air resulting in the generation of buoyancy. If the density is reduced sufficiently then the thermal ascends convectively until it reaches a level of neutral buoyancy in the atmosphere. Depending on the volume of the thermal and the intensity of the cross-wind, an umbrella-type plume may also form. Otherwise the plume may be simply advected downwind and spread out by diffusion.

Small-volume pyroclastic flows may also generate discrete co-ignimbrite plumes that are best described as thermals. As flows run out from source they can become buoyant by heating of entrained air and sedimentation of pyroclasts. Small flows from recent eruptions of Redoubt volcano in Alaska produced co-ignimbrite thermals which ascended to 12 km and spread out to form distinctive umbrella clouds (Woods and Kienle 1994, Figure 1.23).

1.7.5 Surtseyan Eruptions

Surtseyan eruptions are explosive events where magma interacts with large volumes of water. This activity is characterized by numerous discrete explosions that produce steam-rich plumes laden with basaltic pyroclasts. The type example occurred in 1963 during the emergence of Surtsey volcano from beneath the sea just south of Iceland. Seawater mixed with basaltic magma to form a series of spectacular explosions (Figure 1.12). The explosions are described as cockstail, with parcels of tephra and gas following parabolic trajectories. Ejecta from such eruptions are characteristically fine-grained with large proportions of volcanic ash (<2 mm diameter) composed of poorly vesicular blocky glass fragments.

1.7.6 Hawaiian Eruptions

Hawaiian eruptions are characterized by the generally quiescent effusion of basaltic magma as lava flows and fire fountaining. This type of activity has been well documented on the Hawaiian Islands where systematic measurements have been made of magma eruption rates, composition, volatile content and vent geometry. Although this type of volcanism is less explosive than the eruption of more intermediate or silicic magmas, it is

Figure 1.23 Photograph of thermal plume generated above the Redoubt pyroclastic flows. (Courtesy of Juergen Kienle.)

Figure 1.24 Photograph of a fissure eruption taking place at Krafla volcano in Iceland. (Courtesy of Axel Bjornsson.)

possible to develop plumes from fire fountaining activity where a mixture of poorly fragmented magma and gases is ejected up to several hundred metres (Chapter 10). The majority of pyroclasts fall out quickly around the vent area and may form lava flows which then travel away from the vent. However, basaltic ash fragments and the heat transferred to air from large clasts within the fountains can drive convective plumes that rise to higher altitudes than the fire fountain heights. Such plumes have been observed during the Askja 1961 eruption in Iceland and eruptive phases of the 1959–60 activity of Kilauea volcano in Hawaii. An important aspect of Hawaiian-style eruptions is that they can occur from linear vents, or fissures, which extend for several hundred metres to several tens of kilometres (Figure 1.24). The source region of the associated plume is therefore not a localized source. There is a continuous spectrum of eruptive style from Hawaiian to Strombolian in basaltic eruptions and this is discussed in detail in Chapter 10.

1.7.7 Classification Limitations

One problem with the classification of explosive eruptions presented above is that the usage of a limited number of descriptive terms gives the impression that explosive volcanism falls neatly into distinct behavioural categories. In fact there is a continuum between various eruption styles and some eruptions exhibit multiple styles. Furthermore, as more eruptions and pyroclastic deposits are studied in detail, it is becoming increasingly clear that many eruptions do not fit into the traditional categories. Some eruptions show hybrid characteristics and display features where using one of these terms could be misleading. Thus the existing classification of explosive eruptions is somewhat arbitrary and glosses over the real complexity of many explosive events.

1.8 MAGNITUDE AND INTENSITY OF EXPLOSIVE VOLCANISM

Many attempts have been made to characterize the size and energy of explosive eruptions. Walker (1980) has proposed five parameters necessary to describe an explosive eruption. Intensity is the rate at which magma is discharged (kg s^{-1}); magnitude is the total mass of material erupted (km^3 of magma); dispersive power is the area over which the products are spread; violence refers to the distribution of products mainly by momentum; and destructive potential is an assessment of the area over which destruction of buildings, farm land and vegetation occurs. Newhall and Self (1982) developed the Volcanic Explosivity Index (VEI) using some of these parameters in order to provide a semi-quantitative scale for the assessment of historic explosive eruptions. The VEI increments are arbitrarily chosen and assume a fixed interrelationship between the various parameters.

Strombolian and Vulcanian eruptions represent the lower end of the spectrum of explosive volcanism. Strombolian eruptions typically eject very much less than 0.01 km^3 of magma during a single explosion. However, activity may be semi-continuous at a particular volcano for long periods of time. As mentioned previously, the plumes from Strombolian activity are commonly weak and only rise to a few kilometres in the atmosphere. Some violent Strombolian activity such as at Lonquimay volcano in Chile may, however, produce more vigorous plumes (Moreno and Gardeweg 1989). Vulcanian eruptions sometimes produce more ejecta than Strombolian events but the upper range is probably of the order of 0.5 km^3. Because of the more explosive nature of Vulcanian eruptions the plumes can attain substantial heights in the atmosphere, up to 20 km.

Carey and Sigurdsson (1989) have evaluated the range of intensity and magnitude for 45 Pleistocene and Holocene Plinian eruptions. Intensity estimates were based on the application of a theoretical model of pyroclast fallout to the measured dispersal of clasts from the various eruptions (Carey and Sparks 1986 and discussed in Chapter 15). They found that intensity varied over three orders of magnitude from 1.6×10^6 to 1.1×10^9 kg s^{-1}. The highest intensity eruption was the Taupo AD 180 eruption in New Zealand. Walker (1980) has proposed that this event be considered a special class of Plinian eruptions called ultra-Plinian because of its exceptional intensity and wide dispersal of pyroclasts. The magnitude of Plinian eruptions in the data set considered by Carey and Sigurdsson (1989) also varied three orders of magnitude from about 2×10^{11} to 4×10^{14} kg. These values correspond to 0.1–150 km^3 dense rock equivalent of magma.

Among the largest magnitude eruptions known from the geologic record are those that produce large volume pyroclastic flows, typically associated with the formation of calderas (Smith 1979). For example, the eruption which produced the Fish Canyon tuff ejected roughly 3000 km^3 of rhyolitic magma during a single event. The size of such an event is difficult to imagine considering that the largest historic eruption to have occurred was in 1815 at Tambora volcano in Indonesia when 50 km^3 of magma was ejected as tephra fall and pyroclastic flows. Many of these large ignimbrite-forming eruptions also have an initial Plinian phase, although volumetrically the pyroclastic flows are dominant. Unfortunately it is difficult to estimate quantitatively the intensity of such eruptions during flow formation based on the physical characteristics of the deposits, but the peak values certainly exceed the highest intensities shown by Plinian events. Wilson and Walker (1982) have attempted to estimate the intensity of the Taupo eruption in New Zealand based on assumptions about the velocity of flow emplacement and total erupted volume of flows. They arrive at 7×10^9 kg s^{-1} for the average intensity during the flow-generating phase of the eruption.

1.9 FREQUENCY OF EXPLOSIVE ERUPTIONS

In the last section it was shown that a large variation exists in the intensity and magnitude of explosive eruptions, based on historical observations and inferences from the geologic record. For example, the 1815 eruption of Tambora volcano was the largest explosive event in historic times. There are, however, examples of much larger events in the not too distant geologic past. The relationship between the frequency and size of explosive eruptions is now considered. Simkin (1993) has used the Smithsonian database of volcanic eruptions to assess the frequency of explosive eruptions through the use of the VEI. Figure 1.25 is a plot of the number of eruptions per 1000 years versus the VEI. As one might expect, the frequency of small eruptions is much greater than that of very large eruptions. An event similar to the 1980 Mount St Helens eruption is expected to occur about once a decade, whereas one like the Tambora eruption is likely only once or twice every 1000 years.

It becomes increasingly difficult to estimate the frequency of the largest types of ignimbrite-forming events such as the Yellowstone eruption of 2 Ma because of problems with dating and the incompleteness of the geologic record. Decker (1990) has examined

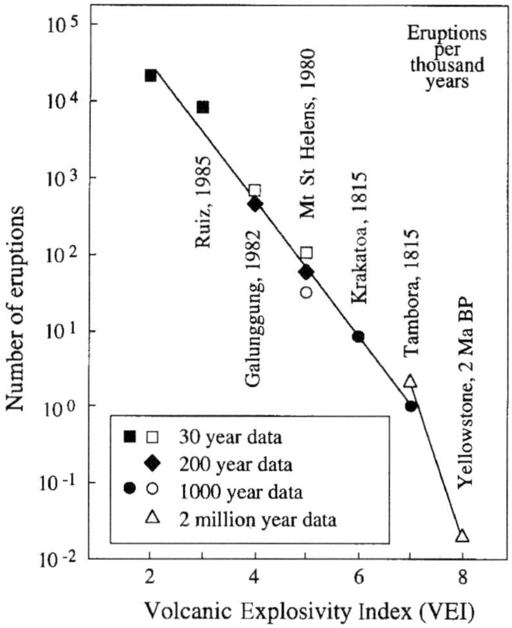

Figure 1.25 Number of explosive eruptions per 1000-year interval versus the Volcanic Explosivity Index (VEI) for Holocene eruptions. Modified from Simkin (1993). Open triangle data points for VEI 7 and 8 are from Decker (1990). Several eruptions are shown according to their VEI rating

the frequency of these large events using data from caldera-forming eruptions which have occurred during the last 2 Ma. His estimate for eruptions of VEI 8 is eight events during a 100 000 year period. The frequency of these types of events is probably a minimum value and will have to be adjusted upwards as more information and dating become available.

Another way to consider the frequency of explosive eruptions is to examine the repose periods that separate eruptions of a particular magnitude. Figure 1.26 shows the distribution of repose periods for eruptions of VEI 0–6 based on historic events. The figure demonstrates that the largest volcanic events at a particular volcano are typically preceded by very long repose periods. The volcanoes which are frequently active actually have a lower probability of experiencing a major explosive event than a more innocuous-looking volcano which may appear to have been dormant for many generations. Such was the case for the recent 1991 explosive eruption of Pinatubo in the Philippines. No historic activity had been recorded at Pinatubo but ^{14}C dating indicates that a large explosive eruption had occurred 400–600 years ago (Wolfe 1992).

1.10 SUMMARY

Volcanic plumes are produced in a myriad of ways during the processes of explosive volcanism and hydrothermal activity. Fundamentally the origin of volcanic plumes can be attributed to the thermal energy of magma which is transferred to the atmosphere or ocean by a variety of mechanisms to drive convective motions.

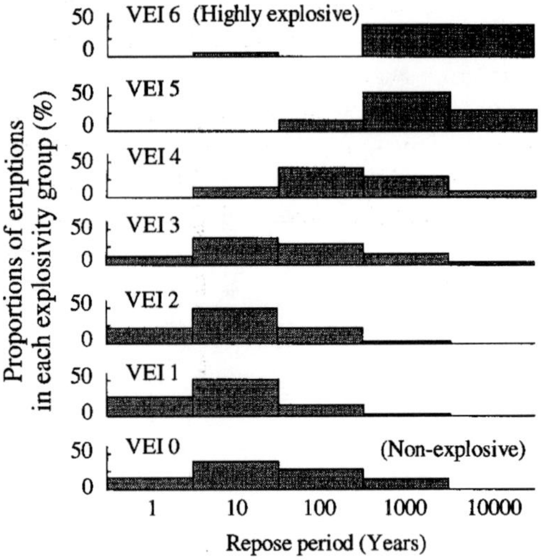

Figure 1.26 Histograms of repose periods between eruptions for VEI classes of eruptions from 0 to 6. Modified from Simkin (1993). Data are for historic eruptions only

In magmatic volcanic eruptions generation of atmospheric plumes involves the fragmentation of magma into small pieces, or tephra, by gas expansion and the consequent rapid acceleration of the gas/particle mixtures. Fragmentation occurs either when dissolved volatile components such as water and CO_2 are exsolved as magma ascends towards the low-pressure conditions of the Earth's surface or when magma interacts with external water such as in the ocean, crater lakes or by the intersection with groundwater. Volcanic plumes can also occur without direct eruption of magma by heating of groundwaters to cause phreatic explosions, fumarolic steam plumes and geysers. On the sea-floor, circulation of water through newly created oceanic crust results in the discharge of mineral-laden hot fluids that rise convectively to form hydrothermal plumes.

Explosive volcanism is typically associated with large-scale tectonic boundaries where crustal plates are consumed back into the Earth's mantle. In these environments recycled volatile components from the descending plate play an important role in magma genesis and the enrichment of water, CO_2 and other components that ultimately drive explosive volcanism at the surface.

The nature of plumes produced by explosive volcanism depends on factors such as magma composition, volatile content, vent configuration and discharge rate. Sustained discharges of highly fragmented silicic magma (Plinian eruptions) generate high-altitude mushroom-shaped columns and widely dispersed plumes. Under certain conditions sustained eruptions do not form convective columns, but instead collapse to produce devastating mixtures of hot gases and particles known as pyroclastic flows. These flows can generate their own secondary plumes (co-ignimbrite) by convective uprise of elutriated fine ash. Eruptions of basic magma by primary degassing usually leads to less energetic plumes (Strombolian and Hawaiian eruptions) as a result of the lower magma viscosity, which facilitates segregation of gas from the magma. Such magmas generally have lower volatile contents and lower degrees of fragmentation. Highly energetic plume production associated with basic magma can occur, however, during magma–water interactions at shallow depths (Surtseyan eruptions).

Large-scale explosive volcanism and the associated plumes pose significant local and global hazards. In the proximal area of erupting volcanoes the local populations are often threatened by pyroclastic flows and building collapse from accumulating tephra fallout. On a more global scale the injection of particulate and volatile species into the atmosphere can have significant impact on the climate system by affecting the amount of solar radiation received at the Earth's surface.

2 General Fluid Dynamical Principles

NOTATION

a	constant in equation (2.45)
A	cross-sectional area of plume (m^2)
B	buoyancy or reduced gravity of the system ($m\ s^{-2}$)
b	radius of plume or jet (m)
b_G	radius where Gaussian plume properties decay by a factor $1/e$ (m)
c	measure of the excess mass or momentum $(kg\ m\ s^{-2})^{5/2}$
C	concentration in plume ($kg\ m^{-3}$)
F	buoyancy flux ($m^4\ s^{-3}$)
F_r	plume Froude number (dimensionless)
g	gravity ($m\ s^{-2}$)
g'	buoyancy or reduced gravity ($m\ s^{-2}$)
G	ratio of the negative buoyancy flux compared with initial momentum flux (dimensionless)
h_a	acceleration height (m)
h_j	jet length (m)
H	height of the plume (m)
k	entrainment coefficient for Gaussian plume model (dimensionless)
L	length scale of the flow (m)
m	mass fraction of MEG in the plume (dimensionless)
M	momentum flux ($kg\ m\ s^{-2}$): also denotes momentum for discrete thermal ($kg\ m\ s^{-1}$)
N	Brunt–Väisälä frequency, characterizing the stratification of the environment (s^{-1})
Q	mass flux ($kg\ s^{-1}$): also denotes mass for discrete thermal (kg)
Q_C	concentration flux ($kg\ s^{-1}$)
r	radial distance from centreline of plume (m)
t	time (s)
u	mean upward velocity ($m\ s^{-1}$)
v	upward velocity as a function of the radius ($m\ s^{-1}$)
w	upward velocity of thermal ($m\ s^{-1}$)
z	height above source (m)
α	density of environment ($kg\ m^{-3}$)
β	bulk density of plume ($kg\ m^{-3}$)
$\delta\rho$	maximum density of MEG–water mixture relative to water ($kg\ m^{-3}$)
ε	entrainment coefficient for top-hat model (dimensionless)
ε_T	entrainment coefficient of discrete thermal plume (dimensionless)

λ dimensionless ratio of the decay scale of velocity and concentration (dimensionless)
ν kinematic viscosity of the fluid ($m^2\ s^{-1}$)
ρ density as a function of radius ($kg\ m^{-3}$)
θ angle of expansion of plume (°)
Ω expansion coefficient for salinity (dimensionless)

Quantities denoted $f(0)$ are evaluated at the plume centreline.

Subscripts

a	acceleration height
d	dimensionless property for non-linear mixing plume
e	property of the environment
G	property of the Gaussian model plume
j	property of a jet or forced plume
l	property of a line plume
m	property evaluated at the point of maximum negative buoyancy
o	property at the source
p	property of plume
T	property of the discrete thermal plume

Superscripts

* dimensionless variables

2.1 INTRODUCTION

In this chapter the motion of fluid released from a source into a second fluid is analysed. If the second fluid is of the same density then the flow is driven by momentum, whereas if the second fluid is of different density then buoyancy-driven flow can ensue. The general principles of these flows provide the essential background to understanding the behaviour of volcanic plumes. If the source fluid is buoyant, then it is driven upwards by gravity and a convecting plume or thermal develops above the source (Figure 2.1). If the motion is sufficiently large or intense that the Reynolds number of the flow is larger than about 10^4, then the flow becomes fully turbulent. The Reynolds number is given by uL/ν where u and L are typical velocity and length scales of the flow and ν is the kinematic viscosity. In many naturally convecting flows the motion is indeed highly turbulent. In particular, volcanic eruption columns and moist convective updraughts can extend over lengths 10–10 000 m with velocities in the range 1–500 m s^{-1}, giving Reynolds numbers in the range 10^3–10^9.

The turbulence manifests itself in the form of energetic eddies which span a range of scales up to the scale of the whole convective structure. In such highly turbulent flows, the molecular properties of the fluid do not limit the rate of mass and heat transfer. Instead, heat and mass transfer is effected by mixing parcels of the surrounding environmental fluid into the convecting fluid. The mixing occurs as large-scale eddies engulf ambient fluid at the periphery of the convecting region. This ambient fluid is then swept into the main body

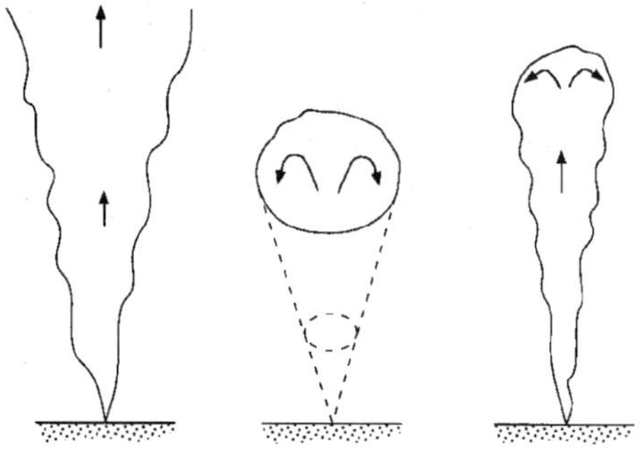

(a) (b) (c)

Figure 2.1 Schematic of (a) a convecting plume, (b) a thermal and (c) a starting plume rising from a source of buoyancy

of the flow where it is thoroughly stirred into the convecting core by smaller scale turbulence. Such mixing is inherently time dependent and, at any instant in time, the boundary of the convecting flow is very irregular. However, on time scales larger than the overturning time of the largest eddies, the convecting fluid may be characterized by its time-averaged mean properties. The time-averaged properties of the flow are quite regular, and the time-averaged boundary of the convecting region is quite smooth (Figure 2.2). Furthermore, although turbulent eddies are extremely vigorous, in the convecting plumes and thermals considered in this chapter, they represent less than 10% of the total kinetic energy of the convecting flow (Papanicolaou and List 1988).

An important experimentally observed feature of the convection which forms above a source of buoyant fluid is self-similarity. This means that the entrainment of ambient fluid, and subsequent spread of the buoyant fluid as it rises, has the same form for all scales of

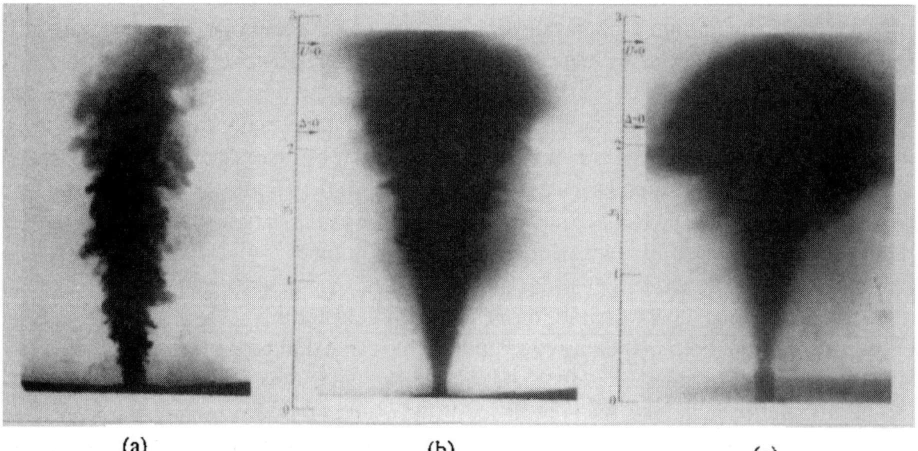

(a) (b) (c)

Figure 2.2 Photographs of a plume, distinguishing between (a) the highly complex shape of the boundary at any particular instant, (b) the smooth time-averaged shape of the boundary and (c) the lateral motion which develops at the plume top in a stratified environment. Reproduced by permission from the Royal Society the motion. The assumption of self-similarity is based upon the premise that the motion is fully turbulent, and so the molecular properties of the fluid do not affect the flow. Such turbulent convective motion contrasts with the much slower, laminar motion of, for example, viscous mantle plumes. Laminar plumes are controlled by molecular viscosity and thermal conductivity. In such flows, fluid can only be added at their edge either by conducting heat to the surrounding fluid, or through viscous coupling with the surrounding fluid.

The detailed morphology of the turbulent flow which develops above a source of buoyant fluid depends upon both the source geometry and the rate of release of the buoyant fluid. However, there are a number of limiting cases in which relatively simple flow structures develop, and these serve as useful paradigms for the interpretation of more complex flow structures. If the fluid is input at high speed and is of the same density as the environment, then a momentum driven *jet* develops. Turbulent eddies mix ambient fluid in from the side of the jet, and as long as the flow is of sufficient scale that molecular dissipation processes are unimportant, the flow spreads out as it decelerates, at a rate proportional to about one-eighth of the distance from the source (Prandtl 1954; Papanicolaou and List 1988). If the fluid released continuously from a point source is buoyant, the classical axisymmetric *buoyant plume* develops (Figure 2.1a). In another limiting case, in which the fluid is released instantaneously, a discrete *thermal* detaches from the source and ascends as a buoyant plume (Figure 2.1b). If the fluid is released from a line source, rather than a point source, then a two-dimensional, rather than axisymmetric, flow develops. In the following sections, analytical models for each of these simple cases are developed. The changes to the ideal model are described for the cases when the buoyant fluid is released from a finite source, when the environment is stably stratified and when the discharge rate is unsteady.

In many naturally occurring buoyant flows, the density varies linearly with the mass of environmental fluid entrained into the flow. However, in a number of situations, including volcanic eruption columns, the density actually varies in a non-linear way with the mass of ambient fluid mixed into the flow. In the final section of the chapter an extension of the simple model of a buoyant plume is described in which the density is modelled as varying in a non-linear fashion with the mass of entrained fluid. A number of further controls upon the motion of buoyant plumes resulting from buoyancy reversal are also considered. This sets the scene for Chapter 4, in which more complex models are developed to describe the detailed motion of volcanic eruption columns.

2.2 JETS

When fluid issues from a nozzle at high speed, such that the Reynolds number of the flow exceeds about 10^4, then the jet is turbulent and a turbulent shear layer at the edge of the jet results in mixing with the ambient fluid. Prandtl modelled the motion of such jets by assuming that the mixing layer is a fraction of the radius of the jet, and so the jet is composed of an inner core upon which a turbulent stress is exerted by the outer edge of the jet. Experimental observations suggest that the radius of the jet, b, increases as $b = z/8$, where z is the distance above the source (Prandtl 1954; Papanicolaou and List 1988). Since the total momentum of the jet is conserved, the mean velocity decreases at a rate inversely proportional to the rate of increase of the radius. Therefore, the velocity of the jet, u, is given by

$$\frac{du}{dz} = \frac{-u}{8b} \tag{2.1}$$

This equation is equivalent to assuming that the volume flux of the jet increases through entrainment of environmental fluid at the rate

$$\frac{d(ub^2)}{dz} = \frac{ub}{8} \tag{2.2}$$

If the vent velocity exceeds the speed of sound, then the material is initially either under- or over-pressured. Therefore, there is an initial transition region in which the flow adjusts to atmospheric pressure. Indeed, only when the flow has slowed to about 0.3 times the speed of sound, do the pressure fluctuations become negligible and the flow asymptotes to that described by equation (2.2).

Equation (2.2) describes the mixing law for a turbulent jet. In the next section an analogous equation for a buoyant plume is derived, in which the motion is driven by the density contrast, rather than the initial momentum. Using this model for a buoyant plume, the entrainment velocity is parameterized as being proportional to the vertical velocity. From equation (2.2) and the observed spreading rate of the jet, one may deduce that the effective entrainment coefficient is 0.065.

Real jets emerge from source nozzles or vents with finite dimensions. Entrainment of surrounding fluid occurs by the development of a shear layer which progressively penetrates into the interior with height (Fischer *et al.* 1979; List 1982). The turbulent flow requires a finite distance for the mixing to penetrate into the centre of the jet and for the flow to become fully developed. For a circular source the distance is approximately 10 times the radius of the source (Fischer *et al.* 1979).

2.3 MAINTAINED BUOYANT PLUMES

The motion of maintained turbulent plumes above a continuous and constant source of buoyancy was first modelled by Morton *et al.* (1956). They introduced time-averaged equations to describe the conservation of mass, momentum and buoyancy. The buoyancy B is defined as

$$B = \frac{g(\alpha - \beta)}{\alpha_o} \tag{2.3}$$

where g is the gravitational acceleration and α and β the densities of the ambient fluid and convecting fluid; α_o is a reference density equal to the density of the ambient fluid beside the source, for example. The buoyancy, B, is also commonly called the reduced gravity, and denoted by the symbol g'. In general, the density varies with the concentration of some conserved quantity C, for example the thermal energy or the salinity. However, the variation in density may depend non-linearly upon C, and so although C is conserved the buoyancy is not. Only in the special case in which the density depends linearly upon the concentration of a conserved quantity C, is the buoyancy conserved on mixing. Most discussions of buoyant plumes assume that this is the case. However, in order to generalize the analysis, for example for application to a volcanic plume, an explicit conservation law for the quantity C is required. This can be combined with a constitutive equation for the density

$$\rho = \rho(C) \tag{2.4}$$

in order to calculate the buoyancy force. In the following model of a plume, it is assumed that the density varies linearly with concentration C. However, in the final section of the chapter, the model is generalized to allow for a more complex constitutive relationship for the density.

The effect of the turbulent eddies is to entrain environmental fluid into the convecting flow. Morton *et al.* (1956) parameterized this effect by assuming that the rate of entrainment of the ambient fluid is directly proportional to the mean upward velocity of the plume at that height. The constant of proportionality is known as the entrainment coefficient. This parameter cannot be predicted theoretically, and therefore must be deduced from experimental measurements. The predictions of the model are in good agreement with experimental observations, and hence they support the notion of an entrainment coefficient (Turner 1979).

Historically, two derivations of the model of a buoyant plume have been described. In the first, the horizontal variation of the properties in the plume are explicitly included in the derivation, while in the second approach, the plume is modelled as being well mixed but of finite lateral extent, with mean values of the plume properties such as velocity, density and temperature assigned at each height. The present discussion follows the second approach, often called the "top-hat" approach. The two models can be related mathematically, so that the variation of properties across the plume can be deduced using results from a top-hat model.

Laboratory measurements of buoyant plumes have found that the time-averaged upward velocity, v, and concentration of the buoyancy producing property, C, are distributed about the centreline of the plume in a Gaussian fashion (Rouse *et al.* 1952; Papanicolaou and List 1988), with the velocity decaying somewhat more rapidly than the concentration C

$$v_G = v_G(0)\exp(-\lambda^2 r^2/b_G^2) \tag{2.5i}$$

$$C_G = C_e + (C_G(0) - C_e)\exp(-r^2/b_G^2) \tag{2.5ii}$$

where r is the radial distance from the centreline, the subscript G refers to the Gaussian description, $v_G(0)$ and $C_G(0)$ are the centreline velocity and concentration, C_e is the concentration in the environment, b_G is the radial distance over which the plume properties decay by a factor $1/e$, $b_G = 0.12z$, with z the distance above the source and $\lambda \approx 1.1$. The horizontally averaged fluxes of mass, Q_G, momentum, M_G, and concentration associated with the plume, QC_G, are defined in terms of the velocity of the plume, $v_G(r)$, the concentration in the plume, $C_G(r)$, and the constitutive relation for the density, equation (2.4).

$$Q_G = \int_0^\infty \rho_G v_G \mathrm{d}A; \quad M_G = \int_0^\infty \rho_G v_G^2 \mathrm{d}A; \quad QC_G = \int_0^\infty \rho_G v_G C_G \mathrm{d}A \tag{2.6i, ii, iii}$$

where dA denotes an integral over a horizontal plane normal to the plume.

Using the entrainment hypothesis described above, it is assumed that the mass of material entrained into the plume, per unit height, at a height z in the plume, is $2\pi k v_G(0) C_e$, where k is the entrainment coefficient. Therefore, the horizontally averaged equations for the conservation of the vertical fluxes of mass, Q_G, momentum, M_G and concentration, QC_G, have the form

$$\frac{dQ_G}{dz} = 2\pi k v_G(0) b_G \alpha; \quad \frac{dM_G}{dz} = \int_0^\infty g(\alpha - \rho_G(r))dA; \quad \frac{dQC_G}{dz} = 2\pi k v_G(0) b_G \alpha C_e \tag{2.7i, ii, iii}$$

In the above conservation relations, α represents the density of the environment and $\rho_G(r)$ represents density of the plume. The model is completed using the constitutive relation for the density, equation (2.4). At any height in the plume, the mass, momentum and concentration fluxes may be evaluated in terms of the value of v_G and C_G on the centreline.

In many situations, to good approximation, the constitutive relation for the density is taken to be linear

$$\rho_G(r) = \alpha + \frac{d\rho_G}{dC}(C_G(r) - C_e) \tag{2.8}$$

In this case, equation (2.7ii) can be written as

$$\frac{dM_G}{dz} = -g\frac{d\rho_G}{dC}(C_G(0) - C_e)\pi b_G^2 \tag{2.9}$$

However, in general, equation (2.7ii) should be left in integral form if the relationship between the density and concentration is non-linear. For ease of notation, the density β_G is defined by

$$\frac{dM_G}{dz} = \int_0^\infty g(\alpha - \rho_G(r))dA = g(\alpha - \beta_G)\pi b_G^2 \tag{2.10}$$

In many naturally convecting flows, the difference between the density of the buoyant fluid and that of the environment is much less than the density of the environment. Therefore, any changes in the density are small in comparison to some reference density. In this limit, the above equations can be simplified by neglecting the variations in the density of the environment and also of the buoyant plume, except in the term on the right-hand side of equation (2.7ii) which explicitly includes a density difference. This term represents the gravitational force resulting from the difference between the density of the ambient fluid and that of the plume, and is therefore responsible for the motion. This simplification is called the Boussinesq approximation (Turner 1979). Using this approximation, and the experimentally observed Gaussian distribution of the velocity and concentration across the plume (equation 2.5), equation (2.7) becomes

$$\frac{d(v_G(0)b_G^2)}{dz} = 2\lambda^2 k v_G(0) b_G \tag{2.11i}$$

$$\frac{d(v_G^2(0)b_G^2)}{dz} = \frac{2\lambda^2(\alpha - \beta_G)b_G^2 g}{\alpha_0} \tag{2.11ii}$$

$$\frac{d(v_G(0)b_G^2(C_G(0) - C_e))}{dz} = -v_G(0)b_G^2\frac{dC_e}{dz}\left(\frac{1+\lambda^2}{\lambda^2}\right) \tag{2.11iii}$$

where the centreline properties $v_G(0)$, $C_G(0)$ and b_G are now functions of the height z and α_o is a reference density. The last equation represents a combination of the mass and concentration equations, and identifies that the concentration anomaly of the plume only varies with height if the background concentration varies with height. This model, based upon the Gaussian distribution of the properties of the plume, can be recast into the top-hat model, which is a simpler bulk description of the plume. In the top-hat model, the plume is assigned one set of averaged properties at a fixed height, and the environment a second set of values. The top-hat velocity, u, radius, b, and concentration, C, are related to the mass, momentum and concentration fluxes as

$$Q = \alpha ub^2; \quad M = \alpha u^2b^2; \quad QC = \alpha ub^2C \qquad \text{(2.12i, ii, iii)}$$

where the density has been approximated with a constant background value. Note that, as is the convention, in the definitions (equation 2.12) and in the remainder of the chapter the factor π is not included, so that, for example, the actual mass flux in a plume would be πQ. The equations governing the top-hat properties can be derived by considering control volumes surrounding the plume, and equating the vertical fluxes into and out of the volume, with the entrainment from the ambient fluid (Morton *et al.* 1956). If the plume is again assumed to entrain at a rate proportional to its mean upward velocity, with the top-hat entrainment coefficient ε, then the top-hat properties are governed by the equations

$$\frac{\mathrm{d}ub^2}{\mathrm{d}z} = 2\varepsilon ub; \quad \frac{\mathrm{d}u^2b^2}{\mathrm{d}z} = \frac{g(\alpha - \beta)b^2}{\alpha_o}; \quad \frac{\mathrm{d}ub^2(C - C_e)}{\mathrm{d}z} = -ub^2\frac{\mathrm{d}C_e}{\mathrm{d}z} \qquad \text{(2.13i, ii, iii)}$$

The top-hat description of the plume is mathematically identical to the Gaussian description of the plume if λ is approximated to be 1 and the top-hat properties are defined in terms of the Gaussian properties according to the following relations:

$$u = \frac{v_G(0)}{2}; \quad b = \sqrt{2}b_G; \quad (C - C_e) = \frac{C_G(0) - C_e}{2} \qquad \text{(2.14i, ii, iii)}$$

where the top-hat entrainment constant, ε, is related to the Gaussian entrainment constant, k, according to

$$\varepsilon = \sqrt{2}k \qquad \text{(2.15)}$$

The top-hat entrainment coefficient has the value of about 0.09 and successfully reproduces experimental measurements. Henceforth the present discussion will focus on the top-hat model, remembering the rigorous relationship with the Gaussian model.

Referring back to section 2.2, it can be deduced that a jet has an entrainment coefficient of about $1/16 = 0.063$. This is considerably smaller than that of a buoyant plume and is a result of the decrease in the efficiency of the entrainment process in a forced momentum-driven jet in comparison to a purely buoyancy-driven plume.

2.4 LINEARLY MIXING PLUMES

The buoyancy can be expressed in terms of the difference in concentration between the plume and environment through the integral equation 2.10. As noted above, in the simplest situation the density depends linearly upon this difference equation 2.8, and so

$$\alpha - \beta = \Omega(C - C_c) \tag{2.16}$$

where Ω is a constant. By combining this linear law with the equation for the conservation of the buoyancy-generating conserved quantity C, equation (2.13iii), the classical equation for the conservation of buoyancy flux can be derived

$$\frac{d(ub^2(\beta - \alpha))}{dz} = -ub^2 \frac{d\alpha}{dz} \tag{2.17}$$

However, as mentioned before, the buoyancy is a derived rather than fundamental variable, and it is actually the quantity C which is conserved. In the more general situation in which the buoyancy varies non-linearly with variations in C, the buoyancy may not be conserved and can in fact change sign, even though C is conserved (Caulfield and Woods 1995). This distinction is of importance in section 2.12 and Chapter 4, where volcanic eruption columns in which the buoyancy can indeed change sign are considered.

In many situations, however, the density does depend linearly upon some conserved property. For example, if the buoyancy is generated by salinity, small variations in temperature, or even small particles in suspension, which only sediment out of the fluid on a much longer time scale than the time of ascent of fluid through the plume. The classical plume model therefore describes the motion of plumes rising from, for example, fires and chimney plumes. Even in many complex situations, including volcanic eruption columns, the density relationship may eventually asymptotically approach a linear mixing law as the density contrast becomes small. Therefore, the motion eventually approximates that given by the linear mixing law.

2.5 THE UNIFORM ENVIRONMENT

In a uniform environment, the motion of a simple buoyant plume is self-similar because the entrainment and mixing are independent of the size of the plume. Therefore, the plume equations may be solved exactly (Turner 1979), and the top-hat buoyancy,

$$B = g' = g\frac{(\alpha - \beta)}{\alpha_o}$$

radius b and upward velocity u can be expressed in terms of the buoyancy flux at the source, $F = QB/\alpha$, and the height above the source, z,

$$g' = \frac{5F}{6\varepsilon}\left(\frac{9\varepsilon F}{10}\right)^{-1/3} z^{-5/3}; \quad b = \frac{6\varepsilon}{5}z; \quad u = \frac{5}{6\varepsilon}\left(\frac{9\varepsilon F}{10}\right)^{1/3} z^{-1/3} \qquad \text{(2.18i, ii, iii)}$$

The velocity of the plume decreases with height and the width increases linearly with height. This similarity solution only applies if the mass and momentum flux at the source are zero, while the buoyancy flux remains finite. Hence, in this ideal (mathematical) solution the density difference at the source tends to infinity. In practice, the model therefore only applies some distance above the point source, when the model density difference is predicted to become finite.

In many real experiments, the initial mass and momentum fluxes are non-zero. For example, if fresh water is released into a tank of saline water. One exception arises when a buoyant plume is generated by a heating unit, which provides a heat flux, but no mass or momentum flux. Therefore, it is the practice to define an effective or virtual source for a real plume. This is located some distance below the actual source. The height of this virtual source may be defined in terms of the buoyancy flux issuing from the real source and either the mass flux or the momentum flux, using the above similarity solution. This lack of uniqueness in defining the location of the virtual source results if the flow conditions in the material issuing from the source do not exactly match the similarity solution.

For a general source, the equations for the conservation of mass and momentum can be combined to obtain the relationship, valid at any height in the plume,

$$Q^2 = \frac{8\varepsilon}{5F\alpha_o^{1/2}}(M^{5/2} + c) \tag{2.19}$$

where c is a constant that relates the mass and momentum fluxes at the source

$$c = \frac{5F\alpha_o^{1/2}Q_o^2}{8\varepsilon} - M_o^{5/2} \tag{2.20}$$

where subscript o denotes properties in the top-hat plume at the source. If $c = 0$, then the similarity solution presented above exactly describes the motion of the ensuing plume. In general, c is non-zero, and the plume behaviour is more complex. If the source from which the plume issues is narrower than that predicted by the similarity solution, for a given mass and buoyancy flux, then the plume has excess momentum flux, and is known as a *forced plume* ($c < 0$). Conversely, if the source is wider, then the plume has less momentum flux than that of the similarity solution, and is known as a *distributed plume* ($c > 0$). As the mass and momentum fluxes increase through entrainment of ambient and acceleration under gravity, the effects of any anomaly in the initial momentum flux decay and the plume rapidly approaches the natural self-similar plume.

In a forced plume, the distance over which the transition to the similarity plume occurs is known as the *jet length*. This represents the distance over which the momentum flux in the plume increases by an amount comparable to the initial momentum flux

$$h_j = \frac{M_o^{3/4}}{\alpha_o^{3/4}F^{1/2}} \tag{2.21}$$

In forced plumes which propagate much further than h_j, the effects of this transient behaviour are localized only near the source.

In the case of a distributed plume, the momentum flux issuing from the source is smaller than in the similarity solution. In this case, the appropriate length-scale over which the plume adjusts to the similarity solution is given by the mass flux issuing from the source. This length can be envisaged as the distance over which the flow in the distributed plume accelerates to a velocity commensurate with that of a similarity plume with the same mass and buoyancy flux as that issuing from the distributed source. This is referred to as the *acceleration length*, h_a, defined as

$$h_a = \frac{Q^{3/5}}{\alpha_o{}^{3/5} F^{1/5}} \tag{2.22}$$

Again, plumes which propagate much further than this length-scale are insensitive to the details of the initial conditions, and except in the initial transition regions, the motion depends primarily upon the source buoyancy flux.

According to the model, the radius of the plume, b, satisfies a relation of the form

$$\frac{\mathrm{d}b}{\mathrm{d}z} = 2\varepsilon - \frac{b(\alpha - \beta)g}{2u^2\alpha} \tag{2.23}$$

The system of equations admits a natural self-similar flow in which the plume spreads with constant angle θ where $\tan\theta = 6\varepsilon/5 = 0.11$ (equation 2.18ii). This prediction of a constant angle of spread is a result of the assumption of self-similarity, and is borne out by laboratory experiments (Morton *et al.* 1956). Measurement of the spreading angle provides an accurate means of estimating the entrainment coefficient in a simple plume. For contrast, a pure, neutrally buoyant jet with smaller entrainment coefficient of $\varepsilon = 0.062$ is predicted to spread with constant angle of $\tan\theta = 2\varepsilon = 0.125$. Therefore, a jet spreads somewhat more rapidly than a buoyant plume (Figure 2.3).

In the above discussion of buoyant plumes, it has been assumed that the entrainment coefficient does not vary as the relationship between the mass and momentum flux diverges from the self-similar solution. However, as the source conditions deviate from those of the similarity plume, so that $c \neq 0$, the entrainment coefficient does in fact change, as in the case of a jet. Although the scalings (equations 2.21 and 2.22) are correct, the detailed adjustment from either forced or distributed plumes to the corresponding similarity plume is not exactly modelled by the above theory. There have been a considerable number of experimental studies to measure the variation of the entrainment coefficient as

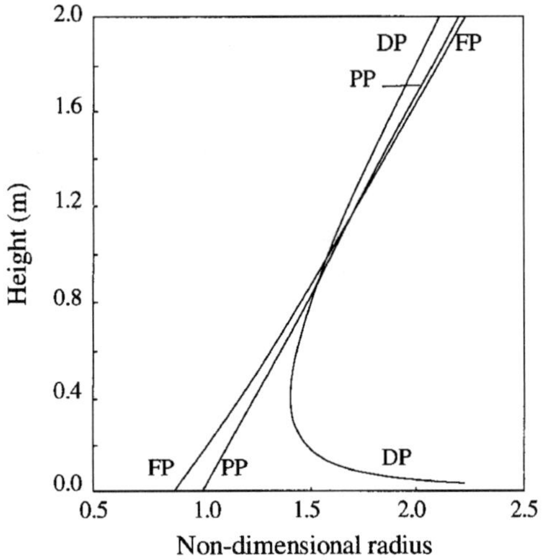

Figure 2.3 Comparison of the spreading of a jet (FP), pure plume (PP) and distributed mass source (DP) as a function of the distance above the source

the plume deviates from similarity form. In the most sophisticated study to date, Papanicolaou and List (1988) have shown that the entrainment coefficient has the value of about 0.056 for a pure jet, while the self-similar plume entrains more efficiently, with an entrainment coefficient of about 0.09.

Some empirical laws have been suggested to describe the variation in entrainment coefficient as the flow changes from jet-like flow to self-similar plume flow, by using the plume Froude number (Schatzmann 1979; List 1982),

$$Fr = \frac{\alpha u^2}{g(\beta - \alpha)b} \tag{2.24}$$

However, there is no one definitive entrainment law as a function of the Froude number which agrees with all the experimental data. Although the effects of the variation in the entrainment coefficient change the details of the transition to the similarity plume from a forced or distributed plume, they do not change the jet or mass length-scales given above.

2.6 THE STRATIFIED ENVIRONMENT

In a stably stratified environment, the density of the ambient fluid decreases with height. Therefore, the fluid in the plume eventually ceases to be buoyant with respect to the surrounding fluid. Since it has entrained the relatively dense fluid from the region just above the source, the plume actually ceases to be buoyant a significant distance below the height at which the environmental density equals the density of the fluid emitted from the source. When the plume reaches the height of neutral buoyancy, it continues rising upwards because of its inertia. However, above this height, it decelerates under gravity and eventually comes to rest. The fluid then falls back under gravity, and spreads laterally, forming a lateral gravitational intrusion in the environment.

Before considering extensions of the plume model described above, it is instructive to use dimensional arguments to deduce the height of the plume. The height of rise of the plume depends firstly upon the buoyancy flux, F, associated with the difference in concentration between the fluid issuing from the vent and the ambient fluid, and secondly upon the density gradient in the environment. The natural time scale associated with the density gradient in the environment is given as the inverse of the Brunt–Väisälä frequency, N, which is defined by the relation

$$N^2 = -\frac{g}{\alpha_o}\frac{d\alpha}{dz} \tag{2.25}$$

Therefore, the height of ascent for the plume is

$$H = 5F^{1/4}N^{-3/4} \tag{2.26}$$

In this relation, used extensively in this book, total buoyancy flux is πF. The constant, of value five in this simple dimensional relationship, is based upon Briggs (1969). Briggs measured the height of rise of a number of different naturally occurring plumes, both in the laboratory and in nature. The result is accurate for motions over a very wide range of scales ranging from centimetres to thousands of metres (Figure 2.4). Note that in the atmosphere, the density gradient used in equation (2.25) should be based on the potential density, i.e. the density relative to the adiabatic value (Gill 1982).

To determine more about the structure of a buoyant plume in a stratified environment, the model presented in the previous section is solved numerically using a uniform value for the stratification, N. Figure 2.5 shows the numerical solution of the equations, as first presented by Morton *et al.* (1956). The solution is plotted here in terms of dimensionless values of the height, radius, velocity and buoyancy defined as

$$z^* = zF^{-1/4}N^{3/4}\varepsilon^{1/2}; \quad b^* = bF^{-1/4}N^{3/4}\varepsilon^{-1/2}$$
$$u^* = uF^{-1/4}N^{-1/4}\varepsilon^{1/2}; \quad B^* = BF^{-1/4}N^{-5/4}\varepsilon^{1/2} \qquad (2.27\text{i, ii, iii, iv})$$

Note that these dimensionless variables are numerically different from those used by Morton *et al.* (1956) who worked with Gaussian plume properties. The graph shows how the velocity of the plume decreases steadily from the source towards zero at the plume top, while the buoyancy of the plume decreases to zero at the neutral height, and then becomes negative until the plume comes to rest.

The radius of the plume increases, almost linearly, until just below the neutral buoyancy height, at which point it begins to increase much more rapidly. This is because the plume behaves almost as if it were in an unstratified environment until just below the neutral height. At this point the buoyancy of the plume then becomes very small, and in fact becomes negative, so that the behaviour diverges significantly from the similarity solution described earlier (equation 2.18). Although formally the model ceases to hold above the neutral height, as the plume diverges from the similarity form, the equations may be integrated up to the point of no motion and, in comparison with experiment, they predict

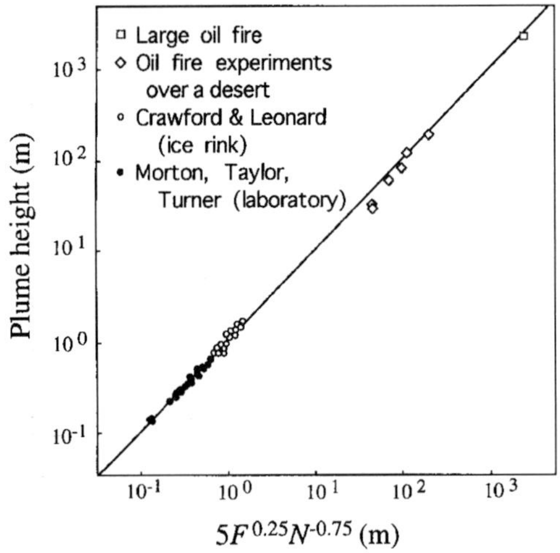

Figure 2.4 Comparison of the observed height of ascent of a number of plumes ranging in scale from the laboratory to large desert fires with the simple scaling law (equation 2.26). Data are from Briggs (1969)

the height of rise of the plume very accurately. Alternatively, the total height of rise may be estimated from the dimensional arguments leading to equation (2.26), in which the constant of proportionality has been determined from laboratory experiments.

As the velocities are very small, the mass entrained above the neutral buoyancy height does not change the buoyancy of the plume significantly, and so any variations in the entrainment constant do not have a large impact upon the accuracy of the model. Some workers have attempted to modify the model in the region above the neutral buoyancy height by introducing more complex entrainment laws (Schatzmann 1979). However, there is no real improvement in the model predictions in comparison with experiment, and such models require the introduction of extra empirical parameters. Furthermore, the plume will form a gravity current intrusion above the level of neutral buoyancy so that the motion and shape of the cloud in this region is not appropriately described by the plume model.

The numerical solution presented in Figure 2.5 represents the plume that develops above a source of finite buoyancy flux, with zero mass and momentum fluxes. This is referred to as the *pure plume solution*, and is analogous to the similarity solution presented in the previous section. In fact, all stratified plumes converge towards this solution as they propagate away from the source. Therefore, they tend to lose memory of any discrepancies between the actual source momentum flux and that associated with the fundamental plume solution at the point where its mass and buoyancy fluxes equal those issuing from the actual source. The motion of a stratified plume is similar to that of an unstratified similarity plume in the region below the neutral height. Therefore, if the neutral buoyancy height of the plume exceeds the jet or acceleration length-scale (equations 2.21 and 2.22), then, as in section 2.3, the length over which any anomalies in the initial momentum flux decay is

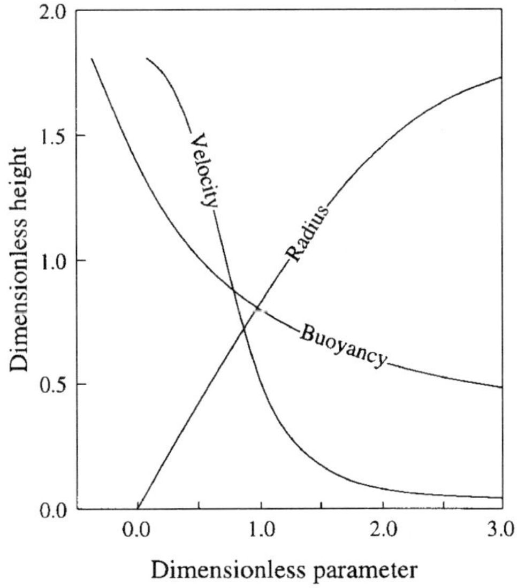

Figure 2.5 Numerical solution of the equations governing the ascent of a pure plume in a linearly stratified environment. The three curves, as labelled, show the variation of the dimensionless radius, buoyancy and velocity as a function of dimensionless height in the plume. The dimensionless plume properties are related to the dimensional values according to equations (2.27i–iv)

given by the jet or acceleration length. In most naturally occurring plumes, any discrepancy in the initial momentum flux typically has a very small jet or mass length compared to the total height of rise of the plume. Therefore, the initial adjustment as the plume converges towards the fundamental plume solution does not influence the subsequent behaviour of the plume significantly (Morton 1959).

If the material issuing from the source has excess momentum flux, then there is a transition region in which this excess momentum flux decays and the plume approaches the pure plume solution. The jet length is small compared to total height of the plume if

$$M_o \ll \frac{F\alpha}{N} \tag{2.28}$$

where M_o is the initial momentum flux. In this case, the transition length-scale is the jet length. In contrast, if the flow issuing from the source has a deficit of momentum flux, then there is a transition region over which the flow accelerates. If

$$Q_o \ll F^{3/4}N^{-5/4}\alpha \tag{2.29}$$

then the acceleration length (equation 2.21) is much smaller than the total height of the plume, and the transition length-scales as the acceleration length.

In the atmosphere, the stratification is given by $N \approx 10^{-2}$ s^{-1}, while in a volcanic eruption column, the buoyancy associated with the thermal energy erupted from the vent is comparable to the gravitational acceleration, 10 m s^{-2} (Chapter 4). A rather simplified calculation, based upon equation (2.28) shows that in order for the jet length associated with the initial momentum of the erupted material to be comparable to the ascent height of the column, the eruption velocities would need to exceed 1000 m s^{-1}. Further, in order for the acceleration length, associated with the initial mass flux, to be comparable to the ascent height of the column, the initial volume flux would need to exceed about 10^{11} m^3 s^{-1} (equation 2.29). These are well beyond the range of realistic source conditions considered in Chapter 3.

Modelling the motion of the material as it intrudes laterally into the environment at the top of the plume forms the subject of Chapter 11.

2.7 TIME-DEPENDENT BUOYANCY FLUXES

In many situations the buoyancy flux issuing from the source may be continuous, but of variable strength. If the time scale of fluctuations in the buoyancy flux is much longer than the time scale of ascent of material in the plume, then the plume behaves essentially as a quasi-steady structure. At any time the properties of this quasi-steady plume scale with the buoyancy flux issuing from the source at that time. The time required for material to ascend to the top of the plume is of the order of $1/N$, the time scale associated with the stratification. In order that the steady-state plume model may be applied to a source of variable strength, the source should only vary on time scales greater than $1/N$. This condition may be expressed in the form

$$\frac{dF}{dt} \ll FN \tag{2.30}$$

If this condition holds then the height of the plume, to leading order, is

$$H(t) = 5F(t)^{1/4}N^{-3/4} \tag{2.31}$$

In the atmosphere, for example, $1/N$ is of the order of 100–200 s, and so to apply steady plume theory the source must remain constant for times of the order of 10 minutes. Otherwise, a steady plume does not become established, and a different model is required. In the limiting case in which the buoyancy is released almost instantaneously relative to the ascent time of the material, then the buoyant material will detach from the source and form a discrete thermal plume. In the next section models of discrete thermal plumes are discussed.

2.8 DISCRETE THERMALS

Discrete thermal plumes develop when the source of buoyant fluid is released from the source instantaneously, or in practice, sufficiently rapidly that the time of release from the source is much shorter than the ascent time to the neutral buoyancy height. When a discrete convecting thermal develops, the flow is highly turbulent and ambient fluid is mixed into the thermal by means of turbulent eddies (Figure 2.1b). Thermals typically take on the structure of a ring vortex (Turner 1979), and so the intensity of the entrainment varies with location on the surface. However, once entrained, the environmental fluid is rapidly mixed throughout the flow. As with plumes, thermals behave self-similarly in a uniform environment because the efficiency of the entrainment is independent of the size and molecular properties of the thermal, and depends only upon the density deficiency of the thermal plume. In his discussion of thermals, Turner (1979) includes a description of the motion of thermals using a simple vortex ring model, in which there is upflow in the centre of the ring, surrounded by a downflow on the periphery.

It is also possible to model the bulk motion of a buoyant thermal plume, in terms of the conservation of mass, momentum and the particular buoyancy generating property, C, such as the salinity or enthalpy. Following Morton *et al.* (1956) and Turner (1979), these three conservation equations have the top-hat, Boussinesq form

$$\frac{\mathrm{d}b_\mathrm{T}^3}{\mathrm{d}z} = 3\varepsilon_\mathrm{T} b_\mathrm{T}^2 \tag{2.32i}$$

$$3\frac{\mathrm{d}(b_\mathrm{T}^3 w)}{\mathrm{d}z} = 2b_\mathrm{T}^3 g\frac{(\alpha - \beta_\mathrm{T})}{\alpha_\mathrm{o} w} \tag{2.32ii}$$

$$\frac{\mathrm{d}\big(b_\mathrm{T}^3(C_\mathrm{T} - C_\mathrm{e})\big)}{\mathrm{d}z} = -b_\mathrm{T}^3\frac{\mathrm{d}C_\mathrm{e}}{\mathrm{d}z} \tag{2.32iii}$$

where b_T is the plume radius, w the upward velocity and other properties are as defined for the top-hat model of the plume. These equations have been evaluated in a frame of reference moving with the thermal. In this frame, the temporal derivatives are equal to the upward velocity multiplied by the vertical gradient of the averaged thermal properties. The model assumes that the shape of the thermal is approximately spherical, and that ambient fluid is entrained at a rate proportional to the surface area of the sphere and the upward velocity of the sphere. The approximation that the changes in the density of the fluid are small in comparison to the background density have also been incorporated, so that only the buoyancy term associated with the acceleration of the thermal need be retained.

In deriving the momentum equation, account has been taken of the motion of the air around the thermal. The air ahead of the thermal must be accelerated around to the rear of the thermal plume. The effect of this displacement of the environmental air around the thermal is equivalent to increasing the mass of the thermal with a mass of air of volume equal to one-half of the thermal (Batchelor 1967). In contrast to the model of the buoyant plume, in which properties were integrated over a control volume fixed in space, the above thermal model actually follows the buoyant fluid as it ascends through space. The entrainment coefficient for thermals has been measured experimentally to be about $\varepsilon_T = 0.25$ (Morton *et al.* 1956). The much larger entrainment coefficient associated with a thermal results from the ring vortex-like structure of the flow, which sweeps fluid into the thermal. This behaviour contrasts with the highly turbulent, but less organized, eddies in a plume, which must engulf the fluid beside the plume.

In an unstratified environment, with a linear dependence of the density upon concentration, the product of the buoyancy and volume of the thermal is constant

$$B_o b_o^3 = g\frac{(\alpha - \beta)}{\alpha_o} b_T^3$$

(see equation 2.32iii) and the equations (2.32) have the similarity solution

$$b_T = \varepsilon_T z \qquad (2.33\text{i})$$

$$w = \left(\frac{B_o b_o^3}{3\varepsilon_T^3}\right)^{1/2} \frac{1}{z} \qquad (2.33\text{ii})$$

$$B_T = g\frac{(\alpha - \beta)}{\alpha_o} = \frac{B_o b_o^3}{(\varepsilon_T z)^3} \qquad (2.33\text{iii})$$

As with the similarity plume, the radius and velocity of the thermal increases linearly with height. In the general situation that the mass or momentum of the thermal is non-zero at the source, the mass and momentum of the thermal are coupled by the simple algebraic relation

$$Q_T^{4/3} = \frac{3\varepsilon}{B_o b_o^3 \alpha^{2/3}}(M_T^2 + c_T) \qquad (2.34)$$

where the constant c_T is evaluated at the source,

$$c_T = \frac{B_o b_o^3 \alpha^{2/3} Q_o^{4/3}}{3\varepsilon} - M_o^2 \qquad (2.35)$$

If the constant c_T is negative, then the thermal initially has more momentum than the self-similar thermal. Conversely, if c_T is positive, the thermal initially has less momentum than the similarity thermal. The latter situation is more common and arises when, for example, a given mass of buoyant fluid is released from rest.

In a stratified environment, a buoyant thermal ascends beyond its neutral height, and comes to rest as the environment becomes less dense. The thermal then oscillates about the neutral height, with a period of oscillation of order $1/N$, until finally coming to rest and intruding into the environment (Morton *et al.* 1956). The height of rise only depends upon the product of the initial volume and buoyancy of the thermal, $B_o b_o^3$, and the stratification of the environment,

$$H = 2.7(B_o b_o^3)^{1/4} N^{-1/2} \tag{2.36}$$

Detailed numerical solutions of the model equations (2.32) for the ascent of a thermal in a uniformly stratified environment, assuming a linear relationship between the density and concentration, are presented in Figure 2.6. In the figure, the dimensionless radius, r^*, velocity, w^* and buoyancy, B^* are shown as functions of dimensionless height, z^*, defined in terms of $F_0 = B_0 b_0^3$ by

$$r^* = \varepsilon^{-1/4} F_0^{-1/4} N^{1/2} r_T; \quad B^* = \varepsilon^{3/4} F_0^{-1/4} N^{-3/2} B_T$$
$$z^* = \varepsilon^{3/4} F_0^{-1/4} N^{1/2} z; \quad w^* = \varepsilon^{3/4} F_0^{-1/4} N^{-1/2} w \qquad (2.37\text{i, ii, iii, iv})$$

Below the neutral height, the motion of the thermal is quite similar to that in an unstratified environment, and the thermal accelerates upwards from the source. However, just below the neutral height, the motion is influenced by the stratification, and the velocity decreases

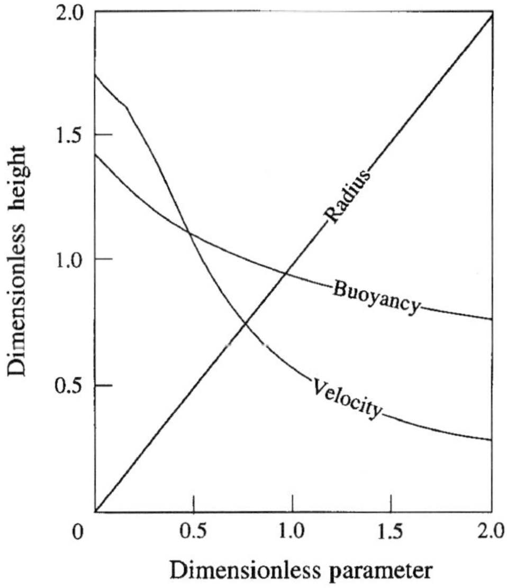

Figure 2.6 Numerical solution of the equations governing the ascent of a thermal emitted from a point source into a linearly stratified environment. The three curves, as labelled, show the variation of the dimensionless radius, buoyancy and velocity as a function of dimensionless height of the centre of the thermal. The dimensionless properties are related to the dimensional values according to equations (2.37i–iv)

towards zero. As a result, the thermal entrains and grows much more slowly. The shape of the velocity profile is quite different from that of a plume, shown in Figure 2.5, in which the velocity decays monotonically.

2.9 STARTING PLUMES

When a plume first rises from a maintained source, a complex thermal-like structure develops at the top of the ascending flow (Figure 2.1c). Behind this structure the motion rapidly adjusts to that of the maintained plume described in section 2.3. Experimental observation suggests that the motion of the whole structure is self-similar (Turner 1979). The thermal is continually supplied with buoyancy from the plume below, and it rises with a velocity equal to about 0.6 times the centreline velocity of the plume (measured using the Gaussian velocity distribution). It is the supply of buoyancy from below which enables the thermal to ascend with the plume. Experimental observations suggest that the thermal at the head of the starting plume spreads with a half angle θ, where $\tan\theta = 0.18$. This contrasts to the similarity thermal, of constant buoyancy, which spreads with a half-angle θ_T where $\tan\theta_T = 0.3$. Further details describing both experiments and models of starting plumes are given by Turner (1962).

The results for this case are particularly significant for volcanic plumes, because most of the data on volcanic plume ascent come from observations of the initial few minutes of the plume development (Chapter 5).

2.10 LINE SOURCES

The geometry of the source can exert an important influence upon the motion of the plume. Thus far, only axisymmetric sources have been considered, and it has been shown that over lengths greater than the jet or acceleration length, the motion approximates that from a point source. However, a more fundamental change in the style of behaviour arises if the source is linear. In this case, a two-dimensional convecting plume develops, and ambient fluid is now entrained from each side of the plume.

Following the simple axisymmetric plume model, a simple top-hat model for line plumes may be derived in which the conservation of mass, momentum and concentration is given by

$$\frac{db_l u_l}{dz} = \varepsilon_l u_l; \quad \frac{db_l u_l^2}{dz} = \frac{g(\alpha - \beta_l)b_l}{\alpha_o}; \quad \frac{db_l u_l(C_l - C_e)}{dz} = b_l u_l \frac{dC_e}{dz} \tag{2.38i, ii, iii}$$

In an unstratified environment, with a linear dependence of the density upon concentration, the line plume has a similarity solution

$$b_l = \varepsilon_l z; \qquad u_l = \left(\frac{F_l}{\varepsilon_l}\right)^{1/3} \tag{2.39}$$

where F_l is the buoyancy flux per unit length, $F_l = Qg'/\alpha$, where Q is the mass flux per unit length and

$$\beta_l = \alpha + \frac{d\rho}{dz}(C_l - C_e)$$

The line plume is also self-similar, and spreads linearly with height. In this self-similar solution, the velocity is constant. The entrainment coefficient is similar to that for an axisymmetric plume, with a value of about $\varepsilon_l = 0.1$ (Linden and Simpson 1990).

In general, the mass and momentum fluxes are related by the simple algebraic relation

$$Q_l^3 = \frac{\varepsilon_l}{F_l}(M_l^3 + c_l) \tag{2.40i}$$

where c_l is a constant defined as

$$c_l = \frac{F_l Q_o^3}{\varepsilon_l} - M_o^3 \tag{2.40ii}$$

As with axisymmetric plumes, even if the initial momentum flux differs from that of the similarity solution, the plume rapidly approaches the similarity solution. Linden and Simpson (1990) found that the entrainment coefficient decreases in forced plumes, and they have estimated that the entrainment decreases to values of about 0.08. Such effects modify equation (2.40), which was derived assuming that the entrainment is fixed. However, equation (2.40) allows one to develop simple scalings for the jet and mass lengths, as illustrated for axisymmetric plumes in section 2.5.

In a stratified environment, the line plume will ascend to a height which may again be found using dimensional arguments based solely upon the buoyancy flux per unit length, F_l, and the ambient stratification. Using experimental measurements of the height (Briggs 1969) to determine the constant of proportionality, the height is given by

$$H = 0.7F_l^{1/3}N^{-1} \tag{2.41}$$

The detailed structure of a line plume rising in a stratified environment is similar to that of an axisymmetric plume. In Chapter 10 this model is extended to model the ascent of fine ash and volatiles above fire fountains issuing from linear fissures.

If a linear source is of finite length, or is elliptical in shape, then the plume tends to become more axisymmetric with height. When the plume has entrained sufficient ambient fluid that the width of the plume equals the greater length of the source, then it behaves as an axisymmetric plume.

2.11 NEGATIVELY BUOYANT JETS

If the material issuing from a source is heavier than the environment, but is driven upwards by its buoyancy, then a *negatively buoyant jet* forms. Negatively buoyant jets rise to a maximum height at which point the buoyancy forces have decelerated the flow to rest. The fluid then collapses back and forms a fountain around the region of upflow. The initial height of rise of a negatively buoyant jet was studied by Turner (1966). Using dimensional arguments, he argued that the height of rise of the jet depends only upon the initial buoyancy flux, F, and momentum flux, M, and therefore has the form

$$H = \frac{1.85M^{3/4}}{F^{1/2}\alpha^{3/4}} \tag{2.42}$$

Once the jet has collapsed, however, the upflow in the centre entrains the relatively dense fluid fountaining around its outer edge, and therefore does not ascend as high, while the fluid fountaining down around the central upflow region entrains some of the ambient fluid. Thus, once the jet has begun to shed fluid which sinks around its outer edge, the simple plume-type model is no longer appropriate. Baines *et al.* (1990) have proposed a model for this flow, based upon two entrainment laws. However, more investigation into the problem is required in order to develop a full understanding of the flow.

2.12 PLUMES WITH NON-LINEAR DENSITY MIXING PROPERTIES

Although the linear relationship between the density and the concentration C of the conserved buoyancy-generating property applies in many situations, there are a number of important cases in which a more complex constitutive relation for the density is required to describe the behaviour of the plume. Most examples of such cases arise when the density depends upon two or more conserved properties, one of which has a non-linear relationship between the density and the concentration.

The primary example of interest in this book arises with dense mixtures of hot ash, water vapour and air. Initially, such a mixture can be very dense relative to the environment. However, as the mixture entrains and heats air, and simultaneously sediments some of the dense ash, the mixture may become buoyant. The influence of the non-linearity in this system will be a central and recurring theme throughout this book. A second example concerns very hot hydrothermal water venting from black smoker plumes at mid-ocean ridges into the deep sea. If the hydrothermal fluid is more saline than the ambient seawater, the difference in density between the plume water and the ambient does not vary monotonically (Turner and Campbell 1987b). Initially, the very hot effluent (>300 °C) is relatively buoyant owing to the large thermal expansion coefficient at high temperature. However, as the plume entrains ambient and cools, the thermal expansion coefficient decreases, the relative density of the plume becomes dominated by the excess salinity and the plume ceases to be buoyant.

In the laboratory, mixtures of methanol and ethylene glycol (MEG) can be used to model this process of reversing buoyancy (Turner 1966). A pure mixture of MEG, with a ratio of about 50 M : 50 EG is less dense than water. However, as it mixes with water, the density of the MEG increases, and becomes greater than the water. Plumes of MEG may therefore be used to model several phenomena particular to plumes with a non-linear constitutive relation for the density. Turner (1966) carried out experiments in which an initially buoyant plume became dense through mixing with the environment, and Woods and Caulfield (1992) have carried out experiments in which the dense jets become buoyant through mixing with the environment. Woods and Caulfield developed a simple analytical model to describe the motion of such MEG plumes by parameterizing the density difference between the MEG/water mixture and water to be a quadratic function of the mass fraction of MEG in the mixture. The density of the mixture relative to the environment is given by

$$\frac{\beta-\alpha}{\beta_{\mathrm{m}}-\alpha}=1-\left(1-\frac{m}{m_{\mathrm{m}}}\right)^{2} \tag{2.43}$$

where m is the mass fraction of MEG in the plume, and m_{m} is the mass fraction when the mixture has its maximum negative buoyancy and density β_{m}. If the initial mass fraction of MEG exceeds the value $2m_{\mathrm{m}}$, then the plume is initially less dense than water, but on mixing with the water it becomes dense. If the initial mass fraction of MEG is less than $2m_{\mathrm{m}}$, then the plume is always dense relative to the water. Using the constitutive relation in equation (2.43), the equations for the conservation of mass and momentum fluxes may be combined to obtain the relation

$$\frac{m_{\mathrm{o}}}{m_{\mathrm{m}}}(Q_{\mathrm{d}}-1)\left(Q_{\mathrm{d}}+1-\frac{m_{\mathrm{o}}}{m_{\mathrm{m}}}\right)=\pm\frac{4\varepsilon}{5G}(M_{\mathrm{d}}^{5/2}-1) \tag{2.44}$$

where Q_{d} and M_{d} are dimensionless mass and momentum fluxes for the jet. This relation captures all the different styles of behaviour which may be exhibited by such plumes of reversing buoyancy. The expression G is a dimensionless constant giving a measure of the initial negative buoyancy flux compared to the initial momentum flux (Woods and Caulfield 1992). If the plume issues downwards into a tank of water then the positive sign is taken, while if the plume issues upwards then the negative sign is taken. Figure 2.7 illustrates how the momentum flux and relative density vary as the mass flux in the plume increases.

For downward-propagating plumes, three situations can arise. In the first case, the plume beomes dense through mixing and continues downwards as a simple buoyant plume. In the second case, the plume comes to rest before it has mixed sufficiently to become dense, and therefore only a small fountain develops. If the initial momentum flux is sufficiently large, then the jet can entrain sufficient environmental fluid to become dense and form a plume

$$M_{\mathrm{d}}^{5/2}>-a\delta\rho\big((1-\delta\rho)^{1/2}-1\big)Q_{\mathrm{d}}^{3} \tag{2.45}$$

where a is a constant and $\delta\rho$ is the initial density of the MEG–water mixture relative to the water. The theoretical criterion in equation (2.45) was successfully compared with laboratory experiments by Woods and Caulfield (1992) (Figure 2.8). In the third case, the plume starts off dense, and therefore remains dense.

For upward-propagating plumes, two different cases arise. In the first, the plume starts off being dense, and therefore only ascends a short distance, essentially as a forced jet, before coming to rest and forming a fountain. In the second case, the plume starts off rising buoyantly, but as it mixes, it becomes dense, and therefore again forms a collapsing fountain. Turner (1966) described some experimental observations in which the plume top oscillates about a mean height, and Baines *et al.* (1990) have presented a model for the re-entrainment of the collapsing fountain into the central plume.

In Chapter 4, it is shown that, as with the jet and acceleration lengths, in most geophysical situations, the length-scale over which the buoyancy reversal occurs is much shorter than the total ascent height of the plume, if the material does become buoyant.

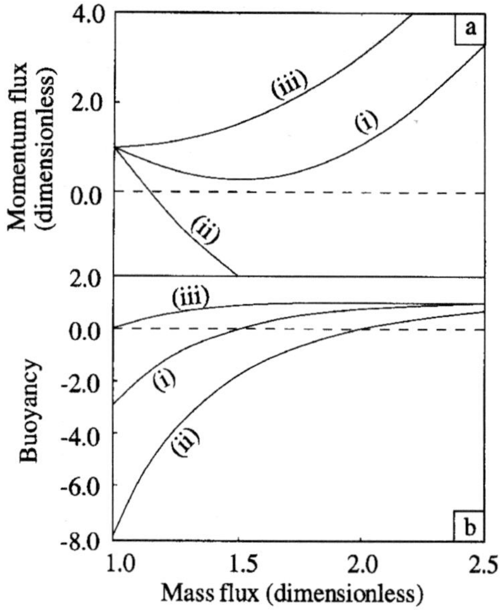

Figure 2.7 Variation of (a) the momentum flux and (b) the buoyancy as a function of the mass flux for the case of a fluid whose density depends non-linearly upon the mass of entrained fluid (section 2.12). In case (i) a buoyant plume develops from an initially dense source; in case (ii) the initially dense source leads to a collapsing fountain; and in case (iii) a neutrally buoyant source leads to a buoyant plume (calculations from Woods and Caulfield 1992)

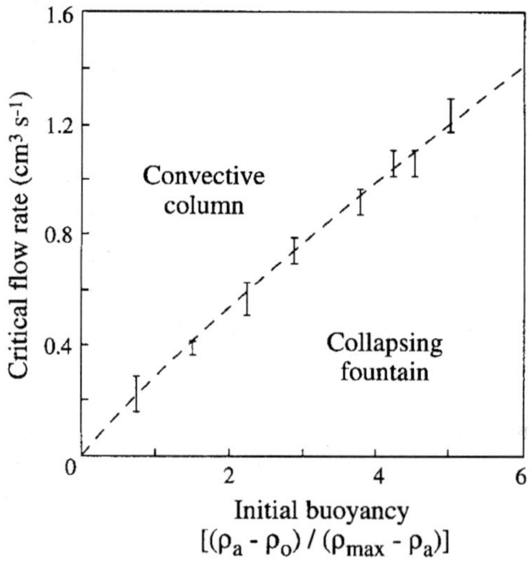

Figure 2.8 Figure illustrating the critical flow rate for the collapse of an initially dense jet which can become buoyant through entrainment, equation (2.45), as a function of the initial density of the jet. The dotted curve represents the theoretical model, and the experimental observations of Woods and Caulfield (1992) are shown as vertical bars

Therefore, the details of the non-linearity are crucial in determining whether or not a buoyant plume may develop. However, they do not affect any subsequent larger scale motion if a buoyant plume develops (Caulfield and Woods 1995). The models and ideas presented herein can then be applied to the modelling of volcanic eruption columns. An important aspect of these models is that the density varies non-linearly, and so many of the ideas of this section are pertinent to understanding the motion of eruption columns.

2.13 SUMMARY

In this chapter, the fundamental fluid dynamics of buoyant plumes, buoyant thermals and jets have been reviewed. The strength of a maintained plume is determined by the buoyancy flux. Motion of an instantaneously released thermal is determined by the initial buoyancy. Jet strength is determined by the momentum flux. Turbulent entrainment of the surrounding fluid is a fundamental feature and has a major influence on the geometry and motion of plumes, thermals and jets. Plumes, thermals and jets engulf surrounding fluid by the formation of large eddies which scale as the half-width of the flow. Instantaneous velocities and flow widths are highly variable due to the turbulence and passage of individual eddies. However, the time-averaged properties can be predicted as functions of position by simple analytical models. Properties such as velocity, particle concentration and temperature show time-averaged Gaussian distributions across the flow.

Pure plumes and jets are self-similar in behaviour, but real flows have finite mass and momentum fluxes as the material issues from the source which do not match the condition of self-similarity. As a consequence the flows have an initial region where they adjust to the self-similar motion of a plume or jet. Also flows issuing from a finite source start entraining ambient fluid in a shear layer at the edge and the flows must reach a height which scales as the characteristic source dimension (the width or radius) for entrainment to penetrate. Plumes and thermals which start with excess momentum are described as *forced*. Those that start out with a deficiency of momentum are known as *distributed*.

Modelling of plume and thermal motion was extended to include the effects of a density stratified environment. Stratification limits the vertical extent of a plume or thermal, because turbulent entrainment transports dense ambient fluid upwards into lower density regions, thus limiting the height to which the plume can ascend. A dimensional argument and model calculation to predict this limiting height compare well with laboratory and field observations of buoyant plumes. The models can be modified to account for a two-dimensional rather than axisymmetric geometry.

Negatively buoyant jets occur when the fluid is denser than the surroundings. These flows ascend under their momentum and then collapse to form a fountain and generate gravity currents. More complex behaviour occurs in systems with non-linear mixing properties. The case where the density of the plume fluid relative to the ambient fluid changes sign as the plume mixes with ambient fluid can produce complex oscillating behaviour between buoyant plume ascent and collapsing fountains. This case represents a simple analogue of an eruption column, in which a relatively dense mixture is ejected from a volcanic vent at high speed. As this jet entrains and heats air, the bulk density falls and can eventually become smaller than the density of the air. At this point, the material is driven upwards by its buoyancy. However, if insufficient air is entrained the flow remains dense and a fountain forms.

In Chapter 4, the models presented in this chapter are extended to examine eruption column dynamics in more detail. First, however, in Chapter 3 the source conditions of the dense jets which issue from volcanic vents and supply eruption columns with a fragmented mixture of tephra and gas are described.

3 Source Conditions in Explosive Volcanic Eruptions

NOTATION

a	speed of sound for Vulcanian eruption ($m\ s^{-1}$)
A	cross-sectional area of conduit (m^2)
c	characteristic speed for two-phase mixture ($m\ s^{-1}$)
C_D	drag coefficient (dimensionless)
C_p	specific heat at constant pressure ($J\ kg^{-1}\ K^{-1}$)
C_v	specific heat at constant volume ($J\ kg^{-1}\ K^{-1}$)
d	clast size (m)
F_R	frictional force per unit volume exerted by the conduit walls on the ascending material ($kg\ m^{-2}\ s^{-2}$)
f	turbulent friction coefficient (dimensionless)
g	gravitational acceleration ($m\ s^{-2}$)
K	permeability of conduit walls (m^2)
L	length-scale of conduit (m)
L_f	length-scale of water flow in fractures (m)
m	mass fraction of lithics in erupting mixture (dimensionless)
n	mass fraction of volatiles exsolved from magma (dimensionless)
n_s	mass fraction of volatiles in solution in magma (dimensionless)
n_o	initial gas mass fraction of magma (dimensionless)
P	pressure in conduit (Pa)
P_1	pressure before adjustment across the shock front (Pa)
P_2	pressure after adjustment across the shock front (Pa)
P_∞	pressure in far field of fracture network (Pa)
Q	mass flux in conduit ($kg\ s^{-1}$)
Q_T	heat transfer per unit mass per unit length of the conduit ($J\ kg^{-1}\ m^{-1}$)
r	radius of conduit (m)
R	gas constant for material in conduit ($J\ kg^{-1}\ K^{-1}$)
s	solubility constant for Henry's law ($Pa^{-1/2}$)
T	temperature of material in conduit, measured in kelvin in the conservation laws and equation of state of the volatiles (K)
T_w	temperature of conduit walls, measured in kelvin (K)
t	time (s)
u	velocity in conduit ($m\ s^{-1}$)
u_c	speed of sound of two-phase mixture ($m\ s^{-1}$)

v free-fall velocity of a clast (m s^{-1})
z distance along conduit (m)
α density of the gas phase (kg m^{-3})
β exponent in Henry's law (dimensionless)
η viscosity of magma (kg m^{-1} s^{-1})
κ thermal diffusivity (m^2 s^{-1})
ϕ void fraction (dimensionless)
ρ bulk density of mixture (kg m^{-3})
σ density of solid material (kg m^{-3})
τ_d time for dynamic equilibration (s)
τ_t time for thermal equilibration (s)
μ_ω viscosity of the volatiles (kg m^{-1} s^{-1})

Subscripts

1 property upstream of shock
2 property downstream of shock
e property at vent
L property of lithics
m property of magma
o initial conditions in magma chamber
w property of water

3.1 INTRODUCTION

In this chapter the processes by which magma ascends from a magma chamber and along a conduit to the vent are discussed quantitatively. This flow results in the eruption of a mixture of pyroclastic particles and gas which then develops into a quasi-steady volcanic plume. Vulcanian explosion dynamics from a lava dome or plugged vent which lead to more instantaneous thermal plumes are also considered. The processes within magma chambers, volcanic conduits and craters determine the properties of the multiphase mixture that emerges from the volcano and the source conditions for plumes.

Eruptive behaviour is strongly dependent on magma viscosity, so for the purpose of discussion low-viscosity (basaltic) systems and high-viscosity (intermediate to silicic) systems are distinguished, while recognizing that there is in reality a continuum in nature. As magma ascends it decompresses and eventually the gas dissolved in the melt becomes supersaturated. Volatiles are then exsolved and the erupting material develops into a magmatic foam. In basaltic eruptions of relatively low-viscosity magma, gas bubbles can move relative to the magma (Vergniolle and Jaupart 1986), and this can lead to bubble merging and separated two-phase flow. Such eruptions exhibit a rich range of dynamical behaviours, including the formation of fire fountains and extensive lava flows. Plumes can be formed in basaltic eruptions, and these are discussed in detail in Chapter 10. In high-viscosity magmas there is negligible flow of the bubbles relative to the liquid and eruptions are generally relatively explosive. This chapter is therefore primarily concerned with explosive eruptions of viscous magmas which are the major source of volcanic plumes.

Provided the conduit walls and degassing magma are sufficiently impermeable, the gas remains within the conduit. As the magma ascends and decompresses it becomes supersaturated and vesiculates. When the gas volume fraction has increased to about 0.4, the mixture forms a foam in which interactions between bubbles become important. Owing to the exsolution of the volatiles, the residual magma becomes much more viscous, thereby continuing to inhibit relative movement of gas and magma. When the volume fraction of gas has increased to a value typically between 0.6 and 0.8 the ascending mixture fragments into a particle-laden gas (Sparks 1978a; Wilson *et al.* 1980; Wilson and Houghton 1990; Thomas *et al.* 1994). In a steady eruption, this transition typically occurs within several hundred metres of the surface and a hot, dusty gas and particle mixture erupts from the vent at the speed of sound. At the vent the material can continue to decompress into a crater and thereby become supersonic. However, if the crater is sufficiently large, then a shock wave can develop, leading to a substantial decrease in the final velocity of the material issuing from the crater. A fluid dynamical model of the steady explosive eruption of silicic magma, which can feed a maintained eruption column is described. The factors that control the discharge rate of magma – exit velocity, tem-perature, bulk density and exit pressure conditions in explosive eruptions – are identified. The problem of determining whether magma erupts explosively or as lava is also con-sidered. Discrete Vulcanian-style eruptions which can generate ash-laden thermal plumes are then considered. Some recent experimental developments directed at understanding the fragmentation process and flow processes in more detail are also described.

3.2 STEADY EQUILIBRIUM ASCENT AND ERUPTION OF MAGMA

In this section a paradigm model for the ascent of magma during sustained quasi-steady explosive eruptions is developed. Since volcanic systems are highly complex it is necessary first to make some simplifications and approximations in order to elucidate the major factors which control eruptive behaviour. Not all the processes in erupting volcanoes are yet well understood and therefore the models should be seen as a first step in developing an understanding of these complex systems.

Eruptions are initiated once a flow path from a magma chamber to the surface develops, and the eruption continues until this conduit has closed or the magma supply is exhausted. The normal stresses in the magma passing through the conduit resist the lithostatic pressure of the surrounding rock, and together with the compressive strength of the rock, they keep the conduit open. If the magma pressure increases to values greater than the tensile strength of the rock plus the lithostatic pressure, then the conduit can expand through wall fracture. Estimates of the tensile and compressive strength of rocks range from 1 to 30 MPa with the lower end of the range corresponding to highly fractured rocks and sediments (Wilson *et al.* 1980; Dobran 1992). These values are consistent with estimates of 10–30 MPa for typical overpressures in magma chambers feeding the conduit (Blake 1984; Tait *et al.* 1989; Stasiuk *et al.* 1993). For comparison, the lithostatic pressure increases by about 25–30 MPa per kilometre below the surface. In this model the conduit walls are normally assumed to be rigid. However, changes in the dimensions of the conduit/vent system can occur by various mechanisms and are clearly evident by the presence of lithic fragments eroded from the conduit, vent and crater walls in the ejecta

(Chapter 13). Models of ascent show that magma can become underpressured which may lead to wall bursting and evolution of the conduit shape (Wilson *et al.* 1980; Macedonio *et al.* 1994). As an eruption develops, conduit erosion might also occur due to the mechanical abrasion of the walls by the erupting magma (Macedonio *et al.* 1994). The walls of craters can fail. Conduit contraction can occur as a result of a decrease in chamber overpressure and the concomitant decrease in flow rate or as a consequence of the plastering of hot ejecta on to the conduit walls. In explosive eruptions, these processes occur over long times compared to the ascent time of the magma, and so in the following analysis, the conduit is assumed to be of fixed geometry.

The following model of the steady-state motion of the erupting material through a conduit builds upon the work of Wilson *et al.* (1980), Giberti and Wilson (1990), Dobran (1992) and Jaupart and Allegre (1991). As the material moves through the conduit, the gas and liquid magma are assumed to move as a homogeneous mixture and the mass fraction of volatiles in solution, n_s, can be related to the pressure according to Henry's law

$$n_s = sP^{\beta} \tag{3.1}$$

where β lies in the range 0.5–1.0, P is the pressure and s the solubility constant. For example, for water-rhyolite, $\beta = 0.5$ and $s = 4.1 \times 10^{-6}$ N m$^{-1/2}$ (Wilson *et al.* 1980). As the magma ascends and decompresses, the pressure falls below the saturation pressure associated with the initial mass of volatiles in solution and volatiles are exsolved. In the paradigm modelling the volatiles are able to exsolve sufficiently fast to remain in equilibrium with the magma, as specified by Henry's law (equation 3.1). Clearly this cannot be strictly correct as exsolution involves nucleation and diffusive growth of bubbles from a supersaturated melt (Sparks 1978a; Toramaru 1989; Proussevitch *et al.* 1993; Hurwitz and Navon 1994). The possible influence of kinetic and disequilibrium effects are discussed in section 3.3.1. If the initial mass fraction of volatiles in solution is n_o say, then in equilibrium, at pressure P, the mass fraction of exsolved volatiles in the mixture, n, is

$$n = n_o - sP^{\beta} \tag{3.2}$$

The density of the mixture of magma and vesicles is defined in terms of the mass weighted averages of the constituent solids, of density σ and mass fraction $1 - n$, and exsolved volatiles of density

$$\alpha = \frac{P}{RT} \tag{3.3}$$

and mass fraction n, giving the mass averaged density

$$\frac{1}{\rho} = \frac{1-n}{\sigma} + \frac{n}{\alpha} \tag{3.4}$$

The void fraction of the mixture ϕ follows from equation (3.4) and has the value

$$\phi = \left(\frac{(1-n)\alpha}{(1-n)\alpha + n\sigma}\right) \tag{3.5}$$

3.2.1 Dynamical Model of Conduit Flow

The conservation of mass in a conduit of cross-sectional area A may be expressed in terms of the mean velocity u and density ρ of the mixture, according to the relation

$$\rho u A = Q \tag{3.6}$$

where Q is the mass eruption rate. The frictional force acting on the flow from the walls of the conduit can be parameterized in the form

$$F_{\mathrm{R}} = \left(\frac{24\eta}{\rho u r} + f\right)\frac{\rho u^2}{2r} \tag{3.7}$$

where f is a turbulent friction coefficient, which depends upon the wall roughness, and has a value in the range 0.001–0.01 for the turbulent flow above the fragmentation region (Schlicting 1968) and η is the viscosity of the magma–volatile mixture. The viscous component corresponds to the drag experienced by laminar Poiseuille flow (Batchelor 1967) and is the dominant effect below the fragmentation region. The mixture viscosity increases with magma crystal content, decreases with magma water content and increases with void fraction. A simple parameterization for the viscosity of vesicular magma as a function of the mass fraction of dissolved water n_{s} is used (Jaupart and Allegre 1991)

$$\eta = \eta_{\mathrm{o}} 10^{5-100 n_{\mathrm{s}}}(1-\phi)^{-5/2} \tag{3.8}$$

where η_{o} is the viscosity of the magma, and ϕ is the volume fraction of gas in the mixture. This parameterization is by no means unique and is strictly only appropriate for rhyolitic magma at 900 °C containing 0.05 mass fraction of water. Other magma compositions and mass fractions of water would have parameterizations which differ in detail from equation (3.8). Here η_{o} has values in the range 10^4–10^6 Pa s for rhyolitic magmas in the temperature range 700–900 °C and mass fractions of dissolved water in the range 0.03–0.06. Above the fragmentation region, the effective viscous friction of the dusty gas, described by the first term on the right-hand side of equation (3.4), becomes negligible and the dominant frictional force is exerted by the turbulent friction coefficient, f. If the equation of motion for the flow along the conduit is averaged across the cross-sectional area of the conduit, the following momentum equation is obtained:

$$\rho u \frac{\mathrm{d}u}{\mathrm{d}z} + \frac{\mathrm{d}P}{\mathrm{d}z} = -F_{\mathrm{R}} - \rho g \tag{3.9}$$

where u is the average velocity, and F_{R} is given by equation (3.7). More detailed analyses include shape factors in this averaging process (Giberti and Wilson 1990), but these have only a small effect upon the quantitative results, particularly in comparison to other uncertainties in the model.

An important ingredient in the model is specification of the conditions at which the fragmentation occurs. This transition represents the point where, on the scale of the conduit, the connected fluid phase changes from a viscous bubbly liquid to a particle-laden gas. The mechanisms of fragmentation are still poorly understood. Some constraints on conditions in the fragmentation region can, however, be made from observations of pumice density which represent the maximum possible state of exsolution of the magma at the time of fragmentation. The vesicularities (void fractions) of pumice mostly fall in the rather narrow range of 0.7–0.8 (Sparks 1978a; Whitham and Sparks 1986; Houghton and Wilson 1990; Thomas *et al.* 1994). Sparks (1978a) interpreted this threshold void fraction as a consequence of the bubbles achieving an optimum packing. However, samples of rhyolite with very low densities (high vesicularities) are known (Thomas *et al.* 1994) and bubble interactions cannot explain the common range of values from 0.7 to 0.8. Models of the expansion of individual bubbles (Barclay *et al.* 1995) and foams (Thomas *et al.* 1994) indicate that it is the increase in viscosity with volatile exsolution which determines the final void fraction in the ejecta. Essentially, as magma decompresses and the gas expands, the liquid films around each bubble spread viscously. During the initial stages of bubble growth, the time scale associated with this spreading is typically much faster than the rate of decompression, and the bubble can grow freely. However, as the volatiles are exsolved from the magma, its viscosity increases (equation 3.8) and bubble expansion is inhibited. Once the viscosity has reached values of about 10^9 Pa s or more, the magma becomes too viscous for the bubbles to expand on the time scale of the decompression (Thomas *et al.* 1994; Barclay *et al.* 1995; Sparks *et al.* 1994). This viscosity threshold corresponds to void fractions in the range 0.7–0.8 consistent with observations of rhyolitic magma. Thomas *et al.* (1994) suggest that fragmentation in viscous magmas occurs at somewhat lower void fractions than observed in the final ejecta. They observe that 0.6 is a common minimum void fraction and take this to be a lower limit.

The void fraction at fragmentation between 0.6 and 0.8 provides an estimate of typical pressures in this region which are several megapascals. For simplicity, in the present model calculations, the fragmentation is assumed to occur when the void fraction of the volatile–magma mixture reaches a value of 0.77.

3.2.2 Dynamic Evolution of the Flow

Combining equations (3.1–3.9), and assuming equilibrium volatile exsolution above the fragmentation region, the equation for the variation of pressure as the mixture rises in the conduit can be derived:

$$\left(1-\frac{u^2}{u_c^2}\right)\frac{dP}{dz} = -g\rho + \frac{2\rho u^2}{r}\frac{dr}{dz} - F_R \tag{3.10}$$

The speed of sound of the mixture, u_c, has value

$$u_c^2 = \frac{1}{\rho^2}\left(\frac{nRT}{P^2} + \frac{s}{2P^{1/2}}\left(\frac{RT}{P} - \frac{1}{\sigma}\right)\right)^{-1} \tag{3.11}$$

where the mixture is assumed to be isothermal and the value $\beta = 0.5$ (equation 3.1) is assumed, corresponding to the exsolution of water. This speed of sound for a two-phase mixture can be considerably lower than that of a pure gas since the material behaves as if it has the compressibility of the gas, but has a much higher density owing to the presence of the dense solid (Kieffer 1977). Figure 3.1 shows the speed of sound from equation (3.11) normalized by the parameter $(n_o RT)^{0.5}$ for different mass fractions of water and pressures

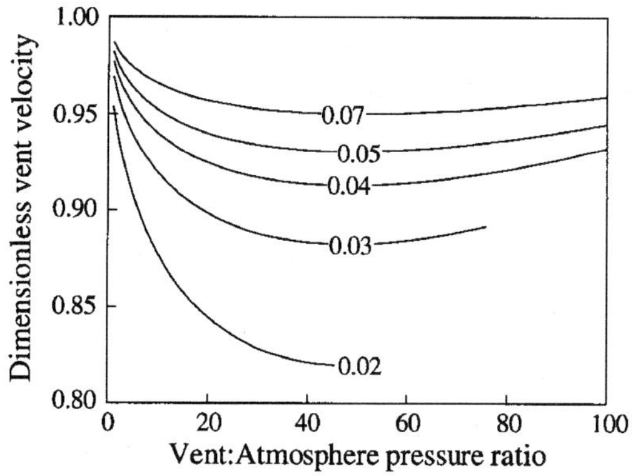

Figure 3.1 Speed of sound of the hot, two-phase mixture as a function of the vent pressure, for several values of the mass fraction of water in the mixture. The speed of sound, which equals the velocity at the vent, has been normalized by the parameter $(n_0RT)^{1/2}$ to define the dimensionless vent velocity. The vent pressure has been normalized by atmospheric pressure

corresponding to the range of conditions typical of explosive eruptions. The normalized velocity is typically just below unity, justifying a useful approximation for the sound speed at pressure above about 1 MPa:

$$u_c = 0.95\sqrt{n_0RT} \tag{3.12}$$

This expression is accurate to within about 2% for water contents in the mass fraction range 0.02–0.07, and is consistent with earlier numerical calculations of Wilson *et al.* (1980). Thus the speed of sound of the mixture is primarily dependent upon the temperature and the gas constant of the volatiles, and only varies weakly for pressures in excess of about 1 MPa. For rhyolite at 800 °C the speed of sound varies from 170 m s^{-1} at 0.06 mass fraction of water to about 100 m s^{-1} for 0.03 mass fraction of water. Note that, if the mixture is heterogeneous or there is significant relative motion of the gas and solid phases, then the problem becomes much more complex, with several different speeds of sound (Wallis 1969; Dobran 1992). Such effects are discussed in section 3.3.4.

The boundary conditions acting on the flow follow directly from equation (3.10). Initially, the mixture moves subsonically, and the coefficient on the left-hand side of equation (3.10) is positive. In this case, the gravitational deceleration and the frictional force on the walls of the conduit (the first and third terms on the right-hand side of equation 3.10) cause the pressure to decrease with distance along the conduit. As the mixture ascends through the conduit and the pressure decreases, volatiles are exsolved, the mixture expands and so the velocity increases. If the speed reaches the speed of sound, then the coefficient on the left-hand side of equation (3.10) becomes zero. Therefore, either the right-hand side of equation (3.10) should fall to zero or the flow must have reached the exit of the conduit, since otherwise a discontinuity in the pressure would develop (Shapiro 1954). However, the right-hand side of equation (3.10) can only fall to zero if the radius increases sufficiently rapidly, as would occur in a flared vent or in a crater-type structure above the conduit. In a conduit of fixed or decreasing radius, the right-hand side of equation (3.10) is always negative, and so the flow remains subsonic, except at the vent.

The flow in a parallel-sided conduit from the magma chamber to the vent is first analysed, and the expansion of this flow into a crater above the vent is then investigated. Since the flow is typically overpressured at the top of the conduit, the flow at the vent is sonic, and then becomes supersonic in the crater. On leaving the crater top, the mixture pressure can be different from that of the atmosphere, and hence there may be a further region in which the mixture decompresses to atmospheric pressure.

3.2.3 Conduit Flow

The above system of equations has been solved in a parallel-sided, cylindrical-shaped conduit. The model calculations assume a magma chamber at 3 km depth with a lithostatic pressure equivalent to that depth. The model also assumes that the magma has the same density as the crust. The consequences of relaxing these constraints are discussed later. Apart from the more detailed parameterization of the viscosity variation with volatile content (equation 3.8), the solutions are analogous to those of Wilson *et al.* (1980) and Giberti and Wilson (1990). As the magma rises and decompresses, the void fraction increases (Figure 3.2b) causing an increase in the speed of the flow (Figure 3.2a). As volatiles are exsolved, the magma becomes more viscous (equation 3.8), and so the frictional resistance increases. The rate of decompression per unit distance then also becomes larger (Figure 3.2c). Eventually, the void fraction reaches the critical value of 0.77, and fragmentation occurs. Above the fragmentation surface, the frictional resistance becomes much smaller, and so the pressure decreases much more slowly with height (Figure 3.2c). Also, as the void fraction increases, the density of the mixture becomes much lower than that of the surrounding country rock (Figure 3.2d). Therefore, because the frictional force is so low above the fragmentation surface, the pressure decreases more slowly than the lithostatic pressure. As a consequence, the material exits the vent with a pressure in excess of atmospheric. The flow is therefore choked at the vent and issues with the speed of sound of the mixture.

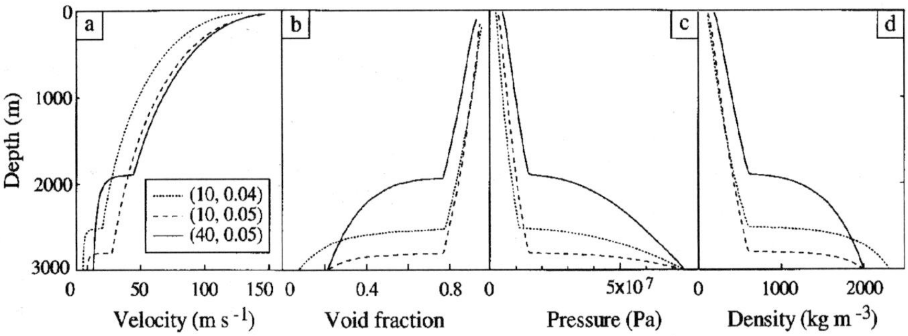

Figure 3.2 Velocity (a), void fraction (b), pressure (c) and density (d) of the magma–volatile mixture as it ascends the conduit. Curves are shown for conduit radii of 10 and 40 m and water mass fractions of 0.04 and 0.05, as indicated in the brackets. The magma chamber is at 3000 m depth, and the magma temperature is 1000 K. The prominent break in slope in each curve represents the fragmentation level

The mass flux erupted from the vent depends only upon the velocity, density and radius at the vent with vent radius being the dominant control (Figure 3.3a). The exit velocity is sonic and nearly independent of the vent pressure (Figure 3.1), but the vent pressure and hence density is strongly coupled to the mass flux and thus to vent radius (Figure 3.3b). As the conduit radius increases, the frictional resistance of the flow decreases and the fragmentation level rises (Figure 3.4). If the initial water content of the magma increases, there is more exsolution, and so the speed of sound and eruption rate increase, although this is a secondary effect compared to variations in the conduit radius.

3.2.4 Decompression into Flared Vents and Craters

The calculations in parallel-sided conduits show that exit velocities are limited by the sound speed and that exit pressures are much greater than the atmospheric pressure. However, volcanic vents are characteristically flared into a funnel shape or crater. In this geometry the speed of sound can be exceeded. This leads to supersonic flow and a rich range of flow regimes within the flared vent or crater. The interactions can be complex and depend sensitively on the exact shape of the near-surface vent. Diverging walls can provide thrust forces on the erupting jet and can suck the jet outwards causing rapid expansion and deceleration. The resulting flows can be underpressured or overpressured and shocks can

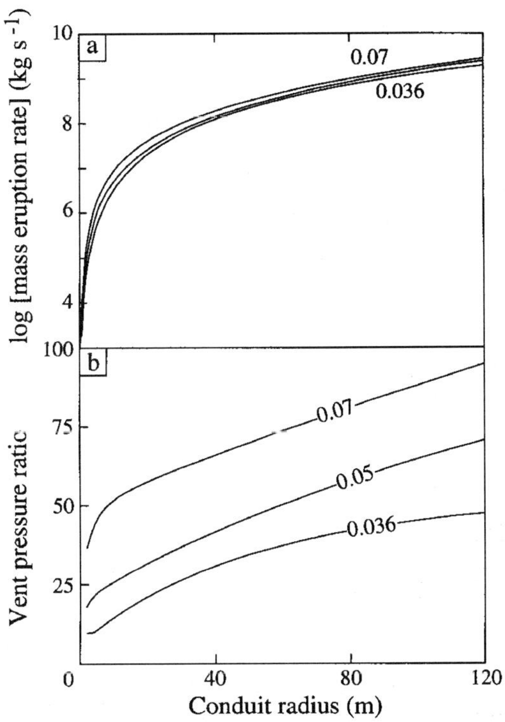

Figure 3.3 Mass flow rate (a) and vent pressure ratio (b) as a function of vent radius for three different water mass fractions (0.036, 0.05 and 0.07). Vent pressure ratio in (b) is defined as the ratio of the mixture pressure to atmospheric pressure

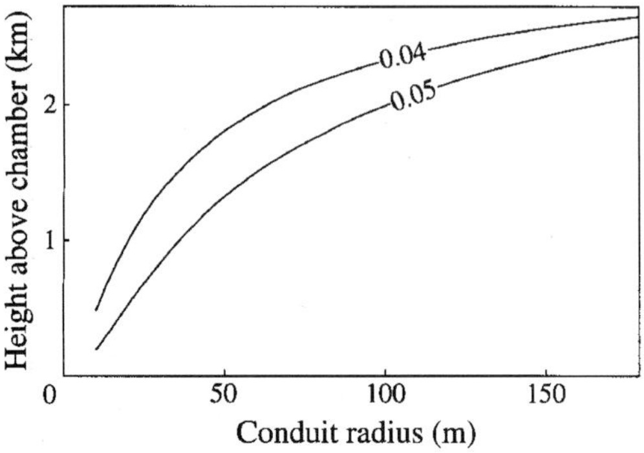

Figure 3.4 Position of the fragmentation surface, expressed as height above a chamber at 3 km depth, as a function of the conduit radius and water mass fraction in the magma (0.04 and 0.05). In this calculation the eruption temperature is 1000 K

develop in craters under some conditions. In the case of rapidly diverging walls, the flow can even separate from the walls, forming a free jet within the crater. Models of these flows are presented based on a two-stage system of a parallel-sided conduit connected to a crater whose walls diverge at a constant angle. The model summarizes the results of Woods and Bower (1995) and is applicable to diverging walls of inclination $<30^\circ$ to the vertical for which flow separation from the walls does not occur.

In a crater, the diverging area dominates the effects of gravity and friction and the right-hand side of equation (3.10) becomes positive. Since the flow at the vent is sonic, the flow becomes supersonic in the crater and the pressure continues to decrease towards the exit. In general, as in classical nozzle theory (Liepmann and Roshko 1957), three different types of flow can issue from the top of the crater (Figure 3.5). If the vent pressure is very large, then the pressure remains greater than atmospheric pressure throughout the crater, and the flow exits the crater top as an overpressured, supersonic flow (curve (i) in Figure 3.5). However, for lower flow rates the vent pressure is lower, and so the material can decompress much more within the crater. For sufficiently low flow rates, the material becomes underpressured in the crater, and issues from the crater top underpressured, but supersonic (curve (ii) in Figure 3.5). At some critical flow rate, the material becomes so underpressured that a shock wave develops in the conduit. The shock wave acts to recompress the flow and increase the density and so the flow becomes subsonic downstream (curve (iii) in Figure 3.5). The material therefore issues from the crater top with a much lower subsonic velocity. The location of the shock is determined by the requirement that the pressure increases to atmospheric at the crater top.

If a shock forms, then in the model, equation (3.10) applies upstream and downstream of the shock. However, the transition in the flow across the shock is given by relations for the conservation of mass and momentum (Liepmann and Roshko 1957)

$$[u_\text{p}] = 0; \quad [u_\text{p}^2 + P] = 0 \qquad (3.13\text{i, ii})$$

where the square bracket notation denotes the difference in value upstream and downstream of the shock. Also, the temperature remains nearly constant across the shock. Curve (iii) in Figure 3.5 was calculated using the shock conditions (equation 3.13) together with the above model of crater flow (equation 3.10).

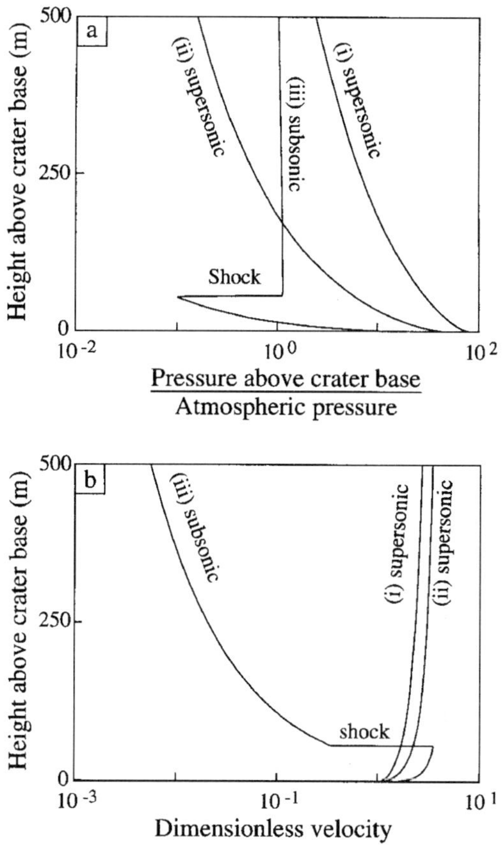

Figure 3.5 Variation of the pressure (a) and velocity (b) in a crater above the vent. Curves are shown for three eruption rates. Curve (i) shows a large eruption rate (vent radius = 108 m) in which the vent pressure is high, and so the flow remains supersonic and overpressured through the crater. Curve (ii) shows an intermediate flow rate (vent radius 34 m) in which the pressure falls below atmospheric, but the flow remains supersonic at the crater top. In case (ii), there is no shock wave which can recompress the material such that the flow is able to vent at atmospheric pressure. Curve (iii) shows the lowest flow rate (vent radius 4 m) in which the pressure in the lower part of the crater falls well below atmospheric, causing the formation of a shock, and a dramatic decrease in the flow speed owing to the much higher pressure and hence density of the flow. In case (iii), the material issues at atmospheric pressure. The velocity is shown normalized to the sonic velocity $(nRT)^{1/2}$, and the pressure is shown normalized to atmospheric pressure

A shock forms if the vent pressure falls below a critical value. In Figure 3.6 model calculations illustrate how the velocity at the crater top varies with the crater shape and the mass flux. For low eruption rates, the development of a shock in the crater leads to very low, subsonic velocities at the crater top, whereas for high eruption rates the thrust provided by the walls of the crater lead to very high supersonic velocities at the crater top. The formation of shocks in a crater can also have an important influence upon the dispersal of large ballistic bombs. Larger particles move through the shock at high speed, and are subsequently decelerated by the relatively slow subsonic mixture ahead of the shock

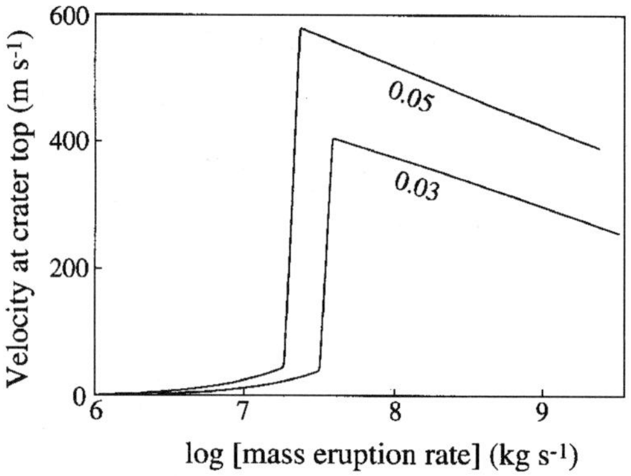

Figure 3.6 Variation in the velocity at the crater rim as a function of the mass eruption rate and mass fraction of water in the magma (0.05 and 0.03). In this calculation the eruption temperature is assumed to be 1000 K

(Chapter 13; Bower and Woods 1996). Below (section 3.2.5) the decompression of a jet is also shown to depend on the crater size and geometry, with shocks forming at higher flow rates as the size of the crater increases.

3.2.5 Pressure Adjustment beyond the Crater

If the flow exits the vent as a supersonic overpressured or underpressured jet then the flow adjusts to atmospheric pressure just beyond the vent. This adjustment is highly complex owing to the turbulent two-phase flow (Wallis 1969). However, some principles of the flow are similar to the discharge of homogeneous fluid from a nozzle (Liepmann and Roshko 1957; Cebeci and Bradshaw 1984). In that case, if the flow is overpressured at the vent, then an expansion fan develops in which the flow decompresses towards ambient pressure (Figure 3.7a). When the expansion fan reflects from the constant pressure boundary a series of compression waves merge to form a shock. In contrast, if the flow is underpressured, then beyond the nozzle, a series of oblique shocks followed by expansion and then compression waves develop (Figure 3.7b).

Kieffer and Sturtevant (1984) carried out a series of experiments to study the motion of sonic and overpressured jets immediately ahead of the source nozzle. They investigated freon jets, issuing from a high-pressure reservoir, through a convergent nozzle. In some experiments, a series of oblique shock waves developed and interacted to form a Mach shock disc ahead of the nozzle, oriented normal to the flow. The flow decompressed across this shock disc, but further oblique shock waves developed, leading to the characteristic diamond pattern (Figure 3.7). In this adjustment region entrainment is relatively small, since large-scale eddy formation is inhibited in high-speed overpressured flows. Valentine and Wohletz (1989) and Dobran *et al.* (1993) have developed numerical super-computer models of the flow ahead of the vent (Chapter 4), and these models predict some features qualitatively similar to the interacting oblique shock waves observed in the experiments of Kieffer and Sturtevant (1984).

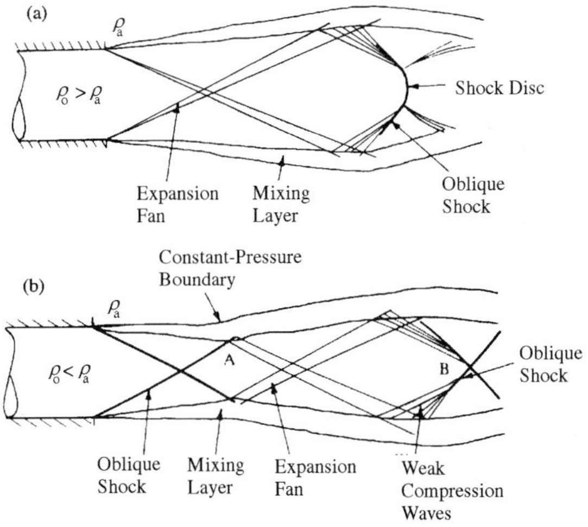

Figure 3.7 Schematic representation of the expansion of (a) an overpressured jet issuing at the speed of sound from a nozzle in which an expansion fan develops through which the flow decompresses, (b) an underpressured jet issuing at the speed of sound in which a series of shockwaves, and expansion and compression waves act to recompress the jet. After Cebeci and Bradshaw (1984)

In order to calculate the flow conditions once the jet has decompressed, a simple model is developed which captures the dominant effects of this pressure adjustment. Assuming there is negligible entrainment of ambient air and that the effects of gravitational deceleration are small during the decompression, the properties of the decompressed jet can be found from the conservation of mass, momentum and enthalpy fluxes across the decompression zone. These assumptions require that the pressure adjustment occurs over a short vertical distance of the order of at most a few hundred metres, because the gravitational deceleration of a jet of velocity 100–300 m s^{-1} occurs over about 1–3 km, and such a jet will double its mass through entrainment over about 1–10 km. The conservation relations across this adjustment region have the form (Woods and Bower 1995)

$$Q = \rho_1 u_1 r_1^2 = \rho_2 u_2 r_2^2 \qquad \text{(mass)} \qquad (3.14)$$

$$u_2 = u_1 + \frac{r_1^2}{Q}(P_1 - P_2) \qquad \text{(momentum)} \qquad (3.15)$$

$$C_v T_1 + \frac{P_1}{\rho_1} + \frac{u_1}{2} = C_v T_2 + \frac{P_2}{\rho_2} + \frac{u_2}{2} \qquad \text{(enthalpy)} \qquad (3.16)$$

where P_2 is the atmospheric pressure, and subscripts 1 and 2 denote properties before and after adjustment. Owing to the large mass fraction of solid in the mixture, the enthalpy conservation relation is approximately equivalent to the condition that the decompression is isothermal. Figure 3.8a shows how the velocity of the decompressed jet varies with the mass eruption rate and mass fraction of dissolved water. Figure 3.8b shows calculations for three different crater sizes and shapes with a fixed mass fraction of water of 0.03. These results for eruption through a crater can be compared with the case of eruption through a parallel-sided conduit which exits straight into the atmosphere. As discussed earlier this latter case involves an overpressured flow which adjusts to atmospheric pressure. Such a

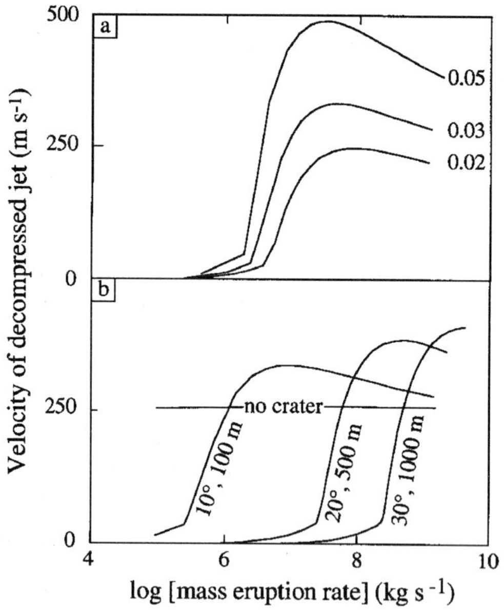

Figure 3.8 Velocity of the jet following decompression to atmospheric pressure as a function of the mass flux. Curves are shown for (a) different mass fractions of water for a fixed crater 500 m deep with an outward-sloping wall having an angle of 10° to the vertical, and (b) for several crater sizes at a fixed water content of 0.03. In (b) the velocities are compared with the case of a freely decompressing jet issuing from a vent directly into the atmosphere (a horizontal line marked as "no crater"). The figures on each curve give the crater depth in metres and the angle between the outward-sloping crater wall and the vertical

freely decompressing jet is found to have an approximate speed of $(1.97 \pm 0.02)\sqrt{n_0 RT}$ which is virtually independent of the eruption rate, but is nearly twice the value at the top of the conduit where the mixture exits at the speed of sound (Woods and Bower 1995). By contrast, the velocities of jets issuing from a crater do not change significantly across the decompression zone. For large mass fluxes, the speeds are very low owing to shock formation in the crater, whereas for high mass fluxes, the speed is very high owing to the additional thrust provided by the crater. Variations in the volatile content of the magma also play an important role in determining the final speed of the material, since mixtures with a higher volatile content propagate much faster.

These results confirm previous studies which showed that velocities at the base of eruption columns are in the range of a few hundred metres per second with velocity typically increasing with the mass fraction of water. Velocities up to 600 m s^{-1} can be achieved. The new calculations, however, show that the vent geometry also has a significant influence, particularly at lower eruption rates of magma.

The decompressed material supplies the eruption column with the dense mixture of ash and gas. The remainder of this book focuses on the motion of the material beyond the region in which the decompression of the jet occurs, so that the flow is pressure adjusted to the atmosphere. In Chapters 4 and 6, variations in the velocity of the pressure-adjusted material, caused by changes in the eruption rate or crater geometry, are shown to be of particular importance in determining whether the erupted material can form a buoyant eruption column or a collapsing fountain.

3.3 CAVEATS AND COMPLICATIONS

Modelling research published thus far has not been able to take into account all the complexities of nature. In this section the various complicating factors which might influence the source parameters for volcanic plumes are assessed in a more qualitative way.

3.3.1 Kinetic Effects of Gas Exsolution

Kinetic and disequilibrium effects are certain to occur in the extreme flow conditions encountered in explosive eruptions. The degassing process requires gas supersaturation. Hurwitz and Navon (1994) have carried out experiments on the decompression of water-saturated rhyolitic melt which indicate that quite large supersaturations can develop in erupting magma. Even with crystal sites for heterogeneous nucleation, supersaturations in the range of 10–70 MPa were needed before substantial nucleation rates were achieved. Their experiments indicate that supersaturations of well over 100 MPa may be necessary for homogeneous nucleation. In such circumstances bubbles would form high in the conduit a short distance below the fragmentation level, rather than over a large depth range in the conduit. Bubble growth is expected to be explosive (Sparks 1978a; Proussevitch *et al.* 1993; Sparks *et al.* 1994) with a large pressure drop over a narrow region. It is also plausible that under rapid flow conditions the melt remains significantly supersaturated throughout the processes of vesiculation, fragmentation and discharge into the atmosphere. While these disequilibrium effects are not expected to introduce any fundamentally new features into the models, the relationships between pressure, viscosity, density and velocity will change in detail. Indeed, using a simple phenomenological model, Woods (1995a) has shown that volatile supersaturations up to 80 MPa can lead to a decrease in the erupting mass flux by a factor of up to about 10.

3.3.2 Fragmentation

The models assume that fragmentation occurs at a single value of void fraction (0.77). However, pumice deposits vary in their mean vesicularities and individual deposits display a significant range of vesicularities around the mean (Thomas *et al.* 1994). The results of calculations will be quite sensitive to the value of vesicularity chosen for the fragmentation level due to the strong dependence of viscosity on vesicularity (e.g. equation 3.8). Woods (1995a) has shown that if the model void fraction at which fragmentation occurs is changed from 0.7 to 0.9, then the predicted eruption rate decreases by a factor of 3–5. The variance in vesicularity may lead to a pressure range over which fragmentation occurs, resulting in fluctuations in fragmentation pressure and various degrees of degassing disequilibria. The conditions in the fragmentation region require further research and are currently being investigated experimentally (e.g. Mader *et al.* 1994, 1996; Phillips *et al.* 1995; Alidibirov and Dingwell 1996).

3.3.3 Controls on Initial Plume Temperature

The temperature of the erupting mixture controls the thermal energy available to carry the erupting mixture high into the atmosphere (Chapter 4). Magmas can have temperatures ranging from 700 to 1200 °C (Chapter 1). Four processes can result in cooling: adiabatic effects, heat transfer to the wall, incorporation of cool lithics into the mixture due to vent and conduit erosion and incorporation of external water during flow through the conduit. These different effects are now evaluated.

As the flow propagates along the conduit, heat is transferred from the hot mixture, of typical temperature 700–1200 °C, to the conduit walls, whose temperature may be as low as 0–300 °C. The efficiency of the heat transfer depends on the processes at the conduit wall. If the conduit wall does not erode then heat is transferred by conduction in the wall which is a very slow process. A simple scaling shows that the maximum thickness of the conduit wall which is affected by this heating is approximately $(\kappa t)^{1/2}$, where κ is the thermal diffusivity of the magma and t is the time of eruption. Assuming a thermal diffusivity of the order of 10^{-7} m^2 s^{-1}, then after the eruption has continued for 100 s, this cooled boundary layer is several millimetres thick. The heat flux conducted through this layer per unit distance along the conduit is about 10^4 J m^{-1} s^{-1}, while the heat flux associated with the flow along the conduit is about 10^{11} J s^{-1}. Therefore, if the conduit is only a few kilometres long, the cooling through the conduit walls has a negligible effect upon the mixture temperature (Wilson and Head 1981). As the eruption proceeds the thermal boundary layer thickness increases and so the heat flux to the wall decreases further.

The equation for the conservation of energy averaged across a conduit can be derived from the second law of thermodynamics:

$$\frac{\mathrm{d}\left(\dfrac{P}{\rho} + C_\mathrm{v}T + \dfrac{u^2}{2}\right)}{\mathrm{d}z} = -g - Q_\mathrm{T} \tag{3.17}$$

where T denotes absolute temperature and Q_T is the heat loss per unit mass per unit length along the conduit, which only becomes important if cold lithic clasts or external water are incorporated into the flow. The first term on the right-hand side of equation (3.17) represents the work done against gravity. Earlier models of conduit flow (Wilson *et al.* 1980; Giberti and Wilson 1990; Dobran 1992) simplified this expression for the conservation of enthalpy (equation 3.17) by assuming that the material in the conduit moves either adiabatically or isothermally. Note that because of the no slip condition on the walls of the conduit, there is no work done on the walls by the flow (Shapiro 1954). Equation (3.17) can be combined with the dynamic model and rearranged to show that the temperature of the mixture changes with pressure and height according to the relation

$$\rho C_\mathrm{v}\frac{\mathrm{d}T}{\mathrm{d}z} = \rho\left(\frac{nRT}{P} + \frac{sP^{1/2}}{2}\left(\frac{RT}{P} - \frac{1}{\sigma}\right)\right)\frac{\mathrm{d}P}{\mathrm{d}z} + F_\mathrm{R} - \rho Q_\mathrm{T} \tag{3.18}$$

The first term on the right-hand side of equation (3.18) has a magnitude which is typically only a few per cent of the term on the left-hand side, since the volatile mass fraction in the ascending magma is only in the range 0.02–0.06. Therefore, adiabatic cooling is typically only a few per cent of the total temperature and the mixture remains nearly isothermal

(Wilson *et al.* 1980). The internal heating due to the frictional dissipation, given by the second term on the right-hand side of equation (3.18), is also small over the length of the conduit, typically resulting in a temperature change of only 1–10 K.

The only process which can cause significant cooling of the mixture is the heat transfer related to incorporation of lithic clasts by erosion of an unstable vent or incorporation of external water. The effects of the addition of external water are considered in Chapter 8, where it is shown that the initial temperature and density change as groundwater is mixed with the erupting material. The effect of adding lithics to the erupting mixture is considered here. Suppose that the mixture erupting at the surface has a lithic mass fraction m_L, and that the initial temperature of the lithics eroded from the walls of the conduit is T_L. Then, if the absolute temperature of the magma–volatile mixture before mixing is T_o, the absolute temperature of the mixture at the vent is

$$T = \frac{(1 - m_L)C_{po}T_o + m_L C_{pL} T_L}{(1 - m_L)C_{po} + m_L C_{pL}} \tag{3.19}$$

where C_p is the specific heat and the subscript L denotes a property of the lithics. Assuming the specific heat of the magma and country rock are comparable, then the temperature decreases approximately linearly from the magma temperature to the lithic temperature as the lithic mass fraction increases from 0 to 1. For example, if the lithic temperature is 400 K and the initial magma temperature is 1000 K, then if the mass fraction of lithics added to the mixture is 0.1, the temperature of the erupting material typically falls from 1000 K to about 940 K.

The simple models presented in this chapter assume that the clasts are always in thermal equilibrium with the surrounding gas and thus the thermal relaxation time of the gas and clasts is rapid compared to the ascent time in the conduit of length L

$$\tau_t \approx \frac{d^2}{\kappa} \ll \frac{L}{u} \tag{3.20}$$

where κ is the thermal diffusivity of the clasts. This condition is typically satisfied for clasts smaller than a few millimetres. This is a reasonable approximation for most, but certainly not all, eruptions. Clearly if the erupting mixture contains a significant proportion of large clasts then a single temperature cannot be assigned to the system and density variations within the flow will be much more complex. Using a simple model of thermal disequilibrium, in which only a fraction of the solid material is assumed to remain in thermal equilibrium, Woods (1995b) has shown that the eruption temperature of the gas may be much smaller. For example, for gas mass fractions in the range 0.04–0.06, and an initial temperature of 1000 K, the eruption temperature lies in the range 800–900 K if 50% of the solid remains in equilibrium but decreases to values in the range 500–600 K if only 10% of the solid material remains in equilibrium.

3.3.4 Unsteady and Heterogeneous Conduit Flow

The models assume that the flow is homogeneous, an approximation which leads to the description of the mixture as a pseudofluid to which single values of temperature, density and velocity can be assigned. If there is significant relative motion of the gas and clasts or there are intrinsic heterogeneities in the relative concentrations of gas and particles then this approximation can become less accurate or even invalid.

Wilson *et al.* (1980) and Buresti and Casarosa (1989) have examined the conditions under which the gas and solid phases can separate. They have shown that under the high-pressure flow conditions typical of the conduit, the product of the density and velocity of the erupting gas is typically sufficient to support most of the clasts, particularly since the mass of the smaller dense particles in the flow contribute to the suspension of the larger clasts (Saffman 1962). The homogeneous flow model is assumed to be accurate if the steady free-fall velocity of the clasts is much smaller than the velocity of the mixture as it ascends to the vent, and also if the time scale over which the clasts accelerate to values near to the mean flow velocity is small compared to the time scale of the motion in the conduit. At such high speeds, most of the clasts in the conduit are supported by the pressure force of the surrounding gas, giving a balance between the quadratic drag force and a free-fall velocity, v, of the form (Suzuki 1983; see Chapter 14)

$$v^2 \approx \frac{dg\sigma}{C_D\beta} \tag{3.21}$$

where d is the clast size, g the gravitational acceleration, C_D the drag coefficient and β the bulk density of the gas and those clasts which are at least an order of magnitude smaller than d. The time scale for the clasts to accelerate to the mean flow speed u, say, is of the order

$$\tau_d \approx \frac{d\sigma}{\beta C_D u} \tag{3.22}$$

Therefore, the homogeneous flow model is valid for clasts which satisfy

$$\tau_d \ll \frac{L}{u} \tag{3.23}$$

where L is a typical length-scale over which the pressure changes by about 10% and $v \ll u$. Wilson *et al.* (1980) and Buresti and Casarosa (1989) have shown that these relations are indeed satisfied except for clasts of size in excess of 10–100 cm. A situation in which phase separation can develop is across the shock waves in the crater. The larger clasts will pass through the shock and then decelerate to the new speed of the gas. However, if the larger clasts only represent a small fraction of the total mass, then the main impact will be upon the dispersal of ballistic clasts from the crater (Chapter 14).

In an isothermal steady-state model, similar to that presented in section 3.2, Dobran (1992) examined the role of relative motion between the liquid bubbles and magma, and also gas and particles, using a parameterized model of two-phase flow. This model shows that, as expected, there is very little motion of the gas bubbles relative to the viscous magma, whereas the gas phase is able to migrate out of the conduit at a somewhat higher speed than the solid particles. However, this latter effect depends upon the grain size distribution of the particles, and is negligible for sufficiently small particles. The results of these more detailed calculations are not significantly different from the homogeneous flow model described in this chapter (Dobran 1992).

Recent experimental studies in shock tubes have indicated that two-phase systems are inherently heterogeneous. Anilkumar *et al.* (1993) have investigated the sudden decompression of an homogeneous static bed of particles held in air at high pressure. The bed expands by developing a series of horizontal partings that distort into cavities in the initial stages of bed expansion (Figure 3.9). As the flow expands further the system evolves into

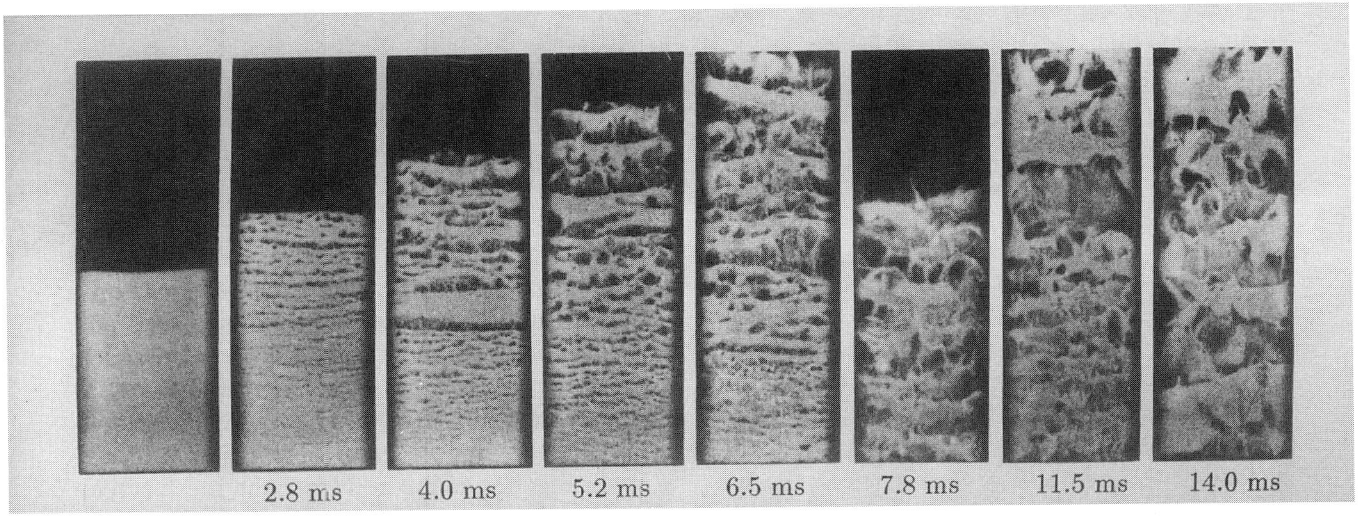

Figure 3.9 Photograph of the decompression of a mixture of particles and high-pressure gas from a shock tube, taken from Anilkumar *et al.* (1993). Sequence of photographs showing the early stages of the decompression of a bed of glass balls initially at rest. Beads are of size 0.125 mm and the photographs are taken at times 0–6.5 ms for the case in which the bed of beads extended to the centre of the window on the side of the apparatus and at times 7–14 ms for the case in which the top of the bed of beads was initially located at the base of the window on the side of the apparatus

irregular regions of high particle concentration and very dilute gas-rich regions. The dilute regions accommodate most of the gas expansion – they have low drag compared with the high concentration regions. Similar heterogeneities have been observed in sprays formed by explosive vaporization and exsolution (Hill and Sturtevant 1989; Mader *et al.* 1994, 1996). These experiments suggest that high-speed two-phase flows are not truly homogeneous. Such heterogeneities will undoubtedly cause complexities in the properties of erupting fluid. For example, the speed of sound will vary from place to place and large fluctuations in density and particle concentration might occur in the vent region.

At low flow rates and pressures, flow instabilities can develop, leading to the periodic formation of either high-density particle-rich "slugs" or low-density gas-rich "bubbles" in the conduit, which erupt periodically. In the intervals between slugs, only the volatile phase and some of the very fine grained particles are erupted (Batchelor 1988; Anilkumar *et al.* 1993). Such instabilities will result in a variation of the local speed of sound, and this affects the exit conditions and the location of the fragmentation front.

There remains a significant amount of work to be done to understand the fundamental physics, as well as the volcanological implications, of such two-phase flow. If two-phase flows are inherently heterogeneous, even for mixtures of fine particles and gas, then the pseudofluid assumption of current models does not provide a complete description of flow dynamics.

3.3.5 Degassing During Magma Ascent: the Lava Problem

One of the most enigmatic features of volcanic systems involving high-viscosity magma is that there can be alternations between lava effusion and explosive activity. The later stages of the eruption of Mount St Helens, Washington, USA, illustrate this kind of activity with several episodes of dome growth and then explosive destruction. There is also good petrological evidence that passively effusing silicic lavas contained as much dissolved gas deep in the conduit as magma which erupted explosively, often in the same eruption (Eichelberger *et al.* 1986). The mechanism of gas loss from high-viscosity magma remains unclear. One explanation is convective cycling of gas-rich and gas-poor magma in the conduit system. Another explanation is that on vesiculation the bubbles become interconnected and permeable to gas flow which escapes from the sides of the conduit. Such gas loss will alter magma properties and may determine whether effusive or explosive eruptions occur.

If the magma and walls of the conduit are permeable (Heiken *et al.* 1988), then gas can escape from the mixture into the fractures. This lowers the void fraction of the mixture and inhibits the fragmentation process (Eichelberger *et al.* 1986; Jaupart and Allegre 1991). If the flow rate is sufficiently low, then there is sufficient time for the volatiles to escape from the conduit walls and vesicular lava slowly effuses from the vent, forming a lava dome. If the volatiles can escape from the magma into the neighbouring fractures, then the rate of loss of volatiles through the fractured, permeable wall rock is governed by Darcy's law, and can be included in the above model by modifying the mass conservation equation to the form (Jaupart and Allegre 1991)

$$\frac{\mathrm{d}u\rho}{\mathrm{d}z} = \frac{-2\alpha K(P - P_{\infty})}{\mu_{\omega} L_{\mathrm{f}} r} \qquad (3.24)$$

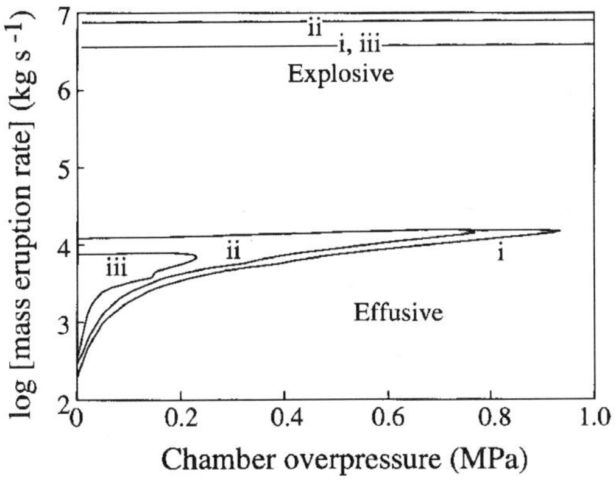

Figure 3.10 Comparison of the effusive and explosive eruption rates as a function of magma chamber overpressure. Either kind of eruption can occur under identical conditions. The explosive eruption is virtually independent of the chamber overpressure, for small overpressures, while the effusive eruption, which requires magma–volatile separation, is very sensitive to chamber overpressure. The model calculations are based on equation (3.24) combined with the conduit flow model of section 3.2 (from Woods and Koyaguchi 1994). Curves labelled (i) correspond to a chamber 2 km below the surface with a conduit radius of 10 m and wall-rock permeability of $10^{-12.5}$ m^2. The magma has an initial volatile content of 0.03 mass fraction. In the curves labelled (ii) the magma has an initial volatile content of 0.05. In curves labelled (iii) the conduit permeability has a value of $10^{-12.75}$ m^2. The explosive eruption regimes for cases (i) and (iii) are indistinguishable. As the volatile content increases, the explosive eruption becomes more energetic (curve ii)

where r is the conduit radius, L_f the scale over which the pressure changes from that of the conduit, P, to the far field hydrostatic value, P_∞, μ_ω the volatile viscosity and K the permeability of the fractures. By including this additional process in the model, the conduit flow in fact ceases to be unique for given conditions in the magma chamber. As well as the slow effusive flow in which the volatiles escape through the conduit walls, the high-speed explosive flow, described above, is also possible. If the flow rate is sufficiently high then only a very small mass fraction of the volatiles actually escape from the magma as it ascends along the conduit, and a high-speed explosive eruption can occur. In Figure 3.10, the different flow rates possible, when the volatiles are able to escape from the magma into the permeable conduit walls, are shown (Woods and Koyaguchi 1994). In fact, owing to the non-linearity of the flow, under identical conditions two slow effusive solutions are possible, as well as the high-speed explosive flow. Such non-uniqueness of eruption rate can lead to instability and a sequence of transitions between effusive and explosive eruptions.

3.4 TRANSIENT VULCANIAN-STYLE ERUPTIONS

As well as the steadily discharging eruptions described above, an important class of eruptions are the short-lived Vulcanian-style explosions. These explosions can lead to the formation of a very fragmented, high-speed ash–gas mixture from which large thermal plumes arise. The models presented earlier in this chapter are now adapted in order to describe such transient explosions, following the work of Wilson *et al.* (1978), Wilson (1980), Turcotte *et al.* (1990) and Woods (1995b).

In a short-lived Vulcanian-style eruption gas pressure builds up in the conduit or in a lava dome. Some workers attribute Vulcanian explosions to the accumulation of magmatic gases and others invoke an important role for external water or hydrothermal fluids. The presence of a resistant plug or cooled cap is often invoked. Here it is proposed that microlite crystallization, triggered by gas loss, creates the excess pressures to cause Vulcanian explosions in lava domes. Eventually, the pressure exceeds the strength of the cap rock plug or crystallized interior of a lava dome, and an explosive eruption develops. When the vent first opens, the pressure at the vent decreases and an expansion wave migrates down into the conduit, lowering the pressure, causing the volatiles and hence mixture to inflate, and possibly causing further volatile exsolution. If the magma contains only a finite depth of a gas-rich region then this wave eventually comes to rest and may be reflected (Figure 3.11). In contrast, if the conduit is connected to a deep magma chamber, then a complex sequence of waves develop in the conduit as the eruption continues (Wohletz *et al.* 1984), until the rate of downward propagation of the decompression wave is balanced by the propagation of magma upwards from the chamber. Steady migration of magma up the conduit then ensues, as described in section 3.2. The time required for the eruption to reach steady state is typically of the order of minutes to tens of minutes.

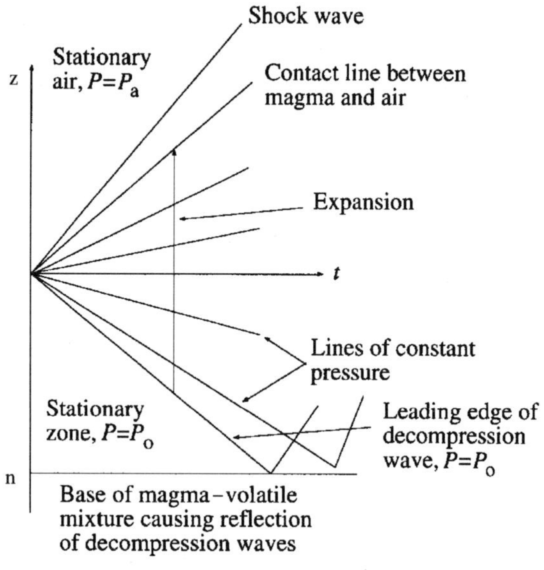

Figure 3.11 Schematic representation of the migration of an expansion wave into the conduit during a Vulcanian eruption. The vertical axis denotes vertical position, with $z > 0$ representing the atmosphere above the conduit, while the horizontal axis denotes time. The eruption starts at $t = 0$ and a decompression wave propagates downwards into the conduit. Ahead of the leading edge of this descending wave, deeper in the conduit, the pressure has its original value and the material is stationary, while above this leading edge the pressure gradually falls. The front of the erupting material advances upwards beyond the vent and forms a "contact line" ahead of which the air is compressed. A shock wave develops in this compressed air and advances into the undisturbed air

As the initial decompression wave propagates into the conduit, an expansion wave moves out of the conduit. This expansion wave is led by a shock wave propagating ahead of the erupting mixture into the air. The motion of such explosion shock waves has been described theoretically by Taylor (1950). Shock waves have been observed in Vulcanian eruptions (e.g. Nairn and Self 1978; Ishihara 1985, 1990). Note that the atmospheric shock wave travels faster than the speed of the erupting mixture because the erupting mixture compresses the air ahead of it and the speed of sound in the atmosphere typically exceeds that in the erupted mixture. Therefore, at any point the initial explosion will be heard before the erupted material arrives. Kieffer (1977) has suggested that any sound waves in the erupted mixture will in fact be dissipated if the speed of sound of the mixture is lower than the actual velocity of the mixture, so that any sound produced in the flow will be trapped in the flow.

Such atmospheric shock waves do not influence the motion of the material in the conduit, because the material issues from the conduit at the speed of sound and so no information can propagate into the conduit from the atmosphere. The rate at which material is erupted at the vent is controlled by the rate of propagation of the decompression wave into the conduit. Turcotte *et al.* (1990) assumed that the material below the hypothesized cap rock or plug was initially volatile-saturated liquid magma, and that on release of the plug, the volatiles were exsolved. Here following Self *et al.* (1979) and Woods (1995b), it is assumed that initially there is a two-phase magma–gas system and that there is no further exsolution of volatiles during the explosion. However, the models lead to very similar predictions for the eruption rate.

The transient conduit flow may be described by the time-dependent relationships for the conservation of mass and momentum. For simplicity, the flow is assumed to be isothermal and homogeneous, and so the conservation of mass and momentum may be written in the form

$$\frac{\partial \rho}{\partial t} + \frac{\partial \rho u}{\partial z} = 0 \tag{3.25}$$

$$\rho\left(\frac{\partial u}{\partial t} + u\frac{\partial u}{\partial z}\right) = -\frac{\partial P}{\partial z} - \rho g - \frac{fu^2\rho}{2r} \tag{3.26}$$

Combining these two equations with the relationship for the density (equation 3.4) two dissipative wave equations are obtained

$$\frac{\partial}{\partial t} + (u \pm a)\frac{\partial}{\partial z}(u \pm c) = -g - \frac{fu^2}{2r} \tag{3.27}$$

where the speed of sound a is defined by

$$\frac{1}{a} = \left(\frac{\partial \rho}{\partial P}\right)^{1/2} = \left[\frac{\rho(n_0 RT)^{1/2}}{P}\right] \tag{3.28}$$

and the characteristic speed c is given by

$$c(P) - c(P_0) = \int_{P_0}^{P} \left(\frac{\partial \rho}{\partial P}\right)^{1/2} \frac{dP}{\rho} \tag{3.29}$$

where P_o is taken to be the initial magma pressure. Assuming the solid density is at least an order of magnitude greater than that of the volatiles, this has the approximate form

$$c(P) = (n_o RT)^{1/2} \log\left(\frac{P}{P_o}\right) + c(P_o) \tag{3.30}$$

At all points in the conduit, the flow is subsonic and so sound waves and information propagate into the conduit from the vent, even though the mixture is moving towards the vent. At the vent, the flow is sonic. The velocity of the magma below the decompressing region is assumed to be zero. Therefore the fragmentation surface propagates into the magma at the speed of sound of the mixture at the original pressure, $a(P_o)$ say, as given by equation (3.28),

$$z = -a(P_o)t \tag{3.31}$$

Friction and gravity are not important during the initial stages of a very shallow eruption (Turcotte *et al.* 1990), and so it follows from equation (3.27) that the quantity $(u + c)$ is constant within the conduit. However, at the leading edge of the decompression wave, the speed $u(P_o) = 0$ and so at any point in the conduit above this front, with pressure P say,

$$u(P) = c(P_o) - c(P) = (n_o RT)^{1/2} \log\left(\frac{P_o}{P}\right) \tag{3.32}$$

where $c(P_o)$ is the speed at the leading edge of the decompression wave and $c(P)$ is given by equation (3.30). Since the flow is sonic at the vent, $z = 0$, the velocity and pressure at the vent is given by the solution of the equation $u(P) = a(P)$. This defines the exit pressure P_e from the relation

$$\frac{1}{\rho} = \frac{n_o RT}{P_e} \log\left(\frac{P_o}{P_e}\right) \tag{3.33}$$

and the speed at which material is discharged from the conduit scales as

$$u = (n_o RT)^{1/2} \log\left(\frac{P_o}{P_e}\right) \tag{3.34}$$

The vent velocity has typical calculated values of 50–150 m s^{-1} depending upon the value of n_o and P_o. The mass flux erupted from the vent during such a transient eruption is shown in Figure 3.12 as a function of the initial pressure and mass fraction of volatiles. Note that the mass flux is larger for smaller initial volatile mass fractions, even though the eruption velocity increases with volatile content. This is because the density of the erupting mixture is much larger for small volatile contents. Since Vulcanian eruptions are short-lived, thermal disequilibrium may develop between the solid and gas phases, leading to lower eruption temperatures and velocities than predicted by the equilibrium model. However, the effects only become significant in very coarse-grained eruptions in which the degree of disequilibrium is high (Woods 1995b).

Once the material has issued from the vent, there will be further decompression of the mixture to atmospheric pressure. This enables the speed of the clasts to increase beyond the speed of sound of the mixture, and thereby attain speeds up to several hundred metres per second, as has been inferred for the eruptions of Ngauruhoe in 1973 (Nairn and Self 1978) and of Arenal in 1968 (Fudali and Melson 1972).

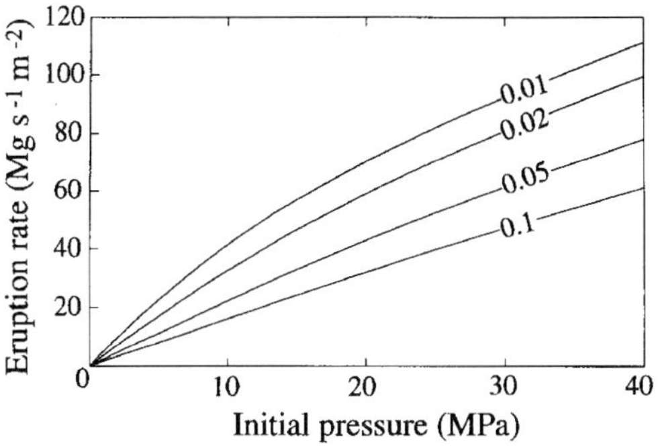

Figure 3.12 Variation of the mass eruption rate as a function of the initial pressure below the cap rock and the initial mass fraction of water (shown on each curve), during a transient Vulcanian eruption

3.5 SUMMARY

Volcanic plumes are generated by explosive eruptions and the source conditions depend on the complex processes of magma ascent, degassing, fragmentation and two-phase flow in conduits and vents. Magma ascending to the Earth's surface becomes supersaturated in dissolved volatiles and exsolves gas to generate explosive flows. The vesiculating magma fragments from a highly viscous magmatic foam to a very low viscosity mixture of gas and particles. This mixture accelerates up the conduit and into the atmosphere to generate high-speed two-phase flows which are the source of volcanic plumes.

Considerable progress has been made in understanding the fundamental controls on flow conditions by computer modelling and experimental investigations in shock tubes. It is clear that the main controls are magma gas content and conduit cross-section which together control mass discharge rates. Ejection velocities are commonly a few hundred metres per second and eruption rates span several orders of magnitude depending principally on conduit size. In Chapter 5 the broad range of predicted flow conditions are shown to occur in nature. Complexities arise, however, because the flows are compressible and they typically reach sonic velocities and a transition can occur to supersonic flows in the near-surface crater and vent system. Overpressured and underpressured flows can develop as well as shock fronts. Modelling shows that the flows can be rather sensitive to near-surface crater or vent geometry. Source conditions for plumes can be influenced by many other factors, including kinetic effects of degassing, thermal effects due to incorporation of lithics eroded from the vent or conduit and external water, incomplete thermal and mechanical coupling of the particles and gas in the two-phase mixture and intrinsic heterogeneities in two-phase flows. Many of these complexities and disequilibrium effects are not yet fully understood and are yet to be included into theoretical models.

4 Eruption Column Models

NOTATION

a_1	2.53×10^8 (kg m^{-1} s^{-2})
a_2	5.42×10^3 K
A	area of plume (m^2)
b	radius of plume (m)
C	average specific heat at constant pressure (J kg^{-1} K^{-1})
C_a	specific heat of air (J kg^{-1} K^{-1})
C_D	drag coefficient (dimensionless)
C_e	specific heat of entrained air (J kg^{-1} K^{-1})
C_o	initial specific heat of eruption mixture (J kg^{-1} K^{-1})
C_s	specific heat of solids (J kg^{-1} K^{-1})
C_v	specific heat of water vapour (J kg^{-1} K^{-1})
C_w	specific heat of liquid water (J kg^{-1} K^{-1})
d	clast size (m)
d_m	maximum clast size at a given height (m)
$e_s(T)$	saturation pressure as a function of temperature (Pa)
E	specific enthalpy (m^2 s^{-2})
f	fraction of solid material in thermal equilibrium (dimensionless)
F_B	effective buoyancy flux of the eruption column (m^4 s^{-3})
F_T	product of the buoyancy and volume of a thermal plume (m^4 s^{-2})
g	gravitational acceleration (m^2 s^{-1})
H	height of rise of an eruption column (m)
H_T	height of rise of a thermal plume (m)
L	latent heat of condensation (J kg^{-1})
m	mass flux in eruption column (kg s^{-1})
m_a	mass flux of air in eruption column (kg s^{-1})
m_s	mass flux of solids in eruption column (kg s^{-1})
m_v	mass flux of vapour in eruption column (kg s^{-1})
n	gas mass fraction in mixture (dimensionless)
N	ambient stratification relative to the adiabat (s^{-1})
P	pressure of mixture (Pa)
$p(d)$	probability density function of clast size d (dimensionless)
q	mass fraction of water in the gas phase (dimensionless)
q_a	mass fraction of water in ambient air (dimensionless)
q_v	mass fraction of water vapour in the eruption column (dimensionless)

q_w	mass fraction of liquid water in the eruption column (dimensionless)
Q	thermal energy flux relative to the environment (kg m^2 s^{-3})
r_T	radius of a spherically symmetric thermal plume (m)
R_H	relative humidity of atmosphere (dimensionless)
R	average gas constant (J kg^{-1} K^{-1})
R_a	gas constant for air (J kg^{-1} K^{-1})
R_c	gas constant for CO_2 (J kg^{-1} K^{-1})
R_v	gas constant for water vapour (J kg^{-1} K^{-1})
T	temperature of mixture (K)
T_1	reference temperature, e.g. ground temperature (K)
u	velocity of mixture (m s^{-1})
w	mass fraction of the gas composed of water vapour (dimensionless)
z	height in plume (m)
z_e	rise height of a particle (m)
z_m	rise height of a particle-laden plume (m)
Z'_o	ratio of particle mass to gas mass (dimensionless)
α	density of environment (kg m^{-3})
β	bulk density of mixture (kg m^{-3})
η_a	ratio of absolute temperature gradient to lapse rate (dimensionless)
ε	entrainment coefficient (dimensionless)
Γ	adiabatic temperature gradient (the lapse rate) (K m^{-1})
ρ	density of the gas (kg m^{-3})
σ	density of solid material (kg m^{-3})
μ_ϕ	logarithmic mean size of clast in plume
σ_ϕ	logarithmic variance of clast size in plume

Subscripts

a	property of environmental air
o	property at vent
e	entrained quantity
s	property of solid material
T	property of a thermal plume
T_o	initial property of thermal plume
v	property of vapour
w	property of liquid water

4.1 INTRODUCTION

In this chapter the physical principles described in Chapter 2, and the source conditions described in Chapter 3, are applied to develop a physical model of the dynamics of volcanic eruption columns. Several eruption styles produce eruption columns, as described in Chapter 1. The principles and understanding developed in the present chapter are subsequently applied to these different eruption styles in the following chapters.

An eruption column is composed of a mixture of particles, volatiles, water, vapour and air. This mixture ascends into the atmosphere as a complex multiphase flow, in which there is heat and momentum transfer between the constituents of the flow, interphase mass transfer resulting from vaporization or condensation, loss of mass through fallout of particles and addition of mass through entrainment of ambient air. The material may be erupted with mass fluxes ranging from 10^4 to 10^9 kg s^{-1} and heat fluxes from 10^{10} to 10^{15} W s^{-1}. Eruption temperatures can be up to 1400 K above the ambient temperature. These values are much greater than industrial plumes, for example from cooling towers, in which the heat flux issuing from the cooling tower is only of order 10^5 W s^{-1}, with exit temperatures only 10–100 K higher than ambient. The dominant processes in operation in a particular eruption column depend upon the particle size distribution, the temperature and the mass of erupted material. Furthermore, the atmosphere into which the column ascends can exert important controls upon the motion of the column, both through the density stratification with height, and also through the initiation of moist atmospheric convection by the eruption column.

In this chapter, a hierarchy of models is developed in which the different processes which influence the motion of an eruption column are identified. The modelling and intuition draws heavily from Chapter 2. A variety of models of maintained eruption columns are considered in some detail, and then the motion of discrete volcanic thermals and starting plumes are analysed. The chapter concludes with a discussion of multiphase supercomputer models. Specific comparisons of the model with observations of volcanic eruptions are left to Chapter 5.

4.2 DENSITY VARIATIONS IN ERUPTING MIXTURES

The density of a well-mixed parcel of hot pyroclasts, magmatic gas and air varies according to the temperature and mass fraction of pyroclasts in the mixture. Before considering dynamical models of eruption columns, the density dependence of the various mixtures of pyroclastic partices and gas is examined. In conjunction with the discussion of non-linear mixing in simple plumes given in Chapter 2, this leads to insight about the possible styles of behaviour of eruption columns.

For pyroclasts of density σ, gas of density ρ and mass fraction of gas n, which includes both entrained air and volcanic gases, the density of the mixture, β, is given by

$$\frac{1}{\beta} = \frac{1-n}{\sigma} + \frac{n}{\rho} \tag{4.1}$$

Assuming that the gas phase behaves as a perfect gas, with gas constant R and temperature T, then the gas density is given by

$$\rho = \frac{P}{RT} \tag{4.2}$$

where P is the pressure of the mixture. The gas constant R is given by the mass average of the gaseous components: for air, $R_a = 285$ J kg^{-1} K^{-1}; for CO_2, $R_c = 185$ J kg^{-1} K^{-1} and for water vapour, $R_v = 460$ J kg^{-1} K^{-1}. The density of an isobaric model mixture of pyroclasts and gas is compared with the density of the environmental air, at pressure P and temperature T_a. It is assumed that the thermal energy of the solids and gas is conserved, that the solid particles initially have a temperature T_o and that the particles in the mixture are sufficiently small that particles and gas tend to thermal equilibrium within seconds (Wilson 1976; Woods and Bursik 1991). Thermal equilibrium is usually a good assumption, but there can be exceptions.

For typical mass fractions of water in the range 0.005–0.07, mixtures of magmatic water, CO_2 and vapour at 0.1 MPa have densities that are considerably greater than the atmosphere. Therefore, the material erupted from the vent is negatively buoyant. Figure 4.1a illustrates the variation of the initial density of an erupting mixture with the mass fraction of exsolved volatiles and with the eruption temperature at sea-level atmospheric pressure (0.1 MPa).

As the erupting mixture entrains and mixes with the ambient atmosphere the density decreases. Conservation of heat between the entrained air and the eruption products may be expressed as

$$(1-n)C_sT_s + n((1-w)C_a + wC_v)T_a = ((1-n)C_s + n((1-w)C_a + wC_v))T \quad (4.3)$$

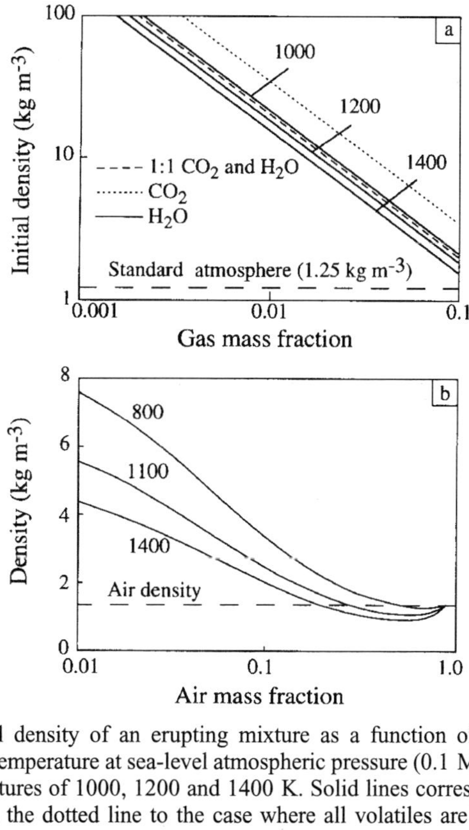

Figure 4.1 (a) Initial density of an erupting mixture as a function of the initial magma water content and the initial temperature at sea-level atmospheric pressure (0.1 MPa). Curves are shown for initial mixture temperatures of 1000, 1200 and 1400 K. Solid lines correspond to the case where all the volatiles are water, the dotted line to the case where all volatiles are CO_2, and the dashed line corresponds to a 1 : 1 mixture of water and CO_2. (b) Density of a mixture of entrained air and erupted pyroclasts and volatiles as a function of the mass fraction of entrained air. The density is shown for three eruption temperatures in kelvin. The mass fraction of water in the erupted mixture is assumed to be 0.03

where T is temperature, C the specific heat, w the mass fraction of water vapour in the gas phase of the mixture and the subscripts s, a and v refer to the solid particles, ambient air and water vapour respectively. Values of specific heat used in this book are $C_s = 1100\ \mathrm{J\ kg^{-1}\ K^{-1}}$, $C_a = 1000\ \mathrm{J\ kg^{-1}\ K^{-1}}$ and $C_v = 1860\ \mathrm{J\ kg^{-1}\ K^{-1}}$.

The density of the mixture is much greater than the ambient air when the gas mass fraction is small, but decreases rapidly during entrainment in the gas thrust region of the column, falling below that of the environment, as the gas mass fraction increases (Figure 4.1b). At high temperatures, the gas is heated by the solids, expands and has a very low density. As a result, even with dense solid particles, the density of the mixture becomes substantially less than the environment. As air is added, the mixture eventually attains a minimum density. As more air is added, the density tends towards the original density of the air, since the mass fraction of ash present is so small that its effect becomes negligible. For a given gas mass fraction, the density decreases as the initial temperature of the solids increases because more heat is available to increase the temperature of the entrained air. Note that if the vapour fraction in the magmatic gas phase increases, the density also increases, owing to the greater gas constant of vapour. However, variations in the vapour content have a relatively small effect in comparison to the variations in the possible range of eruption temperatures.

The non-linear density variations of mixtures of hot pyroclasts and air show that a dense mixture of pyroclasts and volatiles erupted from a vent can become buoyant, and ascend into the atmosphere through heating of entrained ambient air. A number of complications to this picture arise, however, including the effects of thermal disequilibrium between the solids and the gas, loss of solid particles through sedimentation from the mixture and the release of latent heat as water vapour becomes saturated and condenses. These issues are addressed later in the chapter.

The initial temperature of the solid particles in an erupting mixture depends upon the initial magma temperature and also the cooling during degassing (Chapter 3). For example, rhyolitic magmas typically contain about 0.06 mass fraction of water and so cool about 20 K on degassing. In more mafic magmas, there is also additional latent heat released by partial quench crystallization. Some of the ejecta in any eruption is made of accidental lithic clasts which have a low temperature and lower the mean temperature of the mixture (Chapter 3). External water will also produce a lower temperature. This is discussed in Chapter 3 and in more detail in Chapter 8.

4.3 FINE-GRAINED ERUPTION COLUMNS

In the limiting case in which the pyroclasts are very fine-grained, the heat and momentum transfer between the particles and the gas in an eruption column are rapid, and it is appropriate to model the mixture as a quasi-one-phase mixture of uniform temperature and velocity (Wilson 1976; Sparks and Wilson 1976; Woods and Bursik 1991). This is sometimes known as the pseudofluid approximation. In this situation, the simple density mixing relationship described in the previous section is applicable, and processes such as sedimentation of pyroclasts from the ascending column are negligible. A model is therefore developed of a vertical eruption column originating from a single source vent in a similar fashion to the plume models of Chapter 2 and previous studies (Wilson 1976; Wilson *et al.* 1980; Sparks and Wilson 1982; Sparks 1986; Wilson and Walker 1987; Woods 1988; Woods and Bursik 1991; and Woods 1993b).

The initial motion of an ascending eruption column depends upon the source conditions for discharge of pyroclasts and gas. As described in Chapters 1 and 3, material is erupted at high speed from a volcanic vent. Models of magma ascent in conduits, described in Chapter 3, predict that high-speed jets rising from volcanic vents may reach speeds of several hundreds of metres per second. They can erupt either at sonic or supersonic velocities, and can be either pressure adjusted at the vent or overpressured on eruption. However, the flow rapidly expands and decompresses to near ambient pressure. In this chapter the motion of an eruption column is modelled above the height at which it has decompressed.

The material erupted from the vent ascends and forms a large convecting column, with two dynamically distinct regions. In the lower *gas-thrust region* (Figure 4.2a), the column is relatively dense and is driven upwards purely by its inertia, essentially as a jet (Chapter 2). Once the column becomes less dense than the surrounding atmosphere, the material is driven upwards by its buoyancy, and tends to behave as a buoyant plume in the *convection region* (Figure 4.2a). The mixing efficiency of the column with the surrounding air varies between the gas thrust region and the convecting region. In the gas thrust region, the material is dense and can be travelling near the sonic speed, so that entrainment is significantly suppressed in comparison to the convection region, higher in the column (Kieffer and Sturtevant 1984). In the column model developed here, a parameterization is introduced for the entrainment efficiency. As shown in Chapter 2, the details of this parameterization in the lower column are unlikely to have a large influence upon the total height of rise of a buoyant column. This is because the motion of a buoyant column evolves towards that of a buoyant plume with height. However, the particular parameterization chosen has some bearing upon the conditions under which the dense mixture can become buoyant and thereby form a stable plume.

All the material in the column is assumed to ascend as one phase, and so a top-hat model for the conservation of mass and momentum (Woods 1988) in the form shown below can be developed

$$\frac{\mathrm{d}(\beta u b^2)}{\mathrm{d}z} = 2\varepsilon u b \alpha \qquad (4.4)$$

$$\frac{\mathrm{d}(\beta u^2 b^2)}{\mathrm{d}z} = b^2 g(\alpha - \beta) \qquad (4.5)$$

where u is the vertical velocity, b the column radius, z the height, α the density of the atmosphere, ε the entrainment coefficient and g the gravitational acceleration. The definitions of the main parameters are depicted in Figure 4.2b. It is assumed that the interphase momentum transfer is rapid, as justified by Woods and Bursik (1991). In an eruption column, the enthalpy is also conserved as specified by the steady flow energy equation (Shapiro 1954; Woods 1988)

$$\frac{\mathrm{d}\left(\beta u b^2\left(CT + \dfrac{u^2}{2} + gz\right)\right)}{\mathrm{d}z} = 2\varepsilon u b \alpha (C_\mathrm{e} T_\mathrm{e} + gz) \qquad (4.6)$$

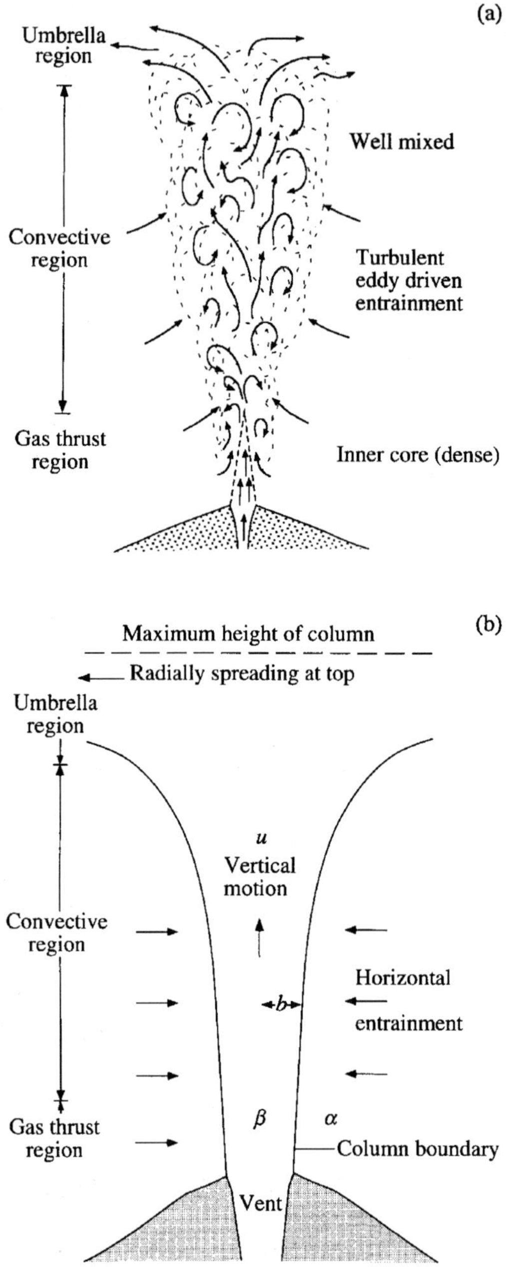

Figure 4.2 (a) Schematic representation of the different regions of a volcanic eruption column, including the lower gas thrust region, the main convective region and the laterally spreading umbrella cloud at the top of the column. (b) Schematic representation of the model eruption column, in which the properties of the plume are assigned one characteristic value at each height, and the entrainment process is modelled with a horizontal inflow velocity

where subscript e denotes a property of the entrained material which is typically composed of air and water vapour. Equation (4.6) follows from the first law of thermodynamics. The left-hand side denotes the increase of the enthalpy, i.e. the total internal energy of the system plus the work done by the system in decompressing the mixture as it ascends to higher altitudes where the pressure is lower. The right-hand side denotes the heat transfer to the system through entrainment of ambient air and the work done by the system as it rises through the atmosphere. If a control volume is considered, fixed in space and enclosing a horizontal slice of the steady column, then the net flux of enthalpy into the system equals the net work done on the system and the heat transfer to the system. In steady state, both the shear work on the control volume due to molecular friction and the heat transfer across the plume boundary are very small in comparison to the turbulent entrainment, leading to the above steady equation for the conservation of enthalpy (Woods 1988).

The final conservation law required for this model of an eruption column is the conservation of solid particles. In this, the simplest model, all the particles are assumed to remain in the column, and so the mass flux of clasts at any height in the column equals the initial mass flux of pyroclasts. This yields the equation

$$\beta u b^2(1 - n) = \beta_o u_o b_o^2(1 - n_o) \tag{4.7}$$

where the subscript o refers to the initial values at the base of the pressure-adjusted flow.

4.3.1 Gas Thrust Region

In the gas thrust region, a number of models have been adopted for the entrainment law (Wilson 1976; Woods 1988). Following the discussion of jets and plumes in Chapter 3, three models of the mixing efficiency in a jet are compared to determine possible differences in the behaviour of the gas thrust region. The simplest model is to assume that the gas thrust region is fully mixed and entrains with the same efficiency as a plume, with entrainment coefficient $\varepsilon = 0.09$ (defined as model B). Another possible model is to assume the entrainment coefficient has a smaller value in the gas thrust region, $\varepsilon = 0.06$ corresponding to that of a jet (Papantoniou and List 1989); and that this increases to 0.09 when the column becomes buoyant (defined as model A). A third, more complex model is to assume that the entrainment depends upon the relative density of the jet and the ambient air (Woods 1988) so that

$$\varepsilon = 0.09\sqrt{\frac{\beta}{\alpha}} \tag{4.8}$$

The third model (defined as model C) is based upon a development of Prandtl's jet theory (1954), including effects of the density variation which produce a buoyancy force (Wilson 1976) and which change the entrainment efficiency. The height of the gas thrust region, defined as the region in which the plume is negatively buoyant, changes with the entrainment parameterization of the three models above (Figure 4.3). For a smaller entrainment coefficient, with the same mass flux and eruption velocity, the gas thrust region extends further before becoming buoyant. There is a somewhat wider range of eruption conditions in which the upward velocity of the erupting mixture decreases to zero before the material becomes buoyant. In these circumstances the column collapses as a

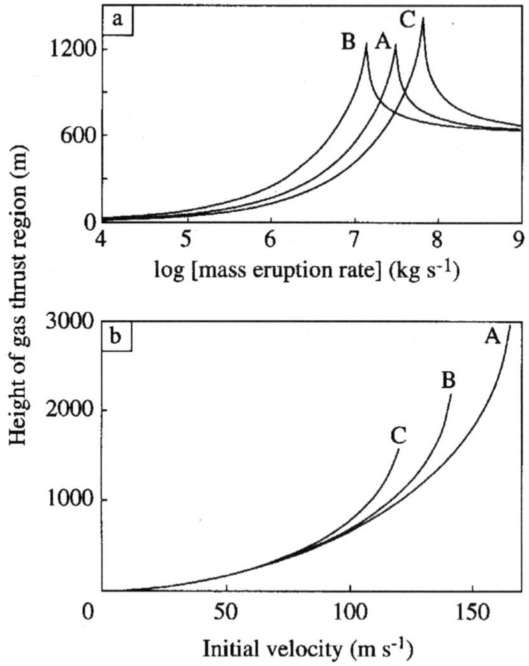

Figure 4.3 Variation of the height of the gas thrust region as a function of the eruption rate for an eruption velocity of 100 m s^{-1}, mass fraction of water of 0.03 and eruption temperature of 1000 K. The eruption velocity is for a fixed volume eruption rate (8×10^6 kg s^{-1}) with a water mass fraction of 0.03 and eruption rate of 1000 K. The figure compares the three entrainment models A, B and C described in the text

fountain as considered in more detail in Chapter 6. Figure 4.3b illustrates that the height of a collapsing fountain increases for a given eruption rate as the eruption velocity decreases and that for a greater entrainment rate, the height also decreases for a given velocity. However, the effects of the different parameterizations of the entrainment rate are relatively small, and in all other calculations the third model is adopted.

The physics of the gas thrust region in fact is far from completely understood. There are numerous complications that have yet to be fully incorporated into the models such as the effects of particle size and concentration (Bursik 1989; see section 4.9) and the entrainment in the overpressured parts of the jet.

4.3.2 Convective Region

In the convective region the entrainment coefficient is taken as a constant, 0.09, as is the case with simple buoyant plumes (Chapter 2).

4.3.3 The Atmosphere

Combination of the conservation equations for the column with the entrainment relation and the thermodynamic model of a well-mixed gas–pyroclast mixture (section 4.2), allows a model for an eruption column to be developed. The only additional complexity is that the atmosphere surrounding the eruption column needs to be parameterized. In order to study particular eruptions, the local atmospheric conditions can be obtained from a radiosonde of the atmosphere. However, in developing a general theoretical model of a column, it is useful to use a standard atmosphere model. The standard atmosphere given by Gill (1982) is adopted in which the temperature is assumed to vary with height according to

$$T = 273 - 6.5z \quad \text{for} \quad 0 < z < 11\,\text{km} \tag{4.9i}$$

$$T = 201.5 \quad \text{for} \quad 11 < z < 20\,\text{km} \tag{4.9ii}$$

$$T = 201.5 + 2(z - 20) \quad \text{for} \quad 20 < z < 40\,\text{km} \tag{4.9iii}$$

The atmosphere is assumed to be in hydrostatic equilibrium, and so the pressure gradient, $\mathrm{d}P/\mathrm{d}z$, is

$$\frac{\mathrm{d}P}{\mathrm{d}z} = -\alpha g \tag{4.10}$$

Finally the atmosphere is assumed to satisfy the equation of state for a mixture of water and vapour

$$P = \alpha RT \tag{4.11}$$

where R is the mass averaged gas constant (section 4.2). In the following sections the basic physical processes controlling the motion of eruption columns are examined and extended to include effects associated with thermal disequilibrium and sedimentation of large pyroclasts. The influence of the condensation and evaporation of atmospheric water vapour is then considered.

4.3.4 The Motion of Dry, Dusty Eruption Columns

Figure 4.4 shows the results of calculations for variations of the temperature, velocity and relative density with height in a model of a typical eruption column. Note that in subsequent calculations in this chapter the initial radius is taken as the height in the eruption column where the erupting mixture has decompressed to 0.1 MPa. This radius only corresponds to the vent radius in special circumstances. The velocity and density of the column rapidly decrease just above the vent in the gas thrust region. The velocity decreases due to the large gravitational deceleration of the dense column; density decreases because air entrained into the column is heated by the solid clasts and expands (section 4.2). Once the column becomes buoyant, the material in the column can accelerate upwards and the momentum flux can increase. The mass flux continually increases with height owing to entrainment of air. The temperature of the column decreases steadily as the column ascends and mixes with the colder ambient air. The column eventually reaches a height at which its density equals that of the atmosphere, referred to as the neutral buoyancy height. As the material ascends above the neutral buoyancy height, it becomes denser than the surrounding air. The column continues entraining ambient air and the temperature and density continue to fall relative to the values in the environment. Since the atmosphere is stably stratified, then near the top of the column, the temperature eventually decreases below that of the surrounding environment.

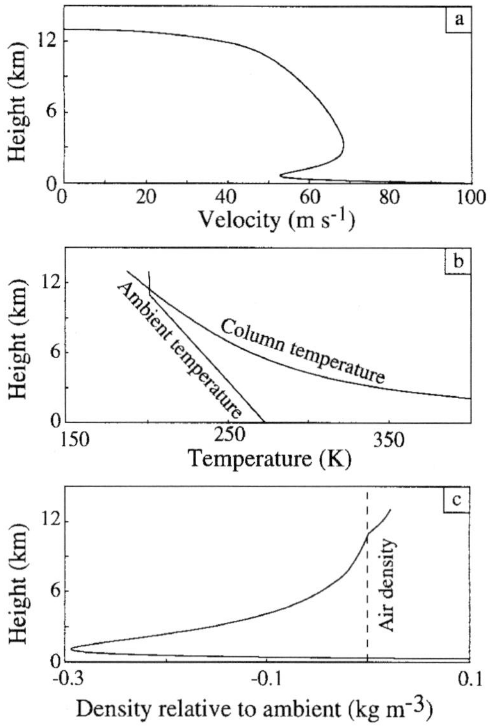

Figure 4.4 Typical variation with height in an eruption column of (a) velocity, (b) temperature and (c) density relative to the ambient atmospheric density. In these calculations the initial velocity is 100 m s^{-1}, the initial radius is 50 m, the eruption temperature is 1000 K and the initial mixture contains 0.03 mass fraction of water

The principal energy transfer in an eruption column is from the thermal energy of the hot material erupted from the vent to the potential energy acquired by this material and the entrained air in ascending to the neutral buoyancy height. The rate of increase of the potential energy at the top of the column is the dominant sink of the enthalpy flux. The eruption column can be regarded as an energy conversion mechanism. The erupted thermal energy is used to heat and expand entrained ambient air, and thereby produce a buoyant mixture which ascends into the atmosphere, imparting potential energy to the erupted material and the entrained air. Indeed the thermal energy flux of a typical eruption column is of the order of 10–100 times greater than the kinetic energy associated with explosive expansion of the magma during decompression.

As the eruption conditions at the vent change, the dynamical balances in the eruption column change, leading to changes in the variation of the velocity and other properties in the column with height. Changes in the initial velocity and radius change the overall velocity profile (Figure 4.5). For large eruption velocities, the velocity decreases monotonically with height in the column. However, for smaller eruption velocities and larger initial radii, the velocity decreases to much smaller values in the gas thrust region. Therefore, on becoming buoyant, the rate of entrainment of ambient air, which is proportional to the velocity of the column, is small, and so the rate of addition of mass is

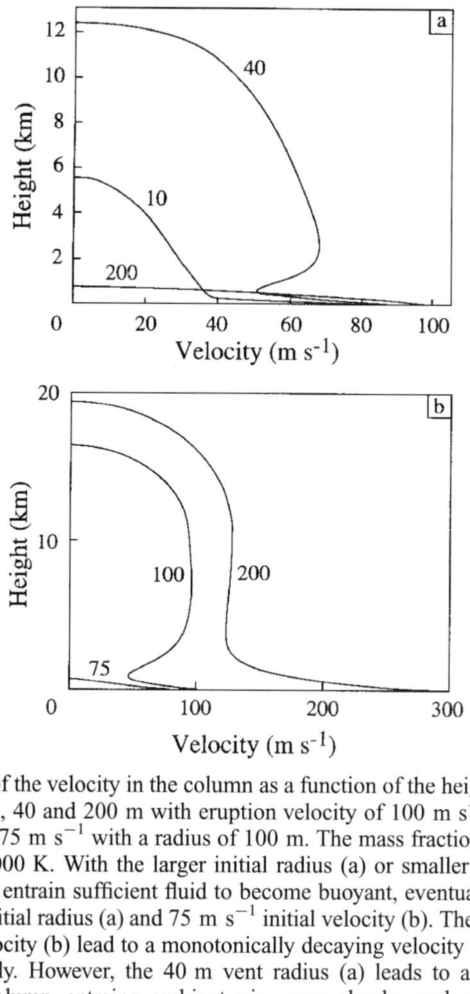

Figure 4.5 Variation of the velocity in the column as a function of the height. Curves are shown for (a) three initial radii, 10, 40 and 200 m with eruption velocity of 100 m s^{-1}, and (b) three eruption velocities 200, 100 and 75 m s^{-1} with a radius of 100 m. The mass fraction of water is 0.03 and the eruption temperature 1000 K. With the larger initial radius (a) or smaller eruption velocity (b) the material takes longer to entrain sufficient fluid to become buoyant, eventually leading to collapse in the case of the 200 m initial radius (a) and 75 m s^{-1} initial velocity (b). The 10 m vent radius (a) and 200 m s^{-1} eruption velocity (b) lead to a monotonically decaying velocity profile, since the material becomes buoyant rapidly. However, the 40 m vent radius (a) leads to a non-monotonic velocity profile, because the column entrains ambient air more slowly, and so the velocity falls off dramatically before the material becomes buoyant. A column with this non-linear velocity profile is referred to as superbuoyant. After Bursik and Woods (1991)

smaller than the rate of increase of momentum due to the upward buoyancy force on the column. Thus the velocity increases with height in the column. This non-linear velocity structure in the eruption column is referred to as *superbuoyancy* (Bursik and Woods 1991). Which profile occurs depends both on velocity and initial radius and therefore mass flux. Figure 4.6 shows a regime diagram where the fields of simple buoyant and superbuoyant plumes are distinguished.

4.3.5 Fountain Collapse

Under some conditions the velocity in the column decreases to zero before the column is able to mix with sufficient ambient air to become buoyant. This leads to the formation of

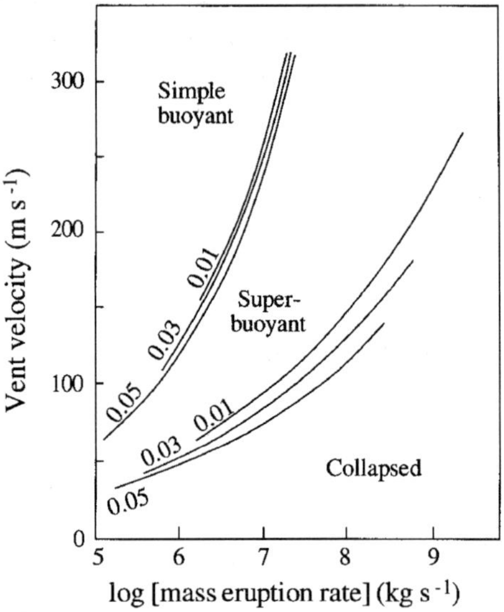

Figure 4.6 Regime diagram of eruption rate and eruption velocity showing the transition from collapsing fountains to superbuoyant columns to simple buoyant columns. Solid curves are labelled with the initial mass fraction of water, and are taken from Bursik and Woods (1991)

a "collapsing fountain" and the subsequent generation of pyroclastic flows, as discussed further in Chapter 6. Collapsing fountains tend to form if the initial upward velocity is small or the vent radius is large (Figure 4.6).

4.3.6 Column Height

A particularly useful prediction is the total height of rise of the column, and its dependence upon the erupted mass and enthalpy fluxes. Such predictions may be used for models of pyroclast fallout, to estimate the climatic impact of eruptions, and to evaluate hazards for aircraft. The total height of rise of an eruption column can be estimated using a simple expression based upon the energy flux supplied to the column and the atmospheric stratification. In order to make such an estimate for the column height, the effective buoyancy flux associated with the energy flux of the column needs to be quantified. Since the material erupted from the vent is in fact denser than the air, the effective buoyancy flux, F_B, is defined in terms of the excess thermal energy associated with the column (Settle 1978; Wilson *et al.* 1978). When the temperature contrast is small relative to the ambient temperature, this has the form

$$F_B = \left[\frac{gub^2\beta(T_o - T_a)}{\alpha T_a}\right] \tag{4.12}$$

where T_o is the initial temperature of the mixture and T_a is the atmospheric temperature.

The stratification in the environment is defined relative to the adiabatic stratification. The stratification is therefore given by

$$N^2 = \frac{g\Gamma}{T_1}(1 + \eta_a) \tag{4.13}$$

where Γ is the adiabatic rate of decrease of temperature in the atmosphere (the lapse rate), η_a is the ratio of the absolute temperature gradient to the lapse rate and T_1 is a reference temperature, for example the ground temperature. In terms of this simple model the height of rise of an eruption column, H, scales as (Morton *et al.* 1956)

$$H = 5F_B^{1/4}N^{-3/4} \tag{4.14}$$

In typical atmospheric conditions, this formula can be simply expressed as

$$H = 5\left(\frac{gQ}{\alpha C_o T_a}\right)^{1/4} N^{-3/4} \tag{4.15}$$

where Q is the total thermal energy flux due to the hot pyroclasts and is related to the mass flux by

$$Q = C_o(T_o - T_a)\beta_o u_o b_o^2 \tag{4.16}$$

where C_o is the average specific heat of the pyroclasts and volatiles at the vent, $\beta_o u_o b_o^2$ is the initial mass flux and T_o the initial temperature.

This model has been shown to predict the height of rise of some historic eruptions quite accurately (Settle 1978; Wilson *et al.* 1978; Sparks 1986), and is discussed in more detail in Chapter 5. However, the scaling approach does not identify the full richness of dynamical behaviour which is exhibited by eruption columns. Also, the model does not include the complexities of the non-uniform stratification in the atmosphere, particularly above the tropopause, where the temperature gradient changes substantially (section 4.3.3).

In Figure 4.7 the rise heights of model eruption columns for a number of eruption temperatures are shown as predicted by the eruption column model and as predicted by the simple scaling argument (equation 4.14). The height of rise of an eruption column increases with eruption temperature for a given enthalpy flux because the maximum buoyancy that can be generated through mixing with ambient air increases with the initial temperature of the solids (section 4.2). However, the buoyancy is only released through mixing with the air, and so is generated above rather than at the source. The simple scaling relationship and the numerical models are in excellent agreement at low heights, but there is an increasing difference as column height increases. Generally the numerical models predict somewhat lower heights for a given mass flux. This is a consequence of the change of stratification above the tropopause which is not taken into account in the simple scaling model, leading to overestimation of column height for large mass fluxes. The details of the variation in column height and the vertical structure of the properties of the plume cannot be resolved by simple scaling laws. Furthermore, non-uniformities in the atmospheric stratification affect the shape of the plume as discussed in section 4.10.

4.4 PARTICLE FALLOUT AND THERMAL DISEQUILIBRIUM

The particle sizes of volcanic ejecta range from very small, submicron-sized ash particles to large bombs, up to several metres in size. Analyses of field data from many fall deposits suggest that total particle-size distributions are approximately log-normal and that the clast sizes vary over several orders of magnitude (Suzuki 1983; Sparks *et al.* 1981). In a coarse-grained eruption, as the larger clasts fall out of the column rapidly, they do not transfer all

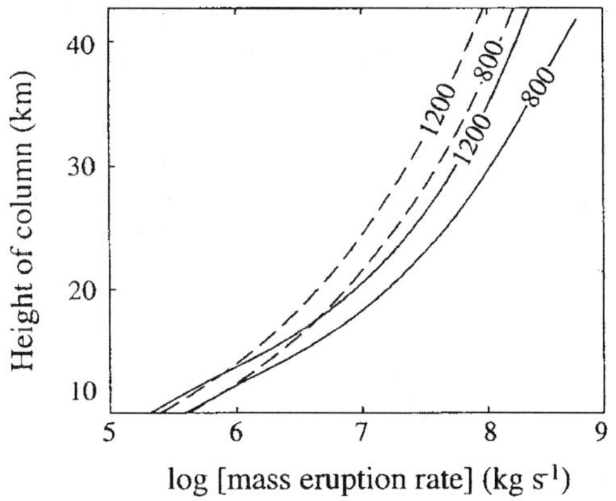

Figure 4.7 Eruption column height as a function of mass eruption rate. Solid lines are calculated from the model of Woods (1988) and dashed lines are calculated from Wilson *et al.* (1978). Curves are shown for eruption temperatures of 800 and 1200 K as labelled in the figure

their heat to the column. Evidence for this inefficiency in heat transfer is provided by welding and high emplacement temperatures in the coarse-grained proximal regions of some pyroclast fall deposits (Thomas and Sparks 1992). The model developed above for fine-grained eruptions can be extended to include these effects (Wilson and Walker 1987; Woods and Bursik 1991). In this extension, fallout is assumed to occur at the height at which the weight of the clast equals the upward drag force of the surrounding flow, of velocity u,

$$4\sigma g d^3 = 3C_D u^2 \rho_a d^2 \tag{4.17}$$

Here $C_D = 0.75$ is the drag coefficient, with d the average diameter of the clasts (Walker *et al.* 1971; Carey and Sparks 1986; Wilson and Walker 1987) and σ, ρ_a are the densities of the solid clasts and the gas phase. The fallout model is a simplification which should give an estimate of the mass lost with height. It is equivalent to assuming that all the pyroclasts at a given locality in the deposit are of the same size. Fallout is assumed to occur as clasts are driven to the edge of the column by the turbulent eddies, at which point they fall into the atmosphere. This phenomenological model allows us to investigate the influence of particle fallout and thermal disequilibrium upon eruption columns. Chapters 14 and 15 consider more detailed models of the sedimentation mechanisms from the plume margins.

The main effect of allowing fallout is that the conservation of mass and enthalpy fluxes must now include a term to represent the particle loss. If d_m denotes the maximum clast size which can remain in the column at height z, then the mass flux of clasts in the column is given by (Woods and Bursik 1991)

$$m_s = \sigma A_o u_o \int_0^{d_m} p(s)\,\mathrm{d}s \tag{4.18}$$

and so this changes according to

$$\frac{\mathrm{d}m_s}{\mathrm{d}z} = \sigma A_o u_o p(d_m) \frac{\mathrm{d}d_m}{\mathrm{d}z} \tag{4.19}$$

where d is the clast diameter, $p(d)$ is the probability distribution function, A_o the area of the base of the plume and u_o the initial velocity; $p(d)$ is defined in terms of the logarithmic mean μ_ϕ and log-normal variance σ_ϕ of the grain size distribution, according to the relation

$$p(d) = \frac{1}{\sqrt{2\pi} d\sigma_\phi} \exp\left(-\frac{(\log_2(d) - \mu_\phi)^2}{2\sigma_\phi^2}\right) \tag{4.20}$$

The mass conservation equation then becomes

$$\frac{d\beta ub^2}{dz} = 2\varepsilon ub\alpha + \frac{dm_s}{dz} \tag{4.21}$$

and the steady-flow energy equation becomes

$$\frac{d\beta ub^2\left(\frac{u^2}{2} + gz + CT\right)}{dz} = 2\varepsilon\alpha ub\left(\frac{u_e^2}{2} + gz + C_e T_e\right) + \frac{dm_s}{dz}\left(CT + gz + \frac{u^2}{2}\right) \tag{4.22}$$

The second term on the right-hand side of these two equations denotes the loss of mass and the loss of enthalpy as a result of fallout. The rate of change gas mass fraction in the column with height may be deduced from the mass flux conservation law, noting that at height z the mass flux of solids in the column m_s is related to the mass of solids in the column $\beta b^2(1 - n)$ by the relation

$$m_s = \beta b^2(1 - n)u \tag{4.23}$$

In this form, the model accounts for fallout of clasts from the column. If it is assumed that the clasts remain in thermal equilibrium with the column, then one can isolate the effect of the fallout of clasts on the motion of the column. The numerical solution of the above set of equations, for realistic column properties suggests that in thermal equilibrium, the column behaviour is insensitive to the fallout. This is because the column moves at such high speed that only very large clasts fall from the lower region of the column. These typically represent only a small fraction of the total solid mass erupted. Therefore, the column is able to entrain a great mass of air before a significant amount of fallout has occurred. When the fallout does occur, higher in the column, the gas mass fraction is very large and most of the thermal energy erupted has been transferred from the solids to the gas. Only in unusually coarse-grained eruptions does the fallout from the column occur sufficiently close to the vent that the thermal energy lost through fallout is significant.

The simple approximation that the column remains in thermal equilibrium is only strictly valid in very fine-grained eruptions. The time for a clast to equilibrate thermally is of the order of $100d^2$ seconds, where d is the clast diameter measured here in centimetres (Woods and Bursik 1991). Therefore, coarser clasts, of size larger than a few millimetres, require a significant fraction of the ascent time to equilibrate thermally with the surrounding gas. As a result, many clasts are hotter than the surrounding air as they fall out, and so the loss of thermal energy from the column may be much greater than if they were in thermal equilibrium (Thomas and Sparks 1992). This effect may be investigated in a very simple heuristic manner by assuming that only a fraction f of the solid material is in thermal equilibrium with the gas (Wilson 1976; Wilson *et al.* 1978; Woods and Bursik 1991). This is incorporated into the model in the definition of the bulk specific heat of the column, as used in the definition of the enthalpy,

$$C = \frac{C_a m_a + fC_s m_s + C_v m_v}{m_a + m_s + m_v} \tag{4.24}$$

The subscripts a, s and v denote the entrained air, solid and water vapour respectively. In the present calculations, the magmatic volatiles are assumed to be entirely composed of water vapour. As the mean particle size of the erupted material increases, the thermal disequilibrium also increases. In Table 4.1, some simple calculations are shown which relate the value of f to the mean particle size (cf. Woods and Bursik 1991). This parameterization allows the investigation of the effect of thermal disequilibrium. However, it does not represent a full solution of the equations, which requires a more complex, fully two-phase model, as discussed later in the chapter.

In Figure 4.8a calculations are presented, based upon this model, of the height of rise of an eruption column, for several values of the thermal disequilibrium parameter f. This parameter decreases as the mean particle size increases because large clasts require much longer to transfer their heat (Table 4.1). As the thermal disequilibrium increases, the column height decreases because less thermal energy is released to heat the entrained air and generate buoyancy. However, if the degree of disequilibrium is arbitrarily fixed, then as the mean particle size of the erupted material changes, the column height does not vary significantly. Differences in the variance of the particle size distribution do not affect the results significantly (Woods and Bursik 1991).

As the degree of thermal disequilibrium increases, the minimum initial velocity required for the erupted material to become buoyant, and hence form a buoyant column, increases. This is because less of the thermal energy in the solids is available to heat the entrained air, and so it must entrain more air in order to become buoyant (Figure 4.8). This has the interesting and important effect that for a given mass eruption rate and eruption temperature, variations in the particle size distribution of the ejecta can change the qualitative behaviour of the column, from a buoyant plume to a collapsing fountain.

In this context attention is drawn to the volcanological literature which recognizes two rather different kinds of fountain. In fountain collapse of columns containing large proportions of fine pyroclasts, pyroclastic flows are formed. However, in eruptions which produce coarse ejecta a "fire fountain" forms and this need not necessarily result in a

Table 4.1

Thermal disequilibrium factor f	Mean grain size $\bar{d}$ (mm)
1.0	$\bar{d} < 2$
0.9–1.0	$2 < \bar{d} < 8$
0.7–0.9	$8 < \bar{d} < 16$
0.3–0.7	$16 < \bar{d} < 32$
0.1–0.3	$32 < \bar{d} < 64$
< 0.1	$64 < \bar{d}$

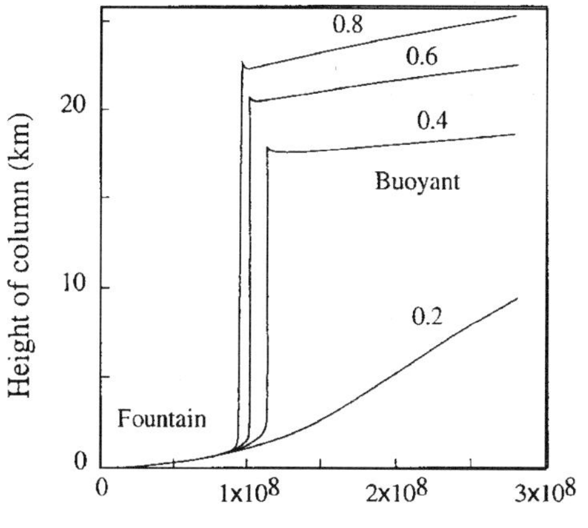

Figure 4.8 Effect of thermal disequilibrium between clasts and air upon the height of rise of an eruption column. Numbers on curves denote the mass fraction of solids in good thermal contact with the gas. With very poor thermal contact, the material does not become buoyant (0.2) and a small collapsing fountain is predicted to develop. The sharp increase in column height which occurs on the curves labelled 0.4, 0.6 and 0.8 corresponds to the point at which the erupted material is able to become buoyant and form a convecting eruption column. After Woods and Bursik (1991)

pyroclastic flow. Basaltic eruptions commonly produce fire fountains (see Chapter 10) of coarse spatter lumps and bombs. The calculations in Figure 4.8 are relevant to both types of fountains.

4.5 ATMOSPHERIC CONTROLS ON COLUMN BEHAVIOUR

4.5.1 Variations in the Environmental Stratification

The model described above considered a standard atmosphere. However, variations in the atmospheric stratification resulting from vertical temperature variations, in particular the location of the tropopause and the intensity of the stratification in the stratosphere, can affect the ascent height of eruption columns. In the troposphere, the atmosphere is only weakly stratified. However, above the tropopause the stronger stratification limits the subsequent ascent of the column. In a typical mid-latitude atmosphere, the temperature decreases steadily to the tropopause, which is located at about 11 km. The temperature then remains nearly constant up to a height of about 20 km where it begins to increase, and the atmosphere becomes even more strongly stratified (e.g. Gill 1982). In contrast, in the tropical atmosphere, the tropopause is located much higher at about 16–18 km. The temperature then increases sharply above this point, producing a very stably stratified atmosphere. In polar latitudes, the tropopause may occur as low as 8 km, and the stratosphere above is then much less strongly stratified than in the tropical or mid-latitude case.

As a result of these variations in the ambient stratification, the height of rise of the column at a fixed eruption rate can vary by a few kilometres. First, in a given atmosphere, such as that considered in the model of section 4.3, the rate of increase of column height as a function of the mass eruption rate decreases for columns that penetrate the tropopause. This can be seen in Figure 4.7 near the tropopause at 11 km by comparing the numerical model and the scale models which do not take account of the stronger stratospheric stratification. Second, as the latitude of the eruption changes from near tropical to near polar, the eruption column penetrates the tropopause at ever-decreasing heights. Therefore, the subsequent ascent height of the plume is smaller for a given mass eruption rate and temperature because the plume is ascending through a deeper region of stratified atmosphere. In Figure 4.9 the ascent heights of eruption columns are compared in model polar, mid-latitude and tropical atmospheres, in which the tropopauses are located at 8, 11 and 17 km respectively and the ground temperatures are chosen to be 263, 273 and 293 K respectively. In the tropical environments the plumes rise significantly higher as expected, owing to the deeper, weakly stratified troposphere.

4.5.2 Wind-blown Plumes

A second environmental control upon the motion of eruption columns arises from the ambient wind. The wind has two main effects: it changes the efficiency of the entrainment and it can bend the plume over. The wind can change the deposition pattern of ash and the influence of the wind is discussed in detail in Chapter 11.

4.5.3 Moist Convection in Eruption Columns

The exsolved volatile contents of magmas are too small to have a significant influence on column behaviour. However, in moist air, a significant mass of water vapour may be entrained into the column from the lower atmosphere. As this water vapour is convected upwards in the column, it is cooled and becomes saturated. Eventually, the vapour can condense into liquid water, releasing latent heat of condensation. This latent heat increases the temperature of the gas in the eruption column, thereby increasing the buoyancy and ascent height of the erupted material. In the limit of very small mass eruption rates, the

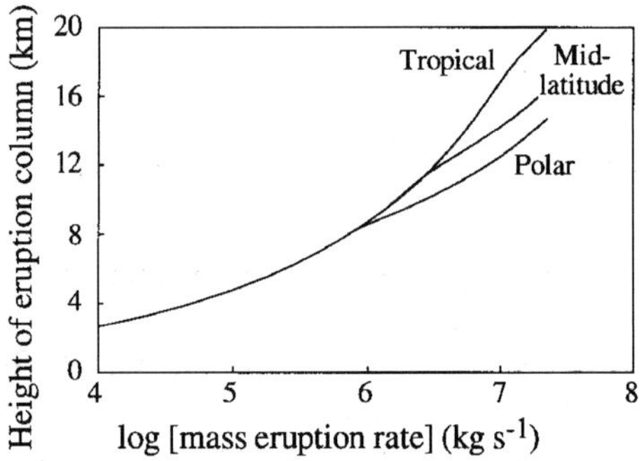

Figure 4.9 Variation in the ascent height of an eruption column with latitude. In the polar case, the ground temperature is assumed to be 263 K and the tropopause is at 8 km. In the mid-latitude case, the ground temperature is 273 K and the tropopause is at 11 km. In the tropical case, the ground temperature is assumed to be 293 K and the tropopause is at 17 km. Differences in the column height are correlated with the differences in the altitude of the tropopause. After Woods (1995a)

moist convection of the atmosphere can provide a dominant control upon the motion of the eruption column. By analogy, sugar-cane fires in Hawaii can heat up the air near to the ground. As the air rises, the vapour becomes saturated and condenses. The latent heat released provides more thermal energy which enables the material to continue rising. The plume ascends much higher than would be predicted from scaling arguments based on the heat released from the ground (Morton 1957; Squires and Turner 1962).

In this section, the eruption column model of section 4.3 is extended to investigate the effects of vapour phase changes upon the column dynamics, following Woods (1993b). In Chapter 8 the model developed here will be applied to eruptions involving significant interaction of external water with the magma. In order to investigate the effects of water vapour on plume dynamics, two new variables are introduced to account for, firstly, the total mass fraction of liquid water and water vapour in the eruption column, q, and, secondly, the mass fraction of water vapour in the gaseous phase of the eruption column, w. As the water vapour condenses, the liquid or solid raindrops that form are relatively small, micrometre to sub-millimetre in size (Rogers and Yau 1989). Therefore, as with the fine ash particles, the condensate can remain in the column up to the umbrella plume where it can then fall out, as described in Chapter 14. However, in some situations the condensate can form larger sized drops, through recycling in the upper parts of the column, as occurs with hailstones, or condensation can occur on the surfaces of much larger particles. Particle aggregation and condensation processes are also likely to complicate the behaviour of water vapour in eruption columns (Chapter 16).

The conservation of the total mass of water in the column is described by the rate of entrainment of ambient water vapour. If the ambient vapour is assumed to constitute a mass fraction $q_a(z)$ of the moist atmosphere, then

$$\frac{\mathrm{d}\beta u b^2 q}{\mathrm{d}z} = 2\varepsilon q_a \alpha u b \tag{4.25}$$

This equation is combined with the mass and momentum conservation equations of section 4.3. The steady-flow energy equation must be modified to include the effects of the release of latent heat as vapour becomes saturated and condenses. If q_v denotes the mass fraction of vapour in the column, then the steady-flow energy equation becomes

$$\frac{\mathrm{d}\beta u b^2 \left(\frac{u^2}{2} + gz + CT\right)}{\mathrm{d}z} = 2\varepsilon\alpha u b\left(\frac{u_e^2}{2} + gz + C_e T_e\right) + L(T)\frac{\mathrm{d}\beta u b^2 (q - q_v)}{\mathrm{d}z} \tag{4.26}$$

When the column is unsaturated, q_v is equal to q, the total mass fraction of water in the column, and the final term in equation (4.26) vanishes. Once the column has become saturated, condensation is assumed to occur sufficiently rapidly for the column to remain just saturated. This is a reasonable approximation, since the high concentration of fine particles provides ample nucleation sites for the condensate. Then the mass fraction of vapour in the gas phase, w, is defined to be exactly that amount such that the partial pressure of the vapour equals the saturation vapour pressure, $e_s(T)$,

$$wR_v P = (R_v w + R_a(1 - w))e_s(T) \tag{4.27}$$

and the mass fraction of water vapour in the column is given by

$$q_v = wn \tag{4.28}$$

The latent heat of condensation is defined as a function of the column temperature

$$L(T) = S + (C_v - C_w)(T - 273) \tag{4.29}$$

where S represents the heat of vaporization at a temperature of 273 K, and the temperature T is measured in kelvin. The density of the gas phase in the column must now be generalized to include both the vapour and the air, according to the relation

$$\frac{1}{\rho} = \frac{w}{\rho_v} + \frac{1-w}{\rho_a} \tag{4.30}$$

where the subscripts v and a denote the density of the vapour and the air. Finally, the eruption is assumed to be fine-grained, such that there is little sedimentation of pyroclasts from the eruption column as it ascends above the neutral buoyancy height. Thus the conservation of the solid material erupted from the vent is simply expressed in the form

$$(1 - n - q_w)ub^2\beta = (1 - n_o)u_o b_o^2 \beta_o \tag{4.31}$$

where $q_w = q - q_v$ is the mass fraction of liquid water in the eruption column.

Liquid water and ice condensed from the vapour are not distinguished, since the latent heat of vaporization is about 10 times larger than that of freezing water. Woods (1993b) has shown that this difference has little effect upon the model predictions in comparison with the effect of the condensation of the vapour. However, the difference between the condensation to form water and to form ice may have implications for the coprecipitation of other volcanic volatiles (Pinto *et al.* 1989; Chapter 16).

The model of a moist eruption column requires that the variation of the atmospheric moisture loading with height be characterized. To study a particular eruption, the precise details of the local moisture loading needs to be specified. However, in this chapter, it is instructive to use a simple model of the moisture loading. The relative humidity with height in the atmosphere is therefore fixed, and the effect of variations in its value investigated. The relative humidity R_H is defined as the ratio of the water vapour loading of the air to the saturated water vapour loading of air with the same pressure and temperature. Therefore, in an atmosphere of relative humidity R_H the mass fraction of water vapour in the air is

$$q_a = R_H \frac{R_a e_s(T)}{R_v P - (R_v - R_a)e_s(T)} \tag{4.32}$$

As the atmosphere varies from humid tropical conditions to drier mid-latitudes, the relative humidity R_H may decrease from about 1.0 to 0.2–0.5. The density and pressure profile in the atmosphere are also modified from section 4.3 to account for the ambient moisture (Woods 1993b). The saturation vapour pressure is given by the Clausius–Clapeyron curve

$$e_s(T) = a_1 \exp\left(-\frac{a_2}{T}\right) \tag{4.33}$$

where the dimensional constants $a_1 = 2.53 \times 10^{11}$ kg m^{-1} s^{-2} and $a_2 = 5.42 \times 10^3$ K (Rogers and Yau 1989).

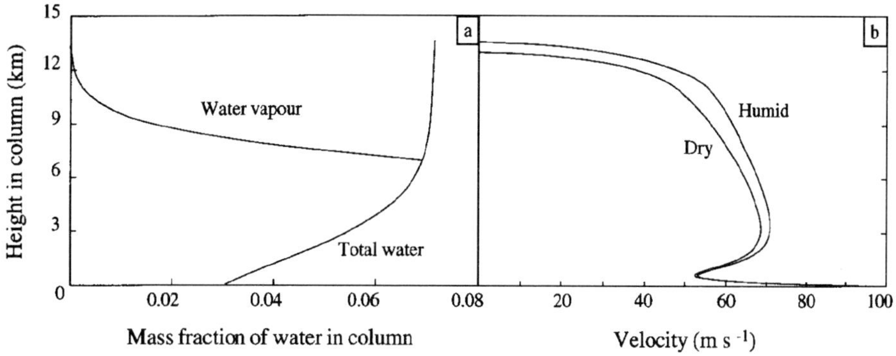

Figure 4.10 (a) Variation of the total mass fraction of water in the column, and the water vapour content of a typical eruption column with height. The atmosphere is assumed to have a relative humidity of 1.0, and the eruption column has an initial temperature of 1000 K, initial velocity of 100 m s^{-1} and initial radius of 50 m. Above about 7.5 km the column becomes saturated and condensation occurs. This lowers the water vapour fraction of the column significantly. (b) Variation of the column velocity with height in an atmosphere with relative humidity 1.0 (humid) and 0.0 (dry). The effect of the ambient water vapour is to increase the column buoyancy, velocity and hence height of rise of the column

Figure 4.10 shows results of a typical model calculation in which the ground level ambient relative humidity is 1.0. Initially, the column is hot and unsaturated, and the water vapour content of the column increases through entrainment of the ambient moisture. However, the column eventually becomes saturated, at about 8 km, because air, entrained from the lower relatively moist atmosphere, becomes saturated at higher altitudes where the pressure is lower. Once saturated, the water vapour content in the column decreases rapidly as the vapour condenses, even though the column is still entraining ambient moisture. The latent heat released by this condensation enables the velocity in the column to decay more slowly than in a comparable dry atmosphere (zero humidity) (Figure 4.10b).

In a moist environment, the height of small columns is considerably greater than in a dry environment (Figure 4.11). This is because the thermal energy released by the condensing vapour is comparable to or even much greater than the thermal energy associated with the erupted material in small eruptions. For small eruptions in relatively dry environments, the column remains unsaturated. In larger eruptions or moister environments, the column does become saturated and therefore the height of rise of the column increases significantly. However, as the erupted mass flux increases beyond about 10^6 kg s^{-1}, the ratio of the erupted thermal energy in the solid clasts begins to exceed significantly the energy associated with the condensation of the water vapour. This is because the ambient moisture is confined to the lower atmosphere. Only smaller columns entrain a sufficient mass of water vapour relative to the erupted mass that the latent heat produced on condensation is comparable to or exceeds the thermal energy erupted in the solids. Larger columns ascend much higher, and so only a small fraction of the entrained air is moist. As a simple criterion, eruption columns whose total ascent height exceeds about 12 km are not significantly affected by environmental moisture.

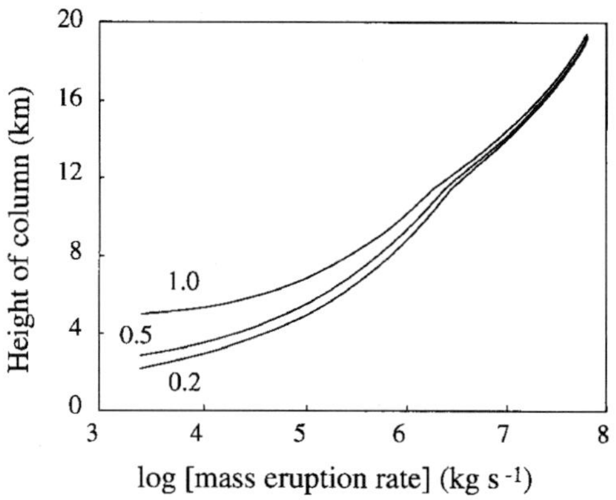

Figure 4.11 Variation of the height of rise of eruption columns in atmospheres of different relative humidities (1.0, 0.5, 0.2). The greater the humidity, the more vapour that is entrained and carried upwards in the column. Thus more latent heat is released leading to a higher column

4.6 SHORT-LIVED ERUPTIONS

In order to model the eruption columns rising from short-lived volcanic explosions, the maintained eruption column model of section 4.3 must be replaced by a model of a discrete plume, in a fashion similar to that of a thermal, as described in Chapter 3. Such a model has recently been proposed by Woods and Kienle (1994), following from the arguments proposed by Morton *et al.* (1956), Wilson *et al.* (1978) and Wilson and Self (1980) in which the relationship between the height of rise of a thermal plume H_T, the product of the initial buoyancy and volume of the thermal, F_T, and the ambient stratification N was given by

$$H_T = 2.7 F_T^{1/4} N^{-1/2} \tag{4.34}$$

where $F_T = V_{T_o} g(T_{T_o} - T_a)/T_a$, V_{T_o} is the initial volume of the thermal plume and T_{T_o} the initial temperature. In order to model the motion of the plume as it ascends into the atmosphere, the volcanic thermal is modelled as being a sphere of radius r, much like a laboratory thermal (Chapter 3). The thermal is assumed to entrain in an analogous fashion to laboratory thermals and so applying the same principles as in section 4.3, the equations for the conservation of the total mass, momentum and enthalpy in the thermal are of the form

$$\frac{d(r_T^3 \beta_T)}{dz} = 3\alpha r_T^2 \varepsilon_T \tag{4.35}$$

$$\frac{d}{dz}\left(r_T^3(2\beta_T + \alpha)\mu_T\right) = \frac{2g(\alpha - \beta_T) r_T^3}{u_T} \tag{4.36}$$

$$\frac{d}{dz}\left(r_T^3 \beta_T \left(E_T + gz + \frac{\mu_T}{2}\right)\right) = 3 r_T^2 \alpha \varepsilon_T (E_e + gz) \tag{4.37}$$

The equation for the density of the mixture of gas and solids is as given in section 4.3. In the limit in which there is negligible fallout from the plume, the conservation of solid mass in the plume has the form

$$(1 - n_{T_o})r_{T_o}^3\beta_{T_o} = (1 - n_T)r_T^3\beta_T \tag{4.38}$$

These equations can be solved using the atmospheric model of section 4.3. The effects of vapour phase change are incorporated in a similar fashion to section 4.6. The plume is assumed to ascend from rest when it is just buoyant. As the plume entrains ambient air, the density begins to fall through the heating and expansion of this air. The plume begins to accelerate and the buoyancy increases further. The radius of the plume grows and eventually the density reaches a minimum, beyond which the acceleration of the plume decreases, and owing to the rapid rate of entrainment of air, the velocity eventually decreases. The plume passes through its neutral buoyancy height, and is driven upwards by

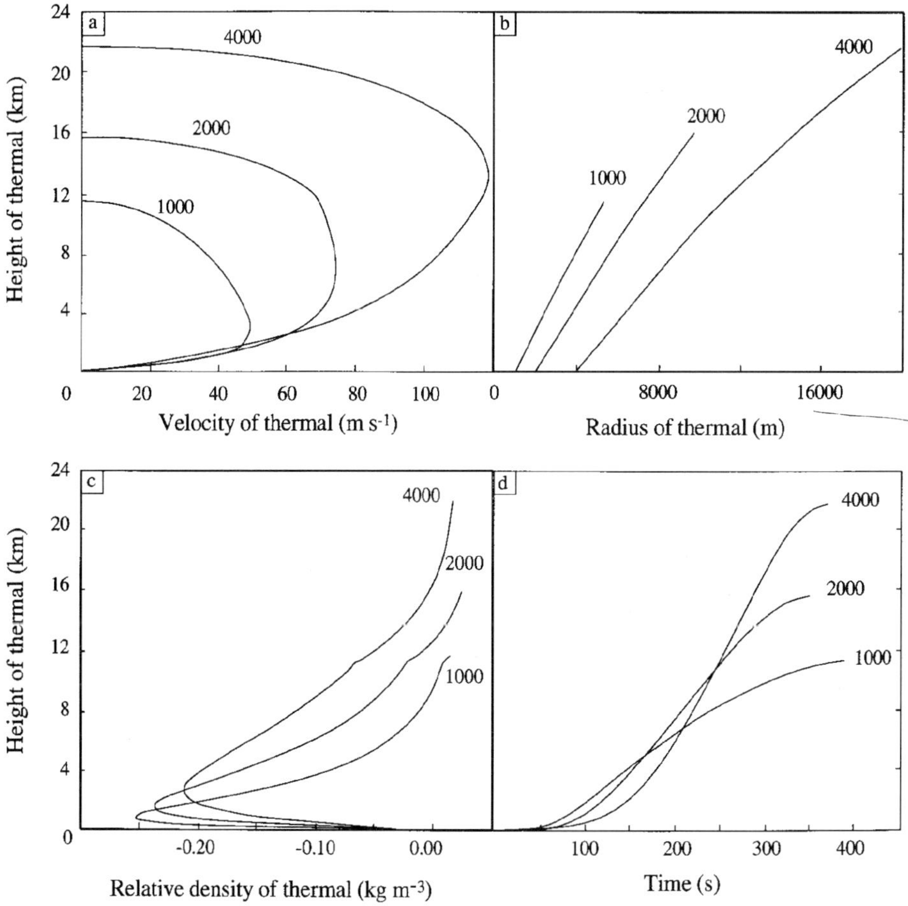

Figure 4.12 Variation of the velocity (a), radius (b), relative density (c) and time after eruption (d) as a function of the height of rise of a discrete volcanic thermal plume. Curves are calculated for initial plume radii of 1000, 2000 and 4000 m

its inertia to its maximum height. A typical set of numerical predictions generated by this model are shown in Figure 4.12. Note that these models assume that the plume starts from rest, whereas Vulcanian explosions have considerable initial momentum. Thus the velocities at low heights are not realistic. As the size of the plume increases, the plume ascends higher and more rapidly, and the plume maintains its buoyancy at higher altitudes. This is because the plume entrains less air per unit mass compared to a smaller plume, since the entrainment is proportional to the surface area whereas the mass is proportional to the volume. In Chapter 5 some further numerical solutions of equations (4.35–4.37) are presented and compared with the ascent of plumes generated from the 1990 eruption of Mount Redoubt (Woods and Kienle 1994).

4.7 STARTING PLUMES

The initial ascent of a steady eruption column can be modelled by appealing to the model and laboratory experiments of starting plumes developed by Turner (1962) (Chapter 3). According to the results of these experiments, the thermal plume atop a developing plume rises with a velocity of about 60% that of the steady plume. This is because the thermal is controlled by the entrainment of fluid from the plume below, rather than by the entrainment of environmental air, as in the isolated thermals considered above (section 4.6). Therefore, the steady-state model of an eruption column can be applied and the velocity scaled by 60% to estimate the time and speed of ascent. Sparks and Wilson (1982) developed such a hybrid model of a starting plume, and applied this to the 1979 eruptions of Soufrière, St Vincent. This is discussed in more detail in Chapter 5.

4.8 ERUPTION COLUMNS ASSOCIATED WITH PYROCLASTIC FLOWS

In all of the above discussions, the eruption column has been modelled as originating from the vent as a high-speed jet with the column having two dynamically distinct regions: a lower, relatively dense gas thrust region and a higher, relatively light convecting region. However, in some situations, part or all of the erupting material does not become buoyant as it rises in the gas thrust region. The material falls back around the vent, forming a fountain and generating pyroclastic flows. For the case of a fine-grained eruption column the condition to form a collapsing fountain occurs when the erupted material fails to mix with sufficient air to form a buoyant column. The erupted mixture runs out of kinetic energy and collapses as a fountain. The boundary between collapsing and convecting fountains is shown in Figure 4.6 and the subject is considered in much greater depth in Chapter 6.

Some of the material in such a flow may subsequently become buoyant and ascend from the flow to form an eruption column commonly referred to as a co-ignimbrite eruption column. Co-ignimbrite eruption columns may be modelled using an adaptation of the models presented in this chapter, and are described in more detail in Chapter 7. The principal difference is in the description of the source conditions and the entrainment law in the lower column, where the radius far exceeds the height of the column, limiting the size of the eddies.

4.9 EFFECTS OF PARTICLES ON LOWER COLUMN DYNAMICS

This section considers the influence of particle size and concentration on the dynamics of the lower parts of eruption columns. Models of the gas thrust region have treated the basal part of an eruption column as a turbulent gas jet containing suspended particles. However, under some circumstances this approach may be inappropriate. Bursik (1989) has applied concepts developed on the fluid dynamics of particle-laden jets to the dynamics of the gas thrust region (Owen 1969; Grace and Mathur 1978; Melville and Bray 1979; Modaress *et al.* 1984). Initially volcanic jets have a high ratio of particle mass to gas mass, Z'_o. In magmatic eruptions Z'_o is approximately 15–30 and can vary from ~ 0.1 to >30 in phreatomagmatic eruptions (Sheridan and Wohletz 1983; Wohletz 1983). Thus the substantial proportion of the initial momentum flux is contributed by the particles. Initially therefore an eruption column is closer in analogy to a spouted particle bed than a gas jet (Grace and Mathur 1978). For the case of a fine-grained eruption which forms a high buoyant eruption column it can be assumed that sufficiently large amounts of surrounding air are entrained so that most of the momentum flux is transferred to the gas phase. However, for collapsing eruption columns or for columns rich in coarse ejecta (such as fire fountains as discussed in Chapter 10) the dynamic interactions between particles and gas in the gas thrust region may become significant.

Bursik (1989) studied the influence of particle size and concentration on the jet dynamics including entrainment of ambient air, but not including heat transfer. The behaviour of a particle-laden jet can be understood in terms of the two dimensionless numbers Z'_o and v/w_o (Melville and Bray 1979; Bursik 1989)

$$\frac{z_m}{z_e} = \frac{2gz_m}{w_o^2} = f\left(Z'_o, \frac{v}{w_o}\right) \tag{4.39}$$

where

$$Z'_o = \frac{1 - n_o}{n_o} \tag{4.40}$$

in which z_m is the rise height of the fountain, z_e the rise height of individual particles at which the kinetic energy has been completely converted to potential energy, n_o the weight fraction of gas in the mixture erupting from the vent, which is the same as the weight fraction of exsolved volatiles in a purely magmatic eruption (Wilson *et al.* 1980).

Figure 4.13 shows results of numerical calculations of the height that a particle-gas jet rises as a function of particle concentration and grain size. To the left (values of $Z'_o < 1.0$) the behaviour is exactly the same as a pure negatively buoyant gas jet for fine suspended particles, reflecting the fact that the particles are locked to the gas motion and the momentum flux is dominated by the gas phase. For coarse particles the particles fall out at heights which are only weakly dependent on the particle concentration. For values of Z'_o (> 1.0) that are appropriate for volcanic eruptions the height of the resulting fountain is dependent on the particle size. Fine-grained fountains reach greater heights than coarse-grained fountains. This is a consequence of the momentum transfer between the particles and gas. As coarse particles fall out they transfer their momentum to the surrounding ascending gas and decelerate the gas by drag.

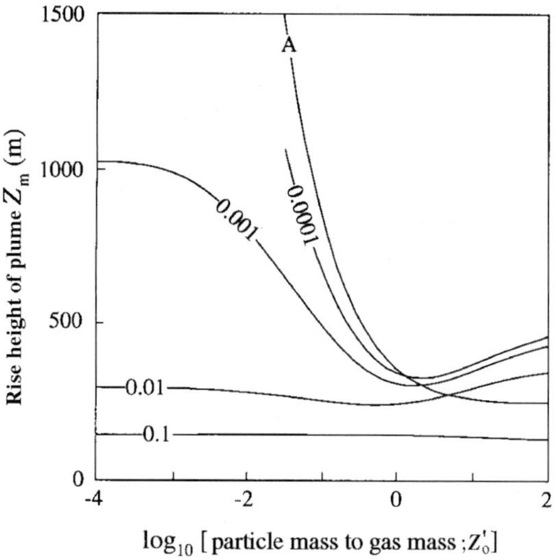

Figure 4.13 Variation of the rise height of particle-laden jets as a function of the mass concentration of particles (measured by the initial ratio of particle to gas mass Z_0'). Curves are shown for a model jet discharged from a vent of 10 m diameter at 100 m s^{-1}. The curve marked A shows the rise height of a negatively buoyant jet with the same bulk density as the particle-laden flows. Curves are for different particle radii in metres

Not only do particles transfer momentum to the gas, but they also weaken the turbulence of the gas motion. Thus at high particle concentrations in the gas thrust region the flow behaves as a dense jet of streaming particles with the relatively small mass fraction of gas dragged upward by the particle motion, as the particles transfer momentum to the flow. The flow may behave like a spouted bed (Grace and Mathur 1978) in that turbulence is suppressed by the streaming motion of the particles.

4.10 MULTI-PHASE NUMERICAL MODELS OF ERUPTION COLUMNS

The models described thus far have treated the column as a coherent plume or jet-like structure, and have modelled the evolution of the horizontally averaged properties in the column as it ascends. Using the analogy between the top-hat and Gaussian properties of a simple buoyant plume, more details about the structure of the eruption column can be deduced. However, volcanic systems are somewhat more complex than is allowed by the top-hat approach. In order to gain further insight about the details of the column motion, a number of multiphase, supercomputer models of the motion of the clasts, volcanic volatiles and ambient moisture have been developed (Wohletz *et al.* 1984; Valentine and Wohletz 1989; Dobran *et al.* 1993). Although these models are still in the early developmental stages, they allow investigation of a number of important processes related to eruption column dynamics, that are not accessible to the simpler steady plume or thermal models.

To date the models have mainly been used to study the initial transient behaviour of an eruption column, during the first few hundred seconds. Some insights have been gained about the initiation of the column collapse process with such computer models, as described more fully in Chapter 6. However, there has been no systematic comparison of these models with the simpler plume theory described in this chapter, which, in contrast to the numerical models, have been successfully tested against analogue laboratory experiments and observations of a wide variety of anthropogenic and natural plumes (Morton *et al.* 1956; Briggs 1969; Turner 1966; Wilson *et al.* 1978; Sparks 1986; Woods and Caulfield 1992). This lack of verification is associated with the fact that the ascent time of an eruption column is comparable to the longest of the numerical simulations of eruption columns carried out with these supercomputer models. The numerical models have not yet been able to simulate a steady, buoyantly convecting column. However, the accuracy of some of the fundamental assumptions implicit in the numerical models, such as the parameterization of mixing by an eddy diffusivity, could be tested in the present context by comparison with the extensive range of published plume experiments (List 1982; Turner 1966). Furthermore, the fundamental processes associated with the sedimentation of particles, even in the absence of heat transfer, have been investigated experimentally (Carey *et al.* 1988; Sparks *et al.* 1991), and it seems that an important step in the progress of numerical modelling is to check the consistency of models with these controlled laboratory experiments. The numerical calculations are themselves sometimes described as "experiments" which generate "data". Use of these terms in this context extends beyond their legitimate meaning. These calculations are in fact theoretical predictions of a sophisticated model which, like all theory, requires verification from real physical experiments. Otherwise their relationship with reality remains unclear.

It is not the purpose of this section to describe the full details of these models. However, significant advances in understanding should follow in the future by the judicious choice of numerical experiments to investigate different aspects of eruption columns. The models so far simulate the motion of axisymmetric eruption columns, and allow for variations in the temporal as well as spatial evolution of the dynamics. Although the models can simulate many of the features associated with column collapse, any asymmetric fountaining which can occur, and which has been observed in the laboratory (Baines *et al.* 1990; Carey *et al.* 1988), cannot be reproduced. This should be remembered when interpreting the physical relevance of numerical results since the numerical code will always produce an axisymmetric flow solution, even if it is unstable.

The numerical models include separate equations for the motion of the solid and the gas phases in the column, and they parameterize the coupling between the motion of the phases with a drag law. Heat transfer is also allowed between the phases, again parameterized by a turbulent heat transfer law. Entrainment and mixing are modelled with a turbulent eddy diffusivity. In a real plume, entrainment occurs through the incorporation of ambient fluid into the plume by vigorous eddies. The model eddy viscosity represents this process as a frictional transfer of momentum, in which the friction has a value much larger than the molecular value. Although this is a useful approximation, the difference in the fundamental physical process of eddy entrainment and frictional entrainment can lead to some limitations of the approach (Papantoniou and List 1989). For example, in regions of higher particle concentration, the intensity of the turbulence can be suppressed (section 4.9).

To date, the supercomputer models only consider eruption columns in which there is one particle size, although this may be extended in the future. Finally, the models do not restrict the column to be of the same pressure as the environment. They can simulate the decompression of the erupted material, as well as supersonic flow and the associated formation of shock waves. This is a useful extension of parameter space beyond the scope of simpler plume models. However, shock waves are dissipated by molecular processes and therefore tend to be very narrow regions. Their proper numerical resolution may require length-scales smaller than the minimum grid size of 10 m, as used in the present models (e.g. Dobran *et al.* 1993). Use of a grid size larger than the phenomenon being modelled may lead to difficulties in interpreting the results.

4.10 SUMMARY

Erupting mixtures emerge from the vent with a density that is typically significantly greater than the atmosphere. A fundamental element of these turbulent flows is the decrease of the density of the erupting mixture as it entrains and heats air. The dynamics of the column can therefore be divided into a lower gas thrust region dominated by momentum and an upper convective region dominated by buoyancy. A buoyant convective volcanic plume forms if the density of the erupting mixture falls below that of the surrounding environment as a consequence of entrainment and heating of air. In this process the initial thermal energy of the erupting pyroclasts is converted to potential energy as both the erupted pyroclasts and the air entrained from the lower atmosphere are convected high into the atmosphere. Variations of the velocity, spreading rate, temperature and buoyancy with height can be estimated from numerical models in fine-grained eruptions for both steady discharges and instantaneous events. Convective eruption columns divide into buoyant plumes, which decrease in velocity with height, and generally more powerful superbuoyant plumes, which accelerate above the gas thrust region, reach a velocity maximum and then decline in velocity.

The height reached by a volcanic plume is a strong function of magma discharge rate, approximately to the fourth power. Height can also be influenced by variations of magma temperature, atmospheric temperature gradients, effects of cross-winds, atmospheric moisture content and particle size distribution. Particle fallout and thermal disequilibrium tend to increase the critical velocity required to produce an eruption column and also reduce the height of rise of the ensuing column. Small eruption columns rising through humid environments can ascend significantly higher than in a dry environment owing to the release of latent heat as water vapour, entrained from the lower atmosphere, is convected upwards and condenses. Although these may be secondary effects in large, fine-grained eruption columns, they can be very important in coarse-grained eruptions such as Hawaiian fire fountains (Chapter 10).

Analysis of the initial dynamics of eruption columns in the gas thrust region has demonstrated that the density can only decrease below that of the air under some circumstances. If part or all of the erupting mixture fails to become buoyant before the flow runs out of kinetic energy, then the column will collapse to form a pyroclastic fountain and will generate pyroclastic flows. The conditions for formation of a buoyant column or a collapsing fountain depend on the initial velocity, initial dimensions of the decompressed flow above the vent and gas content. Numerical calculations show that convective columns are favoured by high exit velocities, high gas contents and narrow vents. Collapsing fountains are favoured by low exit velocities, low gas contents and wide vents. For a given discharge rate there is a critical velocity which needs to be exceeded to form a convective eruption column.

5 Observations and Interpretation of Volcanic Plumes

NOTATION

b	Gaussian length-scale (plume half-radius) (m)
C_c	drag coefficient for cylindrical thermal (dimensionless)
C_s	drag coefficient for spherical thermal (dimensionless)
g	gravitational acceleration (m s^{-2})
H	column height (km)
H_s	scale height of the atmosphere (km)
k	entrainment coefficient (dimensionless)
L	length of cylindrical thermal (m)
M	mass flux of ejecta (kg s^{-1})
n_c	mass fraction of air and volcanic gas in plume (dimensionless)
q	ratio of plume bulk density to atmospheric density (dimensionless)
Q	volumetric magma discharge rate (m^3 s^{-1})
r	radius of spherical thermal (m)
u	velocity of rising thermal (m s^{-1})
u_c	plume centreline velocity (m s^{-1})
z	height above the vent (m)
α	density of air at height z (kg m^{-3})
β_c	bulk density of plume mixture (kg m^{-3})
γ	ratio of specific heats of air (dimensionless)
σ	density of magma (kg m^{-3})
θ_a	temperature of air at height z (K)
θ_{ao}	temperature of air at altitude of vent (K)
θ_c	temperature at centreline of plume (K)
θ_e	temperature of eruption products (K)

5.1 INTRODUCTION

In this chapter observations of the behaviour of plumes in historic volcanic eruptions are presented. Discussion is restricted to observations of column height and shape and to the dynamics of the ascending parts of eruption columns generated at central volcanic vents. Some eruptions will be described as case histories to give further insights into the behaviour of real eruptions and the observations will be compared with the theoretical models of volcanic plumes. The observational data on volcanic plumes are surprisingly meagre and thus this chapter includes some previously unpublished results.

5.2 COLUMN HEIGHT

There are a number of historic eruptions where there is information on the duration of a sustained period of explosive activity, on the eruption column height, on the total volume of ejecta and on magma discharge rates. An original dataset was compared to the predictions of the simple plume theory of Morton *et al.* (1956) by Wilson *et al.* (1978) and by Settle (1978) who found broad agreement. Table 5.1 provides an updated list of data for particular eruptions.

The dataset has a number of uncertainties in the estimates of the parameters. Column height estimates are usually made by trigonometric methods, but there are usually no details on the measurements and possible errors so that it is uncertain how reliable the data are. The Plinian phase of Mount St Helens, Washington, USA, on May 18, 1980 is the only eruption for which there are detailed data on column height variations with time (Harris *et al.* 1981a; Criswell 1987; Carey *et al.* 1990). In this particular eruption the column height fluctuated by over 20% during the nine-hour period of the Plinian phase. The Mount St Helens eruption provides the most detailed data where plume top heights were determined from radar observations, and data on deposit volumes and durations of the four individual phases of the eruption are well determined. Observations are commonly hampered by bad weather, as in the 1991 eruption of Hudson volcano, Chile (Naranjo *et al.* 1994). Increase in column height with time can also be inferred from the common reverse grading of Plinian fall deposits (Pescatore *et al.* 1987). Fluctuations in height will have occurred for all of the eruptions listed in Table 5.1, but for many only a single column height is recorded. It is often not clear whether the height recorded represents a maximum or a mean or a typical height. Uncertainties in the estimates of ejecta volumes, from which mean magma discharge rates can be calculated, are hard to evaluate and depend on the preservation of the deposit (see Chapter 13). Table 5.1 contains ranges where they are available. In the case of the eruption of Vesuvius, Italy, in AD 79, column heights were estimated by the method of Carey and Sparks (1986) (see Chapter 15), but ejecta volumes, durations and discharge rates can be estimated independently from isopach maps, studies of the deposits and historical accounts (Carey and Sigurdsson 1987).

Figure 5.1 plots column height versus discharge rate of magma for the two sets of data. Regression through the data gives the following empirical power law for the dataset:

$$H = 1.67Q^{0.259} \tag{5.1}$$

where H is the column height in kilometres and Q the dense rock equivalent discharge rate of magma in cubic metres per second.

The power exponent is very close to 0.25 as expected from simple plume theory in a stratified atmosphere (Morton *et al.* 1956; Chapter 2). The regression curve through the data is also compared with a number of theoretical curves after Woods (1988) for a standard atmosphere and a variety of eruption temperatures and gas contents (Figure 5.2). The numerical calculations agree well with the regression curve through the observations.

Table 5.1 Data on the mean discharge rates (Q), column heights (H) and eruption durations (T) of historic eruptions

Eruption	Q ($m^3 s^{-1}$)	H (km)	T (hours)
Agung, 1963	650	10	5
Bezymianny, 1956	230 000	36–45	0.5
Cerro Negro, 1971	35	6	144
El Chichon, April A, 1982	27 300	22–30	4
El Chichon, April B, 1982	18 800	22–30	7
El Chichon, March 1982	15 000	22–30	5
Fuego, 1971	640	10	10
Heimaey, 1973	50–150	3–6	—
Hekla, 1970	3 300	14	2.0
Hekla, 1947	17 000–33 000	27.6	0.5
Komagatake, 1928	15 870	13.9	—
Miyake, 1983	570	6.0	—
Mt Pinatubo, June 15, 1991	400 000	34.0	1.3
Mt St Helens, July, 1980	2 000	12.2–13.0	—
Mt St Helens, May 18, 1980, B1	2 520	13.9	2
Mt St Helens, May 18, 1980, B2	5 200	16.0	0.4
Mt St Helens, May 18, 1980, B3	1 560	13.1	6.9
Mt St Helens, May 18, 1980, B4	6 400	17.5	0.75
Nevado del Ruiz, 1985	13 000	24–29	0.3
Ngauruhoe, 1974	10	1.5–3.7	14
Quizapu, 1932	60 000	27–30	18
Santa Maria, 1902	17 000–38 000	27–29	24–36
Soufrière, 1902	11 000–15 000	15.5–17	2.5–3.5
Usu I, 1977	3 375	12.0	—
Usu II, 1977	2 500	10.0	—
Usu III, 1977	3 800	10.0	—
Vesuvius, AD 79 (grey)[a]	60 000	32	11
Vesuvius, AD 79 (white)[a]	34 000	26	7

[a] Column height for Vesuvius inferred from maximum clast size data (Carey and Sparks 1986).
Note: Heights are usually above the volcano although this is not always clear from source reference. Sources for original data can be found in Wilson *et al.* (1978), Settle (1978), Endo *et al.* (1986), Naranjo *et al.* (1986), Carey and Sigurdsson (1986, 1987), Harris and Rose (1987), Carey *et al.* (1990), Hildreth and Drake (1992) and Koyaguchi (1994).

In comparison, the numerical models tend to predict slightly larger heights in the troposphere and slightly lower heights in the stratosphere for a given magma discharge rate. Eruption column height can be used as a good monitor of variations of discharge rate during the course of an eruption because of the very strong dependence of height on discharge rate. Uncertainties in atmospheric conditions, ejecta temperature and ejecta thermal properties are of second order provided the proportion of fine-grained ejecta is high. As discussed in Chapter 4 significant departures might be expected in relatively weak

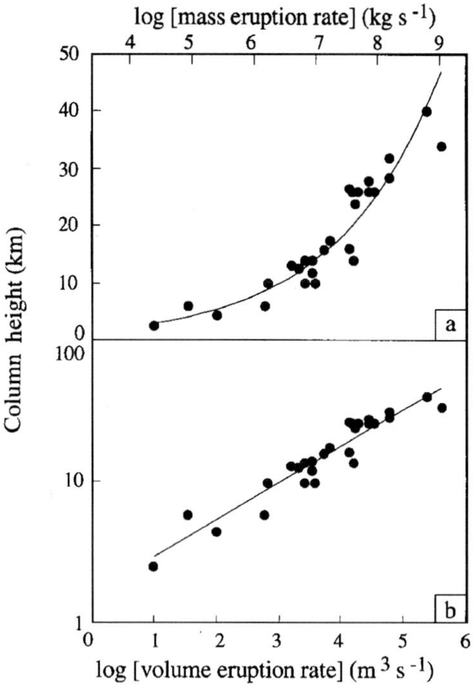

Figure 5.1 Data on column height versus discharge rate in historic eruptions, listed in Table 5.1, plotted as (a) a log-linear plot and (b) a log-log plot. The discharge rate is expressed both as volume eruption rate and mass eruption rate. Solid lines represent a power law regression through the data to yield equation (5.1). The regression has a correlation coefficient $R = 0.921$. Where a range of eruption rates has been given (Table 5.1) the average value has been plotted

eruptions in a moist atmosphere or weak eruptions in a strong cross-wind or in eruptions with a high proportion of coarse ejecta. Ideally, comparisons should be made with numerical models using the precise conditions for the particular eruption. Unfortunately there are no eruptions that have been documented so thoroughly that more exact comparison can be done.

An important consideration when comparing models with observations is the altitude of the vent above sea-level. Figure 5.3 shows numerical calculations of the height above the vent for the case of sea-level and 5 km altitude for a standard atmosphere. For eruption columns less than 10 km above the vent or more than about 16 km above the vent the altitude of the vent is not an important parameter in determining column height. However, for intermediate heights and discharge rates (10^6–2×10^7 kg s^{-1}) the column height can be quite strongly dependent on vent altitude. This result is a consequence of columns from higher altitude vents reaching the stratosphere sooner where the stronger stratification suppresses plume ascent for a given discharge rate. In this intermediate discharge rate region a plume issuing from a 5 km high vent may be 20–30% lower than a plume derived from sea level with the same discharge rate.

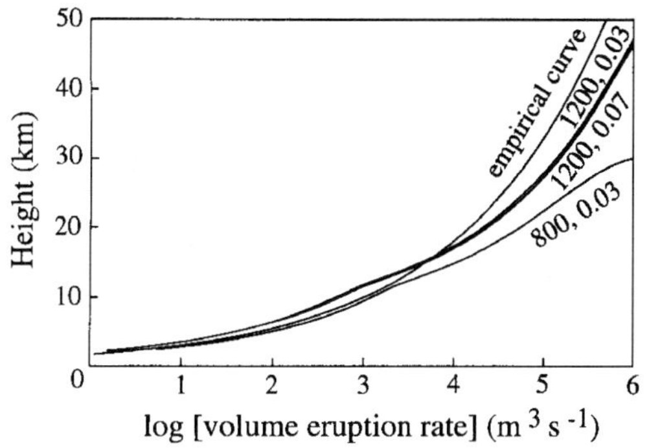

Figure 5.2 Numerical models of column height versus volumetric eruption discharge rate after Woods (1988) are compared with the empirical law given by equation (5.1). Theoretical curves are shown for numerical calculations in a standard atmosphere for the cases of magma temperatures of 1200 and 800 K with mass fractions of water of 0.03 and 0.07

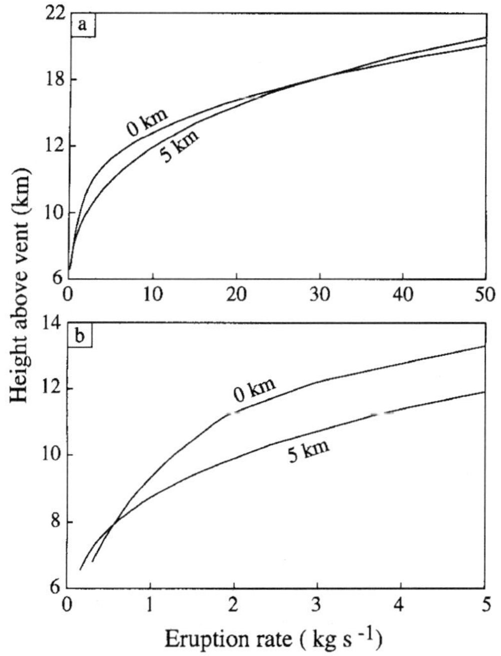

Figure 5.3 The height of an eruption column above the vent is shown as a function of discharge rate for two different values of vent altitude above sea-level (0 and 5 km). The calculations were carried out for a temperature of 1200 K in a standard atmosphere with a mass fraction of water of 0.03. Figure 5.3b is an enlargement of Figure 5.3a in the region where the differences are greatest

5.3 GAS THRUST REGION

There are few systematic sets of observations available on the dynamics of the gas thrust region. Blackburn *et al.* (1976) studied film of the 1973 explosive activity of Heimaey volcano in Iceland and measured gas velocity variations with height by tracking particles. They chose the smallest visible particles (estimated to have been about 5–10 cm diameter) so that the particle motion would reasonably approximate the gas motion. These particles are, however, sufficiently large that some differential velocity must have existed introducing some uncertainty into the gas velocity estimates. Figure 5.4 shows data from several Strombolian explosions in the form of the original measurements (Figure 5.4a) and smoothed curves through the data (Figure 5.4b). There is considerable scatter in the data which may reflect errors in the measurements, turbulence and fluctuations in unsteady flows. Despite these variations the main feature of the data is the rapid deceleration of the mixture due to entrainment of air as predicted by models of the gas thrust region. The decelerations from peak velocities of 150–200 m s^{-1} approach 50*g*. The last few data points show much reduced decelerations and the velocities approach steady values in the range 25–35 m s^{-1}. The columns then continued to rise to heights of between 3 and 6 km. The steady velocities are thought to record the transition region to the convective part of the column where buoyancy becomes dominant.

Wilson and Self (1980) studied small Vulcanian explosions on Fuego volcano in Guatemala during the Febuary 1978 eruptive activity. Two examples of plume ascent profiles are shown in Figure 5.5 and data on velocity versus time are shown for several

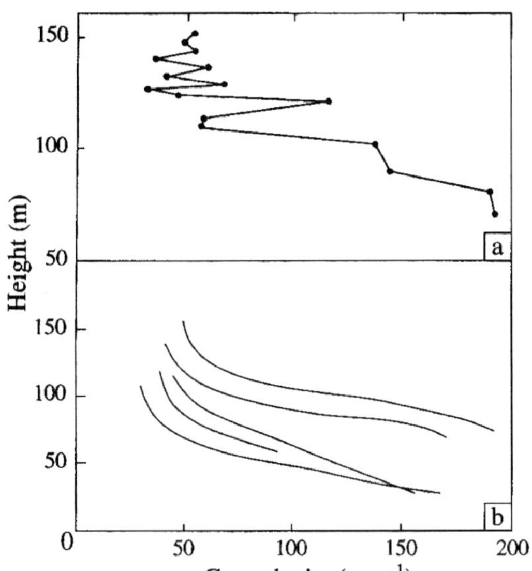

Figure 5.4 (a) Gas velocity estimates with height obtained from 16 mm movie film of an explosive discharge of Heimaey volcano in Iceland in 1973. (b) Smoothed curves of gas velocity versus height from several explosive discharges of Heimaey volcano in Iceland with actual data similarly irregular to that shown in (a). The heights are estimates above the vent with the first few tens of metres obscured by the crater rim

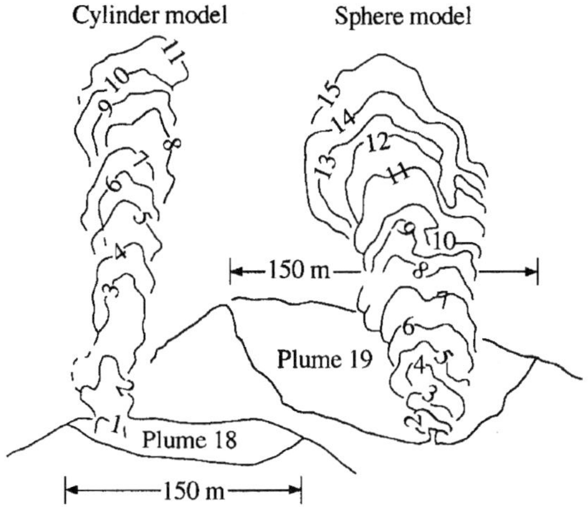

Figure 5.5 Profiles of volcanic thermal shape with time in seconds for two small Vulcanian explosions of Fuego volcano in Guatemala after Wilson and Self (1980). The two volcanic thermals were modelled as a cylinder (thermal 18) and as a sphere (thermal 19) as discussed in section 5.5 of the text. The numbers refer to different eruption events as described in Wilson and Self (1980)

plumes in Figure 5.6. The vent for the eruption is below the crater rim so the initial motions are not observed. However, eruptions 3, 12 and 17 show the initial rapid deceleration characteristic of the gas thrust phase followed by an acceleration phase which is similar to the "superbuoyancy" effect predicted by Bursik and Woods (1991). Thereafter these plumes decelerate within the convective thrust regime.

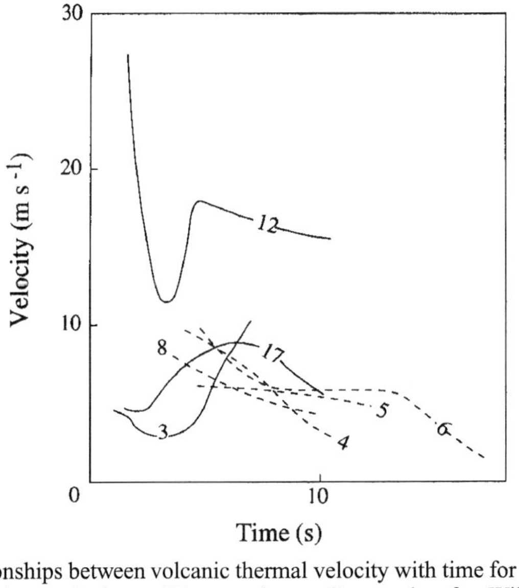

Figure 5.6 Relationships between volcanic thermal velocity with time for several volcanic thermals generated by Vulcanian explosions of Fuego volcano, Guatemala, after Wilson and Self (1980). The curves are regressions through data as presented in Figure 5.5. The numbers refer to different eruption events as described in Wilson and Self (1980)

5.4 STUDIES OF STARTING PLUMES

A small number of explosive eruptions have been photographed or filmed in the first few minutes of activity so that the initial motions can be studied. Three of the cases presented here are examples of the explosive disruption of growing lava domes. The examples described are the April 22, 1979 eruption of the Soufrière volcano, St Vincent, the February 20, 1990 eruption of the Lascar volcano in north Chile and the October 17, 1980 eruption of Mount St Helens. Some photographic data of the 1947 eruption of Hekla, Iceland, provide the only information on a large-magnitude sustained Plinian eruption. The eruptions typically last a few minutes to a few tens of minutes. The first few minutes of activity all show evidence of several closely spaced explosions which merge upwards into a continuous plume. A method has been developed for studying the initial stages of volcanic plume ascent (Sparks and Wilson 1982) based on the turbulent plume theory described in Chapter 4 and the starting plume model of Turner (1962). The activity in the initial stages of plume ascent is far from steady and some caution needs to be applied in comparison with the sustained plume models described in Chapter 4.

There are remarkably few explosive eruptions which have been filmed with the express purpose of making scientific measurements. This book may act as a prompt to colleagues at eruptions to take timed photographs and movies which can later be used for studies of plume motion and dynamics. The precise location of the camera, details on camera conditions, in particular focal length, and angle of the camera should be recorded. A fixed focal length and avoiding zooming in on the plume are recommended. Data on the atmospheric conditions such as temperature, relative humidity, height of tropopause and wind velocities should be recorded. A discussion on how to extract quantitative data from photographs and film is given in Sparks and Wilson (1982).

5.4.1 Starting Plume Model

The model presented here is concerned with the initial stages of volcanic plume ascent and is based on the study of starting plume dynamics of Turner (1962). This work provides an elegant discussion of the relationship between the properties of the plume front with those of the steady plume which follows behind. Turner (1962) treated the plume front as a thermal in which entrainment of surrounding fluid occurs across the top of the plume front and also into the centre of the base. The difference between this and a real thermal is that the fluid entrained at the base consists of material in the developing plume rather than the surrounding fluid (see Chapters 3 and 4). Consequently the buoyancy of the plume front increases with time which is in contrast to the case of an isolated thermal. Turner's experiments indicate that the spreading rate of the plume front is about 20% greater than the following plume. The most significant result of Turner's work is that the front velocity is approximately 60% of the centreline velocity of the following steady plume.

Sparks and Wilson (1982) derived approximate relationships for the plume front motion based on the same theoretical treatment as presented in Chapter 4. An approximate expression for the variation of bulk density with height follows from the equations for the conservation of mass and momentum and has the form

$$(1-q) = \frac{u_c^2}{2g}\left(\frac{1}{6}\frac{\mathrm{d}q}{\mathrm{d}z} + \frac{1}{u_c}\frac{\mathrm{d}u_c}{\mathrm{d}z} + \frac{11.7k}{z}\right) \tag{5.2}$$

where q is the ratio of the plume bulk density to the density of the surrounding atmosphere at height z, u_c the plume centreline velocity and k the entrainment constant. Equation (5.2) is first solved by assuming $dq/dz = 0$ and using the measurements of u_c and du_c/dz from photographs or film. Values of dq/dz are then calculated and are incorporated in a series of iterative calculations to convergence. The bulk density of the plume can thus be calculated using the air density at the measured height.

The bulk density of the plume is a function of both the temperature and the particle content. Sparks and Wilson (1982) adopted a simplified heat conservation relation

$$\theta_c = \frac{n_c}{2}(\theta_{ao} + \theta_a) - \frac{n_c(\gamma - 1)z\theta_{ao}}{2\gamma H_s} + (1 - n_c)\theta_e \qquad (5.3)$$

where θ_c is the centreline temperature, θ_e the temperature of the eruption products at the vent, θ_{ao} the temperature of the air at the altitude of the vent, θ_a the temperature of the air at the height of entrainment, γ the ratio of the specific heats of air at constant temperature and constant volume ($\gamma = 1.4$), n_c the gas mass concentration and H_s the scale height of the atmosphere (approximately 8 km). Note therefore that the particle mass fraction is $(1 - n_c)$. This expression assumes that the temperature of the entrained air is the average over the height interval from the vent to the height of measurement and that the constituents of the plume all have the same specific heat. The expression takes into account the adiabatic cooling of the entrained air. Provided that the height interval is not too large (of the order of the scale height of the atmosphere or less) then equation (5.3) is a fairly accurate and useful approximation.

The relationship between particle mass concentration, plume temperature and bulk density is given by

$$1/q = (\theta_c/\theta_a)n_c + (1 - n_c)(\alpha/\sigma) \qquad (5.4)$$

where α is the density of the air at the measurement height and σ the density of the magma. Equations (5.3) and (5.4) can be combined to estimate unique values of particle content and plume temperature consistent with the estimated bulk density. The total mass flux of solid ejecta, M, at the height of measurement can be calculated as

$$M = \pi(1 - n_c)u_c b^2(\beta_c + \alpha)/2 \qquad (5.5)$$

where β_c is the bulk density of the erupting mixture and b the Gaussian length-scale (equal to half the plume radius). In applying these equations to photographs or movie film of starting plumes some important points need to be made about appropriate values used in the equations. First, if the measurements are of the plume front radius, as is usually the case, then the radius is about 40% larger than that of the following steady plume radius at the same height (Turner 1979; Chapter 2). Thus the observed radius needs to be reduced by 0.72 and then halved to give the correct estimate of the Gaussian length-scale b. Second, u_c is the centreline velocity of the following steady plume which feeds into the plume front and is a factor of 1.62 greater than the observed velocity of the plume front.

The data from eruptions can also be compared with the numerical models presented in Chapter 4. However, as discussed further below, many starting plumes show that discharge rate is unsteady and can increase in the opening moments of an eruption. Thus comparison of steady-state eruption plume models with starting plume motion can be problematic. The method of Sparks and Wilson (1982) therefore provides a simple way of estimating instantaneous plume properties in unsteady starting plumes.

5.4.2 April 22, 1979 Eruption of Soufrière, St Vincent

Explosive activity of the Soufrière volcano, St Vincent, began on April 13, 1979 (Shepherd *et al.* 1979). There were six periods of explosive activity within the first 48 hours, each of which generated a plume which ascended to heights of 17–19 km above sea-level. Further explosive eruptions occurred on April 17, 22 and 25 and these were followed by a period of lava dome extrusion over several months. Each period of explosive activity lasted no more than 30 minutes. The explosive activity has been attributed to phreatomagmatic processes during the ascent of volatile-rich andesitic magma and its interaction with a crater lake (Shepherd and Sigurdsson 1982), with the explosions taking place at a very shallow level.

The April 22 event started at 0635 hours local time. Vigorous seismic tremor indicated that the explosive activity lasted about 14 minutes, during which time a column ascended to a height estimated from the plume-top temperature (Krueger 1982), of about 18 km. However, heights calculated from plume-top temperatures are now thought to be unreliable (Glaze *et al.* 1989; Woods and Self 1992). The plume was dispersed to the south-east. Movie film of the first three minutes of activity and plume rise was analysed by Sparks and Wilson (1982) using the model developed in section 5.4.1. In the first 15 s a pale steam plume (C1) ascended at 8.5 m s^{-1} (Figure 5.7) and was shortly followed by a darker plume (C2) which ascended at a steady velocity of 33.2 m s^{-1}. Plume C1 was observed to accelerate to 25.6 m s^{-1} and then the two plumes merged and became indistinguishable in what was called the main plume. Drift of the plume to the east allowed observation of three further explosion plume fronts which were even darker and more vigorous. These plumes merged with the main plume as they ascended. Peak velocities of 47.7 m s^{-1} for plumes C3 and C4 and of 61.7 m s^{-1} for C5 were observed (Figure 5.7). The main plume had reached a height of 8.9 km three minutes after the eruption started at which point it had an estimated velocity of 58 m s^{-1}.

The Soufrière observations illustrate some of the complexities and observational difficulties in studying eruptions. The plume was fed by a series of closely spaced explosions which differed greatly in ash content and temperature. The observations of plume colour and rise velocity indicate that the explosions became increasingly ash-rich and of higher temperature with time. Observation of each plume rise was partly obscured by earlier plumes and atmospheric clouds. The individual plumes merged into one plume which accelerated with height as a consequence of the increase in temperature and ash content of the explosions feeding the plume (Figures 5.7, 5.8 and 5.9). Thus the early motion of the Soufrière plume is far removed from the steady-state sustained conditions of the models in Chapter 4. However, plume C4 ascended with only minor influence of other plumes (Figure 5.9b) and shows a decrease in velocity with height consistent with numerical models for columns of this strength.

Sparks and Wilson (1982) applied the analysis in section 5.3.2 to estimate variations in plume properties with height. Unfortunately there was an error in the calculation of magma discharge rates presented by Sparks and Wilson (1982) and therefore the rates have been recalculated here. Table 5.2 lists calculations of plume properties for the five plumes at the lowest observed altitude where the model is most reliable and the geometric corrections on plume dimensions are at a minimum. The estimates show the plume temperature and particle content increasing with time and these results are therefore consistent with the

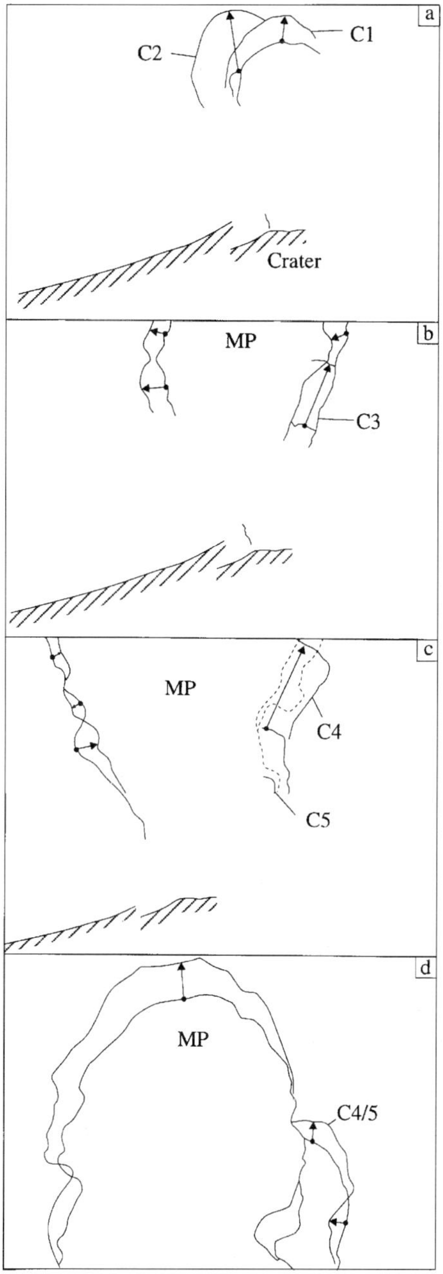

Figure 5.7 Sequence of outlines of the plume of the April 22 eruption of the Soufrière volcano, St Vincent, at different times showing the merging explosion plumes. In each diagram the arrows show the direction of motion of the plumes with two outlines shown in each sketch. C1 to C5 refers to the discrete plume fronts observed as a consequence of individual explosions. MP refers to the main plume. The dashed line on the right-hand side of (c) shows the position of the main plume margins obscured by the motion of plumes C4 and C5

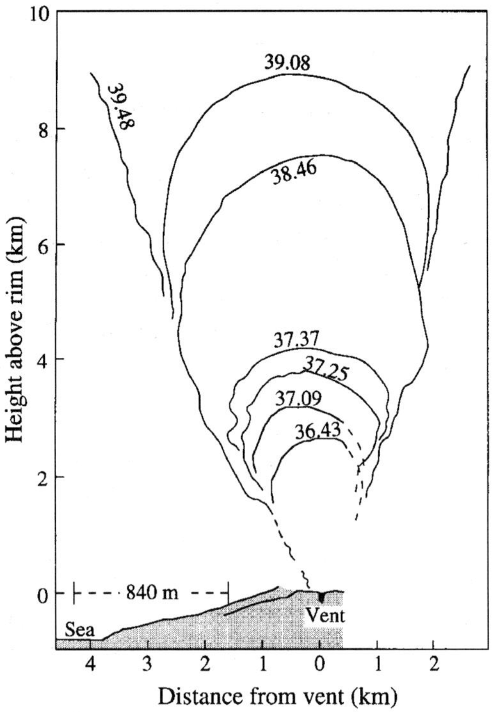

Figure 5.8 Evolving shape of the main plume of the Soufrière volcano, St Vincent, on April 22, 1979. Time is shown in minutes and seconds after 0600 hours local time. 840 m refers to the height of the crater rim above sea-level

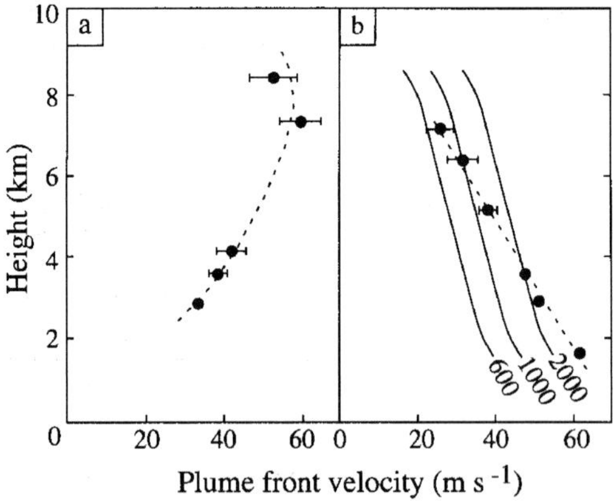

Figure 5.9 Velocity with height for (a) the main Soufrière plume and (b) the C4 plume. Figure 5.9b also shows three theoretical curves for magma discharge rates of 600, 1000 and 2000 $m^3 s^{-1}$. The numerical calculations were carried out for a magma water content of 0.03, initial velocity of 300 m s^{-1} and temperature of 1200 K using the theoretical models described in Chapter 4

qualitative inferences from plume colour and velocities and the acceleration of the main amalgamated plume (Figure 5.9b). The estimates of plume radius are not very reliable for plumes C3 to C5 because these plumes were obscured by the earlier plumes which resulted in calculations of magma discharge rate being only approximate. The first steam plume C1 gives a very low discharge rate of 35 $m^3 s^{-1}$. The magma discharge rates of plumes C2 to C5 increase with time from 465 to 2230 $m^3 s^{-1}$. In Figure 5.9a the main plume accelerates with height and has a maximum velocity of 58 m s^{-1} ($u_c \sim 97$ m s^{-1}) at 7 km height. Application of the methodology gives a magma discharge rate of 2000 $m^3 s^{-1}$ which is in agreement with the discharge rate of 2230 $m^3 s^{-1}$ for plume C5. This result lends further support to the interpretation that the main Soufrière plume was fed by a series of closely spaced explosions which became more intense and hotter with time. The model predicts that a discharge rate of 2000 $m^3 s^{-1}$ should generate a column of approximately 13 km height above sea-level, whereas the satellite observations give an estimate of 18 km for this eruption (Krueger 1982). It is unclear whether the discrepancy is the consequence of an overestimate of column height from plume-top temperatures or is the result of the discharge rate of magma increasing further after the period of observation. The eruption was filmed for the first three minutes, but discharge continued for at least 12 minutes.

The dataset for the motion of plume C4 is the most complete. Estimates of plume properties are listed in Table 5.3. The consistent values of the discharge rate with height indicate that this particular plume involved approximately steady-flow conditions. Figure 5.9b shows numerical models of velocity variation with height using the models in Chapter 4 after Woods (1988) for magma discharge rates of 600–2000 $m^3 s^{-1}$ and data for the plume front velocity versus height for plume C4. The numerical calculations agree reasonably well with the observations.

Observations were also made of the motions of the margins of the Soufrière plumes. Large eddies were observed with circulation velocities in the range 2–15 m s^{-1}. The maximum circulation velocity of large turbulent eddies in plumes typically have values of about one-third of the centreline velocities (Papanicolaou and List 1988; Chapter 4). These observations are therefore consistent with the velocities of the main plume of 40–60 m s^{-1}. The average vertical velocity of marginal eddies in the height range 3–5 km was 2.5 m s^{-1}, only about 3% of the estimated plume centreline velocities. Plumes have Gaussian velocity profiles (Chapter 2) and these observations indicate that the visible edge

Table 5.2 Comparison of plume height (H), particle concentration (1 n_c), mean temperature (q_c), ascent velocity (u_c), and discharge rate of magma (Q) for five starting plumes of the April 22, 1979 eruption of Soufrière, St Vincent

Plume	H (m)	u_c (m s^{-1})	$(1 - n_c)$	q_c (K)	Q ($m^3 s^{-1}$)
C1	2600	13.9	0.011	284	35
C2	3000	54.1	0.032	302	465
C3	2300	78.2	0.077	352	950
C4	3000	82.0	0.04	312	865
C5	1800	101.2	0.257	531	2230

Note: Estimates are shown for similar heights above the vent. The calculations assume the usual exponential decrease in density of the atmosphere with height (scale height 8 km), magma density of 2500 kg m^{-3} and magma temperature of 1273 K.

Table 5.3 Estimates of the plume velocity, u_c, of the ratio of plume density to atmospheric density (q), mass fraction of particles $(1 - n_c)$ and magma discharge rate, Q, at different heights in plume C4

H (m)	u_c (m s^{-1})	q	$(1 - n_c)$	Q (m^3 s^{-1})
3000	82	0.938	0.039	864
3700	77	0.964	0.031	900
5200	59	0.999	0.020	676
6500	49	1.012	0.018	756
7200	39	1.018	0.018	602

of a volcanic plume is approximately twice the characteristic length-scale of the Gaussian profile. This conclusion is in accord with the proposal of Morton *et al.* (1956) that the margin of a plume is taken as the radius at which the time-averaged velocity has decreased to 1% of the centreline velocity.

From the data in Figures 5.7 and 5.8 the spreading angle of the 1979 Soufrière plumes can be measured. The spreading is almost linear with angles in a narrow range from 21° to 23.5° despite wide variations in velocity, ash content and temperature of the different plumes. These values represent the spreading angle of the plume front which is about 20% greater than the following plume according to Turner (1962). The spreading angle of the subsequent main Soufrière plume was 17° which is consistent with Turner's result. This result is equivalent to $b = 0.17z$, where b is the Gaussian length-scale (half the plume radius).

5.4.3 February 20, 1990 Lascar Eruption

Lascar, Chile is a high-altitude volcano on the Andean altiplano at 23°S. The volcano has a large summit crater about 1.2 km in diameter and 300 m deep with a rim approximately 5.5 km above sea-level. An andesite lava dome has been growing in the crater since 1984 (Oppenheimer *et al.* 1993) and has been repeatedly destroyed by explosive eruptions. Several short-lived explosive eruptions have occurred since the new period of activity began in 1984. Each episode partly destroyed the growing lava dome. The dispersal of the eruption plume of September 16, 1986 was described by Glaze *et al.* (1989). The February 20, 1990 Vulcanian eruption is described here. On April 20 and 21, 1993 a relatively large explosive eruption occurred which generated Plinian columns reaching 25 km altitude above the volcano and substantial pyroclastic flows.

The 1990 eruption was observed and photographed by two geologists. J. R. Guerneck took a sequence of 17 photographs at a distance of 52 km from the volcano of which the first 10 photographs were timed with an estimated error of 7 s. S. Foot took four photographs 32 km from the volcano. The first two minutes of the eruption were unrecorded so that the first plume had already ascended to about 3 km above the crater rim in the first photograph. The whole episode involved three closely spaced Vulcanian-type explosions over a period of five minutes and the plumes had risen to their maximum height and were being dispersed downwind only 10 minutes after the eruption began. The ascent of two of the three plumes could be documented from the photographs (Figure 5.10).

Figure 5.10 Photograph of the Lascar plumes of February 20, 1990, at 15:48:30 local time (taken by J. R. Guernick)

The photographs (Figure 5.10) show that the plumes had filled the entire 600 m radius crater on emerging from the crater rim. The source dome was only 200 m in diameter and this gives a maximum constraint on the vent dimension. The erupting mixture must therefore have expanded substantially within the crater. This is attributed to the explosions taking place at a very high level within the dome itself and resulting in lateral expansion of the erupting mixture to fill the crater. Thus the plumes initiated with wide diameters with virtual sources several hundred metres below the crater floor (see Chapter 2 for definition of virtual source). The photographs show that each explosion produced a plume with an expanded plume head and somewhat narrower tail supplying the plume front, which is the characteristic shape of a starting plume (Turner 1962). These observations indicate that material was erupted for some tens of seconds after the initial burst. The supply rate then subsided and the plume began to decrease in height prior to the next explosion taking place.

Height versus time data from the Guerneck photographs for the first two plumes are shown in Figure 5.11 and regressions through the data are used to estimate curves of plume front velocity versus height (Figure 5.12). The heights are estimated from the known field of view of the photographs taken at 51 km from the volcano using the geometrical corrections described in Sparks and Wilson (1982). Both plumes decelerated with height with the most pronounced deceleration at the top where the plumes passed their neutral buoyancy heights. The first plume was much more vigorous than the second and reached a height of 15.3 km above the crater rim, whereas the weaker second plume reached 7.9 km. The maximum radius of the plume front is shown for both plumes in Figure 5.13. Both plumes show the same spreading rate (an angle of 27°) which gives a value of $b = 0.21z$. This result is in good agreement with the numerical spreading rates given by Woods (1988) and in Chapter 4. This estimate assumes that the starting plume fronts have 20% greater radius than the following steady plume and the reader is reminded that the plume radius equals $2b$, where b is the characteristic length-scale for a Gaussian velocity profile.

Variations of magma discharge rate, density relative to the atmosphere, temperature difference and particle content with height in the plumes (Figure 5.14) have been calculated using measurements of the plume front radius and velocity with height, and estimates of the plume density using the method of Sparks and Wilson (1982) and equations (5.2)–(5.5). A temperature of 1200 K was assumed for the andesite ejecta of Lascar. Plume densities are only slightly less than the surrounding atmosphere at a given height so that the results are only weakly dependent on the assumption of the Boussinesq approximation (see Chapter 2) built into the Sparks and Wilson model. The most noticeable feature of the calculations is that the estimated discharge rates of both plumes increase with height. As expected the weaker plume shows the lower particle content, temperature difference and

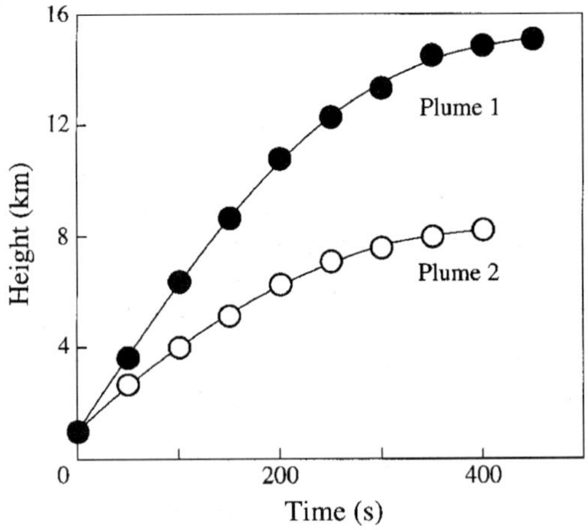

Figure 5.11 Height of the plume top versus time for the first two plumes generated by explosions of Lascar volcano on February 20, 1990. The curves are fourth-order polynomial fits to observations

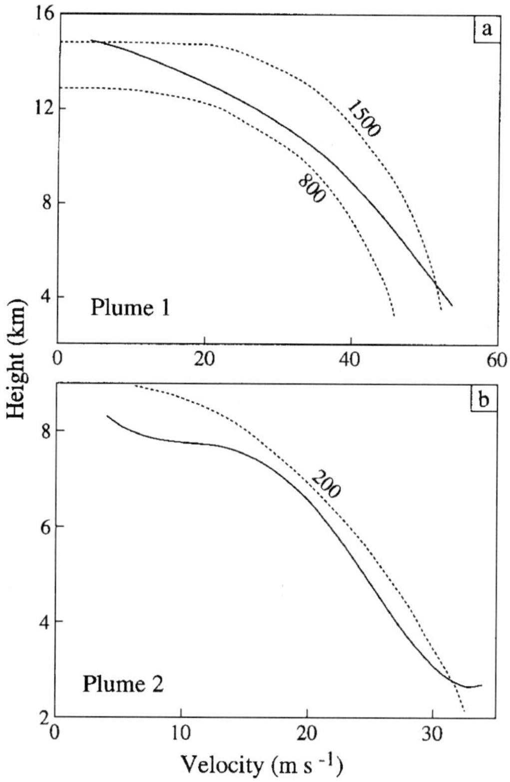

Figure 5.12 Curves of velocity versus height for the Lascar plumes derived by differentiating the polynomial curves in Figure 5.11. Plume 1 is shown in (a) and plume 2 in (b). The curves are compared with numerical calculations (dashed curves) using the theoretical models described in Chapter 4. The calculations assume a vent altitude of 5.5 km and were carried out for a magma water content of 0.03, initial velocity of 300 m s^{-1} and temperature of 1200 K. Plume 1 is compared with discharge rates of 800 and 1500 m^3 s^{-1} and plume 2 is compared with a discharge rate of 200 m^3 s^{-1}

discharge rate at a given height. Application of equation (5.1), using the maximum estimates of magma discharge rate, gives column heights of 7.7 and 11.1 km in comparison with the observations of 7.9 and 14.9 km respectively. The stronger plume also shows a discrepancy between the neutral buoyancy height obtained from the Sparks and Wilson method (9.2 km; Figure 5.14) and the height of 12.3 km expected from simple plume theory. The cause of these differences is not known, but might indicate that the Sparks and Wilson method underestimates discharge rates in stronger plumes. Alternatively there may be a source of extra buoyancy such as condensation or sediment loss which drives the plume to somewhat greater heights than expected.

Figure 5.12 also compares the observations with the numerical models of Chapter 4 after Woods (1988). The weaker plume shows excellent agreement with models for a discharge rate of 200 m^3 s^{-1}. The stronger plume 1 is bounded by the theoretical curves for magma discharge rates of 800 and 1500 m^3 s^{-1}. The observed velocities show a more

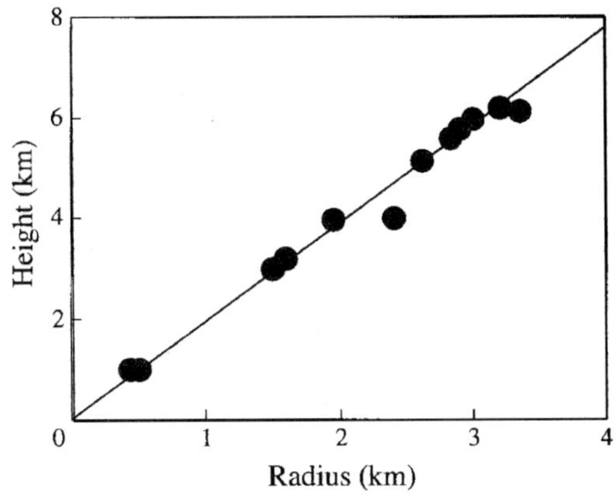

Figure 5.13 The maximum radius of the 1990 Lascar plumes plotted against height. The two plumes are not distinguished because the spreading rates were the same. Note that the height is that at which the maximum plume radius is measured and not the top of the plume

rapid decline with height than the numerical model curves. The numerical results do not suggest an increase in discharge rate with time in comparison to the results of the Sparks and Wilson method.

The apparent increase in magma discharge rate with time could be a consequence of several factors. First, the approximations of the Sparks and Wilson model might generate systematic deviations with height if the entrainment of surrounding air is not perfectly linear. Second, the plumes might not be steady and the results might indicate a plume with either increasing mass flux or temperature of ejecta with time. Third, the plume may not be in complete thermal or mechanical equilibrium as the model assumes. An increasing mass discharge rate with time might be related to the progressive release of heat from the particles (thermal disequilibrium), or to the release of latent heat from water condensation or secondary mineral growth, or to sedimentation.

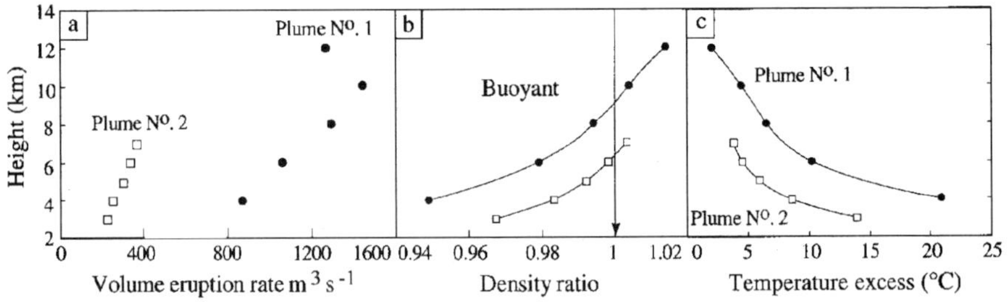

Figure 5.14 Estimates of (a) magma discharge rates, (b) plume buoyancy as measured by the ratio of plume density to atmospheric density and (c) temperature difference between the plume and the surrounding atmosphere versus height for the 1990 Lascar plumes obtained using equations (5.2)–(5.5). The arrow refers to a density value of 1.0 when the plume has the same density as the surrounding atmosphere

5.4.4 October 17, 1980 Eruption of Mount St Helens

A case history of the initial motions of a starting plume is provided by activity of Mount St Helens on October 17, 1980. This plume was recorded by a video camera installed on Harry's ridge about 8.35 km from the volcano (Miller and Hoblitt 1981). The camera was run continuously and captured the initial motions of the activity. The second major explosive event of October 17 began at 0928 hours local time and formed a plume which reached an altitude of 14.3 km above sea-level at 0939 (Markham and Kinoshita 1980).

Figure 5.15 shows profiles of the eruption plumes as they first emerged from the vent in 2 s intervals. Within 5 s it became clear that there must have been at least three almost simultaneous explosions. The initial plume divided into two separate plumes. The plume shown on the left of Figure 5.15 was seen to be fed by two explosions with an interval of about 12 s between each explosion. Only the left plume motion is studied here because the right plume is partly obscured. The plume expanded rapidly within the crater from an initial width of 190 m to 400 m within only 15 s, which implies an average expansion velocity of 14 m s^{-1}. This expansion is attributed to the shallow source of the explosions within the growing lava dome which produced an overpressured flow. Qualitatively the behaviour follows the predictions concerning the effects of overpressured flow issuing into a wide crater as presented in Chapter 3.

Figure 5.16a shows the plume front velocity as a function of height. The initial peak velocity is 73 m s^{-1} and the plume is observed to decelerate rapidly. A second peak velocity is observed between 12 and 16 s and is attributed to the plume from the second explosion overtaking the flow front of the plume from the first explosion. This plume then decelerated rapidly. The gas thrust region is estimated to be approximately 800 m high if 17 s is taken as the time when strong deceleration ceased. After 20 s the plume front velocity steadied and there is an indication of a gradual increase in velocity thereafter. The data are noisy and this is considered a real attribute of the flow with pulses of more

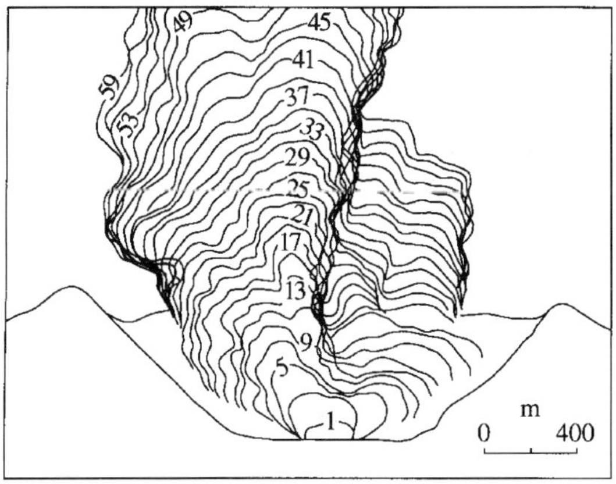

Figure 5.15 Profiles of the plumes generated by the October 17, 1980 eruption of Mount St Helens volcano at 0928. The contours are in 2 s intervals. The profiles were drawn directly from a video screen and therefore the scale varies from place to place. The scale shown is for the central region by the source vent, but should not be used to make measurements

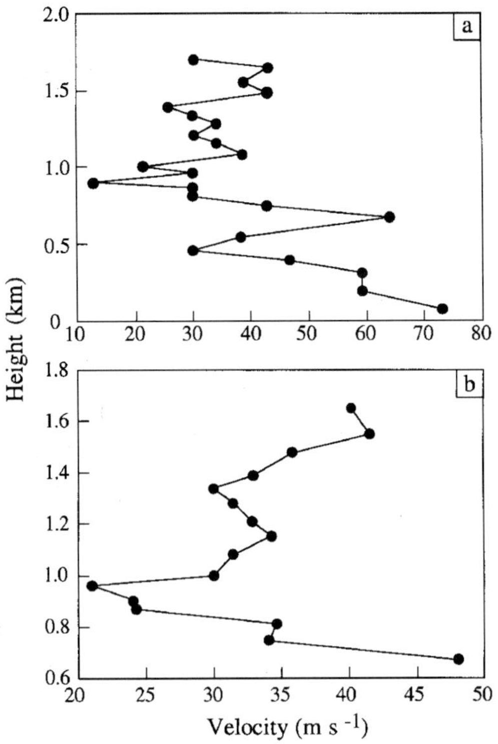

Figure 5.16 Velocity variations with height for data extracted from Figure 5.15. (a) Velocity at 2 s intervals and (b) three-point running averages of velocities at 2 s intervals

buoyant fluid reaching the flow front as described in laboratory plumes (Papanicolaou and List 1988). The increase in velocity is seen more clearly when three-point running averages of the velocity data are plotted (Figure 5.16b) and confirms that the plumes became superbuoyant in these initial stages.

Height versus radius data are shown in Figure 5.17. Note that the height in this figure is at the level where the plume reached its maximum radius. The first 12 s of data are not included in the regression in order to avoid the influence of the rapid expansion in the gas thrust region. The data are consistent with the spreading law $b = 0.23z$, where $2b$ is the radius and z is the height, again on the assumption that the initial plume front had a diameter 20% greater than the following steady plume.

Application of the Sparks and Wilson method to heights of 800 m and above gives the expected decrease in temperature excess and buoyancy with height (Figure 5.18). Magma discharge estimates increase substantially with height (Figure 5.18a). As in the Lascar case it is unclear whether these estimates represent disequilibrium effects in the plume or are the consequence of a real increase in discharge rate. The maximum estimate of 1100 $m^3\ s^{-1}$ at 1600 m height, however, corresponds to an ultimate column height of 10.2 km using equation (5.1) in comparison to an observed height of 12.8 km above the vent which requires a discharge rate of the order of 3000 $m^3\ s^{-1}$. This might suggest that the dis-

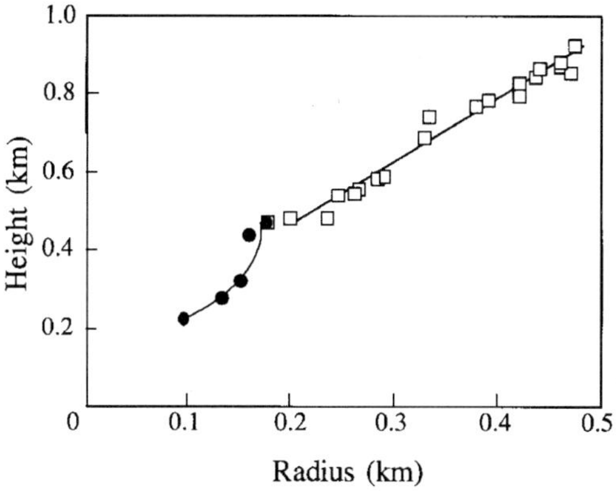

Figure 5.17 Height plotted against radius for the October 17, 1980 Mount St Helens eruption. The height is measured at the height of maximum plume radius. The solid dots are data from the gas thrust region and the open squares are data from the convective region. The solid line through the data in the convective region represents a linear regression and has an intercept at 156 m above the vent

charge rate did increase further with time and that the initial flows were not wholly representative of the average flow conditions over the entire eruption. There were, however, two plumes generated (Figure 5.15) which merged together. Assuming that the other plume had comparable strength then this increases the discharge rate to over 2000 $m^3\ s^{-1}$. Given the many uncertainties the agreement is reasonable.

5.4.5 1947 Hekla Plinian Eruption

The 1947 eruption of Hekla, Iceland, began on the morning of March 29. The initial stages were captured on movie film and Thorarinsson (1967) reports profiles of the plume ascent from the movie. The observations of the height of the plume versus time are shown in Figure 5.19. The average ascent velocity of the plume from 1.8 to 20 km height was

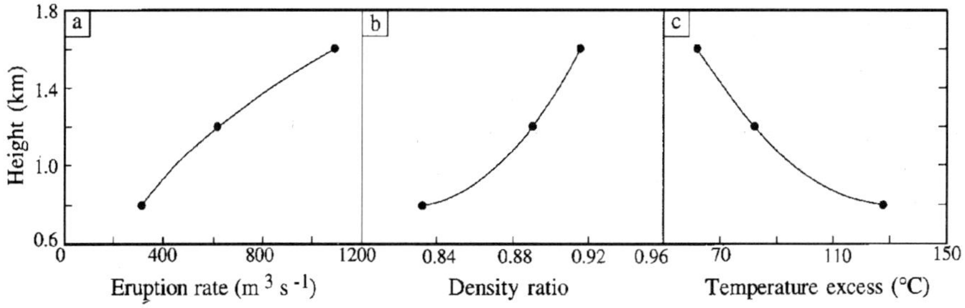

Figure 5.18 Calculated variations of (a) magma discharge rate, (b) buoyancy as measured by the ratio of plume density to atmospheric density and (c) temperature excess with height in the Mount St Helens plume of October 17, 1980

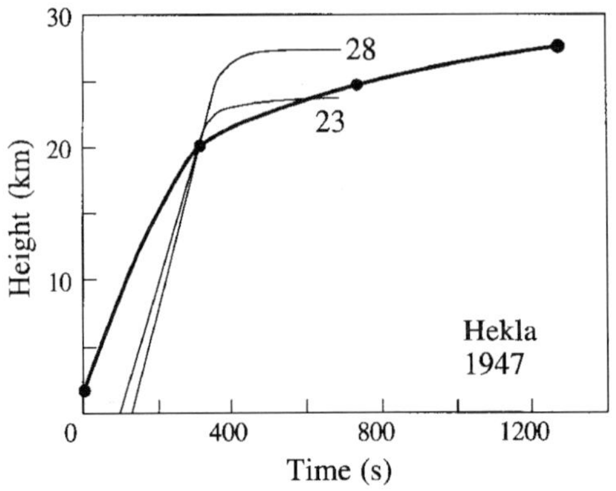

Figure 5.19 Data on plume front height versus time for the 1947 Plinian eruption of Hekla (solid symbols). The heavy line through the observations is hand-drawn to provide a smooth possible height versus time profile through the data. The two thin lines represent theoretical height versus time curves from the numerical model of Woods (1988) for discharge rates of 5×10^4 and 10^5 m^3 s^{-1}, assuming a magma water content of 0.03, an initial vent velocity of 300 m s^{-1}, a temperature of 1200 K and initial vent radii of 149 and 211 m respectively. The figures on the theoretical curves refer to column heights in kilometres

67 m s^{-1}, which is equivalent to the steady plume velocity of 100 m s^{-1}. Above a height of 20 km the data show that the plume reached a region of strong deceleration, indicating that it was above the height of neutral buoyancy.

The results of numerical calculations have been compared with the observations for columns which reach approximately 23 and 28 km (after Woods 1988), equivalent to volumetric magma discharge rates of 5×10^4 and 10^5 m^3 s^{-1} respectively. The results of these calculations are shown in Figure 5.19 in such a way that they pass through the observed time when the plume reached 20 km. The calculated velocities are somewhat higher (140–180 m s^{-1}) than the average velocity to 20 km. This is seen by the steeper slopes of the theoretical curves compared to the linear interpolation between the observations at 1.8 and 20 km. The zero height time intercept of the theoretical curves is required to be significantly later than the observed start of the eruption. This discrepancy is most easily explained by an increase in discharge rate with time in the earliest stages of the eruption as has already been demonstrated for the Soufrière, Lascar and Mount St Helens plumes. Above 20 km the model curves show a very abrupt deceleration close to their maximum height, whereas the observations indicate a more gradual decay of velocity. This may be a consequence of the models not being accurate when the starting plume approaches its ultimate height or could also be evidence that the discharge rate was still increasing with time. The observations are consistent with rather higher discharge rates than given in Table 5.1 based on the estimate of Thorarinsson (1967).

5.5 INSTANTANEOUS EXPLOSIONS

For very short-lived instantaneous events, notably Vulcanian explosions, a sustained plume model is not appropriate. This situation occurs when the time scale of explosive discharge

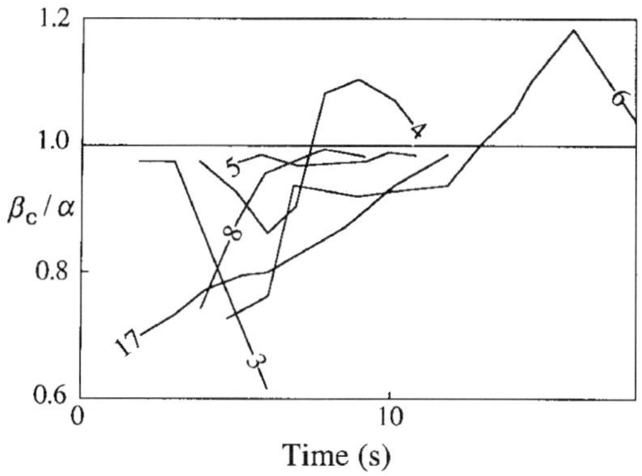

Figure 5.20 Profiles of calculated volcanic thermal density with time for Fuego Vulcanian eruptions after Wilson and Self (1980). The numbers refer to different eruption events as described in Wilson and Self (1980)

is significantly shorter than the ascent time of the plume. Such a buoyant flow can be described as a volcanic thermal rather than a plume. Wilson and Self (1980) developed a method of estimating volcanic thermal properties from their motion. They considered the motion of a sphere and a cylinder, deriving the equations

$$\frac{\beta_c}{\alpha} = \frac{g - \dfrac{3C_s u^2}{8r}}{g + \dfrac{du}{dt}} \tag{5.6}$$

$$\frac{\beta_c}{\alpha} = \frac{g - \dfrac{3C_s u^2}{2L}}{g + \dfrac{du}{dt}} \tag{5.7}$$

where β_c is the bulk density of the volcanic thermal, α the density of surrounding air, u the velocity, r the radius of the sphere, L the length of the cylinder and C_s and C_c the drag coefficients with values of 0.47 and 1.0 for a sphere and a cylinder respectively. All parameters on the right-hand side of these equations can be derived from the kind of data shown in Figures 5.5 and 5.6 by frame by frame analysis of movie or video film. Heat balance calculations can then be used to constrain volcanic thermal temperatures and particle contents. Figure 5.20 shows estimates of the density ratio with time and therefore height, in the volcanic thermals generated in the 1979 eruptions of Fuego volcano, Guatemala. The results indicate that for most volcanic thermals buoyancy decreases with height as a consequence of air entrainment and cooling. Only thermal 3 shows a marked increase in buoyancy and this may be due to heating of entrained air by hot ash dominating over any cooling effects (the superbuoyancy effect).

5.6 SUMMARY

There is a good empirical correlation between volcanic plume heights and discharge rate of magma. When these data are fitted to a power law they give a relationship which is close to that predicted by simple scaling arguments (Morton *et al.* 1956) and numerical models (Woods 1988; Chapter 4). Data on column heights, discharge rates, ejecta volumes and durations are generally not sufficiently accurate to discriminate the subtle effects of atmospheric conditions, source effects and other complexities such as thermal and mechanical disequilibria. Equation (5.1) provides a very simple and useful rule of thumb for monitoring variations in eruption rates from measurements of column heights.

There are unfortunately very few examples of volcanic plumes which have been studied in any detail for comparison with theoretical plume models. All those described here are based on the initial rise of plumes from timed photographs, movie film and video. Observations on the initial ascent of volcanic plumes show broad agreement with plume models. Spreading rates, velocities, temperatures and buoyancies are in the range predicted by numerical models. Features such as the rapid expansion of overpressured vent flows, the rapid deceleration of the gas thrust region, acceleration due to superbuoyancy and more rapid deceleration above the height of neutral buoyancy can be discerned in the observations. An approximate method for analysing starting plumes by Sparks and Wilson (1982) appears to be successful. However, application of the method always gives an increasing discharge rate of magma with time. This may be a characteristic feature of the initial phases of explosive eruptions, but it could also be the result of disequilibrium effects such as incomplete heat transfer between particles and entrained air or condensation effects. There is a tendency for the Sparks and Wilson method to give slightly lower discharge rates than those estimated from observed column heights.

This chapter emphasizes the lack of available data on plume dynamics. It is unlikely that much further progress can be made until methods have been developed for making measurements in the interior of active plumes.

6 Pyroclastic Flows

NOTATION

g	gravitational acceleration (m s^{-2})
H_f	height of pyroclastic fountain (m)
n	mass fraction of volcanic gas (dimensionless)
Q	volumetric flow rate (m^3 s^{-1})
R	gas constant (J K^{-1} kg^{-1})
T	temperature (K)
u	exit velocity from vent (m s^{-1})
ρ	density of fluid (kg m^{-3})
$\Delta\rho$	density difference between fountain and ambient fluid (kg m^{-3})

6.1 INTRODUCTION

Pyroclastic flows consist of mixtures of hot volcanic particles and gas that are usually generated by explosive volcanic eruptions or gravitational avalanching of steep-sided lava domes. They are sometimes called *nuées ardentes*, meaning glowing cloud. Pyroclastic flows that are generated explosively form due to instabilities in the lower parts of eruption columns. Pyroclastic flows mix with the surrounding atmosphere and rising convective plumes can form above them. This chapter is concerned with the dynamical conditions in eruption columns that lead to formation of pyroclastic flows. Pyroclastic flows vary over several orders of magnitude in their volume from less than 0.001 km^3 to those that exceed 1000 km^3. Flow velocities range from a few metres per second in the small historically observed events to estimated velocities of over 200 m s^{-1} in the most violent events. Pyroclastic flows generate plumes with unique characteristics. These plumes vary greatly in size and those generated in the large violent eruptions represent the largest magnitude and most dramatic kind of volcanic plume. For example, the giant plume that formed in the paroxysmal eruption of Pinatubo volcano on June 15, 1991 is believed to have accompanied the generation of pyroclastic flows. Also, the giant umbrella plume which formed during the May 18, 1980 eruption of Mount St Helens rose from the lateral blast flow after it had propagated some distance from the volcano. In most pyroclastic flow eruptions the flow deposits are called ignimbrites (ignis = fire, bris = cloud). The plumes generated from pyroclastic flows are known as co-ignimbrite plumes, discussed in detail in Chapter 7. In order to understand co-ignimbrite plumes it is first necessary to consider the nature and origin of the causative pyroclastic flows.

6.2 THE NATURE OF PYROCLASTIC FLOWS

The geological literature on pyroclastic flows is extensive, complex and sometimes confusing. There are a great many terms and different concepts which result from the geologists' penchant for taxonomy, difficulties in interpretation of the deposits and incomplete understanding of the dynamics of high-velocity multiphase flows. The latter two problems are not surprising because pyroclastic flows by their very nature are difficult to observe and involve poorly understood flow dynamics in complex mixtures of gas, solid particles and liquids. Studies of their deposits have played an important role in developing conceptual models of flow. This book is not the place for a detailed discussion of nomenclature, classification and the geological features of pyroclastic flow deposits, and the reader is referred to thorough accounts in Fisher and Schmincke (1984), Wilson (1986), Cas and Wright (1987) and Carey (1991).

6.2.1 Flows and Surges

Hot particulate volcanic flows or pyroclastic gravity currents have been divided into pyroclastic flows and pyroclastic surges (Carey 1991). The former term is usually equated with flows with a high concentration of particles in which particle interactions play an important role. The latter term refers to highly turbulent flows with a low particle concentration. The archetypal pyroclastic flow is conventionally viewed as a high concentration avalanche of hot particulate ejecta, which moves downslope under the influence of gravity as a granular avalanche or dense particulate dispersion flow (Sparks 1976). Such flows are thought to have a well-defined interface between the high concentration basal avalanche and overlying dilute turbulent surge clouds and convective ash plumes (Figure 6.1). Pyroclastic flows are commonly considered to be partly fluidized (Sparks 1976; Wilson 1986). The archetypal pyroclastic surge is a dilute highly turbulent flow. There are no abrupt boundaries within a pyroclastic surge (Valentine 1987), but it is generally thought that there are vertical gradients of particle concentration. There is uncertainty about whether there is a continuum or a discontinuity between pyroclastic flows and surges. Advocates of the discontinuity point to the tendency of some two-phase systems (such as fluidized beds) to partition into high concentration regions and very dilute gas-rich regions and also draw attention to features of the deposits consistent with a discontinuity (Wilson 1984, 1986; Anilkumar *et al.* 1993). Advocates of a continuous gradation of particle concentration between and within volcanic flows consider that turbulence would generate continuous variations (e.g. Valentine 1987). Certainly most dense pyroclastic flows also generate a dilute turbulent component or ash-cloud surge which can travel beyond the limits of the dense part of the flow. A confusing feature of the nomenclature is that the terms "flow" and "surge" in the volcanological literature are defined in quite different ways to those used in the fluid mechanics literature, where a flow is a general term for sustained movement of fluid and a surge is a pulse of fluid of fixed volume.

The fluid mechanics of pyroclastic flows is still a matter of some uncertainty and controversy. Conceptual models (Figure 6.2) of flow borrow from a variety of analogies with other simpler and better understood flow systems. Ideas developed for single-phase fluid flows are widely employed and flows are often categorized as low Reynolds number

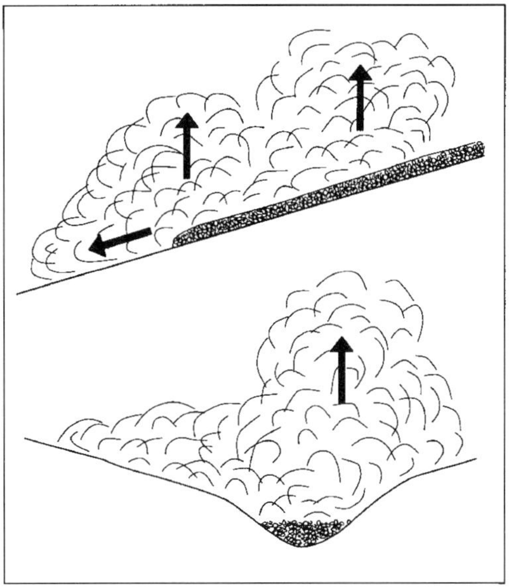

Figure 6.1 Schematic longitudinal section and cross-section of a pyroclastic flow moving down a valley showing the dense basal avalanche and overlying dilute turbulent ash cloud. The arrows imply that one component of the ash cloud is denser than air and can continue on as a gravity current (i.e. a pyroclastic surge) and that the other component is buoyant and can rise as a dilute co-ignimbrite plume

viscous flows or high Reynolds number turbulent flows. However, the concept of a low Reynolds number viscous two-phase flow is problematic and the segregation of particles within a flow can lead to large vertical and lateral variations in particle concentration and flow conditions. Thus a two-phase flow can simultaneously have regions of both strong turbulence and those which have organized internal shear analogous to viscous flows. In these circumstances it is not straightforward to assign a conventional Reynolds number to a pyroclastic flow. Small flows are probably sufficiently "viscous" that the basal avalanche moves in a laminar fashion (Figure 6.2a). However, an analysis of velocity data on a modest-sized pyroclastic flow from Mount St Helens by Levine and Kieffer (1991) indicates that the flow can be modelled by inviscid flow theory using standard hydraulic equations. Wilson and Walker (1980) proposed a model for large high-velocity pyroclastic flows which involves an expanded turbulent flow head analogous to the front of a simple turbulent gravity current (Simpson 1987), followed by a high concentration flow body in which flow is predominantly in organized shear concentrated in a basal zone (Figure 6.2c). Fluidization (Figure 6.2b and 6.2c) is commonly invoked (Sparks 1976; Wilson 1986). The role of fluidization in flow mobility is still unclear, but it probably plays an important role in generation of co-ignimbrite plumes (see Chapter 7).

In very violent flows the distinction between flow and surge based on particle concentration can become problematic. Druitt (1992) demonstrated from studies of the deposits that the violent blast flow generated in the initial phase of the 1980 eruption of Mount St Helens changed from a high concentration flow to a dilute surge as the flow lost sediment and mixed turbulently with its surroundings. In such high-velocity flows it may

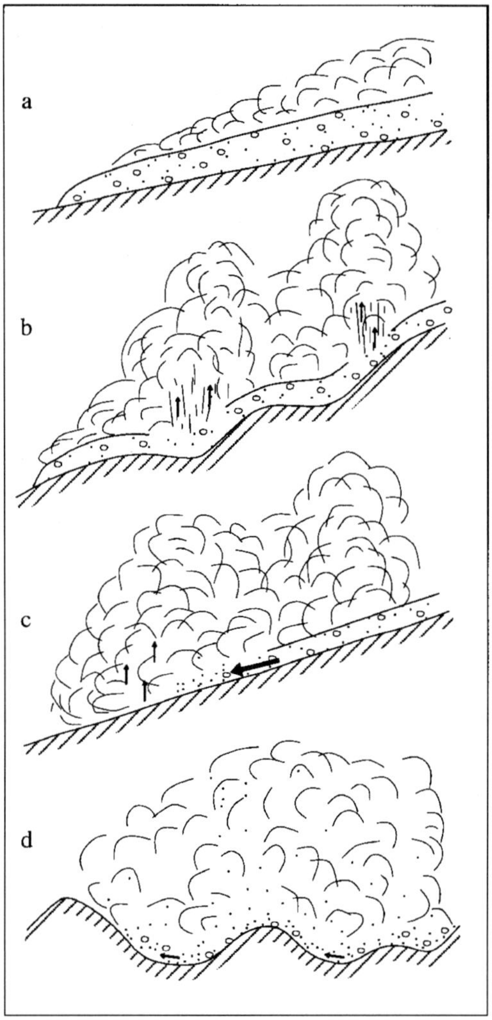

Figure 6.2 Some conceptual models of pyroclastic flows. (a) A small-scale flow predominantly composed of a dense basal granular avalanche moving by organized shear (analogous to viscous flow) and generating a turbulent dilute ash cloud by shear across the upper surface of the avalanche. (b) A similar avalanche which is locally or pervasively fluidized as a result of flowing over rough topography, vegetation or wet ground. (c) A high-velocity pyroclastic flow with strong turbulent mixing and fluidization in the flow head forming a turbulent ash cloud in the wake (analogous to the processes observed in high Reynolds number gravity currents). The body of the flow following the head has a high particle concentration, moves by organized shear and maintains a sharp interface with the overlying ash cloud. (d) A high-velocity pyroclastic flow in which turbulence is pervasive and there is a vertical concentration gradient rather than a sharp division between the basal avalanche and the turbulent cloud. This flow is shown riding over rough topography due to its high velocity

be preferable to envisage a high concentration turbulent sediment gravity current in which there are large local fluctuations in both velocity and density (Batchelor 1988; Anilkumar *et al.* 1993). High-velocity flows may exhibit features of both the archetypal flow and surge (Figures 6.1 and 6.2d). We use the term "pyroclastic flow" for high-temperature volcanic particulate flows. Such flows can have important dilute turbulent surge components (ash-cloud surge) and can evolve into a dilute surge. We use pyroclastic surge in cases where the flow is believed to be entirely dilute and turbulent. Such flows transport particles by a combination of suspension and basal saltation. However, the recent study of Dade and Huppert (1996) models high-velocity pyroclastic flows as turbulent suspensions. A special variety of flow is the base surge (discussed in Chapter 8) generated in hydrovolcanic explosions which can involve abundant steam and low temperatures.

6.2.2 Observations

Most direct observations of pyroclastic flows are of relatively small volume historic eruptions where the flows are generated by explosions or collapse of unstable growing lava domes (e.g. Moore and Melson 1969; Hoblitt 1986; Sato *et al.* 1992). The flows are observed to be confined by topographic depressions (Figures 6.1, 6.3 and 6.4). They generate a billowing dilute turbulent cloud above a basal region (Figure 6.3) which, from observations of the deposits and the effects of the flows, is inferred to be a high concentration avalanche. Flow velocities are typically up to a few tens of metres per second, run-out distances a few kilometres and deposit volumes are fractions of a cubic kilometre. Several historic eruptions have generated more substantial deposit volumes and more extensive flows. Direct observations are, however, limited because of the catastrophic nature of the eruptions. The deaths of several journalists and scientists caused by pyroclastic flows in the 1991 eruption of Mount Unzen in Japan illustrate their tragic consequences. Examples of historic eruptions which exceed a cubic kilometre of pyroclastic flow ejecta include Tambora in 1815, Krakatoa in 1883, Katmai in 1912 and Pinatubo volcano in 1991. In these examples the flows extended to distances of 10 to over 30 km and formed deposits of tens of metres in thickness. It is inferred that the flow velocities were several tens of metres per second. Interpretation of their deposits has led to the view that the flow mechanisms are similar to the smaller avalanche flows, although it is probable that their higher velocities lead to enhanced mixing with the atmosphere and stronger fluidization. In the geological record there is evidence both of much larger and more violent pyroclastic flow eruptions. Such paroxysmal events generate the extensive ignimbrite sheets that are characteristic of large caldera systems. The volume of ejecta can be tens to a few thousand cubic kilometres of ejecta in one individual eruption. One of the largest known ignimbrite eruptions occurred at 75 000 a when Lake Toba caldera in Indonesia disgorged over 2000 km^3 of ignimbrite (Ninkovich *et al.* 1978). These ignimbrites can extend to distances of tens of kilometres and it is inferred from theoretical models (Sparks *et al.* 1978; Valentine and Wohletz 1989; Neri and Dobran 1994) that their velocities were large, in the range 50 to over 200 m s^{-1}.

6.2.3 Range and Aspect Ratio of Deposits

In considering pyroclastic flows it is important to distinguish between the magnitude of an eruption, as measured by the total volume of ejecta, the violence of the flow, as measured

Figure 6.3 Pyroclastic flow moving down a topographic confinement on the flanks of Mount St Helens on August 7, 1980, displaying a turbulent overlying dilute plume

by the dispersal of material by momentum, and the intensity of the flow as measured by the discharge rate of ejecta. Walker *et al.* (1980) defined the aspect ratio as the ratio of the average deposit thickness to the characteristic length of the deposit as measured by the diameter of a circle with the same area as the deposit. They proposed the aspect ratio as a measure of flow violence. A high-aspect ratio ignimbrite (HARI) is interpreted as a deposit formed from a relatively low-velocity flow or sequence of flows. Such deposits tend to be thick, pond in depressions and valleys, and contain many individual flow units. The pyroclastic flows formed in the 1912 eruption of Novarupta, Alaska, have these characteristics (Hildreth 1987). A low aspect ratio ignimbrite (LARI) is interpreted as a deposit formed from a very high velocity flow. Such deposits are thin relative to the area covered and characteristically drape over topography, indicating that the flows were travelling sufficiently fast and were sufficiently thick that they could surmount topographic barriers of several hundreds of metres. The Taupo ignimbrite is the best-documented example of a LARI (Wilson 1985). This was formed by an eruption from Lake Taupo in

Figure 6.4 Generation of a pyroclastic flow during the 1975 eruption of Ngauruhoe volcano in New Zealand. (a) A dense slug of gas and tephra is observed expanding radially from the summit crater 1 s after the initial explosion. The slug of ejecta and gas has an estimated internal pressure of 3×10^5 Pa. (b) After 11 s the dense slug has expanded to a diameter of over 600 m and the external parts of the cloud have become buoyant due to mixing with air. Huge ballistic blocks (up to 30 m diameter) are observed. (c) The dense interior of the column has collapsed to generate a pyroclastic flow after 30 s and the external parts of the column are buoyant, rising to form a high plume. (d) After 60 s the pyroclastic flow has reached the base of the volcano

Figure 6.4 (*continued*)

New Zealand in AD 182. The Taupo ignimbrite (volume 30 km^3) extends to 80 km or more in all radial directions from the source and can be found on the tops of mountain ranges over 1400 m higher than the vent at distances of 50 km. Such a dramatic distribution requires a flow of exceptionally high velocity and Wilson (1985) estimates a velocity exceeding 200 m s^{-1} based on the overtopping of topographic barriers. The initial blast flow of the May 18, 1980 eruption of Mount St Helens is another example of a violent pyroclastic flow of modest volume (0.1 km^3). The flow took only four minutes to spread over 600 km^2 (Hoblitt *et al.* 1981; Sparks *et al.* 1986) and had observed propagation velocities of 90–110 m s^{-1}.

6.2.4 Constituents of Pyroclastic Flows

Pyroclastic flows can be composed of a wide variety of ejecta. The most usual compositions are ejecta with intermediate to high silica contents (andesites, dacites and rhyolites) reflecting the propensity for magmas with high viscosities to erupt explosively. However, basaltic pyroclastic flows are also known (for example the Pucon ignimbrite of Villarrica, Chile). The common constituents of pyroclastic flow deposits are juvenile clasts of pumice or scoria, glass shards, crystal fragments and accidental lithic fragments. Varieties containing volcanic bombs or non-vesicular juvenile blocks are also known. The deposits vary from those rich in volcanic ash (sometimes known as ash flow deposits) to those rich in large clasts of pumice (pumice flow deposits) or scoria (scoria flows) or angular juvenile blocks (block-and-ash flow deposits). The term ignimbrite is conventionally confined to those flow deposits containing a high proportion of juvenile vesicular ejecta (i.e. pumice and pumiceous glass shards). By far the greatest volume of pyroclastic flow deposits are ignimbrites. The temperatures of pyroclastic flows vary greatly. They depend on the temperature of the magma and the cooling processes that occur during their generation and during flow. Flow temperature varies from magmatic (typically 700 to 1000 °C) down to 100 °C. The lower limit is based on the presence of liquid water. Flows below 100 °C are better described as debris flows, mud-flows, lahars or cold avalanches.

6.2.5 Dangers and Hazards

Pyroclastic flows are a major cause of destruction and loss of life in explosive eruptions. Their effects can be devastating as exemplified by the complete destruction of the town of St Pierre and its 30 000 inhabitants in the eruption of Mont Pelée on the island of Martinique in 1902 and by the burial of Pompeii and Herculaneum under several metres of ignimbrite in the AD 79 eruption of Vesuvius. Pyroclastic flows are also responsible for creating huge plumes which have great environmental impact. The 1991 plume of Pinatubo volcano generated voluminous pyroclastic flows and a major associated co-ignimbrite plume. The Pinatubo plume and other co-ignimbrite plumes have caused severe effects in the stratosphere as discussed in more detail in Chapters 7, 11, 12 and 18.

6.3 GENERATION OF PYROCLASTIC FLOWS BY FOUNTAIN COLLAPSE

The principal mechanism of pyroclastic flow formation during explosive eruptions is by a process known as fountain collapse, in which an erupting mixture of tephra, magmatic gas and entrained air forms a collapsing fountain to feed pyroclastic flows. Consideration of the bulk density of the erupting mixture is critical to understanding how collapsing pyroclastic fountains form. In almost all explosive eruptions the volatile contents of magmas are only a few per cent so that the bulk density of an erupting mixture is considerably higher than the density of the Earth's atmosphere. For mass fractions of magmatic water in the range 0.01–0.06, for example, bulk densities of erupting mixtures at 10^5 Pa are in the range 4–18 kg m^{-3}, compared to a standard sea-level atmospheric density of 1.25 kg m^{-3}. If erupting mixtures were to ascend without mixing with the Earth's atmosphere then the mixtures would soon run out of momentum and collapse to form density currents that move along the ground. As outlined in previous chapters, turbulent entrainment of the atmosphere and heating of air results in erupting mixtures becoming buoyant and forming high plumes. The entrainment process and dynamic flow conditions are therefore critical to determining whether a high convecting column or a collapsing fountain forms. Pyroclastic flows are formed when the eruption column fails to entrain sufficient air to become buoyant (Chapter 4).

6.3.1 Observations of Eruptions

Figure 6.4 shows a sequence of photographs taken during a 1975 eruption of Ngauruhoe in New Zealand which beautifully illustrates the key processes of pyroclastic flow formation. The photographs were taken by press photographer R. Foley at timed intervals and the eruption is described in detail by Nairn and Self (1978). In the first photograph (Figure 6.4a), taken about 1 s after the initial explosion, a dense slug of ejecta and gas is observed expanding from the 100 m wide crater. The initial pressure in the slug has been estimated at about 3×10^5 Pa. The second photograph (Figure 6.4b) after 11 s shows the slug expanding in all directions to a diameter of about 600 m. Ballistic blocks of rock up to 30 m in diameter are being ejected from the slug. The velocities at this stage have been estimated at approximately 400 m s^{-1} and a shock wave propagated through the atmosphere. The external surface of the slug displays a structure indicative of highly turbulent motion and atmospheric entrainment. After 15 s the slug had expanded further and the external parts of the developing eruption column had begun to ascend buoyantly. At the same time, material was starting to fall on to the upper flanks of the volcano from the interior of the column. The third photograph (Figure 6.4c), taken after 30 s, clearly shows the interior of the slug collapsing on to the upper flanks of the volcano and feeding pyroclastic flows, while the buoyant exterior of the slug has formed an ascending convecting column. The final photograph (Figure 6.4d), taken after 60 s, shows the pyroclastic flow reaching the base of the volcano and the convective column above the vent which ascended to 8 km altitude.

Figure 6.5 shows the interpretation of events of the Ngauruhoe eruption as a general model for the generation of pyroclastic flows from explosive volcanic eruptions. The erupting mixture has a high pressure and bulk density and initially expands in all directions as the mixture is discharged into an unconfined environment. The turbulent mixture entrains cold atmospheric air at the margins. The air dilutes the ash particles and is heated so that the margins become buoyant and form a vigorous plume. However, the interior of the erupting slug is shielded from significant entrainment and soon runs out of momentum. Because the undiluted mixture has a much higher density than the atmosphere it collapses

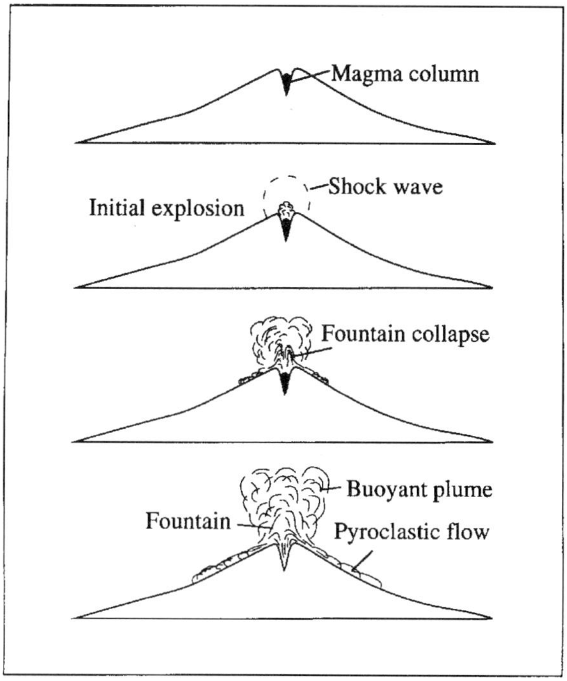

Figure 6.5 Conceptual model for the generation of a pyroclastic flow by a single explosion based on the 1975 Ngauruhoe eruption in New Zealand (see Figure 6.4)

back on to the flanks of the volcano and forms a pyroclastic flow. Nairn and Self (1978) estimated the collapse height in the Ngauruhoe eruption as about 500 m above the vent for an exit velocity of 400 m s^{-1} and bulk density of 20 kg m^{-3}.

The 1980 eruption of Mount St Helens provided further opportunities for observing collapsing fountains. After the main eruption of May 18, 1980 there were several substantial explosive eruptions in the following months related to the explosive destruction of the growing dacite lava dome. These eruptions produced several small volume pyroclastic flows fed by pyroclastic fountains and the eruptions of July 22 and August 7, 1980 were observed in detail (Rowley *et al.* 1981; Hoblitt 1986). Eruptions started suddenly with a sequence of closely spaced explosions which generated plumes rising to heights of 12–18 km. Collapsing ash fountains were observed soon after an eruption began (Figure 6.6). In some eruptions the processes in the vent were obscured by turbulent ash plumes, but some events revealed collapsing fountains reaching 500–800 m above the vent (Hoblitt 1986). Figure 6.6 shows photographs of the July 22 event illustrating collapsing fountains being continuously fed by explosive discharge, rather than by a single discrete explosion as on Ngauruhoe. Later in this eruption a large convective plume formed above the vent and ascended to a height of 12 km (Figure 6.6d).

The initial blast of Mount St Helens on May 18 provides an example of a pyroclastic flow generated with a strong lateral component of motion. Figure 6.7 is a photograph of the early stage of the blast which was triggered by failure of the northern sector of the

Figure 6.6 Photographs of pyroclastic collapsing fountains in the July 22, 1980 eruptions of Mount St Helens (after Hoblitt 1986). (a) and (b) Photographs taken between 19:01:08 and 19:01:38 PDT on July 22, 1980. (c) Photograph taken at 18:25:47 PDT on July 22, 1980. (d) The eruption on July 22, 1980 at about 18:30 PDT shows the two components of a pyroclastic flow moving down the flanks and a turbulent buoyant plume over the crater. For scale the crater of Mount St Helens is approximately 2 km across from rim to rim

Figure 6.6 (*continued*)

Figure 6.7 Photograph of the initial stages of the formation of the Mount St Helens blast flow on May 18, 1980. For scale the field of view is approximately 3 km across

volcano. Prior to failure, intrusion of magma (the cryptodome) deformed the northern flanks and led to catastrophic failure (Christiansen and Peterson 1981). The cryptodome was decompressed by about 12.5 MPa and several closely spaced major explosions formed the high-velocity blast flow. The blast flow then spread over an area of 600 km^2 in about four minutes. This flow is widely regarded as a classic example of a pyroclastic flow generated by a directed blast (Kieffer 1981). While there can be no doubt that there was a strong lateral component in the explosion, photographic evidence also indicates that gravitational slumping of the erupted mixture was a significant factor in the generation of the flow. The northward movement of the flow was partly due to the position of the cryptodome within the volcano and partly attributable to the constraining effect of the newly formed crater wall to the south.

Several features of the Ngauruhoe and Mount St Helens eruptions illustrate important concepts in the fountain collapse process. First, fountain heights are a small fraction of the heights of convective columns and need only be a few hundred metres above a vent in order to generate rapidly moving flows which can extend several kilometres from the vent (Figure 6.6). The fountain heights are controlled by the kinetic energy of the eruption which is only a small percentage of the thermal energy of the ejecta (typically much less than 10%). As discussed in Chapter 4, the heat contained in the ejecta and transferred to the entrained air provides the principal source of energy to form high plumes. Later in this chapter it will be shown quantitatively that the upward momentum at the vent is only capable of generating fountains with heights of a few hundred metres to a few kilometres even in the largest magnitude eruptions. Second, collapsing fountains are almost always associated with simultaneous formation of a substantial buoyant plume. This is because parts of the ejecta mix with sufficient air to become buoyant. The entrainment process works its way into the column from the outside so that it is the external part which tends to become buoyant and the internal region that collapses (Figure 6.4). Third, there may be a component of lateral motion due to decompression of the unconfined erupted mixture and this may contribute significantly to flow formation in some circumstances.

The term "column collapse" has been widely used for the processes described above. This term has sometimes been misunderstood to mean that the whole of a large convective column collapses to form a pyroclastic flow. This is a misconception that arises from the suggestive term column collapse and the emphasis that has been given in the volcanological literature to the common transition between Plinian activity and pyroclastic flow formation. The concept of a collapsing ash fountain can be traced back to the papers of Hay (1959) and Smith (1960) and was given a quantitative theoretical basis by Sparks and Wilson (1976), Sparks *et al.* (1978) and Wilson *et al.* (1980). None of these studies considered the spurious notion of collapse of an entire convective column, but the misconception has persisted. We recommend that the process should now be termed "fountain collapse". Another red herring that persists in the literature is the concept of "boiling over" which is sometimes described as if it were a different mechanism of pyroclastic flow generation. The phrase derives from observations of an eruption of Cotopaxi volcano in Ecuador in 1932 where the ejecta was described as boiling over the crater rim. While this is an evocative description, such flows must still be driven by gravity and should be regarded as an example of a low collapsing fountain.

6.3.2 Experimental Studies

Collapsing fountains have been studied extensively in the laboratory and provide useful insights into fountain dynamics. Turner (1966) and Baines *et al.* (1990) investigated collapsing fountains formed by discharge of jets of saline water into a tank of fresh water. Turner showed that such fountains ascend to a height H_f given as

$$H_f = \frac{1.85 Q^{1/4} u^{3/4}}{(g\Delta\rho/\rho)^{1/2}} \tag{6.1}$$

where Q is the volumetric flow rate from the vent, u the exit velocity from the vent (assumed to be circular in cross-section), g is gravity, $\Delta\rho$ the density difference between the discharging mixture and the surroundings and ρ the density of the discharging mixture (Chapter 2). As an example of a collapsing fountain, Figure 6.8 shows an experiment in which a suspension of fine particles in fresh water was injected into a tank of salty water at a constant flow rate (Carey *et al.* 1988). A fountain is generated which plays around a well-defined but fluctuating height and then feeds a gravity current which spreads radially away from the vent. A buoyant plume is formed due to sedimentation from the current and lofting of a mixture of residual particles and fresh water into the salty water (see Chapter 7 for further discussion of this phenomenon).

Turner (1966) observed that a discharging flow initially reached a height about 40% higher than the steady-state height H_f. This result can be applied to pyroclastic fountains to

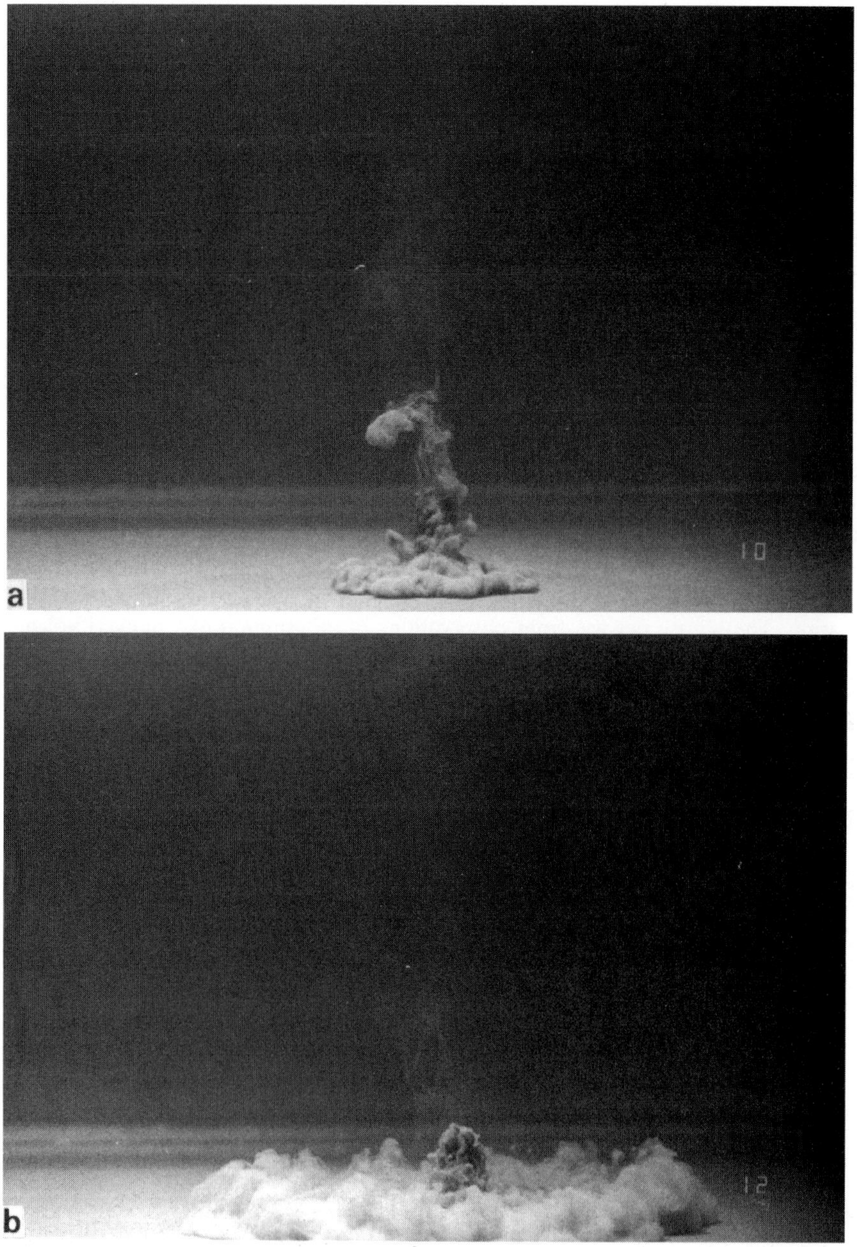

Figure 6.8 Sequence of photographs showing the behaviour of a laboratory collapsing fountain after Carey *et al.* (1988). The jet consists of a mixture of fresh water and suspended fine particles which has a greater bulk density than the salty water in the tank. (a) The jet emerges and collapses to form a radially spreading gravity current. (b) The gravity current advances to a fixed distance. (c) The flow front remains stationary as water and residual suspended sediment forms a plume around the edge of the flow. (d) The plume ascends from the flow front to form a large cloud. The horizontal field of view of the photographs is approximately 30 cm

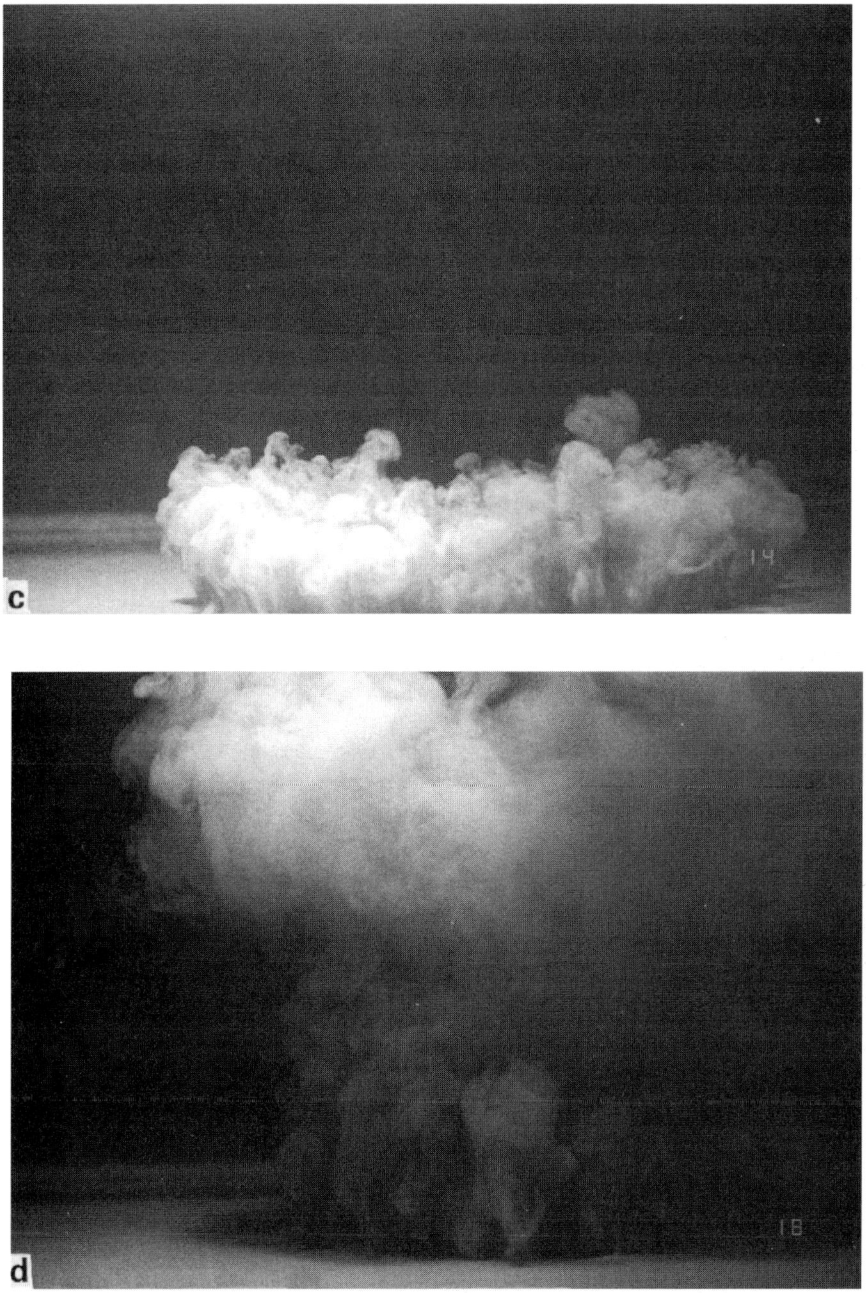

Figure 6.8 (*continued*)

give an approximate indication of the range of heights that might be expected in nature. Table 6.1 gives fountain height estimates obtained from equation (6.1) for a range of mass discharge rates and exit velocities typical of pyroclastic flow eruptions for a representative initial bulk density of 10 kg m^{-3}. The results in Table 6.1 indicate that volcanic fountains should range from a few hundred metres in height, as has been observed in a number of moderate-sized eruptions, to a few kilometres in height in the highest intensity eruptions.

Equation (6.1) does not, however, take into account the non-linear rate of change of buoyancy of the fountain which occurs in a volcanic column. Entrained air is heated and expands and thus generates buoyancy that takes the fountain to greater heights than predicted by equation (6.1) (see section 4.2 in Chapter 4). For a sufficiently large momentum flux the erupted material is able to entrain enough ambient fluid that the mixture becomes buoyant, the fountain does not collapse and a buoyant plume forms (Chapter 4). Woods and Caulfield (1992) have modelled this process experimentally using jets of methanol and ethylene glycol mixtures (MEG). As with the particle experiments of Carey *et al.* (1988), Woods and Caulfield produced collapsing fountains at low flow rates. As the flow rate is increased, the jet eventually became buoyant. As shown in Chapter 2, the critical conditions for fountain collapse to develop in these simple analogue experiments depend upon the erupted mass and momentum fluxes, according to equation (2.34). This criterion has been successfully tested with MEG experiments (Figure 6.9), and shows that if the mass flux is large in comparison with the momentum flux, then column collapse occurs.

Owing to the non-linear increase in the buoyancy of the column as it mixes with surrounding fluid, the height of rise of the fountain, for a given flow rate, will be somewhat larger than that given by equation (6.1). As mentioned in Chapter 4, the deceleration of the MEG flows is not so rapid as in a linearly mixing laboratory plume. The same principle should hold in volcanic flows. This is a consequence of both the heating of entrained air and the tendency of particles to separate from the gas phase. For increases in eruption temperature, the rate of generation of buoyancy also increases, and so the collapsing fountain may be yet higher.

6.3.3 Fluid Dynamical Models

The principles which determine whether a fountain collapses or convects were first described in papers by Sparks and Wilson (1976), Sparks *et al.* (1978) and Wilson *et al.*

Table 6.1 Height of pyroclastic fountains as a function of discharge rate of magma

Discharge rate (kg s^{-1})	H ($u = 100$ m s^{-1})	H ($u = 300$ m s^{-1})
10^6	0.15	0.34
10^7	0.27	0.60
10^8	0.47	1.07
10^9	0.84	1.90
10^{10}	1.49	3.35

Note: Heights in kilometres and estimated from equation (6.1) for a density difference $\Delta\rho = 10$ kg m^{-3}.

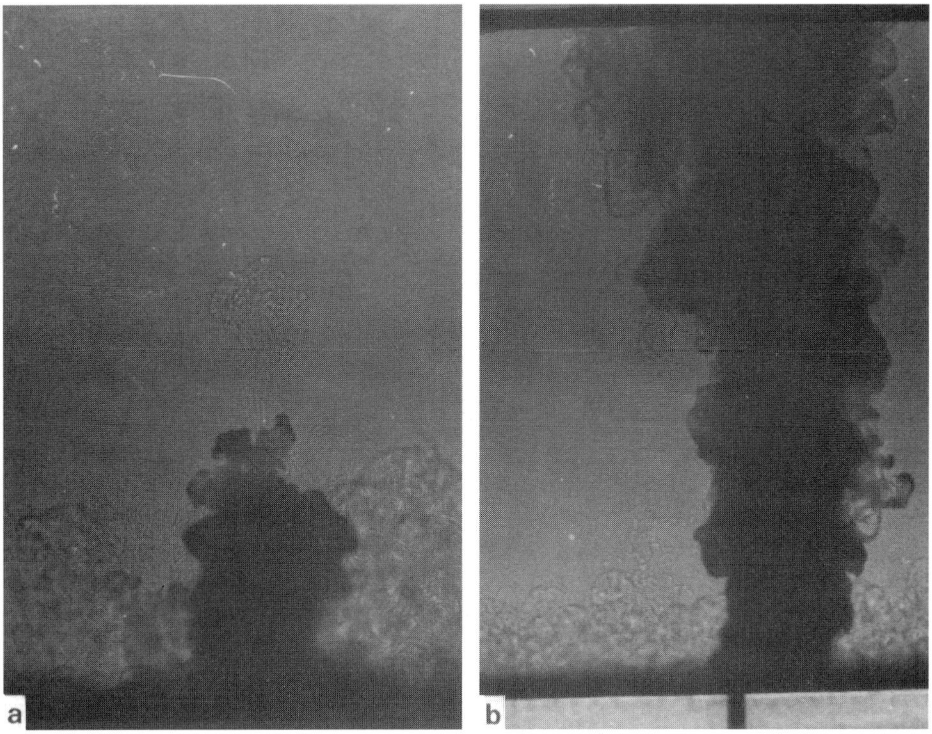

Figure 6.9 Photograph of an experiment in the methanol–ethylene glycol (MEG) system showing fountain collapse due to entrainment in a non-linear mixing system from Woods and Caulfield (1992). The photographs show a jet of MEG is injected downwards from the surface of a tank of water. (a) The MEG is less dense than the water, and therefore decelerates as it sinks into the tank. Eventually, the mixture comes to rest and begins to rise back to the water surface, forming an inverted collapsing fountain. (b) As the MEG spreads out from the fountain along the upper surface of the water, it continues to mix with water and eventually becomes denser than the water. The mixture then separates from the surface and sinks into the tank, forming an inverted analogue co-ignimbrite plume some distance from the original source. The field of view is approximately 7 cm across and 15 cm in the vertical

(1980). In recent years more elaborate computer models and simulations of fountain collapse have been developed (Bursik and Woods 1991; Valentine *et al.* 1992; Dobran *et al.* 1993; Neri and Dobran 1994). The broad principles that control collapse are first described before discussing various quantitative models and the complexities and uncertainties about the conditions in the gas thrust region of an eruption column.

Immediately on exiting the vent, a high-velocity turbulent jet develops a boundary shear layer in which turbulent mixing occurs between the jet and surrounding fluid (Chapter 4). As the jet ascends, mixing causes the jet to expand and the zone of mixing penetrates towards the centre. As depicted in Figure 6.10 there is an inner core which remains undiluted by ambient fluid until the height at which the mixing zone has reached the centre. In a volcanic system the heat transfer between entrained cold air and hot particles results in lowering the bulk density in the mixing zone. Jets issuing from narrow vents

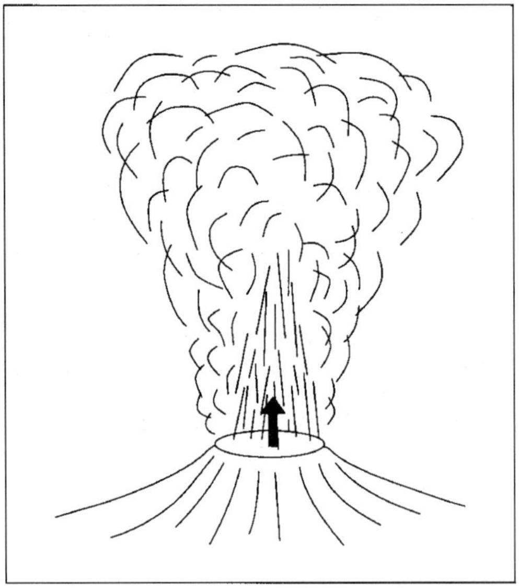

Figure 6.10 Cartoon showing the jet region of a volcanic eruption column and the core of unmixed material with material mixing by turbulent entrainment at the column margin

favour efficient mixing and formation of buoyant plumes, whereas for wider vents the height of the unmixed core is increased as a result of efficient mixing being inhibited. Qualitatively, conditions for fountain collapse are favoured by wide vents (or high exit pressures which cause the column to expand rapidly laterally) and, by implication, high discharge rates. Similarly, high velocities and, usually by implication, high volatile contents lead to effective mixing and the generation of buoyant columns, whereas low exit velocities (and low volatile contents) tend to cause fountain collapse. The key parameters are the initial radius, initial velocity and exsolved gas content. In the above discussion, the radius and velocity refer only to the actual vent if the mixture emerges at atmospheric pressure. In the more general case the initial values refer to the height at which the flow has adjusted to atmospheric pressure. If the exit pressure is high then the initial radius of the pressure-adjusted flow can be much greater than the actual vent radius.

A number of numerical models of the gas thrust region of an eruption column have been developed, as described in Chapter 4. Almost all model calculations assume that the only volatile species is water and that degassing is sufficiently rapid for all the gas to have exsolved on exit, although Wilson *et al.* (1980) also discuss the effects of CO_2–H_2O mixtures. Figure 6.11 shows calculations, using the model of Woods (1988), of plume velocity variation with height which displays the effects of magma volatile content, initial column radius and initial velocity. For the case of a fixed vent radius of 300 m and eruption velocity 100 m s^{-1} (Figure 6.11a), a stable convecting column is formed for water mass fractions of 0.05 and 0.07, while fountain collapse occurs for a water mass fraction of 0.03. Figure 6.11b shows the influence of variable initial radius with fixed initial velocity 100 m s^{-1} and water mass fraction of 0.03. In this case, columns with an

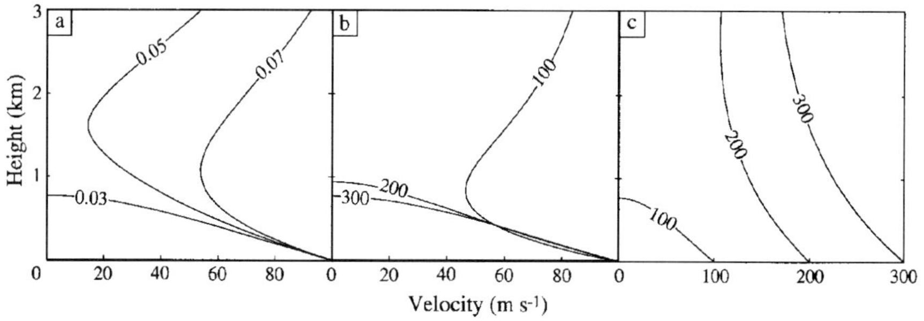

Figure 6.11 Variation of velocity with height in numerical column models: (a) calculations for variable volatile content (mass fraction of water) for a fixed initial radius of 300 m and velocity of 100 m s^{-1}; (b) calculations for constant initial velocity of 100 m s^{-1} and fixed water mass fraction (0.03) with initial radii of 100, 200 and 300 m; (c) calculations for constant initial radius of 300 m, constant water mass fraction (0.03) and variable initial velocity

initial radius of 100 m are able to become buoyant, because the entrainment of the ambient atmosphere is more efficient with a smaller radius. However, with initial radius of 200 and 300 m collapsing fountains about 1 km high are predicted. Figure 6.11c shows calculations for a fixed water mass fraction of 0.03 and an initial radius of 300 m, with initial velocities of 100, 200 and 300 m s^{-1}. The columns with the greater initial velocities are able to form buoyant plumes, whereas the column with the smallest exit velocity forms a collapsing fountain. In many models the velocity shows a minimum in the transition region between the gas thrust region and the buoyant convective region of the column. If volatile content decreases or initial velocity decreases or initial column radius increases, the minimum becomes more pronounced (e.g. Figure 6.11a). For each parameter at a critical value fountain collapse occurs because the vertical velocity of the column falls to zero just before the column density has decreased to that of the surrounding atmosphere.

As described in Chapter 4, a key input to these models of the gas thrust region is the entrainment coefficient. Three models were summarized in Chapter 4, based upon experimental observations of entrainment in jets. The predictions of each of the three entrainment models, as regards the conditions for fountain collapse, are compared. In Figure 6.12, a graph showing the relationship between the height of the collapsing fountain and the eruption velocity, for a given mass flux, is presented. The three curves correspond to three possible entrainment models for the gas thrust region: (a) is the model of Woods (1988), (b) is a model with $k = 0.09$ (where k is the entrainment coefficient) and (c) is a model with $k = 0.06$ in the gas thrust region. As the entrainment coefficient increases for a given initial velocity, the fountain height increases, because the mass of entrained air is greater. The bulk density of the jet decreases and hence the gravitational deceleration is less rapid. This effect dominates the momentum transfer which, when considered in isolation, decelerates the jet. In the remainder of the calculations presented herein, the third of these models has been used, which assumes that the entrainment is related to the density ratio of the volcanic mixture with the atmosphere. Note that if the flow becomes supersonic, then the rate of entrainment of air decreases even further (Kieffer and Sturtevant 1984), although such effects are not yet fully understood. Further details and parametric

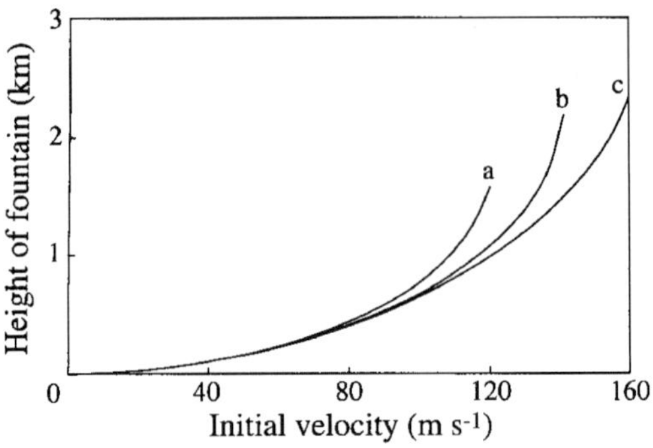

Figure 6.12 Height of collapse fountains as a function of the initial velocity of the column. Curves are for 0.03 water mass fraction, with a constant mass flux of 1.38×10^8 kg s^{-1}. Curves are given for the different entrainment models: (a) Woods (1988) model (Chapter 4); (b) constant entrainment coefficient of 0.09; (c) constant entrainment coefficient of 0.06

studies of the conditions for collapse are given by Wilson *et al.* (1980), Bursik and Woods (1991) and Woods (1995a).

The numerical studies can be summarized in a simple diagram which delineates the fields of stable convective columns and unstable collapsing fountains in terms of initial eruption velocity and mass eruption rate (Figure 6.13). Here the initial eruption velocity is a property of the decompressed jet. Curves are given for three water contents. As inferred qualitatively, the results show that stable convective columns are favoured by low mass fluxes and high eruption velocities and collapsing fountains are favoured by high mass fluxes and low eruption velocities. Figure 6.13 also shows the positions of the transition calculated in two other theoretical studies by Wilson *et al.* (1980) and Bursik and Woods (1991). The differences in the location of these curves are a consequence of detailed differences in the numerical models. Given the uncertainties in factors such as entrainment and heat transfer which affect column stability, there is no basis for preferring one model over another.

In Chapter 3 the eruption velocity was shown to depend upon the magmatic volatile content and the crater geometry, while the mass flux depends mainly upon the conduit radius. In order to understand transitions in eruption style during eruptions, this model of collapse in an eruption column needs to be coupled with a model of the variation of eruption velocity with volatile content and crater geometry. In the case in which there is only a small crater or the vent has precisely the correct diameter for eruption at 10^5 Pa, the velocity of the decompressed jet, u, depends primarily upon the volatile content, and is given by the approximate relation (Chapter 3)

$$u = 1.9(nRT)^{1/2} \tag{6.2}$$

where n is the gas mass fraction, R the gas constant and T the temperature. In this case, for each magmatic volatile content there is a particular eruption rate for transition from collapsing fountains to convecting eruption columns. For a water content of 0.03, this transition occurs when the velocity is about 230 m s^{-1} and the mass flux is about

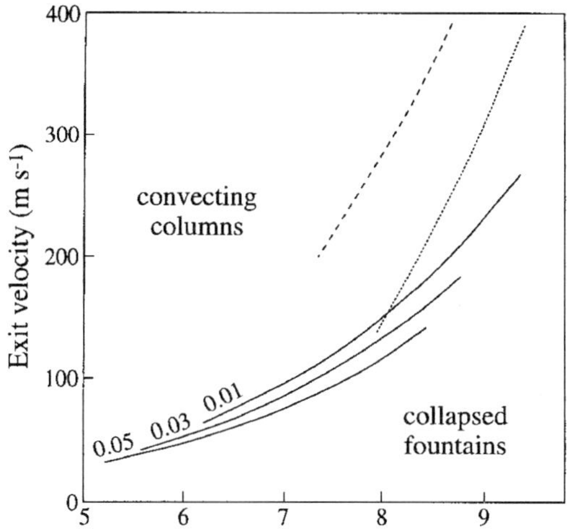

Figure 6.13 Regime diagram of gas exit velocity against mass discharge rate showing fields of convecting columns (above the curves) and collapsing fountains (below the curves) based on numerical calculations. Curves are given for magma water mass fractions of 0.03, 0.05 and 0.07. The dashed line is the transition calculated by Wilson *et al.* (1980) and the dotted line is the transition after Bursik and Woods (1991)

3×10^9 kg s^{-1}, whereas for a water content of 0.05, the transition to collapse only occurs for mass fluxes in excess of 10^{10} kg s^{-1}, which is large in comparison to most explosive eruptions. Figure 6.14 shows the regime diagram after Wilson *et al.* (1980) which assumes 10^5 Pa exit pressure. From the diagram it can be seen that eruptions beginning in the convective field will make the transition to fountain collapse and flow generation either by a decrease in volatile content or an increase in vent radius (and therefore discharge rate). An increase in flow rate unrelated to conduit radius or volatile content would also result in transition to flow generation.

Regime diagrams such as Figure 6.14 are not sufficiently general, as they are based on the restrictive assumption that volatile content and vent velocity are simply correlated. Once a crater has formed, the initial velocity of the pressure-adjusted column depends critically upon the crater shape as well as the volatile content (Chapter 4). Therefore eruption velocity and magma volatile content cannot be linked in a straightforward way. As shown in Chapter 3, at high eruption rates, the material issues from a crater at supersonic speeds, and the velocity of the decompressed jet generally increases with volatile content, with values in excess of velocities predicted by equation (6.2). In contrast, at low discharge rates, a shock wave can form in the crater and this results in subsonic flow and very low exit velocities. In such situations, for a given crater size, shape, and magma volatile content, eruptions with a very low eruption rate have such a small velocity that they are predicted to form collapsing fountains.

Figure 6.15 compares the regime boundaries for volatile mass fractions of 0.03 and 0.05 with the velocity of decompressed material being discharged from craters of different shape. Crater shape is approximated by the angle between the crater wall and the vertical.

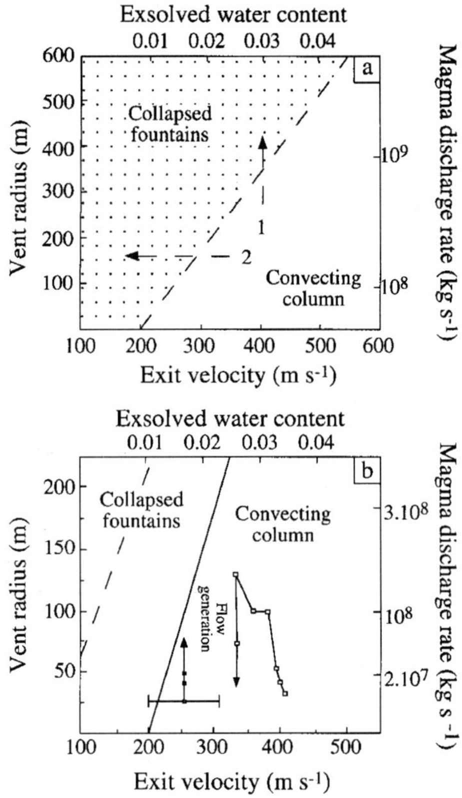

Figure 6.14 Regime diagram after Wilson *et al.* (1980) for an exit pressure of 10^5 Pa where water content and exit velocity are correlated. In (a) the arrows show the paths of eruptions in which vent radius increases with time (1) and gas content decreases with time (2). In (b) parameter estimates from the 1980 Mount St Helens eruption are shown as solid symbols (Carey *et al.* 1990) and the AD 79 Vesuvius eruption are shown as open symbols (Carey and Sigurdsson 1987). The connected points and arrows represent the estimated temporal changes in discharge rate and gas content during eruption and evolution towards pyroclastic flow generation. The dashed line shows the regime boundary after Bursik and Woods (1991) which is strictly only relevant to the velocity axis

Three curves are given: (1) a crater of depth 100 m and crater angle of tilt 10° with magma water content of 0.05, (2 and 3) a crater of 500 m depth and crater angle of 20° with magma water mass fractions of 0.03 and 0.05. The figure shows that for the large crater a decrease in the water content for a relatively small eruption rate can lead to a transition to subsonic flow and hence column collapse. A decrease in water content at a large eruption rate can lead to collapse of the supersonic flow, in a similar fashion to the case where there is no crater. By comparing the small and large crater velocities for a water content of 0.05, it can be seen that for a given eruption rate, crater erosion can cause a transition to subsonic flow and collapse. The velocities of the collapsed subsonic jets are so low that the fountains would have small heights. Some of the pyroclastic flow eruptions that have been described as "boiling over" the crater rim might correspond to this kind of flow condition.

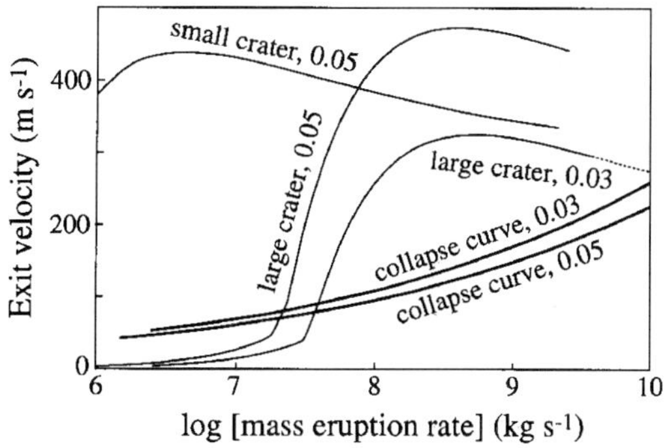

Figure 6.15 Diagram of initial exit velocity against mass discharge rate, comparing the critical velocity required to produce a buoyant plume for magma with both 0.03 and 0.05 water contents with the velocity of a decompressed jet after passing through a crater. Curves are given for two crater sizes, a small crater with depth 100 m, and angle 10°, and a large crater with depth 500 m and angle 20°. The small crater calculation is for magma with 0.05 water content. The large crater calculation is for both 0.03 and 0.05 water content. For 0.05 water content, the velocity of the jet issuing from the larger crater intersects the curve for column collapse at a mass flux of about 2×10^7 kg s^{-1}, suggesting that smaller eruption rates lead to collapse and slow subsonic flow. With 0.05 water content, the smaller crater always leads to supersonic flow for eruption rates greater than about 10^6 kg s^{-1}, suggesting that this always leads to a buoyant column. With 0.03 water content the larger crater intersects the collapse curve at eruption rates of approximately 3×10^7 and 10^{10} kg s^{-1}, corresponding to the two distinct collapse regimes

The collapse process at low eruption rates is distinct from the process at very high eruption rates. This implies that there is a range of eruption rates which lead to buoyant eruption columns, and collapsing fountains develop if the eruption rate is either too small or too large. As the size of a crater increases, the minimum eruption rate for which the flow can issue from the crater supersonically and form a buoyant column also increases. This implies that erosion of a crater during an eruption may lead to column collapse.

The above principles broadly explain the common observation in major explosive eruptions and in pyroclastic sequences that a stable Plinian column grows in height with time and is then followed by instability and pyroclastic flow production. Examples include the AD 79 eruption of Vesuvius (Carey and Sigurdsson 1987), the 1815 eruption of Tambora (Sigurdsson and Carey 1989) and 6000 BP eruption of Mount Mazama in Oregon (Druitt and Bacon 1986). Eruptions commonly initiate in a narrow vent, erupting volatile-rich magma from the top of the magma chamber, and are thus initially in the field of stable convective columns. As the eruption continues the vent widens by erosion and the volatile content decreases, leading to an increase in discharge rate. Consequently the column height increases with time. Two different scenarios may then arise (Figure 6.14a). First, the discharge rate increases and the volatile content decreases such that the eruption rate exceeds the maximum value for the production of a buoyant column and there is an abrupt transition to pyroclastic flow generation. Second, the crater becomes so eroded near the surface that the minimum eruption rate for which supersonic flow is possible eventually exceeds the actual eruption rate, and there is an abrupt transition to subsonic flow and column collapse. Another major area for future research, however, concerns the dynamics of vesiculating magma flows between the chamber and the surface. These flows are undoubtedly complex, because of interactions between vesiculation, viscosity and flow. It may well be that large variations in flow rate can occur unrelated to changes in the conduit shape or volatile content, adding further variation to the criteria for flow generation.

There are few quantitative data with which to compare the model predictions. However, estimates of discharge rates and water contents in the AD 79 eruption of Vesuvius (Carey and Sigurdsson 1987) and the May 18, 1980 Mount St Helens eruption (Carey *et al.* 1990) are shown in Figure 6.14. The May 18, 1980 Plinian eruption of Mount St Helens produced a column which behaved close to the transition with generation of pyroclastic flows through much of the afternoon when discharge rate increased. The estimates of column parameters of Carey *et al.* (1990) lie close to the regime boundary of Wilson *et al.* (1980). The Vesuvius estimates for discharge rate and water content versus time show the transition to collapse significantly to the right of the model boundary of Wilson *et al.* (1980). However, estimates of exsolved water content and the complexities in the relationship between water content and initial velocity make the estimates on the position of the transition boundary curve of the regime diagram subject to considerable error. The important point is that both eruptions show qualitative agreement with increases in discharge rate leading to pyroclastic flow generation.

The models neglect other factors that may influence the details of the position of the boundary and conditions for collapse. A significant source of error in estimating conditions for collapse may arise out of treating the gas thrust region as a negatively buoyant jet and assuming averaged properties across the jet (Chapter 4). The entrainment process must involve substantial gradients of particle concentration, temperature and velocity across the jet (Chapter 2) which have yet to be incorporated into numerical models. Furthermore, exploratory experiments by Kieffer and Sturtevant (1984) show that supersonic jets and jets with large excess densities develop thinner turbulent shear layers and less entrainment than classic subsonic jets. Qualitatively these effects might be expected to expand the field of fountain collapse, as is evident in the above calculations with different entrainment laws. However, this subject awaits further more sophisticated numerical and experimental investigations.

6.3.4 Supercomputer Models

The advent of high-powered computers has led to the development of several numerical models of volcanic explosions and collapsing fountains. Valentine and Wohletz (1989) investigated the first several tens of seconds of an explosive eruption using a supercomputer code originally designed for atomic explosion simulation and adapted for the volcanic case. They investigated the effect of exit pressure on eruption conditions and their results (Figure 6.16) suggest how an overpressured volcanic jet can expand on entering the atmosphere and then collapse. Considerable effort has been invested in developing computer models of collapsing volcanic fountains (Valentine *et al.* 1992; Dobran *et al.* 1993). Another typical result from Dobran *et al.* (1993) in which fountain collapse is reproduced is shown in Figure 6.17. The supercomputer models predict fountain heights which are surprisingly close to the simple analysis given in section 6.3.2 and they have identified a number of additional features. At 3 km from the vent a plume of buoyant material has been

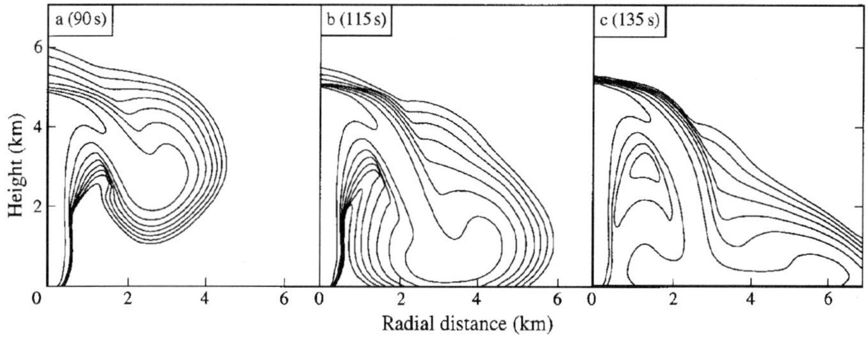

Figure 6.16 Typical numerical results of Valentine and Wohletz (1989) showing the behaviour in the initial stages (first 135 s) of an eruption. Height, in kilometres, versus distance from vent, in kilometres, is plotted. The run is described as having an initial velocity of 300 m s^{-1}, an exit pressure of 0.69 MPa and a vent radius of 300 m. The contours are in units reflecting the volume fraction of solids. The calculations illustrate the large increase in radius of the eruption column as it emerges due to pressure adjustment. The collapsing fountain impinges on the ground at a radial distance comparable to the fountain height. Note that a low concentration region occurs between the ascending column and collapsing fountain. Three times are illustrated

predicted. This has been called a phoenix cloud by Dobran *et al.* (1993). The phenomenon is similar to the convective plumes generated above laboratory flows (Figure 6.8). The models thus reproduce co-ignimbrite convective plumes which are discussed further in Chapter 7. The initial period of collapse involves complex interactions between the ascending jet and descending flow. The ascending fountain causes pressure oscillations in the surrounding atmosphere by converting its kinetic energy both into gravitational potential energy and pressure energy. A maximum pressure is achieved at the top of the fountain and pressure fluctuations in this region result in variations of fountain height. Typical results are shown in Figure 6.18 where the fluctuations of fountain height and maximum pressure variations are shown as functions of time. A steady state is approached after about a minute.

The computer models of Valentine *et al.* (1992) have also investigated the variation of fountain height with time in the first few minutes of an eruption. The dense jet first ascends to a maximum height and collapses. Subsequently the height reduces by about 40% in agreement with the simple analytical result of Turner (1966) and the models of Dobran *et al.* (1993). Their models (Figure 6.19) show the collapsing material reaching the ground at a distance from the vent which scales with the fountain height within a few hundred metres. The rising jet entrains some of this denser collapsing material and there is a further reduction in fountain height several minutes after the start of the eruption.

A noteworthy feature of these studies is that the pyroclastic flows have significantly smaller velocities than those velocities calculated by the model of Sparks *et al.* (1978). This is believed to be because the earlier model neglected segregation of particles in the flows. However, some aspects of the supercomputer models are unrealistic. In the Dobran *et al.* (1993) study, the volatile contents chosen are much lower than typical values of a few per cent expected in explosive eruptions and are hence inconsistent with the high flow velocities used. Furthermore, all the models use only one particle size. Also, the models

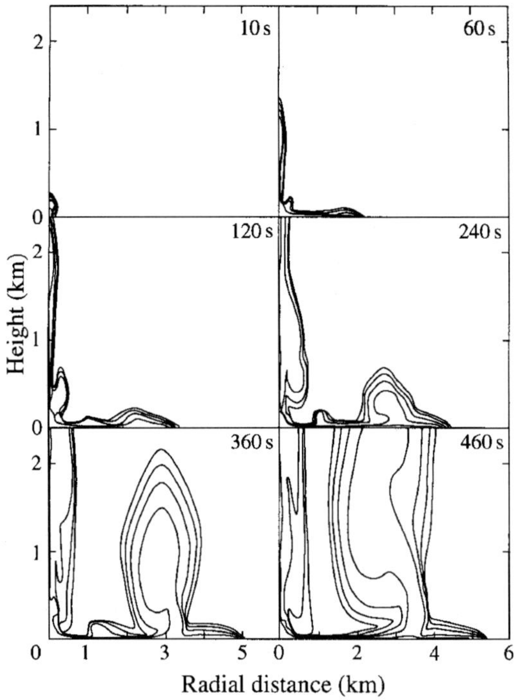

Figure 6.17 Typical numerical results of column dynamics from Dobran *et al.* (1993). Height, in kilometres, versus distance from vent, in kilometres, is plotted. Six time increments are shown with contours representing vapour content to mark the position of the flow. The model is for a 100 m diameter vent, a velocity of 56 m s^{-1} at an exit pressure of 0.1 MPa and a bulk density of 181 kg m^{-3}

implicitly assume that the fountain collapse is symmetric, which is not always the case as demonstrated by observations of eruptions (e.g. Figure 6.7) and shown in the experiments of Woods and Caulfield (1992) (e.g. Figure 6.9). Further discussion of the relevance of such supercomputer models to the actual physical processes is given in Chapter 4.

6.3.5 Influence of Flow Inhomogeneities

One feature of all the modelling work to date is that it is implicitly assumed that the erupting mixtures are homogeneous and that a single bulk density can be assigned to the flow. This is the pseudofluid assumption. Papanicolaou and List (1988) show that in real jets and plumes large inhomogeneities exist due to the complexities of turbulent mixing. Parcels of dense undiluted fluid can survive for substantial periods within jets and plumes. In addition experimental studies in shock tubes indicate that the pseudofluid approximation may not be a good assumption in high-velocity two-phase volcanic flows. Anilkumar *et al.* (1993) described experiments in which a bed of particles in a shock tube was decompressed by rupture of a diaphragm. These experiments involved a wide variety of

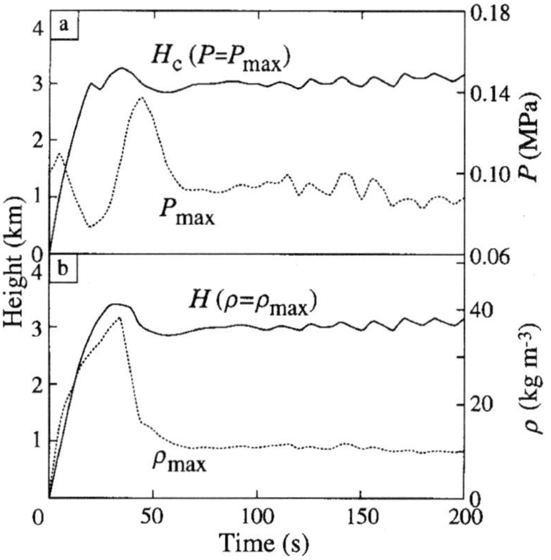

Figure 6.18 Variation of fountain height, H_c, pressure, P, (a) and density, ρ (b) with time from computer calculations of Dobran *et al.* (1993). The heights are calculated at a position in the fountain where there is a pressure maximum (see Dobran *et al.* 1993 for details). Note how the fountain reaches a maximum height after 40 s and subsides to a near steady height after 1 min with marked fluctuations

particle sizes and decompressions up to several atmospheres. Photographs of these experiments are shown in Figure 3.9.

Flow velocities up to 30 m s^{-1} were achieved in the shock tube experiments of Anilkumar *et al.* (1993). The observed phenomena were, however, similar over a wide range of experimental conditions. During initial expansion a number of regularly spaced horizontal cracks formed in the particle beds. These cracks subsequently deformed into bubble-shaped regions of very low particle concentration surrounded by regions of high particle concentration. The low concentration and high concentration regions then deformed into complex shapes as the system expanded and accelerated up the shock tube. The concentrated regions maintained concentrations comparable to the original packed bed for much of the expansion. As a consequence, the mixture emerging from the top of the shock tube was highly inhomogeneous. Comparable inhomogeneity has also been observed in liquid–vapour two-phase systems formed by explosive decompression and vaporization of superheated liquids in a shock tube (Hill and Sturtevant 1989) and in decompression of strongly gas-supersaturated solutions (Mader *et al.* 1994, 1996).

The behaviour of decompressing particle beds has close analogies with the behaviour of gas fluidized beds in which there is a certain gas flow which supports the weight of the particles and excess gas moves through the fluidized particles as dilute bubbles (Wilson 1984). The features in the shock tube experiments can be understood as the same phenomenon, but with added complexities due to the high-velocity turbulent flow. The dilute regions provide low drag pathways for expansion of the gas phase and the much denser high concentration regions lag behind. These results suggest that volcanic mixtures emerging from a vent could be highly inhomogeneous with large density as well as

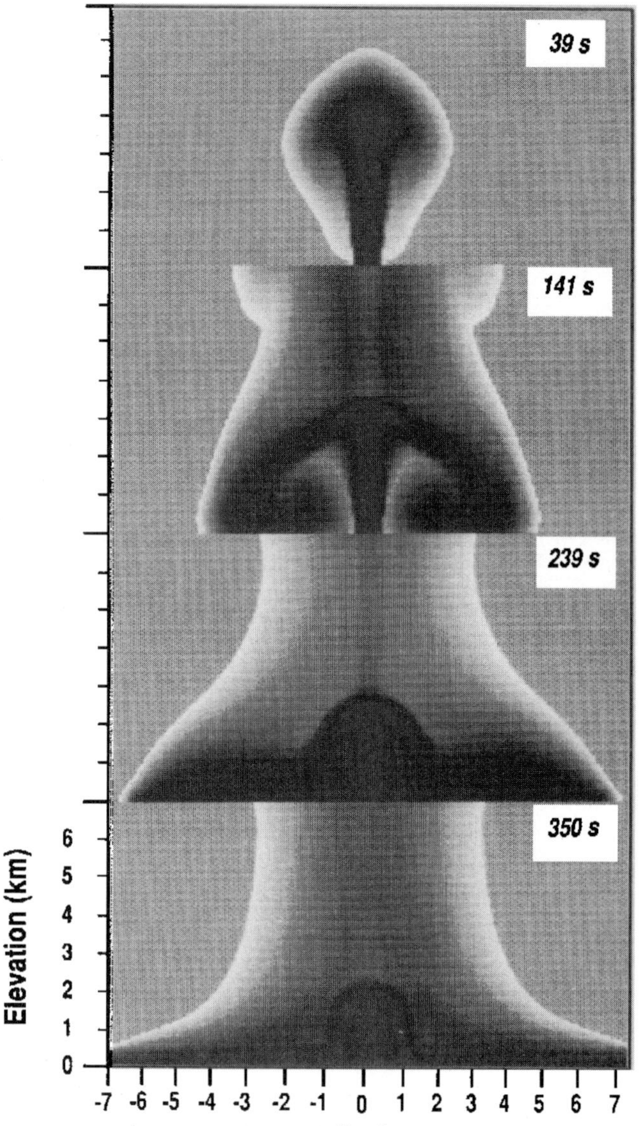

Figure 6.19 Computer graphics illustrating the collapse of a fountain at various times in seconds (after Valentine *et al.* 1992). Colours are relative concentration of particles from red (high) through black to grey (low). The computer run involved an exit velocity of 290 m s^{-1}, 0.017 water content, an exit pressure of 0.1 MPa, a vent radius of 200 m and a discharge rate of 4.6×10^8 kg s^{-1}

velocity fluctuations. The venting of such inhomogeneous materials should have an important influence on the dynamics of the lower parts of eruption columns. The dense regions would not be able to mix as efficiently with entrained air and would have lower velocities than the dilute regions. Situations can be envisaged where dense lumps of two-phase mixtures with high particle concentration fall back to generate pyroclastic flows and the more dilute regions merge to form convective eruption columns. This process is discussed further in Chapter 7 where the formation of co-ignimbrite plumes is considered.

6.3.6 Conclusions on Fountain Collapse

In summary, fluid dynamic models have provided a great many insights into the formation of pyroclastic flows and have identified the broad principles which govern whether entirely stable convective columns or unstable columns that generate collapsed fountains form. The high-velocity two-phase flows that occur above erupting vents are, however, extremely complex and there are undoubtedly many new phenomena and unexpected effects still to be revealed. Supercomputer models such as that by Dobran *et al.* (1993) have already begun to reveal rich physics and novel effects. Laboratory experiments on high-velocity two-phase flows likewise reveal much complexity which has yet to be incorporated into computer models. There is clearly much research to be done in order to understand fully collapsing columns and substantial progress in the next decade can be anticipated.

6.4 OTHER FORMS OF COLUMN INSTABILITY AND FLOW FORMATION

There is geological and experimental evidence for other mechanisms of generating pyroclastic flows and surges from eruption columns. Some occur in the transitional conditions between stable convection and fountain collapse. Some flows can also be generated by fallout of a high proportion of coarse ejecta in the erupting mixture.

6.4.1 Transitional Behaviour

In a series of experiments using MEG plumes, Woods and Caulfield (1992) observed that fountains which collapse just before they are able to become buoyant shed a periodic series of thermal plumes. This phenomenon happens because the collapsing fountain continues mixing with ambient fluid after it has collapsed, while cascading back to the ground. As a result of this mixing, the fluid can become buoyant. If this fluid becomes buoyant sufficiently close to the collapsing fountain, then the collapsing fountain becomes enveloped by the buoyant plume rising around it. The whole collapsing fountain structure may separate from the ground and rises as a thermal plume. Once this plume has risen beyond the original height of the collapsed fountain, a new collapsing fountain becomes established, and eventually this fountain sheds a further thermal plume (Figure 6.20). The periodic nature of the convection results from the interaction of the rising buoyant material with the dense collapsing fountain. At relatively low flow rates, the collapsing fountain material remains dense. Such periodic rise of thermal plumes of ash, with hot pyroclastic flows

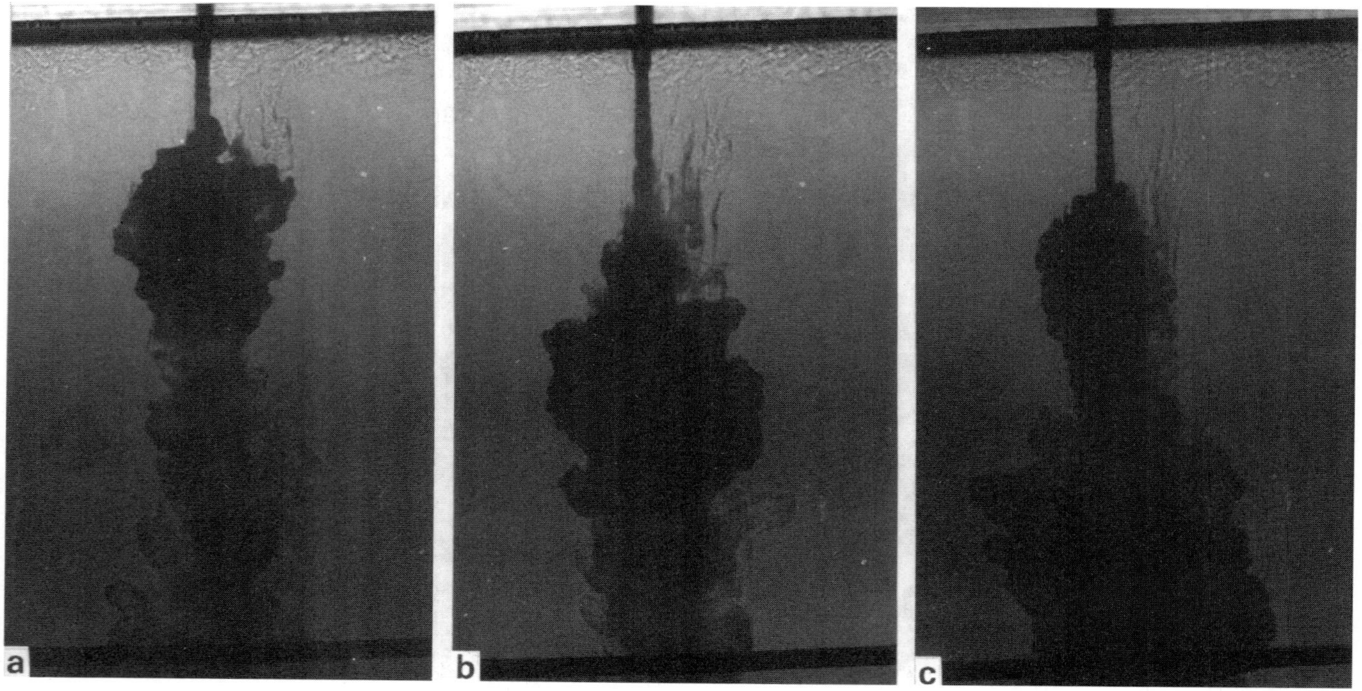

Figure 6.20 Photographs showing periodic collapse of a MEG plume in the transitional regime between fountain collapse and stable convection. The three photographs show, in sequence, the development of discrete MEG–water plumes rising from a maintained source. (a) The relatively light MEG mixture is fired downwards from the source, decelerates and collects just below the source. (b) As this plume grows, it continues to entrain water until eventually it becomes denser than the water and sinks into the underlying layer of water. (c) A new plume of relatively light MEG then begins to collect just below the source, and the process repeats, thereby producing a sequence of discrete dense MEG–water collapse fountains from the continuous supply of a relatively light MEG plume. Field of view of photographs is approximately 7 cm in the horizontal and 15 cm in the vertical

emanating from the vent in the periods between the thermals, should occur under transitional flow conditions.

Unsteady alternations in flow generation and convective column behaviour can occur due to pressure fluctuation in the discharging mixture if conditions are close to the transition between fountain collapse and stable convection. Pressure fluctuations may be caused by a variety of processes such as collapse of the crater walls which temporarily blocks the conduit and creates a higher pressure.

6.4.2 Collapse of Column Margins

Laboratory experiments have revealed other mechanisms of generating gravity currents from particle-laden plumes in addition to full fountain collapse. Carey *et al.* (1988) released mixtures of fresh water and silicon carbide particles into a tank of salty water to form plumes. Various kinds of instability were observed and are shown schematically in Figure 6.21. At low particle concentrations with bulk densities much less than the ambient salty water a stable plume formed. At high particle concentrations with bulk densities much greater than the ambient tank fluid a well-defined collapsing fountain formed (Figures 6.8a, 6.8b and 6.21c).

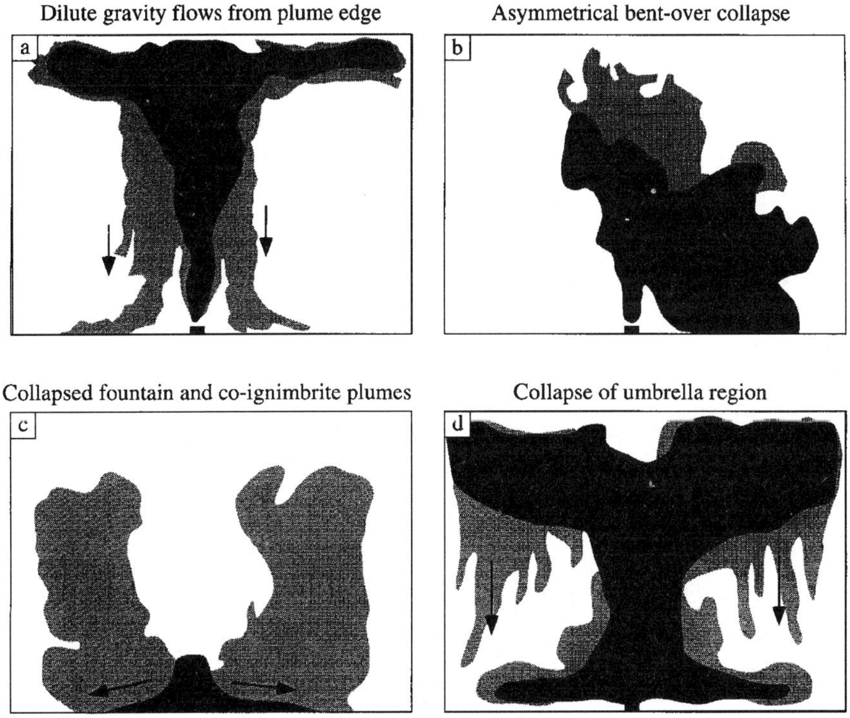

Figure 6.21 Schematic diagram of the different kinds of column stability observed in laboratory experiments (after Carey *et al.* 1988)

The first kind of instability (Figure 6.21a) occurred at bulk densities only slightly less than the ambient tank fluid. A buoyant plume formed, but soon after formation dilute gravity currents were observed to descend along the margin of the plume. These flows always initiated at the top of the plume and gradually spread down the ascending plume margins. Eventually the plume developed into a buoyant ascending interior and descending plume margins which spread dilute gravity currents radially away from the vent. These gravity currents can be explained by re-entrainment of sediment. Fallout of particles creates a veil of sediment around the plume which the plume entrains rather than pure ambient fluid. Carey *et al.* (1988) showed that the experimental plumes developed particle contents that were much higher than contents attributable to the flux of new particles from the vent and demonstrated that these higher contents were a consequence of re-entrainment. Particle concentrations displayed Gaussian distributions across the plume (Carey *et al.* 1988). As a consequence of the time-averaged Gaussian profiles of buoyant plume fluid and particles, the plume developed density maxima at the margins (Figure 6.22) which caused descending gravity currents.

There is observational evidence for this type of instability and the generation of dilute marginal flows in volcanic plumes. In the April 18, 1979 explosive eruption of the Soufrière of St Vincent the column ascended to 18 km height over a period of several minutes (Sparks and Wilson 1982). After approximately 10 minutes a dilute curtain of ash descended from the plume margin and spread down the flanks of the Soufrière and out to sea (Figure 6.23). The dilute ashy gravity current moved at between 7 and 11 m s^{-1} and

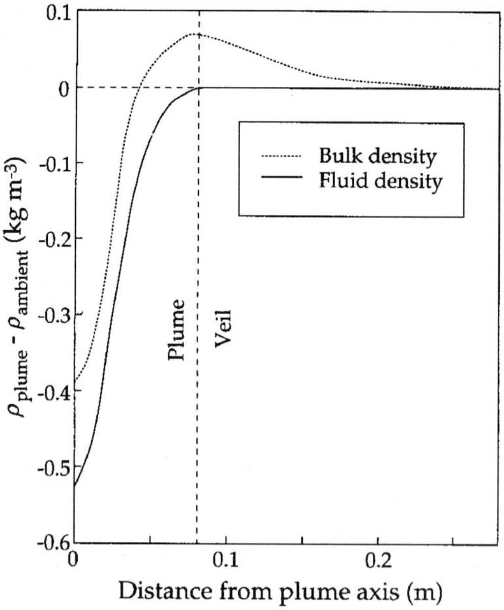

Figure 6.22 Calculations of the density variation across an experimental plume showing density maxima at margins (Carey *et al.* 1988)

Figure 6.23 Photograph of the dilute particle gravity currents generated at the margins of the plume of the 1979 eruption of the Soufrière volcano of St Vincent flowing over the sea to the west of the volcano. The photograph is taken 9 km from the plume front. (Photograph courtesy of R. S. J. Sparks.)

formed a thin (< 1 cm) ash layer. Generation of such dilute flows has also been deduced from geological evidence (Talbot *et al.* 1994).

6.4.3 Coarse Ejecta Fallout

The fountain collapse concept treats the erupting mixture as a pseudofluid in which the particles are sufficiently small that there is no significant segregation of particles from the flow. For large clasts, however, this is not a good assumption and such clasts can behave both independently of the overall flow and, if present in large proportions, can play a significant role in fountain dynamics (Bursik 1989). Very large ballistic projectiles can detach completely from the fountain as exemplified by the Ngauruhoe eruption (Figure 6.4b). Provided large clasts only constitute a small proportion of the ejecta their influence on column behaviour will not be significant. However, in eruptions containing a substantial proportion of large clasts the dynamics of the basal column region can be fundamentally different. The particles detach from the gas motion and are much less efficient at transferring their heat to entrained air than small particles. They also can dominate the momentum flux (Bursik 1989; see Chapter 4). In coarse-grained fountains the falling clasts transfer their momentum to the surrounding gases with the consequence that the fountain height is lowered.

The fallout of a coarse ejecta fountain around the vent can generate a wide range of effects depending on the viscosity of the magma and the ground slope. In Hawaiian eruptions the basaltic ejecta is so fluid that it coalesces to form lava flows or densely welded spatter deposits (agglutinates). In eruptions with somewhat more viscous ejecta the fallout of coarse ejecta on steep slopes of a volcano can avalanche down to form a pyroclastic flow. Very coarse pyroclastic flow deposits with this origin are common on andesitic volcanoes. The Ngauruhoe eruption (Figure 6.4) was probably a hybrid case in which the pyroclastic flows were generated by a combination of fountain collapse and fallout of coarse ejecta from the column interior.

The models of Bursik (1989) and Woods and Bursik (1991) give some theoretical insights into the effects of particle size distribution on column behaviour and the condition when coarse fallout can lead to pyroclastic flow generation (Chapter 4). They established that, for given discharge conditions but variable ejecta particle size distributions, there was a critical distribution below which the column reduced substantially in height forming a low fountain. In general, particles significantly greater than 0.1 m in diameter readily decouple from the gas motion, fall predominantly from the basal region of the column and retain most of their heat. If these large particles are present in large proportions they can also drag back the gas by momentum transfer (Bursik 1989; section 4.9).

6.4.4 Asymmetric Collapse

Another interesting variant of collapse occurred in the experiments of Carey *et al.* (1988) when the bulk density of the erupting mixture was almost equal to that of the tank fluid. In these cases a high unstable fountain formed which collapsed asymmetrically to generate a flow that moved across one sector of the tank floor (Figure 6.21b) and a more dilute stable plume always formed above the vent. These observations show that directed flows do not necessarily require a directed explosion. They also illustrate the hybrid character of fountains where particle–fluid separation at the top of the fountain allows a more dilute buoyant plume to form. Woods and Caulfield (1992) nearly always observed asymmetries in the flow fed by the collapsing fountains in their MEG experiments.

6.4.5 Whole Column Collapse

A third form of collapse was observed (Carey *et al.* 1988) in experiments when the flow feeding the plume was switched off. The whole column structure slowly drifted downwards and outwards to form a thick dilute gravity current (Figure 6.21d). This behaviour was attributed to the sedimentation of particles from the buoyant interstitial fluid of the plume. However, the proportion of ejecta in a column at any one moment is a very small fraction of the total ejecta in a substantial volcanic eruption. If a comparable phenomenon was to occur in eruption columns the flows would be very dilute, of small volume and cool. This mechanism does not, therefore, constitute a promising way of generating hot voluminous pyroclastic flows, but might form dilute drifting ash currents that form thin deposits. Such deposits might be hard to distinguish from fall deposits.

6.4.6 Vent Edge and Decompression Effects

There are non-gravitational effects in the high-speed flow of fluids from vents which may initiate flows along the ground. A very shallow explosion can expand sideways and produce lateral motion on the ground. Symmetric blast flows of exceptional violence have been recorded and may be initiated by explosive decompression in which the initial momentum rather than gravity dominates. The 1951 eruption of Mount Lamington (Taylor 1958) produced a very violent blast flow which wreaked great destruction to distances up to 15 km in all directions from the volcano. The more laterally directed flows of Mont Pelée in 1902 and Mount St Helens in 1980 are also examples where a case can be made for explosive decompression having a significant role (see Kieffer 1981; Fink and Kieffer 1993). However, although decompression may play a significant role in producing an initial impulse, the flow mixtures must also have bulk densities greater than the atmosphere in order to move large distances.

Note also that overpressured mixtures emerging from a vent from a deep source do not in fact expand sideways very much (Kieffer and Sturtevant 1984; Chapter 2) and can even contract. The decompression in such cases mostly creates vertical acceleration.

6.4.7 Geological Observations

Studies of pyroclastic sequences give evidence of eruption column instability, especially in the transitional regime between stable convection and full fountain collapse. In many eruption deposits a reversely graded Plinian fall deposit is overlain by ignimbrite. The upper parts of the Plinian fall deposits commonly contain thin intraformational flow and surge deposits which strongly imply that the column was becoming unstable during the later stages of the Plinian activity and prior to full onset of column collapse. Examples of such sequences include the AD 79 deposits of Vesuvius (Carey and Sigurdsson 1987), the Fogo A Plinian deposit of São Miguel in the Azores (Walker and Croasdale 1971; Bursik *et al.* 1992a) and the Taupo Plinian deposit in New Zealand (Wilson 1985). There are various emplacement mechanisms for these flows and surges, some of which are highlighted by the theoretical and experimental work described above.

6.5 PYROCLASTIC FLOWS GENERATED FROM LAVA DOMES

Pyroclastic flows can be generated by an entirely non-explosive mechanism through the growth of viscous lava domes. Flows of this kind are sometimes known as Merapi-type flows, after eruptions of Merapi volcano in Indonesia. They are also known as block-and-ash flows, because they are characteristically composed of poorly vesicular to non-vesicular lava blocks and ash derived by fragmentation of the dome. The flows contain abundant fine ash and generate ash plumes of the co-ignimbrite type by similar mechanisms to pyroclastic flows generated by explosive processes. Strictly, the plumes cannot be described as co-ignimbrite plumes, because the deposits of Merapi-type flows are not classified as ignimbrite.

Much has been learned about the origin of these pyroclastic flows from the dacite dome eruption of Mount Unzen, Japan (1990–95). These pyroclastic flows were generated by the unstable growth of dacite domes and varied from small rock falls to substantial pyroclastic flows with volumes of up to 10^6 m^3 and extents of over 5 km (Nakada and Fujii 1993). The Unzen flows were associated with characteristic seismic signals that allowed a daily record to be kept. Over the four-year period of the eruption tens to hundreds of pyroclastic flow events occurred daily. Almost all of these were small avalanches and rock falls which formed a scree around the dome and generated small ash plumes.

The lava dome advanced by growth at the top and oversteepening of the margins. Nakada and Fujii (1993) observed that when the margins overhung by more than 12° an avalanche occurred to generate a flow. The largest flows on Unzen (10^5–10^6 m^3 in volume) were formed by substantial sectors of the lava dome avalanching, leaving large horseshoe-shaped scars in the dome side. At Unzen pyroclastic flow deposits make up more than half the volume.

The Unzen flows formed convecting ash plumes which ascended to 2–3 km altitude in the large pyroclastic flow events. The origin of the abundant ash in the flows is problematic. On Unzen it was observed that avalanches of blocks from the unstable dome surface were sometimes simply rock falls and generated little ash or associated plumes. However, when hotter, incandescent parts of the dome margin avalanched, the lava spontaneously disintegrated to produce an ash-rich flow and a substantial convecting plume above the flow. Similar observations of pyroclastic flow generation have been made on Mount St Helens (Mellors *et al.* 1988) and in the andesite dome eruption of the Soufrière Hills volcano, Montserrat (observations by R.S.J.S.). In the Montserrat case, blocks were observed to bounce down the dome margin and spontaneously disintegrate on the second or third bounce. Sato *et al.* (1992) have proposed that the dome lava has the property of autoexplosivity due to high gas pore pressure. Certainly the evidence suggests some mechanism of internal stressing of the lava which causes it to disintegrate and generate substantial plumes containing fine ash.

6.6 SUMMARY

One of the most dangerous phenomenon associated with explosive volcanism is the generation of pyroclastic flows. These are mixtures of hot particles and gases that move down the slope of a volcano under the influence of gravity and can cause widespread destruction and loss of life. Pyroclastic flows are considered to be predominantly high-particle concentration flows within which particle interactions are important. In contrast, pyroclastic surges are low-particle concentration flows where turbulence and saltation play an important role in particle support. High concentration pyroclastic flows can also develop low concentration turbulent surge regions. In pyroclastic flows, partial fluidization contributes to particle support and segregation during transport away from source. The source of gas for this process can be exsolution of primary volatiles, combustion of entrained vegetation, heating of entrained air or fluxing of gas from the base of the flow as a result of particle settling. The deposits of pyroclastic flows that contain vesicular juvenile pyroclasts are known as ignimbrites. Large-scale pyroclastic flows can produce extensive ignimbrite sheets around volcanic centres to distances in excess of 100 km. The volumes of the largest known ignimbrites are of the order of several thousand cubic kilometres.

The principal mechanism for the formation of pyroclastic flows during explosive eruptions is by gravitational collapse of hot particles and gases after they have been ejected from a vent as a high velocity jet. For virtually all explosive eruptions the bulk density of the gas/particle mixture at the vent is substantially greater than the ambient atmosphere. If the jet can mix with, and heat, sufficient quantities of entrained air, it can transform into a buoyant plume and continue to rise convectively. If not, the mixture will reach some critical height and begin to collapse, generating pyroclastic flows. The collapsing fountain that forms over the vent should not be confused with the high-altitude eruption column that can develop prior to pyroclastic flow formation. As flows move away from source, secondary plumes (co-ignimbrite) convectively rise off the flow top, being fed by fine ash elutriated from the main body of the flow (see Chapter 7).

Theoretical and fluid dynamical considerations suggest that the transition from convecting to collapsing conditions during explosive eruptions is predominantly controlled by magmatic volatile content, exit velocity, vent radius and geometry, and magma discharge rate. Collapsing conditions are favoured by large vent radius, high magma discharge rate, and low magma volatile content. Estimation of such parameters for some historic eruptions is in accord with theoretical collapsing/convection regime predictions.

New insights into the behaviour of pyroclastic flow-forming eruptions have been achieved using analogue laboratory experiments and supercomputer models. Experiments with particle-laden plumes and fluids with non-linear density mixing relationships have revealed interesting complexities in collapsing conditions. Fountains may collapse either symmetrically or asymmetrically. In addition, dilute collapse of high-altitude eruption columns can occur as a result of local density increases caused by re-entrainment of particles settling from the umbrella region of eruption plumes. Supercomputer models have been able to effectively simulate the development of a collapsing fountain over the vent and the generation of a co-ignimbrite plume during the early stages of pyroclastic flow runout. Future laboratory and supercomputer simulations are likely to lead to significant improvements in the current understanding of this important eruptive process.

7 Co-ignimbrite Plumes

NOTATION

C_p	heat capacity of ejecta (J kg^{-1} K^{-1})
E_t	total thermal energy (J)
h	column height (km)
r	column radius (km)
U_{mf}	minimum fluidization velocity (m s^{-1})
U_t	terminal fall velocity (m s^{-1})
ΔT	temperature excess in the thermal (K)
ε	entrainment coefficient (dimensionless)

7.1 INTRODUCTION

Volcanic plumes generated during pyroclastic flow eruptions are known as co-ignimbrite plumes. They can form both above the collapsing fountains and from spreading pyroclastic flows. The principal mechanisms for generating co-ignimbrite plumes are by heating of air entrained into the flows or columns and by segregation of particles with the flows. In large eruptions co-ignimbrite plumes can be generated from the entire area covered by the pyroclastic flows and form enormous eruption plumes. They have been responsible for the most historically significant injections of aerosols and ash into the stratosphere. The huge eruption columns associated with the 1815 eruption of Tambora, the 1883 eruption of Krakatoa and the 1991 eruption of Pinatubo are believed to have been co-ignimbrite plumes. The co-ignimbrite plume generated by the blast flow of Mount St Helens in 1980 and from the pyroclastic flows of Mount Pinatubo in 1991 reached altitudes of 30 and 34 km respectively. Ignimbrite eruptions in the geological record have formed co-ignimbrite deposits which have volumes of hundreds to over 1000 km^3 and covered areas of millions of square kilometres. For example, the Toba ignimbrite eruption formed an ash layer which covered much of SE Asia and the Bay of Bengal and estimated to have covered India with a thickness of ash exceeding 10 cm (Ninkovich *et al.* 1978). Co-ignimbrite ash layers therefore represent the products of the most significant kind of plume from the perspective of their global environmental impact. In the following text the term plume is used as a general term to describe both sustained sources and discrete events.

7.2 THE NATURE OF CO-IGNIMBRITE PLUMES

The first scientific observations of pyroclastic flows placed great emphasis on their turbulent billowing character. Indeed it was assumed for some time that *nuées ardentes* were dilute and highly turbulent flows because this is the most conspicuous feature of a *nuée*.

Figure 7.1 shows a classic picture of a 1930 *nuée* generated by an eruption of Mont Pelée in Martinique with a towering buoyant plume rising several kilometres above the flow. Figure 7.2 shows a sequence of photographs of developing co-ignimbrite plumes from a pyroclastic flow generated in the 1980 eruptions of Mount St Helens on August 7. The leading parts of the flow show a tapering geometry. Discrete rising buoyant plumes, generated from the full width of the flow, developed behind the flow front (Figure 7.2b). There were clearly several pulses of plume generation along the length of the flow and the individual plumes merged as they ascended. Figure 7.3 shows a Mount St Helens eruption in which there is a large plume generated above the vent as well as above the flow. Figure 7.4 shows the 30 km high giant umbrella plume formed from the blast flow of Mount St Helens on May 18, 1980 over an area of 600 km^2. Such observations indicate that there is

Figure 7.1 Photograph of a *nuée ardente* of Mont Pelée in 1930 with a towering co-ignimbrite plume. (Photograph by F. Perret.)

Figure 7.2 Photographs of a pyroclastic flow from Mount St Helens erupted on August 7, 1980 (courtesy of Hoblitt 1986). (a) The early flow showed the tapering dilute upper plume which thickened up the length of the flow and from which convecting co-ignimbrite plumes were generated. (b) The flow front stalled as it emerged on to the gentle slopes at the base of the feeding channel and a buoyant plume ascended at the front. (c) Somewhat later at least four individual plumes were generated from the flow which again developed a tapering frontal region. The plumes were distorted to the south-west by winds. (d) The tapering flow front travelled across the pumice plain

Figure 7.2 *(continued)*

Figure 7.3 Photograph of Mount St Helens eruption on July 22, 1980 showing the simultaneous development of a large convective plume above the vent and a co-ignimbrite plume above a pyroclastic flow. (Photograph courtesy of the US Geological Survey.)

a very wide variation in the size, character and behaviour of these co-ignimbrite plumes in different flows.

The existence, significance and characteristics of co-ignimbrite plumes have been deduced from direct observations and geological studies. Hay (1959) observed that the crystal contents of the 1902 pyroclastic flow deposits from the Soufrière of St Vincent were much higher than would have been expected by explosive break-up of the magma. He measured the mass proportions of crystals and volcanic glass in the pumice clasts as representative of these proportions in the original magma. He then observed that the mass ratio of these components was quite different in the ash matrix of the pyroclastic flow deposit, implying that the crystal components had been enriched relative to the glassy components during eruption. Similar studies (Lipman 1967; Walker 1972) established that crystal enrichment is a ubiquitous feature of ignimbrites. The crystal enrichment was attributed to the preferential loss of volcanic glass as a result of its lower density and finer

Figure 7.4 The giant co-ignimbrite plume of the May 18, 1980 eruption of Mount St Helens generated from the entire 600 km^2 area of the blast flow (after Sparks *et al.* 1986). The margin of the rising column on the right is observed to slope slightly inwards. (Photograph courtesy of J. G. Moore.)

grain size, the latter being a consequence of its mechanical weakness and preferential crushing in the flow. The inference is that fine volcanic glass is selectively removed from a flow. Mass balance calculations indicate that up to half the mass of a flow is typically lost, implying that the complementary ash deposits are comparable in volume to the ignimbrites.

The complementary glass-rich ash fall deposits were recognized and described in studies by Sparks *et al.* (1973) and Sparks and Walker (1977). The deposits are typically very fine-grained and depleted in the dense crystal components compared to the original magma. Sparks and Walker (1977) named them co-ignimbrite ash-fall deposits and proposed that they formed from large convective plumes that formed above pyroclastic flows. Co-ignimbrite ash-fall deposits have been recognized as a major component of many ignimbrite eruptions and they cover huge areas. Figure 7.5 shows a map of the co-ignimbrite ash-fall deposit formed during the Tambora eruption of 1815. The ignimbrite has an estimated volume of 20 km^3 (Sigurdsson and Carey 1989). The 1 cm isopach of the co-ignimbrite ash fall covers an area of 850 000 km^2 and has an inferred volume of 30 km^3. In contrast the initial Plinian eruption of 1815 only produced a volume of 1.8 km^3. Figure 7.6 shows log thickness versus distance for the fall deposit of the Mount Mazama 6000 years BP eruption. The plot shows two segments. The inner segment with the steeper slope can be interpreted as the Plinian fall deposit and the outer segment as the

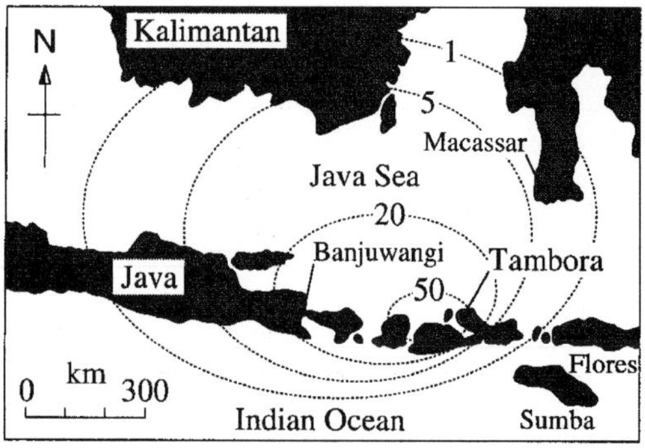

Figure 7.5 Map of the 1815 Tambora co-ignimbrite fall deposit after Carey and Sigurdsson (1989) showing dispersal of over approximately 1 million km^2 across the Java Sea and Indian Ocean. The contours are in millimetres

much more voluminous co-ignimbrite fall deposit. However, note that the break in slope on this plot could also be interpreted in terms of two different sedimentation regimes (see Chapter 14) from the same eruptive phase. As predicted from the crystal concentration studies the volumes of co-ignimbrite ash-fall deposits are typically as great as the associated ignimbrites.

Geological evidence indicates that co-ignimbrite plumes generate the most extensive and voluminous ash-fall deposits known and have the greatest potential for perturbing the global environment. The deposit volumes range from tens to several hundreds of cubic kilometres (Table 7.1) and can cover areas the size of continents. For example, the 600 000-year eruption of Yellowstone caldera generated an ash layer that covered most of the USA. Many co-ignimbrite ash layers are best preserved in deep-sea marine sediments. Eruptions of magnitudes exceeding 100 km^3 have not been experienced in human history although they are common on a geological time scale.

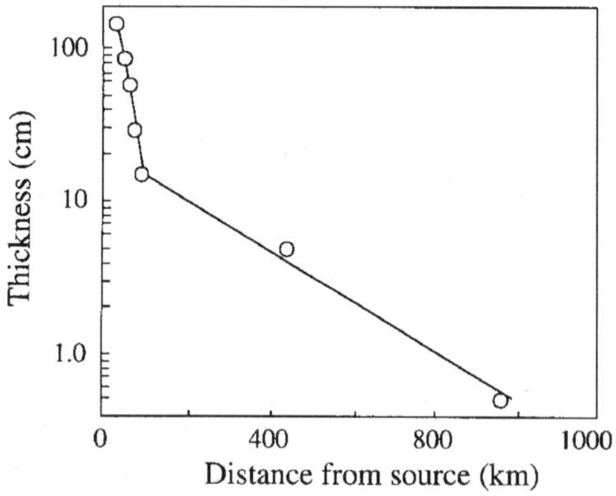

Figure 7.6 Log thickness versus distance for the fall deposit of the Mount Mazama (6000 years BP) eruption (after Williams and Goles 1968)

Table 7.1 Examples of suspected co-ignimbrite ash-fall deposits

	Volume (km^3)	Reference
Tambora (1815), Sumatra	30	Sigurdsson and Carey (1989)
Toba (75 000 a), Sumatra	1000	Ninkovich *et al.* (1978)
Los Chocoyos (84 000 a), Guatemala	500	Drexler *et al.* (1980)
Minoan tephra (3500 a), Greece	32	Watkins *et al.* (1978)
Mount Mazama (6000 a)	37	Williams and Goles (1968)
Bishop ash (0.7 Ma), USA	700	Hildreth (1979)
Campanian (Y-5) ash (33 000 a)	200	Sparks and Huang (1980)

7.3 MECHANISMS OF PLUME FORMATION

7.3.1 Flow-fed Plumes: Fluidization

Co-ignimbrite plumes can be formed by gas fluidization in which fine particles are removed from the flow by escaping gases. When gas flows through a cohesionless bed of particles a condition is met where the drag force due to the flow supports the force exerted by the weight of the bed. At this condition the bed loses its strength and has many of the properties of a liquid and is said to be fluidized. A fluidized particle bed has zero angle of repose, can be characterized by a viscosity (typically in the range of 1–100 Pa s) and can transmit surface waves. At gas velocities greater than the minimum fluidization velocity the bed expands and excess gas moves through the bed as bubbles. At much higher velocities which exceed the particle terminal fall velocity the particles are said to be elutriated. Most pyroclastic flows are thought to be fluidized by a variety of mechanisms (Sparks 1976; Wilson 1984) and the elutriation of fine ash particles and hot gas from fluidized flows is an important factor in the generation of co-ignimbrite plumes.

Pyroclastic flows are typically composed of a wide range of grain sizes from metre-sized blocks to sub-micrometre particles (Chapter 6; Sparks 1976). Theoretical and experimental studies show that it is not possible to fluidize fully such a poorly sorted mixture (Wilson, C. J. N. 1980, 1984). Figure 7.7 shows the variation of minimum fluidization velocity and terminal velocity of particles with a density of 1000 kg m^{-3} in CO_2 at 1000 °C. At any given gas velocity three types of particle can be recognized in a mixture with a wide size range. Particles larger than a certain size cannot be fluidized. Particles above a smaller critical size will be fluidized, but are too large to be removed from the system by elutriation. Particles with terminal velocities less than the gas velocity will be elutriated and swept out of the fluidized bed. Wilson C. J. N. (1980, 1984) investigated the fluidization of poorly sorted mixtures in laboratory experiments. Behaviour is complex, but the results confirm that fine particles are elutriated and coarse particles sink or float in the fluidized bed. Wilson observed that rising gas bubbles form irregular high porosity pipes which focus flow. Fines are swept from the pipes which concentrate coarse and dense particles. Fluidization pipes are common in pyroclastic flow deposits (Walker 1971; Wilson 1984).

There are several sources of gas for fluidization of pyroclastic flows. Hot juvenile particles can continue to exsolve magmatic volatiles within the flow (Sparks 1978b). Pyroclastic flows can incorporate vegetation which is vaporized or incinerated, releasing hot expanding gases. A number of pyroclastic flows have moved through areas of dense

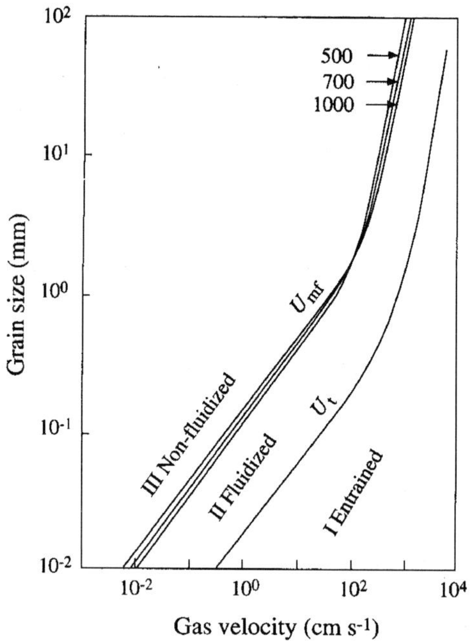

Figure 7.7 Fluidization velocity and elutriation velocity of particles in a flow of CO_2 at 1000, 700 and 500 °C as a function of grain size (after Sparks 1976). The three fields of entrained, fluidized and non-fluidized particles are displayed. U_{mf} is the minimum fluidization velocity in a packed fluidized bed and U_t is the terminal velocity of an individual particle in the gas

vegetation (Walker *et al.* 1980; Wright *et al.* 1984) and the resulting deposits show evidence of depletion of fine particles caused by strong fluidization. Pyroclastic flows can move down rivers and over wet ground and the steam generated can cause fluidization or more violent explosive interactions. An example of this may have been observed in the 1980 blast flow of Mount St Helens where a plume ascended above the flow, 12 km from the volcano in the vicinity of Spirit Lake (Moore and Rice 1984). Another example of interaction with water comes from the 1991 eruptions of Redoubt volcano, Alaska, where pyroclastic flows were observed to generate large co-ignimbrite plumes during emplacement over a glacier (Woods and Kienle 1994).

Entrainment and heating of air can cause fluidization. Abrupt changes in topography can cause mixing with air and therefore fluidization (Calder *et al.* 1997). Strongly fines-depleted deposits have been recognized where flows have spilled over a cliff or have separated from the ground at a sharp break in slope. Mixing with air trapped beneath the flow as it separates from the ground or cascades over a cliff leads to violent gas expansion. The towering plume observed above the 1930 Mont Pelée flow (Figure 7.1) may owe its impressive character to the fact that there is a 200 m high waterfall in the valley down which the flows moved. Mixing between the flow and its surroundings occurs at hydraulic jumps where the flow meets a sudden decrease in slope (Freundt and Schmincke 1985; Levine and Kieffer 1991; Calder *et al.* 1997). In such cases mixing is induced in a standing wave just downstream of the break in slope. Woods and Bursik (1994) have illustrated, in experiments with MEG currents, that the enhanced mixing and entrainment which occurs at hydraulic jumps and positions of flow separation can generate buoyant plumes. As air is entrained, heated and expands, the bulk density of the mixture decreases (Chapter 4), and so a fraction of the flow may rise as a buoyant ash plume above the jump. Levine and Kieffer (1991) have shown that large decelerations in a Mount St Helens flow just beyond a break in slope can be explained by a hydraulic jump. Other evidence for topographic effects is discussed in sections 7.3.3 and 7.4.

Fluidization has also been invoked as a consequence of mixing at the flow front (Wilson and Walker 1982). High Reynolds number gravity currents develop a distinctive flow head which overrides ambient fluid. Air incorporated into the head heats up and expands, fluidizing the flow head.

In summary, internal and external gas sources in pyroclastic flows result in fluidization of the flow and elutriation of ash and hot gas into the overlying dilute turbulent co-ignimbrite plumes.

7.3.2 Flow-fed Plumes: Boundary Shear Mixing

Denlinger (1987) has proposed a mechanism for generation of dilute ash plumes by development of a turbulent boundary layer across the upper surface of a pyroclastic flow (Figure 6.2a). The flow is treated as a dense plate which generates a turbulent shear layer across its upper surface as surrounding fluid is forced over the front of the flow. The turbulent shear layer entrains hot particles from the upper surface which mixes with and heats the air to form a co-ignimbrite plume. The plume thickens upstream producing the tapering geometry observed in some flows (Figures 6.2a and 7.2). Part of this dilute ash plume region can become buoyant and can generate co-ignimbrite plumes. This mechanism should occur in flows where the velocities are modest and Reynolds numbers are such that movement is entirely either laminar or plug-like within a high concentration basal part of the flow.

Entrainment and mixing with a turbulent shear layer is even more pronounced in high Reynolds number pyroclastic flows. In turbulent gravity currents (Figure 6.2b) mixing occurs over the upper surface of the flow head and creates a turbulent wake or mixing layer behind the head (Hallworth *et al.* 1993, 1996). Mixing also takes place within the flow head by overriding buoyant fluid. The rate of mixing increases substantially with slope (Begnin *et al.* 1981; Woods and Bursik 1994). Generation of co-ignimbrite plumes should therefore be more vigorous on the steep slopes of a volcano.

7.3.3 Flow-fed Plumes: Non-linear Mixing Effects

A variety of processes have been described within pyroclastic flows which lead to the mixing of cold air and hot particles. The principles that govern the behaviour of vertical eruption columns are also applicable to mixing in flows as a consequence of the highly non-linear density changes, as described in Chapter 4. Figure 7.8 shows the variation of density, relative to cold air, of mixtures of air and hot particles. At low proportions of particles the mixture density first decreases, reaches a minimum and then increases as the mass fraction of particles increases. Above a certain particle mass fraction, the mixture becomes denser than cold air despite the higher temperature. The upper dilute turbulent

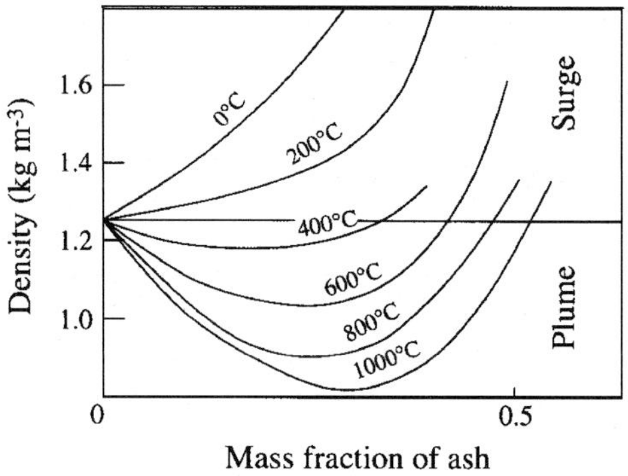

Figure 7.8 Variation of the density of ash–air mixtures for different temperatures as a function of the mass fraction of ash. The horizontal line is the density of the atmosphere at sea-level. The diagram is thus divided into a region of ash-cloud surges which are denser than air and convective co-ignimbrite plumes which are less dense than air

portion of a pyroclastic flow can therefore form both dilute density currents or buoyant convective plumes. Thus the overlying ash plume divides into a dense part and a buoyant part depending on temperature and particle content. The dense part is known as an ash-plume surge (Fisher 1979). The ash-plume surge can move well beyond the margins of the main flow and up valley sides. Particle sedimentation from the ash-plume surge must eventually lead to the formation of a buoyant ash plume.

Experimental studies on gravity currents with fluids with non-linear mixing properties have given insights into how buoyant plumes develop and the way that they influence flow (Huppert *et al.* 1986; Woods and Bursik 1994). The experimental system involved water and methanol–ethylene glycol mixtures (MEG) which produce liquids with a higher density than the pure end members. Experiments were conducted where MEG solutions were released into water and formed a buoyant gravity current along an incline. These "upside-down" experiments are dynamically identical to a dense gravity current generating a buoyant plume by mixing.

Figure 7.9 shows typical results of these experiments. The most prominent feature of the flows is that the flow front propagates in pulses. Relatively undiluted fluid moves into the flow head and then mixes with ambient fluid and decelerates as the density contrast with surrounding fluid declines. The mixture becomes buoyant and rises to form a discrete plume. The flow front momentarily stalls until a fresh pulse of undiluted fluid moves through. Repetition of this sequence of events results in a flow front with strong velocity fluctuations and a series of discrete buoyant plumes above the flow. These flows also differ in their rate of lateral spread down a slope. Normal gravity currents spread linearly with distance, whereas the formation of the buoyant plumes by mixing results in a strong lateral inflow of surrounding fluid towards the rising plumes and thus inhibits the downstream spreading. Similar observations were made by Woods and Caulfield (1992) in the same experimental system in gravity currents generated by collapse of a central axisymmetric fountain. They also observed that the formation of buoyant plumes was inhibited in gravity currents with low Reynolds numbers in which there was only limited mixing.

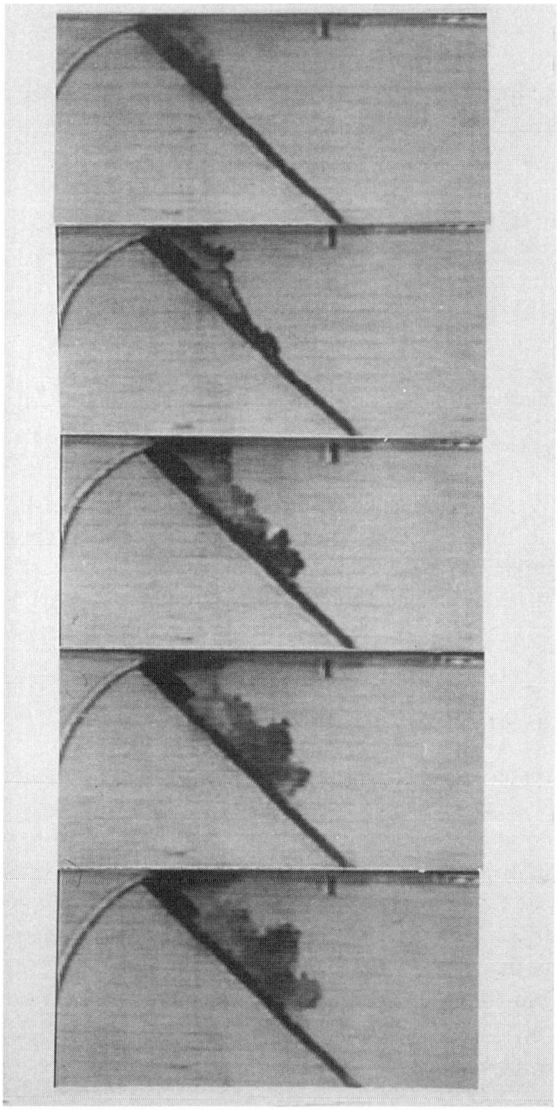

Figure 7.9 A time sequence of a MEG flow on a slope. The fixed volume current propagates along the slope and mixes with ambient fluid. These mixtures become buoyant and separate from the slope. The current generates discrete pulses of buoyant fluid along its length. Note that the photographs are shown upside-down for ease of comparison with pyroclastic flows (after Woods and Bursik 1994)

Woods and Bursik (1994) showed that the run-out distance of discrete currents of MEG decreases as the inclination of the slope increases. This is because the flow becomes more turbulent and can entrain more efficiently on a steeper slope. Therefore, the currents are able to become buoyant over a shorter distance from the source and then separate from the boundary (Figure 7.10). However, in a hot pyroclastic flow, the rate of entrainment of ambient air competes with particle segregation processes, such as sedimentation and

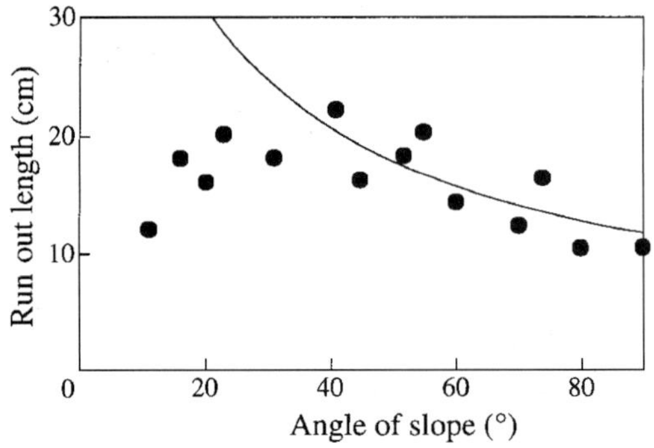

Figure 7.10 Variation of run-out distance (the distance at which buoyant lift-off occurs) of MEG gravity currents as a function of slope is shown for experimental data (dots) and is compared with a theoretical model (after Woods and Bursik 1994)

fluidization within the flow, as mechanisms of buoyancy generation. Experiments with MEG by Woods and Bursik (1994) also show strong effects of topography with abrupt changes of slope producing enhanced mixing and discrete buoyant plumes. Pulse-like motion of the flow front and associated co-ignimbrite plume generation have been observed in flows of Mount St Helens (Hoblitt 1986; Calder *et al.* 1997).

7.3.4 Buoyant Lift-off

There is a much more dramatic mechanism of generating co-ignimbrite plumes from pyroclastic flows in which the whole flow becomes buoyant and lofts. This phenomenon, known as buoyant lift-off, is thought to be the main mechanism of formation of large co-ignimbrite plumes in major eruptions. Buoyant lift-off has been observed in the 1980 blast eruption of Mount St Helens and in laboratory experiments. It is inferred that the giant plumes of eruptions, such as Pinatubo in 1991 and other large historic pyroclastic flow eruptions, formed in this way.

The May 18, 1980 eruption of Mount St Helens was triggered by the failure of the northern flanks of the volcano which generated a large debris avalanche (Christiansen and Peterson 1981). Sudden decompression of the magma that had recently been intruded into the volcano triggered a succession of closely spaced explosions and a laterally directed high-velocity pyroclastic flow known as the blast flow. The flow initiated at 0832 local time and had spread over an area of 600 km^2 by 0837 hours local (PST) time. The flow advanced at velocities between 80 and 110 m s^{-1} (Moore and Rice 1984) and swept over the rugged topography around Mount St Helens climbing ridges over 400 m high. Between 0837 and 0838 hours a giant plume began to ascend and reached an altitude of over 25 km by 0850 hours (Figures 7.4 and 7.11), where it spread out rapidly in the stratosphere between altitudes of 10 and 20 km (Figure 7.12). The giant plume was generated from the entire area covered by the pyroclastic flow and formed an ash-fall deposit containing at least 30% by mass of the flow deposit (Sparks *et al.* 1986).

Several observations show that the blast flow ceased motion rapidly and then lofted. Velocity estimates indicate that the flow was still travelling at about 80 m s^{-1} at 0836

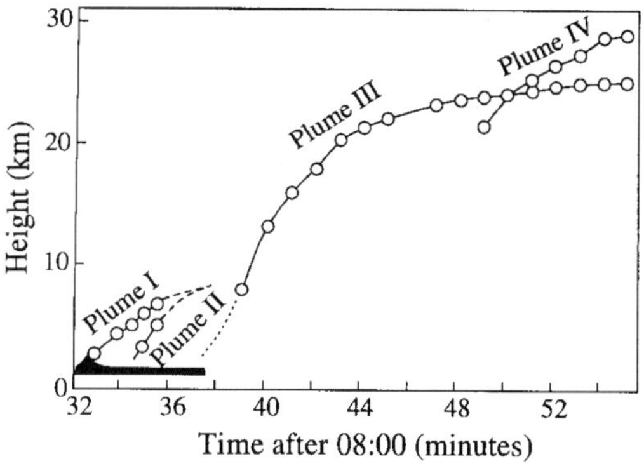

Figure 7.11 Data on the height versus time of the giant co-ignimbrite plume formed from the initial high-velocity blast flow of Mount St Helens on May 18, 1980. Plumes I and II were derived from the initial explosions from the vent. Plumes III and IV were derived from the entire surface of the blast flow (see Figure 7.4). The co-ignimbrite plume shows two discrete pulses of motion (after Sparks *et al.* 1986)

hours when it was approaching its maximum extent. This observation implies a late rapid deceleration. The flow wreaked great destruction and three zones representing different degrees of devastation have been recognized (Figure 7.13). An inner zone is characterized by complete stripping of the forest cover and extends to about 15 km from the volcano. In the next zone all trees were blown down and form spectacular patterns delineating the flow field. In an outer zone only small immature trees were knocked over, and beyond this zone there was no destruction whatsoever. The zone of modest destruction is remarkably narrow, averaging about 500 m, and indicates that the flow went through a rapid transition from predominantly lateral to vertical motion. Witnesses described the flow as rising above them

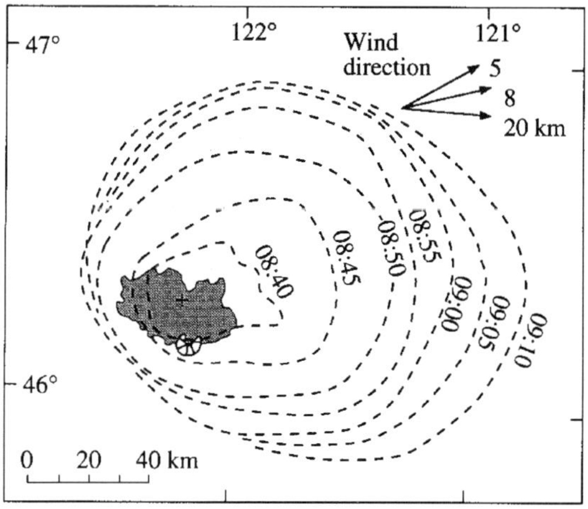

Figure 7.12 The lateral spreading of the giant co-ignimbrite plume of Mount St Helens on May 18, 1980 is shown with contours in five-minute intervals (after Sparks *et al.* 1986). The wind direction at 5, 8 and 20 km altitude is shown by the arrows in top right

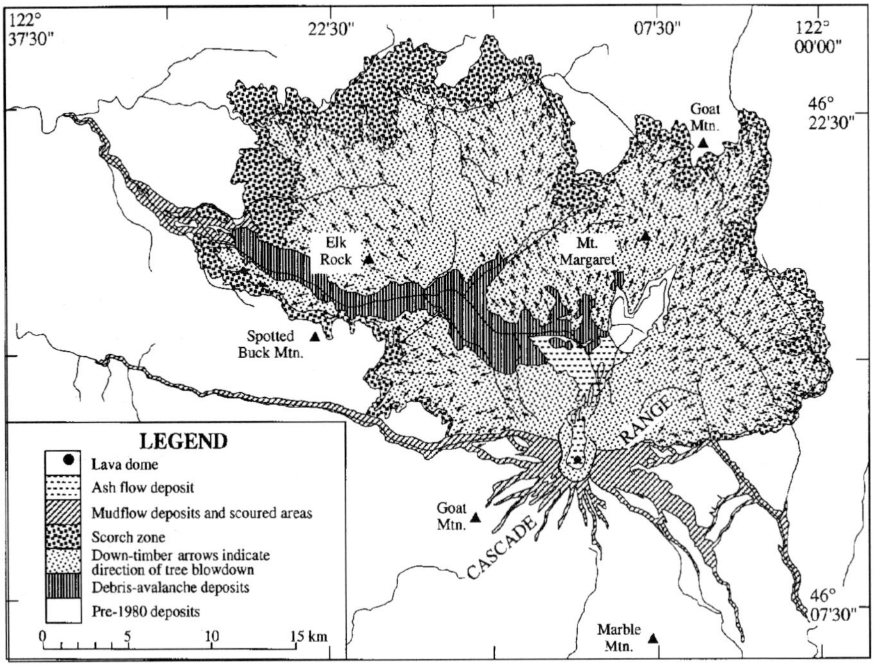

Figure 7.13 Map of the main deposits of the 1980 eruption of Mount St Helens showing the zones of devastation in the blast flow. The inner zone, in which arrows are absent, had all trees swept away. An intermediate zone had all trees knocked down; horizontal tree orientations are shown by arrows. An outer narrow zone occured in which trees were scorched and the flow underwent buoyant uplift. The diagram also shows the distribution of mud-flows, the later pyroclastic flow deposits of 1980 and the debris-avalanche deposit

and photographs of the flow front at this time indicate that much of the flow front was ascending.

Sparks *et al.* (1986) interpreted these observations as evidence that the flow ceased horizontal motion because the whole flow became buoyant and lifted off the ground to form the giant eruption plume. They attributed the lift-off to two main effects. First, the turbulent blast flow entrained and heated cold air as it moved, and second the flow progressively lost particles by sedimentation. They estimated that the flow decreased to the density of the overlying atmosphere after sedimentation of about two-thirds of the mass of the particles. They calculated that the mixture temperature at lift-off was 150 °C and that the mass of ash particles and air in the plume was 1.1×10^{11} and 1.0×10^{11} kg respectively. These calculations are consistent with the mass of particles in the ash-fall deposit formed from the plume and with the emplacement temperatures of the flow deposit (Hoblitt *et al.* 1981). The role of progressive decrease of sediment loading in the case of Mount St Helens is supported by the studies of Druitt (1992), who has shown from studies of the deposits that the flow became progressively more dilute with distance as the flow lost sediment.

Carey *et al.* (1988) observed buoyant lofting in experiments on sediment gravity currents generated by collapse of a particle-laden fountain. In these experiments a mixture of fresh water and particles was injected into a tank of salty water. The bulk density of the mixture was greater than that of the salty water so the flow formed a fountain and fed a radially spreading gravity current. At a well-defined distance from the nozzle the flow front was observed to lift off and formed a ring of buoyant plumes which coalesced to form a large plume which entrained some of the finer particles (Figure 6.8). A remarkable feature of these experiments was that the flow front remained stationary while the fountain continuously fed the gravity currents for several minutes. The distance of lofting increased with increasing flow rate, increasing bulk density of the erupting mixture and decreasing particle size. All the buoyant fluid was generated from the flow front.

Sparks *et al.* (1993) investigated buoyant lofting in flume-tank sediment gravity currents, formed by releasing a fixed volume of fresh water and particles into a 6 m long tank of salty water. Figure 7.14 shows a comparison of a saline gravity current, a particle-laden gravity current with interstitial fluid of the same density as the surroundings and a particle-laden current containing buoyant interstitial fluid. All these currents have the same initial density contrast between the fluid and ambient fluids. The normal particle current decelerates more rapidly than the saline current due to sedimentation. The particle current with buoyant fluid behaves identically to the normal particle current until it becomes neutrally buoyant at which point the current rapidly decelerates and then lofts to form a plume. The late stage rapid deceleration observed in the experiments and in the Mount St Helens flow is a characteristic feature of buoyant lift-off. Sparks *et al.* (1993) presented a quantitative analysis of sedimentation from particle-laden gravity currents based on the theoretical treatment of Bonnecaze *et al.* (1993) which predicts the lofting distance and shows good agreement with experimental results.

Both the processes of sedimentation and entrainment play an important role in the generation of buoyancy in high Reynolds number pyroclastic flows. Each process can cause dilution of the sediment in the flow, and density changes occur both as a result of

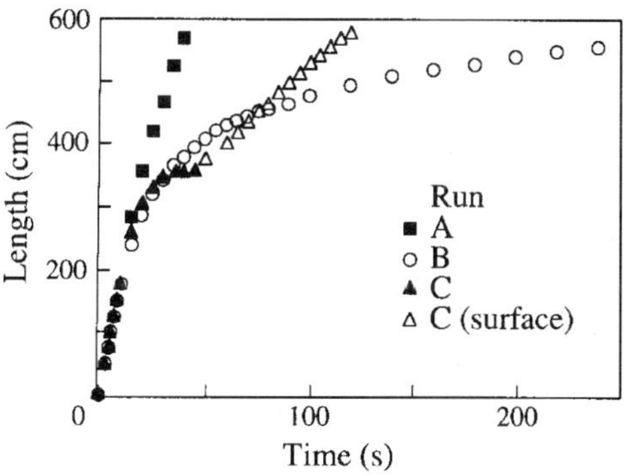

Figure 7.14 The lengths of experimental gravity currents as functions of time for three flows with identical starting conditions (from Sparks *et al.* 1993). Each current started with a fixed volume and had a density excess of 50 kg m^{-3}. Run A is a saline current below fresh water. Run B is a particle-laden current of fresh water below fresh water and slows down faster than the saline current due to sedimentation. Run C is a particle-laden current of fresh water below salty water. In C the sedimentation causes the flow to become buoyant after about 50 s when the flow decelerates rapidly and a plume forms. The surface data are for the gravity current generated at the top of the tank by buoyant lift-off

loss of particles and entrainment with heating of air. Woods and Bursik (1994) investigated the motion of dense mixtures of particles and fresh water, running down a slope into a tank of saline water. As the current propagated along the slope, it entrained ambient fluid and sedimented particles, eventually becoming buoyant through the sedimentation of particles. As the rate of entrainment and dilution of the current increased, the rate of sedimentation from the current decreased. On steeper slopes, the flow entrained more and therefore sedimented less. As a result, the current tended to travel further on steeper slopes when the buoyancy was generated by sedimentation alone. Woods and Bursik (1994) concluded that on shallow slopes, buoyancy will tend to be generated by sedimentation of coarser clasts, yielding a smaller ash plume, while on steeper slopes buoyancy will be generated by entrainment, yielding a greater mass of air in the elutriated plume. Since the mass of solids which may be elutriated in the rising plume tends to increase with slope angle, the mass sedimented may decrease with increasing angle.

The experiments of Sparks *et al.* (1993) illustrated how the lofting plume initially behaves. Figure 7.15 shows that the upward velocity of experimental plumes typically increases. This is due to progressive loss of particles. The flow head first becomes neutrally buoyant so that the flow front stagnates. Further loss of particles results in a gain in buoyancy and the flow begins to rise slowly. The plume accelerates as more particles are lost. The same phase of acceleration can also be inferred from velocity observations of the ascent of the Mount St Helens and Mount Redoubt plumes which are discussed further in sections 7.4 and 7.5.3.

Supercomputer models of pyroclastic flows composed of a single particle size have also predicted buoyant plumes (Dobran *et al.* 1993). These have been termed phoenix plumes.

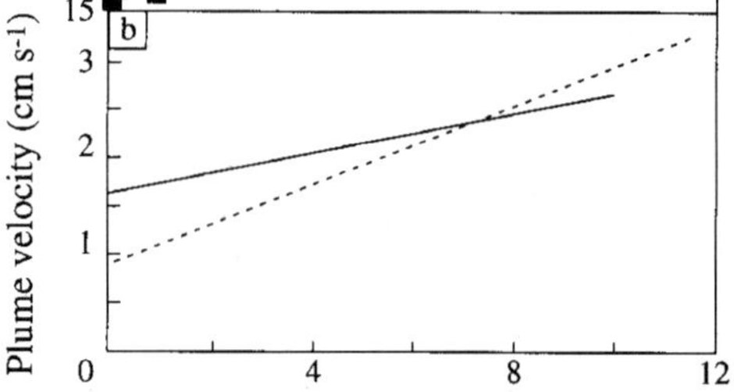

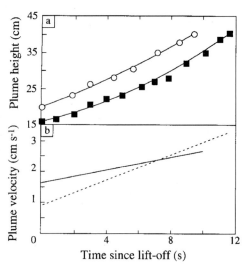

Figure 7.15 Experimental observations (from Sparks *et al.* 1993) of buoyant lift-off in gravity currents of particles and fresh water discharged below salty water. (a) Data for height versus time for two plumes showing acceleration as the lofting plumes lose sediment. (b) Plume velocity versus height for the same experiments as in (a) calculated from best-fit regressions through the height–time data. The dashed line is derived from the experimental data shown with open circles in (a) and the solid line is derived from the experimental data shown with filled squares in (a)

7.3.5 Fountain-fed Plumes

Plumes that form directly above pyroclastic fountains can be understood in terms of the processes of mixing and segregation that have been described in Chapter 6. It is a moot point whether these plumes should be termed co-ignimbrite plumes, but they are included here because they are clearly intimately associated with the production of pyroclastic flows. Studies of co-ignimbrite ash layers have not yet distinguished between the deposits formed from plumes above collapsing fountains and those formed from the pyroclastic flows that are described above.

In fountain-fed plumes mixing with air is clearly a critical factor. The margins of the erupting column first mix with air, can become buoyant and then merge above the fountain to form a plume. This process is well illustrated by the Ngauruhoe explosion (Figure 6.4) and in eruptions of Mount St Helens (Figure 7.3). Laboratory experiments (Woods and Caulfield 1992), using MEG in which buoyancy is generated through entrainment, have identified a mechanism by which discrete thermals may rise above a collapsing fountain. Collapsing fountains were generated by downwards directed turbulent jets of MEG into water. They observed that, if the thermals are shed sufficiently frequently, then they form an almost continuous column. These experiments showed that plumes and thermals can be generated above collapsing fountains in systems with strong non-linear density mixing relationships. In the transitional conditions between fountain collapse and convective plumes some of the batches of fluid at the top of the fountain managed to mix sufficiently to form intermittent buoyant thermals rising from the top of the fountain.

Segregation processes must also play an important role. The decoupling of particles from the hot buoyant gas within and at the top of the fountain can lead to plume formation. This process has been observed in simple laboratory experiments (Carey *et al.* 1988; Sparks *et al.* 1993) where the particles were sufficiently large for decoupling to occur. In laboratory experiments with very small particles the fountain was well defined with no overlying plume (Figure 6.8) and transport of the small particles effectively coupled to the gas motion. Numerical models (Valentine *et al.* 1992; Dobran *et al.* 1993) have also produced plumes overlying the central fountain (Chapter 6) with model flows containing only one size of particle.

Particle-size distribution will have an important influence on plume formation. Eruptions with high proportions of coarse ejecta would be expected to segregate efficiently high density and large particles from low density and small particles within the fountain. Two end-member situations can be envisaged. A fountain composed entirely of very fine particles would not be expected to form a major plume by this mechanism. Likewise a fountain composed entirely of very coarse ejecta (such as a Hawaiian fire fountain) would only form a weak plume (Chapter 10). Qualitatively, therefore, strong fountain-fed co-ignimbrite plumes should be best developed in eruptions with a wide particle size range in which segregation of coarse particles leaves a buoyant mixture of fine particles and hot gas.

The experiments of Anilkumar *et al.* (1993) suggested another mechanism of segregation. Their experiments, on the expansion of two-phase particle beds, indicated that erupting mixtures can be very inhomogeneous with dense and dilute regions (see Chapters 4 and 6). In a collapsing fountain the dense high particle concentration regions would be preferentially fed into the flows and dilute gas-rich regions into the overlying plume.

Segregation processes provide an explanation of the crystal-enriched character of ignimbrites. Fine glass shards should be preferentially partitioned into the plume, and dense and generally coarser crystals into the flows.

7.4 AUGUST 7, 1980 MOUNT ST HELENS FLOW: A CASE STUDY

The eruption of August 7, 1980 has been documented in detail by Hoblitt (1986) and the flow dynamics has been analysed by Levine and Kieffer (1991). New observations of the ascent of the co-ignimbrite plumes from these flows have been presented by Calder *et al.* (1997) from the analysis of video film. The August 7 flow provides a good case history of co-ignimbrite plume generation from a relatively small volume pyroclastic flow. The first seven minutes of this flow was recorded from a video camera at Harry's ridge 8.35 km from the vent. A second film was taken by a hand-held camera from Coldwater peak. Observations from the films were supplemented by still photographs (Hoblitt 1986). Photographs of the flow are presented in Figure 7.2.

Figure 7.16 shows the topographic profile along the valley traversed by the flow, flow front velocity as a function of distance from the vent and the source of four co-ignimbrite plumes. After an initial stage of acceleration the flow front reached a maximum velocity of 34 m s^{-1} at about 3.3 km and then decelerated to a final emplacement distance of 5.6 km. Superimposed on these broad trends are quite substantial fluctuations in flow front velocity. Hoblitt (1986) reported stagnation of the flow front at 1.7, 2.9 and 4.6 km followed by marked accelerations of the flow front. Two of these accelerations coincide with steepening of the channel slope. Levine and Kieffer (1991) have shown that the August 7 flow can be modelled by applying hydraulic theory and assuming an inviscid fluid. Velocity fluctuations can be attributed to slope changes (Figure 7.16). Velocity decreased rapidly from 15 to 5 m s^{-1} at 4.5 km distance where the flow ran out on to the lower slopes of the pumice plain. Levine and Kieffer (1991) have shown that this deceleration can be explained by a hydraulic jump about 600 m downstream of the break in slope.

The August 7 flow generated convective co-ignimbrite plumes (Figure 7.2a). Hoblitt (1986) observed that the plumes started to ascend shortly after the flow front had passed the source of the plume. Hoblitt reported that curves in the flow channel had a significant influence on plume generation. At a bend the overlying ash-plume surge overrode the channel walls and shortly thereafter generated a buoyant plume. Calder *et al.* (1997) examined the video film and have been able to distinguish four separate vigorously rising plumes. Figure 7.17 shows profiles of these four plumes at 2 s intervals and Figure 7.16b shows estimated positions of the source of each plume along the flow path. Figure 7.18 shows data on height versus time for each plume and Figure 7.19 shows velocity–height data derived by differentiation of the best-fit regressions through the height–time data.

Plume B formed at a break in slope. Plume C was clearly associated with a prominent bend in the channel. Plume D formed where the flow expanded and decelerated on to the lower slopes of the pumice plain and may be associated with a hydraulic jump. All the plumes were blown to the south-west by low-level winds. The data show each plume accelerated with height in a similar way to the behaviour of buoyant lift-off plumes in experiments (Carey *et al.* 1988; Sparks *et al.* 1993). The initial velocities of the plumes were in the range 1–3 m s^{-1}. Two of the plumes were observed to reach a maximum velocity

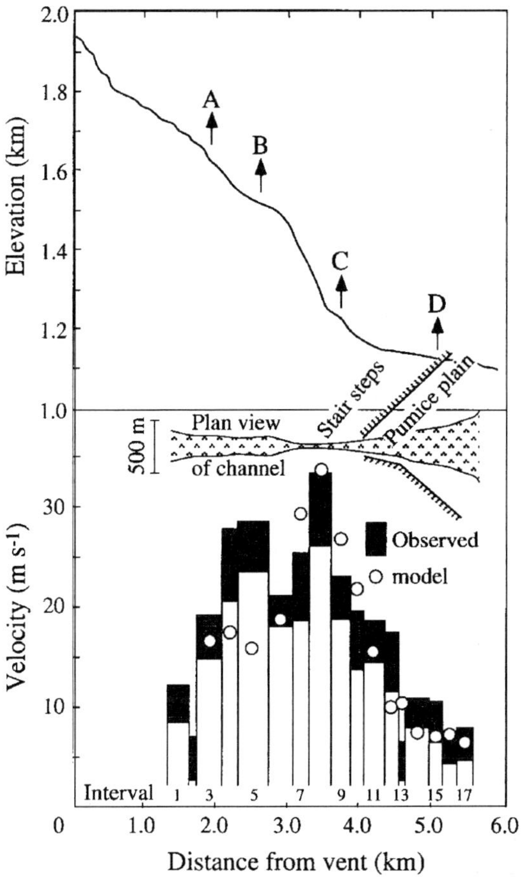

Figure 7.16 Observations of the velocity of the August 7, 1980 pyroclastic flow of Mount St Helens against distance from the vent (after Levine and Kieffer 1991). The filled boxes represent observations and the open circles represent calculated velocities from the model of Levine and Kieffer (1991). The topography of the flow path is shown above by a plan view of the channel and an elevation profile down the centre of the channel. The topographic profile of the channel shows the estimated source position of the four (A–D) co-ignimbrite plumes

and then begin to decelerate. The acceleration is thought to be a consequence of the plumes having been generated from the flow with little initial momentum and then mixing with and heating up entrained air as they rose. A comparison with theoretical predictions is discussed below in section 7.5.

Hoblitt (1986) and Calder *et al.* (1997) interpreted these observations in a manner analogous to the mechanisms discussed in previous sections. They envisaged the flow segregating into a basal avalanche and overriding turbulent surge. The overlying ash-plume surge would shoot ahead momentarily, mix with cold air and deposit particles and then become buoyant, resulting in stagnation of the flow front. The plume would then rise to form a buoyant plume and the denser basal avalanche then pulsates to produce the acceleration. The observations can also be interpreted in terms of a topographic influence on mixing and plume generation. Acceleration down steep parts of channels and hydraulic

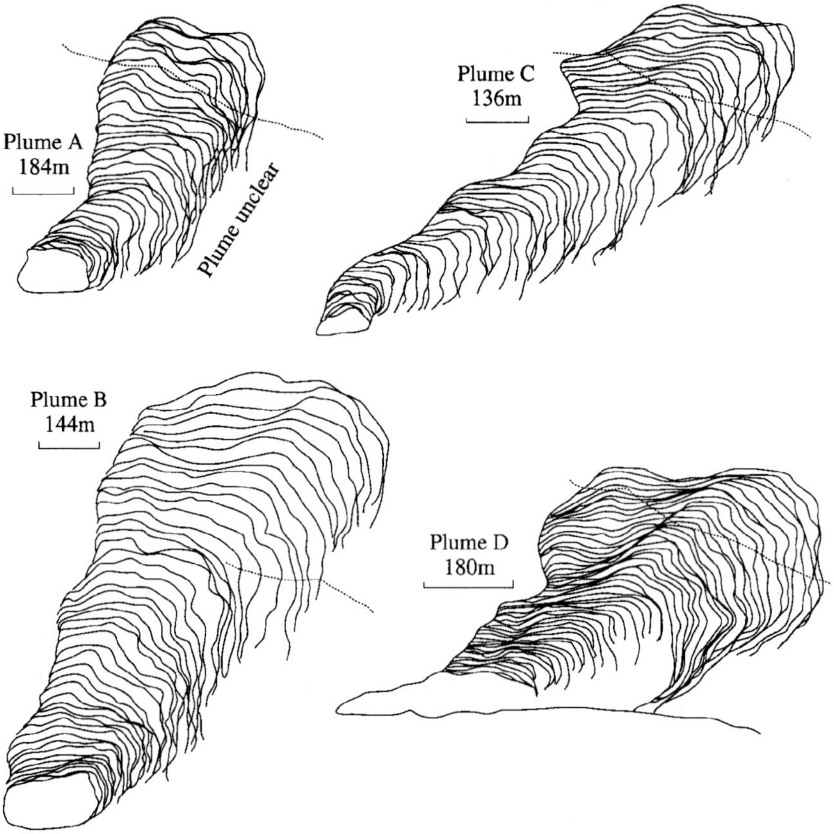

Figure 7.17 Growth profiles of the four co-ignimbrite plumes of the August 7, 1980 pyroclastic flow of Mount St Helens (after Calder *et al.* 1997). The time interval between each profile is 2 s. The dotted line shows the profile of the western flanks of Mount St Helens

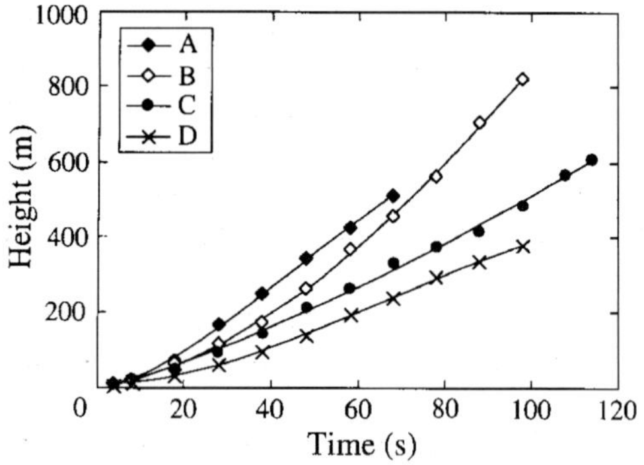

Figure 7.18 Height versus time data for the four co-ignimbrite plumes of the August 7, 1980 pyroclastic flow of Mount St Helens (after Calder *et al.* 1997)

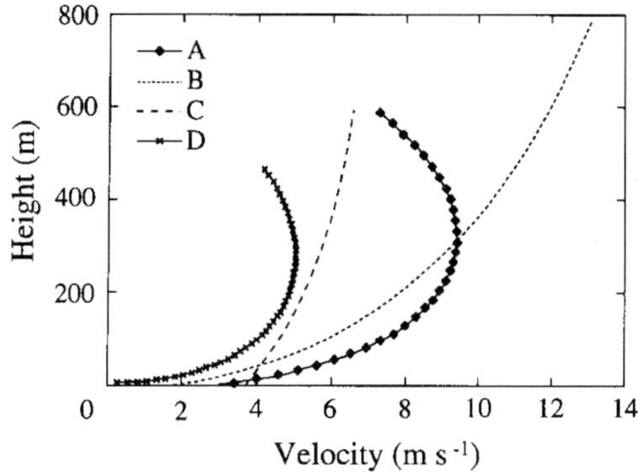

Figure 7.19 Velocity versus height for the four co-ignimbrite plumes of the August 7, 1980 pyroclastic flow of Mount St Helens (after Calder *et al.* 1997)

jumps at breaks in slope enhance mixing and fluidization within flows (Freundt and Schmincke 1985; Huppert *et al.* 1986; Levine and Kieffer 1991; Woods and Bursik 1994) and could therefore explain the occurrence of protuberances associated with changes of slope.

Calder *et al.* (1997) compared theoretical models of co-ignimbrite plume ascent (see sections 7.5.2 and 7.5.4) with the observations. They found good agreement, indicating that the plumes had initial temperatures of 500–600 K and ash masses in the range 3.4×10^5–1.8×10^6 kg.

7.5 THEORETICAL MODELS

7.5.1 A Steady Model

The formation of co-ignimbrite plumes from pyroclastic flows has been modelled as a steady-state column (Woods and Wohletz 1991), using an adaptation of the eruption column model of Woods (1988) and as described in Chapter 4. There are two distinctive aspects of co-ignimbrite plumes. First, they originate from the surface of the flow over a finite and usually substantial area. Second, they have no initial momentum. They are therefore examples of distributed plumes as discussed in Chapter 2. The steady-state model is appropriate for the continuous generation of co-ignimbrite plumes from the surfaces of pyroclastic flows. The model is based on the assumption that the pyroclastic flow acts as a continuous source of ash and hot gases. Furthermore the model assumes that the time over which the pyroclastic flow acts as a source is significantly longer than the ascent time of the plume. The model provides a guide to the likely behaviour of co-ignimbrite plumes, but requires some qualifications as discussed below.

The model assumes that the mixture of fine ash and entrained air above the flow is well mixed, is in thermal equilibrium and is just buoyant in the overlying atmosphere. The flow has some small initial vertical velocity in the range 0.1–10 m s^{-1}, the results being almost independent of the exact value. The mixture of ash and entrained air will only rise from a hot pyroclastic flow when it becomes less dense than the overlying environment. Therefore co-ignimbrite eruption plumes start from rest and have no gas thrust region. They thus rise from a large area, with very small initial momentum.

Figure 7.20 shows how radius, velocity and plume density relative to the density of the atmosphere vary as a function of height in the model calculations. The plumes show a marked acceleration and reach a maximum velocity after which they decelerate. The relative density profiles (Figure 7.20b) show that there is an initial region in which the rising plume becomes increasingly buoyant and a higher region in which the buoyancy of the plume begins to decrease towards zero. The plume starts out in the models marginally buoyant and becomes more buoyant as air is entrained and is heated. Eventually, however, the cooling effect of the entrained air results in the decrease of buoyancy with height. As with other buoyant plumes, the column reaches the neutral buoyancy height, and then decelerates to rest. Another feature of the model is that the plume initially contracts dramatically (Figure 7.20a). In these particular calculations the radius decreases from 5 km to only 1 km over the first 2–3 km of ascent. Above these heights the plume then increases in radius in a similar fashion to plumes from point sources.

The model might be sensitive to the entrainment assumption. Near the ground, the efficiency of entrainment of ambient air into these plumes may be less than that of a fully buoyant plume, because the scale of the eddies and hence rate of mixing initially scales with the height of the column. Only when the height has become equal to the width of the column does the full turbulent entrainment law of section 4.3 hold. For simplicity, Woods and Wohletz (1991) assumed that the entrainment coefficient was constant throughout the column. We extend these calculations here by examining two other entrainment models. The original model (a) of Woods and Wohletz (1991) assumed entrainment to be uniform with height with a constant of proportionality $\varepsilon = 0.09$. Model (b) assumes that the entrainment coefficient has the value $\varepsilon = (0.09h)/r$ for $h < r$ and $\varepsilon = 0.09$ for $h > r$, where h is the height and r the column radius. Model (c) assumes that $\varepsilon = 0$ for $h < r$ and $\varepsilon = 0.09$ for $h > r$. Therefore models (b) and (c) have less entrainment than model (a) in the lower parts of the column. Calculations are presented in Figure 7.20 for the three different assumptions of entrainment. The results are not very different. Model (c), with the least entrainment, shows somewhat more pronounced acceleration and contraction at the base.

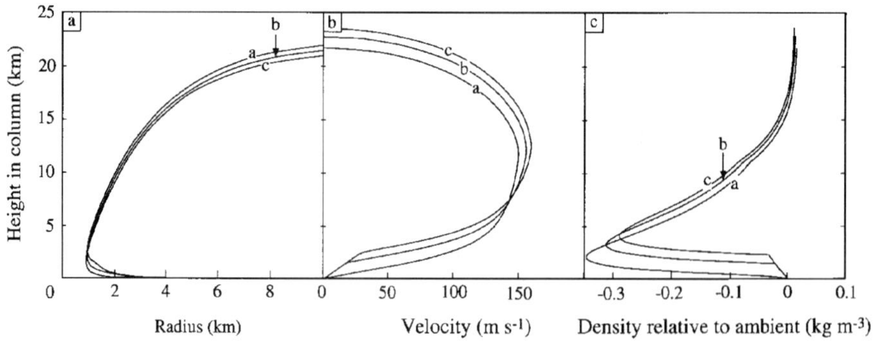

Figure 7.20 Plots to illustrate ascent dynamics of co-ignimbrite plumes, showing variation with height of radius (a), velocity (b) and relative density (c). The curves a, b and c refer to three different assumptions on entrainment as discussed fully in the text. In each calculation initial temperature is 1000 K, initial velocity is 1 m s^{-1} and the Mount St Helens initial radius is 5 km

As in sustained eruption columns from central vents, the maximum height of the plume depends on the mass rate of eruption, the temperature of the ash and atmospheric conditions. For the same eruption rate, the co-ignimbrite plume model predicts a slightly lower height than models of Plinian columns for two reasons. First, only a proportion of the erupted mass and its heat is contributed to driving the plume. In the calculations shown in Figure 7.21 it was assumed that only 35 wt% of the ash was incorporated into the co-ignimbrite plume with the remainder being partitioned into the pyroclastic flow. Second, buoyancy in a co-ignimbrite plume is achieved by mixing with a relatively larger amount of dense air at ground level in comparison to a plume from a point source. This is a consequence of the wide source area of a co-ignimbrite plume. Due to the larger surface area a co-ignimbrite plume incorporates air with a higher density on average for a given mass flux than a vertical Plinian column. The differences are, however, not large. The models demonstrate that very large and high eruption columns can be generated from pyroclastic flows, and this is also clear from observations.

7.5.2 Comparison of Steady Model with Observations

The features of the co-ignimbrite plume model are qualitatively well shown by observations. Data from the August 7, 1980 pyroclastic flows of Mount St Helens display the initial acceleration from rest and the velocity maximum (Figure 7.19). The giant plume of Mount St Helens generated on May 18, 1980 also shows evidence of acceleration from a low velocity (Figure 7.11).

The radius contractions predicted by the model are substantial. Inward contraction has been observed for the Mount St Helens giant plume (Figure 7.4) and the Redoubt plume (Woods and Kienle 1994). However, the observed contractions (see Figure 7.4) are not nearly as pronounced as the model predicts. Calder *et al.* (1997) observed little evidence of contraction in the co-ignimbrite plumes rising off the August 7, 1980 pyroclastic flows of Mount St Helens. The model is only an approximation of nature because, as the plume rises from a large finite area, entrainment and mixing at the margins take some time to mix into the interior. Buoyancy achieved by mixing and heating of air is greatest at the plume margin. Initially, in the central part of the flow, away from the influence of entrainment at

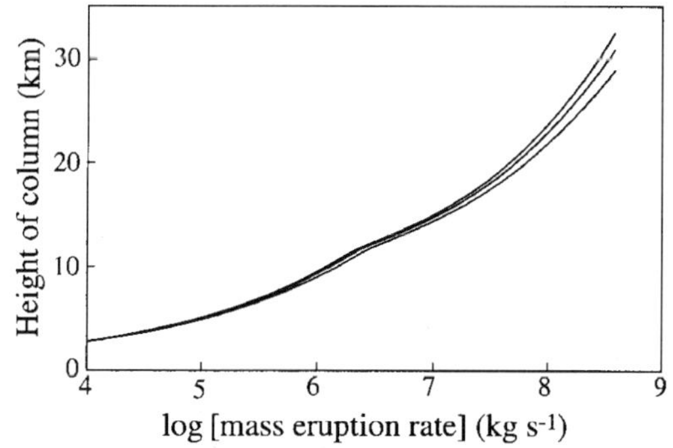

Figure 7.21 Height of steady co-ignimbrite plumes as a function of discharge rate. The three curves are for the different entrainment assumptions and the same initial conditions as in Figure 7.20

the margins, settling of particles will play a more important role than the entrainment in reducing the bulk density. Generation of buoyancy and acceleration due to each of these mechanisms has been observed in experiments (Sparks *et al.* 1993; Woods and Bursik 1994). It is concluded that natural co-ignimbrite plumes take rather longer to develop buoyancy by entrainment and particle loss than the model plumes.

Although the details of the entrainment influence the behaviour in the lower part of the plume, as it rises from the ground, the overall energetics and height of rise are not greatly affected by these complexities. The model (a) of Woods and Wohletz (1991) predicts slightly lower column heights for large, relatively cool co-ignimbrite plumes than the modified models (b) and (c). This is because more of the relatively dense, lower atmospheric air is predicted to be incorporated into the plume in model (a).

A sustained co-ignimbrite plume was generated on the afternoon of the May 18, 1980 eruption of Mount St Helens. Carey *et al.* (1990) have shown that the B_3 phase of the eruption was dominated by pyroclastic flow generation and formation of co-ignimbrite plumes. The mass eruption rate can be estimated from the total mass of the B_3 products and the duration of the B_3 phase. The B_3 phase lasted six hours and the discharge rate is estimated at 4.4×10^7 kg s^{-1} which is three times greater than the mass eruption rate during any of the other Plinian phases of May 18. The model of Woods and Wohletz (1991) predicts a height of 14 km for a co-ignimbrite plume with this discharge rate. This is in excellent agreement with plume heights estimated from radar data (Harris *et al.* 1981a; Carey *et al.* 1990). Photographs of the activity of this period show that the plume was generated from a wide area centred several kilometres from the vent to the north. The B_3 deposit itself is strongly bimodal with a high proportion of fines (Carey *et al.* 1990). A plausible interpretation of these grain size characteristics is that the coarse modes were derived from the plume generated over the fountain at the vent and that the fine mode was generated from the co-ignimbrite plume above the pyroclastic flows.

The ascent of co-ignimbrite plumes from the August 7, 1980 pyroclastic flows of Mount St Helens shows good agreement with the models (Calder *et al.* 1997), but the plumes spread at significantly greater angles (11–12°) than predicted by the models (8°) and do not show the substantial inward contraction.

7.5.3 A Thermal Model

Woods and Kienle (1994) developed a model for the ascent of a co-ignimbrite plume as a discrete thermal. Theoretical models of thermals have already been presented in Chapter 4. Such a model is applicable to situations in which buoyant material is elutriated from a hot ash flow on a time scale which is short compared to the ascent time of the material in the atmosphere. The Woods and Kienle thermal model treats the co-ignimbrite plume as a buoyancy ring vortex and the quantitative aspects of this model have already been described in Chapter 4.

Sudden bursts of buoyant material can develop in a number of ways. Pyroclastic flows can interact with topography where there are sudden changes of slope. Examples have already been given of greatly increased mixing between a flow and the atmosphere where the flow moves over a marked decrease of slope causing a hydraulic jump or over a cliff. Mixing processes in pyroclastic flows and fountains can also lead to generation of pulses of buoyant fluid due to the non-linear mixing properties of ash and air. Pulsations in the generation of buoyant fluid were observed in the MEG experiments of Woods and Caulfield (1992) and in the August 7, 1980 pyroclastic flows of Mount St Helens (Hoblitt 1986).

Figure 7.22 shows how the radius, velocity and relative density of co-ignimbrite plumes vary with height. Results are presented for initial plume radii of 0.5, 1.0 and 2.0 km, with a fixed elutriation temperature of 800 K, and elutriation temperatures of 800, 1000 and 1200 K with a fixed plume radius of 1 km. The results show that there is an initial acceleration of the plume and subsequent deceleration at greater heights as the buoyancy becomes exhausted. This result is similar to the predictions of the steady co-ignimbrite plume model except that the basal contraction is not seen. The plume overshoots the neutral buoyancy height due to its momentum. In contrast to the steady plume models there is no basal contraction of radius and at the base the radius of the plume increases approximately linearly with height. For a given mass, the rise height increases with temperature, because of the greater enthalpy available to create buoyancy. Hotter plumes also accelerate more rapidly. The minimum density in smaller plumes exceeds that for larger plumes with the same initial temperature. This is because smaller plumes entrain more air per unit mass owing to their larger surface area per unit mass. Consequently smaller plumes accelerate upwards more rapidly.

The height of a co-ignimbrite plume, modelled as a thermal, is shown as a function of its mass and temperature in Figure 7.23. The results are close to the simpler theory of Morton *et al.* (1956) which is shown for comparison. Plumes that ascend above the tropopause are

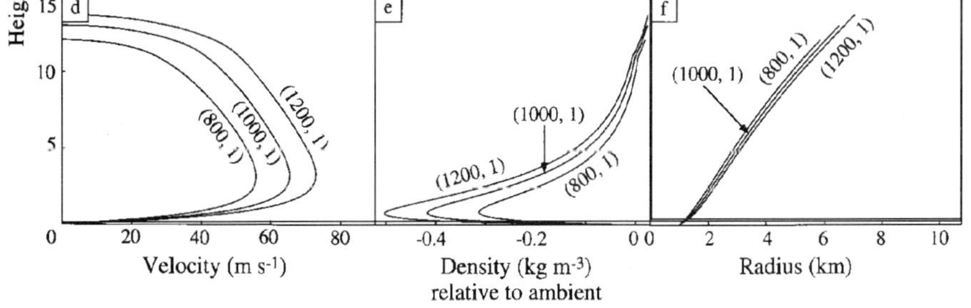

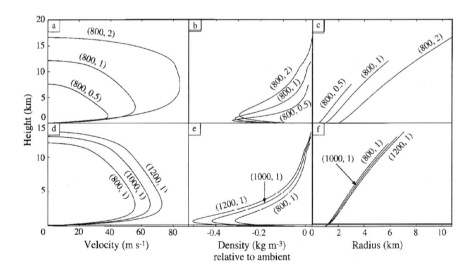

Figure 7.22 Variation of properties of co-ignimbrite plumes with height using the thermal model of Woods and Kienle (1994): (a, d) velocity; (b, e) density and (c, f) radius. Results are presented for initial plume radii of 0.5, 1.0 and 2.0 km, with a fixed elutriation temperature of 800 K, and elutriation temperatures of 800, 1000 and 1200 K with a fixed plume radius of 1 km. In these calculations the initial velocity is 1 m s^{-1} and the values on each curve denote the initial temperature in kelvin and radius of the plume in kilometres respectively. The profiles of density with height show a kink which occurs at the tropopause where the density stratification of the atmosphere changes

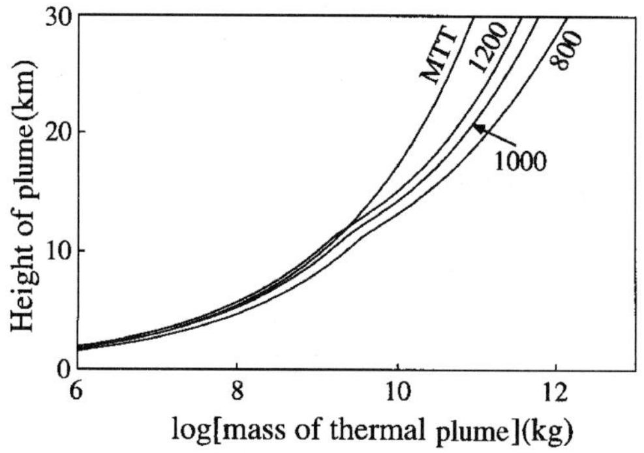

Figure 7.23 Variation of height of co-ignimbrite thermals with total mass values. Model calculations are for initial temperatures of 800, 1000 and 1200 K. The model of Morton *et al.* (1956), plotted as curve MTT, has been converted to the equation for column height, $H = 0.53(E_t/C_p \Delta T)^{0.25}$, where E_t is the total thermal energy, C_p the heat capacity of the ejecta and ΔT the temperature excess in the thermal (after Woods and Kienle 1994)

reduced in height compared to the Morton *et al.* model. This is because the Morton *et al.* model of a thermal assumes a uniform temperature and density stratification in the atmosphere, whereas the Woods and Kienle (1994) model takes account of the increased stratification above the tropopause. The effect of the increased stratification can be seen in the kink in the curves at about 11 km altitude in a standard atmosphere.

7.5.4 Comparisons of Thermal Model with Observations

Only three co-ignimbrite plumes have been described in sufficient detail for comparison with the models. These are the 1980 Mount St Helens giant plume, the small Mount St Helens plumes described in section 7.4 and the 1991 Redoubt plume of Alaska. These events involved emplacement of a pyroclastic flow and generation of a rapidly ascending plume within a period of only a few minutes. For this reason the thermal model should be more appropriate.

The Redoubt eruption of April 15, 1991 involved the emplacement of a pyroclastic flow over a glacier. Seismicity and video recordings provided detailed information on the generation and ascent of the co-ignimbrite plume (Woods and Kienle 1994). A sequence of profiles of the April 15 plume are shown in Figure 7.24. Woods and Kienle (1994) used thermodynamic and observational constraints to estimate that the plume initially contained between 1 and 1.5×10^9 kg of ash, air and vapour with a temperature between 600 and 700 K. The thermal model of a co-ignimbrite plume using the atmospheric temperature gradient recorded in the vicinity of Redoubt on April 15 calculates a plume height of 10 km above the pyroclastic flow source at 2.5 km altitude above sea-level. This prediction agrees well with observations. Figure 7.25 shows the ascent height of the upper surface of the plume as a function of time in comparison to models with elutriation temperatures of 600 and 700 K and an initial mass of ash of 10^9 kg. The models are quite successful in reproducing the observations. In particular the Redoubt plume shows the predicted initial period of acceleration as it lofted. The thermal model predicts that the radius increases

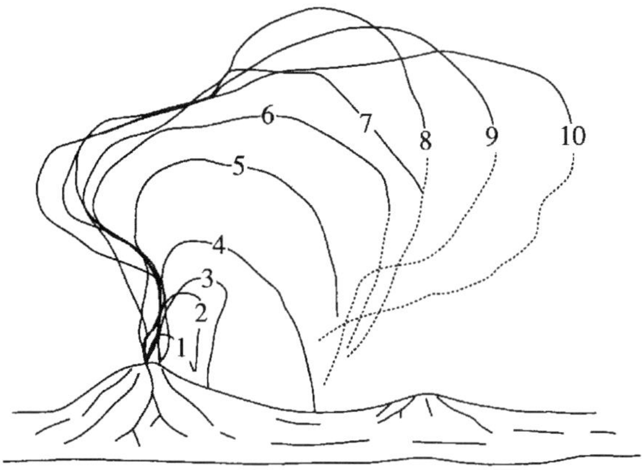

Figure 7.24 Profiles showing the co-ignimbrite plume of the April 15, 1990 eruption of Mount Redoubt, Alaska (after Woods and Kienle 1994). The contours represent plume margins at the following times (in GMT): (1) 22:51:47; (2) 22:52:58; (3) 22:53:35; (4) 22:57:10; (5) 22:58:21; (6) 22:59:33; (7) 23:00:09; (8) 23:01:21; (9) 23:02:21; (10) 23:03:44

linearly with height after the basal contraction of the plume. Applying this model to the observed plume indicates an initial radius of 0.5 km which is consistent with the area of flow from which the plume was generated.

The spreading angle of thermals predicted in the model is 14°, and is slightly greater than observed (Calder *et al.* 1997). The Mount St Helens plumes also show no inward contraction in contrast to the steady plume model in the August 7, 1980 flows of Mount St Helens but close to agreement with the thermal model. The co-ignimbrite plumes of Mount St Helens on August 7, 1980 are interpreted by Calder *et al.* (1997) as thermal plumes sustained for a short while by a flux of diminishing intensity.

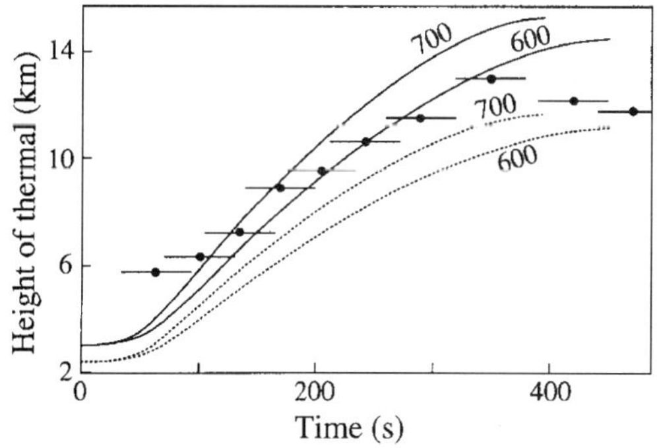

Figure 7.25 The height of the Mount Redoubt co-ignimbrite plume as a function of time (solid dots with error bars) compared with theoretical models (after Woods and Kienle 1994). The solid curves show the theoretical ascent of a thermal plume with 10^9 kg of initial mass for initial temperatures of 600 and 700 K. The dotted lines are inferred because of partial cloud cover

7.6 SUMMARY

Pyroclastic flows generate mixtures of fine ash and gas which become buoyant and form plumes both above the collapsing fountain at the vent and above the flows. The tephra in these plumes is characteristically enriched in very fine glassy ash. Pyroclastic flows can generate co-ignimbrite plumes by shearing across the upper surfaces and by various entrainment processes leading to strong fluidization and elutriation of gas and fines. Topography has a strong control in enhancing mixing at bends, breaks in slopes and where the ground is rough. Large flows can generate very large co-ignimbrite ash plumes by the phenomenon of buoyant lift-off where the whole flow becomes less dense than the overlying atmosphere and rises. Buoyant lift-off is caused by a combination of entrainment and heating of air and sedimentation. Co-ignimbrite plumes can be modelled either as steady plumes fed by a constant flux or as discrete thermals. These theoretical models, experimental studies and observations of co-ignimbrite plumes show that they initiate with little momentum and accelerate to reach a maximum velocity and then decelerate. A theoretical model of a steady plume with no initial momentum predicts that the plume radius contracts substantially in a basal region and then the radius increases approximately linearly with height. In contrast, a theoretical model of a thermal with no initial momentum predicts little inward contraction and plume spreading rates which are approximately twice that of a steady plume. Observation of the May 18, 1980 co-ignimbrite plume formed from the initial blast flow shows contraction at the base. Co-ignimbrite plumes observed at Mount St Helens on August 7, 1980 and Mount Redoubt show linear spreading rates which are more consistent with the thermal model. Co-ignimbrite plumes from large-magnitude ignimbrite eruptions produce the largest volume and most widely dispersed tephra layers of any volcanic process. Co-ignimbrite plumes, therefore, are the most significant kind of volcanic plume from the point of view of global environmental effects and hazards.

8 Geothermal and Hydrovolcanic Plumes

NOTATION

b	plume radius in top-hat model (m)
C_p	specific heat of plume (J kg^{-1} K^{-1})
C_{pa}	specific heat of ambient fluid (J kg^{-1} K^{-1})
e	thermal expansion coefficient of seawater (K^{-1})
F_e	effective buoyancy flux (m^{4}s^{-3})
g	acceleration due to gravity (m s^{-2})
H	maximum plume height (m)
k	entrainment constant (dimensionless)
L	latent heat of condensation (J kg^{-1})
N	Brunt–Väisälä frequency (s^{-1})
n	gas mass fraction in plume (dimensionless)
P	ambient pressure (Pa)
Q	mass flux (kg s^{-1})
Q_e	effective mass flux (kg s^{-1})
R	gas constant for plume material (J kg^{-1} K^{-1})
R_a	gas constant for dry air (J kg^{-1} K^{-1})
R_v	gas constant for water vapour (J kg^{-1} K^{-1})
s	solid mass fraction in plume (dimensionless)
T	temperature of plume material (K)
T_1	temperature 1° above ambient (K)
T_a	ambient temperature (K)
u	plume velocity (m s^{-1})
w	mass mixing ratio of water to dry air in plume (dimensionless)
w_a	mass mixing ratio of water to dry air in ambient air (dimensionless)
z	height above plume source (m)
α	ambient density (kg m^{-3})
β	bulk density of plume material (kg m^{-3})
$d\rho$	density difference between bulk plume and ambient (kg m^{-3})
ρ_s	density of erupted solids (kg m^{-3})
ρ_w	density of liquid water (kg m^{-3})
τ	time (hours)
Ω	condensation rate (s^{-1})

The subscript ‘o’ denotes a quantity at the vent.

8.1 INTRODUCTION

The interaction of hot magma, volcanic gas and external water can lead to a rich variety of behaviour. These interactions may be divided into two distinct categories, those in which water and vapour vent at the surface with no juvenile magma, thereby producing hydrothermal explosions, geysers, hot springs and mud pools, and those in which large quantities of external water and fragmented magma erupt simultaneously in an explosive manner forming "wet" eruption columns.

The first part of this chapter focuses upon venting of water, vapour and gases from geothermal systems where no physical contact occurs between the magma and external water. Geothermal systems are found in many volcanically active regions where cooling magma bodies transfer heat and exsolved gases to subsurface water reservoirs. Vigorous geothermal systems vent fluid at the surface in the form of geysers, hot springs and boiling mud pools. Volatiles exsolving from a cooling intrusion of magma can rise to the surface through an array of fractures and vent at the surface as high-temperature gas fumaroles, or they may mix into the geothermal circulation system. As these hot mixtures of gas and vapour vent at the surface, steam plumes, often rising hundreds of metres into the atmosphere, can form. These plumes are capable of injecting large quantities of volcanic gases into the lower atmosphere (Chapter 18). Well-known examples of large active geothermal fields include the Larderello field in Tuscany, Italy, Yellowstone National Park in Wyoming, USA, and the Taupo volcanic zone, New Zealand (Figure 8.1). The conditions necessary to produce fumarolic vapour plumes can also occur on a smaller scale at the summit of individual active volcanoes. For example, there is almost always a sustained vapour plume composed of a variety of gases at one of the four summit craters on Mount Etna in Italy (Figure 8.2). Chemical analysis of the volatile effluent can provide important information about magma migration and evolution within a volcano (Stoiber *et al.* 1987; De Natale *et al.* 1991; Caltabiano *et al.* 1994) as well as constraining the volcanic contribution of several gas species to the global volatile budget (Stoiber *et al.* 1987; Allard *et al.* 1994). Because these vapour-volatile plumes differ somewhat from other buoyant plumes discussed in this book, a model is presented that describes vapour plumes specifically.

In the second part of this chapter phreatic explosions are considered, which involve the eruption of tephra-laden steam. In such eruptions, the tephra may be hydrothermally altered lithic, mineral and glass fragments, but little or no juvenile material is erupted. A variety of qualitative mechanisms have been proposed to explain such eruptions, including the eruption of vapour from a geothermal system driven by the intrusion of a shallow magmatic heat source. Examples of phreatic activity include the October 1976 eruptions of La Soufrière de Guadeloupe, French West Indies (Heiken *et al.* 1980), and the March 1980 eruptions of Mount St Helens (Christiansen and Peterson 1981). Phreatic eruptions can produce energetic explosions and larger plumes than those associated with fumarolic activity, but the plumes still typically rise only of the order of a few kilometres.

In the final part of this chapter, the physical interactions of external water and erupting magma, which lead to phreatomagmatic or hydrovolcanic eruptions (Thorarinsson *et al.* 1964; Self and Sparks 1978), are discussed. Phreatomagmatic eruptions arise when erupting magma comes into contact with external water. This can occur during ascent, if the magma conduit intersects an aquifer, or at the surface if the vent is overlain by a crater

Figure 8.1 Geyser in the Whakawera wera geothermal fields, New Zealand. (Photograph courtesy of S. Rowland.)

lake or in a shallow submarine environment (Self 1983; Walker *et al.* 1984; Dobran and Papale 1993). Examples of magma interaction with subterranean external water include the explosions in 1977 that produced the Ukinrek Maars in Alaska (Self *et al.* 1980) and the last phase of the AD 79 eruption of Vesuvius, Italy (Sheridan *et al.* 1981). Examples of well-known phreatomagmatic eruptions in a shallow submarine environment include the 1963–64 eruption of Surtsey, Iceland (Figure 8.3), and the 1965 eruption of Taal in the Philippines (Moore *et al.* 1966). The addition of external water to an erupting mixture has important effects on the fragmentation of the magma, the intensity of explosions, the heat budget, variations in eruption rate, transitions from collapsing fountains to buoyant eruption columns, and also upon the rate of formation of aggregates in the eruption column and the subsequent patterns of fallout (Chapter 16). A distinct example of phreatomagmatic eruptions are those volcanoes that erupt entirely under water. The chapter concludes with a brief discussion of these shallow, but completely submarine, eruptions.

Figure 8.2 Mount Etna, Italy, during the September 1989 eruptive phase as seen from the town of Nicolosi. A large ash and water vapour plume generated by phreatic explosions is emanating from one of the four summit craters. (Photograph courtesy of L. S. Glaze.)

Figure 8.3 Formation of the island of Surtsey. (Photograph courtesy of H. Sigurdsson.)

8.2 GEOTHERMAL SYSTEMS

Geothermal systems develop in areas of high regional heat flow, as characterizes young volcanic regions, hot intrusive complexes and high heat-flow tectonic provinces. As external water is heated, it becomes buoyant, and may vaporize, causing the reservoir pressure to build up. Vapour or hot water is then driven along fractures from the reservoir to the surface (Figure 8.4). As the fluid rises to the surface, it passes through progressively cooler rock and its temperature falls. Depending upon the relative rate of cooling and decompression, the vapour can issue superheated at the surface or a fraction of the vapour can condense leading to the relatively passive eruption of a mixture of boiling water and steam (Figure 8.5; Elder 1981). The precise nature of the venting depends upon the relative rate of heating/cooling of the water compared to the mass flow rate through the system. Both continuous discharge of a mixture of water and gas, and periodic jets of water called geysers are possible.

8.2.1 Steady Venting

If there is a large reservoir of geothermal fluid at depth, and a path of relatively low permeability to the surface, then the venting is quasi-steady, and its rate is controlled by the pressure in the reservoir and the frictional resistance of the fluid as it moves to the surface. In this case, the pressure of the external water reservoir only evolves very slowly while fluid circulates through the system. The magma acts mainly as a source of heat to the external water, although some magmatic volatiles exsolved from the magma may seep through the surrounding rock and mix with the external water source. Depending upon the temperature and pressure, the fluid in the reservoir may be liquid, superheated vapour or a two-phase mixture. As the fluid rises to the surface it is typically cooled by the surrounding rock. Both the state of the fluid in the reservoir, and the relative rate of cooling and

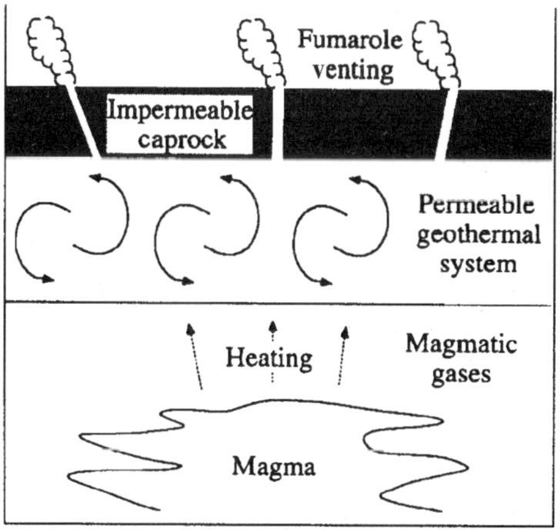

Figure 8.4 A schematic diagram of a geothermal system in which water circulates through the crust and is heated by a cooling body of magma

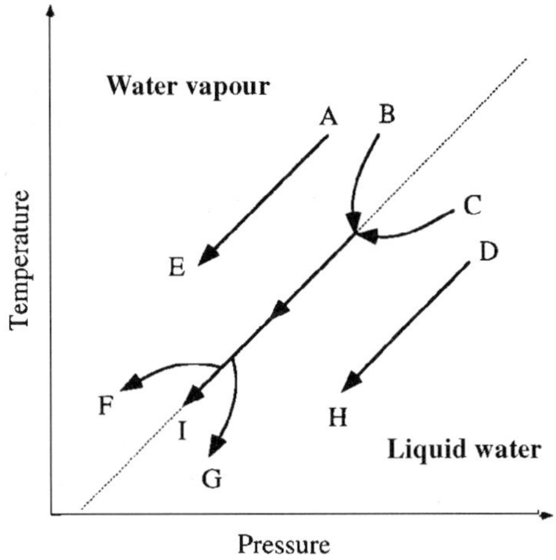

Figure 8.5 Diagram of the behaviour of water showing how, in principle, the initial conditions and the rate of cooling of geothermal fluid during ascent can lead to very different styles of venting. In case AE, the fluid remains superheated and vents as steam. In case B, the fluid starts as superheated vapour, but as it cools and the pressure falls, it becomes saturated. Subsequently the fluid follows the saturation curve (dotted line) as a two-phase mixture. The fluid may remain saturated (case I), eventually become supercooled and condense into a pure liquid flow (case G) or it may become superheated again, at lower pressure and temperature and become vapour again (case F). In case C, the fluid starts off undercooled, as a pure liquid phase, but becomes saturated as the pressure and temperature fall. At this stage a two-phase mixture evolves as described above. In case DH, the fluid is initially supercooled and remains supercooled as it ascends and the pressure and temperature fall. The precise path followed by the fluid depends on the ambient temperature gradient and pressure gradient, which in shallow systems may lie between hydrostatic and vaporstatic

decompression of the ascending fluid, lead to the wide spectrum of venting styles, ranging from warm water to superheated steam (Figure 8.5; Elder 1981).

The chemistry of the hydrothermal fluids is important in this discussion, because mixing of magmatic volatiles as well as the leaching of minerals from the surrounding country rock, can lower the boiling point of the geothermal fluid, resulting in venting of gases and steam at relatively low temperatures. Several models of the discharge of hot water and steam have been developed, based on models of two-phase flow in hot porous layers (Donaldson 1968; Elder 1981). Such models predict the mass fraction and fluxes of water and steam which may issue steadily from geothermal reservoirs as the fluids rise from a quasi-steady reservoir to the surface.

Following this approach, Stevenson (1993) developed a model for determining the depth of a magma body based on the temperature of fumaroles. The model assumes that, in a hot fumarole primarily composed of water vapour, the temperature remains sufficiently high to avoid condensation during the ascent. As the gas rises, it exchanges heat with the surrounding rock. The country rock temperature is assumed to vary either linearly with depth if the country rock is dry or along a hydrothermal geotherm if an active, but independent, geothermal circulation system exists. Using this model, Stevenson (1993) estimated that the magma source at Poas volcano in Costa Rica, which produced fumaroles as hot as

900 °C in 1982, was only 100 m from the surface, while the 300 °C fumaroles found at Vulcano in Italy may be supplied from a magma source as deep as 1500 m below the surface. These estimates provide the source conditions for the buoyant vapour-volatile plumes which are often observed. These plumes differ from other buoyant plumes discussed in this book in their lack of solid particles. We consider the effects that the lack of solids has on the dynamics of vapour-volatile plumes in section 8.3.

8.2.2 Geysers

In contrast to steady degassing, some geothermal systems exhibit intermittent geyser activity, in which hot jets of water issue from a vent at speeds of 20–30 m s^{-1} for intervals of minutes to several tens of minutes. This process injects steam and hot water tens of metres into the atmosphere. Although there are a range of possible mechanisms which result in geysers, the flash boiling model (Williams and McBirney 1979; Elder 1981) appears to be the most successful in explaining observations. Flash boiling can occur under special circumstances in small reservoirs connected to the surface by a conduit. If the rate of supply of relatively cool hydrothermal fluid to the small reservoir exceeds the rate of heating of the fluid in the reservoir, then the fluid is able to fill the reservoir and migrate upwards along the conduit to the surface. During this time, the fluid in the reservoir continues to be heated and may eventually attain the boiling point. At this stage, vapour bubbles nucleate and expand, driving the water in the overlying column upwards. As this water is ejected from the surface, the hydrostatic pressure in the reservoir at depth begins to fall, and the water continues to boil as it decompresses, until the whole column of overlying water has been erupted (Figure 8.6). Subsequently, the reservoir becomes recharged by cooler water from the geothermal circulation system, and the cycle begins again.

Notable examples of geysers occur in Yellowstone National Park (Nicholls and Rinehart 1967). The behaviour of each geyser is somewhat different. Old Faithful is the best known, and tends to produce a series of eruptions in which a mixture of water and steam issue from a vent as jets for periods between one and five minutes, with an eruption rate of the order of 7 m^3 s^{-1} (Kieffer 1984). The erupting jet forms a column of water which rises 10–30 m above the ground, corresponding to exit velocities of 20–30 m s^{-1} (Figure 8.1). After the main eruption of the water–steam jet, there is some more passive venting of steam which continues for several minutes. The system then recharges with water over an interval which ranges from 30 to 90 minutes, before the next eruption. Observations suggest that there is a bimodal distribution of eruption intervals and magnitudes. This observation suggests that there are two connected underground reservoirs at different depths in which water accumulates and is heated. When one of these reservoirs becomes superheated, the water boils and drives the overlying column of liquid from the system. The subsequent eruption can then be triggered in the other reservoir (Nicholls and Rinehart 1967). In a neighbouring Yellowstone geyser named Riverside, the behaviour is somewhat different. Water pours out of the vent for about an hour before the main eruption, which typically lasts for about 20 minutes and produces a jet of water which may rise up to 25 m into the air. Again the eruption ends with a weak steam phase for several minutes before the recharge, which can last up to six hours.

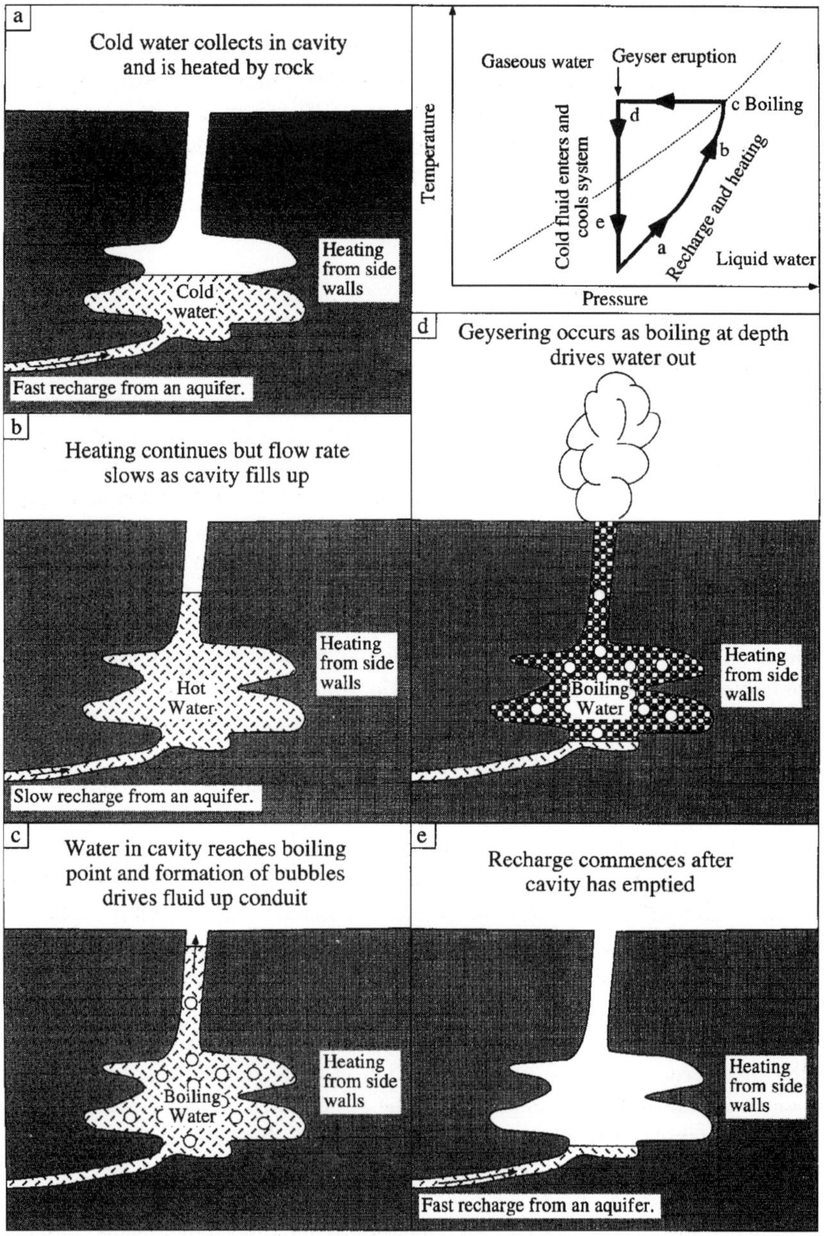

Figure 8.6 A diagram illustrating the sequence of events in a geyser system which leads to flash boiling and eruption of a column of water

A number of models of geysers have been presented in the literature following Bunsen's (1847) pioneering work. One of the most recent contributions of Dowden *et al.* (1991) presents a simple mathematical description of the flash boiling process. In some respects, the dynamics of the fountains produced by geysers resemble basaltic fire-fountaining activity (Chapter 10), although the mechanisms driving the eruptions are somewhat different.

8.3 GEOTHERMAL AND FUMAROLIC VAPOUR PLUMES

While the heating of external water by a magmatic body occasionally results in geysers, it more commonly results in hot springs or gaseous fumaroles, as discussed in section 8.2.1. Fumaroles often carry significant amounts of magmatic gases including, in order of decreasing concentration, H_2O, CO_2, SO_2, HCl, H_2, H_2S, HF, CO, N_2, COS and CH_4 (Faivre-Pierret and Le Guern 1983). Variations in the relative proportions of these gases, in some cases, gives an indication as to the state of activity of a particular volcano. Two examples of detailed studies of the variation of SO_2 with time are those of De Natale *et al.* (1991) for Campi Flegrei, Italy, and Caltabiano *et al.* (1994) for Etna. While sustained fumaroles do not usually produce as much SO_2 in their entire lifetime as a single Pinatubo or El Chichon-type eruption, they do supply a major fraction of exsolved magmatic gases to the atmosphere on an annual basis. Allard *et al.* (1994) have reported that, in a year with no large sulphur-producing eruptions, 10% of the entire global SO_2 budget is derived from gases released from only two Italian volcanoes, Stromboli and Etna, and Stoiber *et al.* (1987) estimated that (exclusive of eruptions such as Pinatubo) 36% of all the SO_2 released by volcanoes is emitted through passively degassing fumarolic vapour plumes. Many measurements of the flux of volatiles issuing from volcanic vents are based upon measurements of the concentrations in the downwind spreading plume. However, information about the height of the vapour plume above the vent can be useful in providing new constraints upon these volatile fluxes. For these reasons, the possible injection heights of geothermal vapour plumes into the atmosphere and their subsequent dispersal by ambient winds are considered in this chapter. This type of steady venting is often capable of sustaining a buoyant plume that differs in several respects to other volcanic plumes discussed in this book.

Geothermal vapour plumes are primarily driven upwards by the heat flux associated with the volatiles erupted at the surface. The heat flux is, in turn, controlled by the temperature and mass flux venting at the surface. A number of field studies of actively degassing volcanic systems have provided constraints upon the total thermal energy associated with the venting. For example, at Poas volcano, the total magmatic volatile flux has been estimated as being about 50 kg s^{-1} with temperatures as high as 900 °C and eruption velocities in the range 6–8 m s^{-1} (Casadevall *et al.* 1984; Stevenson 1993). This suggests that the area covered by the vents is about 10–20 m^2. At Vulcano, steam vents covering an area of 50–350 m^2 have been reported at temperatures of about 300 °C with a mass flux of 2–8 kg s^{-1} (Italiano and Nuccio 1992), suggesting that the material vents with speeds of the order of 0.01–0.1 m s^{-1}. These data provide boundary conditions for estimating the height of rise of volatile plumes rising above fumarolic vents.

8.3.1 Vapour Plume Model

In order to estimate the height to which geothermal vapour plumes rise an approach is used similar to that of Morton *et al.* (1956) (Chapter 2). Equation (2.26) derived by Morton *et al.* for the plume height is based upon the assumption that the temperature difference between the plume and ambient atmosphere is small, and therefore that the density of the plume varies nearly linearly with the temperature. However, if, as in this case, the initial temperature, T, is very large in comparison to the atmosphere, the density of the plume does not vary linearly with temperature as it cools over a large range. Thus, although the initial heat flux is conserved on mixing with air, the initial buoyancy flux changes due to the non-linear change in density with mixing (Chapter 2). Following entrainment, the plume rapidly cools to temperatures close to that of the environment. Therefore, the non-linear mixing can be accounted for by defining an effective buoyancy flux, F_e, which is the buoyancy associated with a plume of the same heat flux relative to ambient, but with an initial temperature, T_1, which is only a small amount (1 K) above ambient (i.e. $T_1 = T_a + 1$ K) (Chapter 9). The mass flux in this equivalent plume, Q_e, is given in terms of the original mass flux, Q, such that

$$Q_e = \frac{QC_p(T - T_a)}{C_{pa}} \tag{8.1}$$

where C_p and C_{pa} are the specific heats for the volatile plume and the air respectively. The density in the equivalent plume, β, now varies nearly linearly with temperature and can be approximated by

$$\beta \approx \frac{P(T_1 - T_a)}{RT_a^2} \tag{8.2}$$

The effective buoyancy flux, F_e, can now be expressed as

$$F_e = \frac{gQ_e}{T_a\beta_o} \tag{8.3}$$

The approximate relation given in Chapter 2 (equation 2.26) can now be rewritten in terms of the effective buoyancy flux:

$$H = 5F_e^{1/4}N^{-3/4} \tag{8.4}$$

where N is the Brunt–Väisälä frequency of the atmosphere defined by equation (2.25).

This simple model is appropriate in dry atmospheric conditions. However, in more humid environments, some of the vapour can condense as it rises in the plume and cools. The latent heat of condensation can then be used to heat the surrounding air, generate more buoyancy and drive the plume higher (Morton 1957; Squires and Turner 1962; Chapter 4). The magnitude of this additional effect, particularly in moist environments, is now considered. The main difference between the condensation in a vapour-volatile plume and a moist eruption column (Chapter 4) is that in the present case there are many fewer solid particles to provide nucleation sites when the vapour becomes saturated. As a result, the plume can become supersaturated and the kinetics of condensate nucleation can become important (Glaze and Baloga 1996). Vapour plumes also differ from particle-laden eruption columns in that there is no source of thermal energy from hot ash.

The dynamics of a vapour plume is modelled using the approach described in Chapter 2, but including equations to account for the conservation of and phase changes between vapour and liquid water (Morton 1957; Squires and Turner 1962; Glaze and Bologa 1996). The plume is assumed to be composed entirely of water, vapour and dry air and, as in Chapter 4, the mass fraction of dry air is denoted by $n(1 - w)$, vapour by nw, and liquid water by $(1 - n)$. Similarly, the mass fraction of ambient dry air is $(1 - w_a)$ and water vapour is w_a. The pressure within the plume is assumed to be equal to the local atmospheric pressure (Chapter 4) and the conservation of the total mass, momentum and energy fluxes may then be written as (Chapters 2 and 4):

$$\frac{\mathrm{d}}{\mathrm{d}z}\left[\beta u b^2\right] = 2k\alpha u b \tag{8.5}$$

$$\frac{\mathrm{d}}{\mathrm{d}z}\left[\beta u^2 b^2\right] = g(\alpha - \beta)b^2 \tag{8.6}$$

$$\frac{\mathrm{d}}{\mathrm{d}z}\left[\beta u b^2 C_p T\right] = 2k\alpha u b C_{pa} T_a + L\Omega b^2 \beta - \alpha u b^2 g\left[1 - \frac{\beta(1-n)}{\rho_w}\right] \tag{8.7}$$

where Ω is the rate of condensation of the water vapour, and L is the latent heat of condensation. Kinetic energy has not been included in the formulation of equation (8.7). To complete the system, the equations for the conservation of vapour and liquid water mass fluxes are

$$\frac{\mathrm{d}}{\mathrm{d}z}\left[\beta n w u b^2\right] = 2k\alpha w_a u b - \Omega b^2 \beta n w \tag{8.8}$$

$$\frac{\mathrm{d}}{\mathrm{d}z}\left[\beta(1-n) u b^2\right] = \Omega b^2 \beta n w \tag{8.9}$$

and the bulk density, β, is defined such that

$$\frac{1}{\beta} = \frac{1-n}{\rho_w} + \frac{nT[R_a(1-w) + R_v w]}{P} \tag{8.10}$$

As a vapour plume rises and cools, both the original vapour and any entrained ambient vapour become saturated and begin to condense (Chapter 4). The condensation rate, Ω, is determined by the availability of condensation nuclei. Specification of the precise value of Ω is a complex problem, depending upon temperature, particle loading of the plume and degree of supersaturation. For simplicity, in the present model, Ω is assumed to have a non-zero constant value. If the time of ascent of the plume is much shorter than the time scale for condensation, $1/\Omega$, then the plume is expected to become very supersaturated as it ascends, with negligible condensation. In contrast, if $1/\Omega$ is very small, then the plume should remain nearly saturated throughout.

8.3.2 Results of Model Calculations

The final height to which a buoyant vapour plume rises is very sensitive to the thermal energy balance within the plume (Morton *et al.* 1956) and there are several factors that can affect this balance. In this section the effects of variations in the initial temperature, the ambient stratification, the rate of condensation and the fallout of condensate from the plume are considered. A brief example is also presented to illustrate the effects of ambient wind on maximum height, followed by a qualitative discussion about the effects on the initial buoyancy of a plume due to variations in its composition.

Initial temperature

Figure 8.7 shows the variation in maximum predicted height of pure water vapour plumes as the initial temperature and mass fluxes vary in the absence of condensation. Curves corresponding to different initial plume temperatures of 100, 300 and 900 °C are shown along with the predictions of equation (8.4). It can be seen that the plume heights predicted by equation (8.4) are in closer agreement with estimates for higher initial temperatures. Although it may seem contradictory to intuition, the "cooler" plumes go higher due to the larger mass flux required to attain the same heat flux value for a lower plume. This result emphasizes the strong influence that the initial mass flux has on the final height of plume rise. Figure 8.7 indicates that the height of rise of the plumes above the Poas fumarolic system ($Q_o C_{po} T_o = 8.9 \times 10^7$ J s^{-1}) would have been 950 m, and for the mass fluxes reported at Vulcano, maximum plume heights are expected to lie within the range 340–480 m.

Ambient stratification

The ambient stratification of the lower troposphere varies both with season and with latitude (Figure 8.8) and may often vary diurnally as a result of heating and cooling. These changes in the stratification can have a large effect on the ascent height of vapour plumes (Chapter 4). For example, the height of the vapour plume at Lascar volcano in Chile has been observed to decrease throughout the day as a result of atmospheric warming (Matthews *et al.* 1997). Figure 8.9 shows how the model plume height varies with ambient temperature gradient (and hence stratification) for a constant initial heat flux of 5.6×10^7 J s^{-1}. It can be seen that plumes rising through atmospheres with positive

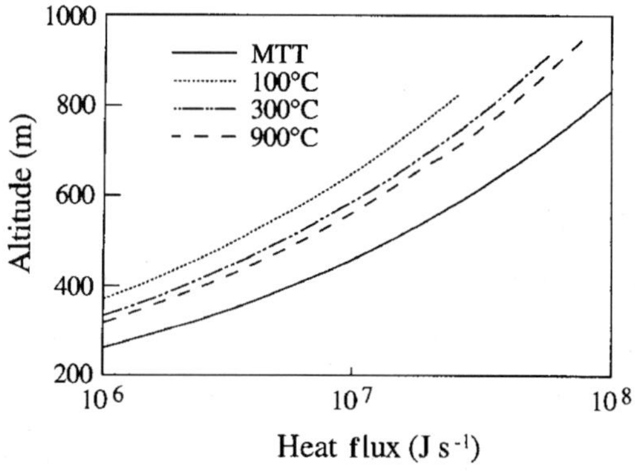

Figure 8.7 Variation in model plume heights as a function of initial heat flux, $Q_o C_{po} T_o$. The three dashed curves correspond to plume heights derived by the model presented here for initial temperatures of 100, 300 and 900 °C, and the solid line represents equation (8.4), after Morton *et al.* (1956)

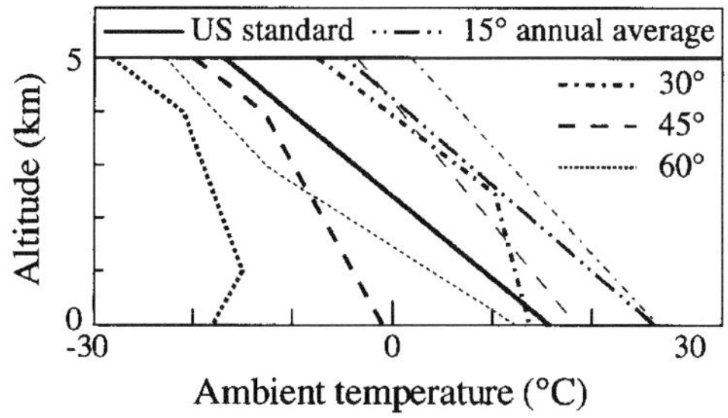

Figure 8.8 Ambient atmospheric temperature profiles between 0 and 5 km taken from Neiburger *et al.* (1973). Curves are shown for several profiles averaged over the months of January and July at specified latitudes. In the above plot, the January and July profiles at a particular latitude are represented by the same line type with the January profiles having a thicker line width. It can be seen that for the first kilometre above the surface, the 60° January profile (far left) actually has a positive temperature gradient and that the 60° July profile has the steepest negative gradient

temperature gradients rise less than half as high as those rising through less stable atmospheres with strongly negative temperature gradients.

Effects of condensation

Entrainment of ambient water vapour, together with the source vapour, can have a strong influence on plume dynamics through release of latent heat if condensation occurs (Chapter 4). This is particularly important for small fumarolic plumes which only rise through the first few kilometres of the atmosphere where the amount of ambient water vapour is large. For example, Figure 8.10 shows several curves representing monthly averages of measured mass fractions of ambient water vapour (in kg kg^{-1}) for January and

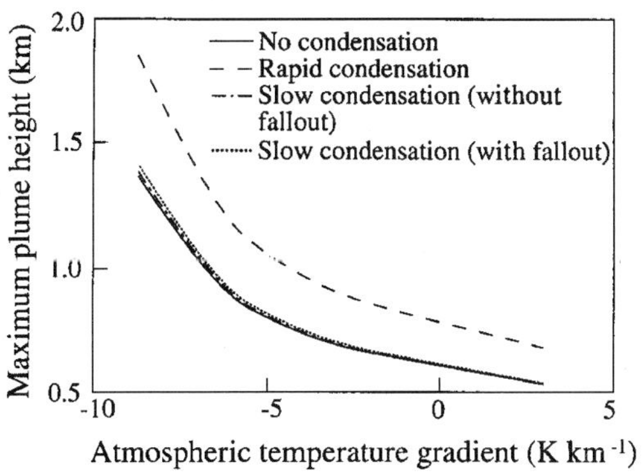

Figure 8.9 Plot of model plume heights as a function of the temperature gradient in the atmosphere. The solid line represents the case when no condensation occurs. The dashed lines represent model calculations when the effects of condensation in the plume are also included. The higher curve corresponds to a rapid condensation rate (10^{-2} s^{-1}) while the lower dashed curve corresponds to a lower condensation rate (10^{-4} s^{-1}). The dotted curve represents a model calculation in which the condensation occurs and the liquid water is allowed to fall out upon formation. In this curve, the lower condensation rate was used

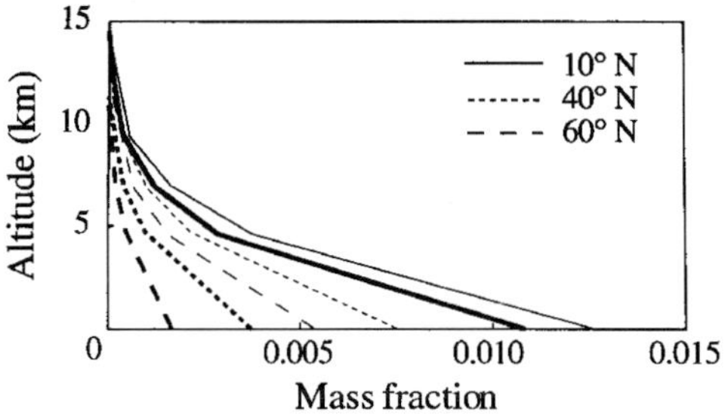

Figure 8.10 Plots of monthly averages of ambient water vapour mass fractions, w_a, for the months of January and July of 1984, for three latitudinal regions (from D. Crisp, unpublished data). The plot depicts January and July profiles for a particular latitude with the same line type where the January profiles are indicated by the thicker line width. The averaged values were derived from actual soundings. It can be seen that the 10°N profile contains the most water vapour and the 60°N profile contains the least

July 1984 (D. Crisp, unpublished data). To illustrate the effects of condensation, model wet and dry atmospheres are defined as the 10°N, July, and the 60°N, January, profiles respectively.

The differences in heights for plumes rising in the wet and dry atmospheres for a given heat flux of 5.6×10^7 J s^{-1} are illustrated in Table 8.1. The case in which the condensation rate is small so that the plume can become supersaturated is compared with the case in which the condensation rate is large so that the plume remains just saturated. For a pure water vapour plume released into the model dry atmosphere, the entrained water vapour comprises between 50 and 70% of the total amount of vapour in the plume. For the plume rising in the model wet environment, the ambient water vapour completely dominates the vapour component of the plume. If the plume is able to condense this vapour, it will ascend much higher owing to the additional buoyancy generated from the release of latent heat. Indeed, for the plume rising in a wet atmosphere with efficient condensation, the plume is able to trigger a moist convective updraft, in a similar fashion to cumulo-nimbus clouds (Squires and Turner 1962), plumes rising from sugar-cane fires (Morton 1957) and some basaltic fissure eruptions (Chapter 10).

When modelling vapour plumes with condensation, an additional complication that must be considered is how much of the condensate mass falls from the plume. Figure 8.9 shows that the effect of fallout of condensate from the vapour plume can lead to a slight increase in the plume height. This is because as the dense liquid water falls from the

Table 8.1 Importance of ambient water vapour on plume rise

	$1/\Omega = 2.8$ h		$1/\Omega = 1.7$ min	
	Dry	Wet	Dry	Wet
Neutral buoyancy height (m)	497	557	806	4465
Maximum height (m)	661	742	1054	5459
Total vapour from source (%)	50	11	30	0.4
Total vapour condensed (%)	1	1.5	83	96

plume, the density of the residual fluid decreases, further enhancing the buoyancy of the plume.

Effects of wind on maximum plume height

Because the plume model presented in section 8.3.1 assumes that there is no ambient wind, the maximum rise heights predicted by the model are higher than would be measured by an observer. The simple assumption is made that in a windy atmosphere, the plume bends over and the motion becomes dominated by the wind when the upward velocity has fallen below the wind speed (Chapter 11). As part of an observation experiment (L. Glaze, unpublished data) the model has been applied to the boundary conditions for a small industrial steam plume (Figure 8.11). From wind measurements made at the observation site and surface weather data, the wind speed several metres above the ground was estimated to be between 1.5 and 2 m s^{-1}. The model velocity in Figure 8.11a falls below these values at an altitude between 13 and 30 m above the pipe opening. The point at which the plume was observed to bend over was estimated to be 27 m, lying within the predicted range. Once the plume has bent over, it is further diluted by the ambient turbulence as described in Chapter 11.

Plume composition

As discussed at the beginning of section 8.3, fumarolic vapour plumes can be composed of a variety of gases. The composition of the gas released at the vent can also affect the buoyancy of a vapour plume. For a plume whose initial composition is entirely water vapour, the initial density of the plume is considerably less than the ambient air, in part because of its higher temperature, but also because the molecular weight of air is greater than that of water vapour. The presence of significant quantities of relatively dense volcanic gases such as SO_2 and CO_2 cause the bulk density to increase, which has the net effect of reducing the plume's buoyancy and consequently the predicted plume height. However, the basic dynamics of such plumes are essentially the same as described above.

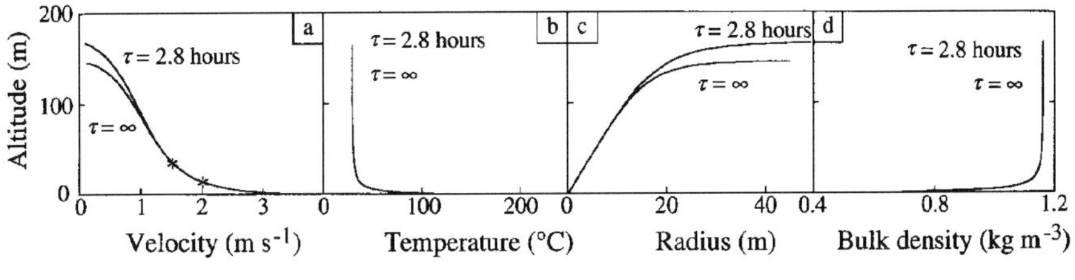

Figure 8.11 Model results for velocity, radius, temperature and bulk density for observed industrial steam plume with boundary conditions: $b_o = 27.9$ cm, $Q_o = 12.6$ kg s^{-1} and $T_o = 232.2$ °C. The gauge pressure of the steam before release was 2.81×10^6 Pa. Ambient air parameters measured locally were $T_a = 27.8$ °C, $P = 1.0154 \times 10^6$ Pa and $w_a = 0.0205$ kg kg^{-1}. Asterisks in (a) represent range of upward velocities between 1.5 and 2 m s^{-1}, equal to the ambient wind speed

8.4 PHREATIC ERUPTIONS

Phreatic eruptions involve the explosive eruption of water (liquid and/or steam) and non-juvenile lithic clasts and altered fine-grained materials. Characteristically, no juvenile magma is erupted. Such events indicate a vigorous hydrothermal circulation system at depth and can occur when the vapour pressure of geothermal fluids exceeds the hydrostatic boiling pressure for a given temperature. The eruptions arise in locations where there is a relatively impermeable caprock which impedes the ascent of the geothermal fluids to the surface. Following failure of the caprock, an eruption can develop with a vaporization wave propagating downwards through the underlying geothermal reservoir. Failure may be induced by the intrusion of magma at depth, hydrofracturing, chemical breakdown of the caprock, which is commonly formed by the precipitation of minerals from geothermal fluids, or unloading of the overlying material, for example, due to lake drainage or an avalanche. This failure enables the vapour to accelerate out through the conduit and up to the surface. Phreatic eruptions can also be triggered when boreholes are emplaced for geothermal power exploration, since this may also create a new pathway for the vapour through the impermeable cap.

Depending upon the resistance of the flow pathway from the caprock to the surface, the erupting steam can be pressure adjusted at the surface and therefore issue at relatively slow speeds, or it may be overpressured and choked at the surface. When overpressured, the steam can issue at very high speed, leading to vent erosion and production of phreatic craters. These craters may range in size from 1 m pits to lake-filled depressions as large as 1 km in diameter. For example, large phreatic explosion craters have been observed in the Eastern Kawerau geothermal field of New Zealand (Wohletz and Heiken 1992) and following the Ukinrek Maars eruption in Alaska, USA, in 1977 (Self *et al.* 1980). Deposits from such eruptions tend to be composed of poorly sorted angular tephra with a range of sizes from sub-millimetre to several metres. The tephra can be emplaced either by a blast flow or as ballistic fallout and many of the lithic clasts are hydrothermally altered.

Two notable phreatic eruptions include the 1976 eruption of Soufrière de Guadeloupe (Heiken *et al.* 1980) and the precursory phreatic eruptions at Mount St Helens volcano in March 1980 (Christiansen and Peterson 1981). During 1976, the background fumarolic activity at Soufrière de Guadeloupe was characterized by the emission of tephra-free steam rising 50–150 m into the atmosphere. However, during September and October 1976, a number of phreatic eruptions occurred, lasting 30–45 minutes each and leading to the injection of tephra-laden steam 350–650 m into the atmosphere, before the plumes were dispersed by the wind. There was no juvenile material in the eruption products and in one of the eruptions, on October 2, a white vapour plume and a medium grey tephra-laden plume were simultaneously erupted. The two plumes were apparently generated from the same geothermal source, but issued from different vents, suggesting that the degree of alteration, strength of the rock or velocity of exit from the two sources were different. The plumes were blown downwind, and then propagated as gravity currents at about 26 m s^{-1} down the valleys of the River Noire and the River des Peres, before spreading over the sea surface. This transport speed is in excess of the prevailing winds speeds of only about 10 m s^{-1}. Using the empirical model of Briggs (1969) for a wind-blown plume, Heiken *et al.* (1980) estimated that exit velocities at the summit were of the order of 20–60 m s^{-1} and that about 10^7 kg of tephra was erupted during the October 2 eruption.

During the period March–May 1980, a number of phreatic eruptions occurred at Mount St Helens volcano prior to the cataclysmic eruption (Christiansen and Peterson 1981). In each of the eruptions, plumes of ash and condensed steam rose as high as 3000 m above the volcano, leading to the formation of a significant crater at the summit. Detailed observations suggest that the ash plumes had distinct regions. Above the crater, finger-like jets of ash propagated upwards several hundred metres. Above the jets of ash a dark brown ash-laden plume developed, overlain by a white steam-like plume. All of the ash was derived from pre-existing rocks.

8.5 PHREATOMAGMATIC ERUPTIONS

Sections 8.2–8.4 have considered the cases where heating of external water and release of magmatic volatiles are the only interactions between magma and water. In the remainder of this chapter, the direct physical interaction of erupting magma with external water, resulting in phreatomagmatic eruptions, is considered. Phreatomagmatic eruptions have a number of distinguishing characteristics, including intense explosive activity (Thorarinsson *et al.* 1964; Houghton and Nairn 1991), the production of anomalously fine-grained tephra (Walker and Croasdale 1972), and the formation of wet base surges from collapsing fountains (Waters and Fisher 1971). For example, during the eruption of Taal 1965, in which lake water had ready access to the vent, Moore *et al.* (1966) reported that "wet turbulent clouds spread with hurricane velocity, transporting ash, mud lapilli and blocks", and that a towering ash plume 15–20 km high developed above the volcano. The deposits of the wet base surges, with relatively cool temperatures of the order of 100 °C, included many accretionary lapilli, and instead of charring the trees in their path, plastered them with mud (Moore *et al.* 1966).

In a number of historical phreatomagmatic eruptions, transitions in eruption style between wet eruption columns and cold dense base surges have been observed. These transitions can result, for example, from changes in the amount of water which has access to the erupting material following erosion of the conduit (Wohletz and Sheridan 1983). Such changes in water content can explain the sequence of activity during the 1963–64 eruption in which the island of Surtsey was formed. As the new island grew, there was easy access of seawater to the vent resulting in pulse-like, high-intensity explosive activity. These eruptions generated turbulent base surge currents which travelled down the side of the volcano and over the sea, as well as large convecting ash plumes over 9 km high (Thorarinsson *et al.* 1964). However, as the eruption proceeded, and the vent rose out of the sea, water access to the vent was restricted and the eruption became largely effusive in character.

Another volcano that has typically exhibited phreatomagmatic behaviour is White Island in New Zealand. White Island is surrounded by the sea, but, in general, the magma conduit has no interaction with the surrounding water or hydrothermal system. During the eruptive phase which lasted from 1976 to 1982, however, erosion and collapse of the conduit walls led to two styles of phreatomagmatic activity (Houghton and Nairn 1991). Near-continuous emission of gas and ash occurred when the walls were relatively stable while collapse of unstable walls resulted in larger discrete explosions followed by prolonged periods of quiescence.

Phreatomagmatic activity can also result from magma interaction with external water in subterranean aquifer systems. Two typical examples include the Ukinrek Maars explosions in 1977 and the last two phases of the Vesuvius eruption in AD 79. In the case of Ukinrek Maars, Self *et al.* (1980) have suggested that the craters that were observed following the explosions were the result of "the collapse of crater and conduit walls, and the blasting-out of debris by phreatomagmatic explosions when the rising magma contacted groundwater beneath the regional water table and a local perched aquifer". Following the initial phase of the AD 79 Vesuvius eruption Sheridan *et al.* (1981) have suggested that cavitation of the roof and walls resulted in rupture of the metamorphic encasement surrounding the conduit and led to interaction of the magma with the regional hydrologic system. They further conjecture that "because the magma level within the emptying chamber was below the principal aquifer... an abundant source of water was available... when the venting pressure dropped below the hydrostatic head".

Magma–water interactions can also take place after silicic magma has been fragmented and discharged from a vent (Walker 1979). Many ignimbrite-forming volcanic centres are near large water bodies or the sea. During explosive eruptions hot pyroclastic flows can be discharged into water (e.g. Cas and Wright 1991). For example, during the 1883 eruptions of Krakatoa volcano in Indonesia, voluminous dacite pyroclastic flows were discharged into the Sunda Straits (Self and Rampino 1981; Sigurdsson *et al.* 1991). Parts of the flow were emplaced subaqueously at high temperature (Mandeville *et al.* 1994), whereas less dense and more turbulent portions continued over the sea surface for distances up to 80 km (Carey *et al.* 1996). The phreatomagmatic interaction of the flows with seawater generated a moisture-rich co-ignimbrite plume and widespread fallout of mud rain. Walker (1979) proposed similar processes of ash plume generation for pyroclastic flows entering the sea in the Taupo Volcanic Zone.

Current modelling efforts suggest that the efficiency of magma–water interaction is controlled by the mass of water mixing into the eruption products, with values of about 0.35 yielding the most intense interactions (Wohletz 1983; Wohletz and Heiken 1992). The following sections discuss the explosive energy and fragmentation that result from magma–water interaction and the properties of erupting water–magma mixtures.

8.5.1 Explosive Energy and Fragmentation

If external water from the sea or a lake mixes with erupting magma, or if the conduit intersects a shallow aquifer, then there may be a very intense interaction which produces a highly fragmented mixture (Walker and Croasdale 1972; Walker 1973; Self and Sparks 1978). The energy driving this fragmentation is derived from the heating and expansion of the water as it comes into contact with the very hot magma. For example, if liquid water is heated to 1200 °C at a fixed volume then the pressure increases to about 500 MPa Thus explosive activity can be much more intense but often characterized by large fluctuations, as typified by the Surtsey example above. Therefore, the total power output from such eruptions is no larger than magmatic eruptions, since in both cases the thermal energy of the magma is the main source of energy.

The rapid chilling, contraction and fracture of the magma when it interacts with the external water can produce blocky ash fragments with planar surfaces (Self and Sparks 1978; Heiken and Wohletz 1985). In some situations, the fragmentation process may be analogous to fuel–coolant interactions which arise in industry when a hot liquid (fuel) comes into contact with a cold fluid (the coolant) (Peckover *et al.* 1973; Wohletz 1983). Vigorous fuel–coolant interactions occur when there is a large surface area between the liquid fuel and the coolant which enables the coolant to be heated and vaporize. The vapour bubbles then expand and collapse, causing the fuel–coolant interface to become convoluted, leading to further boiling of the coolant and break-up of the fuel. Models of fuel–coolant interactions suggest that the rate of break-up and fragmentation of magma depends upon its viscosity and the interfacial surface tension (Wohletz 1983).

If the ascending magma exsolves volatiles and disrupts as it rises along the conduit before coming into contact with the external water (Chapter 3), then there will be a large surface area over which the magma may interact with the water. This represents the most efficient situation for an intense interaction of magma and external water, leading to a highly fragmented vesicular ash (Self and Sparks 1978). However, there is a spectrum of styles of magma–water interaction and associated magma fragmentation which depend upon the efficiency with which the magma and water can come into contact (Kokelaar 1986). There are examples of fine-grained tephras composed almost entirely of angular, poorly vesicular to non-vesicular glass (e.g. Hatepe tephra of Taupo, New Zealand, AD 186 eruption; Wilson and Walker 1985) which suggests that intense phreatomagmatic explosive eruptions can occur without vesiculation of magmatic gases. Wohletz and Sheridan (1983) have proposed that there is an optimal mass ratio of water to magma, in the range 0.1–1.0, for which magma fragmentation can be induced by the water–magma interaction. For higher water contents the magma is chilled and quenched, inhibiting the fragmentation and leading to a less violent style of eruption.

Phreatomagmatic interactions can also occur in more basic magma, such as basaltic andesite and basalt. In such magmas, the volatile content is typically smaller and viscosity lower than in more silica-rich magmas, and so exsolved volatiles are able to move upwards relative to the magma. In the absence of external water there is little fragmentation induced by the volatile exsolution, and the eruptive activity usually consists of fire fountains and lava flows (Chapter 10). However, a number of phreato plinian style basaltic eruptions have been reported, including the 1886 eruption of Tarawera (Walker *et al.* 1984), the 1924 eruption of Kilauea (Decker and Christianson 1984), the 1963–64 eruption of Surtsey (Thorarinsson *et al.* 1964) and the 1983 eruption of Miyakajima (Aramaki *et al.* 1986). In each case, the production of fine ash by external water–magma interaction enabled the magma to transfer its heat to entrained air and produce a buoyant, ash-laden eruption column (Walker and Croasdale 1972).

8.5.2 Properties of Erupting Water–Magma Mixtures

Magma–water interactions can lead to highly fragmented ash. In dry explosive silicic eruptions the ash produced by the exsolution of volatiles in the magma (Chapter 3) is sufficiently fine-grained to enable efficient heat transfer with the entrained air and the formation of a buoyant eruption column so that the additional fragmentation does not affect heat transfer strongly. However, the addition of large quantities of external water to the erupting magma has two other important effects upon the eruption column dynamics. First, as a fraction of the external water vaporizes, the initial density of the material falls, and this tends to promote the formation of a buoyant eruption column (Chapter 4).

However, the external water also causes the temperature of the erupting material to fall and this tends to suppress the formation of a buoyant plume. Figure 8.12 shows how the density, temperature and water content varies with mass fraction of external water, for magma temperatures of 725 and 1125 °C. If small quantities of water (<20%) are added to the mixture, then all the water vaporizes and this lowers the density of the mixture, tending to promote the formation of a buoyant eruption column. However, if larger quantities of water are added to the mixture, then the density increases again and the temperature falls to the boiling point. In this case, some of the liquid does not vaporize, and the erupting mixture becomes relatively cool and wet. The addition of such large quantities of water tends to inhibit the formation of a buoyant eruption column.

To illustrate these effects, a series of calculations are presented showing how the rise height of an eruption column varies with the mass eruption rate and the mass of external water added to the column (Figure 8.13). These calculations were made using the eruption column model described in Chapter 4. For the chosen boundary conditions, an eruption column with no groundwater is predicted to collapse and form pyroclastic flows when the mass flux exceeds about 10^8 kg s^{-1} (Chapter 6). However, as the mass of external water

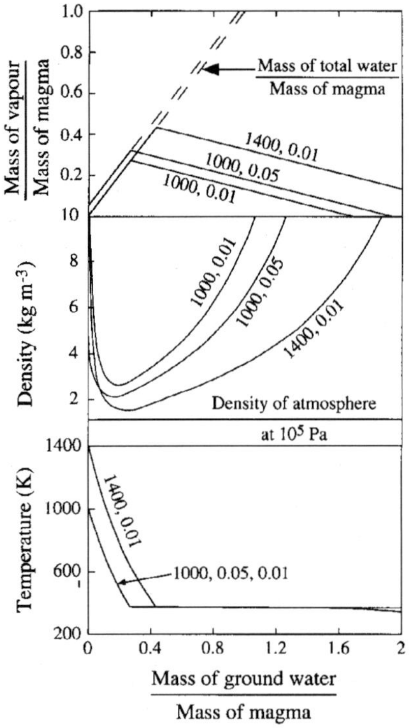

Figure 8.12 Water content, density and temperature for mixtures of ash, volatiles and external water as a function of the external water content. For small amounts of external water, complete vaporization occurs, and the density of the mixture falls. However, as the temperature reaches the saturation value, some of the water remains in the liquid state, and the density begins to increase again. The temperature of the ash in kelvin and the mass fraction of water are shown for each curve, after Koyaguchi and Woods (1996)

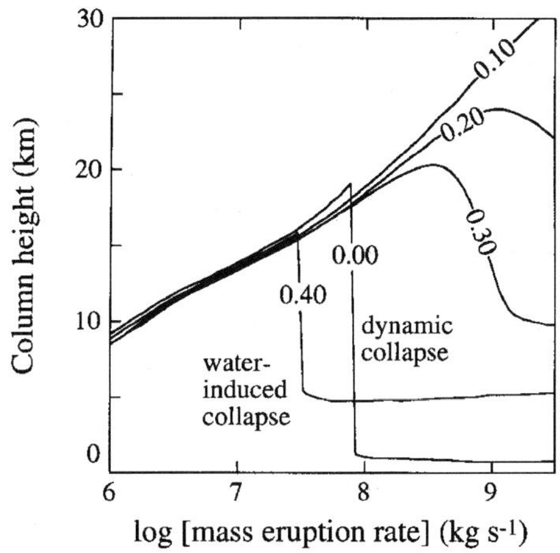

Figure 8.13 Height of an eruption column as a function of the mass flux erupting from the vent. Numbers on each curve denote the mass fraction of external water in the erupting mixture. The abrupt decrease in height to the right represents the change from convecting columns to collapsing fountains, after Koyaguchi and Woods (1996)

added to the erupting mixture increases to a weight fraction of 0.1–0.2, then the material is able to generate a buoyant eruption column directly above the vent for eruption rates of the order of 10^8–10^{10} kg s^{-1}, owing to the lower initial buoyancy of the mixture. Note that for small eruption rates, the rise heights of phreatomagmatic plumes are not very sensitive to the mass of external water added to the erupting mixture. This is because, even though the water is vaporized and initially cools the material, it subsequently condenses, restoring the thermal energy to the system, heating the entrained air and generating buoyancy (Woods 1993b; Koyaguchi and Woods 1996).

As the mass fraction of water in the erupting mixture increases to values of the order of 0.4 or higher, then the initial temperature of the material becomes so low, and the density so high, that collapse occurs for mass fluxes as low as 10^7 kg s^{-1}. The critical velocities

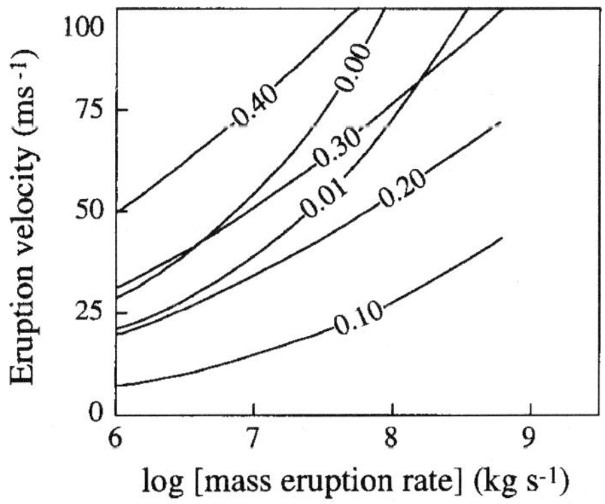

Figure 8.14 Critical velocity required to generate a buoyant eruption column as a function of eruption rate. Curves are labelled with the initial mass fraction of external water, after Koyaguchi and Woods (1996)

for column collapse can be seen in Figure 8.14. These results can account for the frequent observation of cold, wet base surge flows associated with phreatomagmatic behaviour.

8.6 SUBMARINE ERUPTIONS

Explosive eruptions occur in submarine volcanoes, in which the style of eruption is largely controlled by the depth of the vent below the surface, as well as the volatile content and viscosity of the magma, and the efficiency of magma–water mixing. The plumes generated by these eruptions represent an important class of volcanic plumes in which buoyant rise of hot fluids, gases, and particles occurs in an ambient medium of water. There have been few observations of submarine explosive eruptions, and much of the evidence for such activity comes from sightings of floating rafts of pumice, water discoloration and by inference from the geological record. For example, during the eruption of Iwo Jima in 1934, ash-laden steam was erupted from a vent over 300 m below the surface, and large vesicular blocks with incandescent cores rose to the surface and floated for some time before subsequently sinking (Williams and McBirney 1979). Also, during the eruption from a vent at 30 m water depth in the South Sandwich Islands in 1962, pumice rose to the surface and continued to float for periods of up to two years. Indeed, gravel-sized lumps of pumice were washed up in Tasmania 21 months after the eruption (Gass *et al.* 1963; Sutherland 1965). In deep water, evidence for plume formation has been found in the form of bedded hyaloclastite deposits on seamounts at 1240–2500 m depth in the Pacific Ocean (Batiza *et al.* 1984; Smith and Batiza 1989).

Although there has been relatively little detailed dynamical modelling of submarine eruption columns, there is good evidence from submarine fall deposits that such eruption columns can develop (Cashman and Fiske 1991). There are a number of factors that control the formation of submarine eruption columns. In the case of silicic eruptions, the vigour of the eruption is largely influenced by the mass of exsolved volatiles. Hydrostatic pressure increases by about 10^5 Pa for every 10 m of water depth. If magma fragmentation occurs at void fractions of 0.7–0.8 (Chapter 3), then for magmatic water contents of 0.03–0.05, the typical fragmentation pressures are of the order of 5×10^6–2×10^7 Pa, and thus explosive eruptions can only occur from submarine vents shallower than 500–2000 m. For those volcanoes that are able to erupt explosively, the pyroclastic ejecta can form a submarine eruption column or collapse as dense pyroclastic flows (Kokelaar and Busby 1992; Cashman and Fiske 1991). The main controls on these phenomena are the density and temperature of the erupting mixture of gas and vesicular pyroclasts relative to that of the seawater and also the efficiency of mixing of the erupted material with the seawater.

Little is known about the dynamics of such submarine eruption columns, although some simple predictions can be made. During the early stage of the motion above the vent, the material is driven upwards by its momentum, and unless the vent is very deep the mixture will initially be buoyant. As the mixture rises, entrains water and cools, the initial momentum becomes depleted, and some of the larger dense clasts separate from the column. However, assuming the mixture remains homogeneous, the difference in density between this mixture and the surrounding water determines whether a buoyant plume can develop. If the mixture entrains sufficient water, then the hot exsolved volatiles will condense. The density of the material in the plume relative to the ambient seawater is then determined by the excess thermal energy in the plume and the density anomaly of the solid clasts relative to the water. In a relatively weak or deep plume, after mixing with seawater the solid mass fraction, s, in the erupting material eventually becomes relatively small, $s < 0.1$, and the density of the mixture relative to the seawater has the approximate form

$$\mathrm{d}\rho = s\big[(\rho_\mathrm{s} - \alpha) - e\alpha(T - T_\mathrm{a})\big] \tag{8.11}$$

where T is the initial temperature of the eruption products, T_a the seawater temperature, e the thermal expansion coefficient of seawater, and ρ_s and α the densities of the solid clasts and the ambient water. The thermal expansion coefficient of seawater is only approximately 7×10^{-5}, and so unless the density of the pumice lies within the range 900–1100 kg m^{-3}, it is the density of the pumice and not the thermal buoyancy which controls the motion of the eruption column above the vent.

On formation, silicic pumice clasts are typically much less dense than seawater, with densities in the range 200–900 kg m^{-3}. However, pumice collected from subaerial deposits appears to have very well connected vesicles. Whitham and Sparks (1986) have shown that when hot subaerial pumice is dropped into water, it tends to draw in water and rapidly becomes denser than the water. Based on their experimental results, the mean density of hot pumice clasts after immersion into cold water typically lies in the range 1100–1400 kg m^{-3}. Only very vesicular pumice is able to remain less dense than water. Therefore, the data of Whitham and Sparks (1986) suggest that the bulk density of the mixture of pumice and seawater is typically in excess of the seawater (equation 8.11). This suggests that a collapsing fountain will develop, and pumice-laden flows will develop on the sea-floor. As the heavy pumice clasts sediment from this flow, the residual fluid may be sufficiently warm to generate a thermally buoyant column some distance from the volcano which may carry some of the fine-grained solid material to the surface, in a similar fashion to the experiments of Carey *et al.* (1988).

Observations of floating pumice produced by submarine eruptions (Williams and McBirney 1979) suggest that in some situations hot submarine pumice can remain less dense than water after mixing with water. The ability of pumice to draw in water is related to the connectedness of the vesicles. This connectedness is thought to become established through the decompression of the pumice. In submarine eruptions, where the ambient pressure is considerably larger, the vesicles in the pumice may not expand as much, and therefore may remain more isolated. As a result, the pumice may draw in less water and hence remains less dense than suggested by the experiments of Whitham and Sparks (1986). In this case the mixture of pumice and seawater remains buoyant and a plume of pumice and water could develop above the vent, driven upwards primarily by the buoyancy of the pumice. If such a plume becomes established the buoyancy flux is likely to be sufficient for the plume to ascend to the sea surface. Using the expression for the height of rise of a plume in a stratified environment (Morton *et al.* 1956; equation (2.26) in Chapter 2), we find that for eruption rates in the range 10^4–10^8 kg s^{-1}, the associated plumes would rise to heights in the approximate range 500–5000 m.

In the case of eruptions of mafic magma erupted in deep-water environments a somewhat different type of column model is suggested by recent studies in the Pacific Ocean. Based on the recovery of hyaloclastite gravity flow deposits from seamounts near the East Pacific Rise, Smith and Batiza (1989) proposed that vigorous fragmentation of basaltic magma can take place by a form of submarine fire fountaining. They suggest that high discharge rates of magma result in cooling–contraction granulation and perhaps some steam explosivity. A column is developed that consists of a slurry of glass shards, seawater and possibly steam. However, because of the generally low volatile contents of most basaltic magmas and the great water depth (>1200 m) it is likely that the eruption fountains are denser than the surrounding seawater and will rise primarily as a result of their initial momentum. Collapse of these fountains cause shard-rich (hyaloclastite) gravity flows that move away from the source and deposit bedded hyaloclastite layers (Batiza *et al.* 1984). As with silicic submarine columns it is likely that secondary thermal plumes of hot seawater and entrained particles would be developed over the vent from the tops of the gravity flows during the collapse process.

8.7 SUMMARY

This chapter has focused on volcanic water vapour plumes where no physical contact occurs between the magma and external water, phreatic explosions of tephra-laden steam and plumes resulting from the full interaction of external water with magma.

In the case of no physical contact, external water is heated by magma resulting in venting of hot water, vapour and gases from geothermal systems. Vigorous geothermal systems vent fluid at the surface in the form of geysers, hot springs and boiling mud pools. In some cases, hot mixtures of gas and water vapour venting at the surface generate steam plumes rising hundreds of metres into the atmosphere. Chemical analysis of the composition of these steam plumes can provide information about magma migration and evolution within a volcano. A model derived specifically to describe buoyant steam plumes has been presented and the effects of initial steam temperature, ambient stratification, condensation, wind and composition have all been discussed.

Plumes resulting from phreatic explosions sometimes contain tephra comprised of hydrothermally altered lithics and mineral and glass fragments. However, these eruptions involve little or no juvenile magmatic material. Phreatic eruptions generally indicate a vigorous hydrothermal circulation system and occur in locations where there is a relatively impermeable caprock which impedes the ascent of the geothermal fluids to the surface.

Phreatomagmatic eruption plumes result from the physical interactions of external water and erupting magma. This may occur during ascent if the magma conduit intersects an aquifer or at the surface if the vent is overlain by a crater lake or in a shallow submarine environment or where pyroclastic flows enter significant bodies of water. The addition of external water to an erupting mixture has important effects on the fragmentation of the magma, the intensity of explosions, variations in eruption rate, transitions from collapsing fountains to buoyant eruption columns and also upon the rate of formation of ash aggregates in the eruption column. Submarine eruptions can also generate plumes in the ocean as magma is discharged explosively or by its momentum to mix with and heat seawater.

9 Hydrothermal Plumes

NOTATION

a	$(1/V)(\partial V/\partial T)_p$
b	radius of plume (m)
B	buoyancy flux ($m^4\ s^{-3}$)
C_p	specific heat ($J\ kg^{-1}\ K^{-1}$)
d	thickness of neutrally buoyant intrusion (m)
D	depth of neutrally buoyant cloud at the top of the plume (m)
D_v	diameter of diffuse plume source (m)
F	heat flux ($J\ s^{-1}$)
$f(T,S)$	function determining density as a function of T and S
g	gravitational acceleration ($m\ s^{-2}$)
H	height of plume (m)
N	Brunt–Väisälä frequency (s^{-1})
Q	mass flux ($kg\ s^{-1}$)
R	Radius of neutrally buoyant intrusion (m)
S	salinity ($kg\ m^{-3}$)
T	temperature (K)
u	velocity ($m\ s^{-1}$)
U	cross-flow velocity ($m\ s^{-1}$)
V_e	entrainment velocity in the y-direction ($m\ s^{-1}$)
W_e	entrainment velocity in the z-direction ($m\ s^{-1}$)
W_o	vertical velocity of diffuse plume at source ($m\ s^{-1}$)
Y	width of diffuse plume (m)
z	height above source (m)
$Z(x)$	diffuse plume height with distance from source (m)
α	density of ambient atmosphere ($kg\ m^{-3}$)
β	density of plume ($kg\ m^{-3}$)
ε	entrainment coefficient (dimensionless)
λ	ratio of Brunt–Väisälä frequency and cross-flow velocity (m^{-1})
γ_T	thermal expansion coefficient (K^{-1})
γ_c	solutal expansion coefficient (kg^{-1})
ρ	density ($kg\ m^{-3}$)
ρ_u	ambient density at the sea-floor ($kg\ m^{-3}$)
Ω	rotation frequency of Earth (s^{-1})

Subscripts

o property evaluated at the source
e property of environment

9.1 INTRODUCTION

Previous chapters have dealt exclusively with plumes generated by volcanic activity which discharge into the atmosphere. The budget of global volcanism, however, indicates that about three-quarters of the annual volcanic production occurs underwater, mostly along the extensive ocean ridge system. This volcanic activity also generates plumes, but in this case the plumes consist of seawater, with some precipitate particles, driven by the exchange of heat between circulating fluids and hot rocks beneath the sea-floor. This exchange takes place predominantly along the axis of ocean ridges, but has also been documented at subduction zones, back-arc spreading centres as well as rift lakes and volcanic crater lakes.

Early studies of uplifted metalliferous ore deposits associated with rocks of the oceanic crust led to the prediction of vigorous hydrothermal circulation on the sea-floor. However, the first observations of active hydrothermal vents along the East Pacific Rise at 21°N and in the Galapagos were only made about 15 years ago (Edmonds *et al.* 1979; Speiss *et al.* 1980). Since that time the importance of hydrothermal circulation has become increasingly clear (Humphries *et al.* 1995). It has been estimated that submarine hydrothermal systems account for approximately 25% of the total global heat loss (Sclater *et al.* 1981). Along mid-ocean ridges this heat flux greatly exceeds that associated with pure conduction. The circulation of seawater into the oceanic crust and subsequent dispersal as plumes provide an important geochemical exchange mechanism in which both the composition of seawater and the crust are modified. The influx of heat from hydrothermal venting also plays an important role in the generation of abyssal currents (Speer and Helfrich 1995).

Sites of active hydrothermal systems play host to exotic benthic biological communities where chemosynthesis replaces photosynthesis as the basis for a complex hierarchy of organisms (Tunnicliffe 1992). Hydrothermal plumes entrain and disperse the larvae of vent organisms (Kim *et al.* 1994) that colonize new venting areas when their venting area "dies off" and thus ensure the survival of deep-sea life through geological times. Furthermore, the most primitive forms of life on Earth are associated with hydrothermal venting and it is now widely believed that life originated at hydrothermal vents *c.* 4 billion years ago (Holm 1992).

In this chapter the basic features of hydrothermal plumes and models for their dynamic behaviour are described. The application of these models has important implications for understanding the fluxes of heat and chemical components into the ocean, as well as the production of economically significant metalliferous ore deposits.

9.2 GENERATION OF HYDROTHERMAL PLUMES

Along the ocean ridge system new oceanic crust is formed by the injection and extrusion of magma along a relatively narrow zone defined as the spreading axis. The newly formed crust becomes highly permeable owing to extensive fracturing which develops as the magma cools and spreads. Furthermore, seismic tomography experiments have identified that this magma is stored in shallow level chambers, 2–3 km below the surface, providing a long-lived source of heat (Harding *et al.* 1989; Henstock *et al.* 1993). As a result, active spreading centres are ideal locations for the establishment of hydrothermal convection because of the presence of a shallow-level heat source and a relatively permeable boundary separating cold seawater from the underlying hot rocks.

Cold seawater percolates into newly formed fractured oceanic crust over a relatively broad area up to several tens of kilometres from the ridge axis (Figure 9.1). It migrates downward into the crust and begins to react with the surrounding rocks as it is heated. At depths where most of the hydrothermal circulation occurs, 2–4 km below sea-level, the pressure of the overlying seawater is sufficient to suppress the boiling point such that the seawater temperature reaches values as high as 400 °C without vaporization (Tivey 1992). The density of seawater at these temperatures is less than seven-tenths that of cold seawater and thus the hydrothermal fluids are initially extremely buoyant. As a result, a large-scale convective circulation becomes established with the buoyant fluid rising back to the sea-floor in coherent plume-like structures. This leads to the venting of hydrothermal fluids on the sea-floor and formation of plumes. Discharge of the fluid at the sea-floor typically occurs over a much narrower area than the recharge and is generally restricted to about 200 m from the main spreading axis of the ridge.

At such high temperatures the circulating fluid undergoes chemical reactions with the surrounding basaltic rocks. Magnesium and sulphate are nearly totally removed from the hot fluid as clay minerals and sulphates which are precipitated within the crust. On the other hand, Ca, K, Si, S dissolve into the fluid from the basalt, enriching the fluid in sulphide, Fe and Mn. The net result in the oceanic crust is the conversion of relatively fresh basaltic rocks to greenschists and amphibolite facies metamorphic rocks.

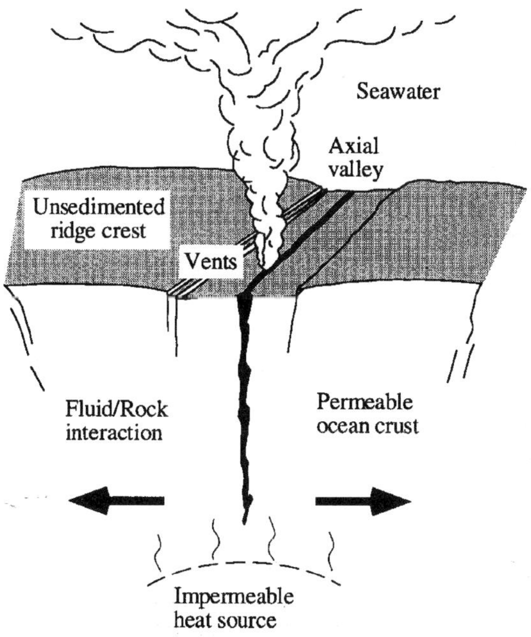

Figure 9.1 Schematic cartoon of the hydrothermal circulation system at a mid-ocean ridge

The exit flow can take a variety of forms. Chemical analysis of fluids venting from these different types of plumes indicates that the spectrum of venting can be attributed to binary mixing between a hot, hydrothermal fluid, which is acidic and reducing, and ambient seawater which is relatively oxidizing (Edmonds *et al.* 1979). Black smokers are the most vigorous type of discharge where high temperature (up to 350 °C) hydrothermal end-member fluid is vented at focused sites with velocities up to 5 m s^{-1} (Figure 9.2). Upon mixing with seawater, the hydrothermal fluid forms the characteristic black precipitate particles, thus giving the name black smokers.

White smokers show many similarities to black smokers except that the fluids are at a lower temperature, the precipitate is white and flow velocities are lower. The least vigorous type of venting occurs over broad areas and involves relatively low-temperature fluids, with a large proportion of ambient seawater (< 5 °C). This warm diffuse flow issues with velocities of the order of 0.1 m s^{-1}, and much of the heat it supplies is carried away by

Figure 9.2 Turbulent black smoker at the Endeavour Segment of the Juan de Fuca Ridge. (Photograph courtesy of A. Schultz.)

currents close to the sea-floor. Hydrothermal plumes generated in shallow environments such as shallow ridges adjacent to Iceland or in island arcs can be rich in gases such as methane and CO_2 so that bubble-rich plumes form.

9.3 HYDROTHERMAL VENTS

9.3.1 Style of Venting

A variety of vent types have been observed and their geometry appears to be related to the intensity of the fluid discharge. Vents range from the classical sulphide chimneys through which there is very rapid venting, to extensive areas and mounds through which there is much slower diffuse flow.

Black smokers are associated with the chimneys, which may be several metres high and typically contain inner conduits from 1 to 10 cm in diameter (Figure 9.3). These chimneys form by the precipitation of minerals from venting fluids. Mixing of hot hydrothermal fluids with cold seawater results in changes in pH and temperature which trigger precipitation. Haymon (1983) developed a model of the formation of such chimneys as a two-stage growth process; a sulphate-dominated stage and a sulphide-replacement stage (Figure 9.4). During the first stage, the venting fluid precipitates an anhydrite collar around the plume and this grows outwards, with some secondary precipitation of pyrrhotite and pyrite occurring at the outer edge of the wall. During the second stage of the growth, once the anhydrite walls of the chimney are in place, mixing of the venting fluid with seawater is suppressed within the chimney, and Cu–Fe sulphides form on the inner edge of the chimney walls. In addition, hydrothermal fluids migrating through the walls of the chimney dissolve and replace the anhydrite with sulphide minerals. The chimney thus grows upwards through precipitation of anhydrite, outwards through anhydrite formation and subsequent replacement by sulphides and inwards by precipitation of Cu–Fe sulphides (Figure 9.4). Tivey and McDuff (1989) have developed a quantitative model of the reaction kinetics associated with these reactions, and by coupling this with a model of the fluid transport through the chimney walls, they were able to predict the rate at which this mineralization sequence occurs.

The growth of black smoker chimneys can also be modelled by analogue laboratory experiments in which plumes of a hot aqueous salt solution are injected into a cooled tank

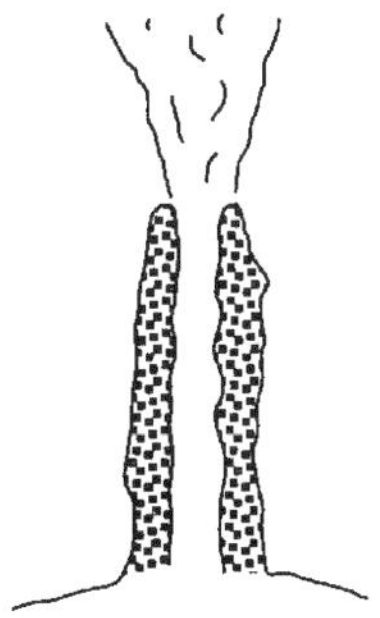

Figure 9.3 Schematic diagram of a black smoker chimney

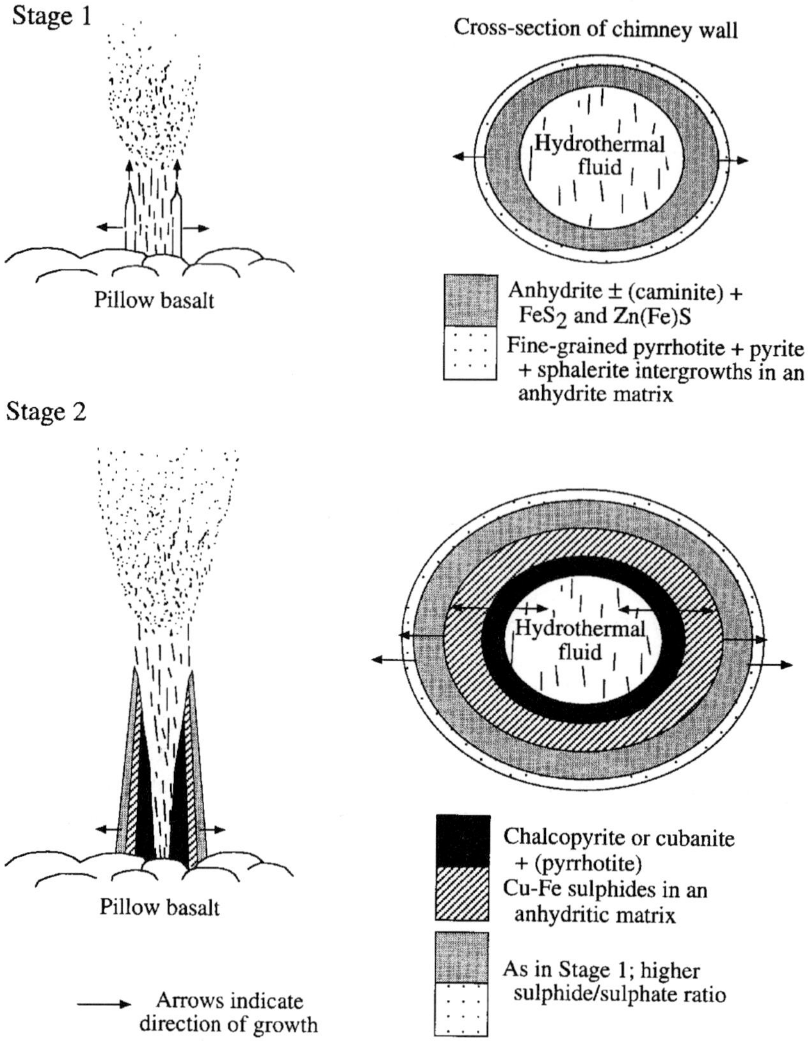

Figure 9.4 The two stages involved in the formation of a smoker chimney (from Haymon 1983)

containing a different aqueous solution (Turner and Campbell 1987a). As the input fluid mixes with the ambient water, the mixture becomes supersaturated and a chimney precipitates around the source. These experimental chimneys were similar to those observed on the sea-floor, and Turner and Campbell demonstrated that their rate of formation depended upon the plume fluid flux and the supersaturation of the mixture.

As the chimney structures grow, some of the pore spaces in the chimney may then become blocked through precipitation. The fluid develops alternative venting channels and can break out through the sides of the chimney. At the Juan de Fuca vent field, these secondary venting sites have been observed to form horizontal flange structures, similar to

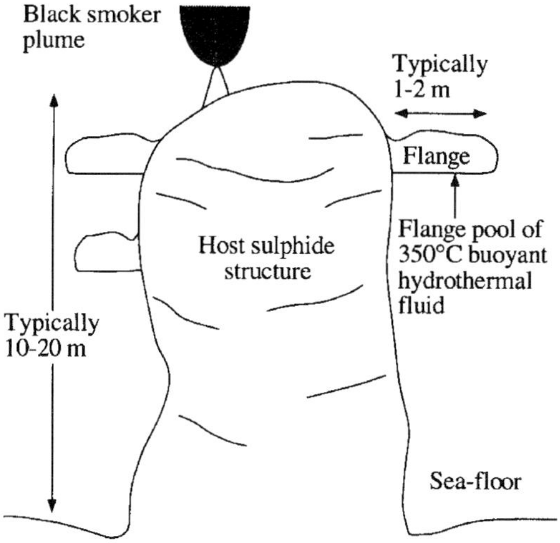

Figure 9.5 Diagram of a flange structure growing from the side of a massive sulphide deposit

porous inverted bowls. Hydrothermal fluid ponds in a pool under the flange, and either percolates through the flange or spills over the edge into the surrounding water, leading to further precipitation of the sulphide structure and the generation of a buoyant plume above the flange (Figure 9.5) (Woods and Delaney 1992). In another vent field, the Guaymas basin, Gulf of California, horizontal plate structures, tens of centimetres in radius have been observed to form on top of tall thin chimneys which are only a few centimetres in radius (Figure 9.6) (Lonsdale and Becker 1985). The mechanism for their formation could be similar to that of the flanges.

9.3.2 Distribution of Venting Sites

As well as the differences in the local venting structures, there is considerable variation in the size and spacing of vents at different spreading centres (Tivey 1992). At 21°N along the East Pacific Rise the vents are located within the axial valley and occur at intervals of 100–1000 metres. Most chimneys are situated on top of low-lying mounds up to tens of metres in diameter. A recent large-scale survey of the East Pacific Rise using ARGO imaging demonstrated a correlation between high-temperature hydrothermal activity and shallowing of the seismic reflector which represents the top of the axial magma chamber (Haymon *et al.* 1991). High-temperature vents are also generally restricted to within 20 m of axial summit calderas. The occurrence of the high-temperature vents is thought to be part of a volcanic–hydrothermal–tectonic cycle that distinct segments of fast-spreading ridges experience over time.

Hydrothermal vent systems along slow spreading ridges have been examined principally from the Mid-Atlantic Ridge in the vicinity of 26°N. The TAG (Trans-Atlantic geo-traverse) vent systems differ significantly from those of the East Pacific Rise (Rona and Speer 1989; Tivey 1992). At TAG the active mound is between 200 and 250 m in

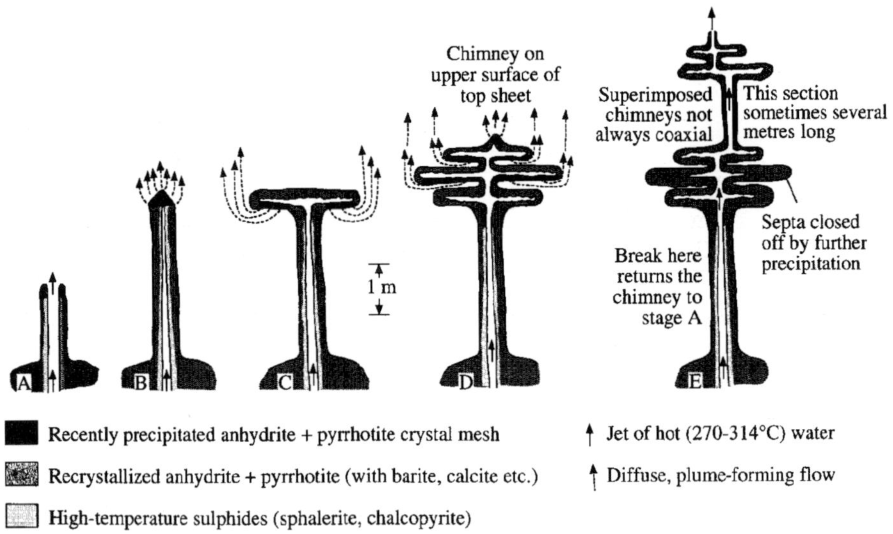

Figure 9.6 Diagram showing development of the complex spires observed in the Guaymas field, Gulf of California (Lonsdale and Becker 1985)

diameter, or about an order of magnitude larger than those observed along the East Pacific Rise. Exploration of the TAG mound by submersibles has shown that all varieties of venting activity can be found over the 200 m extent. Black smokers are found in the central and highest portions of the mound, whereas white smokers develop around a peripheral central zone. At the margins, venting usually takes the form of diffuse shimmering water, with temperatures up to 21 °C. The TAG hydrothermal mound is the only system yet discovered which is similar in size to economic ore bodies described from uplifted oceanic crustal sections.

The vents described in the previous section appear to be associated with well-established hydrothermal circulation systems in which seawater is able to penetrate deeply into the oceanic crust. Hydrothermal plumes can also be generated by more transient geologic events occurring at spreading centres. Haymon *et al.* (1993) reported disorganized venting of high-temperature fluids from the East Pacific Rise crest at 9°45′–52′N. Diffuse black to grey smoke at temperatures up to 400 °C was observed venting from piles of lava rubble on the sea-floor, but was not associated with sulphide mineral chimneys. Other evidence from the area suggested that the site had recently experienced an injection of magma in the form of dikes to <200 m beneath the sea-floor. Haymon *et al.* (1993) proposed that the venting of high-temperature fluid was caused by the circulation of seawater through a hot, permeable surface layer that was invaded by magma. They speculated that, with rapid heat loss and reduction of permeability by mineral precipitation, the system would evolve to a more stable and focused hydrothermal circulation system. Thus, the potential exists for forming large-scale intense hydrothermal circulation of relatively short duration, associated with either direct eruption of magma on to the sea-floor or its injection to very shallow levels in the crust.

9.4 OBSERVATIONS OF SEA-FLOOR VENTING

9.4.1 Submersibles

Most of the visual observations of sea-floor venting have been obtained during the research dives of the deep-submersible vessel *Alvin* and more recently *Nautille* and the remotely controlled *Jason* vehicle. These submersibles are equipped with video recording machines, cameras and on a number of dives infrared sensors have been used to make thermal images of the sea-floor venting sites. In addition to such visual observations, the submersibles are able to collect samples of the sulphide deposits and the venting hydrothermal fluid, for subsequent analysis on board their host research vessels. High- and low-temperature thermometers enable the temperature of the vent fluid and ambient fluid to be measured. As well as the analysis of video recordings, a number of more sophisticated devices have been deployed to measure the velocity of the venting fluid (Schultz *et al.* 1992). One such instrument consists of a flow channel, open at each end, which is placed above a site of venting. As the venting fluid rises through the flow channel, the conductivity of the venting fluid induces an e.m.f. (electromotive force) in an electromagnetic induction coil located in the walls of the channel. The size of this e.m.f. is directly related to the strength of the flow. Velocities of the venting fluid at the Endeavour Segment of the Juan de Fuca Ridge ranged from less than 1 mm s^{-1} in the slow diffuse flow to tens of centimetres to metres per second above the vigorous black smokers.

9.4.2 Remote Surveys of Plume Dispersal

Submersible observations of hydrothermal plumes provide useful information about the near source structure and intensity of fluid venting, but do not allow for tracking of widely dispersed plumes. In order to define the large-scale features of such plumes it is necessary to employ different types of survey techniques. These techniques are based on the ability to detect differences in temperature, composition, or particle content of plume-derived fluids relative to the background water column. One particularly useful tracer of plume activity is the dissolved gas helium. In seawater, helium has two stable isotopes: ^{4}He is produced by radioactive decay of uranium and thorium, while ^{3}He is produced by mantle outgassing during volcanic activity. In hydrothermal fluids ^{3}He/^{4}He ratios are much greater than the background ocean, reflecting the interaction of the ocean with young volcanic rocks at the mid-ocean ridge spreading centres. Water column surveys of ^{3}He^{4}He thus provide a powerful technique for mapping hydrothermal plumes. Figure 9.7 shows a vertical profile of ^{3}He on the East Pacific Rise at 21°N which identifies a hydrothermal plume (Lupton *et al.* 1980).

In order to measure the properties of the black and white smoker plumes as they ascend above the sea-floor, many studies have been made with conductivity–temperature–depth (CTD) instruments. These measure the salinity, temperature and depth of the water column at a single location. By carrying out a series of vertical casts above a site of venting and recording the average properties at each location, and by assuming the venting remains steady over the time scale of such an experiment, a three-dimensional image of the quasi-steady plumes rising from the vents may be obtained (Lunel *et al.* 1990).

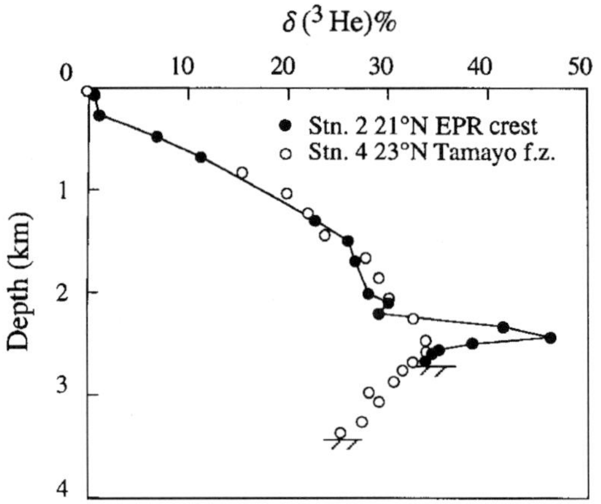

Figure 9.7 Distribution of ^{3}He in the water column on the East Pacific Rise at 21°N and the Tamayo fracture zone at 23°N (Lupton *et al.* 1980)

Baker and Massoth (1987) used deep CTD tows and water bottle sampling to map neutrally buoyant plumes from two major vent fields on the Juan de Fuca Ridge. They found that venting fluids coalesced into single 200 m thick plumes at each vent field that were characterized by temperature anomalies of 0.02–0.05 °C up to 15 km from source. Calculations of heat flux ranged from 580 to 1700 MW and particle fluxes from 92 to 546 g s^{-1}. As an example, Figure 9.8 shows temperature and light attenuation anomalies for surveys of the Southern Symmetrical Segment of the Juan de Fuca Ridge.

9.4.3 Megaplumes

Typically, plumes of diluted hydrothermal fluids above active vent sites are found to rise 100–500 m above source (Lupton *et al.* 1980; Lupton and Craig 1981; Klinkhammer *et al.* 1985). There are examples, however, of especially strong plumes whose geochemical or thermal identity can be recognized at distances up to thousands of kilometres from source (Lupton and Craig 1981; Reid 1982). A particularly large plume was detected by CTD/transmissometer over the Juan de Fuca Ridge in August 1986 (Baker *et al.* 1987). The so-called *megaplume* ascended over 1000 m above the sea-floor with a heat anomaly of about 6.7×10^{16} J. With an average mass discharge of 1.8×10^5 kg s^{-1} and heat flux of 4×10^{11} W, the megaplume represents 1000 times the flux of the entire vent field at 21°N on the East Pacific Rise.

The megaplume fluid contained a high concentration of SiO_2 and Fe, indicating hydrothermal exchange with volcanic rocks. The megaplume also contained mineral grains of anhydrite, chalcopyrite and silica that ranged up to 130 μm in diameter. The large settling velocities of these particles indicates that the plume was emplaced within a few days. By October 1986 the plume had disappeared (Baker *et al.* 1987). A second mega-plume, of about one-half the volume of the first, was discovered about 45 km north of the earlier location in September 1987 (Baker *et al.* 1989). In this plume, the largest particles were only found about 100 m below the thermal anomaly maximum, indicating substantial

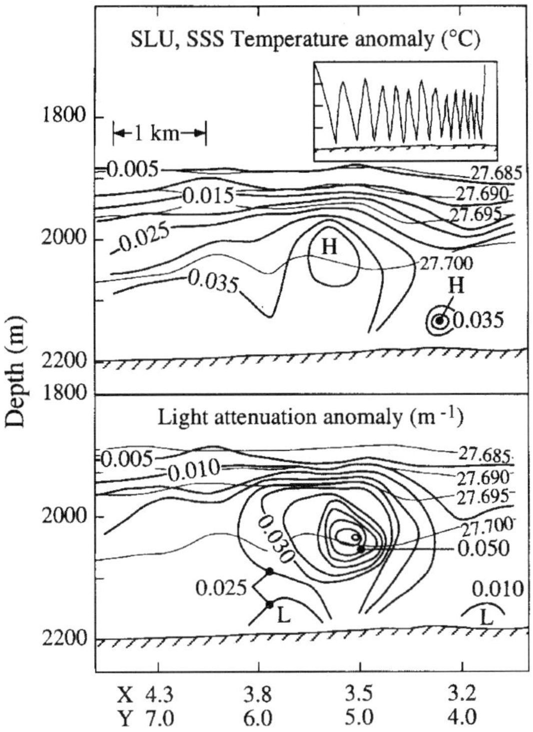

Figure 9.8 Contours of temperature and light attenuation anomalies for Juan de Fuca plumes above the SSS vent field from Baker and Massoth (1987). Temperature (°C) and light attenuation (m^{-1}) are shown as bold contours. Potential density (kg m^{-3}) is shown as thin contours. X and Y are map coordinates

settling of particles, and lending support to the hypothesis that megaplumes are generated by separate short-lived events (several days). The megaplumes have been dated using constraints from scavenging rates of microorganisms and through radiometric dating (D'Asaro *et al.* 1994). These dates support the short-lived nature of these events. The short duration of megaplumes is quite distinct from many other active vent systems which appear to discharge hydrothermal plumes steadily for years or decades (Baker 1994).

Sealing of near surface conduits by mineral precipitation can lead to reduced permeability and the development of a clogged cap to the hydrothermal system. Megaplumes could represent the catastrophic emptying of such a reservoir of hydrothermal fluid following failure of the cap (Baker *et al.* 1989; Cann and Strens 1989). As the temperature of the subsurface fluids increases, their buoyancy increases, thereby exerting an increasing pressure on the cap. Alternatively, a tectonic event may dramatically increase the permeability of the shallow crust (Cann and Strens 1989).

9.4.4 Bubble Plumes

In shallow-water environments (shallow ridges, rift lakes and crater lakes) plumes can contain significant quantities of gas, in particular methane and CO_2. The Steinhaholl plume on the Reykjanes Ridge south of Iceland (German *et al.* 1994) discharges from 350 m water depth and shows vigorous bubbling to the sea surface. This plume intrudes at two different levels in the ocean (Ernst *et al.* 1996b) and this behaviour is a consequence of the bubbles in the plume (Asaeda and Imberger 1993). In volcanic crater lakes, bubble hydrothermal plumes can lead to hazardous eruptive outgassing of CO_2 as modelled by Zhang (1996). For example, at Lake Nyos in Cameroon in 1986, saturation of the dense lower layer of the lake by CO_2 resulted in a violent degassing column that erupted at velocities of 50–100 m s^{-1} through the lake surface and formed a CO_2-rich collapsing jet fountain laden with water drops that was dispersed by prevailing winds. This event killed 1700 people. The dynamics of bubble plumes show significant differences from single-phase plumes (Asaeda and Imberger 1993).

9.5 PARTICLES IN HYDROTHERMAL PLUMES

As with volcanic plumes which discharge into the atmosphere, hydrothermal plumes also carry suspended particles as they rise into the water column. The concentration of the particles is generally insufficient to affect the buoyancy of the plume. The “smoke” of black and white smokers consists of fine-grained particles of sulphides, sulphate and oxides. Some of the particles are entrained from the vent chimneys and the ambient seawater. The majority, however, are formed within the plume fluid by precipitation as the fluid approaches the surface and vents to the sea-floor. Because the plumes are usually turbulent (Chapter 3; Figure 9.2), a key process in the discharge and development of plumes is the entrainment of ambient seawater. This entrainment triggers changes in temperature, pH, and oxidation potential of the fluids. In turn this results in the precipitation of mineral phases from the dissolved components that have been acquired by high-temperature reaction with basaltic rocks at depth.

Direct sampling of plume fluids from hydrothermal vents along the Juan de Fuca Ridge has provided important information about the particles that are transported by plumes (Feely *et al.* 1987). The principal mineral phases in black smokers include sphalerite, wurtzite, pyrite, pyrrhotite, chalcopyrite, cubanite, anhydrite, and hydrous iron oxides. They range in size from <2 μm to as large as 500 μm. White smoker particles, on the other hand, are dominated by amorphous silica, pyrite, barite and anhydrite of roughly the same size range as found in the black smokers (Haymon and Kastner 1981).

An important feature of the particles that are formed in hydrothermal plumes is that they are metastable in ambient seawater at spreading centres and thus begin to dissolve once they leave the plume (Feely *et al.* 1987). Dissolution rates of different mineral phases have been determined by *in situ* and laboratory studies. The rates vary over four orders of magnitude, being lowest for sulphides and highest for sulphates. Anhydrite in particular is rapidly dissolved in ambient seawater. A 50 μm diameter crystal will be completely dissolved in about a day.

9.6 DYNAMICS AND THERMODYNAMICS OF HYDROTHERMAL PLUMES

Hydrothermal plumes are driven by the buoyancy of the hot fluid which issues from the sea-floor. Figure 9.9 shows how the coefficient of thermal expansivity of water varies with

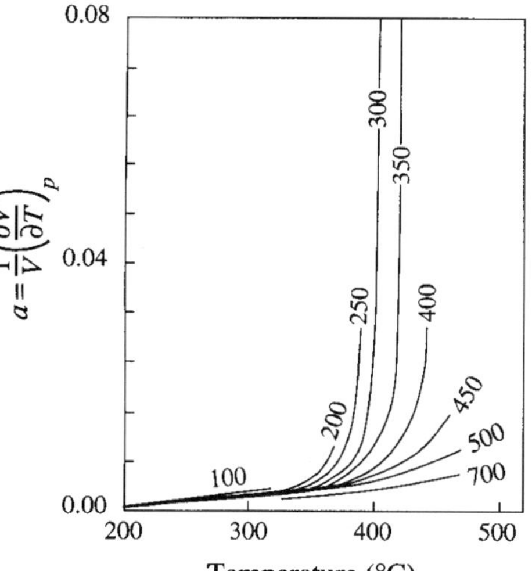

Figure 9.9 Variation of the thermal expansivity of brine γ_T with temperature. Note the rapid increase in expansivity near the critical point (Bischoff and Rosenbauer 1985). Numbers on curves correspond to the pressure in bars

temperature (Bischoff and Rosenbauer 1985). Curves are given for several different pressures. Since the thermal expansion coefficient of water increases rapidly with temperature, especially near the critical point, the density of the hot vent fluid, at temperatures over 300 °C, is much lower than that of the cold overlying seawater. However, higher in the plume, once the vent fluid has mixed with a large volume of seawater, the temperature and coefficient of thermal expansivity decrease significantly. At this stage, differences in salinity between the hydrothermal fluid and the seawater tend to dominate the relative density (Turner and Campbell 1987b). If the vent fluid is more saline, it can result in the plume becoming denser than the environment, even though it was initially much less

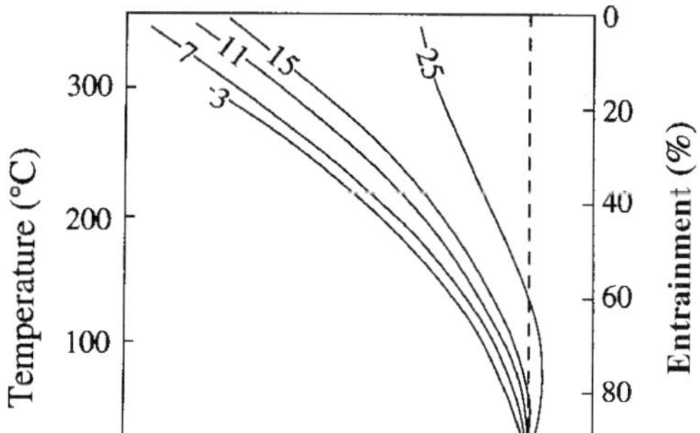

Figure 9.10 Variation of the density of brine with temperature, showing the buoyancy reversal which develops as the hydrothermal fluid entrains seawater of temperature 2 °C and salinity 3–2 wt%. Numbers on curves correspond to the initial salinity of the fluid, after Turner and Campbell (1987b)

dense. As shown in Figure 9.10, the density of hydrothermal fluid increases on mixing with ambient fluid.

The salinity contrast between the vent fluid and the ambient ocean varies between sites of venting and therefore a model is developed to describe both the situations in which the fluid remains buoyant and forms a large convecting plume, and the situation in which the buoyant plume becomes dense on mixing, and therefore forms a collapsing fountain (Caulfield and Woods 1995). The fluid mechanical processes associated with these two situations have been studied in some detail (Chapter 2). Morton *et al.* (1956) developed a model of a simple buoyant plume, and using MEG–water mixtures (Chapter 2), Turner (1966) extended the work to consider plumes in which the density of the plume fluid relative to the environment changes sign, thereby forming a collapsed fountain (Chapter 2). Although it is possible that such buoyancy reversal and collapse occurs in hydrothermal plumes, this has not yet been recognized in the natural environment.

9.6.1 Initial Conditions

Several models of the motion of hydrothermal plumes have been developed from the integral model of Morton *et al.* (1956), which was described in Chapter 2. Just above the smoker, where the thermal expansion coefficient and hence density vary rapidly, the buoyancy flux of the plume is not conserved. However, above an initial transition region in which the plume mixes and cools, the motion of a black smoker plume becomes very similar to the motion of a simple buoyant plume in a stratified environment. Turner and Campbell (1987b) have proposed a simple method of estimating the height of rise of such a plume, incorporating the effect of the increase in the thermal expansion coefficient and associated variation in the buoyancy flux. Since the enthalpy flux of the plume is conserved as it mixes, the height of rise should scale as the buoyancy flux associated with a plume issuing from a source of the same enthalpy flux, but only about 1 °C hotter than the ambient ocean. The effective buoyancy flux, B, can then be calculated using the essentially constant low-temperature thermal expansion coefficient

$$F = [C_p(T_p)T_p - C_p(T_o)T_o]Q \tag{9.1}$$

$$B = \frac{\gamma_T F g}{C_p(T_o)\rho_o} \tag{9.2}$$

Here F is the heat flux, C_p the specific heat, T_p the plume temperature, T_o the ambient water temperature and Q the mass flux in the plume. The buoyancy flux B is linearly related to the thermal expansion coefficient γ_T, and the gravitational acceleration g; ρ_o is a reference density taken to be that of the ambient water at the source. Using this estimate of the effective buoyancy flux of the plume, the height of rise of the equivalent plume H scales as (Chapter 2)

$$H = 5B^{1/4}N^{-3/4} \tag{9.3}$$

where N is the Brunt–Väisälä frequency of the environment (Chapter 2) and H is in metres. Figure 9.11 shows heights in the Pacific as a function of the heat source at the vent, as predicted by this model. The stratification of the environment N is in fact evaluated relative to the adiabat, since as the plume fluid rises, the pressure changes.

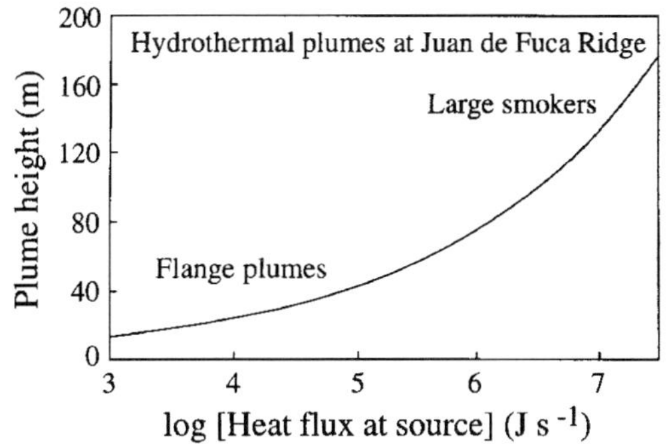

Figure 9.11 Height of rise of hydrothermal plumes in the Pacific as a function of the heat flux, based upon the simple theory of Morton *et al.* (1956) (Woods and Delaney 1992)

9.6.2 Plume Models

A more sophisticated model for hydrothermal plumes, rising in a still environment, is developed by introducing conservation of enthalpy and salinity as fundamental equations (Chapter 2). Hydrothermal plumes are modelled according to the equations

$$\frac{d\beta ub^2}{dz} = 2\varepsilon bu\alpha \tag{9.4}$$

$$\frac{d\beta u^2b^2}{dz} = g(\alpha - \beta)b^2 \tag{9.5}$$

$$\frac{d\left(\beta ub^2(C_pT + \frac{u^2}{2} + gz)\right)}{dz} = 2\varepsilon bu\alpha(C_pT_e + gz) \tag{9.6}$$

$$\frac{d(\beta ub^2S)}{dz} = 2\varepsilon\alpha buS_e \tag{9.7}$$

$$\beta = f(T, S) \tag{9.8}$$

which represent the conservation of mass, momentum, enthalpy and salinity fluxes as a function of height in the plume z and an equation of state for the density as a function of the salinity S and temperature T. As in Chapter 4, u is the velocity in the plume, b the plume radius, β the plume density, while α is the ambient density, ε the entrainment coefficient and T_e the temperature of the environmental fluid. This model is analogous to the model of an eruption column, with the equation for the conservation of salinity replacing that for the conservation of clasts (Chapter 4).

In hydrothermal plumes, the mixture erupting from the smoker can be very hot, and therefore may be of very low density, sometimes as low as 600–700 kg m^{-3}. However, as soon as the mixture has entrained a sufficient mass of seawater that the temperature falls below 30–40 °C, the density of the plume water becomes very similar to the density of the ambient seawater, and the Boussinesq approximation can be made (Chapter 2). If the specific heat remains constant and the kinetic energy is small relative to the thermal and potential energy (since the typical velocities are 0.1 m s^{-1} and the temperature anomalies are greater than 0.01 °C), then the equations simplify to the following form (Speer and Rona 1989):

$$\frac{dub^2}{dz} = 2\varepsilon ub \tag{9.9}$$

$$\frac{du^2b^2}{dz} = \frac{gb^2(\beta - \alpha)}{\rho_o} \tag{9.10}$$

$$\frac{d}{dz}(ub^2(\theta - \theta_e)) = -ub^2\frac{d\theta_e}{dz} \tag{9.11}$$

$$\frac{dub^2(S - S_e)}{dz} = ub^2\frac{dS_e}{dz} \tag{9.12}$$

$$\beta = \alpha - \gamma_T(\theta - \theta_e) + \gamma_S(S - S_e) \tag{9.13}$$

where γ_T and γ_S are the thermal and solutal expansion coefficients (Turner and Campbell 1987b), and the dependence of density upon temperature and salinity has been linearized. The variation of the density with pressure is included in the constitutive relation by working with the potential temperature θ rather than absolute temperature T (Speer and Rona 1989; Gill 1982).

In Figure 9.12, some simple solutions of this model of a hydrothermal plume are presented, as calculated by Speer and Rona (1989), in which the evolution of the salinity and temperature in the plume are shown. By combining the last three equations in this limiting case, the simple equations of Morton *et al.* (1956) for a buoyant plume are derived. McDougall (1990) has analysed this coupled model in some detail to predict various properties of the plume as it mixes with the ambient fluid.

There is some initial adjustment at the source of a black smoker plume in which the temperature decreases rapidly and the buoyancy varies non-linearly with variations in the temperature. However, if the linearized plume model (equations 9.9–9.13) is initialized using Turner and Campbell's (1987b) effective source conditions, then the predictions are virtually indistinguishable from the full model (equations 9.4–9.8). As a simple rule, once a black smoker plume has entrained about 10 times its mass in ambient water, it will behave as a simple buoyant plume. Lunel *et al.* (1990) have measured how the concentration of aluminium, a passive tracer, varies with height in the plume. Their field measurements agree with the simple predictions of the plume model (equations 9.9–9.13), confirming its relevance.

9.6.3 Diffuse Plumes

Recently it has become clear that diffuse hydrothermal flows are as important as the focused flows that form black smoker plumes. Diffuse hydrothermal plumes form by the coalescence of a large number of microplumes that emerge from fractured sea-floor and porous sediments in the vicinity of active hydrothermal systems. Individually, the microplumes exhibit characteristics similar to a large buoyant plume (Converse *et al.* 1984;

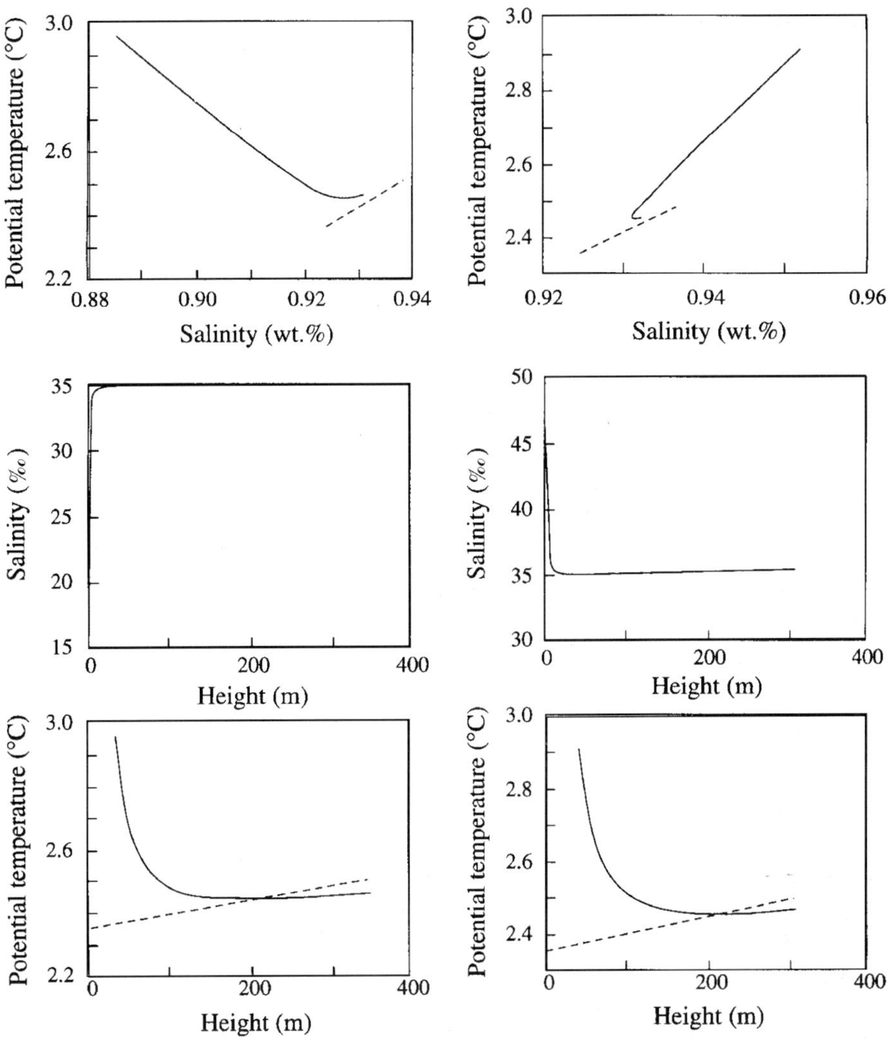

Figure 9.12 Solutions of the plume equations showing the temperature and salinity evolution with height in the plume (from Speer and Rona 1989). The two sets of curves correspond to plumes in the Atlantic, the left-hand one having an initial salinity of 15‰ and the right-hand side being 50‰. The dotted lines correspond to the temperature–salinity relationship in the ocean

Monfort and Schultz 1988). However, the typical spacing of these microplumes in vent fields indicates that they will form a single diffuse plume at heights of 1–10 m above the sea-floor (Trivett 1994). Venting of diffuse plumes may contribute significantly to the total heat flux in hydrothermal areas because although the temperature anomalies are low compared to black smokers, the large areas of diffuse plumes are likely to compensate for this difference (e.g. Rona and Trivett 1992).

Measurements of vertical velocities in diffuse plumes are of the order of a few to tens of centimetres per second and thus the plumes are significantly affected by flow parallel to the sea-floor, such as tidal or other types of bottom currents. Trivett (1994) has developed a simple model for the behaviour of diffuse plumes in a cross-flow at hydrothermal fields. The model assumes that a single diffuse plume forms from the coalescence of micro-plumes and rises into a unidirectional horizontal flow field. Because vertical velocity and buoyancy flux are weak, the plume is rapidly advected downfield. Entrainment of ambient fluid is constrained to occur only on the top and outer sides of the bent-over plumes. Using conservation equations for continuity, momentum and concentration, Trivett (1994) arrived at a relationship for the plume height as a function of distance from source:

$$Z(x) = \frac{-2g}{N^2}\rho_o^*\left(1 - e^{-\omega x \frac{\sin(\lambda x + \phi)}{\sin\phi}}\right) \tag{9.14}$$

where

$$\tan\phi = \frac{2\lambda\omega}{\omega^2 - \lambda^2} \tag{9.15}$$

and

$$\lambda = \frac{N}{U} \tag{9.16}$$

and

$$\omega = \left(\frac{W_e + W_o}{UD_v} + \frac{2V_e}{UD_v}\right) \tag{9.17}$$

with N the buoyancy frequency for ambient conditions, U the cross-flow velocity in the x-direction, W_e the entrainment velocity in the z-direction across the top of the plume, W_o the average vertical velocity of the diffuse plume at source, V_e the entrainment velocity in the y-direction across the side of the plume, and D_v the diameter of the source vent. The density term in equation (9.14) is expressed as a fractional difference compared to the ambient density:

$$\rho_o^* = \frac{\rho_u - \rho_o}{\rho_u} \tag{9.18}$$

where ρ_u is the density on the sea-floor in the undisturbed ambient density profile and ρ_o the density in the plume.

The width of the plume, $Y(x)$, spreads in a linear fashion simply as a result of the entrainment of ambient fluid and is given by

$$Y(x) = \frac{4V_e^2 x}{\omega D_v U^2} + \frac{D_v}{2} \tag{9.19}$$

Trivett (1994) tested equations (9.14) and (9.19) by carrying out laboratory simulations of diffuse plumes with a cross-flow. Very good agreement was found between the predicted plume heights and the observed heights in the experiments. An important conclusion of this work is that entrainment in weak diffuse plumes is not driven by the centreline velocity. A smaller proportion of entrainment occurs per unit distance because entrainment only occurs on the top and sides of the plume as it moves downcurrent over the sea-floor.

9.7 PROPERTIES OF THE PLUME AND NEUTRALLY BUOYANT INTRUSION

The height of rise of hydrothermal plumes depends largely upon the temperature and salinity gradients in the overlying ocean. As in subaerial volcanic plumes, hydrothermal plumes reach a level of neutral density and spread out as a lateral intrusion. In the Atlantic Ocean, the deep bottom water is stably stratified. Relative to the adiabat, both the salinity and temperature increase with height above the sea-floor and therefore the temperature acts as the stabilizing agent. As a result, fluid emplaced as an intrusion above the level of neutral buoyancy will be relatively fresh but cold, since the plume convects the cold fresh deep water upwards. In contrast, in the Pacific Ocean, the salinity decreases and temperature increases relative to the adiabat. Therefore both fields stabilize the density of the water column (Figure 9.13). As a result, the fluid emplaced in the neutrally buoyant intrusion is relatively saline but warm, because both the ambient salinity and temperature tend to decrease the buoyancy of the plume as ambient fluid is entrained. Therefore, the plume is relatively hot at the neutral buoyancy height in order to compensate for the greater salinity of the fluid entrained from below.

In the ocean the dynamics of the neutrally buoyant intrusions can be much influenced by the Earth's rotation. This is important for the dispersal of particulates and larvae. In a series of analogue laboratory experiments in which buoyant plumes were generated in a rotating stratified tank, it was found that as the neutrally buoyant intrusion spreads out, the effects of rotation become increasingly important. Once the intrusion had spread sufficiently that the Rossby number, $Nd/\Omega R$, reached values of the order of 0.5–0.75, the intrusion broke into two rotating clouds (Helfrich and Battisti 1991). Here Ω is the rotation rate of the

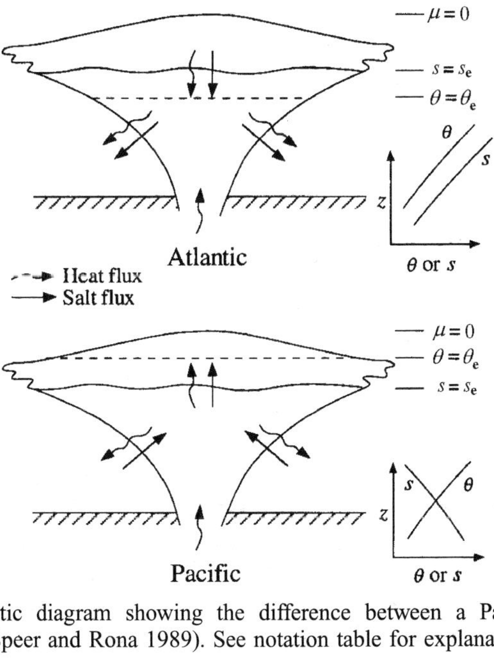

Figure 9.13 Schematic diagram showing the difference between a Pacific and an Atlantic hydrothermal plume (Speer and Rona 1989). See notation table for explanation of symbols

Earth, 10^{-4} s^{-1}, d the thickness and R the radius of the neutrally buoyant intrusion, and N the Brunt–Väisälä frequency of the ocean, 10^{-3} s^{-1}. As these discrete intrusions spiral outwards from the plume and begin to move downstream, a new spreading intrusion can form above the plume, and the process repeats. These large coherent plumes typically have lifetimes of months to years. They can carry particles and larvae in discrete pulses for distances of tens to hundreds of kilometres irrespective of ocean currents. The particles and larvae gradually become mixed into the water column by turbulence and tidal motions.

9.7.1 Effects of Abyssal Cross-flows

A complicating factor in the behaviour of hydrothermal plumes is their response to cross-flows from bottom currents (tidal or thermohaline). In section 9.6.3 the influence of a steady unidirectional cross-flow on the rise and dispersal of diffuse plumes was discussed in the context of the model of Trivett (1994). For more vigorous plumes, such as black smokers, cross-flows can have a number of potential effects. For example, cross-flows can cause bifurcation of plumes to create two lobes that separate in the downcurrent direction (Ernst *et al.* 1994, and Chapter 11). Bifurcation can also be enhanced by cross-currents sweeping around obstacles and by interactions with major density interfaces such as the thermocline (Ernst *et al.* 1994, 1996b). Bifurcation has been recognized in the bubble-rich plume at Steinaholl (Ernst *et al.* 1996b). Complex ocean currents and restriction by topography (e.g. the walls of an axial rift valley) are also expected to influence plume and intrusive behaviour. Rudnicki *et al.* (1994) ascribed an observed 150 m change in the non-buoyant plume height above the TAG hydrothermal vent to variations in the tidal bottom currents over a nine-hour period. They proposed that currents up to 15 cm s^{-1} caused significant bending over of the plume with a subsequent reduction in the neutral buoyancy height.

The effects of a cross-flow on a black smoker plume have been considered on a theoretical basis by Middleton and Thompson (1986). Adapting the treatment of Slawson and Csanady (1971) for bent-over turbulent convective plumes, they arrive at a solution for the plume height given by

$$z = \left(\frac{6B}{\varepsilon^2 UN^2}\right)^{1/3} \tag{9.20}$$

or a stratified environment where B is the buoyancy flux, ε the entrainment coefficient (~ 0.33), U the cross-flow velocity, and N the Brunt–Väisälä frequency. This relationship predicts lower plume heights compared with a buoyant plume in a still environment. Additional discussion about the effects of cross-flows on plumes can be found in Chapter 11.

9.8 FALLOUT OF PARTICLES FROM HYDROTHERMAL PLUMES

Hydrothermal plumes carry particles upwards through the water column. As the plume vertical velocities diminish with height the particles are no longer supported. They are then lost from the plume and begin to settle to the sea-floor. An important difference between hydrothermal plumes and subaerial volcanic eruption columns is that in the former the size of the particles changes as a function of time as a result of precipitation or dissolution and so the settling velocity of particles will be a function not only of their size but of their age.

One of the problems with evaluating the process of particle fallout from hydrothermal plumes is that the resulting deposit is subject to dissolution and thus its grain size and thickness will change subsequent to deposition. In principle, models for the accumulation should take into account the evolving nature of the particles. To date, however, no attempt has been made to integrate fully the time-dependent evolution of particles with a theoretical depositional model to predict the nature and distribution of fall deposits from hydrothermal vents. Feely *et al.* (1987) have developed a model of particle fallout from hydrothermal plumes for particles whose dissolution time is much greater than the particle residence time in the plume. The model assumes particles are released predominantly at the top of the plume and are then advected by time-dependent currents and spread out owing to the turbulent diffusivity. Using input data from a hydrothermal vent on the southern Juan de Fuca Ridge, Feely *et al.* (1987) modelled the depositional pattern for pyrite particles ranging from 25 to 100 μm in diameter (Figure 9.14). The model predicts that large particles ($\sim$ 100 μm) will be restricted to an area within a few hundred metres of the vent in good accord with the sulphide particles collected at various positions around the active vent using sediment traps.

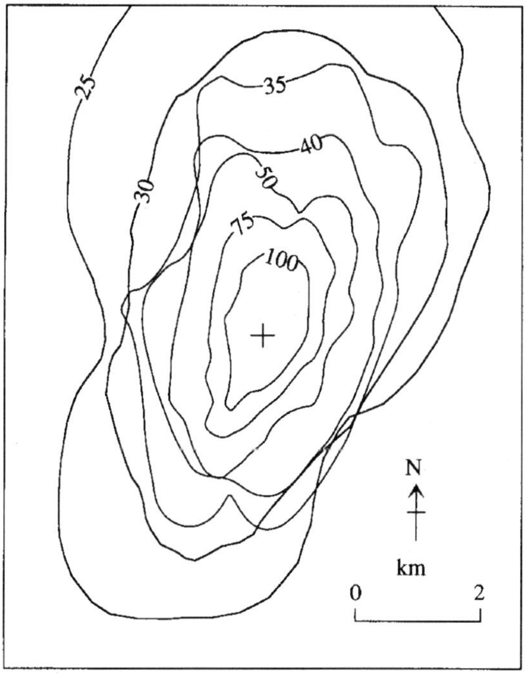

Figure 9.14 Predicted distribution of pyrite crystals around a hydrothermal vent on the Juan de Fuca Ridge based on a particle fallout model of Feely *et al.* (1987). Contours of particle size in micrometres show the areas where 99% deposition would occur. Deposition of material all around the vent and the overlap of some particle contours is the result of a strong tidal component in the bottom current data used for the modelling

A complicating factor in the deposition of particles from hydrothermal plumes, and other particle-laden plumes as well, is the re-entrainment of particles into the plume by inward flow as they settle close to the plume margin (Sparks *et al.* 1991; German and Sparks 1993). Isotopic analysis of plume samples has shown pronounced $^{234}Th/^{230}Th$ isotopic fractionation. German and Sparks (1993) have attributed this fractionation to recycling of old particles within the plume. Using a model of particle sedimentation from turbulent convective plumes they argued that particles within 2–3 km of the vent would be recycled. Lane-Serff (1995) showed that for the conditions of no ambient cross-flow the proportion of fine particles re-entrained into vertical plumes does not depend on particle size. He calculated that 61% of particles should be re-entrained. Ernst *et al.* (1996a) developed a model for fallout from turbulent vertical plumes in a still environment to account for particulate recycling. This model was applied to hydrothermal plumes, the key result being that the majority of hydrothermal plume particulates are too small for much fallout to occur from the vertical column. As a result most of the particles are taken up into the neutrally buoyant plume. An important but unsolved problem concerns the interaction of the effects of currents and of rotation upon this re-entrainment process. Rotation can cause the instability of the neutral buoyant plume and both effects may suppress the long-range re-entrainment.

9.9 SUMMARY

Hydrothermal plumes are formed when circulating fluids exchange heat with a magmatic source and are subsequently discharged into a cooler ambient fluid such as the ocean or a crater lake. The ocean ridge system is an environment where a variety of hydrothermal plumes occur in response to the penetration and heating of seawater into newly formed oceanic crust. Venting of fluids can take many different forms from focused high-temperature (350 °C) black smokers that produce classic turbulent convective plumes to low-energy diffuse plumes that discharge lower temperature water (< 20 °C) over large areas. The vent fields play host to exotic biological communities that are based on chemosynthesis. Remote sensing of plumes indicates that venting can occur on a quasi-steady-state basis in focused discrete plumes and in diffuse flows or be characterized by rapid discharges that produce megaplumes. In shallow water, plumes can contain significant amounts of bubbles.

Some hydrothermal plumes are similar to explosive volcanism plumes in that they carry particles suspended by the vertical velocity of the plume fluid. However, in hydrothermal plumes these particles form *in situ* by chemical precipitation that is triggered by entrainment of ambient fluid. The particles consist mostly of fine-grained sulphides, sulphate and oxides. Such particles are subject to dissolution once they leave the plume and thus the size of particles during transport and deposition evolves with time. An important process affecting the fallout of particles from hydrothermal plumes is the re-entrainment of particles falling from the upper levels of a plume. Over one-half of the particles that escape from rising plumes can be swept back into the plume at lower levels as a result of the net inward flow of ambient fluid by turbulent eddies along the plume margins.

In the ocean, gradients in temperature, density and salinity result in hydrothermal plumes reaching a level of neutral buoyancy and spreading out as a lateral intrusion in much the same way that volcanic plumes in the atmosphere generate an umbrella region. The behaviour and dynamics of "black smoker" hydrothermal plumes can be modelled using the classic treatment of turbulent convective plumes where entrainment is related to the centreline velocity. For diffuse plumes, entrainment of ambient fluid is reduced because the plume is easily advected away from source and entrainment is restricted to only the top and sides of the plume.

Factors that complicate the behaviour of hydrothermal plumes in the ocean include cross-currents and the Earth's rotation. Cross-currents act to distort the rising plume, reducing its maximum rise height and influencing the pattern of particle fallout. If the cross-currents are sufficiently strong bifurcation of the plume can occur. Laboratory experiments have demonstrated that rising plumes in a rotating fluid will form discrete intrusions that spiral outward and separate from the main plume. Coriolis effects will therefore impart rotation to hydrothermal plumes and result in secondary flow fields in the vicinity of active vents. The shedding of large coherent spiralling plumes can transport particles and larvae for distances up to hundreds of kilometres from source.

10 Basaltic Eruptions and Fire Fountains

NOTATION

C_a	specific heat of air ($J\ kg^{-1}\ K^{-1}$)
C_m	specific heat of magma ($J\ kg^{-1}\ K^{-1}$)
C_v	specific heat at constant volume ($J\ kg^{-1}\ K^{-1}$)
D	length of the active fissure (m)
$e_s(T)$	saturation vapour pressure (Pa)
g	gravitational acceleration ($m\ s^{-2}$)
H	height of rise of a line plume above a fire fountain (m)
L	half-width of the fissure (m)
m	mass flux of a fountain ($kg\ s^{-1}$)
m_a	mass flux of air entrained into a fountain ($kg\ s^{-1}$)
n	gas mass fraction in plume above the fire fountain (dimensionless)
P	pressure in the atmosphere (Pa)
q	total water mass fraction in the plume (dimensionless)
q_a	water vapour mass fraction in the air (dimensionless)
q_v	water vapour mass fraction in the plume (dimensionless)
q_w	liquid water mass fraction in the plume (dimensionless)
Q	heat flux per unit length along a fissure ($J\ kg^{-1}\ m^{-1}$)
R	relative humidity of the air (dimensionless)
R_a	gas constant for air ($J\ kg^{-1}\ K^{-1}$)
R_v	gas constant for water vapour ($J\ kg^{-1}\ K^{-1}$)
T_o	initial temperature of the erupting material (K)
T_a	temperature of the air (K)
ΔT	temperature decrease in fire fountain (K)
u	upward speed of the plume ($m\ s^{-1}$)
V	volume eruption rate ($m^3\ s^{-1}$)
w	mass fraction of water vapour in the gas phase (dimensionless)
x	mass fraction of fine ash in fire fountain entrained into the overlying convective plume (dimensionless)
y	mass fraction of volatiles in fire fountain entrained into the overlying convective plume (dimensionless)
z	vertical distance (m)
α	density of the air ($kg\ m^{-3}$)
β	bulk density of the plume ($kg\ m^{-3}$)
ε	entrainment constant (dimensionless)

Γ latent heat of vaporization of water (J kg^{-1})
ρ density of magma (kg m^{-3})

Subscripts

o property evaluated at the vent of the fire fountain

10.1 INTRODUCTION

Plumes associated with explosive basaltic eruptions are of considerable interest, because of the substantial amounts of gas, in particular SO_2, that can be released into the atmosphere. For example in major flood basalt eruptions, such as those that formed the Columbia River basalt group and the Deccan traps (Rampino *et al.* 1988), nearly an order of magnitude more sulphur gas was produced per unit mass erupted than in highly explosive silicic eruptions (Devine *et al.* 1984; Palais and Sigurdsson 1989). Basaltic eruption columns developed above energetic fire fountains can therefore be responsible for the injection of sulphur gases into the upper troposphere or lower stratosphere where sulphuric acid aerosols form. These aerosols increase the optical depth of the atmosphere and can perturb climate (Stothers *et al.* 1986; Thordarson and Self 1993; Chapter 18).

Eruptions of basaltic volcanoes show characteristics which are fundamentally governed by the relatively low magma viscosity. Exsolving volatiles can easily segregate from the magma so that eruptions of lava often dominate volumetrically over eruptions of pyroclastic material. Explosive, magmatic eruptions are often weak to moderate in intensity and the ejecta tend to be coarse-grained. Fire fountains are the most common manifestation of explosive degassing, and much of their ejecta consist of centimetre to metre-sized clasts of ductile material that coalesce on landing to form clastogenic lavas and welded spatter deposits (agglutinates), or accumulate as coarse-grained deposits of volcanic bombs and scoria. Most of these ejecta accumulate within a few hundred metres of the vent to form various landforms such as cinder cones, spatter cones and spatter ramparts.

Volcanic plumes can form above basaltic vents. These plumes are usually much less vigorous than those observed in explosive eruptions of more viscous magmas. The principal reason for this is that most of the heat in a fire fountain is held in the coarse ejecta, which fall from the plume relatively rapidly, so that only a small fraction of the thermal energy flux from the vent is partitioned into the plume (Figure 10.1). Nevertheless, basaltic plumes are an important phenomenon in some circumstances, particularly in the context of the atmospheric effects of volcanism. If the discharge rate of magma is high, as in the case of a major fissure or flood basalt eruption, then even if only a small fraction of the thermal energy held in the large clasts is used, the resulting plume can be sufficiently energetic to pierce the tropopause. There are also some circumstances, notably when magma–water interactions occur, in which the fragmentation of basaltic magma is enhanced so that a more substantial fraction of the thermal flux is utilized to drive the plume. Basaltic magmas also tend to have a higher sulphur content than more evolved magmas.

This chapter describes the major controls on basaltic plumes formed from magmatic eruptions. Basaltic eruptions involving extensive interaction with external sources of water are not considered here, but are described in Chapter 9. There are rare examples of basaltic

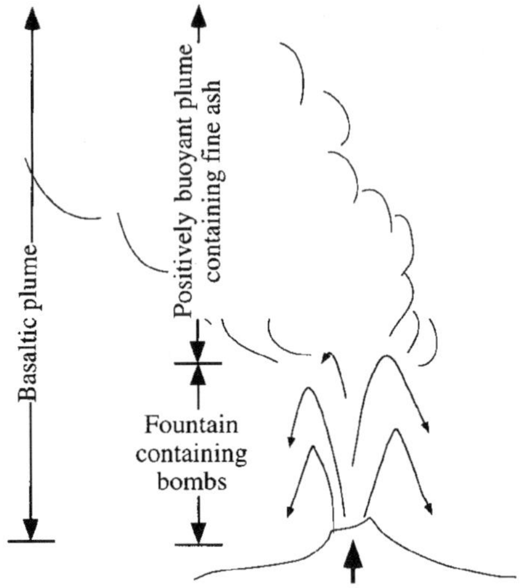

Figure 10.1 Schematic diagram of a fire fountain and fine ash plume rising above the fountain

Plinian and ignimbrite eruptions in which the fragmentation has been anomalously extensive (Williams 1983; Walker *et al.* 1984). There are no intrinsic differences between such eruptions and Plinian eruptions involving more evolved magma, so that the analysis of Plinian eruptions presented in earlier chapters holds for Plinian and ignimbrite eruptions of basaltic magma.

10.2 DEGASSING PHENOMENA IN BASALTIC ERUPTIONS

A fundamental difference between basaltic magma and more viscous intermediate and silicic magmas is that the gas phase can more easily separate from the silicate liquid because of its low viscosity. Indeed, Wood and Cardoso (1997) have shown that such bubble segregation may trigger basaltic eruptions. Segregation can occur in the magma chamber (Vergniolle and Jaupart 1990; Woods and Cardoso 1997), in the conduit (Blackburn *et al.* 1976; Wilson and Head 1981; Vergniolle and Jaupart 1986; Head and Wilson 1987), in lava lakes and fire fountains (Head and Wilson 1989; Wilson *et al.* 1995; Parfitt and Wilson 1995) and may continue in the lava flows (Swanson and Fabbi 1973). The ease of segregation in a particular instance is probably related to several interrelated factors. The values of diffusion coefficients of gases dissolved in basaltic magma are higher than those for more evolved magmas because of the higher temperature. Thus, bubbles can grow to a larger size in a shorter time (Sparks 1978a; Toramoru 1989; Proussevitch *et al.* 1993). Bubbles are also able to coalesce more rapidly (Sahagian *et al.* 1989; Vergniolle and Jaupart 1990; Mangan *et al.* 1993; Parfitt and Wilson 1995), thereby becoming larger and enhancing their mobility since large bubbles are able to rise at velocities much higher than that of the surrounding magmas.

In some cases the magmatic liquid itself may not be rising, and explosive activity will be entirely related to the ascent and bursting of large bubbles. Volcanoes such as Stromboli, Italy, and Villarrica, Chile, have magma-filled conduits that are normally open to the surface. These volcanoes display almost continuous degassing activity as bubbles rise to the surface and burst to cause Strombolian activity and fire fountaining.

10.2.1 Gas Content

Gas contents of basaltic magmas tend to be low. Ocean ridge basalts and Hawaiian lavas have water contents typically less than 0.01 mass fraction, whereas arc basalts can have higher water contents perhaps greater than 0.02 mass fraction (see Chapter 1). Although water is usually the dominant gas, basalts can also contain significant amounts of CO_2 and sulphur gases.

A consequence of the ease of segregation of gas from magma in basalts is that the mass fraction of gas can be highly variable during basaltic eruptions. Some batches of magma can be strongly depleted in gas and so erupt quietly as lava. Other batches of magma can be gas rich because of the physical concentration of exsolved gas bubbles. Vergniolle and Jaupart (1990) have shown that gas bubbles can accumulate at the roof of a magma chamber to form a foam in which the mass fraction of gas is much higher than it was in the original magma. They argued that repeated foam collapse and ascent of the associated gas-rich batches of magma can explain the cyclic patterns of fire fountaining in the Pu'u O'o-Kupaianaha eruption of Kilauea, Hawaii. In the 1973 eruption of Heimaey volcano, Iceland, as in many other basaltic events, the explosive activity was observed to wane with time, and this has been interpreted as indicating that gases were concentrated in the early erupted magma (Blackburn *et al.* 1976). Large gas bubbles rise faster than magma and so tend to be concentrated in the early phases of an eruption. Blackburn *et al.* (1976) estimated that Strombolian explosions on Heimaey in 1973 involved gas mass fractions several times higher than the original dissolved magmatic gas content. In the case of an open conduit system the bursting of bubbles rising to the surface of the lava column can result in fountains dominated by gas to pure gas jets.

10.2.2 Viscosity

Viscosity exerts another important control on the style of basaltic degassing. Basalts vary considerably in rheological properties from very fluid primitive basalts through to higher viscosity evolved basalts and basaltic andesites. There is empirical evidence and some theoretical reasoning to indicate that these variations can have a dominant influence on eruptive style.

Because of the low viscosity and small gas content of primitive basaltic magmas, the gas segregation processes are very effective. Magma is disrupted into fluid lumps of generally large size and the proportion of ash and lapilli-sized ejecta is small to negligible. For example, grain size data for the deposits of the 1959 eruption of Kilauea Iki (Cas and Wright 1987) indicate that only a few per cent by mass of the ejecta were ash grade (<2 mm diameter). As shown in more detail below, Hawaiian fire fountains form very weak plumes because only a very small fraction of the thermal energy flux is partitioned into the plume after the large magma fragments fall out.

In basalts that are more evolved or have higher original gas contents, explosive activity becomes more intense and Strombolian eruptions take place. The higher viscosity of the magma suppresses gas segregation, leading to development of somewhat higher gas pressures and a greater degree of fragmentation. Strombolian tephra deposits typically have mass fractions of ash grade particles in the range 0.1–0.2, while lapilli grade ejecta are in comparable abundance to block grade bombs and spatter (Walker and Croasdale 1972; Chouet *et al.* 1974; McGetchin *et al.* 1974; Self *et al.* 1974; Thordarson and Self 1993). Thus in these eruptions a greater proportion of the thermal flux contributes to the formation of a significant plume. For example, the 1973 hawaiite and mugearite eruption of Heimay in Iceland produced eruption columns 3–6 km high (Self *et al.* 1974; Blackburn *et al.* 1976).

Some highly evolved basalts and basaltic andesite magmas, particularly in island arc settings, are significantly more viscous. The higher viscosity relates to a number of factors including somewhat higher SiO_2 and Al_2O_3 contents of the magmas, lower magmatic temperatures and higher gas contents, which induce quench crystallization in the magma as a consequence of gas loss. Eruptions of these magmas are commonly observed to generate predominantly lapilli-sized scoria and abundant scoriaceous ash. These eruptions show very well-developed plumes as well as intense fire fountaining, and they have been termed violent Strombolian by Walker (1973). The Paricutin eruption in Mexico (1945–53) is a good example of such an event.

10.2.3 Vent Geometry

Vent and conduit geometry impose major controls on degassing behaviour and plume generation. Many basaltic eruptions initiate along fissures and thus the source of the plume is taken to be linear. However, basaltic fissure eruptions rapidly evolve to focus on a limited number of points along the fissure, usually within one or two days of eruption initiation. The geometry of the vent can be significantly modified by a number of processes such as lava drainback, accumulation of ejecta and development of ponds of degassed lava (Wilson *et al.* 1995). Energy must be expended by gas-rich magma for it to force its way through lava lakes and pyroclastic accumulations, and to entrain degassed lava. These effects can have a substantial influence on the energetics of fire fountains and the associated plumes (Wilson *et al.* 1995). Indeed, recent modelling suggests that specific measurements of gas content based on eruption behaviour must be treated with caution because drainback and the subsequent entrainment from lava lakes can substantially reduce fountain heights for a given gas content in the erupting mixture (Wilson *et al.* 1995).

10.3 HAWAIIAN FIRE FOUNTAINS AND STROMBOLIAN ERUPTIONS

Hawaiian and Strombolian eruptions are highly complex, time-dependent processes involving two-phase separated flow effects. As a result, several different models have been developed to investigate and explain their origin.

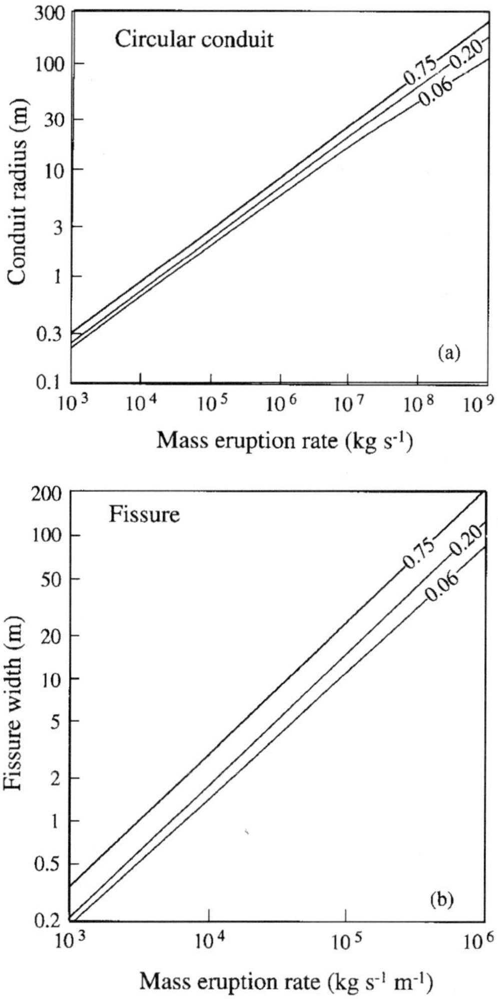

Figure 10.2 Variation of the mass eruption rate as a function of (a) the conduit radius and (b) the fissure width during basaltic eruptions (from Wilson and Head 1981). Curves are shown for different exsolved volatile contents of 0.06, 0.20 and 0.75 wt%

One of the original models of fire-fountaining eruptions, developed by Wilson and Head (1981), is similar to that applied to more viscous Plinian-style eruptions described in Chapter 3. Wilson and Head (1981) assumed that the volatile bubbles are well mixed in the magma reservoir and that, as the melt rises and decompresses in the conduit, further bubbles nucleate and grow. As a result, the mixture inflates and eventually fragments, erupting at the vent as a high-speed homogeneous mixture of gas and magma fragments. The model is most appropriate for fast magma ascent speeds, in which the melt bubble mixture remains homogeneous, and erupts to form a quasi-steady fire fountain. The model describes magma ascent in both thin rectangular fissures and circular conduits. As in Chapter 3, the model is based on the conservation of mass and momentum with the volatile phase gradually exsolving as the pressure falls. Wilson and Head (1981) calculated eruption velocities at the vent as a function of the gas mass fraction of the magma and the eruption rate. In Figure 10.2, model eruption rates are shown as a function of the conduit radius or fissure width for gas mass fractions of 0.75, 0.2 and 0.06. For fissures of width 1–10 m, the eruption rates are calculated to lie in the range 10^3–10^5 kg s^{-1} m^{-1}.

This homogeneous flow model is only valid for fast flow with velocities greater than about 1 m s^{-1}. When the magma has a relatively slow ascent speed, < 0.1 m s^{-1}, the gas bubbles have large rise velocities through the magma. Using the model of Sparks (1978a), Wilson and Head (1981) and Parfitt and Wilson (1995) have shown that such relative motion leads to the production of very large bubbles in the conduit through bubble merging. Figure 10.3 shows the numerical predictions of bubble size as a function of magma rise speed. For speeds smaller than about 0.1 m s^{-1}, the bubbles are typically a few metres in diameter and as a result, separated two-phase flow develops. Parfitt and Wilson (1995) have suggested that the primary difference between Hawaiian and Strombolian events lies in the ability of the bubbles to coalesce and grow. They argue that in Hawaiian eruptions, there is little coalescence and eruptive activity is controlled by the bursting of many small bubbles at the fragmentation surface (see also Head and Wilson 1987). In contrast, they proposed that in Strombolian eruptions, the eruptive activity is manifested by the bursting of large bubbles at the upper surface of the magma (Blackburn *et al.* 1976; Vergniolle and Brandeis 1994).

This picture can be compared with the concepts developed by Vergniolle and Jaupart (1986) who applied the two-phase flow model of Wallis (1969) to examine two-phase flow in magmatic conduits in more detail. They described a series of possible conduit flow regimes, which depend on gas content, bubble size and the melt viscosity. This suggested that, rather than remaining as a homogeneous flow, some Hawaiian fire-fountaining eruptions may involve transitions from bubbly flow to annular flow, in which there is a central stream of gas bounded by liquid moving along the conduit walls. However, they also argued that Strombolian eruptions involve transitions from bubbly flow to slug flow, in

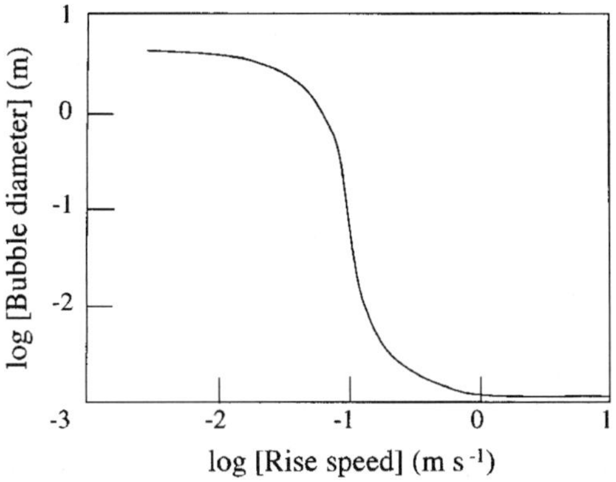

Figure 10.3 Variation of the bubble diameter as a function of magma rise speed as calculated by Parfitt and Wilson (1995) using the bubble growth model of Sparks (1978a). The calculation is for magma of viscosity 30 Pa s with an initial volatile content 0.5 wt%

which large bubbles of gas develop and rise through the residual bubble-poor liquid (Figure 10.4).

Jaupart and Vergniolle (1988) noted that there was a relatively large mass fraction of gas in the eruption products of the 1969–71 Mauna Ulu eruption and the 1983–89 Pu'u O'okilavea eruption compared to the volatile content of the parent magma. They argued that this resulted from volatile–magma separation in the magma reservoir beneath the conduit. Through a combination of theoretical models and laboratory experiments, they showed that a volatile-rich foam layer can develop at the roof of a magma reservoir as bubbles separate from the underlying body of melt. If the supply rate of volatile bubbles to this foam from the underlying body of magma is sufficiently fast, then the foam tends to thicken since it is unable to migrate into the conduit and erupt as rapidly as it is deepening. If the foam becomes sufficiently deep, then it becomes unstable, bubbles coalesce, and a large gas pocket forms. This gas pocket then rises along the conduit as a slug, and bursts at the surface. After some time, a new layer of foam builds up and a further slug develops thereby repeating the cycle. As the bubbles accumulate at the chamber roof, the liquid ascends the conduit. However, once the gas pocket has formed, ascended the conduit and burst, the liquid level in the conduit falls. This process is similar to the so-called gas-piston activity associated with fire-fountaining eruptions in which large variations in the depth of lava lakes are observed. Jaupart and Vergniolle (1988) also found that for small bubble supply rates in the chamber, the foam layer is able to flow into the conduit sufficiently rapidly that the foam remains thin and there is little bubble coalescence. They showed that this leads to a more continuous stream of bubble-rich magma along the conduit and produces a more continuous style of eruption at the vent.

Both mechanisms probably operate, to differing degrees in different systems, with volatile–magma separation occurring in both the chamber and the conduit. In Hawaiian-style eruptions, in which Parfitt and Wilson (1995) argue that there is relatively little bubble merger, the homogeneous conduit flow model of Wilson and Head (1981) provides

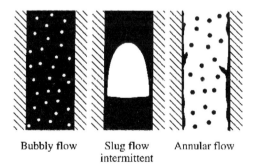

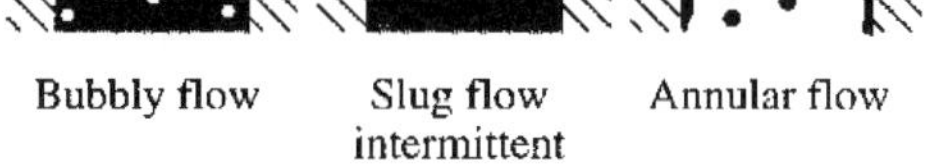

Figure 10.4 The three main flow regimes of separated two-phase flow, as described by Wallis (1969). For relatively small gas fluxes, the bubbly flow regime develops in which unconnected bubbles ascend through the liquid. As the gas volume flux increases, bubbles merge, leading to the intermittent formation of gas slugs. Finally, at even higher gas fluxes, there is a transition to annular flow in which there is a central core of gas surrounded by a film of liquid on the walls of the conduit. Vergniolle and Jaupart (1986) have suggested that Strombolian eruptions involve slug flow, while Hawaiian eruptions may involve annular flow

a first approximation for the flow. In the following discussion, this model is used to calculate the height of rise of the ensuing fire fountain.

10.3.1 Fire Fountain Activity

The presence of a distinct fountain-like jet structure in Hawaiian plumes distinguishes these eruption types from the others considered in this book. The fountain structure arises because of the large pyroclast size, which inhibits dynamic and thermodynamic coupling between the pyroclasts and the volcanic and entrained gases (Woods and Bursik 1991). The pyroclasts are unable to transfer much of their heat to the gases during flight and the particle trajectories are less affected by the gas motion (Chapter 11).

Head and Wilson (1989) have described the classic Hawaiian fire fountain as consisting of several parts. The inner fountain is composed of the hottest clasts which are ejected nearly vertically from the vent. This part of the fountain is nearly isothermal and remains incandescent yellow to a great height. Most of the pyroclasts in the inner fountain region are deposited within a cone surrounding the vent. They can therefore accumulate in and contribute to a lava lake over the vent. Such a lake can either deepen sufficiently to overtop the cone or can undermine and breach the cone to produce lava flows. The intermediate part of the fountain is composed of clasts which are ejected at slightly lower angles and which land on and near the crater rim. Pyroclasts in this region remain sufficiently hot that they can flow upon impact. Depending upon their initial landing position and mobility, these flows can spread down the inner crater wall into the lava lake, remain on the rim to form an agglutinated spatter, or flow down the outer crater wall as rootless flows. The outer fountain is composed of cinders that cool sufficiently from contact with entrained air that they are brittle upon impact. The cooled cinders produce pyroclastic deposits that are composed of a high proportion of broken clasts.

The landform that results from Hawaiian activity typically has a core consisting of a welded spatter cone. Above the spatter cone is a cinder cone consisting of interlayered lava, often of clastogenic origin, and pyroclastic deposits, with an agglutinate rim. With stratigraphic height the agglutinate rim commonly migrates away from the vent and then moves back towards the vent. This variation in position is caused by the waxing and waning of the eruptive phase responsible for the landform. Finally, lavas originating from the lava lake or by coalescence of spatter from the fire fountain extend beyond the cinder cone.

10.3.2 Height of Rise of Fire Fountains

The flight of the largest pyroclasts that comprise a Hawaiian fire fountain is governed by ballistic behaviour, and can be modelled using the well-established ballistic equations of motion (see Chapter 11). The maximum height reached by the pyroclasts, H, may be calculated to first order from a simple dynamical balance between the eruption velocity of the magma at the vent, u_o, and the gravitational deceleration, g:

$$H = u_o^2/2g \tag{10.1}$$

Wilson and Head (1981) have calculated the height of rise of fountains using equation (10.1), together with the eruption velocity associated with a given mass eruption rate and exsolved volatile fraction as predicted using the conduit flow model. Their results are shown in Figure 10.5 in which it is seen that the maximum fountain heights are of the order of 1 km. The height increases rapidly with the mass fraction of exsolved volatiles, essentially owing to the lower density of the vesiculated magma and the greater speed required to achieve a given mass flux.

Because of the dependence of the eruption velocity, u_o, on conduit and vent conditions, it is possible to predict the clast velocity and fountain rise height given the conduit conditions (Wilson 1980; Wilson and Head 1981), or to perform the inverse problem, and deduce information about conduit conditions from observed fountain heights. Head and Wilson (1987) have applied the model to interpret variations in the fountain heights observed at Pu'u O'o Kilauea in terms of variations in the mass of exsolved volatiles. However, more recent modelling has identified that some caution should be exercised in such comparisons since fountain heights also decrease as the degree of coupling between the bubbles and magma decreases (Parfitt and Wilson 1995), and as the degree of magma recycling by entrainment from a lava lake at the vent increases (Wilson *et al.* 1995). In each of these cases, the fundamental control on the fountain height is the mass of exsolved volatiles available to drive the magma clots from the vent (Chapter 3). Typical maximum Hawaiian fountain heights of 500 m contrast with Strombolian fountains which display

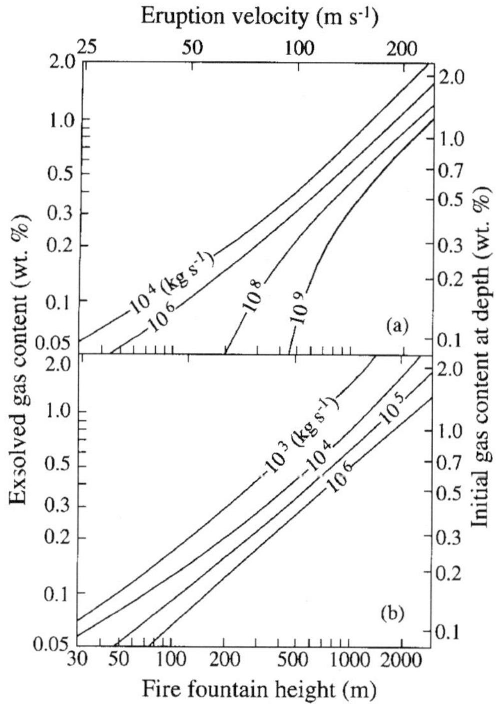

Figure 10.5 The variation of the height of rise of fire fountains and the eruption velocity as a function of the magma volatile content (from Wilson and Head 1981). Curves are shown for a range of eruption rates: (a) corresponds to eruption in a cylindrical conduit while (b) corresponds to eruption in a rectangular fissure. The associated conduit radii and fissure widths may be deduced from Figure 10.2

heights which are generally less than 100 m (Parfitt and Wilson 1995). This difference in characteristic height primarily results from the much lower driving pressure differentials of Strombolian-style behaviour (Wilson 1980; Wilson and Head 1981).

10.3.3 Variations in Eruptive Activity

The above model captures some of the important features of Hawaiian eruptions and gives fountain height estimates which are comparable with observations. Wolfe *et al.* (1987) observed a number of eruptions of Kilauea volcano from January 1983 until June 1984 and found that fountain heights were typically of the order of 400 m, suggesting that eruption velocities were of the order of 100 m s^{-1} (Wilson and Head 1981). However, the fountains exhibited large variations in height (50–800 m) over time scales of as little as two hours. Several theories have been proposed to explain the variations in fountain height that have been observed to occur through time during a single eruption.

Vergniolle and Jaupart (1986) and Jaupart and Vergniolle (1988) have attributed fluctuations in the eruption rate, and hence fountain height, to the result of a dynamic instability in the magma chamber feeding the conduit. As explained earlier, a natural eruption cycle can develop due to foam accumulation below the chamber roof.

Following the calculations of Wilson and Head (1981) (see Figure 10.5), Head and Wilson (1987) and Parfitt and Wilson (1995) argued that fountain height is relatively insensitive to volume flux, and is more sensitive to the degassing style which is controlled by the magma rise speed and apparent exsolved gas content. They reasoned that high magma rise speeds results in fire fountaining producing fountains with heights of several hundred metres. In contrast, at low magma rise speed, bubbles rise relative to the magma, coalesce extensively, and burst at a shallow magmatic surface in Strombolian explosions. In the transition between these two cases, characterized by intermediate magma rise speeds of the order of 0.02 m s^{-1}, they argued that fragmentation occurs at some depth because the close packing of bubbles is exceeded, yet bubbles also have an opportunity to coalesce. A relatively high fire fountain therefore develops because of the release of high-pressure gas. However, its height is unstable, because of the bursting of the occasional large, coalesced bubble.

Wilson *et al.* (1995) suggested that the drainback of substantial amounts of magma from the fountain into the vent is responsible for changes in fountain height, since energy must be expended to drive the degassed backflow upward. Temporary clearing of the backflow accumulation results in the resumption of maximum fountain heights. These then decrease as the drainback accumulates, thus starting a new cycle.

10.3.4 Strombolian Activity

The structure of a Strombolian burst differs considerably from that of a Hawaiian fountain. A Strombolian burst consists of a discrete blast of gas and pyroclasts which together comprise the walls and gaseous contents of a bubble that has burst at the upper surface of the magma column. For example, during the eruption of Heimay, 1973, the bubbles were inferred to have initial radii as large as 10 m and volumes of 30–50 m^3 (Self *et al.* 1974). Strombolian events are relatively instantaneous and separated by periods in the range 0.1 s to several hours (McGetchin *et al.* 1974; Blackburn *et al.* 1976). They also typically include a greater fraction of fine pyroclasts than Hawaiian-style fire fountains.

As the bubble bursts, there is an initial, upward gas thrust caused by the sudden release and decompression of the gas within the bursting bubble. This strongly impulsive motion can eject material 100–150 m above the vent in the gas thrust region where the erupted mixture decelerates. In some explosions, near the upper part of the gas thrust phase, heat from the smaller, fragmented pyroclasts is transferred to the air and thereby drives a convective thermal (see Chapter 4). At Heimaey, this convective phase extended to heights of 3–6 km. Within the eruption column, the largest pyroclasts (> 0.2 m) primarily follow ballistic trajectories (Chouet *et al.* 1974; Blackburn *et al.* 1976), while the motion of the smallest pyroclasts (1–50 mm) is coupled to the gas motion (Self *et al.* 1974). Intermediate-sized pyroclasts (50–200 mm) seem to be released from the plume near the top of its initial, gas thrust rise.

Observations of Strombolian eruptions at Heimaey, Etna and Stromboli have provided constraints on the velocities of the gas thrust phase of such columns. For example Blackburn *et al.* (1976) observed that the mean rise speed of the initial, gas thrust phase of explosion plumes at Heimaey was 157 m s^{-1}, with the maximum observed speed being 230 m s^{-1}, while Strombolian explosions produced from two vents at Stromboli had velocities of the order of 31 ± 12 and 56 m s^{-1}. The average initial speeds of ballistic pyroclasts observed at Stromboli were 25 m s^{-1} (Chouet *et al.* 1974), 51 m s^{-1} at Northeast Crater at Etna (McGetchin *et al.* 1974) and 75–110 m s^{-1} at Heimaey (Self *et al.* 1974). These ballistic speeds are lower than the associated gas speeds, because of the slip between large clasts and gas. Indeed, smaller ballistic clasts were observed to have higher initial speeds (McGetchin *et al.* 1974) and the largest ballistic clasts barely surmount the crater rim (Self *et al.* 1974). Clast trajectories range from being nearly vertical to subtending angles of 45° from the horizontal. This wide range of angles of ejection, and the relatively fine grain sizes of the incandescent pyroclasts, lends Strombolian bursts their typical "fireworks" appearance in night photographs (Chouet *et al.* 1974).

The landform most typical of Strombolian activity is the cinder cone (see Figure 13.10). Progressive growth of Northeast Crater at Mount Etna allowed for the interpretation of the growth sequence (McGetchin *et al.* 1974). In contrast to Hawaiian spatter/cinder cones, Strombolian cones are composed predominantly of cinder and ash that were brittle solids, although often incandescent, upon impact. Cone growth occurs as a series of accumulation and avalanche events. Cinder fallout is concentrated at a point within the topographic rim, thus fallout on the outer wall of a cone is greatest at the rim itself. Because of the consequent oversteepening of the top of the cone, the cinders frequently avalanche to bring the slope of the outer wall to values in the range 31–35°, representing the angle of repose for unconsolidated cinder. In contrast to the intermediate-sized pyroclasts that comprise the majority of the cone, the largest pyroclasts often roll to the base of the cone upon impact and accumulate in an apron. The smallest pyroclasts are generally carried downwind and accumulate in a fines-rich deposit on the downwind side of the cone. The vent within a cinder cone is often choked with talus originating from avalanched cinders and debris derived from small, transient lava ponds.

10.4 THE PLUMES ABOVE FIRE FOUNTAINS

Fountain eruptions produce large quantities of heat, as well as variable but usually minor amounts of fine pyroclasts (see section 10.2.2), which can drive a convective column above a fire fountain. These columns differ from typical eruption columns in that a much smaller fraction of their heat is supplied by the small clasts carried within them. Instead, much of the heat is supplied from the larger clasts in the fountain which heat up the air around and within the fountain. Since little thermal energy is provided by hot, small pyroclasts, Hawaiian and Strombolian plumes are typically much lower than Plinian columns, at comparable total eruption rates (Woods and Bursik 1991). Water, CO_2, sulphur compounds and other volatiles can be released in large quantities into the plume and the atmosphere. Basaltic magma degasses very efficiently either just before it enters or as it rises through the fire fountain (Gerlach and Graeber 1985; Stothers *et al.* 1986; Stothers 1989).

For basaltic eruptions there are three major sources of thermal energy driving the buoyant material upward. First, there is the flux of hot ash from the fire fountain which constitutes a fraction, x, of the mass in the fountain. The fraction of basaltic magma which can be fragmented into ash-sized particles and entrained into the plume from the fire fountain depends upon the viscosity of the magma, the intensity of the eruption and volatile content (see section 10.2.2). In basaltic fire fountains estimates of the mass fraction of fine-grained pyroclasts that are entrained into the plume vary from 0.0 to 0.17 (Walker 1973; Stothers *et al.* 1986). For example, grain-size data for the Kilauea Iki 1959 scoria suggest that the fines constitute only a few per cent of the erupted material (Cas and Wright 1987). In contrast, in more energetic fissure eruptions, the fraction of fine ash can have values of the order of 0.1–0.2, as in the Strombolian tephra of the 1783 eruptions of Laki (Thordarson and Self 1993). Second, there is a flux of volatiles issuing from the vent, which constitutes a fraction, y, of the mass in the fountain. The exsolved volatiles, typically a mixture of water and CO_2, constitute a mass fraction of the erupted material which is of the order of 0.005–0.02, with the relative fraction of water to CO_2 in the range 0.3–1 (Stothers *et al.* 1986; Gerlach and Graeber 1985; Vergniolle and Jaupart 1990). Finally, the material in the fountain cools by an amount ΔT, transferring its thermal energy to the surrounding air which rises into the plume. The fountain cooling, ΔT, is thought to be of the order of 1–30 K (Stothers *et al.* 1986). The ash, magmatic volatiles and entrained air rise from the fountain with a small initial momentum. However, as in a co-ignimbrite column, the plume can accelerate upwards as it entrains more air and increases its buoyancy (cf. Chapter 6).

10.4.1 Height of Rise for a Line Plume

Depending upon the intervent spacing along an erupting fissure and the total fissure length, the plumes above fountains behave as if issuing from either a point or a line source (Walker *et al.* 1984). Fissure eruptions invariably evolve to become localized to central vents (Thordarson and Self 1993). Since the largest basaltic eruptions initially issue from highly elongated fissures, and since axisymmetric sources have been treated extensively in preceding chapters, an eruption plume model for line sources is developed in this and the following section.

Stothers *et al.* (1986) applied the dry-plume theory of Morton *et al.* (1956) (Chapter 2) to argue that the height of a plume issuing from a line source (a line plume) is

$$H = 9.1Q^{1/3} \tag{10.2}$$

where the heat flux released per unit length along the fissure,

$$Q = \rho V((1 - x - y)C_m \Delta T + xC_m(T_o - T_a) + yC_a(T_o - T_a))/D \qquad (10.3)$$

in which V is the volumetric eruption rate, D the active fissure length, T_o the initial temperature of the erupting material, T_a the air temperature, C_m and C_a the specific heats of the magma and air respectively, x is typically 0.01–0.2 of the total mass erupted, and y is typically 0.005–0.02. They have compared this model with historical observations of the plumes which developed during fissure eruptions at Askja, 1961, Mauna Loa, Hawaii, 1984 and Laki, 1783. The results (Figure 10.6) show that the model can reproduce these historical observations, although the model requires fairly large values for x, the mass fraction of fines produced in the eruption.

The heat flux associated with the volatiles and fine ash rising from fire fountains is typically small compared to the total magma heat flux. Rise heights are correspondingly low relative to Plinian eruption columns, for which about 95 wt% of the mass flux issuing from the vent is carried by pyroclasts. Furthermore, because the total amount of ash generated in fountain eruptions is small, relatively minor variations in this amount result in rather large changes in plume height. Indeed, plume height can increase by 1–2 km if the mass fraction of fines in the eruption column increases from 0.01 to 0.2 (Stothers *et al.* 1986) (Figure 10.6). Such variations in the fine ash fraction do arise, at least in Strombolian-style eruptions. Self *et al.* (1974) have noted that grain-size variations occurred during the 1973 Heimaey activity as the result of partial blocking of the vent by fallback during especially vigorous eruptive phases. The dense black plumes of fine ash generated during clearing of the fallback from the vent indicate that the fallback became more finely fragmented than non-recirculated ejecta. On the other hand such recycled material may already have cooled significantly.

The cooling of the material in the fountain is the direct result of the heat transfer to the air entrained into the fountain. Therefore the fountain cooling can be used to estimate the amount of air entrained into the fountain itself. If the fountain cools by an amount ΔT, of

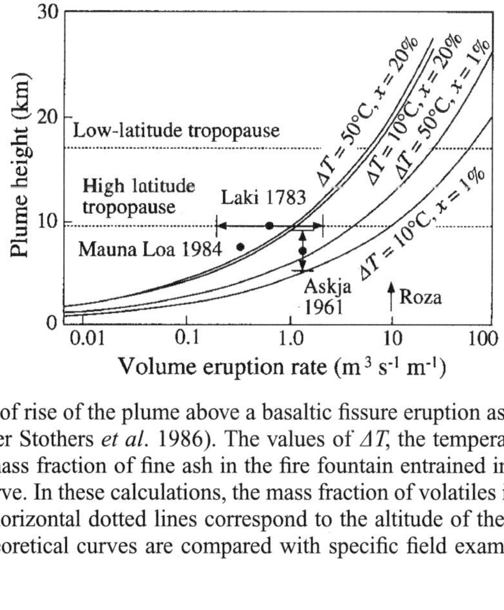

Figure 10.6 Height of rise of the plume above a basaltic fissure eruption as calculated by equations (10.2) and (10.3) (after Stothers *et al.* 1986). The values of ΔT, the temperature decrease in the fire fountain, and x, the mass fraction of fine ash in the fire fountain entrained into the overlying plume, are shown on each curve. In these calculations, the mass fraction of volatiles in the plume, y, has been set to 0.5 wt%. The horizontal dotted lines correspond to the altitude of the tropopause in high and low latitudes. The theoretical curves are compared with specific field examples

the order of 1–30 K, then in thermodynamic equilibrium, the mass flux of air entrained into the fountain, m_a, is given in terms of the mass flux in the fountain, m, according to

$$m_a = m\frac{C_m \Delta T}{C_a(T_o - \Delta T)} \tag{10.4}$$

Figure 10.7 shows typical values m_a/m as a function of the cooling of the fountain, ΔT. Typically the mass of entrained air constitutes about 1–2% of the total mass erupted from the vent. This is about a factor of one-tenth of the initial mass of volatiles. Thus, even at the top of the fountain, the gases in the column may include a significant mass of air.

Since most of the plume material by weight is gas, the height of rise of the plume should be sensitive to the environmental conditions, in particular to the atmospheric water vapour loading (Woods 1993b; Glaze and Baloga 1996; cf. Chapters 4 and 8). The effects of water vapour will be greatest in the small plumes typical of fountain eruptions since water vapour is mostly confined to the lower few kilometres of the atmosphere (Barry and Chorley 1976; Rogers and Yau 1989). Hawaiian and Strombolian plumes thus tend to entrain large quantities of ambient vapour relative to their initial mass flux. As the entrained water vapour is carried upwards to regions of lower temperature, it approaches saturation. Eventually, it becomes saturated, condenses, and releases latent heat. The release of latent heat decreases the density of air in the plume, and thereby increases the buoyancy and height of rise by a substantial amount (cf. Chapter 4). In the next section, an integral model for line source plumes is developed in which latent heat plays an important role in adding thermal energy to the plume.

10.4.2 A Dynamical Model of a Hawaiian Plume

The one-dimensional "top-hat" model described in Chapter 4 (Morton *et al.* 1956) can be used to describe convecting plumes rising from a fire fountain and entraining water vapour. The plume rising from a fissure is defined to have half-width L, velocity u, bulk density β, temperature T, total mass fraction of water plus water vapour q and mass fraction of water vapour alone q_v, and that the mass fraction of solid material in the plume is $1 - n$. Here n

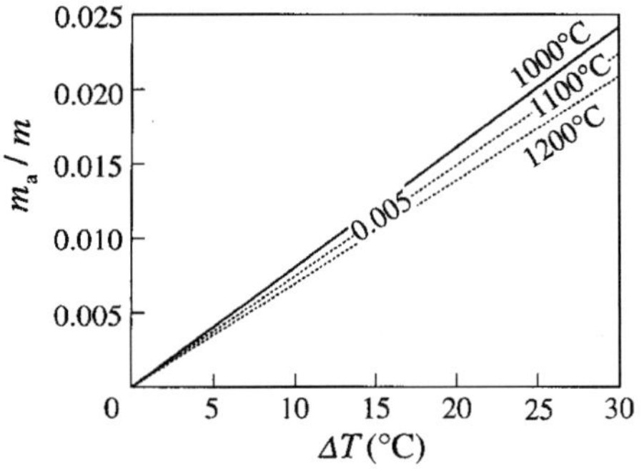

Figure 10.7 The variation of the mass flux of the entrained air, m_a, compared to the initial mass flux, m, as a function of the amount of cooling within the fire fountain (equation 10.4). Curves are shown for a magmatic volatile mass fraction of 0.005 with initial magma temperatures of 1000, 1100 and 1200 °C

is the gas mass fraction at the top of the fire fountain and so can contain air entrained into the fountain as well as magmatic gas. The plume is assumed to entrain ambient air at a rate proportional to the upward velocity, with constant of proportionality $\varepsilon = 0.09$ (Morton *et al.* 1956). If the ambient density is α, and z is the vertical coordinate, then conservation of mass requires

$$\frac{\mathrm{d}\beta uL}{\mathrm{d}z} = 2\varepsilon\alpha u \tag{10.5}$$

and if $q_a(z)$ is the mass fraction of vapour in the atmosphere, then the conservation of water requires that

$$\frac{\mathrm{d}q\beta uL}{\mathrm{d}z} = 2\varepsilon\alpha u q_a \tag{10.6}$$

The conservation of momentum has the form

$$\frac{\mathrm{d}\beta u^2 L}{\mathrm{d}z} = -g(\beta - \alpha)L \tag{10.7}$$

and the conservation of total enthalpy is given by

$$\frac{\mathrm{d}\beta uL\left(gz + \frac{u^2}{2} + C_v T + \frac{P}{\beta}\right)}{\mathrm{d}z} = 2\varepsilon u\alpha(gz + C_a T_a) + \Gamma\frac{\mathrm{d}uL\beta q_w}{\mathrm{d}z} \tag{10.8}$$

where C_v is the specific heat at constant volume, q_w denotes the mass fraction of liquid water in the column, and Γ is the latent heat of condensation. The saturation vapour pressure is parameterized as in Chapter 4. If it is assumed that once the plume becomes saturated, exactly that amount of vapour condenses so that the plume remains saturated, then the mass fraction of the gas phase composed of vapour, w, is exactly that amount such that the partial pressure of the vapour equals the saturation vapour pressure, $e_s(T)$,

$$w = \frac{R_a e_s(T)}{R_v P - (R_v - R_a)e_s(T)} \tag{10.9}$$

where R_v and R_a are the gas constants for vapour and air respectively. The total mass fraction of vapour in the plume is therefore

$$q_v = wn \tag{10.10}$$

and the mass fraction of liquid water in the plume is

$$q_w = q - q_v \tag{10.11}$$

In contrast, when the column is unsaturated, all the water in the plume remains as vapour and

$$q_v = q \tag{10.12}$$

The model is completed by the equation for the conservation of ash in the plume. Since the ascent velocities are of the order of tens of metres per second (Figure 10.5) and only fine ash of the order of 1 mm or smaller ascends above the fire fountain (Thordarson and

Self 1993), all the solid particles and condensate are assumed to be carried to the top of the plume before falling out (Morton 1957; Squires and Turner 1962). This yields the relation

$$(1 - n - q_w)\beta u L = (1 - n_o)\beta_o U_o L_o \tag{10.13}$$

More complex models, allowing for the fallout of ash (Woods and Bursik 1991) and precipitation (e.g. Schlesinger 1978) have been developed in other contexts (Chapter 4). However, in the present situation such effects are negligible.

The importance of the condensation of water vapour upon the motion of line plumes above fissure eruptions is primarily a function of the relative humidity of the atmosphere, because the amount of entrained water vapour is much greater than the amount of juvenile water vapour. This effect is demonstrated in Figure 10.8 in which the water content, the density of the plume relative to the environment and the plume velocity are shown as functions of height in atmospheres of relative humidity 0.50, 0.75 and 1.0. The relative humidity of the atmosphere, R, is defined as the fraction of the water vapour that would be contained in a fully saturated atmosphere, so that the mass fraction of water vapour, q_a, in an atmosphere of relative humidity R is

$$q_a = R\frac{R_a e_s(T)}{R_v P - (R_v - R_a)e_s(T)} \tag{10.14}$$

In an environment of high relative humidity, $R = 1.0$ (vapour saturated), the mass of vapour entrained into the column and hence total water content of the column can be two to three times larger than in less humid environments, where $R = 0.5$ or 0.75 (Figure 10.8a). Also, in the vapour-saturated atmosphere, the mass flux of vapour entrained into the column is nearly 10 times the vapour flux rising from the fire fountain. This increases

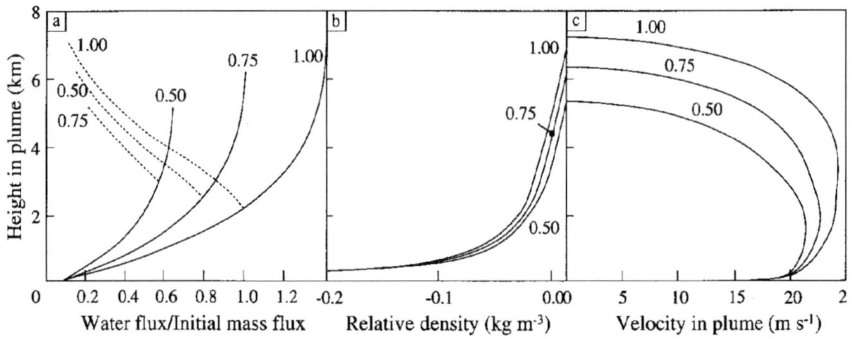

Figure 10.8 Model calculations of the ascent of the plume above a fire fountain during a line fissure eruption (Woods 1993a). The eruption temperature is 1400 K, the fountain is assumed to cool by 10 K, the initial velocity of the plume is 2 m s^{-1}, the initial volatile fraction is 0.01 and mass fraction of fine ash is 0.1. Curves are shown for relative humidities of 1.0, 0.75 and 0.5. The figure shows the variation with height of (a) the total H_2O content as a fraction of the mass erupted into the plume (solid line); the water vapour as a fraction of the total mass erupted into the plume (dashed line); (b) plume density relative to that of the environment; and (c) top-hat velocity of the plume. In this and subsequent calculations, the specific heats of water, vapour, magma and air are taken as 4167, 1400, 1100 and 1400 J kg^{-1} K^{-1}. The gas constants of vapour and air are 467 and 285 (Rogers and Yau 1989), and the latent heat of condensation at 273 K is 2.5 × 10 J kg^{-1} (Rogers and Yau 1989)

the thermal energy flux of the buoyant plume nearly 10-fold through the release of latent heat. As a result, the plume in a vapour-saturated atmosphere ascends several kilometres, or up to 40% higher than in a dry atmosphere. Initially all the water in the plume is present as vapour. However, once the material has risen above the height at which the water is saturated, about 2–3 km for the conditions illustrated in Figure 10.8a, some of the vapour condenses into liquid drops, and the vapour content of the plume decreases. The height at which condensation first occurs decreases with relative humidity, because of the greater water content of the plume in the humid atmosphere. The latent heat released on condensation warms the gas within the plume, thereby slowing the decrease in buoyancy (Figure 10.8b) and increasing the vertical velocity (Figure 10.8c). Since the atmosphere is stably stratified, and the vapour loading of the atmosphere decreases with height (Chapter 4), the plume eventually ceases to be buoyant (Figure 10.8b).

The rise height of the plumes above Hawaiian fire fountains increases with erupted mass flux, fraction of fines and exsolved volatile content as well as with relative humidity (Figure 10.9). The effect of relative humidity is most pronounced at lower eruption rates. Larger plumes entrain less air per unit of mass flux and also penetrate higher into the atmosphere, where absolute humidity is lower. The release of latent heat per unit mass is therefore less, and the dominant heat source is the ash. Because fine particles are transported into the buoyant plume, increasing the fraction of fines increases plume heat flux and therefore rise height. Rise height increases by 1–2 km as the fine ash fraction increases from 0.01 to 0.2 (Figure 10.9b, curves a, b, e). This effect increases with eruption rate because the ash is cooled by less entrained air per unit mass in larger plumes. If the exsolved volatile content increases, the heat flux supplied to the plume is again larger for a given eruption rate since all volatiles are transported into the plume, and rise height increases. However, the measured range of volatile contents of basaltic magma is small, typically 0.005–0.02 mass fraction. Thus any natural variation in the initial volatile content only leads to a small change in the height of the plume for a given eruption rate (curves b, d, Figure 10.9b). Similarly, for ash fractions greater than 0.1, variations in the heat transfer from the cooling of the fountain, ΔT (equation 10.4), have only a small effect upon column height (curves b, c, Figure 10.9b).

As fissure eruptions proceed, the venting tends to organize itself into a series of localized sources along the fissure (e.g. Thordarson and Self 1993), and the plumes then appear to rise from separate point sources. The localization process has been explained as a dynamic instability of the ascending magma. If local fluctuations in the ascent speed develop along the fissure, then in the regions of slower flow, the magma cools against the conduit walls. This increases the viscosity of the magma locally, thereby further lowering the local flow rate. Subsequently, magma tends to flow around these regions and a series of flow channels develop in the conduit (Bruce and Huppert 1989).

The above model of the eruption plume can be modified for application to such localized axisymmetric sources and the predictions of the axisymmetric model are shown in Figure 10.10. As expected the qualitative variation of rise height and velocity as a function of relative humidity, fines fraction and volatile fraction still occurs (Figure 10.10). It is useful to define the conditions under which the plume will act as if issuing from a point or a line source to determine which model may be appropriate in a given situation. A simple criterion can be found by finding the ratio of vent length, D, to plume height, H. If $D/H \ll 1$, then the plume will coalesce to a nearly axisymmetric shape far below its final rise

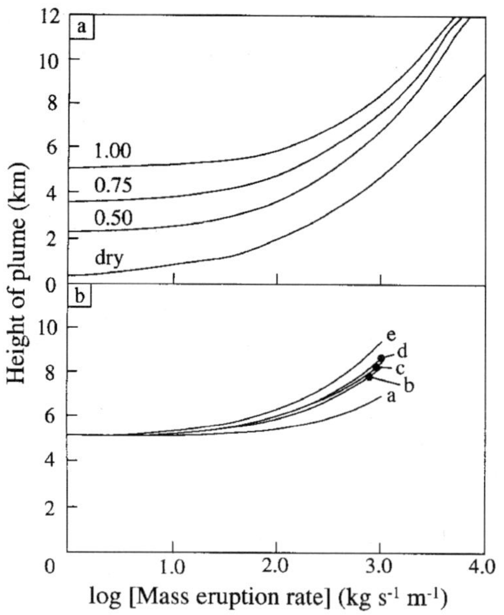

Figure 10.9 Model calculations of the height of rise of linear eruption plumes as a function of the eruption rate of magma (Woods 1993a). Figure 10.9a illustrates the effect of a variation in the ambient humidity (1.0, 0.75 and 0.5). As in Figure 10.8, the fountain is assumed to cool by 10 K, the initial volatile fraction is taken to be 0.01 and the initial fines mass fraction is 0.01. Figure 10.9b illustrates the effect of a variation in the mass fraction of fine ash, x, volatile content, y and fountain cooling, z K. (Curve a) $x=0.01$, $y=0.01$, $z=10$; (curve b) $x=0.1$, $y=0.01$, $z=10$; (curve c) $x=0.1$, $y=0.01$, $z=30$; (curve d) $x=0.1$, $y=0.02$, $z=10$; (curve e) $x=0.2$, $y=0.001$, $z=30$. In these calculations the relative humidity is assumed to be 1.0

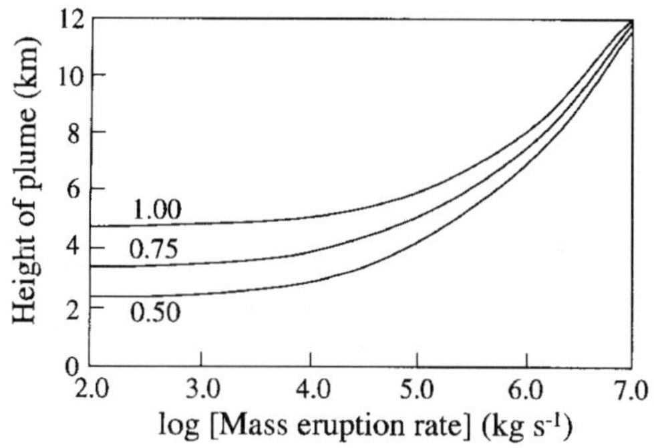

Figure 10.10 The variation of the height of an axisymmetric plume issuing from a localized fire fountain as a function of the mass eruption rate. Curves are shown for relative ambient humidities 1.0, 0.75 and 0.5

height and will therefore act as if issuing from an axisymmetric source. If $D/H \gg 1$, then the plume will reach its rise height with its elongate geometry intact. In this case, the line source model should provide a better representation. In situations where D/H is of the order of 1, the behaviour may well be intermediate between the line and axisymmetric source models.

10.4.3 Comparison with Observations

There are few detailed quantitative observations that can be used to test the models of plumes above Hawaiian fire fountains, partly because it is difficult to estimate the fraction of fine ash or the eruption rate of the plume. Furthermore, the dependence of plume rise height on atmospheric conditions prevents detailed analysis for many historical eruptions because accurate meteorological data are not available. However, the observations that do exist support the theoretical predictions and suggest that relative humidity and source geometry have an important effect on plume rise.

Qualitative visual observations support an important role for water vapour in Hawaiian plumes. For example, during the 1959–60 eruption of Kilauea Iki, Hawaii, hot, turbulent draughts of fine ash, air and condensate were observed to rise high above the fire fountains, and, at least in one instance, large cloud banks were pushed up and formed high above the fire fountains (Richter *et al.* 1984).

During the eruption of Mauna Loa, 1984 (Lockwood *et al.* 1984), a tall cumulus-like volcanic plume was observed to ascend 7.3–7.6 km above the erupting fissure on March 25. The total eruption rate during the event has been estimated to be 0.8–1.5×10^6 kg s^{-1}. The length of the active fissure was about 2 km. If it is assumed that the mass fraction of fines was in the range 0.01–0.1, and the relative humidity was 75–100%, then a corresponding point source would produce a model plume 6–8 km high (Figure 10.10). The equivalent line source, of strength 400–700 kg s^{-1} m^{-1}, would produce a plume of height 5.0–6.5 km (Figures 10.8 and 10.9). In a dry atmosphere, the Stothers *et al.* model (1986) given by equations (10.2) and (10.3) predicts that the maximum rise height for a model fissure event is 5.5 km, assuming the most favourable conditions (volume eruption rate, $V = 600$ m^3 s^{-1}, fissure length, $D = 2000$ m, ash mass fraction, $x = 0.2$, volatile mass fraction, $y = 0.02$, cooling of the fountain, $\Delta T = 50$ °C). This suggests that some other process causes an increase in the rise height, and is consistent with the presence of a large quantity of condensing atmospheric moisture.

Askja volcano in Iceland erupted on October 26, 1961, and generated a Hawaiian plume with an approximate maximum rise height of 9 km above a 700 m long active fissure. The maximum mass eruption rate has been estimated as 2.9×10^6 kg s^{-1} (Thorarinsson and Sigvaldasson 1962). Including uncertainties in the source conditions, and assuming the plume behaved as if issuing from a point source, the model predicts a plume of height 8–9 km, in good agreement with the observation. In this case, similar heights can also be calculated using the simpler model (equations 10.2 and 10.3) of plumes rising in a dry atmosphere (Stothers *et al.* 1986).

The Skaftar Fires, Iceland, eruption of 1783 produced one of the most voluminous volcanic deposits of historical time (Thorarinsson 1969). In August 1783, sightings by a distant observer of the plume rising above the fissure during Phase VI of the eruption were consistent with a plume height greater than 12 km (Thordarson and Self 1993). The mass eruption rate calculated from the deposit volume and the duration of the eruption phase is 5.8×10^6 kg s^{-1} from a 4.5 km long active fissure (Thordarson and Self 1993), or 1290 kg m^{-1} s^{-1}. Using these estimated source conditions and equation (10.2) or its axisymmetric analogue, the plume height is predicted to be about 8.8 km. Thus it is difficult to explain the observed plume height of 12 km in terms of the standard plume models for a dry atmosphere. However, if instead it is assumed that the relative humidity of the atmosphere was high, then the plume could have risen into the stratosphere, attaining heights in the range 10–15 km, which are more consistent with the observations.

10.5 BASALTIC PLINIAN AND IGNIMBRITE ERUPTIONS

There are rare situations where basaltic magma can be involved in much larger explosive eruptions similar to the Plinian and ignimbrite eruptions of more silicic magmas. Examples of basaltic Plinian eruptions are the 1886 eruption of Tarawera, New Zealand (Walker *et al.* 1984) and the scoria fall deposits of the Fontana lapilli and San Judas Formation from Masaya Caldera Complex, Nicaragua (Williams 1983). In the case of the Tarawera eruption the Plinian eruption can be attributed to enhanced fragmentation due to interaction of the erupting magma with water along a major fissure. However, the Masaya deposits do not appear to be particularly fine-grained. Large-volume basaltic ignimbrites have also been recognized. Examples include the basaltic parts of the strongly welded P1 ignimbrite on Gran Canaria (Schmincke 1969; Freundt and Schmincke 1992) and post-glacial ignimbrites of Llaima and Villarrica volcanoes in Chile (H. Moreno and J. Naranjo, personal communication). These examples have volumes of a few to tens of cubic kilometres, showing that basaltic magma can under some circumstances produce large magnitude ignimbrites.

The common causative factors in these examples of Plinian deposits and ignimbrites are likely to be very high eruption rates, higher than normal fragmentation and high gas contents. High discharge rates of magma may be required to inhibit the efficient segregation of gas from magma. As in violent Strombolian eruptions the high gas contents and perhaps associated higher magma viscosities may lead to enhanced fragmentation. Interaction with water could also enhance fragmentation. From the point of view of modelling, basaltic Plinian plumes and co-ignimbrite plumes associated with major basaltic pyroclastic flow eruptions can be treated in the same way as more silicic eruptions.

10.6 SUMMARY

Basaltic magma has a low viscosity and gas can segregate rapidly from the magma during eruption. Thus the ratio of gas to lava can vary from almost entirely degassed lava to almost entirely gas. The efficiency of segregation varies greatly and leads to a great variety of surface activity, including Hawaiian fire fountains, Strombolian explosions, discharge of almost pure gas jets, development of lava lakes, and formation of clastogenic lavas, spatter ramparts, cinder cones and agglutinates. Activity is often highly variable as a consequence of gas segregation mechanisms within the magma chamber and conduit and complex processes at the surface, where gas-rich magma interacts with gas-poor magma and pyroclastic accumulations.

Hawaiian fire fountains and explosive Strombolian bursts differ in their degassing behaviour, pyroclast generation and dynamics. Although several models have been proposed to explain the difference in behaviour, the most recent model of Parfitt and Wilson (1995) suggests that fire fountaining occurs when there is relatively little bubble coalescence in the conduit, and this occurs when the magma ascent speed is high. In contrast, they suggest that Strombolian eruptions occur when there is significant bubble merging, leading to the explosion of gas slugs at the vent. Because many of the clasts land while still plastic, fire fountains typically generate spatter cones with a high proportion of flow as well as pyroclastic material, whereas Strombolian bursts result in cinder cones that are primarily pyroclastic.

There is typically a small fraction of fine-grained material in the ejecta and this can separate from the fire fountain and form a weak convecting plume. Since basaltic eruptions commonly occur in rift environments, Hawaiian and Strombolian plumes are often formed from fissure (line) sources. If the fissure is long relative to plume rise height, the plume acts as if generated from a line source, and this requires some modification to the standard integral equations of plume dynamics. In addition, there can be a significant mass of water vapour entrained into the plume. As this condenses, the associated latent heat released can contribute a significant fraction of the total heat flux of the plume, possibly enabling the erupted volatile gases to penetrate the tropopause.

Basaltic plumes have particular importance for environmental effects on climate, because of the high sulphur content of basaltic magma. The plumes above large flood basalt vents can reach the stratosphere and result in substantial aerosol loadings. There are examples of large-magnitude basaltic Plinian and ignimbrite eruptions which can also cause massive loadings of sulphur aerosols.

11 Atmospheric Dispersal

NOTATION

b_o	vent radius (m)
b	exponent that describes umbrella cloud growth
g	gravitational acceleration (m s^{-2})
g'	reduced gravity, $g(\bar{\rho} - \alpha)/\bar{\rho}$, (m s^{-2})
g'_o	reduced gravity at vent level (m s^{-2})
h	volcano height (m)
k_1, k_2, k_3, k_4	empirical constants
r	radial coordinate (m)
t	time (s)
u	mean centreline upward component of velocity in plume (m s^{-1})
w	mean horizontal windspeed (m s^{-1})
x	coordinate in the downwind direction (m)
y	downwind plume width (m)
z	vertical coordinate, positive upward (m)
B_o	buoyancy flux at vent (m^4 s^{-3})
Fr	topographic Froude number (dimensionless)
H_b	height at umbrella cloud base (italicized in text) (m)
H_t	height of top of umbrella cloud (italicized in text) (m)
N	atmospheric buoyancy frequency (s^{-1})
P	pressure (kg m s^{-2})
Q	volumetric flow rate (m^3 s^{-1})
Q_{H_b}	flow rate of material from vent into umbrella cloud (m^3 s^{-1})
Q_o	flow rate at vent (m^3 s^{-1})
R	umbrella cloud radius (m)
R_{H_b}	initial umbrella cloud radius (m)
R_{H_s}	radius of thermal (m)
R_o	umbrella cloud radius at $t = 0$ (m)
T_a	ambient atmospheric temperature (K)
T_o	eruption temperature (K)
V	umbrella cloud volume, constant for an instantaneous source (m^3)
V_o	initial umbrella cloud volume
v_R	radial velocity (m s^{-1})
α	atmospheric density (kg m^{-3})
$\dfrac{d\alpha_R}{dz}$	atmopsheric density gradient relative to the ambient atmosphere

β	parameter describing the temporal change in flow rate into an umbrella cloud ($m^3\ s^{-2}$)
K_h	horizontal eddy diffusivity ($m^2\ s^{-1}$)
λ	umbrella shape factor (constant)
ρ	mean density of umbrella cloud ($kg\ m^{-3}$)
ΔH	umbrella cloud thickness (m)
ΔH_o	umbrella cloud depth at $t = 0$ (m)
Λ	wavelength of volcano lee wave (m)

11.1 INTRODUCTION

The spreading and dispersal of volcanic plumes within the atmosphere results from interaction with atmospheric motions and stratification. Strong plumes form horizontal intrusions around the height where they become neutrally buoyant, whereas weaker plumes are significantly distorted by the wind as they rise (Figure 11.1). Plumes interact with ambient atmospheric motions throughout their entire height of rise. These plume types (those that form umbrella clouds and those that are bent over) represent end-members in a continuum that depends on relative plume and wind strength. Locally, the atmospheric motions can be broken into two components: a steady component, the wind, and an unsteady component, turbulence. On a larger scale, the motion of plumes shows that the wind is merely a part of a larger-scale atmospheric turbulence (e.g. Barton *et al.* 1992). The wind affects plumes in different ways, depending on plume size and strength.

Large, vigorous plumes reach their maximum height of rise almost undistorted by wind (Figure 11.1). They then spread out as gravity intrusions (umbrella clouds) and start to interact with the wind as their lateral velocity diminishes. Dispersal is a consequence of the motion of the intrusion, the wind and atmospheric turbulence. Small, weak plumes are bent over like industrial plumes as they rise (Figure 11.1), and can still be buoyant when plume motion is primarily horizontal. They gain momentum by entrainment of moving air and develop complex vortices and circulation patterns due to interaction with the airflow. Hemispheric and global plume transport are also affected by the winds. Only since the

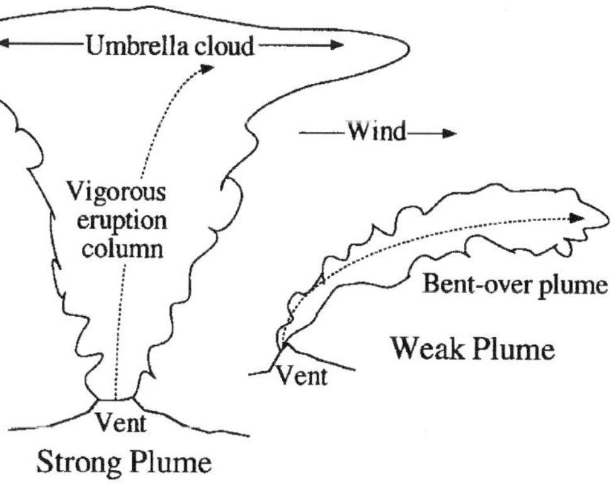

Figure 11.1 Schematic diagram illustrating the contrasting behaviour of strong and weak plumes in a wind field. A strong plume is little affected by the wind. Weak plumes can be bent over to a considerable degree

advent of satellite imagery has such large-scale transport been observed and analysed. Satellite imagery of the long-distance transport of plumes has greatly increased our understanding of the Earth's atmospheric circulation system (see Chapter 12).

In this chapter theoretical models of radially spreading umbrella clouds and the effects of the atmospheric circulation on plume motion are described. The models are compared with observations of umbrella clouds that formed during the May 18, 1980 eruption at Mount St Helens volcano, Washington, USA, and of April 21, 1990 at Redoubt volcano, Alaska, USA, and June 15, 1991 at Pinatubo volcano, Philippines, which demonstrate some of the principles of umbrella cloud growth. The chapter also describes the effects of wind on plumes of varying strength. Wind-blown plume trajectories for the May 18 and July 22 eruptions of Mount St Helens are described and compared with the models. Some of the major results of global plume tracking are discussed.

11.2 DYNAMICS OF UMBRELLA CLOUDS

When the material in a convecting eruption column reaches its neutral buoyancy height, it is carried further upward by its inertia. Since the column is now more dense than its surroundings, it decelerates under gravity. Eventually, the material in the column reaches a maximum rise height, and then begins to flow downward and outward to form an intrusive, particle-laden layer around its neutral buoyancy height. This radially driven intrusion has been called the umbrella cloud (Sparks 1986) and it is a characteristic feature of large volcanic eruption columns. Similar gravity driven intrusions develop as anvil-shaped intrusions at the top of thunderstorm updraughts (Houze 1993) and hydrothermal and effluent plumes. Their motion has been described by Simpson (1987).

11.2.1 Models of Umbrella Cloud Growth

Volcanic plumes spread horizontally in the atmosphere because the atmosphere is stratified. In unstratified environments, such as laboratory tanks, plumes will continue until contacting the upper surface of their immediate environment or until they are completely diluted in the ambient fluid. This occurs because their density cannot rise above that of the ambient fluid. In stratified environments, however, plume density eventually climbs above ambient (Chapters 2 and 4), and the plume thereafter is subjected to a downward gravitational force. The plume's momentum will carry it to a final height, H_t, from which it will slump back gravitationally toward the height at which its density equals that of the ambient fluid (Figure 11.2). The downflow will encounter the upflowing, trailing plume material resulting in the formation of a wedge-shaped, intrusive gravity current, the umbrella cloud. In general, these currents have an extremely sharp leading edge and a smooth outer appearance relative to the plume, suggestive of a lower degree of turbulence (e.g. Abraham and Eysink 1969; Cardoso and Woods 1994). Analysis of the laboratory photographs of Morton *et al.* (1956) and Abraham and Eysink (1969) suggests that, in general, the height of the base of the umbrella cloud, H_b, lies below the neutral buoyancy height (Figure 11.2). However, since these heights are not greatly different, in the following discussion it is assumed that H_b is equal to the neutral buoyancy height.

A useful approximation for the initial depth of volcanic umbrella clouds produced by continuous sources as a function of plume top height and mass eruption rate has been

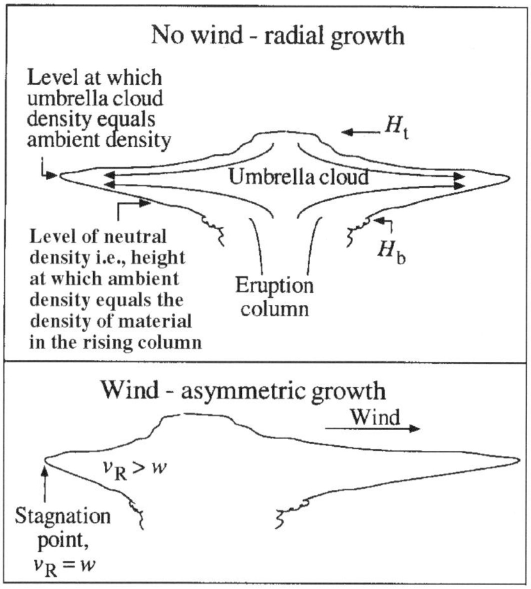

Figure 11.2 Schematic diagram illustrating the important features of a volcanic umbrella cloud and developing downwind plume; v_R is the radial velocity and w the wind velocity

presented by Sparks (1986) and Bursik *et al.* (1992b). An approximation for ΔH, umbrella cloud thickness, as a function of eruption size as parametrized by H_t can be derived from the relationship

$$H_t = 1.32(H_b + 8b_o) \tag{11.1}$$

where H_t is the final height and b_o is the vent radius. Hence

$$\Delta H = 0.24H_t - 8b_o \tag{11.2}$$

For a continuous eruption, H_t can in turn be found as a function of the volumetric flow rate of material into the plume, Q_{H_b} and is given approximately by

$$H_t = 0.287(Q_{H_b})^{0.19} \tag{11.3}$$

These relationships will be used in the following discussion to parametrize the motion of umbrella clouds.

Sparks *et al.* (1986) recognized that flow in the umbrella region is predominantly radial. They suggested that entrainment within the umbrella cloud is negligible and therefore that the rate of supply of mass to the plume from the eruption column equals the outward rate of spread of the plume. Assuming that the umbrella cloud spreads as a cylinder of depth ΔH, and radius R, the conservation of mass equation, or continuity, takes the form

$$\frac{\mathrm{d}\pi R^2 \Delta H \bar{\rho}}{\mathrm{d}t} = Q_{H_b} \tag{11.4}$$

where $\bar{\rho}$ is the mean density of material in the umbrella cloud.

In an umbrella cloud, the density differs from the environment, hence the hydrostatic pressure in the plume is different from that outside. As a result, there is a horizontal pressure gradient between the plume and the environment, which drives the intrusion radially outwards. This pressure gradient scales as

$$\mathrm{d}P/\mathrm{d}r = \bar{\rho} g'(\Delta H/R) \tag{11.5}$$

where $g' = g(\bar{\rho} - \alpha)/\bar{\rho}$ is the reduced gravity, and α the atmospheric density. Note that the Boussinesq approximation is used in equation (11.5). This approximation is used when density variations from a reference density are small, and therefore density can be cancelled from the momentum equation except where it affects the magnitude of the gravitational driving force. Since umbrella clouds are of such vast scale, more than a few kilometres in depth, and tens of kilometres in radius, the frictional forces acting upon the motion are small, hence the (gravitational) pressure gradient driving the outward flow is balanced by the inertia of the plume. The momentum equation for the radial flow may be expressed as (this also holds for instantaneous releases)

$$\frac{\mathrm{d}R}{\mathrm{d}t} = \lambda(g'\Delta H)^{1/2} \tag{11.6}$$

where λ is a constant of order unity, which accounts for the detailed shape of the plume. In reality, as clasts fall from a plume, the density contrast will increase with time. At the same time, however, the density contrast will decrease because of entrainment, hence the motion will be slightly more complicated than described. In general, however, the wind will begin to transport a plume before these processes have significant effect (Bursik *et al.* 1992a; Woods *et al.* 1995).

Since the plume spreads at its neutral buoyancy height, the reduced gravity, g', can be approximated by noting that locally the atmosphere is nearly linearly stratified. Therefore,

$$g' \simeq N^2 \Delta H \tag{11.7}$$

where N is the buoyancy or Brunt–Väisälä frequency of the atmosphere, defined as

$$\left(\frac{-g}{\alpha}\frac{\partial \alpha_{\mathrm{R}}}{\partial z}\right)^{1/2}$$

where $\partial\alpha_{\mathrm{R}}/\partial_Z$ is the density gradient relative to the adiabat (Chapter 4).

Integrating equation (11.4), then substituting from equation (11.7) into equation (11.6), then into equation (11.4) to take out ΔH, and integrating again yields a relationship for the growth of a plume fed by a steady column

$$R^3 = R^3_{H_{\mathrm{b}}} + \frac{3\lambda}{2\pi} N Q_{H_{\mathrm{b}}} t^2 + \frac{3\lambda N V_{\mathrm{o}} t}{\pi} \tag{11.8}$$

Therefore, the radius of the plume may be found at any time given the initial plume radius, $R_{H_{\mathrm{b}}}$, the initial cloud volume V_{o}, and the steady flow rate, $Q_{H_{\mathrm{b}}}$, into the plume at height H_{b}.

In discrete volcanic explosions, the umbrella cloud can be emplaced rapidly, relative to the duration of spreading. In this case, the plume is assumed to be emplaced instantaneously, and the mass and volume of the plume are assumed to be conserved as it intrudes laterally. Under these conditions, assuming that the entrainment is negligible, the volume of the cloud relates the radius and depth:

$$V = \pi R^2 \Delta H = \text{constant} \tag{11.9}$$

Substituting from equations (11.6) and (11.7) into (11.9), and integrating yields:

$$R^3 = R_{H_s}^3 + (3\lambda VN/\pi)t \tag{11.10}$$

Here, R_{H_s} is the radius of the thermal when its top reaches H_t. R_{H_s} can be evaluated using the relationships in Chapter 2, and to a first approximation, $R_{H_s} = 0.2H_t$. A comparison of equations (11.8) and (11.10) reveals that cloud radius should increase more rapidly for a plume fed by a steady column than for a rapidly emplaced plume (Figure 11.3). This occurs because the depth, and hence the pressure force on the cloud is greater at a given radius for a continuously supplied column than it is for a thermal.

A more general situation can be analysed in which the mass flux supplied to the plume may vary as $Q(t) = Q_o t^b$. In this relationship, $b = 0$ corresponds to continuous emplacement of the cloud, $b > 0$ implies the supply rate of material increases with time and $b < 0$ implies the supply rate decreases with time. Following the same procedure as that used to derive equation (11.8), the radius then varies according the relationship

$$R^3 = R_o^3 + \frac{3\lambda NQt^{b+2}}{\pi(b+2)(b+1)} + \frac{3\lambda NV_o t}{\pi} \tag{11.11}$$

where the plume has radius R_o and volume V_o at time $t = 0$, due to the emplacement of an initial pulse of material. This pulse would be supplied from a starting plume, for example, before the maintained supply from the eruption column begins. The second term on the right-hand side shows that material emplaced after the initial pulse only causes the cloud to

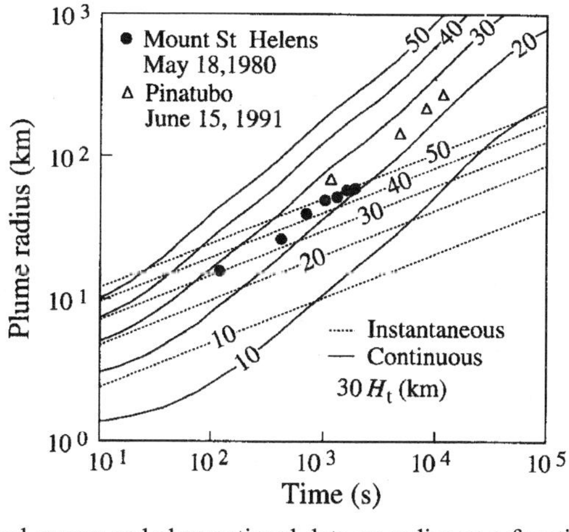

Figure 11.3 Theoretical curves and observational data on radius as a function of time for umbrella clouds. The radius grows more rapidly as a function of time for a continuous (solid lines) as opposed to an instantaneous source (dotted lines). The figures on the curves are for different plume heights, H_t. In the calculations: $N = 0.035\ \text{s}^{-1}$, $b_o = 100$ m, $\lambda = 0.8$ (Bursik *et al.* 1992b; Woods and Kienle 1994), and for thermals, when $H = H_t$, $R = 0.2H_t$ (Chapter 2). The growth of the May 18, 1980 Mount St Helens and June 15, 1991 Pinatubo plumes are also plotted

spread more rapidly if $b > -1$; otherwise, the leading edge of the intrusion is driven by the initial pulse, and the subsequent supply from the eruption column will not be observed in the spreading rate of the umbrella cloud.

11.2.2 Entrainment

Although models have been developed for the case of no entrainment into umbrella clouds, there is experimental evidence that some entrainment occurs at the cloud top, where the column reaches its maximum rise height, and at the transition region between the vertically rising column and the horizontally expanding cloud. Experimental evidence shows that as a column begins to flow outward to form the cloud, air is engulfed in the rising and falling eddies that develop at the top of the cloud just above the column (Turner and Yang 1963; Cardoso and Woods 1994). Where the column impinges on the spreading current from below, and vertical velocities are transferred rapidly to horizontal velocities, the flow goes through a hydraulic jump that generates turbulence and entrains extra atmosphere (Chen 1980). Instability and entrainment also occur in laboratory experiments where processes internal to an initially stable intrusion or layer destabilize it (Shy and Breidenthal 1990; Kerr 1991). Such processes include sedimentation, which releases light unstable fluid (Holasek *et al.* 1996) and evaporative cooling, which causes an increase in density (Srivastava 1987). Entrainment at the base and top of a continuous eruption column could result in an increase in plume depth, but a decrease in the density contrast between the laterally spreading current and the ambient atmosphere. To a first approximation, however, these effects are negligible in assessing the growth rates of volcanic umbrella clouds.

11.3 PLUME–WIND INTERACTION

All volcanic plumes are eventually distorted by the wind to some degree. Distortion of plumes arises because either air with horizontal momentum is entrained into the plume or horizontal momentum is gained from the airflow around the plume. The largest plumes and those erupted into a still atmosphere are able to initially spread to form umbrella clouds. In contrast, less vigorous plumes rise buoyantly in the wind field even as they are advected by it (Figure 11.1). These plumes do not form umbrella clouds. These differences are analysed by investigating the large-scale interactions to establish the criteria for determining which kind of plume will form. The effect of wind on weak to moderate strength plumes is then considered in terms of local interactions.

The large-scale features of the plume–wind interaction can be usefully categorized by a comparison of the relative strength of the plume and the wind. If some characteristic velocity of a plume, say the centreline velocity, u, is compared to the mean horizontal wind speed, w, then the two modes of behaviour (Figure 11.1) arise. For $w \ll u$, plumes are little bent over in the wind field, and plume trajectories are subvertical. For $w \gg u$, plumes are bent over in the wind field a substantial amount. These relationships can be formalized further in terms of plume source or initial conditions rather than local conditions, which are necessarily more difficult to characterize.

The difference between vigorous and weak plumes in their interaction with the wind can thus be parametrized in terms of the ratio of the wind speed, w, to the characteristic plume speed (Wright 1984), where the latter is given by

$$k_1(B_oN)^{1/4} = k_1(Q_og'_oN)^{1/4} \tag{11.12}$$

where k_1 is a constant equal to 4.38 (Turner 1979), B_o the buoyancy flux, Q_o the volume flux and g'_o is reduced gravity at the source. Chapter 4 describes how B_o and g'_o are calculated for volcanic plumes. The parameter $w/(B_oN)^{1/4}$ increases as wind speed increases and decreases as eruption size increases. For $w/k_1(B_oN)^{1/4} \gg 1$, a weak plume will form, which will be much affected by wind, and will be severely bent over in the wind field. For $w/k_1(B_o\mathrm{N})^{1/4} \ll 1$ (a strong plume) column rise will dominate the plume trajectory, and the column will be little affected by the wind as it rises. These relationships are illustrated for typical volcanic buoyancy fluxes and atmospheric windspeeds in Figure 11.4. For most plumes with a total rise height greater than 20 km, no typical values of wind speed are able to affect plume rise substantially, and all eruption columns above this height should develop umbrella clouds. Plumes less than 10 km in height will often be strongly affected by the wind except on quiescent days.

11.3.1 Strong Plumes

If $w/k_1(B_oN)^{1/4} \ll 1$, as in the case of a strong tropospheric plume in a weak wind field, or most plumes that reach the stratosphere, then the plume will rise relatively unmodified

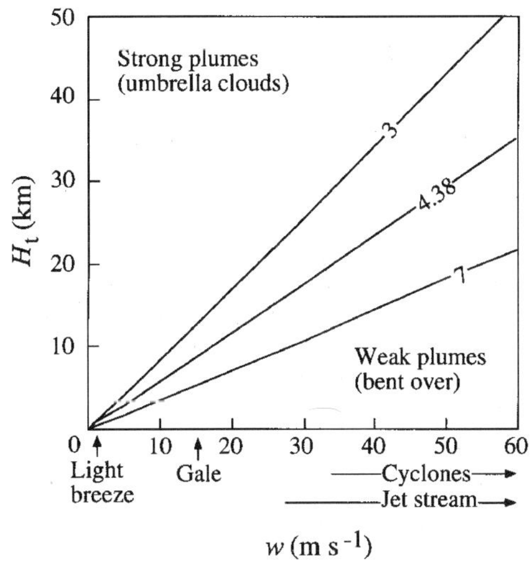

Figure 11.4 Plume height, H_t, as a function of wind speed, w, for the transition from weak, bent-over, to strong plumes. The diagonal lines represent the case where wind speed equals characteristic plume velocity (see text). For typical tropospheric wind speeds of < 10 m s^{-1}, all but the smallest volcanic plumes form umbrella clouds. For typical stratospheric jet stream velocities columns greater than 20 km height will normally form umbrella clouds. In the calculations also: $N = 0.035$ s^{-1}, eruption temperature, $T_o = 1173$ K, ambient temperature, $T_a = 293$ K. Numbers next to curves are values of the proportionality constant, k_1, with values of 3 and 7 shown to illustrate a plausible range. From Turner (1979), the best estimate of the value of this parameter obtained in the laboratory is 4.38

through the atmosphere to form a typical umbrella cloud at its top. Figure 11.5a shows an outline of the vigorous starting plume from the July 22, 1980 eruption of Mount St Helens. The eruption column is bent only slightly by the wind as it follows the ring vortex at the plume head. Wright (1984) suggested that the trajectory for a vigorous plume in a stratified, windy environment should be similar to that in an unstratified environment up to the neutral buoyancy height, based on the observations of Morton *et al.* (1956) that plumes in both environments rise and grow in similar fashion below this height. Wright assumed therefore that the variation of vertical velocity with height scaled as (Chapter 2)

$$u = w\left(\frac{\mathrm{d}z}{\mathrm{d}x}\right) = k_2 B_{\mathrm{o}}^{1/3} z^{-1/3} \qquad (11.13)$$

where x and z are downwind (horizontal) and vertical coordinates, u the centreline column velocity and k_2 is a constant. Upon rearrangement and integration, equation (11.13) yields a relationship for the trajectory, $z(x)$, of the column. The centreline trajectory for a vigorous, buoyant plume in a cross-flow (wind) and uniform environment is therefore

$$z = k_3 B_{\mathrm{o}}^{1/4} w^{-3/4} x^{3/4} \qquad (11.14)$$

where k_3 is a constant with an empirically determined value of ~ 2.5 in laboratory experiments.

As a vigorous column passes above its neutral buoyancy height, it continues upwards due to its momentum. It reaches its maximum rise height, then spreads radially as a gravity current because of its excess density to form an umbrella cloud. As the gravity current spreads upwind, it encounters resistance. At the stagnation point (Carey and Sparks 1986), the wind speed and the current speed are equal, and the upwind progress of the current is halted (Figure 11.2). For strong plumes the stagnation point can occur at substantial distances upwind of the vent. For example, the umbrella cloud from the May 18 Mount St Helens blast plume spread upwind to distances of 15 km before it was balanced by stratospheric winds (Sparks *et al.* 1986). Downwind, the plume speed approaches the wind speed as it loses particles and slows, while in the crosswind direction, the plume can continue to spread as a gravity current (Bursik *et al.* 1992a). The plume will thus spread in an oval, rather than a concentric, pattern. As plume strength decreases, the situation is approached where the displacement of the plume centreline is roughly equal to the upwind spreading in the umbrella region. This results in a plume that has a nearly vertical upwind profile and a downwind umbrella region.

With distance, the wind eventually dominates the downwind plume motion. Friction between the upper and lower interfaces of the downwind plume and the atmosphere results in downwind transport at the wind speed (Chen 1980). In the crosswind direction, the lenticular plume continues to spread because it is more dense at its top and less dense at its base than the surrounding stratified atmosphere. Behaviour in this regime is thus dominated by conservation of volume flux downwind and gravity flow crosswind (Bursik *et al.* 1992a). Assuming that negligible amounts of air are entrained and that crosswind plume shape can be approximated by an equivalent rectangle, the downwind continuity equation becomes

$$Q = y\Delta H w = \text{constant} \qquad (11.15)$$

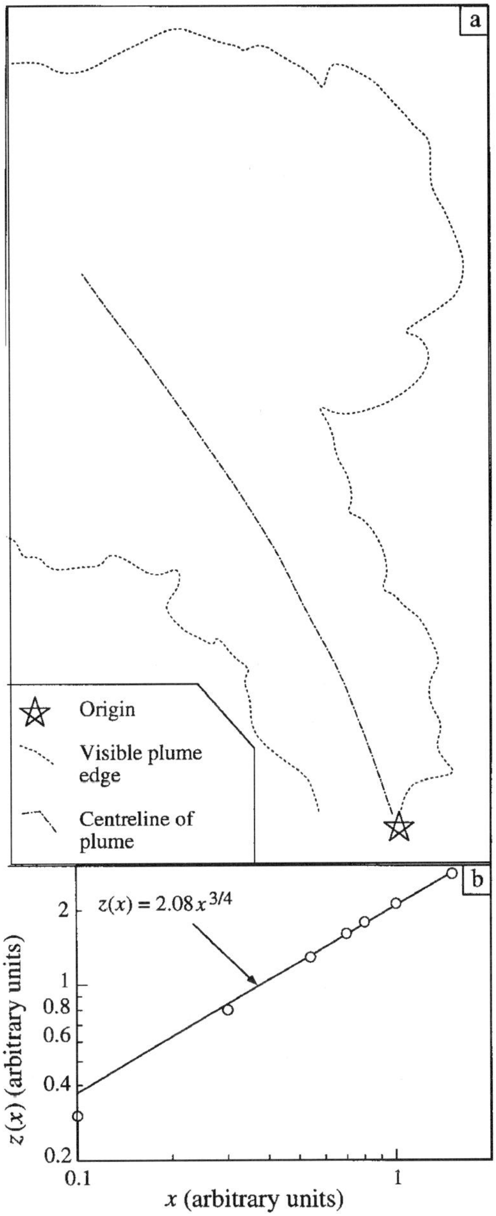

Figure 11.5 The starting plume of the July 22, 1980 eruption of Mount St Helens volcano. (a) Tracing showing only slight bending over in the wind. (b) The trajectory data, $z(x)$, are consistent with the power law relationship expressed in equation (11.14) for a strong plume. Distances are in arbitrary units

where y is plume width. Crosswind, the plume front advances with a speed

$$\frac{dy}{dt} = w\left(\frac{dy}{dx}\right)\lambda N \Delta H \tag{11.16}$$

due to gravity spreading, where λ is again a constant of order unity.

Assuming $Q = Q_{H_b}$, and substituting from equation (11.15) into (11.16) and integrating, the width as a function of distance can be expressed as

$$y = \frac{\left(2\lambda N Q_{H_b} x\right)^{1/2}}{w} \tag{11.17}$$

Because of their corresponding greater depth, plumes from higher eruption columns spread more rapidly crosswind than do those from lower eruption columns (Figure 11.6a), given the same wind. Conversely, given separate plumes of the same height, those erupted into a region of higher winds will be distorted into a more elongated shape.

Further downwind, the plume loses its density contrast with the atmosphere and can be advected as a lens of aerosol and gas with nearly constant width (Sarna-Wojcicki *et al.* 1981; Robock and Matson 1983). It thins, spreads and disperses slowly as shearing and small-scale atmospheric turbulence act at its margins and as very fine ash settles out. The elongated shape assumed by such plumes is illustrated by the aerosol band generated during the El Chichon eruptions of 1982 (Figure 11.7). This band maintained nearly constant width and became gradually more dilute due to shearing at its margins as it completely encircled the globe (Pollack *et al.* 1983; Robock and Matson 1983). Recent results suggest that plume properties over large distances are dominated by the clast settling process (Carey and Sigurdsson 1982; Armienti *et al.* 1988; Glaze and Self 1991; Heffter and Stunder 1993). In the Mount St Helens case, the shear of the horizontal wind components acted to disperse the plume through segregation of the different grain size fractions (Danielson 1981; Woods *et al.* 1986). As the plume was transported downwind, the dominant particle size remaining within the plume, and height of origin changed with time due to differential settling speeds.

11.3.2 Weak Plumes

In the case of $w/k_1(B_o N)^{1/4} \gg 1$, a subvertical eruption column cannot develop. Instead, a weak, bent-over plume whose motion is dominated by the wind rises from the vent. Later in the day of July 22, 1980, at Mount St Helens, as the eruption lost its vigour, the plume was modified into the typical bent-over shape (Figure 11.8a), quite different from its earlier form (Figure 11.5a).

Some of the major features of weak to moderate strength plumes that are affected by the wind must be understood in terms of the local interactions of the plume and wind. Earlier work such as McMahon *et al.* (1971), Chassaing *et al.* (1974) and Moussa *et al.* (1977) suggested that jets acted as solid cylindrical objects in their interaction with wind. The wind/plume system can be divided into three regions as the wind is followed around the plume: the freesteam, plume and the wake (Figures 11.9 and 11.10). Within the freestream the wind velocity remains constant. Beginning slightly upstream of the plume, the wind streamlines are diverted and accelerated around the plume. Vortices are shed in the wake

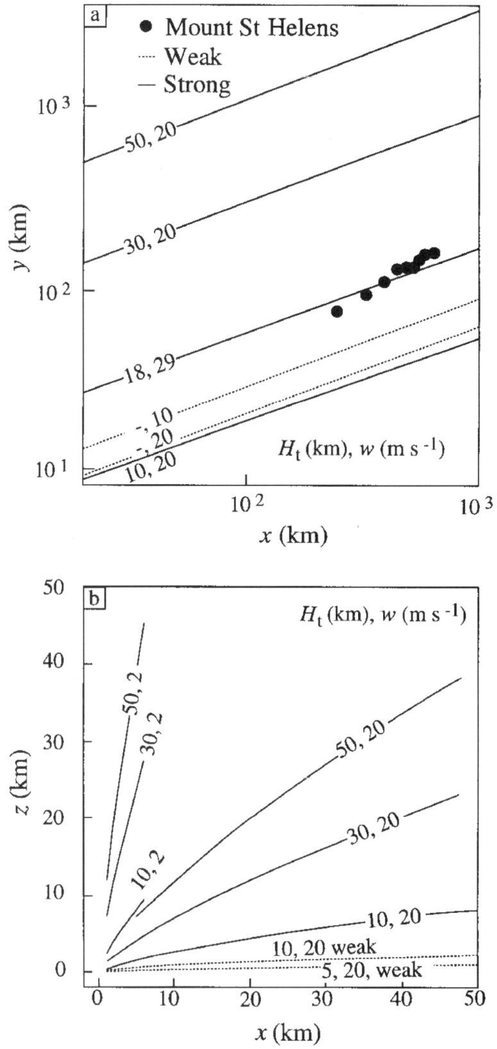

Figure 11.6 (a) Horizontal growth of the plume from continuous eruption columns for strong and weak plumes showing plume width, y, as a function of downwind distance, x. For strong plumes, numbers next to curves are values of column height, H_t, in kilometres and windspeed, w, in metres per second. For weak plumes, $k_h = 5000$ m^2 s^{-1} (Armienti *et al.* 1988) and numbers next to curves are windspeeds. Downwind data for the May 18, 1980, Mount St Helens plume are consistent with a column height of 18 km and wind speed of 29 m s^{-1}. (b) Contrast of trajectories for strong and weak plumes. Numbers next to curves are column heights and wind speeds, as in part (a) for strong plumes showing plume width, x, against height, z. The 10 km high column in a 20 m s^{-1} wind has been calculated by both methods as it falls near the boundary between the two plume types

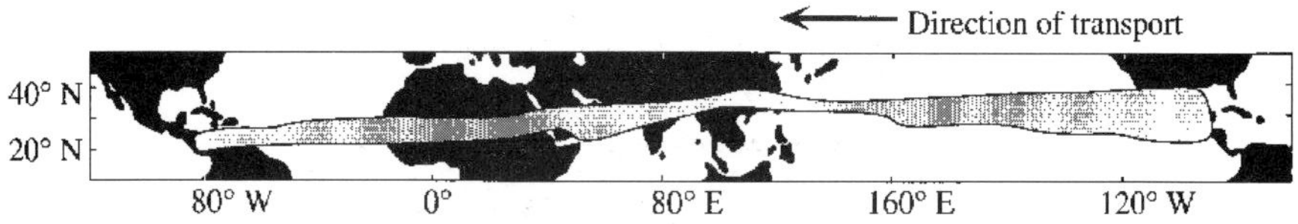

Figure 11.7 The circumglobal aerosol band generated by the El Chichon eruptions of 1982 maintained a nearly constant width with distance (redrawn from Robock and Matson 1983)

region. However, recent research (Korthapalli *et al.* 1990; Fric and Roshko 1994) has shown that the compliant and dynamic nature of a plume or jet results in rather more complex interactions than the analogy with flow round solid cylindrical objects allows. Four kinds of interaction can be recognized (Figure 11.10).

Korthapalli *et al.* (1990) and Fric and Roshko (1994) have shown that the vortices shed downstream of a plume or jet have a fundamentally different origin to those shed from a solid cylinder. In contrast to a solid object, the flow around a compliant jet or plume does not separate from the jet itself and does not shed vorticity into the wake. Instead the ground boundary layer develops vortices in the wake that are then convected downstream. A series of separation events lead to an alternating sequence (known as a street) of regularly spaced vortices in the boundary layer region near the ground. This flow produces horseshoe vortices around the base of the jet or plume (Figure 11.10).

In the wake region the separation of the horseshoe vortices around the plume (Figure 11.10) creates vorticity. During the separation events in the wake region this vorticity is tilted and stretched in the entrainment field of the rising bent-over plume overhead. Because angular momentum is conserved, vorticity stretching is responsible for the development of tornado-like structures (Figure 11.10) or whirlwinds as the fluid in the boundary layer of the wake region is drawn towards the underside of the bent-over plume. These structures have been observed in the wake of plumes in the 1963–67 Surtsey

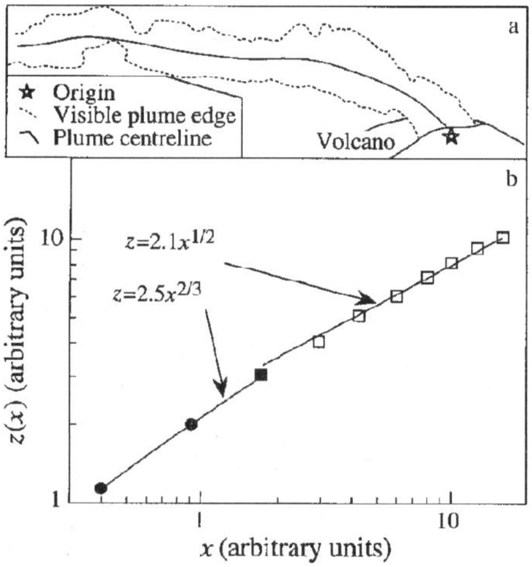

Figure 11.8 The weak plume that developed late in the day on July 22, 1980 at Mount St Helens volcano. (a) Tracing of the eruption column showing the bent-over structure. (b) The power-law relationship for the trajectory, $z(x)$, is close to that predicted for a weak plume by equation (11.19). Note that the distal downwind regime wherein $z \sim x^{1/2}$ also occurs for the July 22 plume

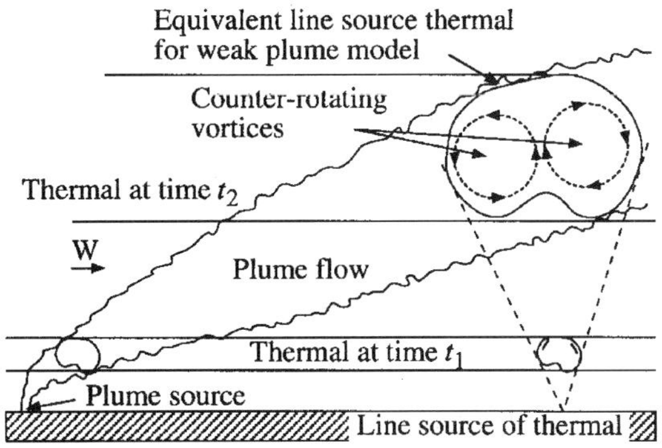

Figure 11.9 Schematic diagram showing the local features of the interaction of a plume with the wind (after Figure 5a of Ernst *et al.* 1994). When the plume is bent over into a subhorizontal orientation, it resembles a thermal in cross-section

eruption (Thorarinsson and Vonnegut 1964). Rotational velocities of the order of 90 m s^{-1} are expected in these structures.

Two other kinds of interaction can be recognized (Figure 11.10). Further up the plume the effect of bending over is to develop a pair of counter-rotating vortices when the plume is bent over (Figures 11.9 and 11.10). A vertical cross-section through the plume is thus dominated by the presence of the attached vortex tubes and therefore tends to resemble a cross-section through a thermal emanating from a line source (line thermal) (Figures 11.9 and 11.10), with a buoyancy per unit length of B/w. This structure is the consequence of the large plume half-width-scale eddies that develop in an ascending vertical plume and are bent over by the wind. The result of the bending over of the plume is to produce two counter-rotating vortices which form tubes with horizontal axes in the downwind direction. These vortices have a characteristic kidney shape in cross-section (Figure 11.9) with a strong updraft in the middle. Yet another interaction occurs as the ambient fluid moves across the upper surface of the bent-over plume or jet. Shear occurs between the ambient wind and the plume or jet-creating vortices on the upwind side (Figure 11.10).

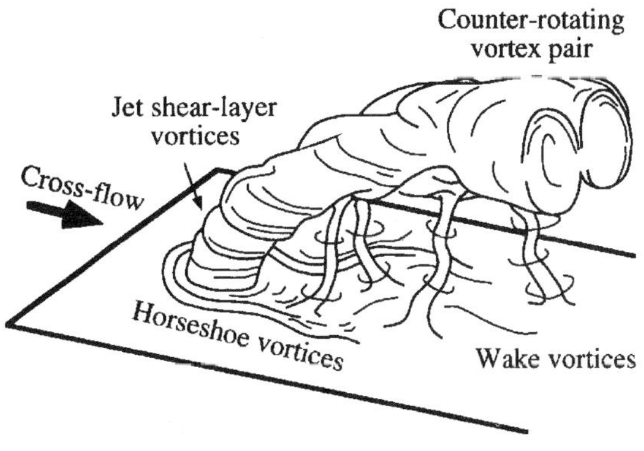

Figure 11.10 Diagram showing four types of vertical vortical structures developed by interaction of a plume or jet with a cross-flow (after Fric and Roshko 1994)

The overall variation of plume velocity with height (Wright 1984) is

$$u = w\left(\frac{\mathrm{d}z}{\mathrm{d}x}\right) = k_4 B_\mathrm{o}^{1/2} w^{-1/2} z^{-1/2} \tag{11.18}$$

Upon rearrangement and integration, the centreline plume trajectory, $z(x)$, follows the power-law relationship (Richards 1963; Wright 1984):

$$z = k_5 B_\mathrm{o}^{1/3} w^{-1} x^{2/3} \tag{11.19}$$

where k_5 has been determined to have a value of ~ 1.7 (Wright 1984). Thus for weak or bent-over plumes, trajectories will be much closer to subhorizontal than for vigorous plumes in the same wind field (Figure 11.9). Wright (1984) notes that a further downwind regime exists. Once a bent-over plume reaches the height at which its density is approximately equal to that of the surrounding atmosphere, it no longer rises by buoyancy but by its momentum, and the resulting trajectory follows the relationship $z \sim x^{1/2}$.

Unlike strong plumes which form umbrella clouds with a turbulence level below that of the eruption column, weak plumes do not lose their puff-like structure as they reach their maximum rise height, since their flow characteristics are not reorganized in the upper atmosphere (Figures 11.11 and 11.12). The plume mixes with the turbulent atmosphere as it is advected, and the plume width should increase with downwind distance by turbulent diffusion (Taylor 1921; Csanady 1980; Rose *et al.* 1988) such that in a steady mean flow:

$$y = 4\sqrt{\frac{\kappa_\mathrm{h} x}{w}} + b_\mathrm{o} \tag{11.20}$$

where κ_h is the atmospheric horizontal eddy diffusivity, and it is assumed that Fickian diffusion occurs (Bursik *et al.* 1992a) and that the plume margin lies at twice the characteristic crosswind length-scale from the plume centreline (Morton *et al.* 1956). Note that the equation (11.20) should also apply in situations where a plume is just able to rise slowly into the jet stream near the tropopause and has insufficient energy to form an umbrella cloud.

Once a plume is completely bent over in the wind field, it begins to move with the wind, approximately parallel to the wind vectors, and travels as a downwind plume. The transfer of momentum from the wind to the plume can apparently be accomplished with little transfer of mass (Chen 1980). Although there are few data on the shapes of downwind plumes, in general they seem to be oval lenses in cross-section, with their long axes horizontal (Hobbs *et al.* 1991). When looked at in planimetric view, they often consist of elongated lenses of material that become detached from the volcano when the eruption ceases (e.g. Glaze *et al.* 1989). Sometimes, however, smaller plumes can have rather irregular outlines because of their susceptibility to local atmospheric motions (Sawada 1983a). They can even have several lobes at different heights due to wind shear effects (Sarna-Wojcicki *et al.* 1981).

Some plumes develop a highly organized structure in the wind field. The attached counter-rotating vortex tubes that form as the subvertical eruption column is bent over can become strong enough to dominate downwind plume motion, resulting in splitting, or bifurcation of a plume in the far field (Scorer 1959; Turner 1960; Crabb *et al.* 1981; Ernst *et al.* 1994). Plume bifurcation develops most markedly when the ratio of wind velocity to characteristic plume velocity is in the range 2–8 (Ernst *et al.* 1994). Some examples of

Figure 11.11 A Landsat image showing the plume of April 3, 1986 issuing from the crater of Augustine volcano, Alaska. (Photograph provided by W. I. Rose.)

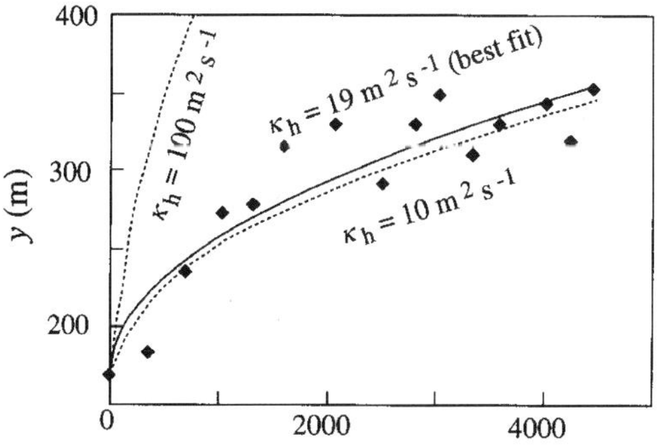

Figure 11.12 Data for the April 3, 1986 plume of Augustine showing downwind spreading. The observations for $y(x)$ are consistent with the prediction of equation (11.20), assuming a very low but reasonable value for atmospheric turbulent diffusivity. Wind speed is assumed to be 23 m s^{-1} (Rose *et al.* 1988)

plume bifurcation are illustrated in laboratory experiments (Figure 11.6) and also can be seen in satellite imagery of volcanic plumes (section 11.4.2). The origin of bifurcation has been explained as the consequence of the lateral pressure gradients that develop as the cross-flow moves around the margins of a plume or jet (Ernst *et al.* 1994). The resulting pressure distribution forces the vortex pair formed in these flows to move apart. Turner (1960) modelled a bent-over plume in uniform surroundings as a buoyant vortex pair. He argued that buoyancy should be conserved as the plume ascends, but momentum should increase. As the momentum of a vortex pair is directly proportional to the separation distance, Turner deduced that there should be a linear increase in separation with distance, as observed in experiments (Turner 1960; Ernst *et al.* 1994; see Figure 11.13). Bifurcation can also be enhanced by cooling and evaporation at the plume margins (Scorer 1958) and by interaction with a major density interface (Hayashi 1972). Two-dimensional line source plumes can split into two vortices of different strengths spreading at different heights if the cross-flow is oblique to the line source (Wu *et al.* 1988).

11.3.3 Topographic Effects

Because the volcano itself can represent a disturbance to the wind field of the same scale as a weak plume, the flow of air around the volcano itself has been observed to affect the motion of weak plumes. A well-documented example of this occurred during the April 3, 1986 eruption of Augustine volcano, Alaska (Rose *et al.* 1988). In this case, the plume was so weak that it was unable to rise from the vent more than 10–100 m (Figure 11.14). In the downwind direction, the material in the plume provided an excellent marker to delineate a strong lee wave in the wake of the volcano, similar to those generated by airflow over steep mountain fronts which often generate lenticular plumes (Long 1953; Scorer 1959). The wavelength, L, of such a wave is given by

$$L = 2\pi w/N \tag{11.21}$$

and the local Froude number for the flow will be

$$Fr = w/Nh = L/2\pi h \tag{11.22}$$

where h is the volcano height.

The wavelength at Augustine was measured to be 8 ± 2 km. For a volcano height of 1.2 km, $Fr \sim 1.1$, suggesting slightly supercritical flow. The airflow does resemble that obtained in laboratory experiments for stratified flow over obstacles at similar Fr (Long 1953; Hunt *et al.* 1978), in which an atmospheric hydraulic jump is created downstream of the flow disturbance.

11.3.4 Regional and Global Transport

When a downwind volcanic plume has persisted for a sufficiently long time, it becomes fully subjected to atmospheric motions. These motions fall into a spectrum of characteristic scales, from microscopic molecular agitation, to hemispheric geostrophic flow (e.g. Holton 1992). The effect of these scales of motion varies greatly despite the fact that all atmospheric agitation has its origin in the incoming solar radiation, and as Richardson (1922) so succinctly put it: "Big whorls have little whorls that feed on their velocity and a

b

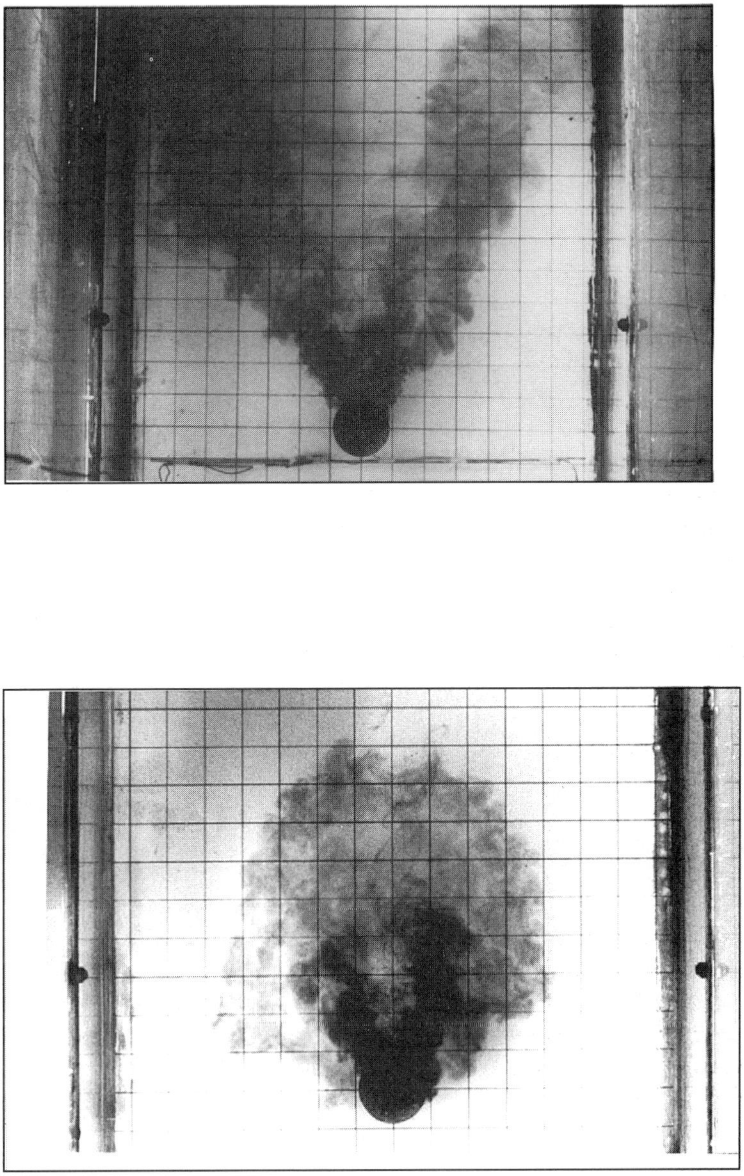

Figure 11.13 Plan views of plume bifurcation in laboratory experiments. The cross-current moves from base to top and splits the ascending plumes into two lobes. The bifurcation is more prominent in the plume (a) with the faster cross-current (after Ernst *et al.* 1994)

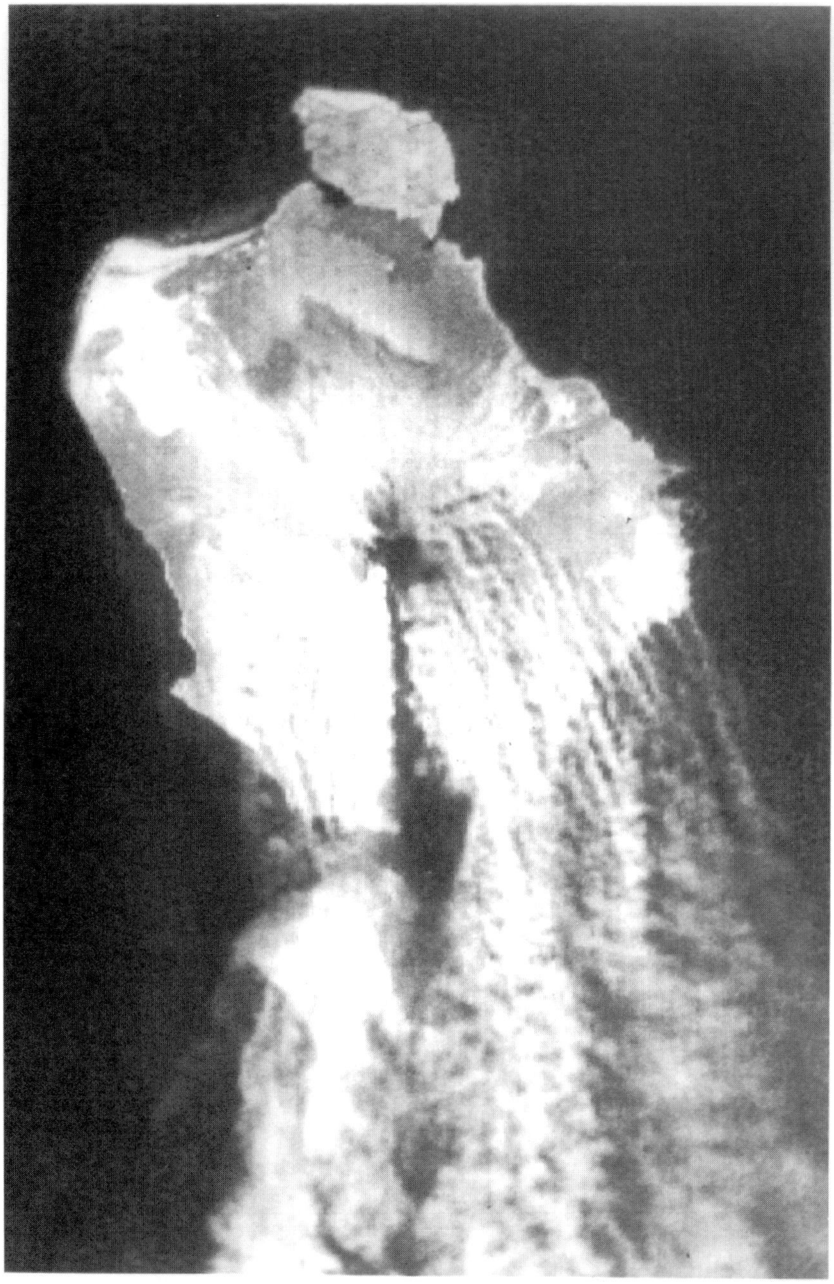

Figure 11.14 Photograph showing lee waves visualized by the April 3, 1986 plume of Augustine in the wake of Augustine volcano. (Photograph provided by W. I. Rose.)

little whirls have littler whirls and so on to viscosity." The atmospheric motions can be categorized by referring to the energy cascade, or the power spectrum of atmospheric motions (Figure 11.15). Range I eddies, or enstrophy-cascade scale motions, are of the order of 10^3–10^4 km in size and quasi-horizontal because their characteristic lengths are greater than the effective depth of the atmosphere (Gifford *et al.* 1988). Range I eddies include the large-scale zonal and geostrophic flows in the atmosphere, which because of their size are capable of distorting air parcels, but themselves cause little relative diffusion. Because of the shear between range I eddies moving in different directions and between the eddies and the Earth's surface, energy is transferred from them to smaller atmospheric motions, the range II eddies. Range II motions span a vast range in sizes from 10^3 to 10^{-5} km, and include features such as regional scale frontal disturbances, through motions generated by frictional effects within the planetary boundary layer (PBL – within a kilometre of the Earth's solid surface), down to gusts and other highly localized disturbances. The smallest motions are those responsible for the viscous dissipation of the atmosphere's turbulent energy. These range III eddies are those primarily related to the molecular (Brownian) agitation of air molecules and aerosols.

The effects of the different scales of atmospheric motion on volcanic plumes are very diverse. Range I eddies primarily distort moderate-sized plumes, but cause little growth or diffusion. Observations of plumes distorted in the geostrophic and zonal winds suggest that distortion primarily occurs as simple stretching and bending, although higher more complex modes of modification have also been observed. Range I mixing is observed when plumes have grown to sizes at least as large as the lower bound of range I motions, approximately 10^3 km. Range II eddies are primarily responsible for relative diffusion of plumes and turbulent dispersion of ash particles that have fallen from plumes. Range III eddies do not affect the movement and dispersal of plumes directly. However, their agitation is partly responsible for bringing ash, gas and aerosol particles into contact so that aggregation becomes possible (Chapter 16).

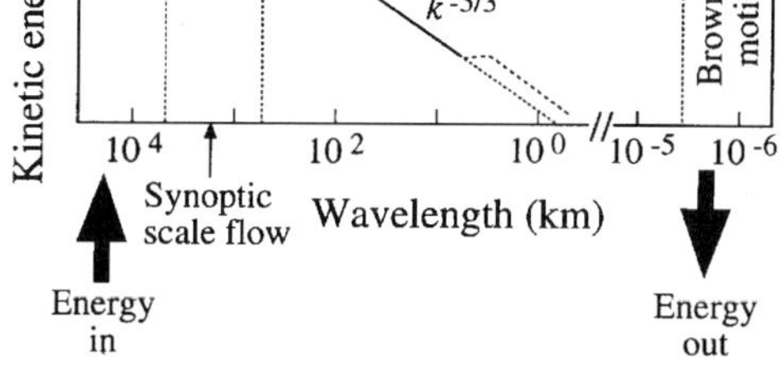

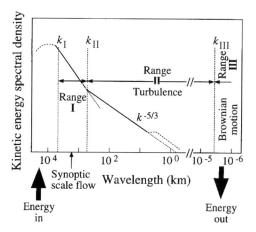

Figure 11.15 The atmospheric energy cascade showing the different scales of atmospheric motion. k_I, k_{II} and k_{III} are the wavenumbers at which energy is transferred from one scale of motion to the next scale down. (Redrawn from Gifford *et al.* 1988.)

Long-range plume dispersal can be estimated with atmospheric trajectory models (Heffter 1983; Kahl *et al.* 1991; Schoeberl *et al.* 1993; Heffter and Stunder 1993). Trajectory models track the motions of air parcels transported in the stratosphere or troposphere. They can be derived from either global circulation models (GCMs; e.g. Barton *et al.* 1992) or weather balloon sounding data (Heffter 1983), by using these sources to define the wind field and other properties at different heights in the atmosphere.

11.4 COMPARISON WITH OBSERVATIONS

11.4.1 Umbrella Clouds

Mount St Helens, USA, May 18, 1980

Sparks *et al.* (1986) studied the giant umbrella cloud that rose from the devastated area during the May 18 1980 eruption of Mount St Helens (Figure 11.16). Using column speeds, initial plume area and thickness measurements from photographs, and atmospheric values for the density at the boundary between column and plume, α, and at the mid-depth of the plume, $\bar{\rho}$, the radial expansion of the plume as a function of time can be compared, with the simple models described by equations (11.8) and (11.10). Using values for the parameters from Sparks *et al.* (1986), a reasonable fit to the Mount St Helens data can be obtained for a rapidly emplaced umbrella cloud (Figure 11.17). Note that there are some discrepancies in the model fit at smaller radii, perhaps related to the emplacement process. A check on these results can be performed by considering the data presented in Figure 11.3. Because the measurements in Figure 11.3 were not made using the interpolation procedure used in Figure 11.17, the data may be more representative. These data indicate that the plume grew at a rate intermediate between the instantaneous thermal and steady-state plume models, consistent with the suggestion of Woods and Wohletz (1991) that the emplacement time for the plume was comparable to the column rise time of about 10 minutes.

Pinatubo, Philippines, June 15, 1991

The sequence of gigantic eruptions of Pinatubo, Philippines on June 15, 1991, produced numerous umbrella clouds that were observed by the Japanese Geostationary Meteorological Satellite (GMS; Figure 11.18 (Plate I)). Koyaguchi and Tokuno (1993) produced timed tracings of the largest of the plumes which were generated by the paroxysmal event. Here the accuracy of the measurements has been independently corroborated. The tracings can be used to plot the growth of the plume as a function of time (Figure 11.3). The data are consistent with the growth of an umbrella cloud from a continuous source with $H_t = 25$ km. This result is in agreement with the measurements from the imagery of the height of the outer edge of the plume of 25 km (Koyaguchi and Tokuno 1993) and with SAGE II measurements of an aerosol concentration between 20 and 25 km, and maximum altitude of 29 km (McCormick and Veiga 1992). The eruption duration was at least as long as the time of plume growth. The estimated plume top height from this method is considerably below the single measurement of 34 km made of the central swell (Tanaka *et al.* 1991), but it is not known how long this height was maintained.

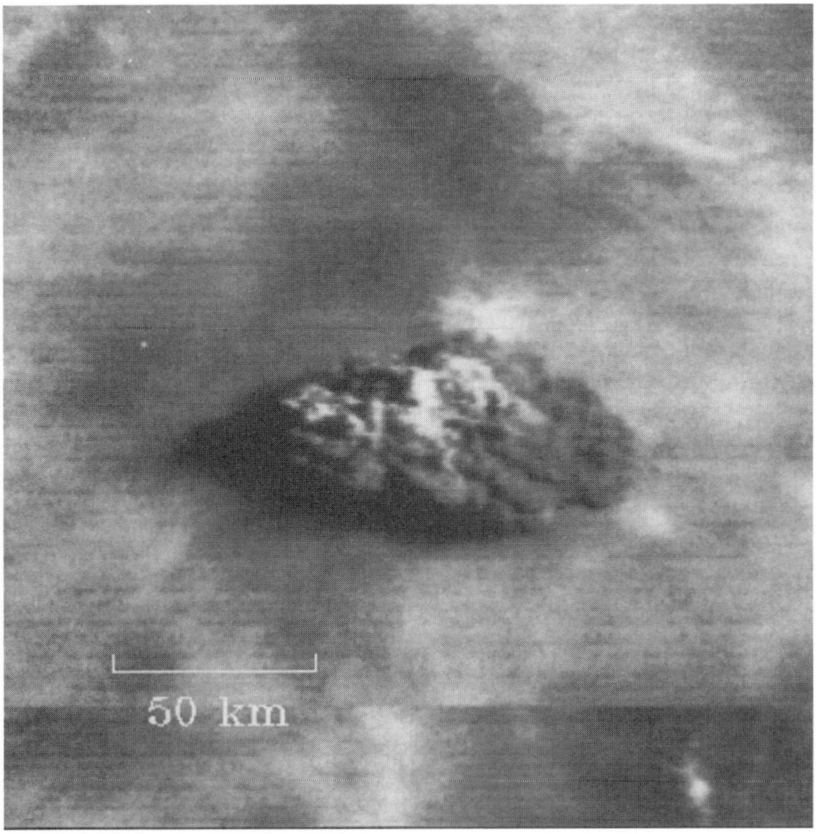

Figure 11.16 Image acquired by the VISSR instrument on board the GOES satellite (see Table 12.1 for a list of satellite acronyms) in the visible channel of the giant umbrella cloud generated by the May 18, 1980 eruption of Mount St Helens showing the radial growth and central peak of the cloud

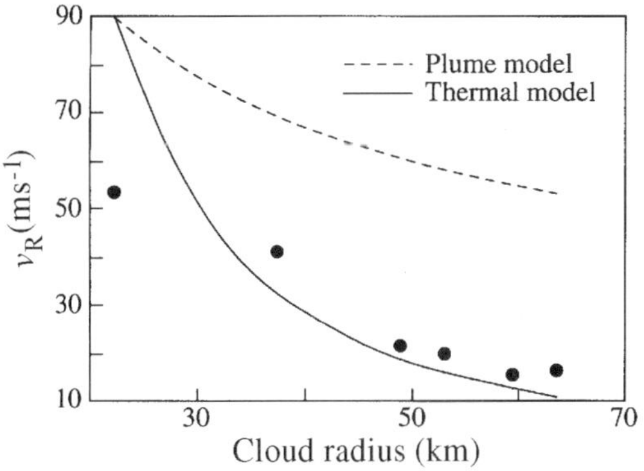

Figure 11.17 The growth of the giant umbrella cloud from the May 18, 1980 eruption of Mount St Helens is consistent with emplacement from an isolated thermal. Values of the parameters used in equations (11.8) (plume model) and (11.10) (thermal model) are $N \approx 0.035$ s^{-1}, $l = 0.8$ (Bursik *et al.* 1992b), $Q_{H_b} = 4 \times 10^{10}$ m^3 s^{-1} and $V = 5 \times 10^{12}$ m^3 (Sparks *et al.* 1986)

Redoubt volcano, USA, April 21, 1990

The expansion of the umbrella cloud which formed from the April 21, 1990, co-ignimbrite plume of Redoubt was recorded in a sequence of timed photographs (Woods and Kienle 1994). Woods and Kienle successfully compared the radial expansion of the upper intrusion with the theoretical prediction for the rate of expansion of an instantaneously emplaced intrusion, equation (11.10) (Figure 11.19). The agreement of the model and the observations suggests that the dominant dynamical processes assumed by the model are valid, and that the entrainment of air or fallout of ash during this initial motion are negligible. The results of this effort suggest that ground-based photography as well as satellite imagery can be used to study umbrella cloud dynamics, providing appropriate geometric corrections are made.

11.4.2 Downwind Spreading and Plume Dispersal Patterns

Strong and weak plumes

A comparison of equations (11.14) and (11.19) suggests that if the trajectories of the plumes are plotted on logarithmic diagrams, they should display different slopes of ~3/4

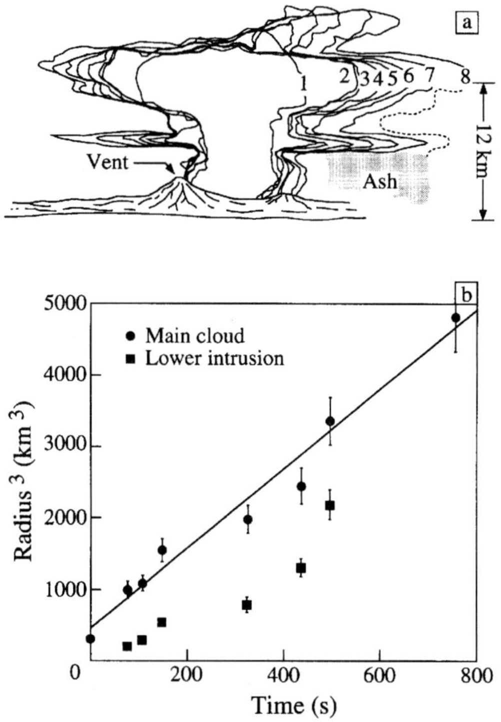

Figure 11.19 The Redoubt cloud of April 21, 1990 grew as an instantaneous release. (a) Timed tracings of the outline of the cloud showing lateral growth and thinning. (b) R increases with t according to equation (11.10). As measured from photographs, the volume of the cloud was 2×10^8 m^3. Values of other parameters are $N \approx 0.035$ s^{-1} and $\lambda = 0.8$ (redrawn from Woods and Kienle 1994)

in the case of a vigorous plume, and $\sim 2/3$ in the case of a weak plume. The data for the different times of the July 22, 1980 Mount St Helens plume (Figures 11.5b, 11.8b) fit these trajectory relationships reasonably well, with the starting plume (Figure 11.5a) being strong (Figure 11.5b), and the later plume (Figure 11.8a) being weak (Figure 11.8b). Thus the simple relationships determined for plumes in an unstratified ambient atmosphere seem to work for volcanic plumes in a stratified atmosphere.

Downwind plumes

With the development of remote sensing techniques (Chapter 12) it has become possible to test models of downwind plume spreading and dispersal. The May 18, 1980 eruption of Mount St Helens provides the best documented example of the development of a downwind plume from a vigorous eruption column (Sparks *et al.* 1986; Bursik *et al.* 1992b). Initially, the giant umbrella cloud of this eruption evolved into a slightly wind distorted oval pattern (Figures 11.16 and 11.20). As the gravity flow weakened upwind, the plume front reached a stagnation point (Sparks *et al.* 1986) at a radial distance of 15 km from the vent, where timed satellite observations and windspeed measurements from balloons showed both plume and wind travelling at 20–25 m s^{-1}. Downwind, the plume spread in a parabolic pattern with vertex facing upstream consistent with equation (11.6) (Figure 11.9)

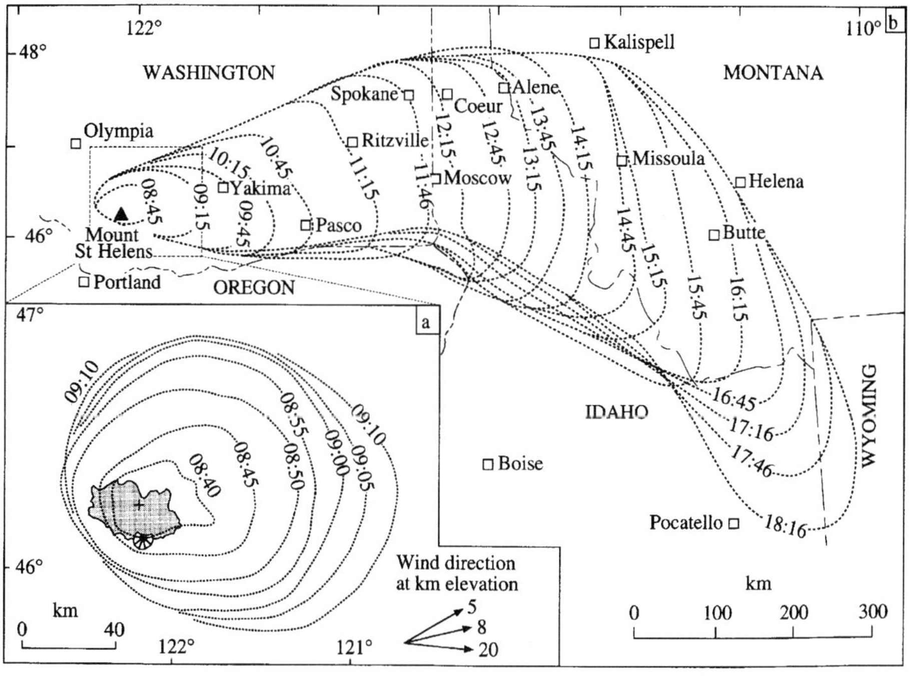

Figure 11.20 (a) Map traced from satellite imagery showing the initial growth of the giant umbrella cloud of Mount St Helens (from Sparks *et al.* 1986). Contours are in five minute intervals labelled with local time in hours and minutes. (b) Map traced from satellite imagery showing the later growth of the May 18, 1980 Mount St Helens plume (redrawn from Sarna-Wojcicki *et al.* 1981)

while the velocity of the downwind plume front approached the maximum stratospheric wind speed of 28 m s^{-1} (Figure 11.21).

Frequent small eruptions of Augustine volcano in Cook Inlet, Alaska, USA, provide examples of the dispersal of a weak downwind plume (Figure 11.11) (see references by Rose *et al.* 1988, and Holasek and Rose 1991). The puff-like structure of a typical Augustine plume is maintained downwind, and the plume disperses as predicted by equation (11.20) under the assumption of constant diffusivity (Figure 11.12). For the plume of April 3, 1986, that is illustrated, the effective horizontal eddy diffusivity is found to be 19.2 m^2 s^{-1}, which is within the wide range of values measured for diffusivity over brief time intervals of 10–10^4 m^2 s^{-1} (Heffter 1965). The low value of diffusivity could be related to the strong wind shear noted during this eruption (Rose *et al.* 1988).

Bifurcation of volcanic plumes up to 16 km high has been observed in many instances (Figure 11.22) (Ernst *et al.* 1994), although the pattern of bifurcation can be complicated by a number of factors including wind shear (section 11.3.2). Ernst *et al.* (1994) reviewed controls on bifurcation and carried out laboratory experiments (Figure 11.13) suggesting bifurcation of volcanic plumes should be common. In fact they recognized bifurcation for many plumes described in the literature (e.g. Sawada 1985; Holasek and Rose 1991) and drew attention to many examples of bilobate fall deposits consistent with fallout from bifurcating plumes. Ernst *et al.* (1994) showed that bifurcation should best be developed in plumes of weak to moderate strength in a strong cross-flow. Bifurcation is also enhanced for volcanic plumes in moderate winds that interact with the tropopause.

11.4.3 Examples of Hemispheric to Global Transport

Large-scale to global transport patterns of vigorous plumes are seen in the dispersal of the plumes of El Chichon, 1982, and more recently Pinatubo, 1991, and Hudson, 1991. In each case, long-distance and global transport are affected by the large-scale wind field of the atmosphere. In general, plumes subjected to long-distance transport travel as elongated lenses of gas and aerosol (Figure 11.23 (Plate II)). Slow fallout of aggregating aerosol and very fine ash particles, and shearing and turbulence acting at the plume margins account

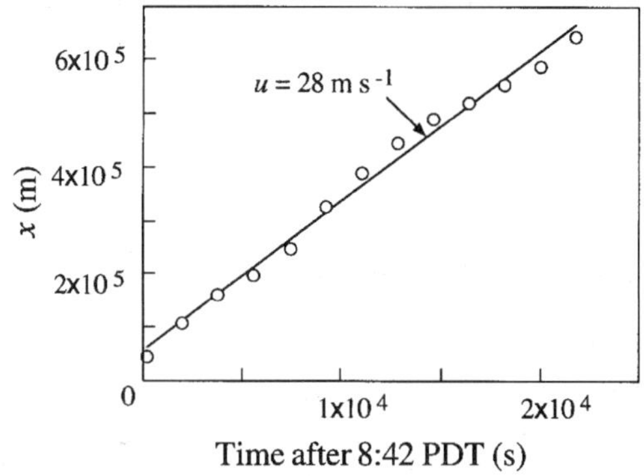

Figure 11.21 Measurements taken from the data shown in Figure 11.20b. The front of the plume advanced at the (maximum) wind speed measured at the base of the stratosphere on May 18, 1980 of ~28–29 m s^{-1} (from Bursik *et al.* 1992a)

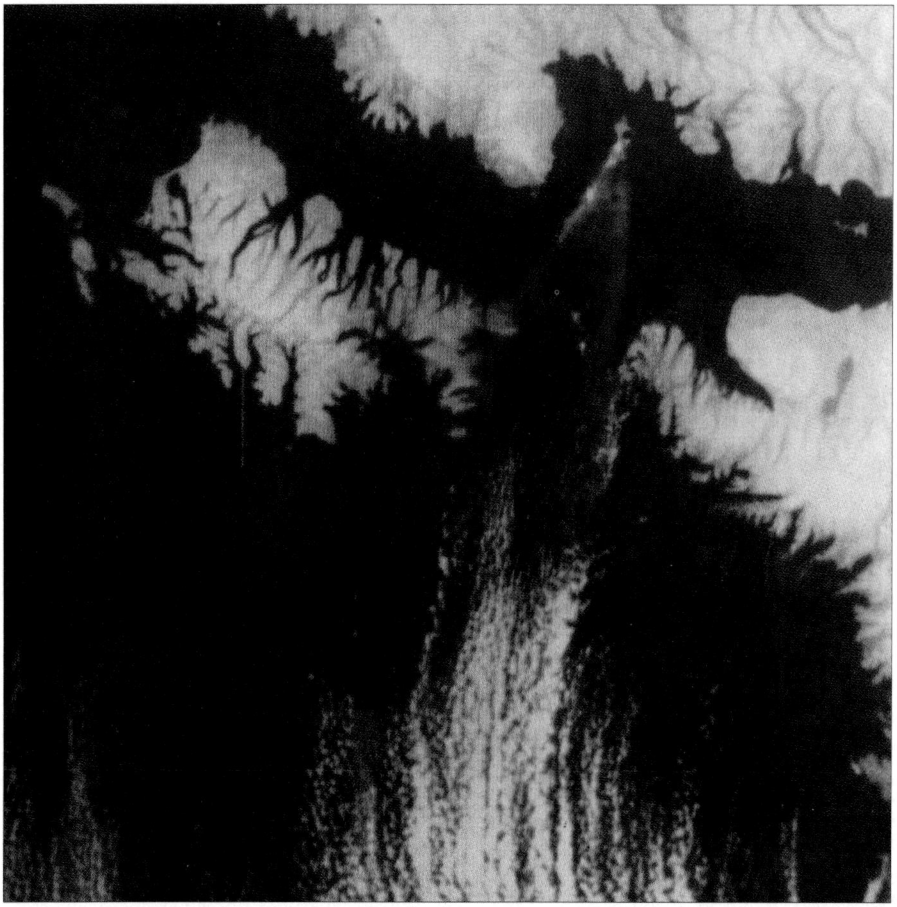

Figure 11.22 Image acquired during the 1986 eruptions of Augustine, Alaska, by the AVHRR satellite instrument, showing apparent plume bifurcation. (See Table 12.1 for a list of satellite acronyms.) The two thermal infrared channels have been ratioed in order to differentiate between the volcanic plume and the surrounding meteorologic clouds. With the ratioed image displayed in red, one of the unprocessed thermal infrared channels displayed in green and the mid-infrared channel displayed in blue the volcanic plume appears orange and the meteorologic cloud appears white and blue. Plume differentiation is discussed in Chapter 12

for gradual spreading, stretching, break-up and dissipation. Monitoring of the spread of these plumes has recently allowed atmospheric scientists to better understand transport processes within the stratosphere.

The eruption of Mount Hudson in Chile during August 1991 generated a plume that reached the lower stratosphere, where the circulation carried the plume eastward in a region of strong zonal flow (Figure 11.24 (Plate III)). Reports showed that the plume passed over Australia five days after the eruption, and again eight days later. The plume was carried poleward and dispersed during September 1991 by disturbances in the circulation north of the polar vortex (Pitts and Thomason 1993).

The plume interacted strongly with the range I scale of motion in the atmospheric circulation, and its transport provides a qualitative comparison with the dispersion model of Gifford *et al.* (1988). For the first several days, both the plume and the tracer gas in the experiment reported by Gifford *et al.* were primarily transported and deformed by large-scale eddies, which fall in the range I (enstrophy-cascade) scale of motion. Few differences can be discerned when comparing the degree of convolution of the tracer gas release simulation to the natural Mount Hudson plume. This similarity indicates that the circulation in the troposphere at 5 km (the level of the tracer gas simulation) was comparable to the circulation in the lower stratosphere at the level of the Mount Hudson plume. Barton *et al.* (1992) and Schoeberl *et al.* (1993) were able to show that the dispersal pattern of the plume during its first seven days was the result of the wind field measured within the polar night jetstream, just north of the region of the southern polar vortex. Its lack of elongation during this time is thought to be the result of weak vertical wind shear (Schoeberl *et al.* 1993). The plume apparently thus acted as a passive tracer of the atmospheric motion. A comparison of the Hudson transport pattern with the experimental results of Behringer *et al.* (1991) and the lower stratospheric modelling of Pierce and Fairlie (1993) suggests that the plume was trapped in a zonal flow across which mixing into the polar vortex is achieved only with great difficulty.

The spread of the June 1991 eruption plume of Mount Pinatubo in the Philippines has recently lent further insight into mixing into the stratospheric polar vortex. The eruption plume circled the Earth in about 22 days at a speed of 21 m s^{-1} (Bluth *et al.* 1992) and an altitude of 15–30 km (Trepte *et al.* 1993). The plume initially stayed relatively confined in latitude, but eventually did show significant dispersal within two months of the eruption (Figure 11.25). Unlike the plume from Mount Hudson, the Pinatubo aerosol did not enter the south polar region until late November 1991 because the aerosol was unable to penetrate the polar vortex that was present at the plume's altitude of residence (Pitts and Thomason 1993).

Tracking of the Pinatubo plume from its initiation near the equator into the polar vortex allowed visualization of the flow that brings equatorial air to the poles (Trepte *et al.* 1993). The flow of air in the equatorial middle stratosphere is dominated by the quasi-biennial oscillation (QBO), which is an oscillation approximately every two years between an easterly and a westerly flow. During its easterly phase, the QBO has the same directional sense as the zonal flows to north and south, and stirring of equatorial air is inhibited. During its westerly phase, however, mixing of equatorial air into the poles is possible, as the planetary waves generated by the opposing flows break in a hemispheric "surf zone" (McIntyre and Palmer 1983). The Pinatubo aerosol was seen to be drawn out in filaments from the equatorial reservoir, across the mid-latitudes and finally into the polar vortex by energetic activity in the stratospheric surf zone of the Southern Hemisphere. In contrast, in the Northern Hemisphere, most of the poleward transport of air resulted from circulation patterns probably related to tropospheric anticyclones, consequently meridional transport towards the North Pole occurred at lower altitudes.

Another eruption that penetrated into the equatorial middle stratosphere was that of April 4, 1982 from El Chichon, Mexico. The transport of the El Chichon plume shows that the atmosphere can sometimes inhibit plume spreading, as well as cause plume spreading. Although a short-lived ash plume spread eastward to Haiti with upper tropospheric winds (Rampino and Self 1984), the main plume spread rapidly westward because the QBO and

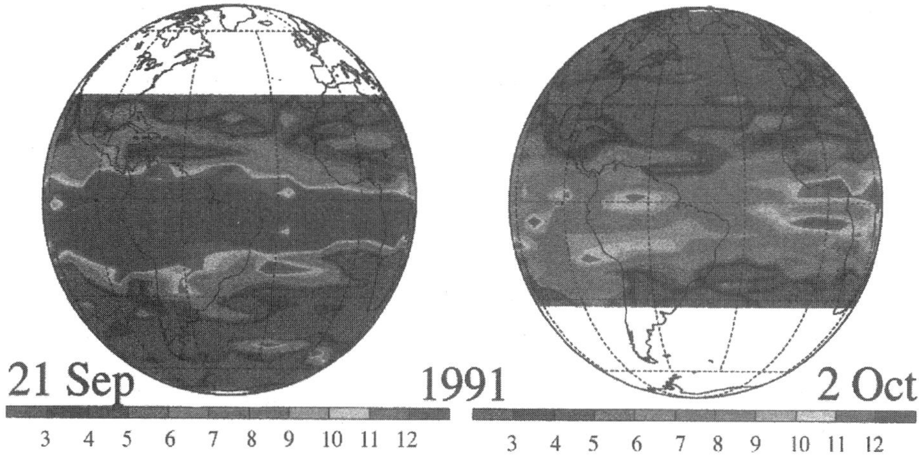

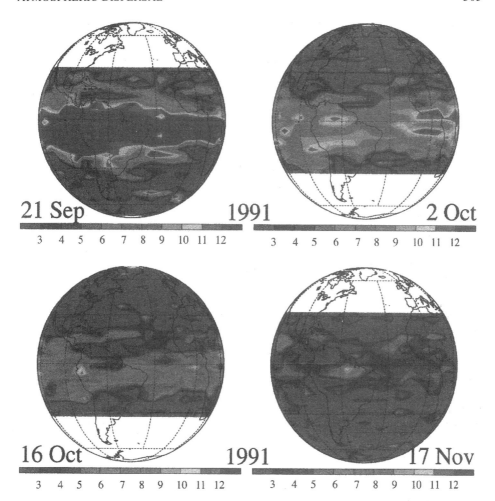

Figure 11.25 The decay over time of the Pinatubo SO_2 plume as measured by the MLS satellite instrument. (See Table 12.1 for a list of satellite acronyms.) The plume showed considerable hemispheric asymmetry. (The image was processed and provided by W. G. Read and J. W. Waters.)

the Northern Hemispheric stratospheric summer winds were both in an easterly phase (Robock and Matson 1983). Satellite imagery showed the main portion of the plume travelling westward until it had circled the globe in about 21 days at an average velocity of 22 m s^{-1}. The plume initially extended from about 16 to 30 km in altitude, with a peak concentration at 27 km that descended because of aerosol sedimentation to 21 km by the end of 1982 (Pollack *et al.* 1983). The majority of the plume remained confined to a small latitude range for more than six months, with little movement past 30°N and 10°S latitude. Late in the summer of 1982 as the summer northern polar vortex weakened, meridional transport of the plume mass to the northern mid-latitudes occurred as material “fingered” from the tropics, in a manner similar to the southward transport of the Pinatubo plume. Approximately five times less material was transported into the Southern Hemisphere, however, because the plume's southward advance was hindered by a horizontal wind shear (Pollack *et al.* 1983; Trepte *et al.* 1993).

11.5 SUMMARY

Volcanic eruption plumes and thermals are dispersed in the atmosphere at and below their maximum rise height. Horizontal momentum is gained by the plume either by entrainment of moving air or by momentum transfer from the wind that moves around the plume. Strong volcanic plumes can be defined as those with characteristic velocities greater than the atmospheric wind velocities that affect them. Such plumes are not significantly distorted during ascent and spread as umbrella clouds above their height of neutral buoyancy under gravity. The umbrella cloud can gradually gain horizontal momentum and develops into a horizontal advected plume which continues to spread sideways under gravity. The largest umbrella clouds can spread to radial distances of hundreds of kilometres and can move large distances upwind. Weak plumes are defined as plumes with characteristic velocities that are smaller than the atmospheric wind velocities that affect them. Such plumes are bent over as they rise and can undergo complex interactions. Phenomena such as plume bifurcation into two lobes and formation of downwind tornadoes can occur. Characteristically weak plumes are distorted into two counter-rotating vortex tubes with their long axes in the downwind direction. Volcanic plumes which attain heights greater than 20 km classify as strong plumes. Volcanic plumes in the range 10–20 km can be strong or weak depending on the strength of the wind. Most plumes less than 10 km high are weak and are blown over in typical atmospheric conditions under all but the most tranquil conditions. Large-scale atmospheric circulation, together with settling processes and turbulent diffusion, can eventually transport volcanic material around the globe.

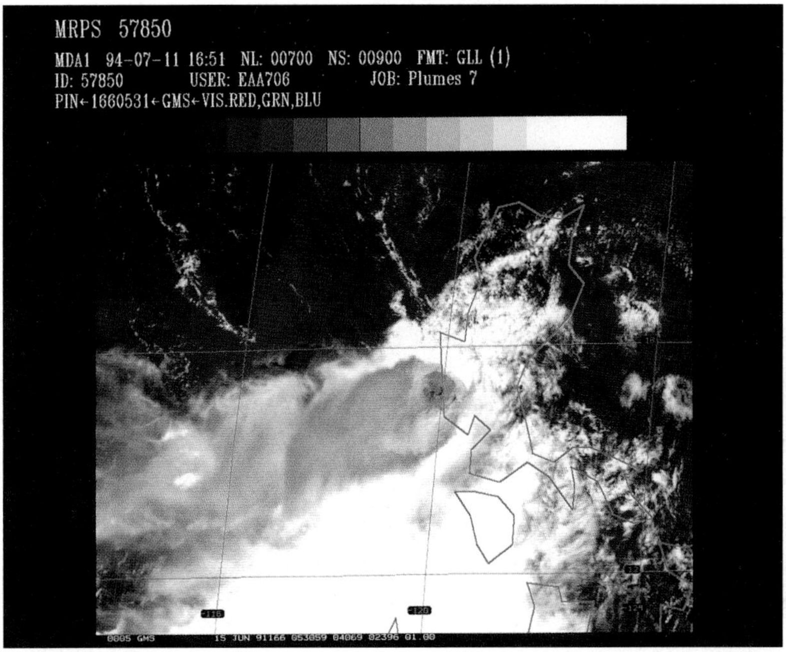

Plate Ia For caption see Plate 1c.

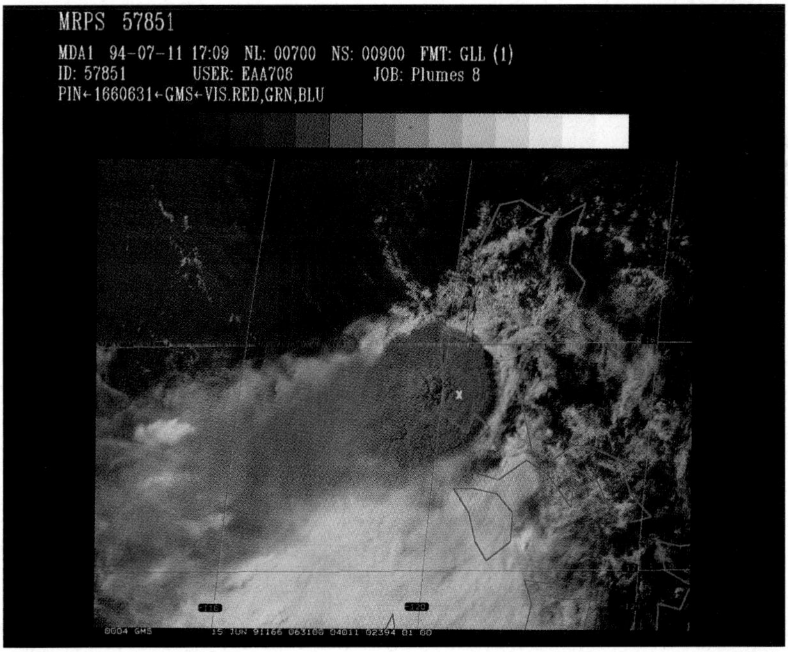

Plate Ib For caption see Plate 1c.

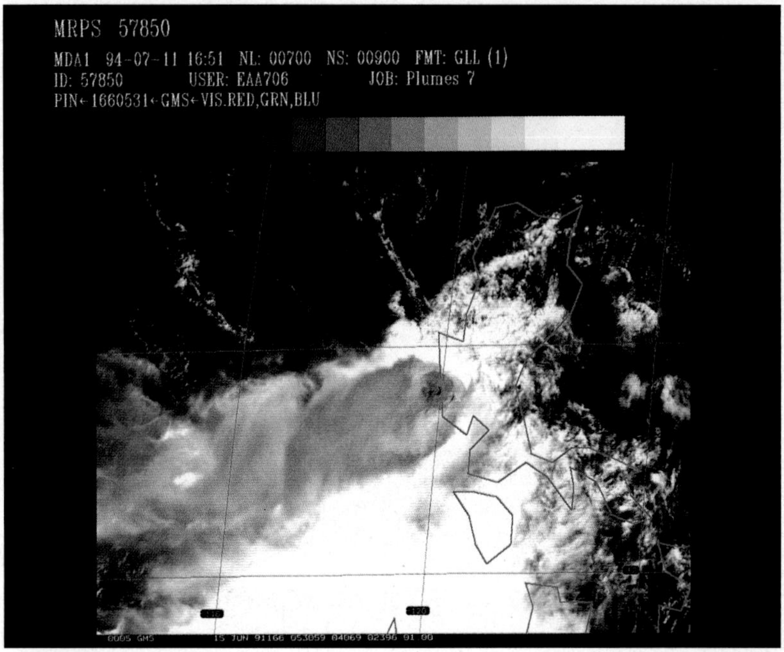

Plate I Figure 11.18 Japanese Geostationary Meteorological Satellite (GMS) images of the gigantic June 15, 1991 umbrella cloud of Pinatubo, Philippines. Outline of the Philippines and latitudes and longitudes provided for scale. Central upswelling of plume is apparently offset from volcano (5), due to geometric distortion. Images were acquired at approximately 0540, 0640 and 0740 hours UT. Images are at visible wavelengths with a ground resolution of 1.25 km at nadir. (Images processed and provided by R. Holasek.)

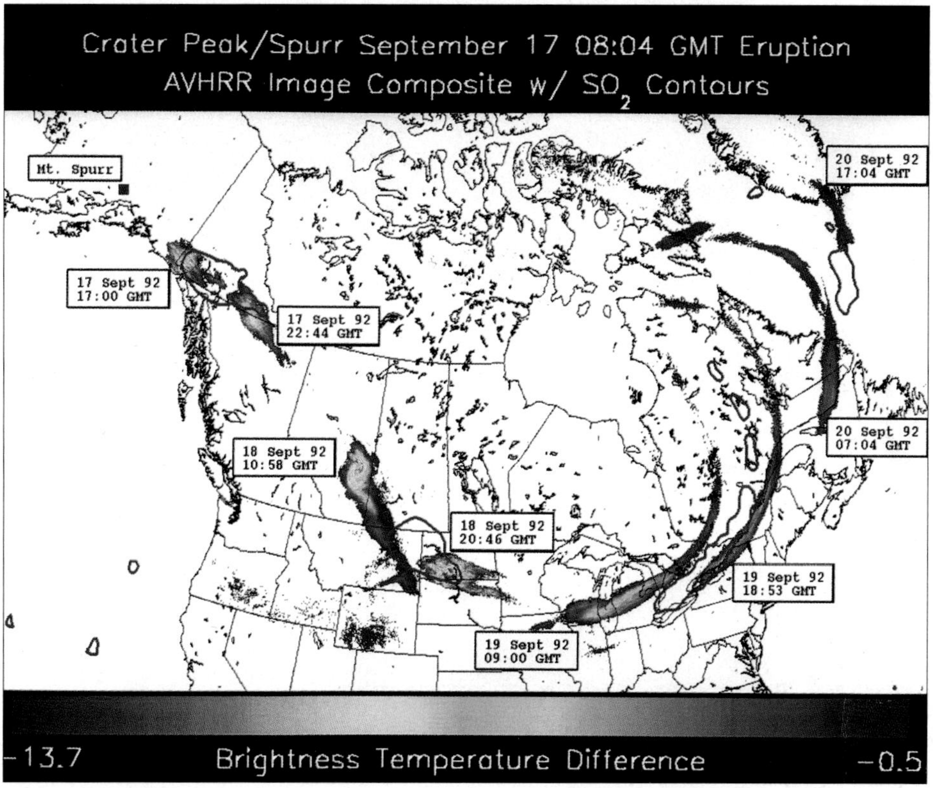

Plate II Figure 11.23 Image showing the transport of the September 17, 1992 Spurr eruption plume over a period of four days. The plume was primarily advected and distorted in the Range I atmospheric motions, with little Range II diffusion until being disarticulated in the polar jet. The image is a compilation of several AVHRR scenes (shown in colour), and several TOMS images (red outlines) showing the transport of volcanic $S0_2$. (See Table 12.1 for a list of satellite acronyms.) The TOMS images were not acquired simultaneously with the AVHRR images. A discussion of these data types and their interpretation is given in Chapter 12. (These data were processed and provided by D. Schneider and W. I. Rose.)

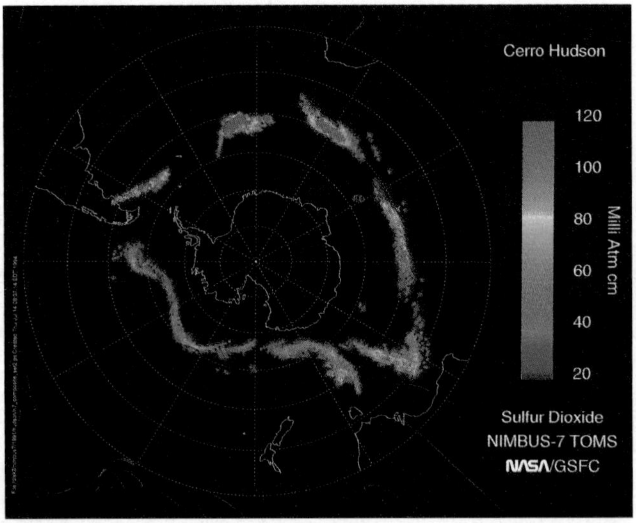

Plate III Figure 11.24 Composite image of TOMS SO_2 data for a period of eight days following the August 15, 1991 eruption of Hudson volcano in Chile. (See Table 12.1 for a list of satellite acronyms.) The eruption cloud travelled eastward around the pole, showing little evidence of diffusion by Range II eddies, but considerable Range I distortion. The images of the plume on days 7 and 8 overlap. (The image was processed and provided by I. Sprod and A. J. Krueger.)

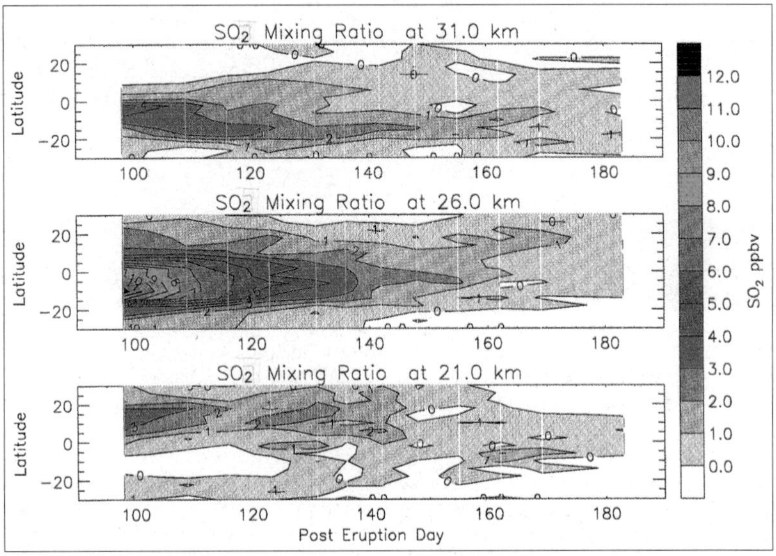

Plate IV Figure 12.11 Plot of mixing ratios for SO_2 at three different levels in the atmosphere as measured by MLS following the eruption of Pinatubo. Each of the plots shows the variations in SO_2 mixing ratio as functions of latitude and time. Most of the Pinatubo SO_2 was concentrated in a region 10°S of the equator and 26 km altitude. Note that the MLS began operation almost 100 days after the June 15, 1991 eruption and continued to measure anomalously high concentrations of SO_2 for almost two more months. (This figure is reproduced with the permission of W. G. Read from Read *et al.* (1993), copyright by the American Geophysical Union.)

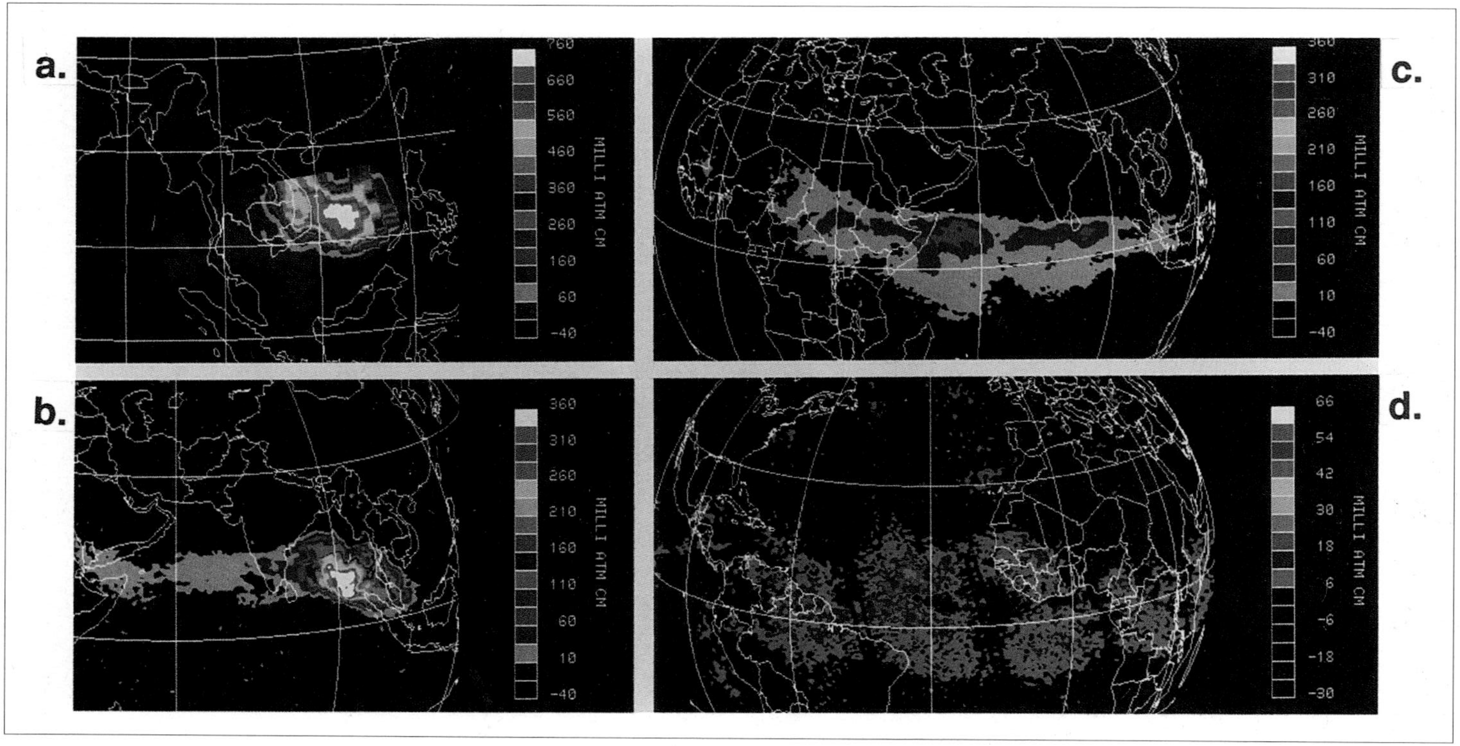

Plate V Figure 18.9 The westerly drift of the Pinatubo sulphur dioxide cloud on (a) June 16, (b) June 18, (c) June 23 and (d) June 30, 1991 (Bluth *et al.* 1992). These false-colour images are produced from TOMS satellite data, and show the sulphur dioxide volumes over column areas in units of milliatmosphere-centimetres (the amount of gas affecting the reflection of ultraviolet light through a scanning column from the satellite to the Earth's surface, given in terms of the one-dimensional thickness of the pure gas layer at STP). During this two-week period the sulphur dioxide mass of the cloud decreased by about a third, due to gas-to-particle conversion, with formation of a stratospheric aerosol

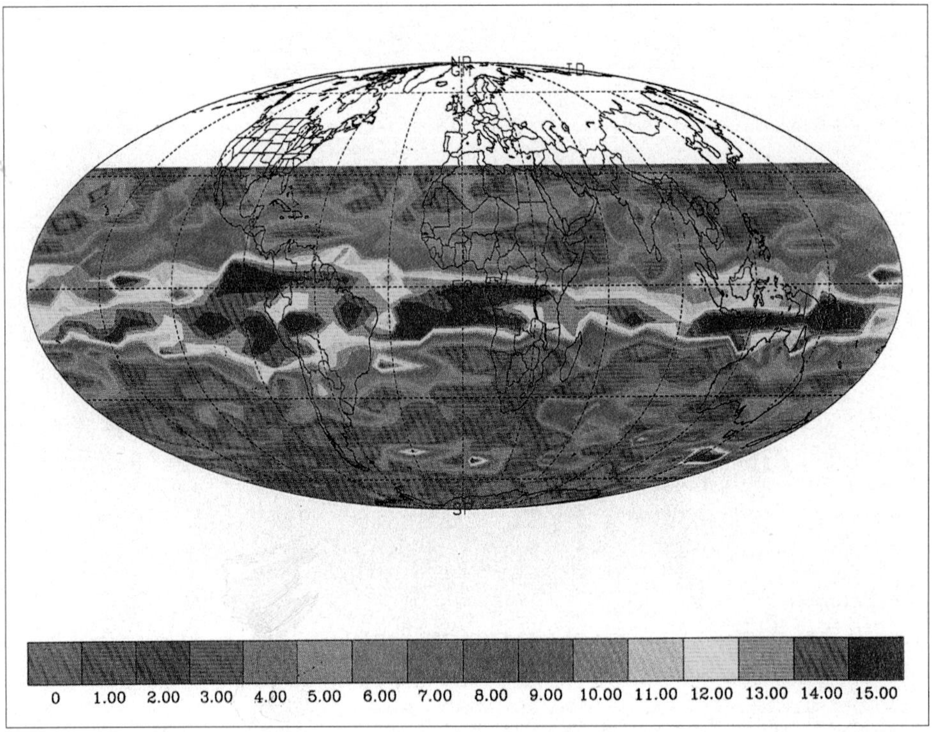

Plate VI Figure 18.10 The global distribution of sulphur dioxide at the 26 km level on September 21, 1991, approximately three months after the Pinatubo eruption (Read *et al.* 1993). The colour bar units are in parts per billion by volume

12 Remote Sensing of Volcanic Plumes

NOTATION

A	fractional area of pixel occupied by plume
C	speed of sound (m s^{-1})
c	speed of light in vacuum (3×10^8 m s^{-1})
D	declination (°)
D_f	Doppler shift (Hz)
f	frequency (Hz)
h	Planck's constant (6.266×10^{-34} J s)
I	radiance received at the sensor (W m^{-2} μm^{-1} $ster^{-1}$)
k	Boltzmann's constant (1.38×10^{-23} J K^{-1})
L	latitude (°)
R	spectral radiance (W m^{-2} μm^{-1} $ster^{-1}$)
T	temperature (K)
T_p	temperature of plume material (K)
T_s	surface temperature (K)
t	transmissivity coefficient for plume material (dimensionless)
u	vertical velocity (m s^{-1})
α	angle used in acoustic measurements (°)
ε	emissivity (dimensionless)
ϕ	azimuthal angle of sun from the north (°)
λ	wavelength (m)
θ	angle of sun above the horizon (°)
ξ	angle used in deriving azimuthal sun angle (°)

12.1 INTRODUCTION

Volcanic plumes are best observed from a distance. Quiescent fumaroles and small eruption plumes can be observed from the ground or from aircraft. Large-magnitude explosive eruptions are often only observable in their entire extent by remote sensing satellites which have recorded unique information on many of the most powerful eruptions, including Mount St Helens in the USA, El Chichon in Mexico and Pinatubo in the Philippines. A wide variety of remote sensing data are currently available for use in observing volcanic plumes. The term *remote sensing* applies to all observations where direct physical contact with the object of interest is not made and includes measurements of such quantities as acoustic vibrations, electric fields and electromagnetic radiation (a photograph is considered remotely sensed information).

The first five sections of this chapter will concentrate on measurements of variations in electromagnetic radiation. Electromagnetic remote sensing data can be acquired over a broad range of spectral wavelengths at a variety of spectral, spatial, radiometric and temporal scales. Each of these factors, as well as the continuity of the observations, determines the information that can be derived. Ground- and aircraft-based instruments can often provide detailed spectral and/or spatial information on portions of a plume which can then be interpolated to estimate a given property for the entire plume. Satellite-borne instruments, on the other hand, cannot usually offer this kind of spectrally or spatially detailed information, but are well suited for making regular observations of an entire plume in a single synoptic view. Satellite instruments designed to monitor weather patterns are useful for observing volcanic plumes and mitigating the risks posed by ash plumes to aircraft. Measurements can be made of several physical plume parameters such as rise rate, radius, radial spreading rate, downwind spreading rate, temperature structure and height of the plume top. This information can constrain models describing plume behaviour. Another important parameter that can be measured by remote sensing is the amount of volcanic SO_2 output, considered by many to be climatically important (see Chapter 18).

Volcanic plumes can also be remotely monitored by studies of physical parameters other than electromagnetic radiation, including electric fields and acoustic vibrations. Sections 12.6 and 12.7 discuss principles and present data for remote electric potential gradient measurements and acoustic monitoring of volcanic plumes. While few studies have been made, results suggest that electric fields can detect and track plumes, and that acoustic vibrations can estimate particle velocities within an eruption column.

12.2 PRINCIPLES OF ELECTROMAGNETIC THEORY

In order to interpret the information acquired by remote sensing instruments the characteristics of electromagnetic energy should be understood. The electromagnetic spectrum is divided into several regions based on wavelength (Figure 12.1). Energy throughout the spectrum is emitted both from the Sun and from any material at a temperature above absolute zero (0 K, −273 °C), although some materials are more efficient emitters than others. The most efficient emitter is called a *blackbody* which emits all of the energy that it absorbs. A perfect blackbody has an emissivity, ε, of 1 at all wavelengths, while less efficient emitters have emissivities less than 1 that can vary with wavelength. The amount of spectral radiance, R, emitted by an object at a temperature, T, can be determined as a function of wavelength, λ, by using Planck's law, defined as

$$R(\lambda) = \varepsilon \frac{2\pi h c^2 \lambda^{-5}}{\pi \left[\exp\left(\frac{hc}{\lambda k T} \right) - 1 \right]} \qquad (12.1)$$

where h is Planck's constant $= 6.266 \times 10^{-34}$ J s, c the speed of light in a vacuum $= 3 \times 10^{8}$ m s^{-1}, k is Boltzmann's constant $= 1.38 \times 10^{-23}$ J K^{-1}, T is in kelvin, and λ is in metres. Spectral radiance is usually given in watts per metre squared per

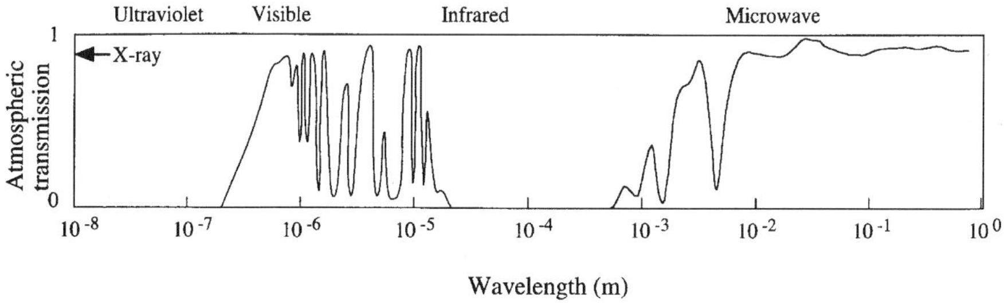

Figure 12.1 Plot of atmospheric transmission across the electromagnetic spectrum. A transmission of 1 (100%) indicates an atmospheric window where radiant energy of a specified wavelength is able to pass through the atmosphere without being absorbed or scattered. The names of the various wavelength regions of the spectrum are indicated

micrometre per steradian (W m^{-2} μm^{-1} $ster^{-1}$). Figure 12.2 shows several blackbody Planck curves for different temperatures. As the temperature of the emitting blackbody increases, the wavelength of maximum emitted radiance decreases. Thus, if the peak of the blackbody curve can be ascertained, the temperature of the blackbody is then uniquely known. Temperatures within the range experienced daily on Earth (-50 to 100 °C) have their peaks in the thermal infrared (hence the name). The peak for higher temperatures (400–500 °C) occurs in the short-wavelength infrared, and for very high temperatures (>600 °C) in the red end of the visible range, explaining the glow that can be seen from objects that are red-hot.

The radiation received by a detector travels a complicated path. Interpretation of the measured radiation intensity requires an understanding of how the energy got there. Transmission, absorption, scattering and reflection all occur when radiation emitted from one source travels through space or encounters an object in its path (Figure 12.3). The occurrence of these processes depends upon the wavelength of the radiation, the composition of the transmitting medium, and the size, shape and composition of objects in the radiation path. In the simplest case, electromagnetic waves travel through a vacuum with 100% transmission (no attenuation). If, however, a gas is present that absorbs 10% of the incident radiation at a particular wavelength, only 90% will be transmitted. Radiation can also be diverted by scattering or reflectance. The total amount of radiant energy, however, is always conserved (often referred to as Kirchhoff's law). Energy radiated by the sun is first transmitted through the Earth's atmosphere where it may be scattered by aerosols, or absorbed by chemical constituents (e.g. ozone has very strong absorptions in the ultraviolet region). If a measuring instrument is directed upwards, the radiation received at the detector is simply the radiation from the sun that has survived transmission through the atmosphere. If a measuring instrument is directed downwards (from an aeroplane or satellite) the solar radiation is reflected from the ground and/or plume and then transmitted again back to the detector. Some portion of the radiation can be absorbed and scattered during both trips through the atmosphere or may be absorbed by the surface and re-emitted at a longer wavelength (e.g. the Earth's surface heated through absorption of radiation at visible wavelengths subsequently emits radiation at a new warmer temperature). The radiation received at the detector is a combination of the partially transmitted and scattered

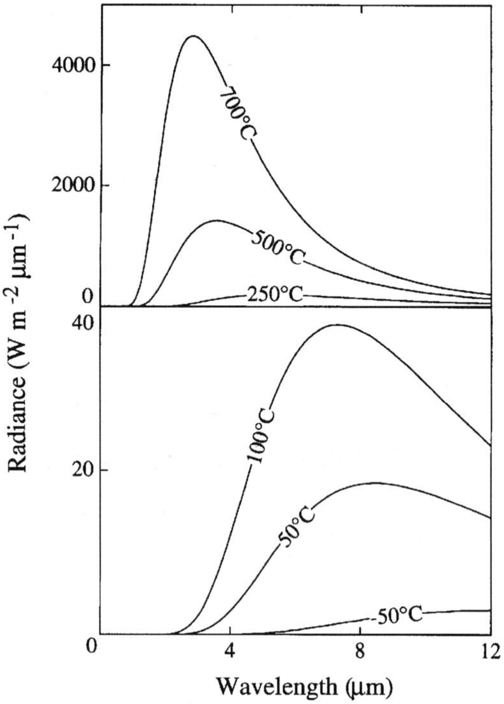

Figure 12.2 Emitted radiance as defined by the Planck function (equation 12.1) in the visible and infrared spectral regions for blackbodies with temperatures of −50 °C, 50 °C, 100 °C, 250 °C, 500 °C and 700 °C. Note that wavelengths are given in micrometres (10^{-6} m) and that the top and bottom plots have different radiance scales. Note also that peak emitted radiances occur at lower wavelengths for higher temperatures

solar radiation and any emitted radiation from atmospheric constituents or the surface itself.

The instruments described in sections 12.3, 12.4 and 12.5 operate on the principle that all materials reflect, emit, and transmit electromagnetic radiation to varying degrees. While all of the instruments measure electromagnetic radiation, the ways in which their detectors function vary. Unless otherwise stated, the instruments use electro-optical detectors that measure the intensity of radiation over a certain wavelength range. The radiation received at an electro-optical detector is transformed to electrical energy, the intensity of which is assigned a digital number within the detection limits of the instrument. Digital numbers, more commonly called *DN*, are integer values that usually lie between 0 and 255, where 0 corresponds to the lower detection limit of the instrument and 255 corresponds to the upper limit. These limits define the dynamic range of an instrument at the prescribed wavelength range. The process of converting the *DN* value into a real radiance value requires instrument calibration, which is instrument specific. Many instruments have a linear relationship between *DN* and radiance such that

$$R(DN) = (\text{gain}) \times (DN) + \text{offset} \tag{12.2}$$

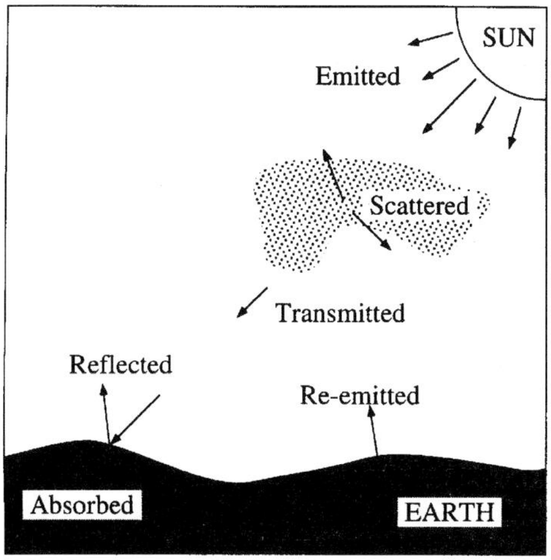

Figure 12.3 Diagram illustrating one possible progression of electromagnetic radiation through the atmosphere. In this figure, the radiation is originally emitted by the sun. As it passes through a gaseous cloud, some of the radiation is scattered, and some is transmitted along the original path. At the ground, some of the radiation is reflected while the remainder is absorbed and then re-emitted at a longer wavelength

where the offset is the minimum radiance detectable by the instrument (corresponding to $DN = 0$) and the gain defines the range over which the instrument is sensitive. The gain and offset values for each image are usually provided with the digital data. Some instruments have more complicated algorithms for radiance retrieval.

The reasons why particular materials possess certain spectral properties is explained by complex theories of the wave nature and the quantum nature of electromagnetic radiation. For thorough discussions of these topics as well as the mechanics of remote sensing instruments, the reader is directed to the books by Siegal and Gillespie (1980), Colwell (1983), Elachi (1987), and Rees (1990).

12.3 ELECTROMAGNETIC REMOTE SENSING BASICS

Many remote sensing instruments measure the amount of electromagnetic radiation. In order to interpret such data, it is necessary to understand several basic concepts. In this section some of the information used in remote sensing is introduced and terms used throughout this chapter are defined. The three end-member instrument types are radiometers, spectrometers and imagers. A radiometer measures the intensity of radiation over a broad spectral region for a single location (not necessarily small). A radiometer is used when radiation intensity is important in a certain wavelength range and when information on spatial variability is not needed. A spectrometer looks only at a single point, but has much higher spectral resolution meaning that it will have two or more, typically hundreds, of narrow (spanning a few nanometres or less) spectral bands over a particular region of the spectrum. Spectrometers are often used to identify the unique spectrum of a particular material on a surface or along the path of observation. An imager creates a two-dimensional image of spectral information. For example, a photographic camera is a two-dimensional imager that operates over a broad spectral range in the visible region. In many cases instruments are a combination of these end-members. For example, an imaging spectrometer may acquire a two-dimensional image in as many as 200 spectral channels.

12.3.1 Spectral Region and Resolution

The spectral region in which data are acquired determines their interpretation. For example, data collected in the visible region are interpreted as reflected sunlight, while data in the thermal infrared are interpreted as temperature variations. The number of channels and the wavelength range over which each channel is sensitive (referred to as the channel width, or *spectral resolution*) are also important. Radiometers have one or two very broad spectral channels that receive radiance averaged over an entire wavelength range, while spectrometers have many narrow channels that can detect specific spectral features that would be lost in a broad channel. The radiation emitted or reflected from the object can be attenuated by the atmospheric constituents. Water vapour is a major component of the atmosphere and absorbs radiation at many wavelengths. For this reason the wavelength ranges of many channels are designed to correspond to atmospheric *windows* (see Figure 12.1), or regions where absorption by water vapour is low (with the exception of channels specifically designed to measure water vapour). The dynamic range for a particular channel is also determined by the intended application. In this section several volcanic plume observations that can be made in each spectral region are described. More detailed discussions and references for each of the measurement techniques are given later in the chapter.

Ultraviolet region

Instruments that measure ultraviolet (UV) radiation can detect and monitor volcanic SO_2, which has several distinct absorption features between 0.297 and 0.315 μm. Some of these instruments were originally designed to measure ozone, which also has several absorption features in the UV region between 0.3 and 0.4 μm. Upward-looking ground- and aircraft-based systems measure the amount of incoming radiation in each channel. Downward-directed satellite UV instruments measure the amount of upwelling backscattered UV radiation. The principal difficulty with using UV instruments is that the signal is small and so optimal measurements are made when the largest magnitude signal is received. This occurs at local noon when the solar radiation takes the shortest possible path to the instrument. Because the detection of UV radiation requires an emitting UV source such as the Sun, measurements cannot be made at night.

Visible region

Visible (VIS) and short-wavelength infrared (SWIR) imagers and spectrometers are primarily intended for regional mapping of the Earth's surface or for observing weather patterns. The VIS channels usually measure reflected radiance in one broad or several narrow bands between 0.4 and 0.7 μm. Visible data acquired from a satellite platform can be very useful for simple plume tracking and for making measurements of physical plume properties. Hand-held still, movie and video cameras all operate in a single broad visible channel. In contrast to the digital, electro-optical instruments discussed previously (section 12.2), cameras are analogue instruments that function by exposing film sensitive to visible radiation. Chapter 5 discusses the studies that have used cameras to observe plumes in more detail.

The SWIR channels measure solar reflectivity just beyond the visible range from 0.7 to 3.5 μm. Similar to the UV region, VIS and SWIR measurements must be made during daylight hours because the energy being measured is all reflected radiation (which requires original emitted radiation from the Sun). Very high temperature phenomena (e.g. lava flows) can also be observed in the SWIR region. High-temperature measurements, however, are not relevant to volcanic plume observations. Rothery *et al.* (1988) and Oppenheimer *et al.* (1993) discuss the theory behind measurements of high-temperature volcanic phenomena.

Mid- and thermal infrared regions

The mid-infrared (MIR) region between 3.5 and 8 μm is dominated by broad bands of water vapour absorption. Several weather satellites have channels in the MIR specifically to determine abundances of water. Thermal infrared (TIR) instruments generally measure the thermal emittance of the target between 8 and 12 μm. These instruments have determined the two-dimensional temperature variations over volcanic plume tops. Useful data in the TIR can be recorded both day and night.

Another type of instrument that measures electromagnetic radiation in the VIS and infrared regions is lidar. All of the other instruments described thus far derive information on an object by measuring the attenuated radiation reaching the detector over a finite range of wavelengths. This type of remote sensing is referred to as *passive* remote sensing because the measured radiation is either emitted by the observed object, or emitted by an external source (e.g. the Sun) and then reflected off the observed object. Lidar operates differently. Lidar provides its own radiation source by emitting a laser beam at a specific wavelength chosen for looking at a particular material. This type of remote sensing, where the instrument provides the radiation source is called *active* remote sensing. The beam emitted by the lidar instrument is absorbed, reflected and scattered as it progresses through the atmosphere. Radiation scattered back to the detector is then measured to obtain concentration estimates for the atmospheric constituent of interest. Volcanic SO_2 and aerosols are often measured by these instruments.

Microwave region

Instruments that collect data in the microwave (MW) region (1 mm–1 m) can also operate as thermal emission radiometers and are usually used to measure atmospheric temperatures and gas concentrations such as H_2O, ClO, NO_2 and HCl as functions of altitude, often referred to as atmospheric sounding. Most radar systems also operate in the MW or radio wavelength region and are often used to derive topography and to measure surface roughness. Radar has been used in a few studies of volcanic plumes to estimate particle concentrations and to determine maximum plume heights. Radar systems are similar to lidar in that a signal is sent out and the amount of radiation scattered back to the detector is measured. Two advantages of radar systems are that data can be collected day or night, and there is very little water vapour absorption at such long wavelengths, so that observations can be made through clouds.

12.3.2 Spatial Resolution

Spatial resolution of an instrument influences the kinds of measurements that can be made. In general, an instrument's spatial resolution is controlled by the instantaneous field of view (IFOV) of the detector (defined as the smallest area that can be resolved by a detector element) and the distance to the target. The IFOV is usually given in radians and refers to the viewing angle of the instrument detector. Radiometers and spectrometers often have small IFOVs. A field of view of a few microradians results in a *footprint* of several square metres for a target a few hundred metres away. If the observed object is not homogeneous in both composition and temperature within the field of view, the instrument will detect an area weighted average of the radiance emitted and reflected from the components. Imagers operate in much the same way as radiometers, except that instead of having a single detector, they generally have a two-dimensional array of detectors. Each of the detectors behaves as the single radiometer detector with a finite field of view. The information recorded by each detector is referred to as a picture element, or *pixel*, and the sampling interval of the detector is called the pixel size.

Because most of these instruments have similar instantaneous fields of view, the difference in spatial resolution is primarily dependent on the distance between the instrument and the target. In most cases this distance is determined by the instrument *platform*. Three types of observation platforms are considered: ground, aircraft and satellite. Ground-based instruments range from hand-held, field portable cameras or imaging radiometers, to spectrometers mounted on cars or boats, to permanently installed lidar facilities. Instruments carried on aircraft platforms range from those that are easily installed on small Cessnas, to those permanent on much larger aircraft such as the DC-8 or ER-2 (Rothery and Pieri 1993). Satellite-based instruments are, of course, permanently mounted on their platforms, and can operate for several years.

12.3.3 Observation Opportunities

A major factor determining the usefulness of remote sensing measurements is the timing of, and the repeat interval between, the observations. Portable ground-based instruments are usually taken into the field and operated by individuals. The advantage of field instruments is that they often have much higher spatial resolution than similar instruments on other platforms. In general, measurements made in this manner are costly and have not led to routine sources of data. There are, however, two notable exceptions. The first is permanent ground-based lidar facilities. These collect data on atmospheric gases and aerosol concentrations on a consistent and continual basis. They have provided vast databases of information on the location and size of several large volcanic gas and aerosol plumes. The second exception includes measurements of volcanic SO_2 emissions collected at quiescently degassing volcanoes. Most of these data have been collected using the same instrument, called a correlation spectrometer (COSPEC).

Portable instruments can also be installed on small aircraft. These types of measurements are often restricted by the ability to get the instrument to an eruption and to find an appropriate and available aircraft. Several aircraft around the world are permanently equipped with a variety of remote sensing instrument packages. These aircraft are in general owned and operated by their respective governments and the data collected by the instruments are often available to the ardent investigator willing to sort through extensive archives. The primary difficulty with data collected by government operated aircraft is that they are often scheduled for deployments months or years in advance. For this reason data for volcanic eruption plumes acquired from these platforms are rare. In a few instances, however, larger organized efforts have been made to get appropriate instruments to useful locations in order to make measurements. For example, following the eruption of Pinatubo in June 1991, a US National Aeronautics and Space Administration (NASA) Electra aircraft equipped with several instruments including a depolarization lidar and a COSPEC was quickly dispatched to the Caribbean, over which the Pinatubo plume was passing.

Much useful information for studying volcanic plumes has been derived from satellite-based remote sensing data. In some cases, where a volcano has erupted in a region that is not easily accessible, for geographical or perhaps political reasons, the information gained from satellite images may have been difficult to derive by other more conventional means. For well documented eruptions, satellite remote sensing data can complement field observations of physical parameters, and may provide the only means for estimating parameters such as the temperature of the plume top. There are currently dozens of satellites in orbit about the Earth carrying a multitude of instruments controlled and operated by a variety of countries (Curtis 1994). Many more are planned for launch by the beginning of the twenty-first century as part of an international project known as the Earth Observing System (EOS) which is headed by NASA. Unfortunately, the list of acronyms that accompanies the lists of instruments and satellites is extremely long! The acronyms used in this chapter are given in Table 12.1. Table 12.2 lists the instruments mentioned in this chapter as well as their satellite platform, spectral ranges, spatial resolution and the most common applications to volcanic plume studies. Instruments currently in operation cover a wide range of spectral and spatial resolutions. The frequency of observation opportunities ranges from a few minutes to over two weeks. Some geological mapping satellites only collect data on an advance request basis and are, therefore, not optimal for observations of unpredicted and short-lived phenomena. In contrast, weather satellites gather consistent data on a regular basis. This not only greatly increases the chances of catching an eruption in progress, or of measuring the effects of an eruption, but also allows for comparisons between measurements separated in time and space. Lists of volcanic plumes observed from satellites are given in Sawada (1983) and Holasek and Rose (1991).

Another restriction on satellite observation frequency is the presence of meteoric cloud which obscures features of interest. Only instruments with detectors in the microwave region are able to see through meteoric clouds. All other instruments are virtually useless for seeing beneath opaque clouds. Fortunately, volcanic plumes often rise into the stratosphere, above the cloud layer.

While many satellites acquire data continuously or on a regular basis, there are several factors inherent to satellite platforms that determine how often an instrument will actually be able to see a particular spot on the Earth's surface. Because many of the studies

Table 12.1 Acronym definitions for satellites and satellite-borne instruments

Acronym	Definition
AVHRR	Advanced very high resolution radiometer
ERBS	Earth radiation budget satellite
GMS	Geostationary meteorological satellite
GOES	Geostationary operational environmental satellite
HIRS 2	High-resolution infrared sounder
MLS	Microwave limb sounder
MSS	Multi-spectral scanner
NOAA	National Oceanographic and Atmospheric Administration
SAGE II	Stratospheric aerosol and gas experiment II
SPOT	Système probatoire d'observation de la terre
TM	Thematic mapper
TOMS	Total ozone mapping spectrometer
UARS	Upper atmosphere research satellite
VISSR	Visible and infrared spin scan radiometer

Table 12.2 Satellites and their instruments (all acronyms are defined in Table 12.1)

Satellite	Instruments	Applications to volcanic plumes	Spectral ranges of channels used for volcanic studies	Resolution
ERBS	SAGE II	Detect volcanic aerosols	VIS, SWIR	Low
GOES (-East, -West)	VISSR	Regional plume tracking, plume temperature and height	VIS, TIR	Moderate
GMS	GMS	Regional plume tracking, plume temperature and height	VIS, TIR	Moderate
Insat	Insat	Regional plume tracking, plume temperature and height	VIS, TIR	Moderate
Landsat	MSS	Observations of persistent plumes	VIS, SWIR	High
	TM	Observations of persistent plumes	VIS, SWIR, TIR	High
Meteosat	Meteosat	Regional plume tracking, plume temperature and height	VIS, TIR	Moderate
Nimbus	TOMS	SO_2 detection and global tracking	UV	Low
NOAA	AVHRR	Global plume tracking, plume temperature and height	VIS, MIR, TIR	Moderate
	HIRS 2	SO_2 detection and global tracking	VIS, MIR, TIR	Low
SPOT	SPOT	Observations of persistent plumes	VIS, SWIR	High
UARS	MLS	SO_2 detection and global tracking	MW	Low

presented in sections 12.4 and 12.5 have used satellite data, a more detailed discussion of constraints on satellite observation frequency is given here. The observation frequency of a satellite instrument depends primarily on the type of orbit the satellite is in and the acquisition mode. The orbit determines how often the satellite has an observation opportunity, and the acquisition mode determines how often the instrument actually takes data when given the opportunity.

Satellites in geostationary orbit are placed in an orbit synchronous with the Earth's rotation such that they remain above a given point at all times. These satellites can image approximately one-fifth of the Earth's surface area with each data acquisition. Most geostationary instruments are capable of acquiring a "full Earth disc" image (a complete image of the surface under the orbit) in about 20 minutes (the newest generation of geostationary satellites are even quicker, capable of tracking a hurricane every few minutes). It then takes approximately 10 minutes before the instrument can begin to acquire data again. Not all geostationary instruments, however, acquire data at every 30-minute opportunity.

Because geostationary satellites have the most frequent repeat viewing opportunities of all satellites, the majority of images of the onset and development of explosive activity have been acquired by these instruments. Examples include La Soufrière of St Vincent, 1979 (Krueger 1982), Mount St Helens, 1980 (Sparks *et al.* 1986; Holasek and Self 1995), Pagan, 1981 (Sawada 1983), Galunggung, 1982 (Sawada 1983b), El Chichon, 1982 (Matson 1984, 1985), Mayon, 1984 (Sawada 1985), Lascar, 1986 (Glaze *et al.* 1989) and Pinatubo, 1991 (Koyaguchi and Tokuno 1993). There are currently five geostationary satellites in operation. They are all in orbits centred on the equator at different longitudes, two GOES at 135°W and 75°W (named -West and -East, respectively), GMS at 140°E, Insat at 94°E and Meteosat at 0°. The instruments on these five satellites cover the majority of land masses at mid-latitudes, but are able to see the higher latitudes only at oblique angles. All of the examples of volcanic plume studies listed above were acquired by either GMS or one of the GOES satellites. In 1994 a new series of GOES satellites were launched carrying multichannel instruments capable of acquiring images very rapidly. At the time of publication of this book, however, no volcanic plume observations using the new GOES data have been published in the open literature. The GMS and older GOES carry instruments that have one broad band visible and one broad band TIR channel. The GOES and GMS instruments have square pixel sizes at the equator, directly beneath the instrument (referred to as nadir looking), of 1 and 1.25 km on a side in the visible channel respectively, and 4 and 5 km on a side in the TIR channel. To date, studies using either Meteosat or Insat for volcanic plume observations have not been published in the general literature. For more detailed information on the GOES satellites, see Glaze *et al.* (1989) and for the GMS satellite.

Other satellites orbit the Earth at varying angles of inclination to the equator. Their instruments are continually looking at a different location on the surface. An instrument's ability to see the highest latitudes depends upon the inclination of the orbit, where an inclination of 90° is a truly polar orbit, and an inclination of 0° is an equatorial orbit. The frequency of repeat observation opportunities by an instrument on a polar orbiting satellite depends upon the *swath width* of the instrument. The swath width is the actual width of the image measured on the ground. This width is a function of the number of pixels in a scan line, and the area on the ground represented by each pixel. Low spatial resolution instruments have large pixel sizes of the order of 10–50 km on a side and corresponding swath widths of the order of several thousand kilometres. These instruments can observe the same spot on the Earth at least once a day, and sometimes twice (once during the day and once at night). Instruments with higher resolution pixel sizes of 10–1000 m on a side have much narrower swath widths and, consequently, have much longer repeat cycles. The repeat observation time for some high spatial resolution instruments is more than a week between day and night observations, or over two weeks between successive daytime observations.

Polar-orbiting instruments generally acquire much more spectral information than do the geostationary instruments. Many also have much higher spatial resolution in their TIR channels. However, their long repeat viewing times have resulted in very few images of the initial phases of an eruption and a complete lack of time series images during the first several hours of an eruption. TOMS, SPOT, MSS, TM, AVHRR, HIRS 2 and MLS are all examples of instruments carried on polar-orbiting satellites that have been used for volcanic plume observations (see Table 12.1 for definitions). A few eruptions that have been observed from polar orbit include Mount St Helens (Danielsen 1981; Sparks *et al.* 1986), El Chichon (Stowe and Schwedfeger 1982; Bandeen and Fraser 1982; Krueger 1983; Matson 1984), Nevado del Ruiz (Krueger *et al.* 1990), Augustine (Kienle and Shaw 1979; Holasek and Rose 1991), Colo (Malingreau and Kaswanda 1986), Pinatubo (Stowe *et al.* 1992; McCormick and Veiga 1992; Bluth *et al.* 1992; Read *et al.* 1993) and Rabaul (Rose *et al.* 1995).

The SPOT, MSS and TM instruments all have relatively high spatial resolution (Figure 12.4 shows several satellite instruments along with their respective spatial resolutions and spectral ranges). The TM has a pixel size of 30×30 m in the visible and short wavelength infrared channels, and 120×120 m in the TIR channel. TM's predecessor, MSS, has a pixel size of approximately 80×80 m in all four of its channels. SPOT can operate in two modes, a multispectral mode with two visible and one very short wavelength infrared channel all with a pixel size of 20×20 m, or a panchromatic mode which has one broadband visible channel with a pixel size of 10×10 m. A consequence of high spatial resolution is a narrow swath width and a lower frequency of observation opportunities. As a result, high spatial resolution instruments such as TM, MSS and SPOT are not generally useful for observations of transient events such as volcanic eruptions because of their infrequent coverage. In addition, requests to acquire data by TM, MSS and SPOT must be placed months in advance, thus further decreasing the possibility of observing volcanic activity.

Moderate spatial resolution, AVHRR weather data have been used very successfully to monitor the distribution of ash plumes over periods of hours to days. There are often two AVHRR instruments in operation concurrently that record data in five channels in two modes, the local area coverage mode (LAC) with 1 km resolution, and the global area coverage mode (GAC) with 4 km resolution. The AVHRR also transmits 1 km resolution data directly to receiving stations in a mode called the high resolution picture transmission (HRPT) mode. The swath width for the AVHRR is 2400 km. Thus, in GAC mode the AVHRR can image every spot on the Earth at least twice a day (more often at high latitudes where adjacent swaths overlap), once during daylight hours, and once at night. Some LAC data are acquired every day on a regular basis (over most of the USA, for example) and permanently archived. Other LAC data must be requested in advance.

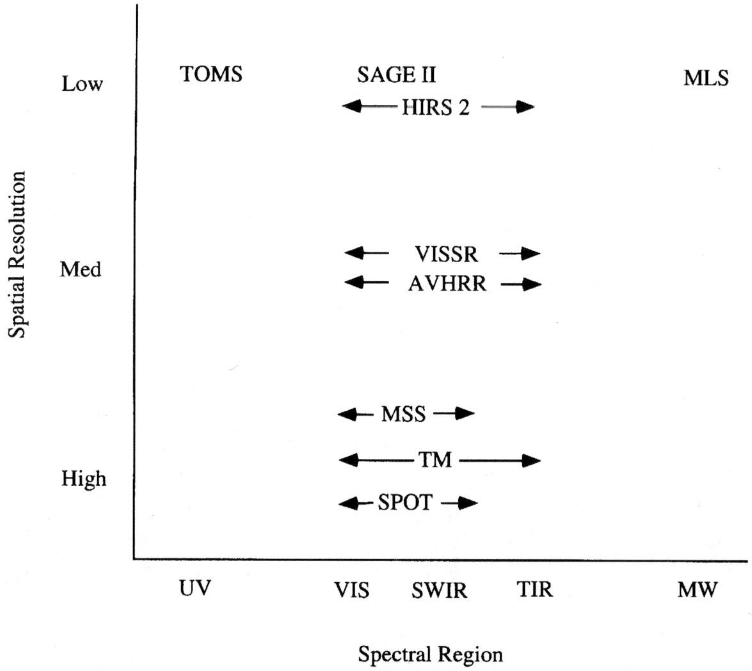

Figure 12.4 Diagram illustrating the relative spatial resolutions and operational spectral ranges of several instruments discussed in this chapter

Therefore, when conducting a retrospective study of archived data, the highest resolution LAC data may not have always been acquired over the area of interest on the necessary date. In addition to the continuation of the AVHRR series of instruments in the future, the Earth Observing System also intends to launch an instrument with many of the same capabilities, called the moderate resolution imaging spectroradiometer (MODIS). MODIS will have 36 spectral channels with varying spatial resolutions between 250 and 1000 m.

TOMS, MLS and HIRS 2 have very low spatial resolution and high spectral resolution. They were designed to map several atmospheric constituents on a daily basis. The TOMS instrument has six channels in the UV region with a pixel size of approximately 50 × 50 km. Because TOMS measures UV radiation, it is only effective during daylight hours and is at its best near local noon. TOMS was originally designed to measure O_3, but the fact that SO_2 has several absorption bands in the same region has enabled its use for tracking SO_2-rich eruption plumes. The MLS is a thermal emission radiometer that collects data in the microwave region with a spatial resolution of several hundred kilometres. It has the ability to measure several atmospheric gases including O_3, H_2O, ClO, NO_2, SO_2 and HCl, as well as temperature profiles through the atmosphere. The MLS instrument is capable of distinguishing varying amounts of each constituent at different altitudes with a vertical resolution of 3 km. The HIRS 2 instrument has one visible channel and 19 channels in the TIR region between 3.6 and 15 μm. It was designed to measure atmospheric gas constituents but can also measure volcanic SO_2.

Despite the difficulty in acquiring data from polar-orbiting instruments on the first few minutes of a volcanic eruption, a few examples do exist, including the initial giant umbrella clouds of Mount St Helens (May 18, 1980) and Pinatubo (June 15, 1991), which were both fortuitously captured by AVHRR. Unfortunately, the 1 km LAC scene of Mount St Helens has been permanently lost, leaving only 4 km GAC data for all channels. Nothing has been published on this dataset. The only TM image of a major plume in existence is of the eruption of Augustine volcano, Alaska, on March 27, 1986 (EOSAT Landsat Data User Notes, 1986 and Figure 12.5). Several images acquired by a polar orbiting military satellite during the first hour of the May 18, 1980 Mount St Helens eruption have been studied by Moore and Rice (1984) and Sparks *et al.* (1986). However, these are the only cases where time series data over such a short scale from a polar-orbiting

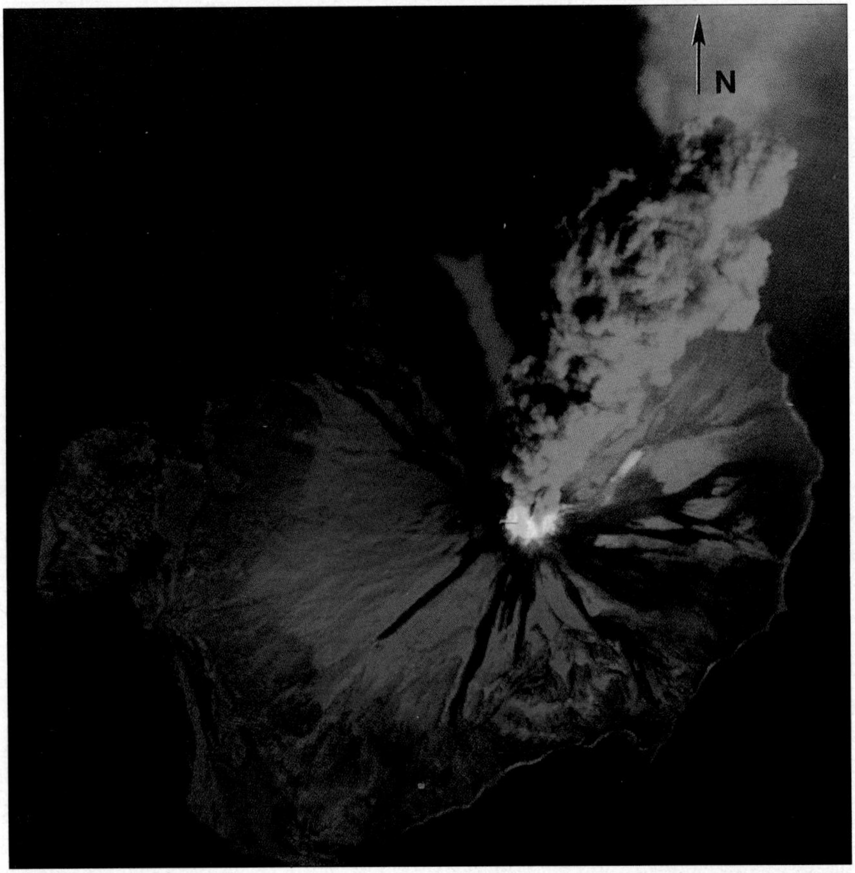

Figure 12.5 Landsat thematic mapper image of Augustine volcano, Alaska, during an explosive eruption in 1986. The image shows an area approximately 15.4 km on each side with north at the top. This false colour composite was created by displaying the thermal infrared (TIR) channel in red, one of the short wavelength infrared (SWIR) channels in green and a visible channel in blue. Note the pyroclastic flow (red in this image) that is still sufficiently warm to be detected in the TIR channel. The crater of Augustine was sufficiently hot that emitted radiance was detected by both the TIR and SWIR channels, appearing yellow in this image

satellite have been used to observe a volcanic plume. In general, data from high spatial resolution, fast snapping military satellites are not available to the general public.

12.4 DETERMINATION OF PLUME PROPERTIES

The theoretical models discussed in this book predict the behaviour of several plume properties. Two of these, plume height and plume top temperature, can be estimated using remote sensing instruments. A third measurement is the amount of SO_2 released into the stratosphere during an eruption. H_2SO_4 aerosols evolve from SO_2 through several chemical reactions and can dramatically increase the absorption of solar radiation resulting in a warming of the stratosphere and cooling of the Earth's surface (Labitzke *et al.* 1983; Chapter 18). In this section several methods for estimating each of these parameters using remote sensing data are presented.

12.4.1 Plume Height

Ground-based radar has been used to measure plume height. Two examples are the eruptions of Mount St Helens in 1980 (Harris and Rose 1987; Harris *et al.* 1981a) and in 1982 (Harris and Rose 1983). The National Weather Service (NWS) radar in Portland, Oregon, operating at a 5 cm wavelength, monitored the height of the May 18, 1980 column. The smaller March 19, 1982 ash eruption of Mount St Helens was also observed with the Portland radar. The plume height and volume were determined. Also, using an estimate of the particle size distribution, the concentration of ash in the plume and the total mass erupted were calculated. The method by which these parameters are extracted from radar data is described by Harris *et al.* (1981a).

Satellite images have been used to estimate eruption column height. Four methods for estimating plume height using satellite imagery are described here. The first method is known as the shadow method, and calculates the maximum plume height from its shadow. A second method compares measured values for horizontal plume velocity with the wind velocity at different altitudes. A third method estimates the plume top temperature and compares this with the temperature of the surrounding atmosphere. The fourth method uses a photogrammetric technique to derive plume height and plume top surface topography.

The shadow method (Glaze *et al.* 1989) uses basic geometry to estimate the column height by measuring the length of the shadow cast on the Earth's surface by the top of the plume (Figure 12.6). A plume shadow can be seen in Figure 11.16, which is an image of the Mount St Helens umbrella cloud on May 18, 1980. Measurement of a plume shadow is not always simple because it is sometimes difficult to match a shadow feature on the ground to a corresponding feature on the plume. The edge of the shadow does not necessarily relate to the highest point of the plume. Lower sun angles (closer to the horizon) are more likely to cast shadows of the highest point. Most data collected by weather satellite instruments are not corrected for geometric distortions before distribution. Data can be corrected by selecting easily recognizable control points, or "tie" points, and referencing them to a map projection or previously corrected image. Most image-

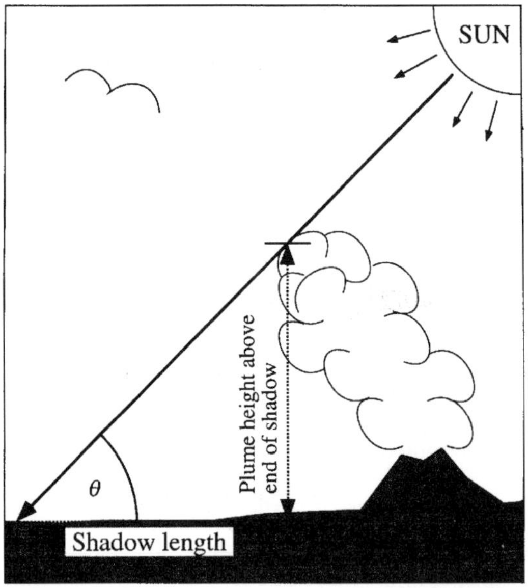

Figure 12.6 Diagram illustrating the shadow method for determining the height of an eruption column. The angle, θ, is derived using astronomical ephemeris data and the shadow length is measured using the satellite image

processing software packages provide algorithms for reshaping the image to the required map projection.

It is then important to determine the pixel size. For polar orbiting instruments, the centre of a swath is generally nadir viewing (unless the instrument is pointable) and the pixels become wider toward the edges of the swath. Geostationary instruments have a wide range of pixel sizes. For the visible channel of the visible and infrared spin scan radiometer (VISSR) instrument on the GOES satellite, the range is from 1×1 km square pixels at nadir (the centre of the full Earth disc) to rectangular pixels several kilometres on a side towards the edges of the image. By resampling an image such that each pixel is roughly square and then selecting two points of known distance from each other along a roughly horizontal or vertical line, the average pixel size can be determined.

Once the distance between the edge of the shadow to the point directly below the plume feature has been estimated, the Sun's angle may be determined by locating the sun's position from ephemeris data. The Greenwich hour angle (GHA) of the sun, as well as its declination, D, are tabulated in the *Nautical Almanac*, published jointly by the US and British governments, for every minute after 1960. The *Astronomical Almanac* is another source of ephemeris data for dates after 1981. In addition to the traditional hardcopy format, the *Astronomical Almanac* provides the information on CD-ROM along with software for determining the appropriate angles. For a point on the Earth west of the prime meridian the longitude is subtracted from the GHA (GHA $\approx$ Greenwich mean sidereal time-apparent right ascension angle), and added if it is to the east, to give the local hour angle (LHA). The angle of the sun above the horizon, θ, at latitude, L, is given by

$$\theta = \arcsin[\sin D \sin L + \cos D \cos L \cos (LHA)] \quad (12.3)$$

The azimuthal angle, ϕ, of the sun measured eastward from the north is then found by

$$\phi = \begin{cases} \zeta & \text{if} \quad \sin(LHA) < 0 \\ 360 - \zeta & \text{if} \quad \sin(LHA) \geqslant 0 \end{cases}$$

where

$$\zeta = \arccos\left[\frac{\sin D - \sin L \sin \theta}{\cos L \cos \theta}\right] \tag{12.4}$$

The shadow method is most accurate for morning and evening images when the sun angle is lowest.

A second method of determining plume height requires a series of images acquired relatively closely in time. The distance the plume front has travelled between images is measured and an average transport speed is determined. The direction and speed of transport are then compared with available wind data for the local area. This method can constrain the plume altitudes but is often not conclusive.

A third method is to determine the coldest temperature of the plume top (temperature determination is discussed in section 12.4.2) and compare this temperature with local atmospheric measurements (Kienle and Shaw 1979; Sawada 1983; Malingreau and Kaswanda 1986; Glaze *et al.* 1989). This method assumes that the top of the plume has equilibrated to ambient temperatures. While this assumption should hold for plumes some distance from the vent (Woods *et al.* 1996), it does not hold for the initial column above the vent due to undercooling (described in section 4.3.4) that occurs in the last few kilometres of plume rise (Woods and Self 1992). For an undercooled plume below the tropopause, this method results in an overestimate of the plume height, and for a plume above the tropopause, an underestimate of plume height.

The fourth method uses shading techniques adapted from planetary photogrammetry applications (Wilson *et al.* 1996). This technique utilizes a single visible wavelength channel and assumes that, for a uniform material, variations in the grey value of a pixel are due to changes in the tilt of the material contained within that pixel. Thus the amount of tilt can be quantified by assuming a direct relationship between the amount of tilt and the ratio of *DN* values (defined in section 12.2). To find the maximum plume height, the altitude of the plume edge is found by the shadow method, and then the topography of the plume itself is found using the method just described. This technique provides an entire topographic surface. For a plume such as the Pinatubo giant umbrella cloud, such data should provide important information on the spreading dynamics. While in theory, stereo imagery could also produce plume top topography, images acquired simultaneously are required. The only instrument in operation that can gather such data is the along track scanning radiometer (ATSR) on board the European Space Agency's European remote sensing satellites, ERS-1 and ERS-2. To date, no accounts of ATSR applications to volcanic plumes have been published. The year 1998 should see the launch of another multi-angle instrument called the multi-angle imaging spectroradiometer (MISR) as part of the US Earth Observing System. Both ATSR and MISR can potentially observe a volcanic plume from several angles within a matter of a few seconds. The main problem with viewing volcanic plumes with these instruments will be their long repeat viewing period due to their narrow swath widths.

12.4.2 Plume Temperature

Plume temperature can be measured by a variety of instruments that operate in the TIR. A radiometer can be used on the ground for volcanic plume observations in a single broad TIR band. Detailed temperature structure can be determined by pointing at several areas. The temperature measured by a radiometer pointed at a diffuse source such as a gas or ash plume corresponds to the radiance emitted by the plume constituents between the margin and the point at which the plume becomes opaque at that wavelength. This point may be several centimetres inside the plume, or even metres for plumes with diffuse margins.

Estimates of the two-dimensional plume top temperature surface can be made using TIR image data. All published studies of plume top temperature have used satellite imagery. Commonly, geostationary data have been used because they provide the highest probability of seeing the beginning of an eruption. However, TIR channels have poor spatial resolution (4–5 km, at best) which makes interpretation of any *structure* in the two-dimensional temperature surface difficult. For large umbrella clouds, even poor spatial resolution images will indicate structure (Figure 12.7), but higher resolution data are certainly preferable. Volcanoes that maintain an eruption column for several hours are more likely to be captured by polar orbiting weather satellites such as AVHRR (Figure 12.8), but because AVHRR does not always acquire data in the LAC mode, the best available resolution may still be only 4 km.

When using satellite weather data for quantitative studies that rely on measured radiance values it is important to be aware of whether or not the *DN* (defined in section 12.2) values have been reversed. Typically, the coldest parts of an image will have the lowest *DN* values (appearing darker) and the warmest parts will have the highest *DN* values (appearing brighter). Meteorologic clouds and volcanic plumes are generally much colder than the ground beneath them and appear very dark compared to the surface. Some data distributors, however, invert the TIR data so that the meteorologic clouds appear white when displayed. For temperature determinations the image should not be geometrically corrected. Geometric correction of an image results in the resampling of pixel values in order to stretch, shrink, or rotate the image. Depending on the algorithm used to execute the resampling, the actual measured *DN* value for each pixel may be averaged with other pixels in order to arrive at the new pixel values. The original *DN* value for each pixel may be converted to an actual radiance value by means of calibration information provided in the documentation for the data (see section 12.2).

In general, the transmissivity of the atmosphere in TIR windows is high (75–80% from surface to satellite, increasing to near 100% from the stratosphere) which means that rough temperature estimates can be made from the raw radiance value. For more accurate measurements of low-altitude plumes, application of an atmospheric modelling program may be necessary (e.g. LOWTRAN, Kneizys *et al.* 1988). These programs predict path radiance and the percentage of radiance that should be transmitted to the detector. The boundary conditions for the models can be chosen to reflect specific conditions such as temperature and pressure profiles obtained from radiosonde, or estimates of increased aerosol loading. For volcanic plume problems, the models can be used to determine the transmissivity of the atmosphere for the relevant channels and for any unusual components (e.g. SO_2). Once the transmissivity has been determined the actual radiance of the plume

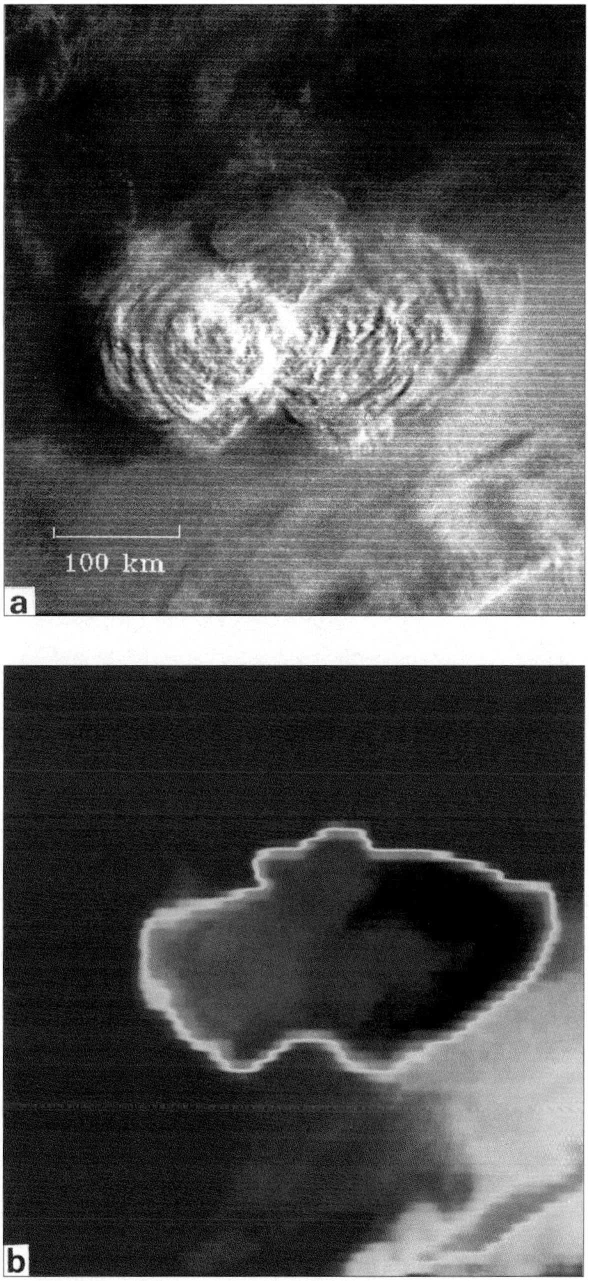

Figure 12.7 VISSR images of El Chichon, Mexico, acquired during the April 4, 1982 eruption in the (a) visible and (b) thermal infrared (TIR) channels. The TIR image has been processed such that cooler temperatures are displayed in blue and warmer temperatures are displayed in red. The scale is the same in both images and north is at the top. The visible image has a much higher spatial resolution than the TIR image where individual pixels can be resolved

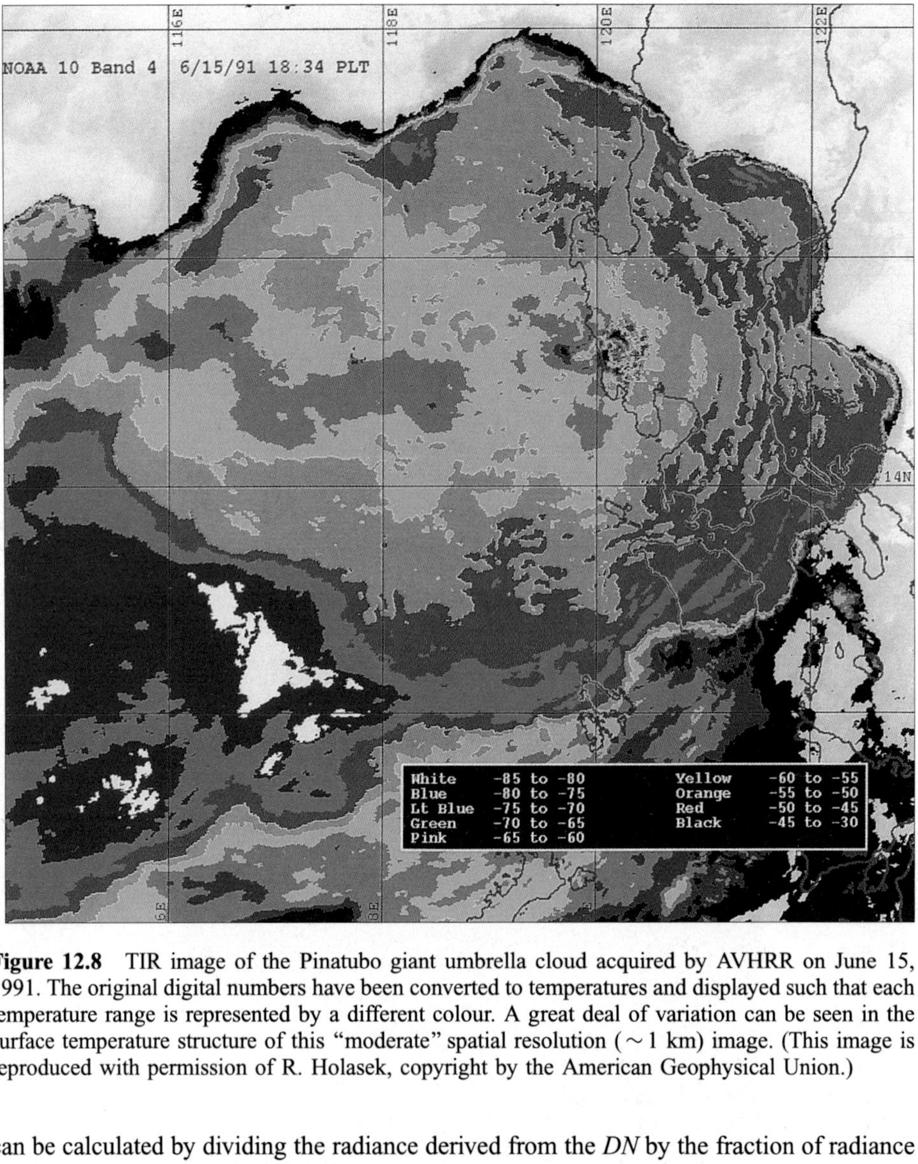

Figure 12.8 TIR image of the Pinatubo giant umbrella cloud acquired by AVHRR on June 15, 1991. The original digital numbers have been converted to temperatures and displayed such that each temperature range is represented by a different colour. A great deal of variation can be seen in the surface temperature structure of this "moderate" spatial resolution (~ 1 km) image. (This image is reproduced with permission of R. Holasek, copyright by the American Geophysical Union.)

can be calculated by dividing the radiance derived from the *DN* by the fraction of radiance reaching the instrument platform.

A *brightness* temperature, T, can be calculated from the measured radiance by inverting Planck's distribution law (equation 12.1) such that

$$T = \frac{(k/hc)\lambda}{\ln\left[\dfrac{\varepsilon(2hc^2)\lambda^{-5}}{R(\lambda)} + 1\right]} \tag{12.5}$$

where λ is the central wavelength of observation and $R(\lambda)$ is the measured spectral radiance measured at the satellite. The amount of radiation emitted from the plume top depends upon the temperature and emissivity of the plume components, as well as the scattering effects of the diffuse boundary layer. These factors are not well understood, and the question of how far into the plume one can see has not yet been answered. Most studies have used an emissivity of 1 in the TIR region, with the assumption that the plume top is a fairly dense body, such that radiance occurs from a very thin boundary layer (of the order of a few millimetres). The laboratory TIR reflectance curves shown in Figure 12.9 indicate that volcanic particles have a very low reflectance and, therefore, an emissivity very near to 1 in the TIR region between 8 and 14 μm. However, the validity of these assumptions remains to be tested.

The principal problem with the above method for temperature determination is that it is not clear what the measured temperature actually corresponds to. Assuming that the plume is very dense, the temperature should correspond to the temperature of the material on the well-defined plume surface. However, in most cases there is a region of lower particle concentration at the top of the plume and, in such instances, the temperature measured corresponds to a depth within the plume at the point where the plume becomes opaque.

12.4.3 Output of SO_2

Volcanoes are an important source of sulphur. Castleman *et al.* (1974) first recognized volcanoes as significant contributors to sulphuric acid aerosols in the atmosphere. Volcanic SO_2 reacts with H_2O to create H_2SO_4 aerosols that have very long stratospheric residence times, of the order of years. These aerosols have a significant impact on global climate (Labitzke *et al.* 1983; Chapter 18). Several instruments have been used to measure levels of H_2SO_4 aerosols and amounts of SO_2 injected into the stratosphere by eruptions. Ground- and aircraft-based instruments that measure SO_2 or H_2SO_4 aerosols include COSPEC, lidar instruments and the NASA thermal infrared multispectral scanner (TIMS). The satellite instruments that have been used to determine SO_2 and H_2SO_4 aerosol loadings include TOMS, MLS, SAGE II, and HIRS2. These instruments each operate in different spectral ranges and have had varying success. The methods for retrieving SO_2 or

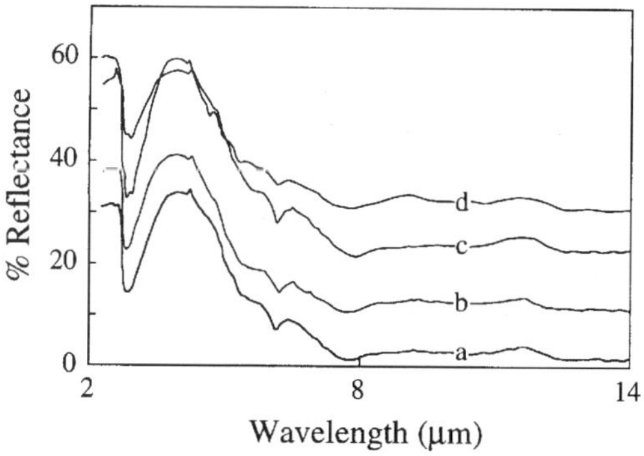

Figure 12.9 Infrared reflectance spectra for (a) dacitic ash from the May 18, 1980 Mount St Helens, (b) dacite pumice from the July 22, 1980 Mount St Helens, (c) trachyandesitic ash from the April 3, 1982 El Chichon and (d) ash from the April 2, 1986 Augustine eruptions. Each plot has been offset by 10% (i.e. between 8 and 14 μm all of the data are coincident with line (a)). Note that the spectral signatures for each of the samples show several similarities

H_2SO_4 amounts from these different data types are complex and most are beyond the scope of this chapter. However, the science teams for some of the SO_2 measuring satellite instruments produce maps of SO_2 concentrations following eruptions as special data products.

The COSPEC measures SO_2 concentrations by comparing incoming radiation in seven channels between 0.297 and 0.315 μm to a known concentration of SO_2 in a small cell in the instrument. The COSPEC has been used to estimate the amount of SO_2 outgassing at active volcanoes (Stoiber and Bratton 1978; Malinconico 1979; Hoff and Gallant 1980; Millan *et al.* 1985; Stoiber *et al.* 1986; Rose *et al.* 1986, 1988; Kyle *et al.* 1990; Symonds *et al.* 1990; Andres *et al.* 1991; Caltabiano *et al.* 1994). The concentration of SO_2 in either a vertical column or horizontal (bent-over or wind-blown) plume is determined by *scanning* across the plume perpendicular to the direction of transport. The scan is started by measuring incoming radiation from a clear sky to measure background SO_2. The plume is scanned at a steady rate until the background concentration is again reached on the other side. By knowing the time interval for each scan segment, and using an estimate of the rise rate (for a vertical plume) or transport rate (for a horizontal plume), the volume flux of SO_2 is determined. Daily SO_2 output can be estimated for volcanoes that steadily release SO_2. Errors occur when physical barriers restrict perpendicular plume observations or when the relative locations of the instrument and plume are not known precisely. Another source of error is the estimation of plume rise rate for vertical plumes, or transport velocity for horizontal plumes. Errors can be reduced when the instrument is flown. For example, when flying under a plume, there are no obstacles that restrict flight perpendicular to the direction of its transport, and the estimate of wind speed measured by the aircraft at the altitude of the plume is generally more accurate.

Another ground-based system popular for studying volcanic plumes is lidar, which may be used to measure abundances (defined in section 12.3.1) of H_2SO_4 aerosols for a dispersing plume. The permanent lidar systems at Mauna Loa Observatory and in Boulder, Colorado, use lasers in the visible and TIR spectral regions. These systems, among others in the world, have produced useful data on the large SO_2-producing eruptions of El Chichon (Swissler *et al.* 1983; Oberbeck *et al.* 1983; McCormick *et al.* 1984) and Pinatubo (Goldman *et al.* 1992; Post *et al.* 1992). The world-wide system of stations has been able to track the volcanic plumes for a month or two after the eruptions, recording when the maximum concentration passes overhead. Other ground-based systems, not necessarily lidar, were used following the Pinatubo eruption to measure depletions in stratospheric constituents, such as NO_2 (Johnston *et al.* 1992).

NASA has several aircraft that are permanently fitted with remote sensing instruments operating in the VIS, SWIR, TIR and MW regions (radar). The NASA Electra aircraft, deployed to the Caribbean following the 1991 Pinatubo eruption, carried several instruments to study SO_2 and other volcanic gases including a total-direct-diffuse spectral radiometer, and a Fourier transform infrared (FTIR) spectrometer (Winker and Osborn 1992a). In contrast to many ground-based lidar systems, the airborne lidar used in the Pinatubo study had only one laser at 0.532 μm. Preliminary results of the lidar measurements are given in Winker and Osborn (1992b) and estimates of SO_2 using COSPEC are given in Hoff (1992). Other measurements resulting from the deployment included spectral optical depths, and derived aerosol size distributions using the total-direct-diffuse spectral radiometer (Valero and Pilewskie 1992). The infrared channels of the FTIR were used to measure SO_2, HCl and O_3 (Mankin *et al.* 1992). The thermal infrared multispectral scanner (TIMS), carried on board the NASA C-130B aircraft, has derived information from the TIR spectral range about volcanic SO_2 plumes. The TIMS has six relatively narrow channels in the TIR between 8 and 12 μm and uses the SO_2 absorption feature between 8 and 9 μm to quantify the amount of SO_2 emitted from a volcano (Realmuto *et al.* 1994). Figure 12.10 shows an SO_2 plume at Mount Etna as derived from TIMS data by Realmuto *et al.* (1994). The US Earth Observing System (due to launch in 1998) will include an instrument called the advanced spaceborne thermal emission and reflection radiometer (ASTER), similar in capabilities to the TIMS. This instrument can measure SO_2 output from steady emitters such as Etna, but will have trouble capturing short-duration eruptions due to its long repeat viewing cycle.

The satellite instrument that has had the most success detecting and measuring volcanic SO_2 is TOMS (Krueger *et al.* 1995). The TOMS instrument primarily functions as an ozone mapper. Following the 1982 eruption of El Chichon, it was discovered that TOMS was also capable of mapping SO_2 (Krueger 1983). Since then, the TOMS archive has been searched for volcanic plumes, and several have been found (Krueger 1985). An SO_2 retrieval algorithm is applied to the data to remove the O_3 from the profile and to estimate the mass loading of SO_2 for the entire vertical column of atmosphere in a pixel. These abundances (defined in section 12.3.1) are added together to estimate the total SO_2 in the TOMS scene, and assimilated with other TOMS scenes to estimate the total output of SO_2. Estimates are complicated by the removal of SO_2 with time in chemical reactions forming H_2SO_4. The TOMS data have also been used to track SO_2 plumes during dispersal. Plumes produced by the El Chichon (Krueger 1983; Bluth *et al.* 1992) and Pinatubo (Bluth *et al.* 1992) eruptions were tracked for several weeks. A major problem with using TOMS to measure erupted SO_2 is that, due to its large pixel size ($\sim 50 \times 50$ km), a minimum of 10 kilotonnes of SO_2 per pixel is required for TOMS detection (Bluth *et al.* 1992). Therefore some of the world's largest sulphur producers that slowly and continuously release SO_2 through fumarolic activity (e.g. Mount Etna) cannot be detected.

The MLS instrument is an atmospheric sounding device that began operation in the autumn of 1991 ($\sim$100 days after the June 15, 1991 Pinatubo eruption). MLS sounds the atmosphere by measuring thermal emissions in the microwave region on the limb of the Earth (the curved horizon as seen from space). It was designed to measure abundances and distributions of several atmospheric chemical constituents including ClO, O_3, H_2O, and O_2. The MLS can also measure SO_2. When first activated, MLS observed anomalously high concentrations of SO_2 from Pinatubo in a belt near the equator (Read *et al.* 1993). Maps of the SO_2 distribution were created every two weeks until SO_2 concentrations had returned to background levels (Figure 11.24; Plate III). In contrast to TOMS, MLS is able to measure the profile of SO_2 with a vertical resolution of 3 km corresponding to an area approximately 300×300 km on the ground, which can then be integrated to give total column abundance. Figure 12.11 (Plate IV) shows SO_2 concentrations for three levels in the atmosphere as functions of latitude and time. Because MLS measures thermal emission in the microwave region, it functions at night as well as during the day.

The SAGE II instrument also looks at the Earth's limb in order to measure stratospheric aerosols. The instrument acquires 15 profiles at sunrise and 15 at sunset every day with a 1 km vertical resolution. These measurements provide information on the global distribution of all aerosols, including small ash particles and sulphuric acid droplets. From

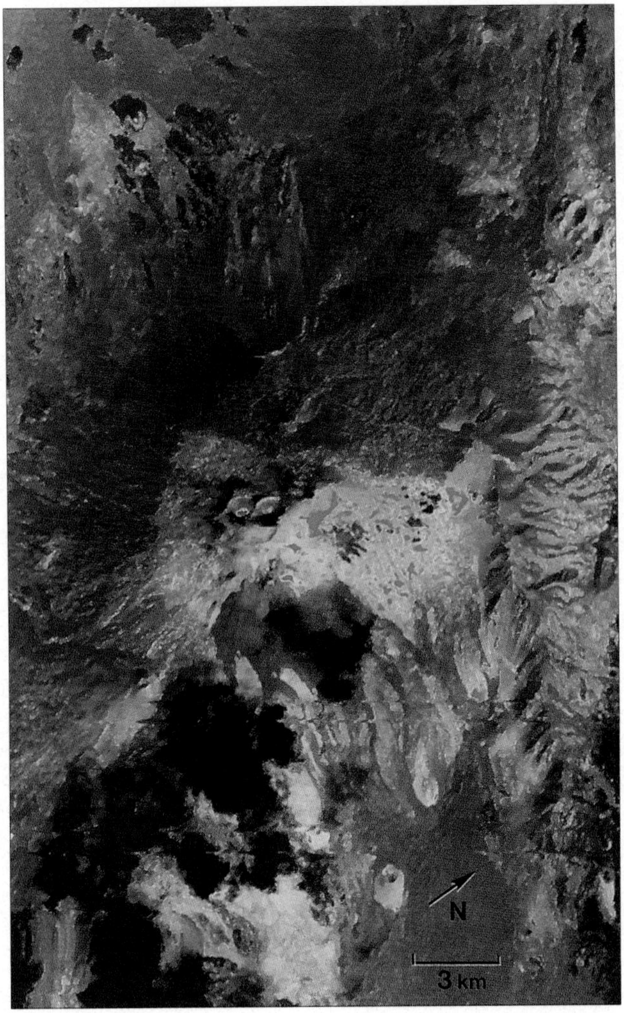

Figure 12.10 TIR image of Mount Etna, Italy, acquired by the TIMS instrument flown on board the NASA C-130 aircraft. The image has been processed to show an SO_2 plume (yellow) emanating from one of the craters at the summit. North is indicated by the arrow and the 3 km scale bar is valid for the summit region of the volcano in the centre of the image. Due to the large change in elevation on Etna, pixels on the flanks of the volcano (greater distance between the sensor and the surface) represent larger areas than those at the summit. (This image was processed and provided by V. J. Realmuto.)

estimates of aerosol loading, calculations of optical depth may be made. SAGE II tracked the temporal evolution of the Pinatubo aerosol cloud as it migrated towards the poles (McCormick and Veiga 1992; Post *et al.* 1992). SAGE II measures aerosol concentration, but is unable to discriminate between different aerosol types. This limits comparison between SAGE II results and the estimates of SO_2 loading, measured by instruments such as TOMS and MLS.

The HIRS2 operates in the TIR. SO_2 and H_2S both have absorption features in the TIR. The SO_2 detection limit for HIRS 2 is five times higher than for TOMS and observations of the May 18, 1980 Mount St Helens plume indicated that satellite-borne instruments operating in the TIR are capable of detecting volcanic SO_2 (J. Crisp, unpublished data). However, retrieval of actual concentrations of SO_2 and/or H_2S in the TIR is an unreliable process. To date, no quantitative measurements of SO_2 or H_2S from HIRS 2 data have been published.

12.5 SATELLITE PLUME DIFFERENTIATION AND ERUPTION MONITORING

There are several types of remote sensing studies of volcanic plumes that can only be conducted using data from a satellite platform. These studies fall into the categories of plume differentiation and eruption monitoring. The phrases "plume differentiation" and "eruption monitoring" from a remote sensing point of view often refer to retrospective studies of images, conducted months or years afterwards, involving searching through a data archive for images of plumes. Exceptionally, scientists have had access to polar orbiting weather data in very near real time. For example, the Alaska Volcano Observatory has access to an AVHRR HRPT (high resolution picture transmission, 1 × 1 km pixel size) data receiving station which was used during the 1989/90 Redoubt eruption to monitor plume movement and alert aviation officials (*SEAN* **14** (11), 1989). So far no eruption has been detected by a satellite, and certainly not in real time. There are, however, several studies currently under way aimed at detecting volcanic plumes in real time. Three of these are presented here as examples of applications of remote sensing data. Eruption monitoring studies from satellite are primarily interested in synoptic aspects of volcanic plumes. These studies have used satellite images to look at plume transport, ash dispersal, atmospheric dynamics and SO_2 and aerosol dispersal. In this section results from several of these monitoring studies are presented as further examples of applications of satellite remote sensing data.

12.5.1 Volcanic Plume Distinction

The use of satellite data to distinguish between volcanic plumes and meteorologic clouds has been encouraged by many encounters between aircraft and volcanic plumes since the early 1970s (section 17.7) that have resulted in temporary jet engine failure. There have been several methods proposed for differentiating volcanic plumes from meteorologic clouds and eventually these methods may be implemented by all commercial passenger aircraft. So far, however, none have been shown to be completely reliable. Two such methods use the multispectral information acquired in the two AVHRR TIR channels (Prata 1989; Holasek and Rose 1991), while a third uses the ability of TOMS to detect SO_2 (Krueger 1983). Each of these is effective under certain conditions, but each has problems that make them unreliable as a routine mitigation tool.

Prata (1989) first attempted to distinguish volcanic plumes from meteoric clouds by using the fact that silicate ash has absorption bands in the TIR due to vibrations of the SiO bonds (so-called "restrahlen bands"). Prata (1989) determined brightness temperatures for each pixel in the two infrared channels. The temperature derived from the 10.3 to the 11.3 μm channel was then subtracted from the temperature derived from the 11.5 to the 12.5 μm channel. A new image was created of the resulting differences. If the targets behave as blackbodies, with no wavelength dependence, the two temperatures will be equal and the difference is zero. Prata (1989) found that for thin water–ice clouds, the difference is positive, and for eruption plumes containing ash the difference is negative. Because this method uses data acquired in the TIR region of the spectrum, it can be applied to data acquired during the day or at night. There are, however, some difficulties. If there is a large amount of sulphuric acid present in the plume, or if the ash particles are coated with sulphuric acid, the brightness temperature difference may be positive, suggestive of a water–ice cloud. While this method is not likely to identify a meteorologic cloud falsely as a volcanic plume, it may fail to identify plumes of volcanic origin.

Holasek and Rose (1991) also distinguished volcanic plumes from meteoric clouds from the TIR absorption of volcanic ash using the two AVHRR TIR channels by ratioing the 10.3–11.3 μm channel with the 11.5–12.5 μm channel. ([*DN* in 10.3–11.3 μm channel]/[*DN* in 11.5–12.5 μm channel].) The operation is applied to every pixel in the image with the result being a new image of the ratioed values. This new image, when displayed in red and combined with the 11.5–12.5 μm channel displayed in green and the 3.5–3.9 μm channel displayed in blue, results in a yellow or orange volcanic plume in contrast to blue or white meteoric clouds (Figure 11.22). While the Holasek and Rose method has worked very consistently for plumes from the Alaskan volcanoes, Augustine, Redoubt and Spurr, it has not worked well for others, including Mount St Helens (L. Glaze, unpublished data), and one of the smaller Pinatubo plumes of 1991 (R. Holasek, personal communication). A problem with this method is that no corrections for attenuation by the atmosphere are made before performing the ratio operation. Thus, data toward the edges of a swath are affected by the increased thickness of atmosphere. The sensitivity of this technique to variations in particle size and the presence of H_2SO_4 or water–ice coatings on ash particles is also not known.

In several of the images processed by Holasek and Rose, the mid-infrared (3.5–3.9 μm) channel is displayed in blue. The contrast between the volcanic plumes (orange or yellow) and meteorologic clouds (white) in these images is partly due to the lower reflectance of volcanic plumes in the 3.5–3.9 μm channel (Figure 12.12). Instances can be found, however, where the reflectance of the volcanic plume and nearby meteoric clouds are similar in the mid-infrared channel. No published information explains this feature, but this channel may also be useful for distinguishing volcanic plumes.

The ability of TOMS to detect SO_2 provides a possible method for distinguishing volcanic plumes. There are several difficulties with this approach. The principal problem with the TOMS data is that they do not see every eruption. Many eruptions that are particle-rich (and dangerous to aircraft) are not large producers of SO_2 (e.g. the 1986 eruption of Augustine, Alaska). Furthermore, TOMS has difficulty seeing below the stratosphere due to strong absorption by ozone. Consequently, sulphur-poor plumes, or low-altitude sulphur-rich plumes are difficult for TOMS to identify. Ultraviolet instruments such as TOMS operate only during daylight hours and are most effective near to local noon, making it difficult to catch smaller eruptions in progress. The large pixel size ($\sim 50 \times 50$ km) exacerbates the ability of TOMS to detect smaller eruptions. If a plume is sulphur-rich but does not fill a significant area of a pixel, it will not be detected above the background.

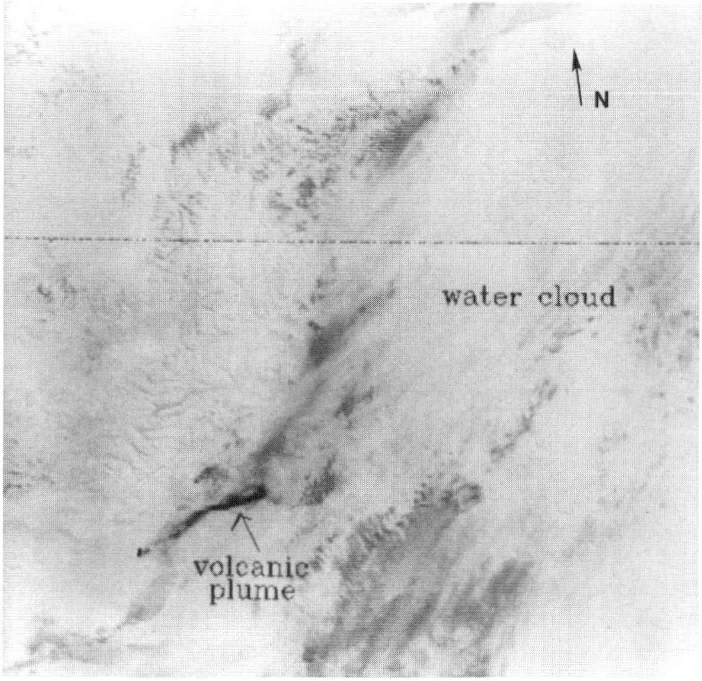

Figure 12.12 Plume generated during the eruption of Augustine in 1986, captured by the AVHRR mid-infrared channel in HRPT (high resolution picture transmission, 1 km spatial resolution) mode. In this image both the snow on the ground and the water/ice clouds have a very high reflectivity, thus appearing bright. In contrast, the volcanic plume has a much lower reflectivity, thus appearing dark. The image is approximately 512 km across and north is at the top. (Reproduced by permission of Springer-Verlag, New York.)

An interesting question is whether or not the particle and gas constituents of volcanic plumes travel together. Based on a comparison of AVHRR and TOMS data for the Redoubt and Pinatubo eruptions, Schneider and Rose (1992) suggested that, in at least some cases, ash and gas plumes do not separate. Similar comparisons for El Chichon (D. J. Schneider, unpublished data), however, exhibit a dramatic separation of the plume into what appears to be ash-rich and gas-rich plumes moving in opposite directions. Lane and Gilbert (1992) have also invoked separation of gas from particles at the top of eruption columns and ahead of spreading plumes (section 12.6) in order to explain electric potential gradient measurements at Sakurajima volcano in Japan. The process of ash and gas separation has been simulated in analogue laboratory experiments (Holasek *et al.* 1996).

12.5.2 Plume Dispersal Observations from Satellite

Once a plume has been injected into the atmosphere, it is transported downwind and dispersed. Chapter 11 discusses the dynamics of the dispersal processes. Data from satellite instruments provide a unique perspective for monitoring the dispersal of plumes. In this section several case studies are presented where satellite images have been used to study volcanic plumes.

A potential use of satellite images is for estimations of particle loading in volcanic plumes. For a series of images an estimate of particle sedimentation rates from plumes could constrain theoretical models. Recent work of Wen and Rose (1994), based on work by Yamanouchi *et al.* (1987), Prata (1989) and Lin and Coakley (1993), has shown this to be possible. The theory relates a plume's opacity (the ratio of intensity of transmitted light to the intensity of incident light) to the constituent particle properties such as particle size distribution, density, mass, concentration and refractive index. The temperature difference method (Wen and Rose 1994) assumes that the observed radiance in each of the two thermal infrared AVHRR channels is the sum of the radiance emitted from the plume and from the underlying surface, where the surface radiance is transmitted without attenuation when no plume is present. If the area within a pixel that contains plume material is expressed as A, then the measured radiance for each AVHRR channel can be written as:

$$I_\lambda = (1 - A)R(T_s) + A[\varepsilon_\lambda R(T_p) + t_\lambda R(T_s)] \qquad (12.6)$$

where T_p is the temperature of the plume material, T_s the temperature of the underlying surface, R the radiance defined by Planck's function (equation 12.1), and ε_λ and t_λ the emissivity and transmissivity of the plume at wavelength λ. The emissivity and transmissivity are functions of the particle size distribution and the effective radius of the particles. Wen and Rose have conducted numerous radiative transfer calculations in order to estimate the mass of particles in a given volcanic plume. The approach is sensitive to the assumption that the volcanic particles are spherical and that the particle size distribution is uniform and monodisperse within each pixel. Bredow *et al.* (1995) have begun to test some of these assumptions in the laboratory.

Much can be learnt about explosive volcanic eruptions and the dispersal of plumes using broad-band visible, moderate spatial resolution (of the order of 1–5 km) geostationary satellite images. In remote areas of the world the images can allow evaluation of the style of activity and time of eruption. Examination of several images in a time series (e.g. Figure 12.13) provides clues to the type of eruption and its duration (Glaze *et al.* 1989). For GOES satellites, the longest interval between images is 30 minutes. From the first two images, it is usually easy to discern whether or not an eruption plume has disassociated from the source and therefore the approximate time scale for activity. Observations can be made quickly from a series of raw images.

Glaze *et al.* (1989) have used geometrically corrected images (see section 12.4.1) to estimate the start time of an eruption of Lascar, Chile, in 1986. By measuring the distance the plume front had travelled between images, they determined the average plume speed and then extrapolated back to the volcano to find the onset time.

Depending on the mass erupted and the column height, a plume may either quickly disperse, or be caught up in the stratosphere and tracked for days or weeks. The type of material erupted, particularly whether the eruption has a low or high sulphur content, can affect the ability of the plume to be tracked over large distances. The types of measurements include estimates of dispersal area, height of the dispersing plume, transport velocity, plume thickness, and total output of SO_2. For small short-lived eruptions, geostationary instruments are the best choice for following their dispersal. Observations of plume dispersal using geostationary data include studies of Mount St Helens, 1980 (Holasek and Self 1995), Alaid and Pagan, 1981 (Sawada 1983), Mayon, 1984 (Sawada

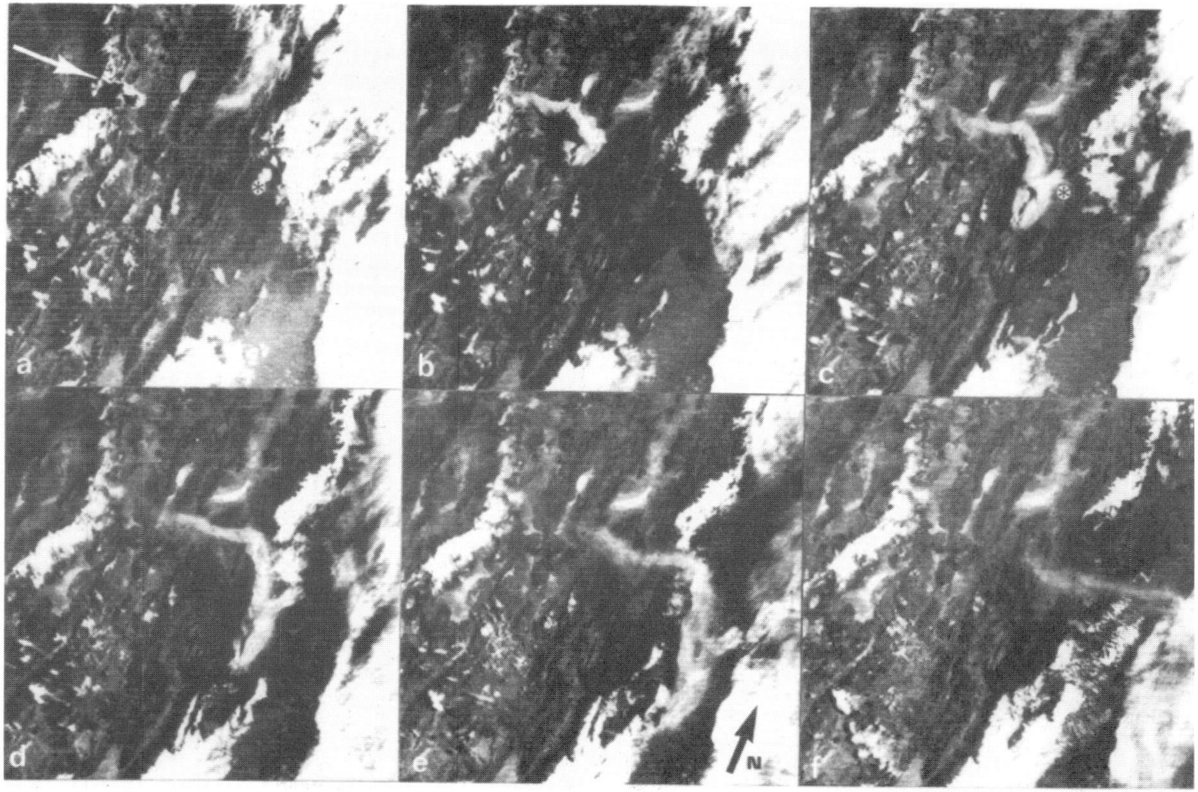

Figure 12.13 Series of VISSR VIS images of the September 16, 1986 eruption of Lascar, Chile. Each image is approximately 600 km on a side. The arrow in (e) points to north, the arrow in (a) points to Lascar and the asterisk (*) in (a) and (c) denotes the town of Salta, Argentina. (This figure is reproduced by permission of Springer-Verlag, New York.)

1985), Lascar, 1986 (Glaze *et al.* 1989), and Pinatubo, 1991 (Koyaguchi and Tokuno 1993). Each of these studies used a series of images to create a map of the dispersal area and influence of ashfall. In the case of Lascar, data on the actual fallout area were not obtainable due to its isolation in the central Andes. For Alaid, Pagan, Mayon and Pinatubo data were not available due to the large amount of ash which fell over the ocean. The study of the Lascar plume by Glaze *et al.* (1989) mapped the vertical trajectory of the plume through time using the shadow method. The plume appeared to fluctuate vertically, possibly due to rising and falling draughts on the downwind side of the Andes.

In some instances, particularly at high latitudes, geostationary data are not available for an eruption. Therefore, one or two polar orbiting AVHRR images can be used. Two such studies using AVHRR LAC images are by Kienle and Shaw (1979) of the July 23, 1976 Augustine eruption, and by Malingreau and Kaswanda (1986) of the July 28, 1983 Colo, Indonesia, eruption. A single AVHRR image was used by Keinle and Shaw (1979) to determine the height of the eruption plume above the vent using the shadow method. It was also possible to distinguish several plumes from individual eruptions. These individual plumes were correlated to seismic events associated with eruptions earlier in the day. Malingreau and Kaswanda (1986) used several AVHRR images to demonstrate the usefulness of multispectral satellite data for monitoring and hazard mitigation in remote locations. Colo had to be monitored from neighbouring islands due to the danger posed by pyroclastic flows and ashfall. Satellite images provided information on eruption events that were otherwise unattainable. Malingreau and Kaswanda used the AVHRR infrared data to determine brightness temperatures (temperatures associated with the measured radiance, without any kind of correction). They then compared the derived temperatures to a nearby radiosonde profile to estimate the plume height. The problems inherent in the method, however, render the results unreliable, as discussed in section 12.4.1.

The aerosols, particles and gases in stratospheric plumes generated by large eruptions can often be tracked for weeks or months due to much longer fallout times and the slow rate of transfer between the stratosphere and troposphere. Large plumes are best observed by polar-orbiting satellite instruments that have near global coverage. Long-lived eruptions can also be observed by polar-orbiting instruments such as the AVHRR and the TOMS. These two instruments have vastly different spatial resolutions and each operates in different spectral regions, providing very different types of information. Studies with AVHRR have successfully mapped the global transport of volcanic ash plumes. For example, the study by Robock and Matson (1983) mapped the progress of the April 3, 1982 plume from El Chichon as it circumnavigated the Earth. TOMS followed the El Chichon plume by tracking the high concentrations of SO_2 produced in the eruption (Krueger 1983). In 1991, TOMS followed the sulphur plume from Pinatubo as it circled the equator in 22 days (Bluth *et al.* 1992), and in 1991 tracked the Hudson volcano, Chile, plume around the South Pole (Figure 11.24 (Plate III)).

The Pinatubo eruption in 1991 provided excellent opportunities for tracking a globally distributed plume, in part because more instruments were in place than ever before. In addition to the TOMS SO_2 measurements, SAGE II and AVHRR were able to track the aerosol cloud produced by the Pinatubo eruption. Preliminary results from more than two months of measurements following the eruption can be found in McCormick and Veiga (1992) for SAGE II, and Stowe *et al.* (1992) for AVHRR. McCormick and Veiga determined that the Pinatubo aerosol cloud reached a maximum of 29 km altitude with most of the material between 20 and 25 km. They also determined that, after two months, the optical depth in the Southern Hemisphere had increased 10 times compared to measurements shortly after the eruption. SAGE II estimates of the total amount of aerosol produced by Pinatubo were reported to be 20–30 × 10^9 kg (20–30 megatonnes), approximately twice the amount produced by El Chichon (12 × 10^9 kg). The increased aerosol concentration is thought to have contributed to a 2.5 °C temperature increase in the stratosphere (at 30 mb, or ~25 km at mid-latitudes) over the 26-year mean (Labitzke and McCormick 1992). The AVHRR channel centred at 0.5 μm was also used to compare the change in reflected solar radiation which could then be related to the aerosol optical depth (Stowe *et al.* 1992). The maximum aerosol optical depth observed by the AVHRR was 0.31 which corresponds to 13.6 × 10^9 kg of SO_2.

12.6 MONITORING OF ELECTRIC POTENTIAL GRADIENTS AND LIGHTNING GENERATED BY PLUMES

Only a small number of exploratory studies of electric potential gradients, or electric fields, and lightning associated with volcanic plumes have been made. This work has revealed that plumes have electric properties substantially different from the ambient atmosphere. Therefore monitoring of these electrical characteristics has the potential to identify and track plumes. In this section results of field studies of this nature are presented.

12.6.1 Background

The potential difference between two points is the mechanical work per coulomb needed to move a small positive charge from one point to another (Chalmers 1967). If a positively charged body is moved from a point at a lower potential to one at a higher potential then the body is moved against a force. The ratio of the force acting on the body to the charge on the body is known as the electric field. The direction of the electric field is opposite to that of the rate of change of electric potential, or electric potential gradient (Figure 12.14). The convention used is that the positive direction of the potential gradient is upwards from the surface of the Earth (Chalmers 1967). Because the surface of the Earth is a conductor, it follows that the lines of force must reach the surface in a normal direction. Thus, where the Earth's surface is horizontal, the lines of force are vertical and the potential gradient is vertical (Figure 12.14).

Atmospheric electric fields are routinely measured and, in fine weather when meteoric clouds are absent, there is a weak electric field of approximately -120 V m^{-1} (i.e. an electric potential gradient of 120 V m^{-1}) at the surface of the Earth (Iribarne and Cho 1980). Soundings made from balloons and aircraft show that the fine-weather atmospheric electric field decreases in absolute value with altitude until it disappears, approaching the highly conducting part of the atmosphere known as the ionosphere. In general, it is not possible to recognize the electric fields of meteoric clouds against the normal variations of electric field in the surrounding cloudless atmosphere. The exception to this is the thundercloud, which is capable of producing fields sufficiently large to result in lightning. Thunderclouds are characterized by convective instability with strong updraughts and downdraughts. Measurements beneath and within thunderclouds suggest that most have a

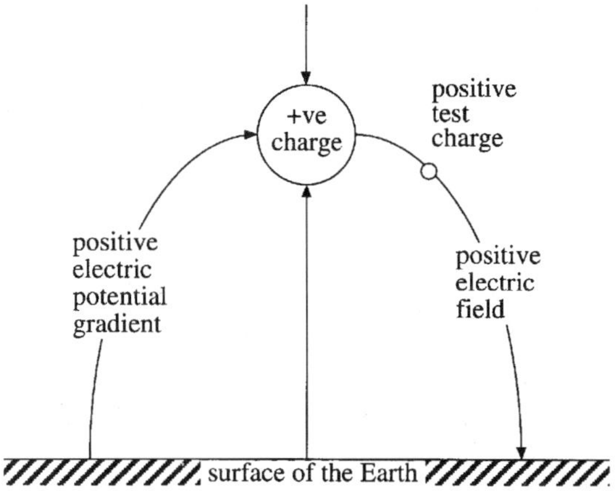

Figure 12.14 Explanation of electric field and electric potential gradient. A small positive test charge in the vicinity of a positively charged body will seek to reduce its energy by travelling along an electric field line to regions of lower potential. The potential gradient is measured along lines of increasing potential. The electric field and potential gradient have the same magnitude but opposite directions

concentration of positive charge in their upper regions, with a lower region of negative charge (Chalmers 1967). In many cases there is a further concentration of positive charge in a limited zone at the base of the cloud. This arrangement is thought to result from gravitationally driven charge-separation processes operating at different levels within the cloud (Chalmers 1967). Charge distributions have also been described in non-raining clouds (Whitlock and Chalmers 1956) and in dust devils (Freier 1960; Crozier 1964). Electric potential gradient data collected during fair weather, beneath clouds, during rainfall and snowfall, and at the coast (where breaking waves produce space charges which affect the fine-weather potential) are relatively abundant (Chalmers 1967). Electric field and potential gradient measurements of thunderclouds have been made by launching instrumented rockets (e.g. Winn and Moore 1971), balloons and kites into clouds, but these types of studies have not been carried out for volcanic plumes.

Spectacular displays of lightning are common in particle-laden volcanic plumes indicating that locally the air ionization limit (Blythe and Reddish 1979) has been exceeded. Among many other examples, lightning has been observed at Vesuvius in 1872 (Palmieri 1873), Krakatoa in 1883 (Simkin and Fiske 1983), Vesuvius in 1906 (Perret 1924), Stromboli in 1907 (Perret 1924), Anak Krakatoa in 1933 (Simkin and Fiske 1983), Surtsey in 1964 (Anderson *et al.* 1965), Heimaey in 1973 (Brook *et al.* 1974), Usu in 1977 (Kikuchi and Endoh 1982), Mount St Helens in 1980 (Rosenbaum and Waitt 1981), Redoubt in 1989–90 (Hoblitt and Murray 1990), Hudson in 1991 (José Naranjo, personal communication) and Sakurajima. Therefore, high degrees of charge are present in volcanic plumes and, as well as generating magnificent visible displays, may play a pivotal role in the formation of ash aggregates (Chapter 16). Characterization of charge levels within eruption plumes will allow plume electrical properties to be used for remote sensing.

12.6.2 Field Measurements at Volcanoes

Electrical studies of volcanic plumes fall into two groups. First, those which detect charged particles and aerosols within plumes (e.g. Lane and Gilbert 1992; Lane *et al.* 1995) and second, those which detect lightning associated with plumes (e.g. Hoblitt and Murray 1990; Hoblitt 1994).

Detection of charged particles and aerosols in plumes

Data on charged material within eruption plumes have been collected at a small number of volcanoes. In most cases, measurements have been made by workers on the ground near the plumes by means of upward-looking commercial electrostatic fieldmeters.

Electric field data collected during eruptions of Aso volcano, Japan, in 1950 (Hatakeyama and Uchikawa 1952) led to the interpretation that plumes are dominated by positive charge in their lower and negative charge in their upper regions. The authors suggested that charge of one sign was carried by relatively large particles and charge of the other sign by small particles. Electric potential gradient measurements made during the eruption of Surtsey in 1964 (Anderson *et al.* 1965) found the plume to be dominated by positive charge in its upper and negative charge in its lower region, the converse of the Japanese work at Aso. Lane and Gilbert (1992) and Lane *et al.* (1995) have explained their potential gradient data, collected at Sakurajima volcano, in Japan, by assigning an average negative charge to particles, which dominate the lower regions of the plume, and an average positive charge to liquid drops formed by condensation of plume gases.

The relative movements of a plume dipole structure (Lane and Gilbert 1992) inferred from electric potential gradient data collected at Sakurajima volcano (Figures 12.15 and 12.16) are shown in Figure 12.17. The idea is that in the early stages of an eruption (Figure 12.17a), positively charged gases separate from the particles at the top of the column due to gravitational forces. Particle aggregation may enhance this process by increasing terminal fall velocities. At Sakurajima, this charged gas plume emerges first from the 300 m deep crater and is detected by the measuring equipment (electrostatic fieldmeters) prior to the ash column. It is this gas which generates a small positive potential gradient recorded at all fieldmeters regardless of their location, relative to the plume, on the flanks of the volcano (Figures 12.15a, b, 12.16b). As the column ascends (Figure 12.17b) the negatively charged ash-rich lower regions of the column dominate the measured potential gradient at which point all fieldmeters show a negative potential gradient (Figures 12.15a, b, 12.16a, b). The plume is then distorted by the prevailing wind into a laterally spreading gravity current (Figure 12.17c). The potential gradient measured during this period depends on the relative positions of the fieldmeter and the plume. For example, higher wind speeds at relatively high altitudes carry the positively charged gases over any downwind fieldmeters and this produces large positive potential gradients (Figures 12.15a, 12.16a, b). Fieldmeters upwind of the plume record a reducing negative potential gradient (Figure 12.16b) as the plume recedes from the fieldmeter in the prevailing wind. As the plume travels over the downwind fieldmeters, with time, they detect the average negative charge on the ash-laden part of the plume (Figure 12.17d), particularly when ashfall occurs (Figure 12.15a). As the eruption switches off and the plume recedes from the vent, all fieldmeters return to measuring the pre-eruption levels of atmospheric potential gradient (Figures 12.15 and 12.16).

Lane *et al.* (1995) suggested that only a few to a few tens of coulombs of charge imbalance in the concentration of positive and negative charges in the separating ash and gas plume are required to produce the observed potential gradient fluctuations. Lane and Gilbert (1992) estimated the mass of particles that would require separation (by gravity) from the condensed gases to yield the observed potential gradients to be carried by approximately 0.1 wt% of the ash. These authors suggested that, because charge resides on the surface of particles, the amount of charge generated per unit mass of erupted material is likely to be a function of particle size (surface area) and that consequently, explosions which generate relatively large quantities of fine ash (such as during phreatomagmatic eruptions) create more charge per unit mass than dry eruptions. Therefore, high levels of charge accompanied by small degrees of separation of particles from condensing gases

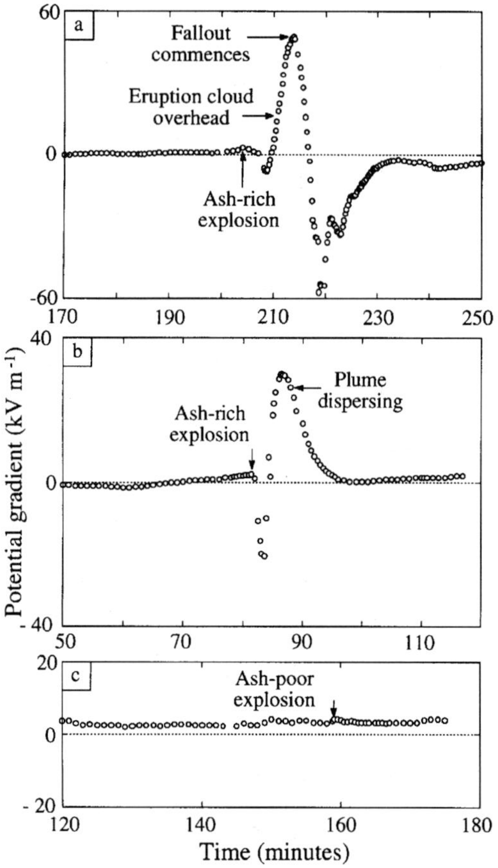

Figure 12.15 Potential gradient versus time for three eruptions at Sakurajima volcano, in Japan. (a) Data collected at a locality 2.75 km downwind of the vent for an explosion on May 3, 1991. (b) Data collected at a locality 2 km from the vent off the dispersal axis for an explosion on April 30, 1991. (c) Data collected at a locality 2.75 km downwind from the vent for a gaseous plume on April 30, 1991. (After Lane and Gilbert 1992, reproduced with permission of *Bulletin of Volcanology*.)

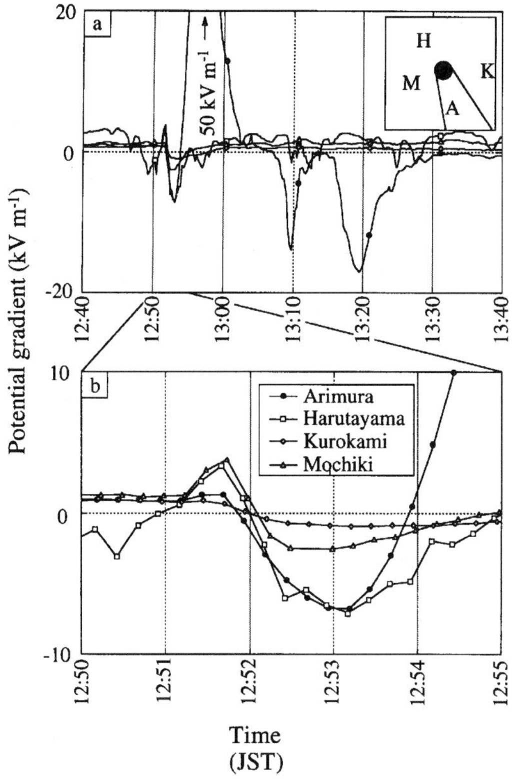

Figure 12.16 Potential gradient versus time for an explosive ash-rich eruption at Sakurajima volcano, in Japan, at 12:52 hours JST on December 27, 1993. (a) Data collected simultaneously by four fieldmeters. The inset indicates the relative locations of fieldmeters, vent and plume. (b) Expanded version of Figure 12.16a showing details of changes in potential gradient at the onset of eruption measured at all four locations. (After Lane *et al.* 1995, reproduced with permission of Sakurajima Volcano Research Center.)

will produce substantial perturbations of the ambient potential gradient. This type of monitoring at several sites around a volcano simultaneously may be used, in the absence of wind directional data and visual information, to indicate (1) the travel direction of a plume, and (2) the area of ash or mud–rain fallout. Electric potential gradient monitoring may therefore be used as a remote sensor of volcanic eruptions.

Detection of lightning associated with plumes

During activity at Redoubt volcano, in Alaska, in late 1989 and early 1990, inclement weather and restricted daylight hours resulted in few of the eruptions being observed clearly. Lightning accompanied all of the ash-producing events (Hoblitt and Murray 1990; Barker and Vonnegut 1990; Hoblitt 1994). This prompted deployment of a commercial lightning-detection system near Cook Inlet, Alaska, in an attempt to temporarily remotely monitor plume-to-ground volcanogenic lightning associated with the eruptions. The sys-

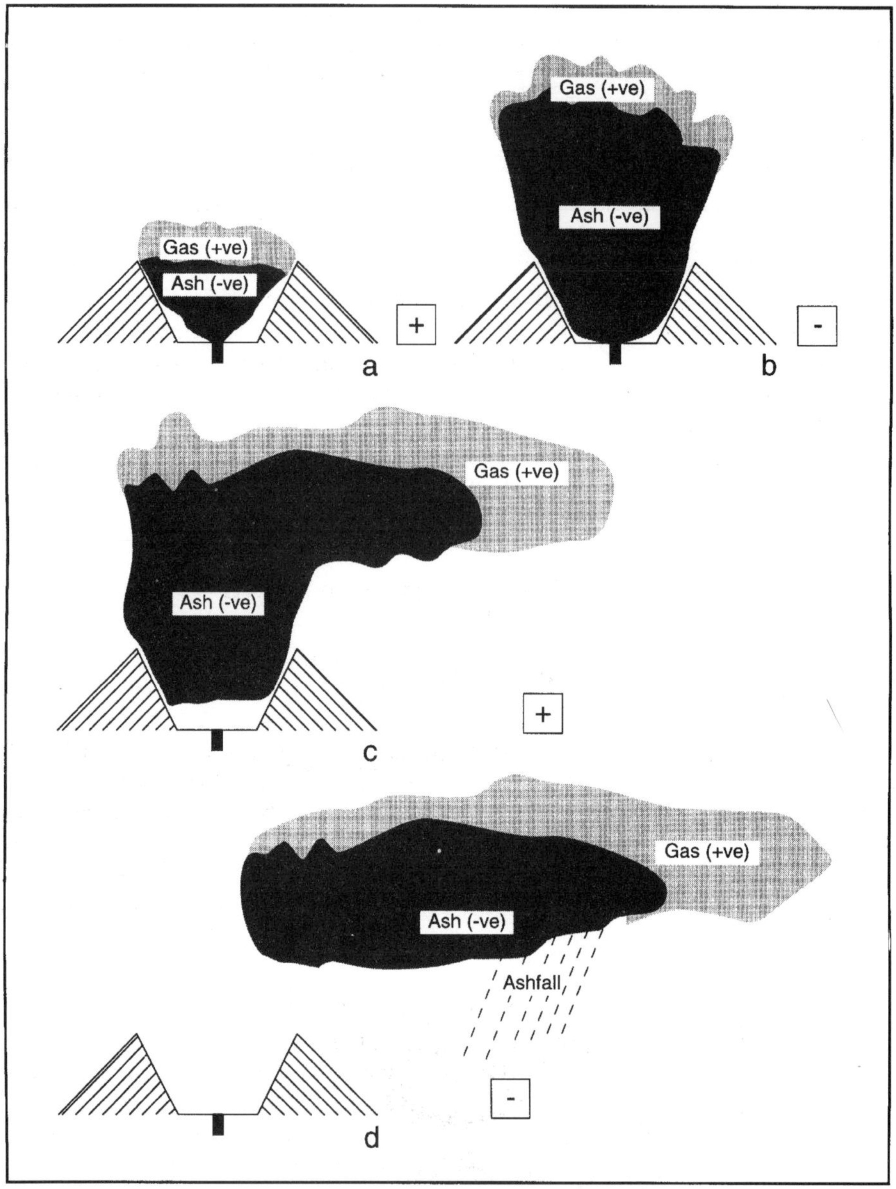

Figure 12.17 Schematic representation of the growth and dissipation of an eruption plume to explain changes in potential gradient measured at ground level. See text for details. (After Lane *et al.* 1995, reproduced with permission of Sakurajima Volcano Research Center.)

tem works on the basis of detection of broad-band radio waves that are radiated from lightning. The lightning-detection equipment became operational on February 14, 1990 and lightning was detected in 11 and located in 9 of the 13 subsequent eruptions. The lightning was generated by ash-laden plumes rising from pyroclastic flows produced by collapse of a lava dome emplaced near the summit of Redoubt. In individual eruptions, early flashes tended to have a negative polarity (when negative charge is lowered to the ground) in contrast with late flashes which tended to have a positive polarity (when positive charge is lowered to the ground). Hoblitt (1994) suggested that the quantity of lightning produced increases with the quantity of material in the resulting plume, in agreement with the suggestion of Lane and Gilbert (1992) that explosions which generate relatively large quantities of fine ash create relatively high levels of charge per unit mass. Hoblitt (1994) showed that lightning detection and location is a useful technique for volcano monitoring, particularly when poor weather or darkness prevents visual observation. Analysis of the frequency of lightning flashes provides a crude but rapid and remote measure of eruption magnitude, and monitoring of the location of lightning should indicate the direction of plume movement.

12.7 ACOUSTIC MEASUREMENTS OF VOLCANIC PLUMES

The acoustics of volcanic plumes provide another means for remote sensing. Acoustic measurements have been used (Weill *et al.* 1992) to obtain vertical velocities of volcanic jets. Such velocities have traditionally been retrieved by photography of ballistics (Chouet *et al.* 1974). Photographic studies are generally not continuous and the advantage of the acoustic technique lies in the fact that it has potential for continuous monitoring. This method generates real-time vertical velocities and is well suited to remote monitoring of temporal variations in velocity during an eruption. However, the technique cannot be used successfully in a strong wind. In this section the basic principles involved in acoustic remote sensing and measurements carried out at Stromboli volcano in the Aeolian islands, Italy, are discussed.

12.7.1 Principles

Sodar (sound detection and ranging) can characterize velocities of atmospheric plumes (Weill *et al.* 1992), industrial plumes and experimental jets (e.g. Brown and Hall 1978). The method is based on the Doppler effect for particles in motion and uses a monostatic sodar. This is an instrument which sends an acoustic wave and receives an acoustic backscattered wave at the same location (similar to the way in which the lidar instruments measure backscattered electromagnetic waves as described in section 12.3.1). The Doppler shift, D_f, between the transmitted wave of frequency, f, and the backscattered wave is related to the vertical velocity, u, inside the jet. In the absence of any horizontal component

$$D_f/f = -(2u \sin \alpha)/C \tag{12.7}$$

where α is the angle between the antenna axis and the horizontal, and C the speed of sound in the jet; C is a function of both temperature, T, and composition. Assuming $T = 1000$ °C, $C \sim 718$ m s^{-1} and the acoustic wave frequency is 2000 Hz. With an angle $a = 45°$, and a vertical velocity of 50 m s^{-1} (Chouet *et al.* 1974) the Doppler shift is equal to 197 Hz. This is sufficient to allow the determination of vertical velocities. The reader is referred to Weill *et al.* (1992) for further information.

12.7.2 Measurements at Stromboli

Stromboli has shown remarkably steady-state activity during the last 2000 years. Its eruptive pattern consists of a series of explosions (typically 10–15 s long) at two vents, which propel ejecta into the atmosphere, interspersed with periods of quiescence. The explosions are thought to be due to large gas pockets bursting at the vent (Blackburn *et al.* 1976; Jaupart and Vergniolle 1988).

A slanting monostatic acoustic antenna was used to detect the vertical component related to the gas vertical velocities in the Doppler spectrum (Weill 1991). Velocities were measured as close to the vent as possible (at ~20 m above the vent) in order to obtain the most representative estimate of the gas velocity. During five days, more than 117 explosions were identified and analysed. Figure 12.18 shows a continuous measurement of the average vertical velocity over a period of 35 minutes during which time there were two explosions. The velocities for the explosions were 40 and 45 m s^{-1}.

Weill *et al.* (1992) took $C = 700$ m s^{-1} and, assuming a temperature of 900 °C, estimated that the uncertainty of C was 10%. With the mass ratio of particles to gas < 0.5, the sound speed in the jet was close to that measured in a pure gas at the frequency of the transmitted wave (Soo 1967). The authors estimated an uncertainty of 4 m s^{-1} in their measurements and found that they were of the same order of magnitude as those measured by Chouet *et al.* (1974) and Blackburn *et al.* (1976) from photographic studies. Therefore, this method has been demonstrated to be useful as a remote sensing technique for monitoring velocities of volcanic ejecta.

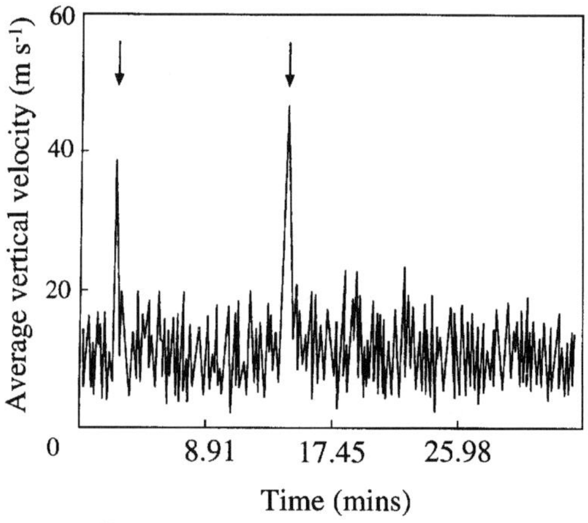

Figure 12.18 Example of average vertical velocity estimates for ejecta from Stromboli, Italy, calculated over a 35 minute period on September 24, 1991. Two explosions are indicated by arrows and correspond to visual observations. (After Weill *et al.* 1992, reproduced with permission of the American Geophysical Union.)

12.8 SUMMARY

This chapter has discussed a variety of remote sensing tools that can be used to observe volcanic plumes and to measure critical plume characteristics. These tools include instruments that measure electromagnetic radiation, acoustic vibrations and electric fields.

Electromagnetic remote sensing data can be acquired over a broad range of spectral wavelengths at a variety of spectral, spatial and temporal scales. Each of these factors determines the types of information that can be derived from a particular data set. Emittance, scattering, absorption and transmission of electromagnetic energy all affect the remote sensing measurements. The spectral resolution and the spatial resolution also affect the type of measurements that can be made. The electromagnetic spectrum can be divided into several regions: ultraviolet, visible and microwave. The fact that materials respond differently at different wavelengths allows measurement of a variety of volcanic plume phenomena. The positioning of the remote sensing instrumentation, whether it is on the ground, flown on an aircraft or in orbit on board a satellite, affects the spatial resolution and observation frequency.

Many instruments for measuring electromagnetic radiation currently exist and can be used from the ground, aircraft or satellite. Measurements that can be made using electromagnetic remote sensing methods include plume rise rate, plume radius, radial spreading rate of an umbrella cloud, downwind spreading rate, temperature structure and height of the plume top, and the mass of gases generated during an eruption. Satellite remote sensing data are also often used to monitor ongoing eruptions. Studies are currently under way to develop methods for differentiating volcanic plumes from meteoric clouds for real-time eruption detection.

Particle-laden volcanic plumes commonly generate lightning, indicating that the air ionization limit has been exceeded locally, and that volcanic plumes are highly charged. In addition to playing an important role in the formation of ash aggregates, this high level of charge allows the use of electric field measurements for monitoring purposes. Several studies have detected charged particles and aerosols, and lightning within volcanic plumes.

Acoustic measurements have recently been used to obtain vertical velocities of volcanic jets, such as those commonly observed at Stromboli volcano, Italy. The method is based on the Doppler effect for particles in motion and measures the Doppler shift between transmitted and backscattered acoustic waves.

13 Tephra Fall Deposits

NOTATION

A	area of isopach thickness T (m^2)
b_t	distance over which deposit thickness decreases by one-half (m)
b_c	distance over which particle size decreases by one-half (m)
D	area enclosed by 0.01 T_{max} isopach (m^2)
F	percentage < 1 mm at the 0.1 T_{max} isopach (dimensionless)
k	decay constant (m^{-1})
Md_ϕ	median grain size (mm)
σ_ϕ	sorting coefficient (dimensionless)
σ_v	standard deviation for settling velocity histograms (dimensionless)
T	isopach thickness (m)
T_{max}	maximum thickness of fall deposits at source (m)
V	volume of fall deposits (km^3)

13.1 INTRODUCTION

Tephra fall deposits from volcanic plumes provide a record of explosive volcanism in the geological past. From this record much can be learned about the frequency, style, magnitude and intensity (eruption rate) of past eruptions. Tephra fall deposits also provide a rich source of information on the dynamics and behaviour of volcanic plumes by inversion of the observations on the grain size and thickness variations, and interpretation of their depositional structures. Tephra deposits can be very widespread, covering areas of hundreds to millions of square kilometres. These layers provide excellent time horizons because they represent a virtually instantaneous geologic event that can be correlated over wide geographic areas. The deposits can be dated by a variety of techniques and thus they have been utilized as a key stratigraphic and geochronological tool in such diverse fields as volcanic hazards studies, palaeo-oceanography, palaeoecology, hominid evolution and archaeology.

Obtaining direct measurements and information on explosive eruptions is difficult because of their violent nature. A major explosive eruption can only be observed from afar and information is usually restricted to the size and shape of the eruption column, and occasionally information on their velocity during the initial period of ascent (Chapter 5). Remote sensed observations can also provide data on plume heights, temperature and the effects of wind dispersal (Chapter 12). Direct measurements of the internal variations in velocity, temperature and particle concentrations, of the overall grain size distribution of the ejecta and discharge rates in volcanic eruption columns have, thus far, not been attempted. Methods for so doing are not yet available. Tephra fall deposits contain potential information on these plume properties. The petrology of juvenile ejecta also provides important information on the conditions within magma chambers during explosive eruptions.

In this chapter the principal features of tephra fall deposits are described. The most significant characteristics of fall deposits are thickness variations, grain size variations, the physical and chemical properties of the ejecta and the sedimentary structures that form during deposition. This chapter and the following Chapters 15 and 16 describe how these geological materials and features can be interpreted to give qualitative and quantitative information about the conditions in the magma chamber, volcanic conduit, eruption column and plume during explosive eruptions. Their study can yield information on the total volume of ejecta, the total grain size distribution of ejecta, temperature and gas content of the erupting magma, variations in eruptive conditions, gas content of the plume and whether or not the explosive eruption involved substantial amounts of external water. Tephra fall deposits can also be used to reconstruct the history of explosive volcanism at a volcano or in a volcanic region.

13.2 EJECTA COMPONENTS

Fall deposits consist of material which is fragmented during explosive eruptions. The fragmentation process can result from degassing of volatiles dissolved in magma at high pressure, quenching or shattering of magma by thermal shock due to contact with external water, or the disruption of a volcanic edifice by steam explosions. These fragmentation processes generate a wide variety of particle sizes and shapes which are ejected during the eruption and eventually settle out to create a fall deposit. The principal components of fall deposits are juvenile fragments, crystals and lithics. Juvenile fragments consist of quenched pieces of magma. They can range from blocks or bombs (>64 mm in diameter) to ash size fragments (<2 mm in diameter). In most cases the juvenile fragments are vesicular quenched magma (e.g. pumice, scoria and glass shards) which consist of glass and sometimes crystals. Free juvenile crystals, which were present in the magma prior to eruption, are commonly present (Figure 13.1). Dense non-vesicular juvenile clasts of volcanic glass or obsidian can occur. The density of juvenile clasts can vary considerably from those which approach magmatic density (~ 2600 kg m^{-3}) to highly vesicular pumice (~ 500 kg m^{-3}).

Studies of fall deposits have shown that there is tremendous diversity in the morphology of juvenile fragments (Heiken and Wohletz 1985). This diversity results from differences in magma rheology, volatile content and eruption mechanism. Much effort has been devoted to the use of particle morphology, as ascertained from SEM images, for determining the style of the source eruptions. The reader is referred to Heiken and Wohletz (1985) for a thorough treatment of this topic. Free crystals are liberated from the magma during fragmentation or are derived from the comminution of wall rock in the volcanic edifice (Figure 13.1b). The principal crystals include quartz, plagioclase feldspar, potassium feldspar, orthopyroxene, clinopyroxene, hornblende, mica, and Fe–Ti oxide. The

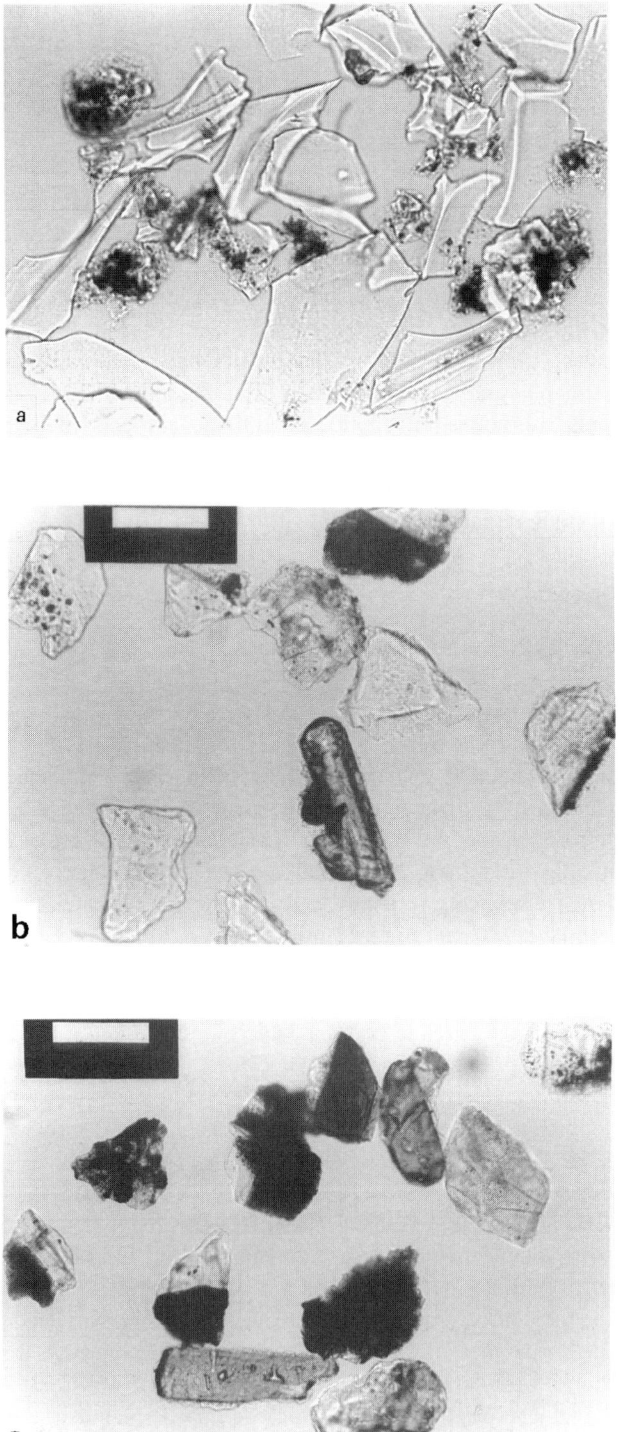
a
b
c

ferromagnesian-rich minerals such as the pyroxenes and hornblende are known as mafics. Crystals typically range in size from tens of micrometres to a few millimetres in diameter. In some instances, they can contain a thin coating of glass, indicating that they were present in the magma prior to eruption. They also exhibit a range in densities (2400–5000 kg m^{-3}) which is a function of their composition.

Another significant component of tephra fall deposits are lithics (Figure 13.1c). These are fragments of pre-existing rock from the conduit and vent rocks that are broken up during the eruptive process and incorporated into the turbulently erupted mixture of gases and juvenile particles. Lithics generally exhibit a range of sizes similar to juvenile fragments (blocks to ash), but are usually of a more restricted density range. Their lithology can be igneous, sedimentary or metamorphic depending on the structure and composition of the particular source volcano. Lithic clasts are usually easy to distinguish from juvenile material. However, fragments of pre-existing pumice, ash, free crystals and obsidian can occur and more elaborate analytical techniques have to be applied to determine whether such materials are juvenile or not.

An important aspect of tephra fall deposits is that the source plumes consist of a heterogeneous mixture of particles with greatly contrasting sizes, shapes, and densities. Transport of this material through the atmosphere will therefore result in strong fractionation of components relative to one another because of the difference in particle terminal settling velocities. This fractionation results in the concentration of certain components at various distances downwind from source. A good example of this process can be seen in the May 18, 1980 fall deposit of Mount St Helens, USA. This is an excellent deposit in which to study the abundances of different components and lateral variations in the composition of a fall deposit, because virtually all deposition occurred on land and extensive sampling was carried out before erosion could affect the primary depositional relationships. Figure 13.2 shows the variation of pumice and glass shards, felsic crystals, mafic crystals and lithics as a function of distance from Mount St Helens. Within 10 km of the source, the deposit consists mostly of pumice and lithics. With increasing distance downwind these components decrease rapidly, being replaced by feldspar and mafic crystals. The maximum crystal abundance occurs at about 140 km from source, where the total abundance is about 70 wt% of the deposit. Beyond this distance the abundance of feldspar and mafic crystals along with lithics decreases significantly with a corresponding increase in pumice and glass shards. At 400 km from source the deposit consists of over 80 wt% of the latter two components. The variation of component abundance with distance demonstrates that the bulk composition of fall deposits depends strongly on distance from the vent. It is generally not possible to use the bulk chemical composition of a fall unit for correlation purposes over large distances. It is more reliable to use the composition of individual components, such as glass shards or specific mineral phases.

Figure 13.1 Photomicrographs of (a) silicic glass shards from a tephra layer in ODP Site 999 sediments (full field of view = 500 μm), (b) plagioclase and orthopyroxene crystals from the 1980 eruption of Mount St Helens volcano (scale bar = 200 μm), and (c) lithic particles (opaque) and crystals from the 1980 eruption of Mount St Helens volcano (scale bar = 200 μm)

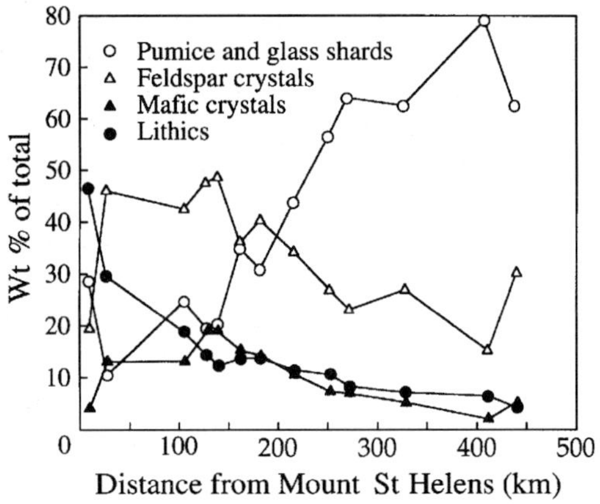

Figure 13.2 Variation in pumice and glass shards, feldspar crystals, mafic crystals and lithics as a function of distance along the main dispersal axis of the May 18, 1980 Mount St Helens fall deposit modified from Carey and Sigurdsson (1982). Each component has a peak in abundance at a specific location downwind from source as a result of the aeolian fractionation of particles according to settling velocity

13.3 PETROLOGY OF EJECTA

The study of the juvenile components of pyroclastic ejecta can provide important information on the basic conditions occurring during explosive eruptions. Previous chapters have shown how properties such as magmatic gas content, magma chamber pressure, magma viscosity and temperature exert a fundamental control on the source conditions for a volcanic plume. A number of techniques have been developed for estimating these properties from studies of the petrology of the ejecta.

The composition and abundance of minerals crystallizing in magma chambers is determined by the principles of phase equilibria. Specifically, the combination of minerals (the mineral assemblage) and the composition of each mineral is controlled by the bulk chemical composition of the melt, the volatile components in the melt and gas phase, and the thermodynamic properties of the system including temperature, volatile partial pressures, total pressure and oxidation state of the magma. Many of the minerals in magmas are solid solutions in which chemical composition varies due to substitution of different cations on particular sites in the mineral structure. Knowledge of phase equilibria, the thermodynamic controls on the substitutions and experimental data allow estimates of these parameters to be made from the mineralogy of tephra. The reader is referred to Wood and Fraser (1976) for more detailed discussion of the general principles of phase equilibria. However, in order to apply the principles of phase equilibria it is important to demonstrate that crystals present in the juvenile component of a tephra deposit formed at equilibrium. In some cases processes such as mixing of different magmas, assimilation of crust by magma and incorporation of refractory minerals (restite) from the source region result in more complex disequilibrium crystal assemblages where the principles cannot be applied in a straightforward way.

The temperature of magma is an important parameter to determine. This is because the thermal flux plays a critical role in the dynamics of volcanic plumes. It is often not feasible to measure directly the temperature of volcanic ejecta during an explosive eruption. However, the composition of coexisting minerals can be used to calculate the temperature at which equilibration last occurred. The values obtained are assumed to be the pre-eruption temperature of the magma within the chamber. A number of mineral geothermometers and geobarometers have been developed and applied to volcanic ejecta. The iron-titanium oxide phases magnetite and ilmenite are commonly found in silicic tephra fall deposits. Their compositions are a function of temperature and the fugacity of oxygen (Buddington and Lindsley 1964) and they have been widely used to calculate eruption temperatures (e.g. Hildreth 1981; Bacon and Druitt 1988). The accuracy of the determinations can be less than 15–25 °C for very fresh minerals, but alteration and more complex thermal histories (e.g. mixing of magmas of different temperature during the eruption) can reduce the accuracy considerably and can even make the calculated temperatures meaningless. Other examples of commonly used geothermometers include the assemblage clinopyroxene and orthopyroxene (Wood and Banno 1973) and the assemblage hornblende, plagioclase and quartz (Holland and Blundy 1994).

Accurate estimates of total pressure are less easy because many common igneous minerals are stable over a wide range of pressure and their compositions tend to be more sensitive to temperature than total pressure. A more complicated approach, which has been successful in recent years, is to carry out high-pressure laboratory experiments on natural ejecta compositions in order to determine the often unique combination of physicochemical conditions which reproduce the observed mineral assemblage, volatile content, and coexisting melt phase observed in the natural samples. In these types of experiments several variables can be adjusted independently to map out the phase equilibria of a particular system. The first step in the methodology is to select natural samples which are representative of the magma and to identify the natural mineral assemblage. A basic requirement is that the sample contains an equilibrium assemblage of minerals and melt. Magma mixing is a common phenomenon in explosive eruptions and investigations of hybrid material with disequilibrium assemblages can be difficult or even meaningless. The samples are then studied petrologically and estimates of temperature, volatile partial pressures and oxidation state are made if suitable minerals are present.

This approach is exemplified by the study by Rutherford *et al.* (1985) on the ejecta of the May 18, 1980 Plinian eruption of Mount St Helens. The 1980 Mount St Helens magma consisted of a dacitic melt phase (60 wt% SO_2) in equilibrium with plagioclase, orthopyroxene, amphibole, titaniferous magnetite and ilmenite. Ejecta temperatures were constrained quite accurately by Fe–Ti oxide geothermometry (920–940 °C), but pressures could not be determined with any precision. Experiments were carried out under a variety of conditions guided by the petrological studies. The main unknowns were total pressure and volatile composition because temperature was well determined. Figure 13.3 shows a phase diagram for the Mount St Helens dacite based on experiments carried out at various degrees of water undersaturation, but at a constant total pressure of 220 MPa. The products of these experiments were then analysed and compared to the natural sample. By equilibrating melts with a mixed H_2O/CO_2 gas phase it is possible to vary H_2O activity at fixed total pressure and to investigate situations where magma may have been water undersaturated within the chamber. Based on these results Rutherford *et al.* (1985) esti-

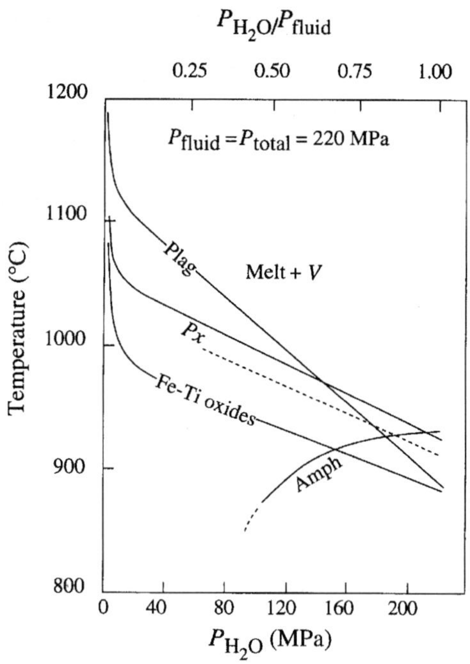

Figure 13.3 Phase diagram of the 1980 Mount St Helens dacite based on experiments carried out at 220 MPa with variations in the partial pressure of H_2O, modified from Rutherford *et al.* (1985). Solid curves denote stability boundaries for various mineral phases in equilibrium with melt. The calculated temperature of the dacite is 920–940 °C based on Fe–Ti oxide geothermometry

mated that the Mount St Helens dacite was derived from a magma chamber with a temperature of 920–940 °C, a water content of about 4.5 wt% and a total pressure of 250 MPa. The results suggested that the magma was not water saturated at depth prior to the eruption, although a separate CO_2-rich gas phase may have been present. The total pressure inferred from the experiments is equivalent to a chamber at a depth of 6–7 km which is in excellent agreement with geophysical estimates based on earthquake locations.

Volatile content is a particularly important factor which influences the nature of explosive volcanism. The experimental technique described previously provides constraints on volatile content and degree of volatile saturation, but is time consuming and requires extensive petrological analysis. Another more direct way of determining pre-eruption volatile contents of magma is by the study of glass inclusions trapped in phenocrysts. These inclusions may represent small aliquots of melt that were incorporated into crystals as the crystal grew in the chamber. Analytical techniques such as Fourier transform infrared spectroscopy (FTIR), electron microprobe and ion probe mass spectrometry allow determinations of the content of major volatile constituents such as H_2O, CO_2, SO_2, Cl and F in melt inclusions that are typically only a few tens of micrometres in size. The assumption is that the melt in such samples was quenched by explosive eruption and has not leaked volatiles as the pressure was reduced to atmospheric. Difficulties arise if crystallization of the inclusion has taken place because this tends to increase the apparent concentration of volatile components. Such crystallization modifies the composition of the melt inclusion relative to the coexisting melt. Using melt inclusion analysis Anderson *et al.* (1989) estimated the water content of the Plinian phase of the Bishop Tuff eruption to be about 6 wt%, whereas magma from the pyroclastic flow phase had a water content of about 4 wt%. A similar result was found by Dunbar and Hervig (1992) for the Lower Bandelier tuff with between 4 and 5 wt% water in the Plinian and water decreasing from 4 to 2 wt% during the ignimbrite phase of the eruption. In contrast, Dunbar and Kyle (1993) found that there was no difference in water content in the transition from the Plinian to pyroclastic flow phase of the Taupo AD 180 eruption of New Zealand. Other examples of such studies include Lowenstein and Mahood (1991), Webster *et al.* (1993) and Gardner *et al.* (1995). Some explosive eruptions eject fragments of obsidian derived from depth and these can contain elevated volatile contents. For example, Newman *et al.* (1988) were able to make minimum estimates of the H_2O and CO_2 contents of magma involved in the AD 1340 sub-Plinian eruptions of Mono Craters, California, USA.

Some minerals are sensitive to volatile partial pressure. For example, equilibria between plagioclase feldspar and glass is sensitive to water partial pressure (e.g. Arculus and Wills 1980) and hornblende will only precipitate from magmas of a given composition once melt water contents reach some minimum level (>3 wt%, Burnham 1979). Application of these types of equilibria will provide information about the partial pressure of H_2O but cannot constrain the total pressure of the system. Minerals such as anhydrite ($CaSO_4$) only form at elevated sulphur contents under oxidizing conditions (Carroll and Rutherford 1987).

13.4 GENERAL DESCRIPTION OF FALLOUT

Some of the general qualitative features of deposition from volcanic plumes are now considered. Quantitative aspects will be discussed in Chapters 14, 15 and 16. Figure 13.4 shows the general features of particle fallout from an eruption column. The figure emphasizes the division of the column into an upward rising convective region and a horizontally expanding umbrella region with a corner at the boundary of the two regions. Four categories of fallout ejecta are defined for convenience (Bursik *et al.* 1992b), while recognizing that these categories are in practice gradational.

The coarsest particles are sufficiently large that their motion is only weakly influenced by the flow in the eruption column. These clasts are accelerated up the conduit in a sustained eruption and ejected into the atmosphere in a discrete explosion. Their subsequent motion depends on their size and launch velocity (Wilson 1972). They are described as *ballistic* ejecta because their motion is comparable to the behaviour of projectiles. These clasts (typically called volcanic bombs or ballistic blocks) can leave the column margins and fall from low heights within the gas thrust region. The size of clast that can be regarded as ballistic in behaviour depends on the velocities in the basal parts of the eruption column. For typical vent velocities of 100–300 m s^{-1} clasts significantly larger than 10 cm in diameter can be regarded as ballistic. The proximal parts of many fall deposits can contain large proportions of ballistically emplaced ejecta. Self *et al.* (1980) recognized two categories of ballistic: those sufficiently large to be unaffected by gas motions and somewhat smaller ballistics that are affected by turbulent motions in the column. The dynamics of ballistic clasts is discussed further in section 14.3 of Chapter 14.

A second category of fallout takes place along the ascending margins of a volcanic plume. Theoretical models and experimental studies (Chapter 14) show that marginal fallout is restricted predominantly to relatively large clasts. A lower limit on clast size is imposed by the strong inward flow of air into the margins of plumes due to lateral entrainment. For columns rising at typical convective velocities of 50–200 m s^{-1} the inward velocity at the plume margins is approximately 10 wt% of the average upward velocity. Particles much less than a centimetre in diameter cannot easily escape from the margins of plumes with heights exceeding 10 km. This is discussed quantitatively by Ernst *et al.* (1996a) and in Chapters 14 and 15. The fallout of coarse ejecta from the column margins, predominantly in the size range 1–10 cm diameter, will contribute to the formation of proximal parts of fall deposits.

The bulk of ejecta from a volcanic eruption column falls out of suspension from the umbrella region and the wind-blown plume. The vertical velocities and vigorous turbulent motions are sufficient to transport most particles into the umbrella region. Fallout of suspended particles takes place at the base of the spreading umbrella cloud and from the plume as it is advected downwind. The quantitative aspects of this process are given in Chapter 14. Particles are transported in two stages before reaching the ground: first within the highly turbulent plume itself and then by the wind after they leave the base of the plume. Smaller and/or lower density particles are in general transported further than larger denser particles so that complementary lateral changes in grain size distribution of the deposit are expected. Particle fallout in the size range of a few centimetres diameter down to fractions of a millimetre can be described in terms of transport by the turbulent plume and advection by wind. Particles falling out near the vent are also influenced by the inflow of air towards the eruption column. Suspended particles can be further classified into high Reynolds number coarser particles that have fall velocities proportional to the square root of the particle diameter, and low Reynolds number finer particles that have fall velocities proportional to the square of the particle diameter (section 14.2 in Chapter 14). The threshold between the two fall regimes occurs typically between 100 and 500 μm diameter depending on altitude in the atmosphere.

The final category of fallout includes the smallest particles in an eruption plume. Their fallout is influenced by additional factors. First, atmospheric turbulent diffusion can have a major influence on their dispersal. The importance of atmospheric diffusion in dispersal of particles is still uncertain and is discussed further in Chapter 14. However, all models indicate that submillimetre particles will be increasingly influenced by diffusion as particle size decreases. Second, fine particles can aggregate by a variety of processes (Chapter 16) and fall much faster than individual particles. Aggregation is known to strongly affect particles of the order of 200 μm or less.

These general considerations alone imply that the characteristics of tephra fall deposits are strongly influenced by the grain size distribution of the ejecta and the relative proportions of different particle types. The implication is that the variation in deposit thickness from the vent can be divided into at least three distinct zones. In the proximal zone the fallout will be dominated by ballistic ejecta and coarse clasts from the column margins. In Chapter 14 it will be demonstrated that thickness decreases rapidly in this zone which extends approximately to the column corner (Figure 13.4). Beyond the corner another zone is defined by fallout from the umbrella cloud or wind-blown plume and thickness decreases less rapidly. Empirical observations and theoretical considerations show that

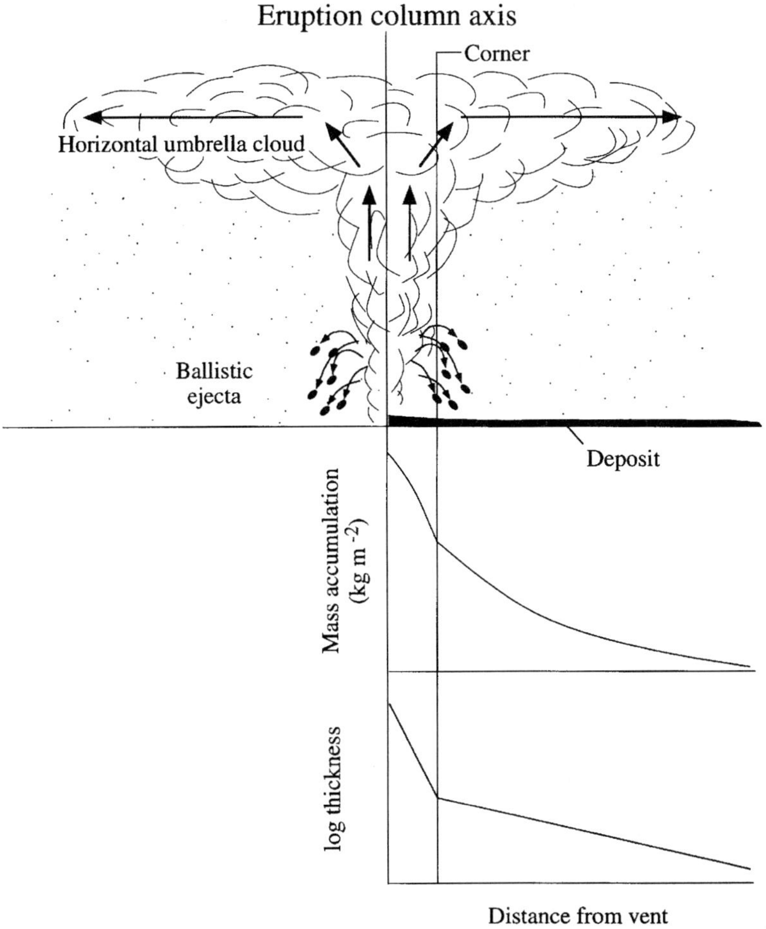

Figure 13.4 Schematic representation of fallout from different regions of a sustained convective plume. At the base of the plume large clasts are ejected ballistically and deposited close to the vent. Within the rising convective region moderate to large clasts are lost from the column margin and accumulate in the proximal area. The majority of particles fall from the base of the horizontal umbrella cloud. In the lower part of the figure idealized plots of mass accumulation and log thickness are shown as a function of distance from the vent. The inflections in both plots correspond to changes in the nature of particle fallout at the corner of the plume where motion shifts from dominantly vertical to horizontal

thickness decreases approximately exponentially with distance (Pyle 1989; Sparks *et al.* 1992). However, there may also be a change of sedimentation rate that relates to the change from high Reynolds number coarse suspended particles to low Reynolds number fine suspended particles. This change in rate could explain the slower exponential thinning observed in distal parts of some tephra fall deposits (section 13.5.1). A third zone of secondary thickening can occur at large distances from the volcano due to fallout of fine ash as aggregates (Chapter 16) as exemplified by the May 18, 1980 Plinian deposit of Mount St Helens (Carey and Sigurdsson 1982).

The different categories of fallout are representative of deposition from a sustained eruption plume from a point source. The extent to which these different categories are developed during a particular eruption will depend on the character of the event. For example, deposition from a Strombolian type of eruption will be dominated by ballistic fallout because a large convective plume with a well-defined umbrella region and a high proportion of ash are not produced during this type of activity (Chapter 10). A co-ignimbrite plume is likely to develop an umbrella region, but will lack ballistic ejection of material from its basal region and will be dominated by low Reynolds number particles (Chapter 7).

13.5 CHARACTERISTICS OF FALL DEPOSITS

13.5.1 Thickness

Fallout of tephra occurs from heights of several kilometres to a few tens of kilometres. The resulting deposit drapes the topography and has a uniform thickness on the local scale of an outcrop. Individual beds from particular explosive events show characteristic mantle bedding (Figure 13.5), a feature which helps discriminate fall deposits from other kinds of pyroclastic deposits formed by flow processes. Individual fall deposits cover areas of hundreds to over 1 million km^2 in the largest eruptions. Thickness data are typically

Figure 13.5 Plinian fall deposit from Vesuvius volcano showing characteristic mantling of underlying topography

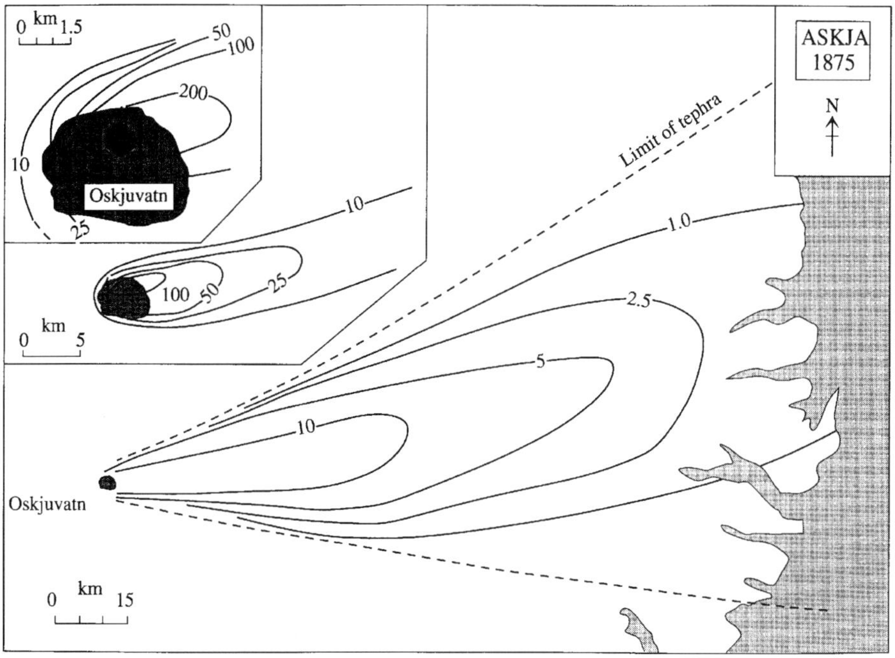

Figure 13.6 Isopach map of the 1875 Plinian fall deposit of Askja volcano, Iceland, modified from Sparks *et al.* (1981). Stippled area corresponds to the caldera lake known as Oskjuvatn which formed as a consequence of the eruption. Isopach contours are in centimetres. The elongation of isopach contours is a reflection of the strong winds that affected the plume during the eruption

plotted on a map and contoured to produce an isopach map which shows the large-scale regional variations. Such maps provide a good qualitative indication of the height of an eruption column and hence the intensity of the eruption as well as information on the strength and direction of prevailing winds and the location of volcanic vents (Walker 1971). Figures 13.6–13.9 and 13.11–13.13 show some examples of isopach maps. These

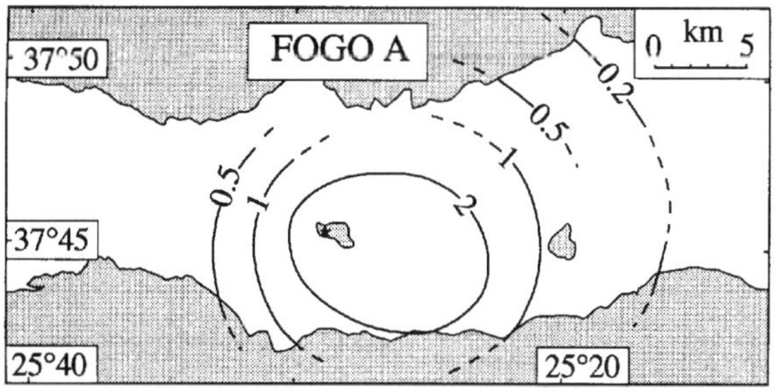

Figure 13.7 Isopach map of the Fogo A Plinian fall deposit from São Miguel in the Azores, modified from Walker and Croasdale (1971). Isopach contours are in metres. The circularity of the isopach contours suggests a strong eruption plume with only minor wind influence. The star shows the source vent

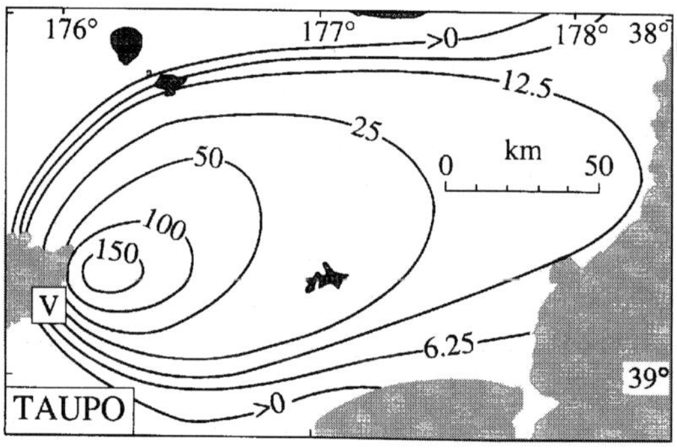

Figure 13.8 Isopach map of the AD 180 Taupo ultra-Plinian fall deposit in New Zealand, modified from Walker (1980). Isopach contours in centimetres. The eruption is inferred to have had an eruption column height of 50 km which produced a roughly circular accumulation pattern in the proximal area with substantial deposition on the upwind side of the vent which is indicated by V

illustrate some of the more common features and the qualitative inferences that may be drawn from the distribution patterns.

The isopach map of the 1875 eruption of Askja in Iceland (Figure 13.6) is an example of a deposit formed from a powerful Plinian event. The eruption was short-lived (6.5 hours) and generated an eruption column with an estimated height of approximately 30 km. The plume was advected by jet-stream winds with a mean velocity of 30 m s^{-1} (Sparks *et al.* 1981). The isopach map is quite symmetrical and displays strongly elliptical contours as a consequence of the strong wind which did not change direction significantly during the short period of activity. The isopach map shows the characteristic thinning of fall deposits with distance from the vent. The isopach contours become more elliptical as the thickness decreases, reflecting the increasing influence of the wind on dispersing finer particles. The axis of symmetry of the deposit is known as the dispersal axis and defines the main direction of wind transport during deposition. Normals to the dispersal axis are the cross-wind directions.

The Fogo A Plinian deposit of Agua de Pau volcano on São Miguel in the Azores resulted from a very powerful eruption which took place approximately 5000 years BP, with an estimated column height of 20–27 km in a very weak wind (Walker and Croasdale 1971; Bursik *et al.* 1992b). The isopach map (Figure 13.7) is quite symmetrical and the contours are almost circular. In detail the Fogo deposit is rather more complex and shows greater thicknesses to the south and east. The column height increased with time and the early periods of the eruption involved relatively small columns which were blown to the south (Bursik *et al.* 1992b).

The Plinian deposit of the AD 180 eruption of Taupo in New Zealand is the most powerful known eruption of this type with an estimated column height of over 50 km (Walker 1980). The isopach map (Figure 13.8) shows features typical of very powerful eruptions. For example, the inner contours are quite circular with significant deposition upwind of the vent as a consequence of the strong upwind transport of tephra in the umbrella region of very powerful columns. The contours become increasingly elliptical and distorted downwind as thickness decreases. The maximum thickness is located some distance downwind.

The eruption of Lonquimay volcano in southern Chile in 1988–90 resulted from deposition from modest Strombolian eruption columns during a long-lived eruption (Figures 13.9 and 13.10). The explosive activity was most vigorous in the first few months when the ascending magma was actively degassing (Moreno and Gardeweg 1989). The column heights reached a maximum of 12 km in January 1989 and gradually declined over the following several weeks as the eruption intensity decreased. Much of the ejecta was deposited close to the vent to build up the Navidad cinder cone which reached a diameter of 1100 m and height of 600 m by April 1989 (Figure 13.10). The conical shape of such cinder cones is determined by fallout close to the vent from relatively weak columns and by the attainment of the critical angle of repose of loose particles (approximately 30°). Beyond the base of the cone such deposits become distributed in all directions (Figure 13.9). The Lonquimay deposit has elongated lobes to the south and east. Deposits of this kind are strongly influenced by variations in wind directions which can vary greatly over an eruption lasting several months. The isopach pattern is thus controlled by regional weather patterns. Such deposits can be quite symmetrical even though the eruptions are weak and can display complex lobate patterns (see also Self *et al.* 1974).

Complex tephra fallout patterns can also develop where eruptions occur in areas with significant variations of wind velocity with height or with time. For example, in low latitudes wind directions can be opposite at different heights. The fall deposit of the 1902 Plinian eruption of Santa Maria volcano, Guatemala, had tephra dispersed to the east by low-altitude anti-trade winds and tephra dispersed to the north and west by higher level trade winds (Williams and Self 1983) causing a bilobate dispersal pattern (Figure 13.11a). Changes in wind direction during an eruption can also cause asymmetrical contour patterns. The deposit formed from the 1104 eruption of Hekla in Iceland shows non-elliptical contours (Figure 13.11b). Thorarinsson (1967) showed, by a systematic study of the dispersal of the layers of different chemical composition, that this pattern was the result of

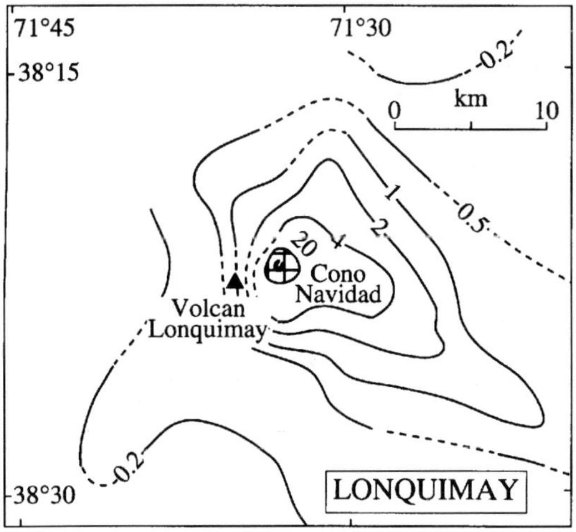

Figure 13.9 Isopach map of the fall deposit formed by the 1988–90 violent Strombolian eruptions of Lonquimay volcano, Chile. Contours are in centimetres. (Modified from Moreno and Gardeweg 1989.)

Figure 13.10 (a) Cinder cone produced by the 1988–90 eruption of Lonquimay volcano, Chile. The cone is breached by effusion of lava on the right-hand side. The yellow colour on the rim is due to precipitation of $FeCl_2$ from fumaroles. (Photograph courtesy of R. S. J. Sparks.) (b) The eruption of Lonquimay volcano, Chile, in 1989 showing a plume from Strombolian activity and the cinder cone. The cone is breached to the right by effusion of lava. (Photograph courtesy of M. C. Gardeweg.)

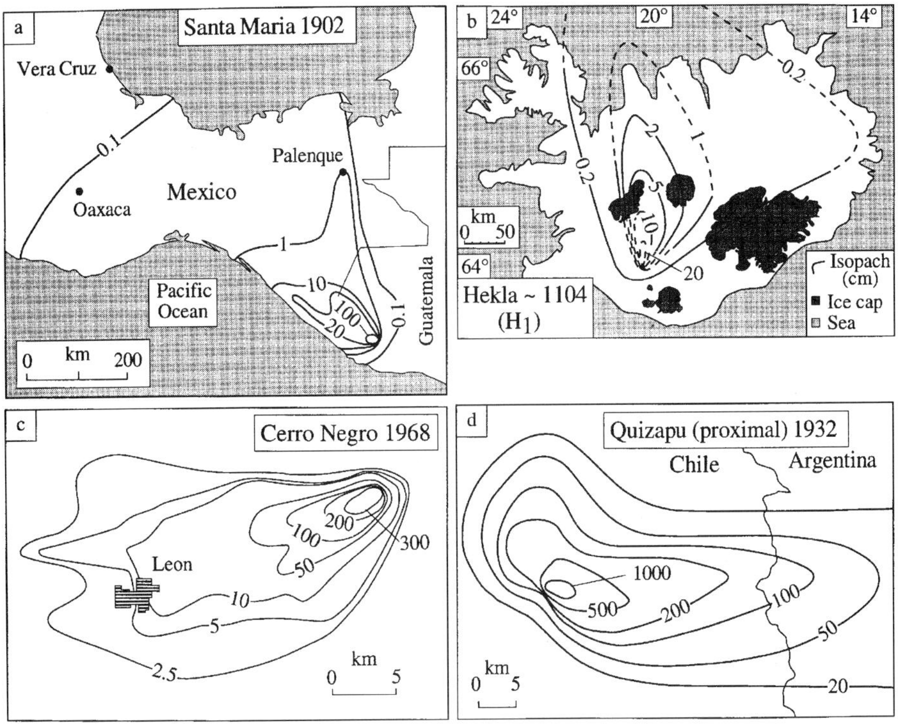

Figure 13.11 Examples of isopach maps showing bilobate dispersal patterns: (a) 1902 Plinian fall deposit of Santa Maria volcano in Guatemala, modified from Williams and Self (1983); (b) AD 1104 Plinian fall deposit of Hekla volcano (H_1), Iceland, modified from Thorarinsson (1971); (c) 1968 violent Strombolian fall deposit of Cerro Negro, Nicaragua, modified from Rose *et al.* (1973); (d) 1932 Plinian fall deposit of Quizapu, Chile, modified from Hildreth and Drake (1992). Isopach contours are in centimetres for all maps. The bilobate dispersal patterns may be attributed to variations in wind direction at different altitudes, variations in wind direction during an eruption and interaction of the plume with a crosswind, causing plume bifurcation (Ernst *et al.* 1994)

a clockwise rotation of the wind direction during the eruption. Another cause of a bilobate pattern is plume bifurcation due to interaction with a cross-wind. This phenomenon has been observed in laboratory experiments and in satellite images of volcanic plumes (Ernst *et al.* 1994; see Chapter 11). Possible examples of bilobate isopach maps attributable to bifurcation include the 1968 Cerro Negro deposit, Nicaragua, and the 1932 Quizapu deposit, Chile (Figure 13.11c and 13.11d). These more complex patterns show that it should not be automatically assumed that isopach contours are elliptical.

Figure 13.12 shows the isopach map of the Plinian deposit of the May 18, 1980 eruption of Mount St Helens (Sarna-Wojcicki *et al.* 1981). This phase of the eruption lasted approximately nine hours and involved discharge of 0.2 km^3 of dacitic ejecta (Criswell 1987; Carey *et al.* 1990). The eruption column height varied between 16 and 18 km and the plume was dispersed by stratospheric winds with a mean velocity of 28 m s^{-1} at 12–14 km altitude (Bursik *et al.* 1992a). At distances closer than 250 km from the vent the pattern is as expected for a short-lived moderately high intensity eruption in a strong wind. However, a well-defined zone of secondary thickening occurs between 250 and 350 km

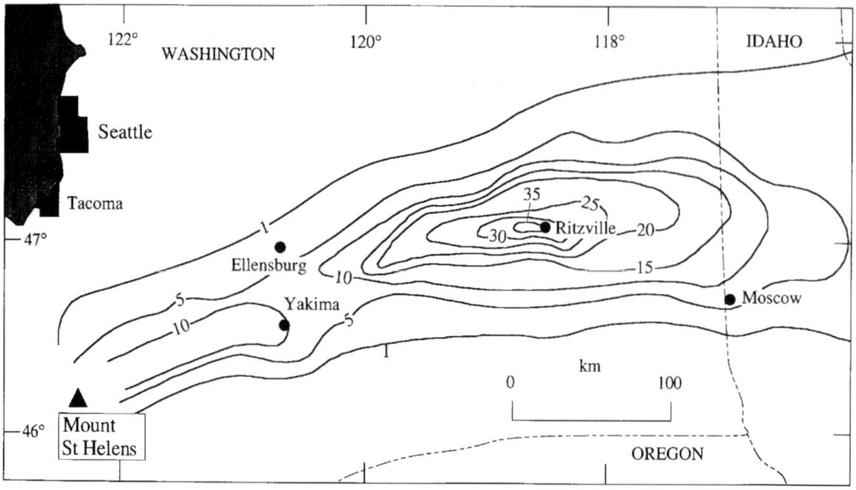

Figure 13.12 Isopach map of the Plinian fall deposit from the May 18, 1980 eruption of Mount St Helens, modified from Sarna-Wojcicki *et al.* (1981). A second thickness maximum is present in the vicinity of Ritzville, Washington, and has been attributed to premature settling of fine ash by aggregation. Contours are in millimetres

near Ritzville, Washington (Sarna-Wojcicki *et al.* 1981). The secondary thickening has been attributed to fallout of aggregated fine particles (Carey and Sigurdsson 1982). Other examples of secondary thickening in tephra fall deposits are given in Chapter 16 (section 16.2.1).

Figure 13.13a shows the isopach map of the deposit of the giant cloud that lofted off the initial blast flow in the 1980 eruption of Mount St Helens (Sparks *et al.* 1986). The proximal part of the Plinian deposit is shown for comparison (Figure 13.13b). The tephra fall deposit of the giant cloud is a good example of a co-ignimbrite fall deposit and is compared with the near source pattern of the subsequent Plinian deposit. The co-ignimbrite cloud deposit had a source which was the entire area of the flow which was dispersed asymmetrically to the north (Sparks *et al.* 1986; Sisson 1995). The deposit therefore has a centre of symmetry which is displaced 17 km north of the volcano. Sisson (1995) suggested that the maximum thickness was controlled by topographic effects.

Isopach maps can be used to identify source vents. The maximum deposit thickness usually occurs at the vent and so the isopach contours enclose the vent like a bulls-eye. However, there are a few examples of deposits formed from very powerful eruptions, such as the Taupo ultra-Plinian deposit, where the maximum thickness is displaced a few kilometres downwind of the vent (Figure 13.8). Another technique for locating the vent is to exploit changes in wind direction during an eruption and the fact that the vent should lie on the dispersal axis. If different fall units of the same eruption have different dispersal directions, then the intersection of the dispersal axes will give the vent. Of course this method only works if the two eruption phases come from a common vent. During some eruptions the vent position changes with time (e.g. the Askja 1875 eruption; Sparks *et al.* 1981).

Thorarinsson (1967) observed that thickness decreases exponentially with distance along the dispersal axis of Plinian fall deposits of Hekla in Iceland. Pyle (1989) proposed

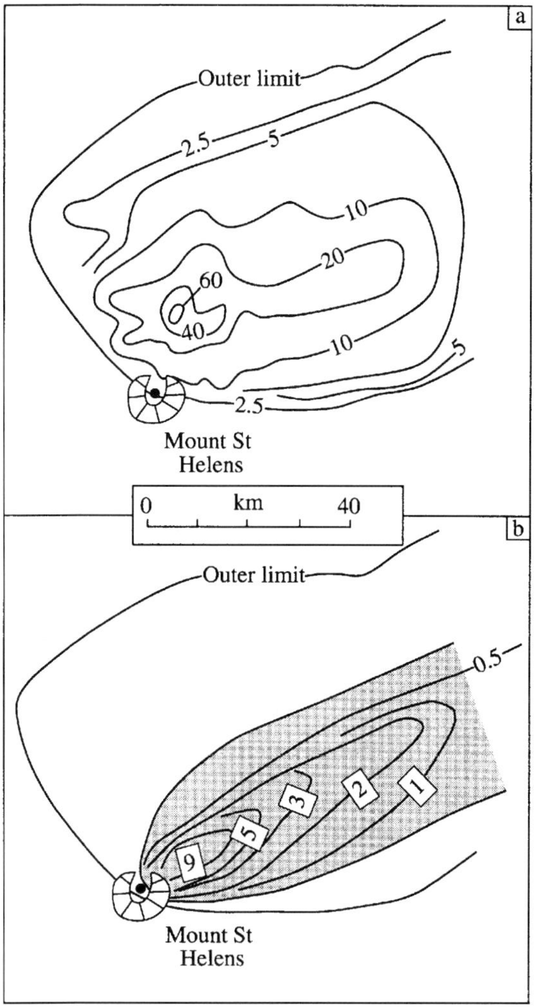

Figure 13.13 Isopach maps of (a) the co-ignimbrite fall deposit from the initial May 18, 1980 blast/surge of Mount St Helens, and (b) the Plinian fall deposit of the May 18, 1980 eruption of Mount St Helens. In (b) the shaded area represents the extent of the Plinian fall deposit and the line labelled "outer limit" refers to the fall deposits from the blast/surge. Isopach contours are in millimetres (a) and centimetres (b). Modified from Sparks *et al.* (1986).

that thickness variations in fall deposits can be described by a more general exponential relationship

$$T = T_{max} \exp(-kA^{0.5}) \tag{13.1}$$

where A is the area enclosed by the isopach contour of thickness T, T_{max} the maximum thickness of the deposit and k the decay constant. Data from isopach maps are plotted in the form of equation (13.1) in Figure 13.14 to illustrate the good agreement with this empirical formulation. Many other deposits also show this exponential behaviour (Pyle

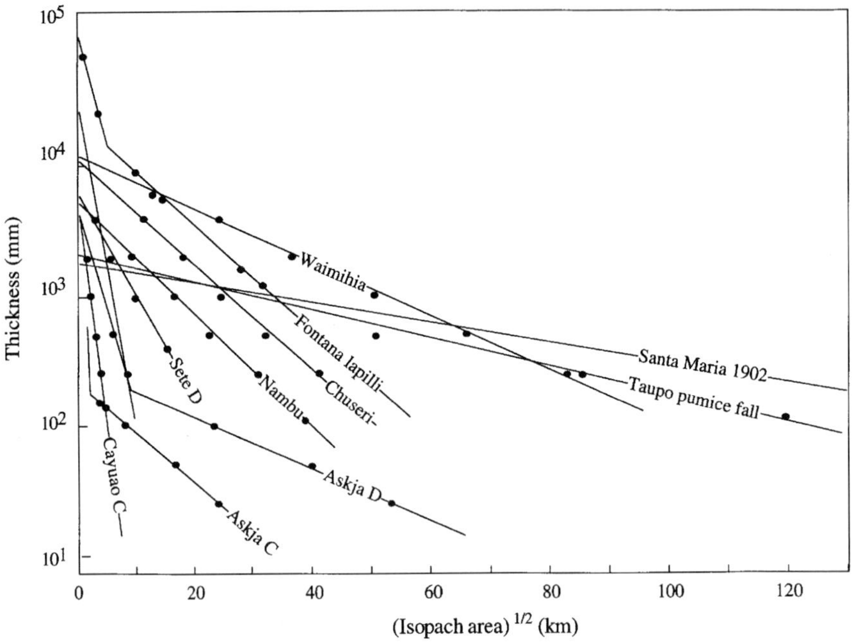

Figure 13.14 Plot of thickness versus square root of isopach area for a variety of tephra fall deposits, modified from Pyle (1989). Most deposits fall along straight-line segments on this plot indicating an exponential decay thinning behaviour and many also exhibit two line segments. The inflection point of the line segments has been interpreted to reflect the corner of an eruption plume where motion of gas and particles changes from dominantly vertical to horizontal

1989; Fierstein and Nathensen 1992). Each deposit can therefore be characterized by a decay constant. Pyle (1989) proposed the thickness half distance, b_t, as an appropriate parameter ($b_t = 1/k$) which is the characteristic distance over which the thickness halves. Deposits with complex lobate shapes and with thickness maxima away from the vent still fit this exponential relationship; see for example data for the May 18, 1980 Mount St Helens deposit (Figure 13.15).

Some deposits show two or more distinct straight-line segments on this diagram (Figure 13.14) with an inner segment with a steeper gradient. In some examples (Figure 13.14) the break-in-slope occurs between two segments within a few kilometres of the vent. In these cases the two segments have been interpreted as the consequence of sedimentation from the two zones shown in Figure 13.4, i.e. the rising plume region and the region beyond the plume corner where transport is predominantly horizontal (Sparks *et al.* 1992; Ernst *et al.* 1996a). In other examples the change in slope occurs much further from the vent: for example the 1980 Mount St Helens Plinian deposit (Figure 13.15), the Mazama tephra fall deposit (Figure 7.6) and the 1991 Hudson tephra (Scasso *et al.* 1994). These latter cases cannot be explained by the column corner model and require an alternative explanation. Physical explanations of the exponential thinning and the origin of the segments are presented in more detail in Chapter 15.

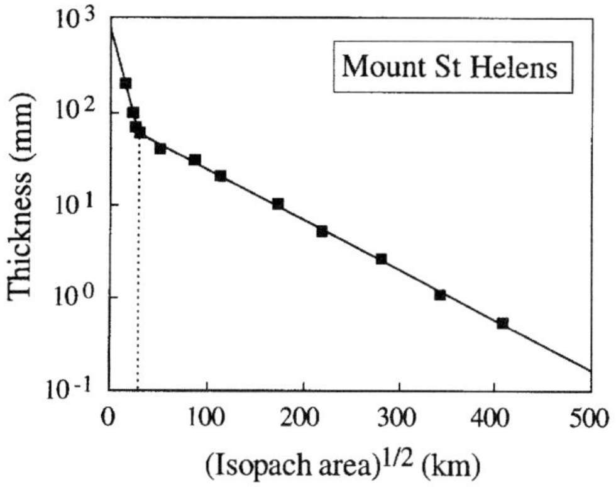

Figure 13.15 Plot of thickness versus square root of isopach area for the May 18, 1980 Mount St Helens Plinian fall deposit. The deposit follows the exponential decay relationship despite the occurrence of a secondary thickness maximum. There is a marked change of slope in the plot at about 27 km

13.5.2 Volumes

Thickness data can be used to estimate the total volume of ejecta from an eruption and, if the duration of the event is known, the discharge rate of ejecta feeding the plume. Such data are important in evaluating and refining the application of theoretical models of eruption plume dynamics.

A variety of methods have been developed to calculate deposit volumes (see Pyle 1989, 1995 and Fierstein and Nathensen 1992 for detailed reviews). All are empirical and should yield approximately the same result within the preserved area of the deposit irrespective of the function used to parameterize the isopach map (e.g. polynomial, power law or exponential). The main difficulty is extrapolating beyond the area that is preserved. The thin distal areas of fall deposits are quickly eroded or made unrecognizable by formation of soils. However, the distal parts may contain significant volumes of ash. There are very few deposits with extensive and reliable data beyond 100 km from the vent except for deposits, such as the 1980 deposit of Mount St Helens and the 1991 Hudson tephra (Scasso *et al.* 1994), which were studied immediately after eruption. Various commonly used empirical formulations have no physical basis (Pyle 1989, 1995) so that their extrapolation may give spurious results. The exponential relationship (equation 13.1) has the merits that it is simple to extrapolate and most, if not all, preserved parts of fall deposits display exponential behaviour. Recent modelling of sedimentation from plumes indicate that approximate exponential thinning is predicted (see Sparks *et al.* 1992 and discussion in Chapter 15) which gives a physical basis for this formulation. The volume, V, of the deposit with exponential thinning is given as

$$V = 13.08 T_{\max} b_t^2 \tag{13.2}$$

Application of equation (13.2), however, may only give a minimum estimate of volume. Examples such as the 1991 Hudson tephra show markedly different thinning of rates of fine-grained distal parts. Many older tephra deposits may not preserve their fine distal parts and there is often no way of knowing whether there might have been large amounts of fine ash which was distributed with a different value of the decay constant in equations (3.1) and (3.2). Pyle (1989) gives formulae for calculating volumes with more than one decay constant.

Another way of estimating deposit volumes is the crystal concentration method (Walker 1980). The crystal to glass mass ratio can be measured in pumice and then compared with the free crystal/glass mass ratio in the fine-grained components of the preserved deposit. If the latter material is the result of the fragmentation of the pumice then the ratios should be equal if all of the deposit is preserved. However, if the heavier and coarser crystals are preferentially concentrated in the preserved part of the deposit and the finer grained glassy component is preferentially concentrated in the distal regions beyond the limits of preservation, then the crystal/glass mass ratios in the deposit should be enhanced. Mass balance considerations then allow the missing volume to be calculated. This method requires a great deal of painstaking laboratory work in order to establish an overall mass ratio. Consequently there are few examples of such work. In the case of the Taupo Plinian deposit there is an unexplained discrepancy between the exponential method (Walker 1980; Pyle 1989) and the crystal concentration method. The latter gives much larger volumes. The crystal concentration method involves certain assumptions which must be tested. For example, some crystals may be non-magmatic xenocrysts and preferential sampling of crystal-rich pumice clasts in heterogeneous deposits. Alternatively the distal fine part of the Taupo deposit beyond the preserved limits may be much more voluminous than calculated by extrapolation of the more proximal exponential thinning rate.

It is preferable to express volumes of ejecta as the equivalent volume of unvesiculated dense magma. This volume is referred to as the *dense rock equivalent* (DRE) volume. In order to calculate DRE volumes it is necessary to estimate the average bulk density of the deposit which must take into account both the intraparticle voidage (i.e. the vesicularity) and the interparticle pore space. For individual deposits the bulk density can vary significantly from proximal to distal areas. Coarse-grained Plinian deposits exhibit a range of bulk density from about 500 to 1500 kg m^{-3}, with the main control being the abundance and density of juvenile pumice fragments. The lowest bulk densities are attained by freshly fallen fine ash. For example, the distal component of the May 18, 1980 Mount St Helens fall had a bulk density of only 300 kg m^{-3} two days after it was deposited. Conversion of deposit volume to DRE can therefore involve large adjustments in density which can translate to substantial uncertainties in the accuracy of the estimate of total mass.

13.5.3 Particle Size

The particle size of a fall deposit is determined by the initial fragmentation properties of magma, the height of the eruption column and the strength of the prevailing wind. Conventionally grain size is described by a grain size analysis expressed in terms of mass percentage of each sieve interval. Grain size is expressed as the logarithm of the particle diameter in millimetres to the base 2. For cumulative frequency curves, statistical parameters such as median diameter (Md_ϕ) and sorting coefficient (σ_ϕ), which is the standard deviation, are derived (see Fisher and Schmincke 1984 for more details). Measures of grain size such as median size and maximum size behave in a similar way to thickness in that they generally decrease exponentially away from source (Pyle 1989). Figure 13.16

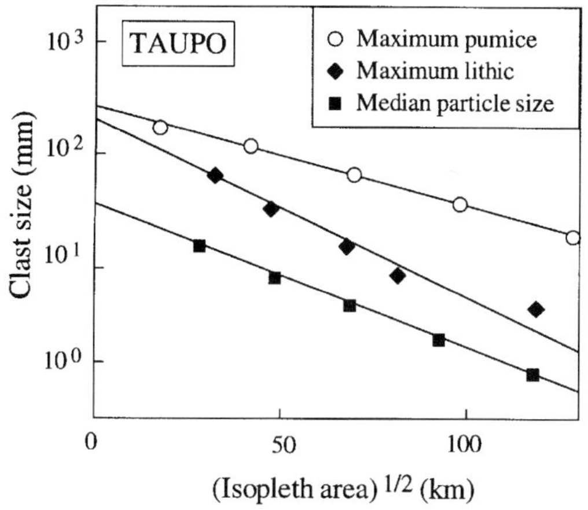

Figure 13.16 Variation in clast size as a function of square root of isopleth area for the deposit of the AD 180 Taupo ultra-Plinian eruption in New Zealand, modified from Pyle (1989)

shows the exponential decay of particle size for the Taupo ultra-Plinian deposit. A distance, b_c, can be defined as the distance at which some measure of the grain size decreases to one-half of the initial maximum value. This parameter is analogous to the thickness half distance, b_t, that is defined from the change in thickness as a function of distance from source. In Chapter 15 the relationship between these parameters and the source eruption is described in greater detail.

Another important particle size characteristic of fall deposits is the sorting coefficient (σ_ϕ). This is a measure of the range of particle sizes at a particular depositional location and is typically derived from a cumulative grain size plot. Deposits with a narrow range of particle size are considered to be well sorted, whereas a broad range is characteristic of poor sorting. Most fall deposits become better sorted with increasing distance from source as a result of the progressive fractionation of components during aeolian transport. It is possible, however, for fall deposits to contain a relatively broad range of particle sizes yet still be considered to be well sorted. Particles are separated from one another during transport in an eruption plume, and during their fall through the atmosphere by differences in their settling velocity. Settling velocity is a function of both particle size and density. Thus it is possible for two particles of contrasting size to have the same settling velocities if their densities are inversely related. For example, a 2 cm pumice with a density of 1000 kg m^{-3} would fall at the same rate as a 1 cm diameter lithic with a density of 2400 kg m^{-3}. Both of these particles could be transported a similar distance from source and deposited simultaneously. The two particles are considered to be aerodynamically equivalent. To illustrate this point Figure 13.17 shows a comparison of a particle size histogram and settling velocity histogram for the Fogo A pumice fall deposit in the Azores. Fogo A contains both low-density pumice and abundant high-density lithics and crystals. Some samples have strongly bimodal particle size distributions. The distribution of particle sizes can be expressed in terms of the distribution of settling velocities in air at sea-level. These distributions are much narrower and are unimodal. The apparent "sorting" (as measured by one standard deviation) improves and the bimodality disappears.

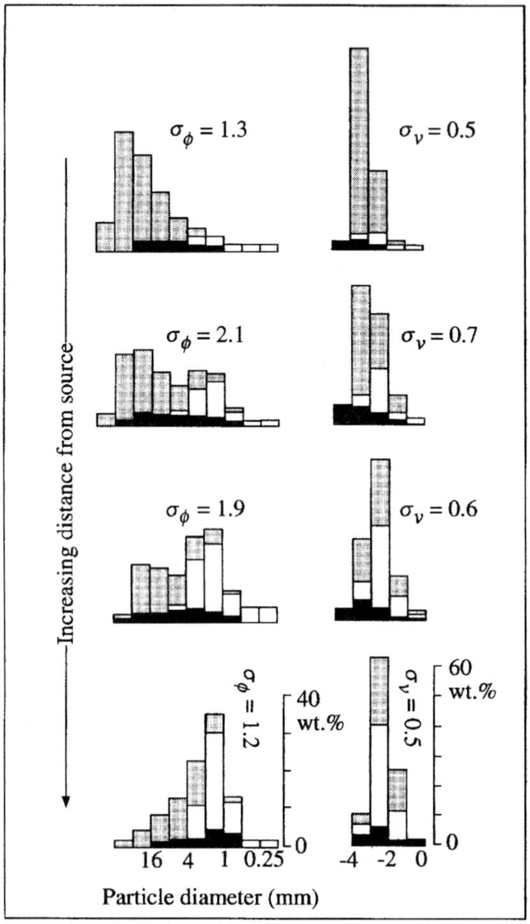

Figure 13.17 Particle size histograms of the Fogo A Plinian fall deposit are shown on the left-hand side. Stippled pattern = pumice, white = crystals and black = lithics. To the right the grain size histograms have been converted to histograms of the distribution of particle settling velocities, calculated at sea-level. The improvement in the standard deviation (measured by σ_ϕ for grain size and σ_v for the settling velocity histograms) reflects the aerodynamic equivalency of particles with contrasting size and density

Another factor which influences sorting is the increase in density of pumice which occurs as particle size decreases (Walker 1971). This acts to diminish the density contrast between particles in small particle size intervals. The fractionation of particles according to their settling velocity more closely approaches that of particles of similar size. This probably contributes to the observed improvement in sorting with increasing distance from source. Walker (1971) recognized that a combination of particle size measures such as sorting and median size could be useful for discriminating fall deposits from other types of pyroclastic deposits. Figure 13.18 shows a plot of sorting coefficient (σ_ϕ) versus median diameter for tephra fall deposits, pyroclastic flow deposits and pyroclastic surge deposits. In general, fall deposits are better sorted (i.e. they have a lower sorting coefficient) and

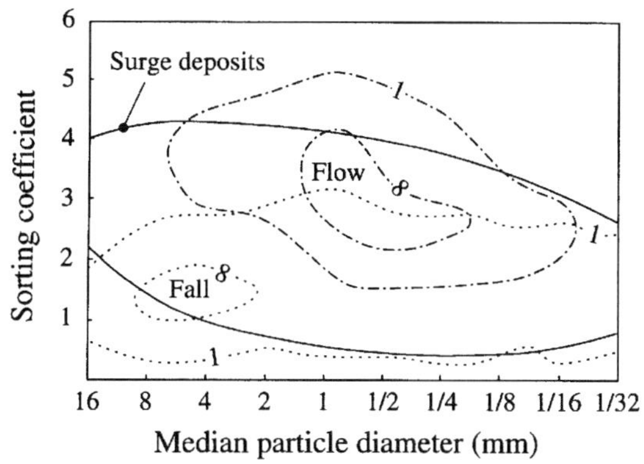

Figure 13.18 Plot of sorting coefficient (σ_ϕ) versus median particle size (Md_ϕ) for pyroclastic fall, flow and surge deposits, modified from Walker (1971). This diagram can be used in conjunction with other observations to discriminate between deposit types

coarser grained than pyroclastic flow deposits and can be effectively separated on this type of diagram. There are, however, exceptions to these general relationships. In particular, the sorting and grain size relationships of deposits which involve external water can differ substantially from deposits originating from purely magmatic eruptions. Magma–water interactions commonly result in increased fragmentation, the production of abundant fine-grained ash and wet depositional environments. Phreatomagmatic eruptions also generate abundant steam in the eruption plume which can act to flush out fine ash by aggregation processes (Chapter 16). In this way the fine ash is deposited much closer to source than if it had settled out slowly as individual particles. This can produce a deposit which is poorly sorted in terms of total grain size distribution (section 16.2.2).

Figure 13.19 compares particle size relationships of the Plinian and phreato-Plinian deposits of the 1875 eruption of Askja volcano in Iceland. The Plinian deposit shows a systematic migration of the particle size population with increasing distance from source. In contrast, the phreato-Plinian distributions are characterized by a decrease in only the coarse part of the distribution with distance from source. The fine-grained part of the distribution is present at all locations, indicating that deposition of both coarse and fine particles occurred simultaneously in proximal areas. This behaviour is referred to as coarse tail grading and is typical of deposits formed from eruptions that involve substantial interactions with external water.

13.6 CLASSIFICATION OF FALL DEPOSITS

Walker (1973) introduced a classification of fall deposits that was based on the extent to which material was dispersed and a measure of the overall grain size distribution. He introduced two parameters to quantify these characteristics. First, the dispersal index, D, is defined as the area enclosed by the $0.01T_{max}$ isopach, where T_{max} is the maximum thickness. For most deposits T_{max} is estimated by extrapolating thickness versus distance curves back to the vent. The dispersal index serves as a measure of overall dispersal. The

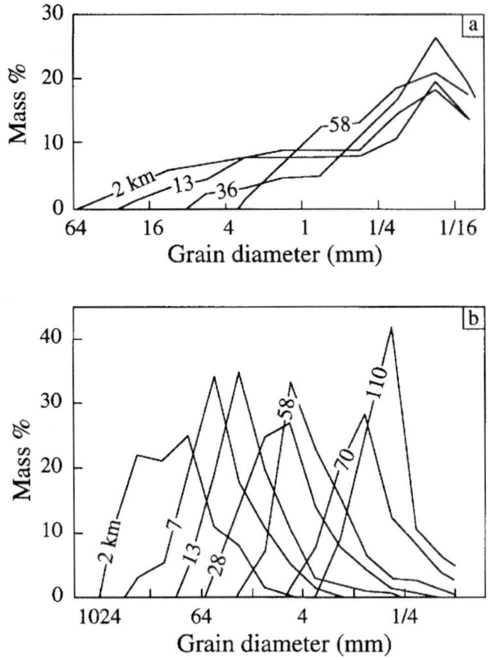

Figure 13.19 Particle size frequency plots for the (a) phreato-Plinian and (b) Plinian deposits of the 1875 eruption of Askja volcano, Iceland, modified from Sparks *et al.* (1981). Numbers on the curves indicate distance from source in kilometres. The phreato-Plinian deposit is coarse tail graded with distance from vent, whereas the Plinian exhibits distribution grading

second parameter is the fragmentation index, *F*, defined as the percentage of a deposit finer than 1 mm at the point on the axis of dispersal where it is crossed by the $0.1T_{max}$ contour. Figure 13.20 shows a plot of *F* and *D* and boundaries proposed to subdivide different types of fall deposits. In a qualitative sense the dispersal axis can be correlated with eruption column height and eruption intensity. The fragmentation index is partly related to the extent to which magma is fragmented during an eruption and is primarily related to the amount of juvenile gas and the extent to which magma has interacted with external water.

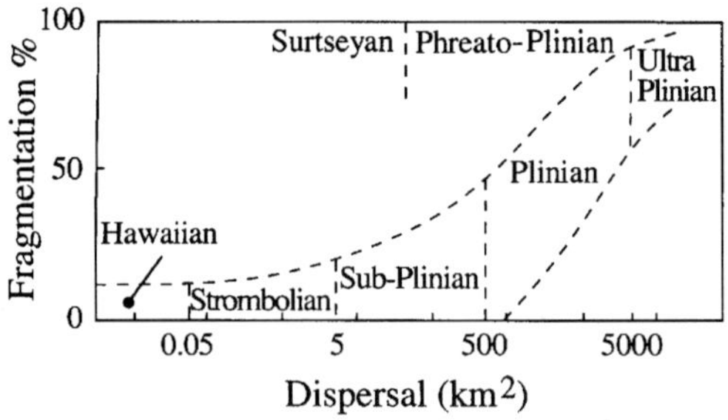

Figure 13.20 Walker (1973) classification of fall deposits based on fragmentation (% of deposit < 1 mm in size at the $0.1T_{max}$ position along the dispersal axis) and dispersal (area enclosed by $0.01T_{max}$ isopach). T_{max} is the maximum thickness at the vent

The degree of fragmentation often increases substantially in eruptions where external water has interacted with magma. The fragmentation index can also, however, be controlled by premature fallout of fine particles due to aggregation processes (Chapter 16) and therefore cannot be simply used as a measure of overall ejecta fragmentation. Deposits which plot on the right-hand side of Figure 13.20 have sheet-like morphology, whereas deposits on the left are best characterized as cone-shaped.

Pyle (1989) provided new insights into the analysis of fall deposits by focusing on the exponential decay behaviour of thickness and particle size characteristics commonly observed in fall deposits. He defined two parameters, b_t and b_c, which refer to the distance over which a deposit decreases its thickness and particle size respectively by one-half from that at the source. These parameters are thus analogous to the half-life of a radioactive nuclide. The thickness half-distance, b_t, is a measure of the dispersal of a deposit and it thus is somewhat analogous to the Walker dispersal index. There are important differences, however, and Pyle (1989) points out that the dispersal index does not necessarily relate exclusively to column height. Similar values of dispersal can be achieved from greatly contrasting column heights if the bulk particle size distribution of the eruption is different. The clast half-distance, b_c, can be measured for either the maximum clast or the median diameter of the deposit.

Pyle proposed a new classification diagram that uses the ratio b_c/b_t versus b_t (Figure 13.21). The ratio b_c/b_t was interpreted as a measure of the degree of fragmentation of a deposit. Low ratio values correspond to a fines-rich total size population that would exhibit wider dispersal for a given column height compared to a coarser grained distribution. In Pyle's classification scheme, b_c corresponds to the half-distance of the maximum clast size. This can be directly related to column height using the model of Carey and Sparks (1986). In Figure 13.21 curves of constant b_c, or column height, are used to subdivide fields into various eruption styles.

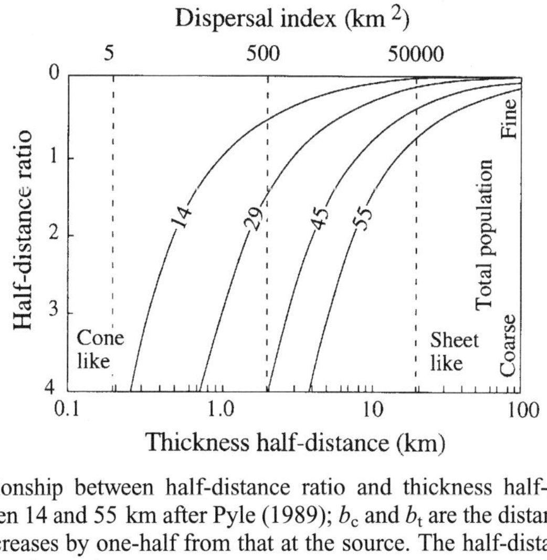

Figure 13.21 Relationship between half-distance ratio and thickness half-distance for eruption column heights between 14 and 55 km after Pyle (1989); b_c and b_t are the distances at which the clast size and thickness decreases by one-half from that at the source. The half-distance ratio $= b_c/b_t$. The column height is shown for each curve.

This classification scheme is an important step towards a more quantitative basis for fall deposit classification because its defining parameters are based more closely on the physical processes of sedimentation from volcanic plumes. There are, however, some limitations and caveats that must be considered. The existence of two or more segments on plots of log thickness versus the square root of distance means that some deposits are characterized by two or more values of b_t. Until the origin of the different segments is better understood the appropriate choice of b_t will be uncertain. Although b_c (maximum clast) is likely to be primarily related to column height, simple modelling by Sparks *et al.* (1992) has suggested that b_c may also show some dependence on total grain size distribution. Pyle chose a b_c value of 3 km (29 km column height) as the division between sub-Plinian and Plinian deposits. Sparks *et al.* (1992) have pointed out that many deposits that have been classified as Plinian in the literature would be classified inappropriately as sub-Plinian on the new diagram. They proposed that the sub-Plinian–Plinian boundary be placed at a b_c value of 1 km (14 km column height). This would normally correspond to a dispersal index of > 500 km^2.

13.7 CO-IGNIMBRITE FALL DEPOSITS

The most widespread fall deposits in the geologic record are derived from large-scale pyroclastic flows. Fallout occurs from plumes that rise from the top of moving flows and carry large quantities of fine-grained ash into the atmosphere (Chapter 7). An important feature of these plumes is that their source areas can be laterally extensive, unlike plumes that are generated from a single vent. Some pyroclastic flows cover areas up to 10^5 km^2 and thus the area of plume generation is likely to have been equally as large. These types of deposits are important to study because they represent the upper limit of the environmental hazard associated with ash fallout. There are, however, relatively few thorough studies of these types of deposits.

A relatively recent example of a large co-ignimbrite ash fall was caused by the 75 000 a eruption of the Toba caldera in northern Sumatra (Ninkovich *et al.* 1978; Rose and Chesner 1987). Ash from this eruption has been found on Sumatra, Malaysia, throughout the Bay of Bengal and in northern India (Figure 13.22). The total fallout area is at least 4×10^6 km^2 and may have been even larger. Rose and Chesner (1987) have studied the grain size of the ash which fell at distances greater than 500 km and found that there is no systematic change in median or maximum size at up to 3100 km from source. Most of the ash in this area consists of thin, bubble-wall shards which Rose and Chesner (1987) argue would have reduced terminal settling velocities because of their highly elongated nature. No explanation has been offered as to why the deposit shows little size variation over such a large distance. It may be that the fragmentation process generated an enormous amount of ash that corresponds to the mean grain size. A restriction in the size of the smaller particles generated during the eruption may have been controlled by the dominant bubble size produced by vesiculation or the abundance and size of crystals in the magma and the way in which they influenced the breakup of magma during the fragmentation process. Alternatively, the lack of variation may reflect particle aggregation processes.

The volume of the widely dispersed Toba ash has been estimated to be between 800 and 840 km^3 DRE based on an extrapolation of the area versus thickness relationship and an

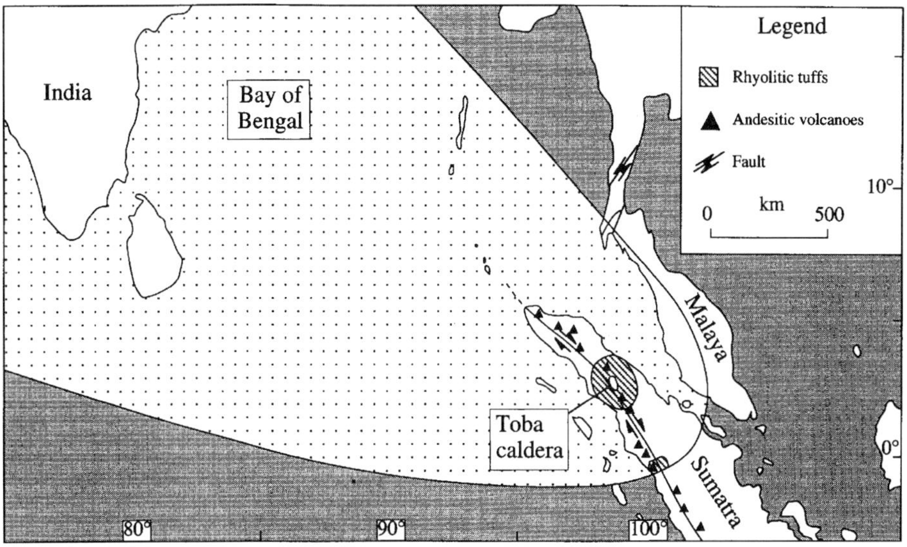

Figure 13.22 Map of the distribution of volcanic ash from the 75 000 a Toba eruption in Indonesia, modified from Ninkovich *et al.* (1978)

assessment of the crystal/glass fractionation. In addition, the volume of pyroclastic flow deposits in the vicinity of Toba caldera has been estimated to be > 2000 km^3 DRE. Thus the total erupted volume is at least 2800 km^3 DRE. This eruption is the largest known to have taken place during the Quaternary period and it has been suggested that it may have played a role in initiation of the major glaciation which began 75 000 years ago (Rampino and Self 1992).

A deposit of similar scale, but even larger than the Toba, has recently been correlated between rocks in North America and Europe (Huff *et al.* 1992). The Millbrig K-bentonite (altered volcanic ash) bed, found in eastern North America, corresponds in age and composition to the so-called "Big Bentonite" bed in Baltoscandia. This deposit was produced by an enormous eruption which took place at 454 Ma. Ash was spread over an area of several million square kilometres, which included the pre-existing Iapetus Ocean which separated North America from Europe at that time. The total volume of ash has been estimated to be 1140 km^3 DRE, or about 35% larger than the Toba fall deposit (Huff *et al.* 1992). As yet the source of this ash has not been identified, but judging from its magnitude it is likely to be associated with a major pyroclastic flow-forming event. Geochemical evidence suggests that the magma was generated in a convergent margin setting by partial melting of continental crust.

13.8 TEPHROCHRONOLOGY

In geologic terms the deposition of volcanic ash from widely dispersed eruption plumes is a virtually instantaneous event. Tephra fall layers thus form ideal chronostratigraphic markers which can be used to correlate between different sedimentary environments and establish well-constrained chronologies. The usefulness of these layers lies in their wide geographic dispersal, distinct lithology, and the ability to date their constituent material. The study of tephra layers for correlative and dating purposes is known as tephrochronology, a field that was defined by the Icelandic geologist, Thorarinsson. A collection of papers that illustrate the diversity of tephrochronologic work can be found in a book edited by Self and Sparks (1981).

13.8.1 Correlation and Dating

Widely dispersed tephra fall layers commonly form distinctive lithologic units compared to their surrounding sediments or rock units. In distal locations the layers are usually a few to tens of centimetres thick and typically have sharp lower contacts. The upper contacts vary from sharp to diffuse. Preliminary correlation of these beds can often be accomplished on the basis of colour, mineral assemblage and stratigraphic position. However, these features may not be constant over a large geographic area. As plumes transport tephra there is significant aeolian fractionation of components according to their settling velocity (Figure 13.2). This results in marked changes in the lithology of a layer at various distances downwind from source. For example, a bulk chemical analysis of the May 18, 1980 Mount St Helens fall deposit at 100 and 300 km distance from the vent would yield completely different compositions owing to the large variation in crystal abundance at the two locations.

A more useful correlation technique involves the analysis of individual components from fall deposits such as volcanic glass or minerals. Ejecta produced during individual eruptions is often chemically distinct and thus it is possible to "fingerprint" the material by geochemical analysis. The major element composition of glass shards, as analysed by electron microprobe, can serve as a powerful tool for correlating tephra fall layers (Smith and Westgate 1969). Similarly, the composition of specific minerals such as Fe–Ti oxides can be distinctive for individual fall units (Smith and Leeman 1982). The caveat with such techniques is that multiple eruptions from the same volcano or closely related volcanoes can potentially erupt ejecta which may not be sufficiently chemically distinct to use major elements simply as a discriminant.

One of the most precise techniques for correlation involves the use of trace elements in glass separates. Sarna-Wojcicki *et al.* (1987) showed that the rare earth element content of tephra fall deposits from several major eruptions of the western USA can be used to discriminate and correlate layers which are not compositionally distinct in terms of their major element content. This technique works well for correlating deposits from some major ignimbrite-forming eruptions which produced large, chemically homogeneous volumes of magma. Some eruptions, however, tap magma chambers that are compositionally zoned (e.g. Hildreth 1981) and thus the deposits from a single eruption can contain a range of glass and mineral compositions. In such cases the compositional diversity may actually be used as a distinctive criteria for correlative purposes.

The scale at which correlations can be accomplished varies according to the intensity of the source eruptions, the environment of deposition and the age of the enclosing sedimentary sequence. For large ignimbrite-forming eruptions the potential area of correlation can be quite large. For example, the tephra fall deposit from the Bishop Tuff eruption in

California can be correlated over 10^6 km^2. The most extensive correlation of an individual tephra layer has been proposed for the Millbrig K-bentonite bed of eastern North America and the "Big Bentonite" bed in Baltoscandia (Huff *et al.* 1992). The total depositional area including the Iapetus Ocean was at least several million square kilometres.

Dating of tephra fall layers is critical to their use as chronostratigraphic horizons and a variety of techniques can be applied. Glass shards and zircon crystals can be dated by the fission track method, but its use is restricted to material older than 100 000 years. The conventional K–Ar dating technique can be used on sanidines and plagioclase phenocrysts with lower limits of about 70 000 and 200 000 years respectively (Naeser *et al.* 1981). For younger deposits, ^{14}C techniques can be applied to various types of carbon found associated with the fall layers. In proximal areas some fall deposits are sufficiently hot to carbonize vegetation and incorporate this material directly into the deposit. Dating of such material provides the best opportunity to date the fall phase of an eruption. More commonly the source eruption can be dated if the event also produced pyroclastic flows which carbonized and incorporated vegetation. Another approach to carbon dating is to date organic material in the level just below the fall deposit. High precision accelerator mass spectrometer (AMS) dating of carbon from soil horizons below fall layers has been used to date eruptions of Santorini, Vesuvius and Mount St Helens (Vogel *et al.* 1990). The dating of young rocks is currently being revolutionized by improvements in mass spectrometry and single crystal studies using laser techniques, in particular for application of $^{39}Ar/^{40}Ar$ dating. There are also new dating methods being developed such as ^{3}He exposure techniques.

Tephra fall layers can also be dated by their position within a sedimentary chronostratigraphic framework. In marine sediments chronostratigraphy for relatively recent sediments can be established from high-precision oxygen isotope analysis of carbonate microfossils. The widespread Toba tephra fall deposit from Indonesia has been dated by this technique in sediment cores from the Indian Ocean at 75 000 a (Figure 13.22). A somewhat less precise method in sedimentary sequences is to use palaeomagnetic chronostratigraphy.

13.8.2 Archaeological Applications

Tephra fall layers have several different types of application in the field of archaeology. One involves the burial and preservation of archaeological sites. In the proximal area of active volcanoes the accumulation of tephra by fallout may often exceed several metres in thickness and thus powerful eruptions can bury sites of human habitation. Sigurdur Thorarinsson and archaeological colleagues in Iceland pioneered such studies when they discovered and excavated a medieval farmhouse buried by the AD 1104 eruption of Hekla in Iceland (Thorarinsson 1967).

One of the best known examples of this is the burial of the city of Pompeii by fallout from the AD 79 eruption of Vesuvius volcano. The eruption began with a period of coarse, Plinian fallout which lasted for approximately 12 hours (Sigurdsson *et al.* 1985). This activity deposited 2.5 m of pumice fall on Pompeii before the eruption shifted to the production of deadly pyroclastic flows and surges. The complete burial of the city was the result of the accumulation of fall, flow and surge deposits, but the fall component was 50% of the total thickness. In the span of only 20 hours the entire city was entombed and remained so until it was accidentally discovered in 1594 by digging for a water works project. Excavation of the Pompeii site provided an unparalleled opportunity to learn about life in a Roman city.

In the Late Bronze Age a catastrophic explosive eruption occurred on the island of Santorini in the Aegean. At least 30 km^3 of rhyodacitic ejecta was erupted and a large caldera formed (Bond and Sparks 1976; Sparks and Wilson 1989). The island was already populated by a sophisticated culture with strong links to the Minoan civilization on Crete. The town of Akrotiri was buried beneath several metres of fallout tephra and ignimbrite, preserving spectacular frescoes and an elaborate city. The Santorini tephra was dispersed over much of the eastern Mediterranean (Watkins *et al.* 1978) and has been used to resolve a controversial archaeological problem. At one time there was a widespread view that the eruption had been responsible for an abrupt decline in the Minoan civilization. However, the tephra layer has now been found below the pottery stage that preceded the destruction levels on Crete and in the eastern Aegean. This exemplifies how tephra can be used to improve stratigraphic resolution in archaeological sites and the potential for large volcanic eruptions to influence the history of ancient peoples.

The Minoan tephra is currently associated with another controversy concerning the chronology of the Late Bronze Age. Frost-ring damage in Irish oak at 1639 BC (Baille and Munro 1988) and a sulphate-rich level at approximately 1645 BC in Greenland ice cores (Hammer *et al.* 1987) have been correlated with the Minoan eruption. However, conventional archaeological chronology places the eruption closer to 1500 BC. Unfortunately radiocarbon dates are too imprecise and scattered to resolve the problem. The tree-ring and ice-core correlations with Santorini are speculative. Recently, large explosive eruptions comparable in scale to Santorini and with the same radiocarbon age have been recognized in Alaska and on Villarrica volcano in Chile, providing alternative sources for the seventeenth-century BC anomalies.

13.8.3 Marine Tephrochronology

Tephra fall layers are important components of some deep-sea sediments adjacent to areas of explosive volcanism (Kennett 1981). Figure 13.23 shows the distribution of tephra fall layers, based on studies of sediment cores, in the world's oceans. The distribution patterns reflect the direction of the prevailing atmospheric circulation as it affects eruption plumes from volcanically active areas.

The study of these layers is useful for understanding volcanic activity on both a regional and local scale. Kennett *et al.* (1977) compiled the frequency of tephra fall layers in marine sediments recovered by the Ocean Drilling Program and proposed that the tempo of global explosive volcanism had varied considerably over the past 20 Ma and that there were distinct peaks during the Middle Miocene (14–17 Ma), Lower Pliocene (3 Ma) and Quaternary (<2 Ma). The cause of such episodes is still under debate. Some of the mechanisms proposed include variations in sea-floor spreading rates or hot-spot activity. However, the frequency of tephra layers correlate well with the influx of aeolian terrigeneous material to deep-sea sediments. Thus the frequency of tephra layers may be saying more about palaeoclimatology than palaeovolcanology.

In island arcs the terrestrial record of explosive volcanism is often poorly preserved because of extensive erosion and widespread dispersal of ejecta into the surrounding

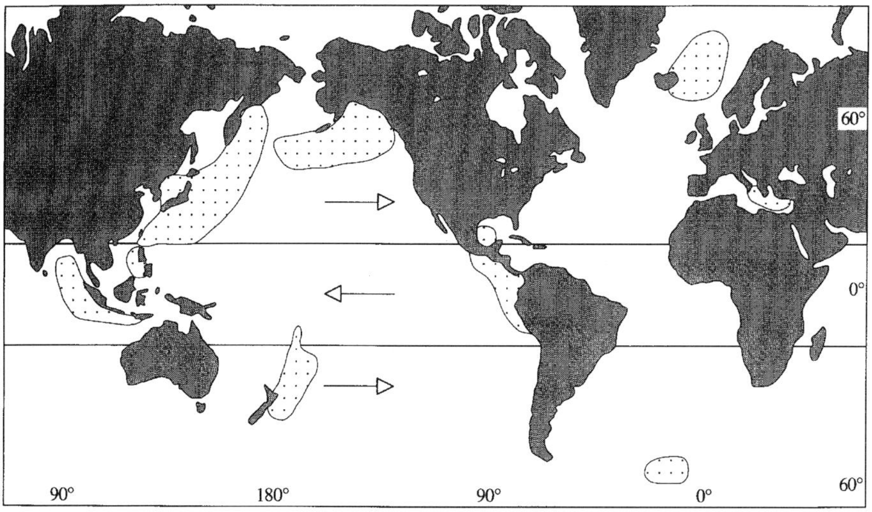

Figure 13.23 Main areas of tephra fall deposition in the world's oceans based on the presence of volcanic glass in marine sediments. Arrows represent main wind direction in stratosphere

oceans. The sedimentary record in the adjacent marine environment thus provides the opportunity to study the nature and frequency of explosive volcanism in specific volcanic areas. The Lesser Antilles arc provides a good example of how marine tephrochronology can be used to elucidate the history of volcanic activity on a local scale. Subduction of western Atlantic oceanic crust beneath the Caribbean plate has created the Lesser Antilles

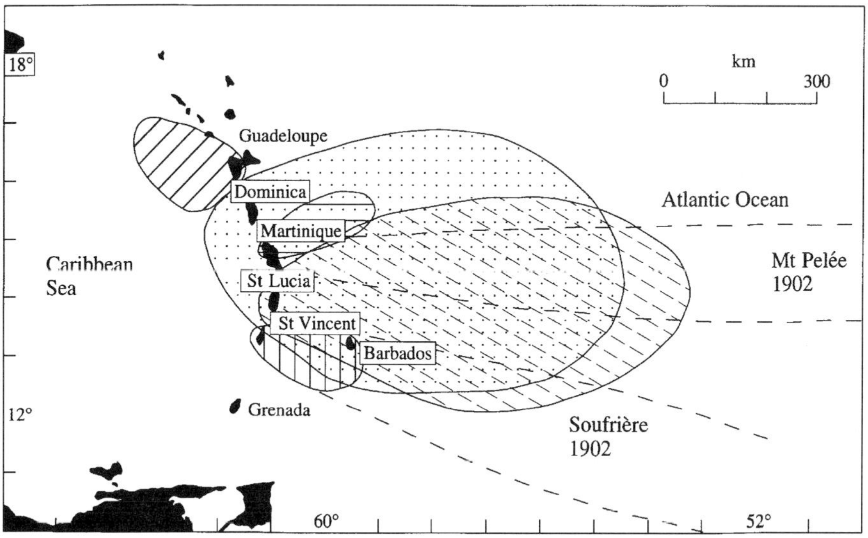

Figure 13.24 Tephra fallout patterns from Quaternary explosive eruptions in the Lesser Antilles arc as reconstructed from the distribution of volcanic glass in western Atlantic and eastern Caribbean sediments

volcanic arc where explosive volcanism has occurred frequently in the past. Tephra fall has occurred primarily to the east of the arc (Figure 13.24), controlled by the prevailing anti-trade winds. Sediment cores from the western equatorial Atlantic have recovered numerous tephra fall layers interbedded with carbonate and terrigeneous marine sediment (Sigurdsson *et al.* 1980). These layers have been correlated to specific source volcanoes within the arc by geochemical analysis of glass shards and mineral phases (Sigurdsson and Carey 1981). Dating of these layers has been accomplished by oxygen isotope stratigraphy of the carbonate sediments, thus providing a reconstruction of the source and timing of major explosive eruptions which have taken place in the Lesser Antilles during the last 1 million years. Such studies provide important constraints on the frequency and scale of large-magnitude eruptions in these types of volcanic systems.

13.9 SUMMARY

Tephra fall deposits provide a record of explosive volcanism from which important information can be extracted on the frequency, style, intensity and magnitude of past eruptions. The deposits are composed of juvenile fragments, crystals and lithics and can vary widely in grain size from deposits composed of large blocks and bombs to fine layers of dust. Density of the particles also varies considerably as a consequence of vesiculation of juvenile particles and density variations in minerals. Particles are sorted aerodynamically with respect to size and density with distance from source by sedimentation from the plume and by the wind. Grain size characteristically decreases exponentially away from source. Aggregation can cause premature fallout of fine particles, however (Chapter 16).

The ejecta in tephra fall deposits contains information on the eruptive conditions which can be extracted by petrological and geochemical studies. The mineral and glass compositions can be analysed or interpreted to give estimates of magma temperatures, volatile compositions, contents and partial pressures, chamber depth and oxidation state of the magma. Mineral assemblages in tephra clasts can also be reproduced experimentally to provide unique constraints on eruption and magma chamber parameters.

Tephra fall deposits are formed by sedimentation from volcanic plumes. The processes involved depend on particle aerodynamics and plume structure. The coarsest particles are ejected as ballistic clasts in the basal region of columns. Coarse suspended particles (typically 1–10 cm) are sedimented from the plume margins. Most particles are sedimented, however, from the umbrella cloud in powerful eruptions and from advected downwind plumes. Very fine ash particles may have more complex sedimentation due to aggregation.

Tephra fall deposits thin exponentially with distance. Isopach maps show a variety of patterns which depend principally on column height, wind strength and direction, and eruption duration. Vent locations and directions can be identified from isopach maps. Qualitative inferences on eruption strength can be made from grain size and whether the deposits form localized cones or tuff-rings or extensive thin sheets. Other factors such as vent geometry, shifting of the vent location, plume bifurcation and particle aggregation can cause complexities to isopach map patterns. For example, thickness secondary maxima can occur due to particle aggregation. Co-ignimbrite tephra fall deposits are characteristically very fine-grained and show patterns that can have a thickness maximum well away from the source volcano due to their origin from a distributed source (Chapter 7).

The volume of tephra fall deposits can be determined from an isopach map. Advantage can be taken of exponential thinning relationships as many deposits can be characterized by a thickness decay constant when thickness data are plotted on a diagram of log thickness against the square root of the area enclosed by isopach contours. Examples of tephra fall deposits with two or more segments with different exponential decay rates suggest, however, that caution is necessary in extrapolating thinning relationships to distances beyond the preserved deposit. Thickness and grain size decay constants can be used as a basis for the quantitative classification of tephra fall deposits.

Tephra layers are excellent time markers. They provide extensive isochronous horizons which can be dated and correlated with important applications to stratigraphy, volcanic geology and archaeology.

14 Sedimentation from Volcanic Plumes

NOTATION

a	acceleration (m s^{-2})
a_p, b_p, c_p	principal axes of a pyroclast, such that $a_p > b_p > c_p$ (m)
b	characteristic plume radius (m)
d	equivalent particle diameter (m)
g	gravitational acceleration (m s^{-2})
h	thickness of control volume (m)
k	re-entrainment coefficient (dimensionless)
m	pyroclast mass (kg)
m_s	mass of pyroclasts deposited per unit distance (kg)
p	probability parameter (dimensionless)
q	momentum flux (kg m s^{-2})
q_o	initial momentum flux (kg m s^{-2})
r	radial coordinate (m)
r_{bf}	radial backflow distance (m)
r_o	initial value of radial coordinate (m)
t	time (s)
t_m	time of flight (s)
t_o	initial time (s)
u	horizontal component of wind velocity (m s^{-1})
u	vector velocity (m s^{-1})
u_o	inflow speed evaluated at b (m s^{-1})
u_r	radial inflow speed (m s^{-1})
V_t	terminal fall velocity (m s^{-1})
w	upward component of velocity (m s^{-1})
$\bar{w}$	centreline upward plume speed (m s^{-1})
x	horizontal coordinate (m)
x_m	pyroclast range (m)
x_o	initial horizontal distance (m)
y	difference in radius at top and bottom of control volume for re-entrainment calculation (m)
z	vertical coordinate, positive upward (m)
A	cross-sectional area of pyroclast (m^2)
B	$\pi v/Q$ (m^{-2})
C	concentration of particles in plume (kg m^{-3})
C_D	coefficient of drag (dimensionless)

D	$5/6\alpha$ (dimensionless)
F	pyroclast shape factor, $(b+c)/2a$ (dimensionless)
F_D	drag force (kg m s^{-2})
F_G	gravitational force (kg m s^{-2})
F_I	inertial force (kg m s^{-2})
F_o	volumetric buoyancy flux at vent (m^4 s^{-3})
F_V	viscous force (kg m s^{-2})
$J(r)$	stratification parameter for ambient medium
K	$0.1\ (9\alpha F_o/10)^{1/3}$
M	particulate mass flux (kg s^{-1})
M_o	initial value of particulate mass flux (kg s^{-1})
M_S	total mass of pyroclasts in a plume through time (kg)
N	buoyancy frequency (s^{-1})
Q	volumetric flow rate (m^3 s^{-1})
Q_o	volumetric flow rate at vent (m^3 s^{-1})
Re	Reynolds number (dimensionless)
S	mass of pyroclasts deposited per unit area (kg m^{-2})
S_o	initial mass of pyroclasts deposited per unit area (kg m^{-2})
T	eruption duration (s)
X	$p\ (5/6\alpha)^{1/3}(9\alpha F_o/10)^{-1/3}$
α	entrainment coefficient (dimensionless)
ϕ	ejection angle, measured from the horizontal (°)
ξ	difference in altitude between crater rim and site of deposition (m)
κ	tensor coefficient of eddy diffusivity (m^2 s^{-1})
λ	empirical shape factor (dimensionless)
μ	drag parameter, $\rho C_D A/2m$; dynamic viscosity (Pa s)
ν	kinematic viscosity (m^2 s^{-1})
ρ	density of ambient medium (kg m^{-3})
σ	particle density (kg m^{-3})
ΔH	thickness of umbrella cloud (m)
Ξ	$(2\nu/3u)(2\lambda N/Q)^{1/2}$
Φ	source term for advection/diffusion model (kg m^{-3} s^{-1})
Ω	control volume perimeter (m)

Subscript

1	evaluated from distance at which eruption column flow becomes plume-like

14.1 INTRODUCTION

Volcanic plumes consist of several distinct regions with different flow characteristics. Plumes carry tephra and aerosols of a great variety of sizes, densities and shapes. Because of complex plume structure and variations in particle size, shape and density, several different sedimentation regimes occur within plumes as described in Chapter 13. The most massive, ballistic particles (metre scale) interact little with the eruption column, and are ejected as projectiles from the vent. Intermediate-sized particles of the order of 0.01–1 m are able to escape re-entrainment by the strong inflow at the plume margins and fall from the steep margins of the eruption column. In this intermediate category larger particles are ejected as ballistic projectiles in the lower parts of a column, but interact with the large-scale turbulent motions within the column. Somewhat smaller particles in this intermediate category are carried in suspension and sediment out of the sides of the rising column. Smaller particles cannot escape from the column margins and are carried into the umbrella cloud and downwind plume, from which they gradually settle according to their fall velocity. Particle aggregation can cause enhanced sedimentation of some of the finest grains (Chapter 16). This chapter develops a theoretical understanding of sedimentation from plumes and compares the results with laboratory experiments and field data. Chapter 15 uses some of the results discussed in this chapter for the quantitative interpretation of tephra fall deposits.

14.2 PARTICLE SETTLING

The terminal fall speed of a particle is an important control on its behaviour in an eruption column as well as in the atmosphere during fallout. The fall speed can be determined by balancing the atmospheric drag force on a particle against gravity

$$ma = F_G - F_D \tag{14.1}$$

where m and a are particle mass and acceleration, F_G the gravitational force on the particle and F_D the drag force acting on the particle due to the resistance of the ambient atmosphere. At the fall speed, V_t, where $ma = 0$, the balance in forces gives the condition

$$mg = F_D = \tfrac{1}{2} A C_D \rho v^2 \tag{14.2}$$

where A is the cross-sectional area of the particle, C_D the drag coefficient, which is a dimensionless number that is not necessarily a constant, ρ the ambient fluid density and g is acceleration due to gravity.

In reality, there are two resistance forces provided by the ambient medium that can control the form of the drag force, F_D, and determine the value of the drag coefficient (Figure 14.1): F_V, the viscous force resisting particle motion, is caused by the friction acting on the surface of the particle (skin friction or friction drag), and F_I, the inertial force resisting particle motion, is caused by having to accelerate and move air from the particle's path. These forces, related to pressure variations around the particle, are known as form drag. The Reynolds number, Re, for a falling particle is the ratio of the inertial force per unit mass to the viscous force per unit mass

$$Re = F_I/F_V = \frac{V_t d}{\nu} \tag{14.3}$$

where d is the particle diameter, V_t the terminal velocity of the particle and ν the fluid kinematic viscosity. Re is the only dimensionless parameter that can be formed from the basic flow variables. Because of this, C_D must be a function of Re alone, since it too is dimensionless. The functional relationship between C_D and Re has been worked out in

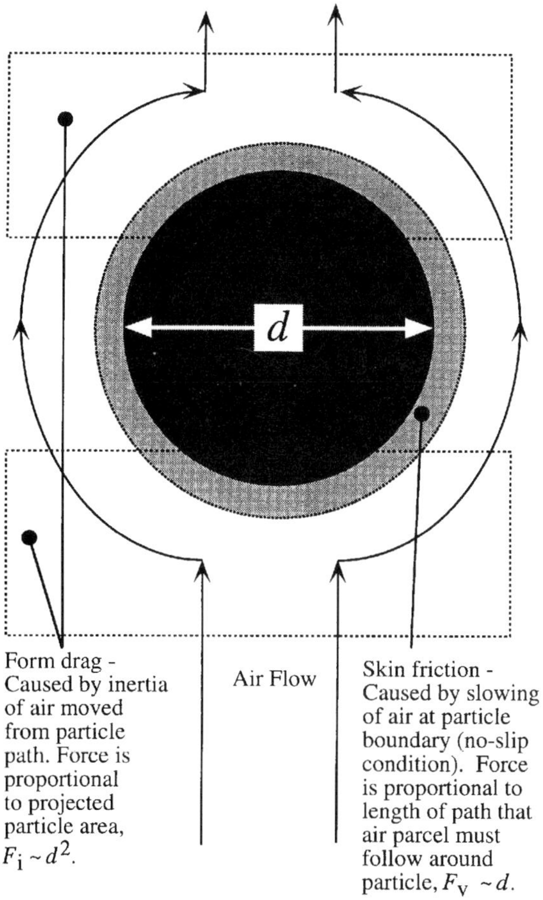
d
Form drag - Caused by inertia of air moved from particle path. Force is proportional to projected particle area, $F_i \sim d^2$.
Air Flow
Skin friction - Caused by slowing of air at particle boundary (no-slip condition). Force is proportional to length of path that air parcel must follow around particle, $F_V \sim d$.

Figure 14.1 Schematic diagram showing the flow around a particle resulting in form drag and viscous drag

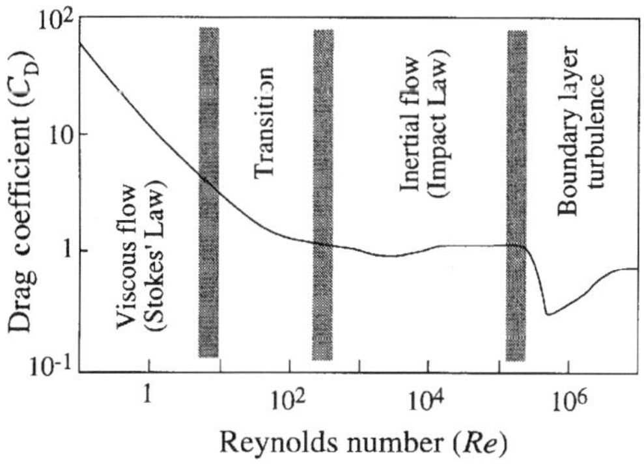
Drag coefficient (C_D)
10^2
10
1
10^{-1}
Viscous flow (Stokes' Law)
Transition
Inertial flow (Impact Law)
Boundary layer turbulence
1
10^2
10^4
10^6
Reynolds number (*Re*)

Figure 14.2 The drag coefficient, C_D, varies little as a function of Reynolds number, *Re*, over a wide range of *Re* in the range of inertial flow. Smaller particles have lower *Re*, larger particles higher *Re*

many numerical calculations and laboratory experiments for a variety of objects, including pyroclasts (Figure 14.2).

For large particles, the skin friction is small compared to the inertial force required to accelerate air to accommodate the particle, and the fall velocity is therefore controlled by inertia. In this case, C_D approaches a constant value of ~ 1 (Figure 14.2), since the drag force, on the right-hand side of equation (14.2), is already defined to be proportional to the inertial force on the particle (equation 14.3). After substituting for A in equation (14.2) and rearranging, the terminal speed, for a sphere of diameter d, is approximated by

$$V_t = \left(\frac{4}{3} \frac{d(\sigma - \rho)g}{C_D \rho} \right)^{1/2} \tag{14.4}$$

where σ is particle density. For volcanic particles in the atmosphere, $\sigma \gg \rho$ and therefore ρ can be ignored in the numerator on the right-hand side of equation (14.4) which holds provided the Reynolds number exceeds 500. At $Re \sim 2\text{–}3 \times 10^5$, C_D decreases suddenly (Figure 14.2). This effect results from the onset of boundary layer turbulence, which causes flow separation and a consequent reduction in the size of the low-pressure region on the downstream side of the particle. Flow separation decreases the total pressure difference developed in the flow around the particle, hence the form drag is reduced. In air, and for typical column speeds of 50 m s^{-1}, this corresponds to a spherical particle diameter of 10 cm. Because this size is greater than most of the particles usually studied in field deposits, this change in C_D is usually not considered in attempts to derive information regarding eruption columns from deposits, except in the most proximal deposits (McGetchin *et al.* 1974; Thomas and Sparks 1992).

For small particles, the skin friction is large relative to the form drag. In this case, dimensional arguments lead to the relationship that $C_D \propto 1/Re$. For smooth spheres the constant of proportionality is equal to 6π (e.g. Tritton 1977) leading to the result

$$mg = 3\pi\mu d V_t \tag{14.5}$$

and therefore

$$V_t = \frac{\sigma g d^2}{18\nu} \tag{14.6}$$

which is the well-known relationship for the Stokes settling velocity of a sphere. In the case of irregular objects such as pyroclasts, d must be defined carefully and effects of shape on the drag coefficient must be taken into account if the Stokes relationship is to be useful. When the Reynolds number is between 1 and 500 the viscous and inertial forces are of similar magnitude, and therefore C_D becomes a more complex function of Re. Numerical solutions and laboratory measurements have been used to determine the relationship in this transitional range of Reynolds numbers (Figure 14.3).

Typical terminal fall velocities for pyroclasts at sea-level illustrate the transition from low to high Reynolds number behaviour (Figure 14.4). At sea-level pyroclasts smaller than about 100 μm exhibit viscous drag consistent with the importance of skin friction on their small mass, whereas particles above about 0.5 mm exhibit behaviour consistent with high Reynolds numbers and constant C_D. The grain size boundaries between these regimes increases in diameter with altitude in the atmosphere, because of the exponential decrease of density with height.

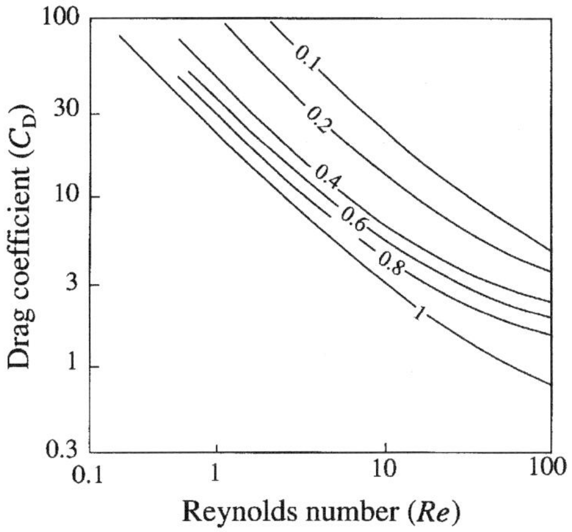

Figure 14.3 The best fit model of Wilson and Huang (1979) to their data for pyroclast fall speed showing the variation of C_D with Re for a range of shape parameters, F (numbers next to curves; redrawn from Wilson and Huang 1979)

14.2.1 The Influence of Particle Shape

The influence of particle shape on the fall velocity of real pyroclasts is now considered. The diameter can be difficult to define. The shape of a pyroclast can be quite irregular and is often not easily approximated by an equivalent sphere (see the discussion in Chapter 13, and Heiken and Wohletz 1985). Walker *et al.* (1971) and Wilson and Huang (1979) measured the fall speeds of a large number of small pyroclasts and found that, for these irregular objects, C_D can be quantified by the use of only one parameter in addition to Re. Wilson and Huang (1979) defined this parameter as the shape factor, F, where

$$F = (b_p + c_p)/2a_p \tag{14.7}$$

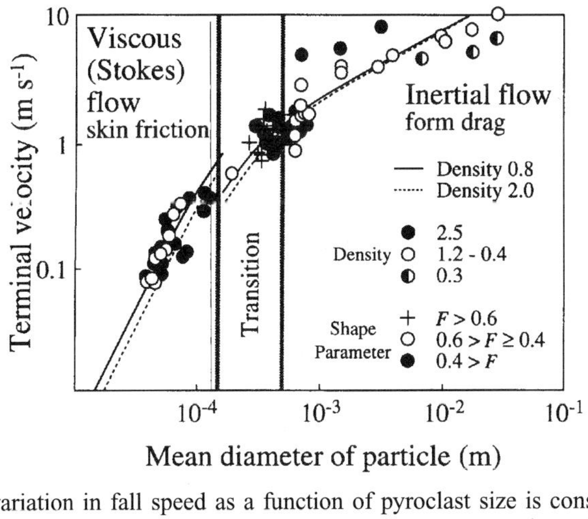

Figure 14.4 The variation in fall speed as a function of pyroclast size is consistent with Stokes' flow for the smallest pyroclasts and inertial behaviour for the largest pyroclasts. The dashed curve is that derived from equation (14.8a), the solid from 14.8b (redrawn from Suzuki 1983). Data are from Walker *et al.* (1971), Wilson and Huang (1979) and Suzuki (1983)

in which a_p, b_p and c_p are the principal axes of a particle such that $a_p > b_p > c_p$. Their equation for C_D for a number of particles < 1 mm on the long (a) axis was

$$C_D = \frac{24}{Re} F^{-0.828} + 2\sqrt{(1.07 - F)} \tag{14.8a}$$

which is illustrated in Figure 14.4. Suzuki (1983) proposed a modification to this relationship that seemed to provide a better fit to the experimental data, including those of Walker *et al.* (1971) (Figure 14.4):

$$C_D = \frac{24}{Re} F^{-0.32} + 2\sqrt{(1.07 - F)} \tag{14.8b}$$

The advection/diffusion models of particle fallout (Suzuki 1983; Armienti *et al.* 1988; Glaze and Self 1991) use equation (14.8a) or a similar relationship for the drag coefficient for the small particles that are typically considered in such calculations. For larger pyroclasts ($> \sim 1$ mm at sea-level) that are frequently measured in the field, it has generally been assumed that C_D is a constant with a value near unity (Carey and Sparks 1986; Wilson and Walker 1987). The influence of shape on the fall velocity is taken into consideration by either measuring three axes of roughly equant pyroclasts, or by using tabulations of average pyroclast shapes to derive effective clast diameters from field values.

14.2.2 Variation of Fall Velocity with Altitude

Wilson (1972) calculated the variation of fall velocity with altitude and presented results in terms of the variation of fall time with height of release (Figure 14.5). His results show that because of the great increase in fall velocity with height (McDonald 1959), a tenfold increase in release height results in only a three- to fivefold increase in fall time. This effect must be carefully considered when calculating sedimentation patterns for clasts released from a vertical eruption column, because as clasts reach greater heights within the column, their fall velocity increases accordingly, and this has ramifications for their fallout behaviour (section 14.4.3).

14.3 BALLISTIC PARTICLES

The largest particles ejected from the vent during an explosive eruption can have fall velocities similar to or even greater than the initial gas velocity, in which case they can travel as ballistics (Figure 14.6). There have been a number of studies of the trajectories of ballistic pyroclasts (Wilson 1972; Chouet *et al.* 1974; McGetchin *et al.* 1974; Self *et al.* 1974, 1980; Blackburn *et al.* 1976; Steinberg and Lorenz 1983; Fagents and Wilson 1993; Bower and Woods 1996. Distinction can be made between purely ballistic pyroclasts that are so large that they are not affected by the motions in the eruption column (Figure 14.6), and somewhat smaller ballistic pyroclasts that are influenced by the turbulent plume motion (Self *et al.* 1980). In the first type of ballistic (type I of Self *et al.* 1980) the momentum of the largest particles can carry them further from the vent than somewhat smaller particles. In the second type (type II of Self *et al.* 1980) the smaller ones will be carried higher than the larger ones before leaving the column, and therefore a fallout

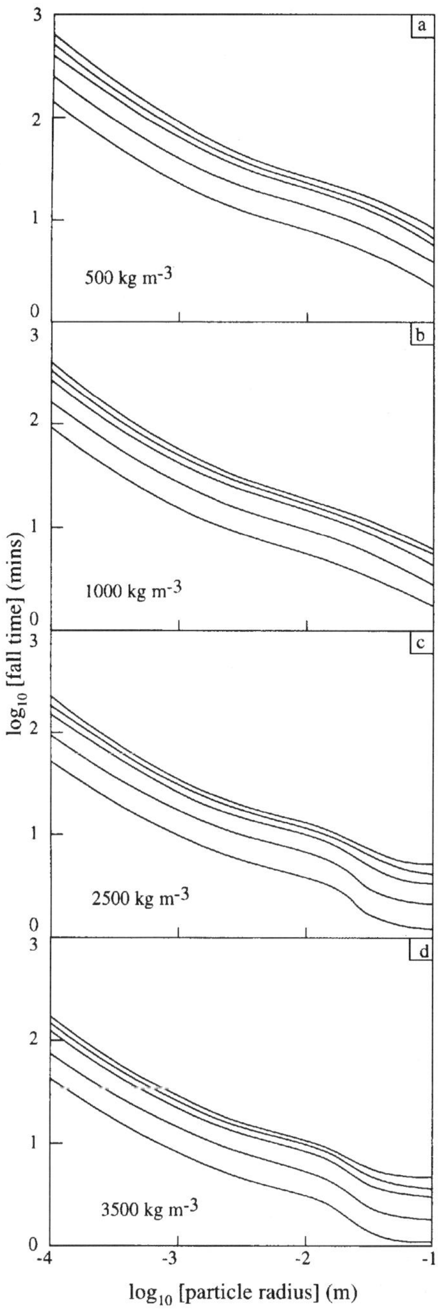

Figure 14.5 Variation of fall time with particle size and height of release for four different particle densities (redrawn from Wilson 1972). The curves (from bottom to top) are for particles released from heights of 5, 10, 20, 30 and 50 km

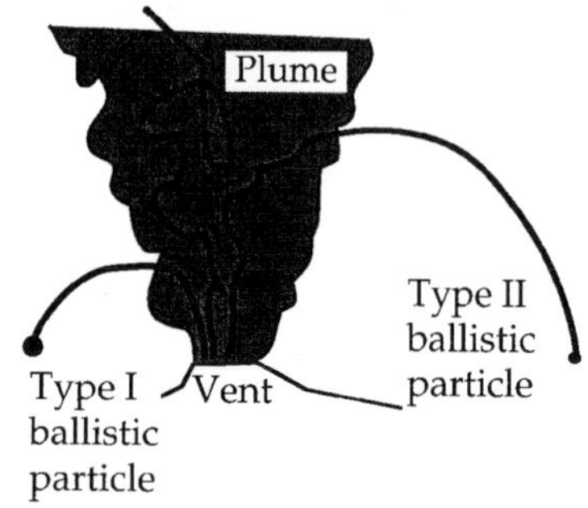

Figure 14.6 Schematic diagram of the two major types of particle transport and sedimentation from volcanic plumes – ballistic and suspended

pattern of decreasing ballistic size with distance will develop for a sustained, powerful plume.

For the largest pyroclasts that spend their entire flight as ballistic projectiles, the equations of motion can be written as

$$m\frac{dz}{dt} = mg - \frac{1}{2}\rho A C_D \frac{dz}{dt}\left[\left(\frac{dz}{dt}\right)^2 + \left(\frac{dx}{dt}\right)^2\right]^{1/2} \tag{14.9}$$

and

$$m\frac{dx}{dt} = -\frac{1}{2}\rho A C_D \frac{dx}{dt}\left[\left(\frac{dz}{dt}\right)^2 + \left(\frac{dx}{dt}\right)^2\right]^{1/2} \tag{14.10}$$

These equations require numerical solution in general, but Self *et al.* (1980) presented a rather simplified version of the model which has analytical solutions for range, x_m of:

$$x_m = \frac{1}{\mu}\ln(\mu(w_o \cos\theta - u)t_m + 1) + ut_m \tag{14.11}$$

with w_o being vent velocity, θ ejection angle, and u wind speed. Here μ is a factor dependent solely on the properties of the particle and ambient atmosphere, such that $\mu = \rho C_D A/2m = 3\rho C_D/\pi\sigma d$. Flight time, t_m, is given by

$$t_m = \frac{1}{\sqrt{\mu g}}\left(\cosh^{-1}\left(\sqrt{1 + \frac{\mu}{g}w_o^2 \sin^2\theta \cdot \exp(\xi\mu)}\right) + \tan^{-1}\sqrt{\frac{\mu}{g}} \cdot w_o \sin\theta\right) \tag{14.12}$$

where ξ is the difference in altitude between the crater rim and the deposition site. Notice that the range of the largest ballistic projectiles will be greater than the range of smaller projectiles, since x_m increases as μ decreases or d increases.

Figure 14.7 is a plot of the ranges for these large projectiles. McGetchin *et al.* (1974) calculated typical ranges for ballistic pyroclasts on the Earth, Moon and Mars. They found that at the highest reasonable ejection speeds of 300 m s^{-1} (Wilson L. 1980) 10 cm diameter pyroclasts could be ejected to ranges of 0.5 km, whereas 1 m diameter pyroclasts could reach up to 2.7 km. Fagents and Wilson (1993) made calculations of ballistic clast trajectories, taking account of the movement of air ahead of the explosion. They demonstrated that clasts can travel further due to the reduced drag of the moving air and large clasts approach the vacuum range. Projectile calculations by Bowers and Wood (1996) showed that there is an optimum size for ballistic clasts in explosive eruptions, with

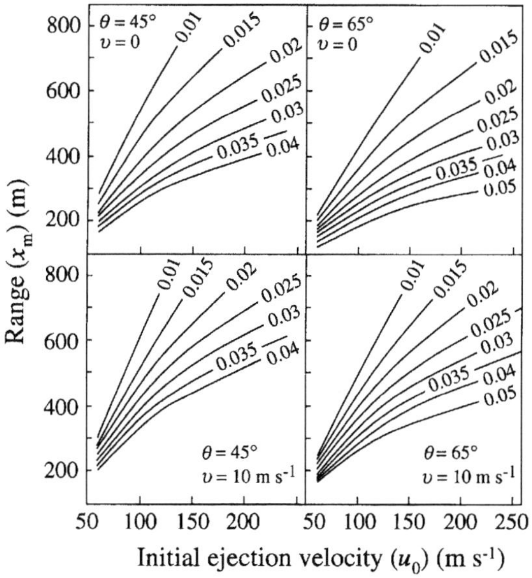

Figure 14.7 Range, x_m, as a function of size and ejection angle, θ, for ballistic pyroclasts. Values next to curves are for the size parameter $4C_DA/m$, where A is particle cross-sectional area and m is particle mass. This parameter decreases with increasing particle size (redrawn from Self *et al.* 1980). The top two graphs are for still air ($v = 0$) and the bottom two graphs are for a moderate cross-wind ($v = 10$ m s^{-1})

clasts of approximately 1 m in diameter having the farthest range. Clasts much larger than 1 m are increasingly affected by air drag and so range decreases as size decreases.

The motion of smaller ballistic clasts that are influenced by the turbulent motions of the column are less well understood. Some insight can be gained from the experimental study of strong particle-laden jets by Longmire and Eaton (1992). They observed how large particles become concentrated between the largest eddies in the flow and are accelerated and ejected outwards by the eddy motions. This behaviour was observed to heights of approximately five times the nozzle diameter for particles where the ratio of the particle time constant to fluid time constant was in the range 3–12. This ratio is known as the Stokes number. The particle time constant is the time it takes particles to adjust to the terminal fall velocity due to changes in surrounding fluid velocity and can be deduced from equations (14.4) and (14.6). The fluid time constant is the ratio of eddy length-scale to characteristic eddy velocity. Stokes numbers greater than 12 were not influenced by the jet motion and therefore correspond to type I ballistics.

Bursik (1989) has shown that particle behaviour is strongly affected by interactions between the gas and particles, with transfer of momentum from particles to gas and vice versa (Chapter 4). For volcanic jets the ratio of particle mass to gas mass is high and typically greater than 1. Most of the momentum flux is in the particles. Bursik (1989) produced models of jets which showed that particle release heights are influenced by momentum transfer between entrained fluid, the gas phase and the particles. Not only do ballistic particles transfer momentum to the gas, but they also weaken the turbulence of the gas motion (see Chapter 4).

As discussed in Chapter 4, the physics of the lower gas thrust parts of columns where ballistic particles originate is complex and incompletely understood. Thus much remains to be done to understand the nature of the interactions between large particles and the column.

14.4 SEDIMENTATION FROM TURBULENT SUSPENSIONS

14.4.1 Basic Principles

Most pyroclasts within a plume are retained as suspended particles because of their low fall velocities relative to plume velocity. A simple model for sedimentation of suspended particles from dilute suspensions is based on the idea of a box filled with a turbulent fluid, convecting in a sufficiently vigorous fashion to suspend particles uniformly (Martin and Nokes 1988). In the boundary layer at the base of the box, velocities approach zero because of the no-slip condition, hence turbulence will not keep particles in suspension, and they will fall out. The mass of particles that fall out will equal the mass per unit height multiplied by the thickness of the zone at the base from which particles leave the system. This concept can be expressed as

$$\frac{\mathrm{d}M}{\mathrm{d}t} = -\frac{V_\mathrm{t}M}{h} \tag{14.13}$$

where M is the mass flux of particles through the control volume of fluid, V_t the particle fall velocity and h the depth of the box. This result is readily extended to the case of steady one-dimensional horizontal flow of a suspension layer to investigate the sedimentation as a function of distance

$$\frac{\mathrm{d}M}{\mathrm{d}x} = -\frac{V_\mathrm{t}}{uh}M \tag{14.14}$$

where u is horizontal flow speed. The mass that passes beyond a distance x from the origin over the entire flow duration, T, is $M_\mathrm{s} = MT$. Note that M_s is also the amount of material deposited beyond a given distance if the particles fall vertically from the base of the layer. Thus as fall speed increases or flow speed decreases, the sedimentation rate per distance increases. Equation (14.14) has solutions of the form

$$M = M_\mathrm{o} \exp\left(-\int_{x_\mathrm{o}}^{x} \frac{V_\mathrm{t}}{hu} \mathrm{d}x \right) \tag{14.15}$$

Deposition can be measured in a number of ways. If it is measured as the mass of material deposited per unit distance in the flow direction, m_s in kilograms per metre, then

$$m_\mathrm{s} = \frac{\mathrm{d}M_\mathrm{s}}{\mathrm{d}x} = \frac{V_\mathrm{t}M_\mathrm{s}}{uh} \tag{14.16}$$

Alternatively, if deposition is measured in unit area, S (kg m^{-2}), and if deposition is uniform perpendicular to the flow direction, then

$$S = -\frac{1}{\Omega}\frac{\mathrm{d}M_s}{\mathrm{d}x} = \frac{V_t M_s}{\Omega u h} \tag{14.17}$$

where Ω is the perimeter of a control volume whose surface is a surface of equal flow time.

The model can be applied to different flow geometries to deduce the amount of sediment released from a steady, moving fluid as a function of distance. This concept has been tested in the laboratory for a variety of configurations and flows and predicts sedimentation rates for a convecting particle layer (Martin and Nokes 1988), a radially spreading gravity current (Sparks *et al.* 1991), gravity currents (Bonnecaze *et al.* 1993) and vertical plumes and jets (Ernst *et al.* 1996a).

14.4.2 Radial Gravity Currents

Sedimentation from radially spreading gravity currents can be modelled using the continuity condition with no entrainment (Sparks *et al.* 1991), in which volumetric flow rate, Q, is defined as

$$Q = 2\pi r \Delta H u_r = \text{constant} \tag{14.18}$$

where u_r is the outward radial speed. If $x \equiv r$ and $h \equiv \Delta H$, then $1/\Delta H u_r = 2\pi r/Q$ and

$$M_s = M_{s_0} \exp(-\pi v(r^2 - r_0^2)/Q) \tag{14.19}$$

where M_s is the total mass of pyroclasts of a given grain size fraction deposited beyond the distance r, r_0 and M_{s_0} are the initial radius and mass. This model can describe the sedimentation from a volcanic umbrella cloud in a quiescent atmosphere or the lateral intrusion of a neutrally buoyant hydrothermal plume in the ocean. In cases where grain size isopleths and isopachs are only slightly distorted by wind into ovals, a geometric correction factor can be substituted for π (Bursik *et al.* 1992b). The total sedimentation pattern can be calculated by summation of all the grain size fractions within the cloud.

14.4.3 Plumes and Jets

Understanding the instantaneous structure of turbulent plumes and jets enables modelling of fallout from column margins. Simple scaling relationships (Chapter 2) can be used to show how plume width and speed vary with height. These relationships can be combined with the sedimentation model (equation 14.16), to determine how deposition should vary with radial distance from the vent.

The sedimentation model can be modified to consider fallout from the sloping margins of the plume, and to exploit the features of the plume scaling laws (Bursik *et al.* 1992b; Ernst *et al.* 1996a). The plume is assumed to be sufficiently turbulent to retain a mass flux, M, of particles in homogeneous suspension, although particles are lost from the plume margins. Then the sedimentation can be written

$$\frac{\mathrm{d}M}{\mathrm{d}t} = -\frac{pM}{b} V_t \tag{14.20}$$

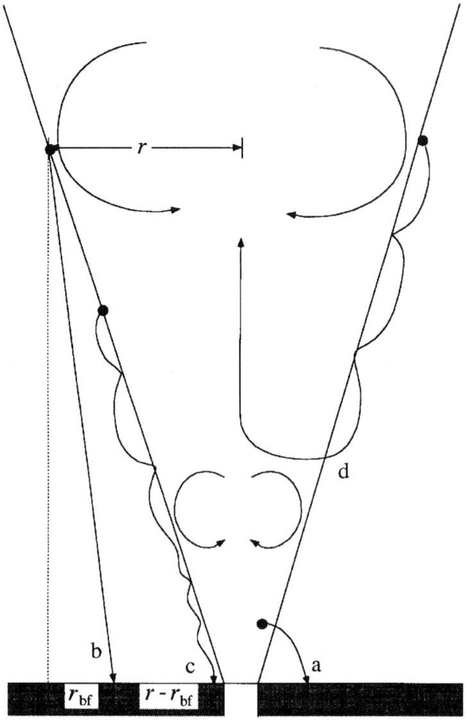

Figure 14.8 Diagram to illustrate sedimentation processes from the margins of a plume or jet. Particles can follow a variety of paths once they fall from a plume, here shown schematically (from Ernst *et al.* 1996a). Particle "a" follows a ballistic path to the ground. Particle "b" hits the ground at a distance, $r - r_{bf}$, that is closer to the plume centreline than is the distance at which it fell from the plume, because of the inward flow of ambient fluid into the plume. The effect of the backflow on particle "c" is sufficiently strong that the particle is almost re-entrained, although the particle is too large to remain in the plume. Particle "d" is simply re-entrained

where p/b is a geometric probability factor that represents the fraction of particles per unit height which are in the sloping margins of the plume and can therefore fall out, rather than remain within the plume (see Figure 14.8). For a plume in a homogeneous atmosphere, p has an estimated value of 0.226 (Bursik *et al.* 1992b). In a later section, a theory of re-entrainment is developed that accounts for the recapture of some of the particles by the inflow of ambient fluid to the plume and results in a modification of equation (14.20).

For steady plume flow, equation (14.20) reduces to

$$\frac{\mathrm{d}M}{\mathrm{d}z} = -\frac{pM}{bw} V_\mathrm{t} \tag{14.21}$$

where w is upward plume speed. Substituting from the scaling relations in Chapter 2, the solution to equation (14.21) for a plume is (Bursik *et al.* 1992b; Ernst *et al.* 1996a)

$$M = M_\mathrm{o} \exp(-Xv(r^{1/3} - r_\mathrm{o}^{1/3})J(r)) \tag{14.22a}$$

where $X = p(5/6\alpha)^{1/3}(9\alpha F_\mathrm{o}/10)^{-1/3}$, α is the entrainment coefficient (Chapter 2), and $J(r)$ is a function of the atmospheric stratification (Bursik *et al.* 1992a); $J(r) = 1$ in a uniform environment (Ernst *et al.* 1996a) and, for the atmosphere, $J(r)$ can be approximated by the polynomial expansion

$$J(r) = 3.069 + 3.466 \times 10^{-5} Br + 1.5315 \times 10^{-9} B^2 r^2 \tag{14.22b}$$

where $B = 5/6\alpha$.

14.4.4 Backflow

Not all particles that fall from the plume margins reach the ground. A certain fraction will be re-entrained because of the inflow of ambient fluid towards the plume. The fall paths of other particles will be deflected by this same inflow to produce a backflow of pyroclasts towards the plume (Figure 14.8). The average trajectories of backflow pyroclasts can be calculated from the relationships:

$$\frac{\mathrm{d}z}{\mathrm{d}t} = -V_\mathrm{t} \tag{14.23}$$

and

$$\frac{\mathrm{d}r}{\mathrm{d}t} = u_\mathrm{r} = bu_\mathrm{b}/r = -0.1\bar{w}b/r \tag{14.24}$$

where u_b is the inflow speed evaluated at b, and $\bar{w}$ the centreline plume speed. The trajectory relationship for a plume can be obtained from the above two equations and the proper scaling relationships (Chapter 2). For a plume in an unstratified ambient fluid, which is representative of one in a stratified fluid up to the neutral buoyancy height (see Chapter 11 and Morton *et al.* 1956), equation (14.24) becomes

$$\frac{\mathrm{d}r}{\mathrm{d}t} = \frac{\mathrm{d}r}{\mathrm{d}z}\frac{\mathrm{d}z}{\mathrm{d}t} = -\frac{\mathrm{d}r}{\mathrm{d}z}V_\mathrm{t} = -Kz^{2/3}/r \tag{14.25}$$

where $K = 0.1\,(9\alpha F_\mathrm{o}/10)^{1/3}$. Rearrangement and integration yields for the distance of radial backflow, r_bf

$$r(\mathrm{o}) = [r^2(2) - (6Kz^{5/3}/5V_\mathrm{t})]^{1/2} \tag{14.26}$$

Thus particles falling from a height z will not strike the ground at the corresponding distance $r(2)$, but at $r(\mathrm{o})$ (Figure 14.8). Measurements of particle fallout at different radii on the ground will not correspond directly with measurements at different radii along the plume margin. The backflow thus represents a transformation of coordinates between the plume margins and the ground. Theoretical curves of deposition can therefore be plotted in the coordinate system of the ground to model the effect of the backflow.

14.4.5 Re-entrainment

Particles that are sufficiently small can be swept back by the backflow into the plume after falling out (Figure 14.8). For the case of particles sedimented from an umbrella cloud or neutrally buoyant intrusion fed by the plume, particles released from the base of the flow closer than a critical distance cannot reach the ground and are re-entrained. Sediment re-entrainment for the case of a steady radial gravity current atop a plume in a quiescent

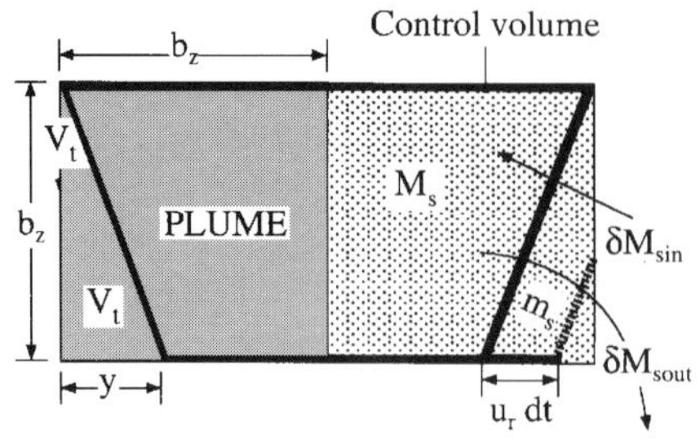

Figure 14.9 A schematic diagram of the assumed configuration for the particle re-entrainment model starts with a control volume scaled to the plume half-width, b_z, containing a mass, M_s, of particles. In time, dt, a mass, m_s, of particles with fall speed, V_t, falls from the plume into a triangular region of width, y, at the base of the control volume. Of these particles, a mass δM_{sin} are within a distance $u_r dt$ of the plume margin, where u_r is the radial inflow speed, and are therefore re-entrained. A mass δM_{sout} successfully escapes the plume

environment has been considered by Sparks *et al.* (1991), German and Sparks (1993) and Lane-Serff (1995). Although the critical distance increases as the particle size decreases, Lane-Serff (1995) showed that the proportion of particles re-entrained is in fact independent of both plume strength and particle size.

For plumes and jets the process of particle re-entrainment can be approximated by using a control volume surrounding the fluid directly adjacent to the plume (Figure 14.9). In general, the number of particles that fall from a fluid parcel within the plume and that are re-entrained by the plume will follow the relationship

$$\mathrm{d}M = \frac{-MpV_t}{b}\mathrm{d}t + \frac{Mkpu_r}{y}\mathrm{d}t \tag{14.27}$$

or

$$\frac{\mathrm{d}M}{\mathrm{d}t} = \frac{-MpV_t}{b} + \frac{Mkpu_r}{y} \tag{14.28}$$

where u_r is the radial inflow speed, y the difference in plume radius at the top and bottom of a control volume and k is the re-entrainment coefficient (Ernst *et al.* 1996a) which is related to the fraction of particles that have fallen from the plume parcel adjacent to the control volume that are in the re-entrainment zone. The solution to equation (14.28) is

$$M = M_o \exp\left[-\int_{t_o}^{t}\left[\frac{pV_t}{b} - \frac{pku_r}{y}\right]\mathrm{d}t\right] \tag{14.29}$$

For a simple jet, the scaling relationships from Chapter 2 can be used to change variables on the right-hand side of equation (14.29) to obtain

$$M = M_o \exp\left[-\int_{z_o}^{z}\left(\frac{pV_t}{2\alpha z} - \frac{pkw}{2\alpha z}\right)\frac{\mathrm{d}z}{w}\right] \tag{14.30}$$

or, in terms of r

$$M = M_o \exp\left[-\int_{r_o}^{r} \left(V_t X - \frac{pk}{2\alpha r} \right) dr \right] \tag{14.31}$$

where

$$X = [0.267p/4\alpha^2 M_o^{1/2}]$$

so that

$$M = M_o \exp[-V_t X(r - r_o) + (r/r_o)^{(pk/2\alpha)}] \tag{14.32}$$

where the subscript o denotes the point where the jet becomes fully developed.

Thus, if the rate of accumulation on the ground is controlled by the rate of the re-entrainment process, then M will vary as a power law function of r.

For a plume (Ernst *et al.* 1996a)

$$S = S_1 \left\{ \exp\left(-Xv(r^{1/3} - r_1^{1/3})\right) \left(\frac{r}{r_1}\right)^{\frac{5pk}{6\alpha}} \left(\frac{XV_t/3r^{5/3} - 5pk/6\alpha r^2}{XV_t/3r_1^{5/3} - 5pk/6\alpha r_1^2} \right) \right\} \tag{14.33}$$

where the subscript 1 refers to the distance at which flow in the column becomes plume-like, given by

$$r_1 \approx \frac{6}{5} \alpha Q_o^{3/5} F_o^{-1/5} \tag{14.34}$$

where the subscript o refers to values at the source.

No study has yet been made of re-entrainment by plumes affected by a cross-flow, but the same process should happen for small particles.

14.4.6 Effects of Wind

Wind can affect the motion of pyroclasts falling from a volcanic plume as well as the motion of the plume itself (Chapter 11). The wind-modified shape of a downwind plume results in sedimentation patterns that differ from those produced by a radially expanding umbrella cloud. The most basic observation of the sedimentation patterns resulting from deposition from downwind plumes is that they are elongated rather than axisymmetric. The amount by which sedimentation changes with distance also differs from that associated with axisymmetric plumes. If the model for the gravitational spreading of the downwind plume is adopted from Chapter 11 (equation 11.16), and applied to the relationship for sedimentation from a steady flow (equation 14.17), then the sedimentation along the plume axis will take the form:

$$S = S_o \left(\frac{8M_o V_t}{\sqrt{\pi} Q} \right) \exp(-\Xi(x^{3/2} - x_o^{3/2})) \tag{14.35}$$

where S is sedimentation (mass per unit area), $\Xi = (2V_t/3u)(2\lambda N/Q)^{1/2}$, u the wind speed, λ an empirical constant and N the Brunt–Väisälä or buoyancy frequency, x the downwind distance along the wind-blown plume centreline, and x_o the distance at which particles first reach the ground with deposition of S_o. Such particles are derived from the column corner between the vertical rising column and the horizontal wind-advected plume.

A complete description of sedimentation from wind-advected plumes has not been developed. This is a major gap in development of plume sedimentation models.

14.4.7 Atmospheric Advection/Diffusion Models

Once a downwind volcanic plume has spread and thinned sufficiently, the particles within it are no longer subjected to the motions of the plume itself. Instead, atmospheric motions determine particle dispersal. The downwind distance at which the transition between these two types of particle transport occurs is a complex function that depends on the relative magnitudes of a number of parameters, including the atmospheric turbulent diffusivity, wind shear, grain size of particles in the plume, degree of particle aggregation within the plume and the gravitational force on the plume. However, in an atmosphere lacking significant vertical wind shear, and for an umbrella cloud with such fine particles that few fall out over the time that it takes the cloud to be transported, the transition distance, or umbrella cloud thickness, can be parametrized by comparing the horizontal turbulent diffusivity, i.e. the rate at which the plume is being dispersed by turbulence, with the driving gravitational force. Hence, a number that expresses the ratio of these two driving forces should be of the order of unity at the transition distance

$$\frac{\sqrt{\overline{v'^2}}}{\lambda N \, \Delta H} \approx 1 \tag{14.36}$$

where $\sqrt{\overline{v'^2}}$ is the RMS velocity of turbulent motions in the atmosphere and ΔH the plume thickness. Using the value for $\overline{v'^2}$ of 2–3 $m^2\ s^{-2}$ which was measured in the troposphere (Nastrom and Gage 1985), an umbrella cloud would need to thin to 10–100 m before the parcels of material that comprise it are spread significantly by atmospheric diffusion rather than gravitational spreading of the cloud as a whole. Plume breakup and prevailing winds may cause diffusion and advection to predominate over gravitational spreading before such cloud thicknesses are attained, so 10–100 m should be regarded as the lower bound at which gravity current behaviour and the related settling can occur.

A number of studies of sedimentation from volcanic plumes have been based on the principles of turbulent advection and diffusion (Richardson and Proctor 1926; Carey and Sigurdsson 1982; Suzuki 1983; Armienti *et al.* 1988; Glaze and Self 1991; Heffter and Stunder 1993). The models have thus far considered the contribution of the turbulence structure of the atmosphere in determining the motions of particles once they leave the plume, by using simplified vertical line diffusers as particle sources, and then allowing the particles to be advected and dispersed in a turbulent wind field given by

$$\frac{\partial C}{\partial t} + u_x \frac{\partial C}{\partial x} + u_y \frac{\partial C}{\partial y} + \frac{\partial (vC)}{\partial z} = \kappa_x \frac{\partial^2 C}{\partial x^2} + \kappa_y \frac{\partial^2 C}{\partial y^2} + \kappa_z \frac{\partial^2 C}{\partial z^2} + \Phi \tag{14.37}$$

where C is the concentration of particles in a given grain-size class, κ an empirically determined eddy diffusivity for the atmosphere that is not necessarily isotropic, and Φ a source term that describes the change in particle concentration through time at the origin. The second and third terms on the left-hand side represent the advection of particles by the steady wind, and the fourth is the settling term. The equation (14.20) that has been used for sedimentation from a turbulent suspension is a special case of the diffusion equation.

There are considerable uncertainties in the application of advection–diffusion models (see also section 14.5.2). One of the major characteristics of the turbulence structure of the atmosphere is its inhomogeneity. That is, the turbulence does not look the same at different spatial scales. Much of the turbulent energy is contained in eddies of 1–100 m formed as the wind interacts with the planetary boundary layer (the Earth's solid surface and oceans), and in much larger eddies of 100–10 000 km related to the general atmospheric circulation. Energy is transferred into the small eddies from the larger planetary flow. The result of the inhomogeneities is that κ in equation (14.37) cannot be taken as a constant if the spatial scale of an atmospheric disturbance, such as a volcanic plume, is large. The eddy diffusivity can be modified to take this into consideration by equating the size of the disturbance with an equivalent time scale (to ease solution to equation (14.37); see Crank 1956) and modifying κ to be a function of time. Such a modification, however, requires numerous measurements of the turbulence structure at many atmospheric heights, or a complex turbulence model. Pasquill and Smith (1983) and others have summarized and modelled such data, but these have yet to be fully incorporated into the volcanic diffusion models. An approach that would be applicable to given wind patterns, either measured or estimated from a global circulation model (GCM), would be through the use of atmospheric trajectory models, which have been recently applied to the study of the long-range transport of volcanic plumes (e.g. Heffter and Stunder 1993). The trajectory models could be modified to predict ground-level deposition of volcanic particles, much as they have already been used to predict deposition of radioactive and chemical contaminants (Apsimon *et al.* 1988; Stunder *et al.* 1986).

14.5 OBSERVATIONS

Numerous measurements have been made of sedimentation patterns from plumes both in the laboratory and in nature. As visual observations have confirmed that larger pyroclasts do travel as ballistic projectiles (McGetchin *et al.* 1974; Chouet *et al.* 1974), the sedimentological observations have primarily been directed towards testing whether the models for sedimentation from turbulent suspensions, given by equations (14.19), (14.22) and (14.35), adequately describe observed sedimentation patterns. For the most part, these studies have found that the models are consistent with the data.

14.5.1 Laboratory Experiments

Data collected from laboratory experiments are consistent with the models for sedimentation from a radially spreading gravity current and from vertical plumes and jets as outlined in the previous section (Sparks *et al.* 1991; Ernst *et al.* 1996a). Sparks *et al.* (1991) investigated the sedimentation of suspended particles from the radially spreading current at the top of buoyant jets. Although these were interface flows, trapped between the ambient water in the experimental tank and the overlying atmosphere, and therefore differing somewhat in mechanism from intrusive gravity currents, such as umbrella clouds, the conservation equations were similar to those for umbrella clouds. Their data exhibit the variations with distance that can be expected from equation (14.18) (Figures 14.10 and 14.11).

Ernst *et al.* (1996a) studied the somewhat less accessible problem of plume and jet sedimentation. For sufficiently coarse particles they found good agreement between experimental results and theoretical models of sedimentation from plumes (equation 14.22) and jets in uniform environments. However, for finer particles re-entrainment

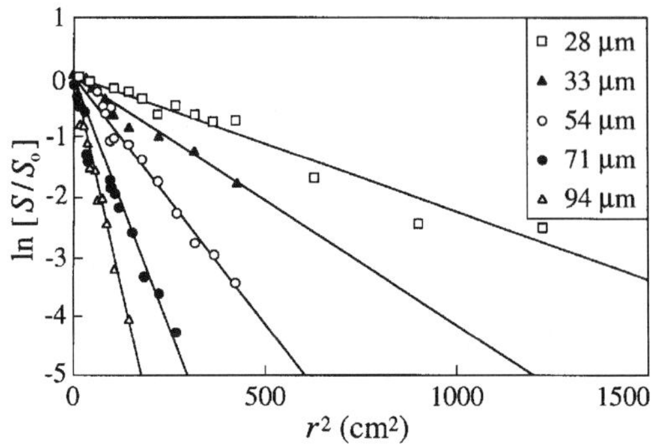

Figure 14.10 Particles falling from an "umbrella cloud" generated in a laboratory tank exhibit the predicted relationship for sedimentation with distance (from Sparks *et al.* 1991). The normalized deposition per unit area, S/S_o, decreases exponentially with distance squared, r^2, from the plume centreline. The decrease in sedimentation with distance is more rapid for larger particles, as predicted by equation (14.19)

became important. They found good agreement with experimental results by adjusting the re-entrainment coefficient, k (Figure 14.12). They found that if the ratio of the characteristic plume velocity to particle settling velocity was greater than 1 then experimental results gave a value of $k = 0.4$, independent of particle size and plume strength. A value of $k = 0.1$ was found for jets. The identification of a constant re-entrainment coefficient suggests that for small suspended particles re-entrainment reaches a steady state as found for the case of lateral intrusions above plumes (Lane-Serff 1995).

14.5.2 Volcanic Deposits

In studying eruption deposits, it has proved essential to consider sedimentation in the separate regions of plume, umbrella cloud and downwind plume, as end-member situa-

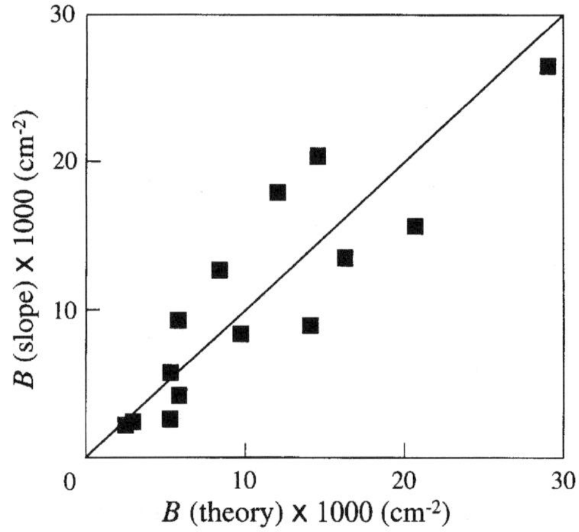

Figure 14.11 The value of the exponent, $B = \pi v/Q$, calculated from the slope of the curves shown in Figure 14.10 compares well with the value estimated by calculating the terminal fall velocity of the particle, V_t, and measuring the flow rate, Q

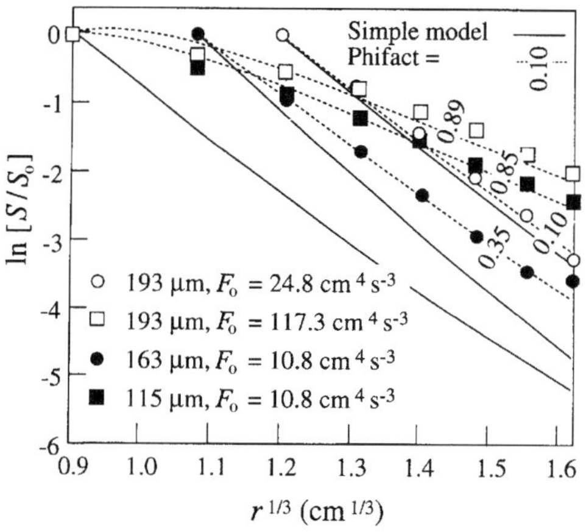

Figure 14.12 The normalized sedimentation per unit area from a vertically directed laboratory buoyant jet is in agreement with the model, equation (14.38), if backflow and re-entrainment, equations (14.28) and (14.32), are accounted for (from Ernst *et al.* 1996a). Four experiments are shown with particle diameter in micrometres and buoyancy flux, F_0, shown for each experiment. The solid lines show the predictions of the simple theory neglecting re-entrainment. The dashed lines labelled "phifact" are for the theory including re-entrainment and show the value of the re-entrainment coefficient

tions. Moreover, ballistic deposits of large particles have been treated separately from deposits resulting from sedimentation of intermediate-sized pyroclasts from turbulent suspensions, which have in turn been modelled separately from the deposition of small pyroclasts that are affected by the atmospheric turbulence structure.

Ballistics

Data on the distribution of ballistic clasts confirm qualitatively the model predictions. Ballistic clasts of 1–2 m diameter have been observed at distances of up to 5 km at Arenal volcano, Costa Rica (Fudali and Melson 1972) and at Lascar volcano, Chile (Matthews *et al.* 1997). Ballistic fallout patterns for pyroclasts ejected during the eruption at the Ukinrek maars in Alaska have been studied by Self *et al.* (1980). They found that the largest ballistic pyroclasts increased in size with increasing distance from the vent, as predicted by equation (14.11) for type I ballistic pyroclasts. Somewhat smaller ballistics decreased in size with increasing distance, suggesting that they were originally transported within the turbulent plume from which they were ejected as type II ballistic projectiles. Observations of ballistic clasts erupted in the September 17, 1996 explosion of the Soufrière Hills volcano, Montserrat, show that blocks approximately 1 m in diameter have the greatest range in agreement with the theoretical predictions of Bower and Woods (1996).

Suspended particles

Granulometric data from several volcanic deposits are consistent with fallout and deposition from turbulent suspension in the vertical eruption column, the nearly radial gravity current and the windblown plume. The Fogo A deposit has provided data that can be compared with the radial umbrella cloud current and the eruption column (Bursik *et al.*

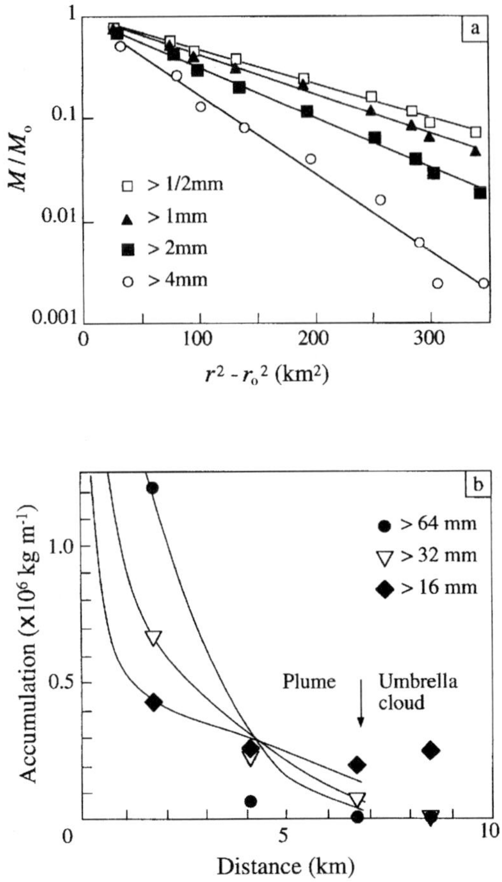

Figure 14.13 The fallout, as measured by M/M_o is shown as a function of the radial distance squared. Note that here M and M_o are the total mass deposited rather than mass sedimentation rate. Fallout from both the umbrella cloud (a) and the eruption column (b) for the Fogo A deposit are consistent with model predictions (equations 14.19 and 14.22 respectively; from Bursik *et al.* 1992b)

1992a). Fogo A was a powerful Plinian eruption which occurred in a weak cross-wind so that the dispersal pattern is nearly symmetric around the vent with only minor wind distortion. Grain size variations in the Fogo A deposits show excellent agreement with the model for the sedimentation from a radial umbrella cloud (equation 14.19; Figure 14.13) and suggested that the eruption column model may also be reasonable (equation 14.22). The laboratory work of Ernst *et al.* (1996a) corroborated that the eruption column model is reasonable, but also suggested that re-entrainment can sometimes be an important process. The Mount St Helens deposit from the May 18, 1980 eruption has provided data on deposition from a windblown plume (Bursik *et al.* 1992a). The data for the downwind plume of Mount St Helens suggested that, to distances of at least 500 km, the source flow characteristics rather than the atmospheric turbulence structure dominated the sedimentation pattern (Figure 14.14).

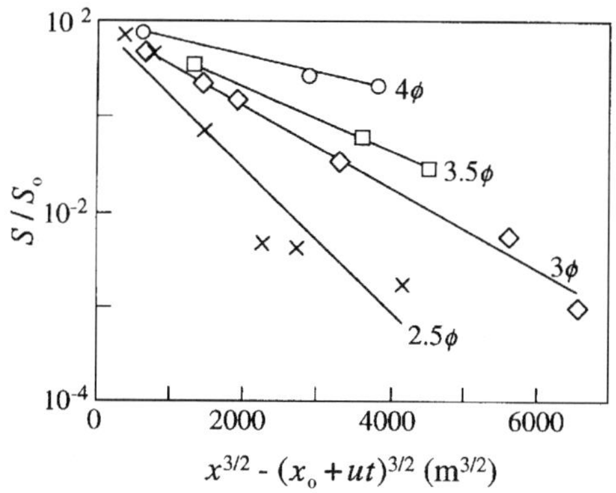

Figure 14.14 Sedimentation data collected along the downwind axis of the fall deposit from the May 18, 1980 eruption of Mount St Helens. The data are consistent with model predictions for a gravity current being transported downwind (equation 14.35; from Bursik *et al.* 1992a)

A major change in sedimentation pattern is anticipated to be associated with the eruption column corner where the flow changes from predominantly vertical to horizontal flow in the umbrella cloud. On plots of thickness and grain size against a characteristic distance (Pyle 1989) several deposits show a marked break-in-slope that can be interpreted as related to the influence of the column corner (Sparks *et al.* 1992). For the Fogo A deposit, a maximum in radial accumulation occurs for many grain size classes below the position where the plume begins lateral flow as an umbrella cloud or downwind plume (Bursik *et al.* 1992b). This observation suggests that the loss of vertical column speed and turbulence intensity triggers enhanced fallout for the grain sizes that dominate the deposits of volcanic eruption columns, from approximately 100 μm to 10 cm for these intermediate-sized eruptions with column heights of 20–30 km.

Advection/diffusion models

General results of the application of the advection/diffusion relationship have been presented by Suzuki (1983), specific examples by Armienti *et al.* (1988) and Glaze and Self (1991), and forecasts of deposition from hypothetical future eruptions by Barberi *et al.* (1990). The results resemble deposition patterns found in nature (Figure 14.15).

Such modelling probably adequately describes the deposition patterns from small plumes dominated by wind dispersion (Rose *et al.* 1988; Glaze and Self 1991). However, for many even moderately powerful plumes there are problems in applying diffusion models even at considerable distances from the vent. The analysis in section 14.4.7 (equation 14.36) and the indication that gravity controls lateral spreading of the Mount St Helens downwind plume (Bursik *et al.* 1992a) indicate that atmospheric diffusion may not be a dominant influence on the sedimentation of most particles. Diffusion only becomes important at distances where all but the very finest particles have been deposited. Thus, this kind of approach may be best suited to modelling of the long-distance dispersal of plumes over hundreds to thousands of kilometres and to understand the dispersal of very fine ash and aerosols.

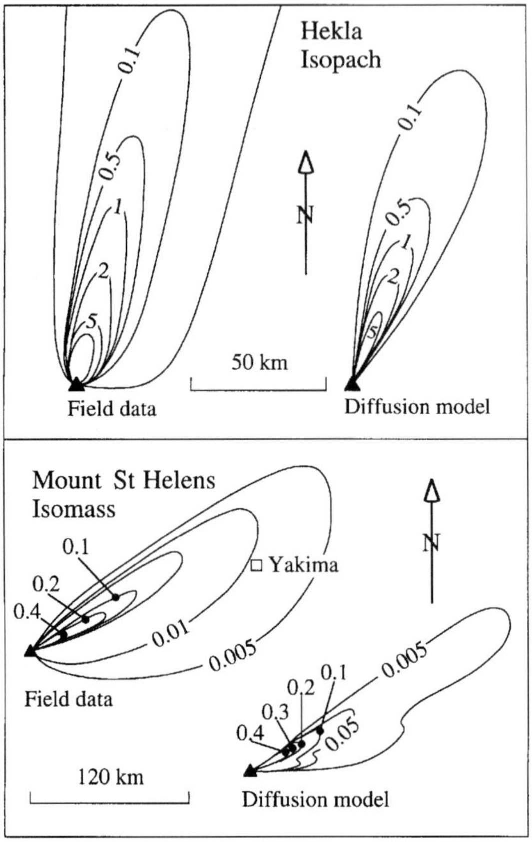

Figure 14.15 Predicted deposition patterns compared to isopach data for the August 17, 1980 eruption of Hekla, Iceland, and for the July 22, 1980 eruption of Mount St Helens, Washington, USA. The results from the advection/diffusion modelling are in reasonable agreement with field data for these relatively small eruption plumes (from Glaze and Self 1991)

The major modifications to be made in the future development of diffusion models are to use more realistic characterizations of the source and of the atmospheric turbulence structure. Up to the present, source models have considered the eruption column and umbrella cloud as simple linear sources of diffusion, centred on the volcano. Not only do these models bear an unknown relationship to reality, but also for distances of over 500 km from the vent in eruptions of the size of the May 18, 1980 Mount St Helens event, the source shape and structure play important roles in determining sedimentation patterns (Bursik *et al.* 1992a, b). Only at considerable distances, where the details of the source become unimportant, might current diffusion models describe average particle motions. However, at these distances, the unknown structure of turbulence in the upper atmosphere has frequently made it difficult to model distal sedimentation patterns (Glaze and Self 1991). Until this structure is better characterized, application of advection/diffusion models will be problematic.

14.5.3 Re-entrainment

Particle re-entrainment has been observed in laboratory experiments (Sparks *et al.* 1991; Ernst *et al.* 1996a). This process also accounts for the excess particle concentrations observed in profiles across laboratory plumes where sedimenting particles were re-entrained into plumes (Carey *et al.* 1988). Geochemical observations of uranium series disequilibria in hydrothermal plumes on the ocean ridges show that the particles have significantly longer average residence times in the plume systems than expected. These observations can be explained by re-entrainment of particles (German and Sparks 1993). Large accretionary lapilli with multiple concentric layers of fine ash are also best explained by re-entrainment (Chapter 16).

14.6 SUMMARY

The terminal fall velocity of particles is determined by a balance between the gravitational force and the drag force of the surrounding fluid. Terminal fall velocities of particles at high Reynolds numbers are controlled by inertia and the fall velocity is proportional to the square root of the particle diameter. Viscosity controls the fall of particles at low Reynolds numbers and the velocity is proportional to the square of the particle diameter (Stokes' law). The drag coefficient is characteristically of the order of unity at high Reynolds number, but varies in detail according to particle shape and roughness. Flow separation above *Re* of $2–3 \times 10^5$ results in a marked drop of drag coefficient.

There are two major types of pyroclast transported by volcanic plumes: ballistic and suspended. The largest ballistic pyroclasts follow paths that are determined by equations of motion similar to those used in studies of projectiles. Larger clasts can be projected further from the vent than smaller ballistic clasts because of their greater momentum. Smaller ballistic pyroclasts are influenced by fluid motions in lower parts of the eruption column. They can be "fired" along ballistic trajectories from large eddies. Suspended particles fall from the steep margins of plumes, umbrella clouds and downwind plumes. Sedimentation rates are dependent upon their concentration within the transporting flow, the flow rate and time. They can be adequately described by a simple modelling concept of a uniform turbulent suspension. Sedimenting particles can be affected by backflow towards the plume, re-entrainment and atmospheric motions subsequent to fallout. Models of sedimentation from turbulent gravity currents can describe the dispersal of tephra from umbrella clouds and windblown plumes even at large distances from source. Models based on atmospheric diffusion can describe tephra dispersal adequately, but are largely empirical with adjustable source functions and diffusivities. Such models are not based on physically appropriate descriptions of the near-field dynamics of plumes. However, distal transport and deposition of very fine ash and aerosols can be affected by the motions of the atmosphere, and are best modelled using advection/diffusion equations modified from those used in contaminant transport.

15 Quantitative Interpretation of Tephra Fall Deposits

NOTATION

b	distance where centreline velocity decreases by $1/e$ (m)
β	gas density (kg m^{-3})
C_d	drag coefficient (dimensionless)
d	particle diameter (m)
g	gravitational acceleration (m s^{-2})
H_b	height of plume neutral buoyancy (m)
H_t	maximum plume height (m)
M	mass flux of air and particles at H_b (kg s^{-1})
M_o	mass eruption rate of magma (kg s^{-1})
R	radial distance in umbrella region (m)
R_w	cross-wind width of umbrella cloud across vent (m)
S	accumulated mass per unit area (kg m^{-2})
σ	particle density (kg m^{-3})
S_o	initial accumulated mass per unit area (kg m^{-2})
U_c	plume centreline velocity (m s^{-1})
U_v	time-averaged vertical velocity (m s^{-1})
x	radial distance from plume centreline (m)

15.1 INTRODUCTION

The geologic record of volcanic plumes consists of tephra fall deposits (Chapter 13). For historic eruptions observations of plume height, duration, geometry and dispersal can be related directly to the physical characteristic of the resultant fall deposits. However, for ancient deposits it is necessary to develop models for the dispersal and deposition of tephra that can reconstruct the characteristics of the source plume from measurements of the associated fall deposits. Application and development of such models falls within the nascent field of quantitative palaeovolcanology. Much of the research to date has focused on dispersal of coarse ejecta in the regions relatively close to the volcano.

Development of such models is important in the field of volcanic hazard assessment (Chapter 17). One of the objectives of a hazard assessment is to understand the past behaviour of a volcano so that future activity can be predicted. The different types of pyroclastic deposit produced by a volcano, such as falls, flows and surge, are identified to provide information about the frequency and styles of past eruptions. The distribution, thickness and grain size characteristics of the deposits provide information about the strength and duration of an eruption and environmental conditions. Together these elements form the basis for evaluating how a volcano has behaved in the past and thus what its behaviour will likely be in the future.

The particle size characteristics and dispersal patterns of tephra fall deposits are controlled principally by a combination of eruption column height, the local wind field, and the particle size distribution of the ejecta. Atmospheric conditions can also influence the dispersal of fine ash particles (Chapter 11). In order to quantify the source plume characteristics the effects of the wind must be decoupled from these other factors. In this chapter models are described which use the distribution of large pyroclasts to infer eruption column height and the strength of a cross-wind. The estimation of column height from ancient fall deposits can elucidate the dynamic range of explosive volcanism for which the historic record of observations is incomplete. The predictions of the models compare well with observations from historic eruptions. Important information is also provided about the height to which gas and particles are injected into the atmosphere, the volume of ejecta erupted, the rate of eruption and the duration of an eruption. The results of such palaeovolcanological studies can be applied to investigation of many aspects of eruption dynamics.

The thickness and bulk particle size variations of tephra fall deposits are discussed. Tephra fall deposits typically show a general exponential decay in thickness and particle size away from source (Chapter 13). This behaviour is explained by sedimentation processes from the umbrella region of an eruption plume. Modelling can also investigate the influence of particle size distribution and column height on the thickness, mean grain size and sorting features of fall deposits.

15.2 MAXIMUM GRAIN SIZE DATA

A very useful field measurement is the size of the largest clast within a tephra fall deposit. Qualitatively these data have been used to compare the relative intensity of volcanic eruptions (Walker 1981c).

There is no standard method for the collection of maximum clast size data. Some workers use the average of the three largest clasts, others use the average of five. Irregularities in the particle shapes introduce problems in data collection as well as variations in the terminal fall velocity of a clast of given mass and density (see Chapter 14). Clasts are often elongate and thus some workers average the three principal axes to obtain a mean diameter, whereas others use only the largest diameter. The size of the largest clast found will tend to increase as the size of the search area is enlarged, but should eventually approach some upper limit.

Maximum clast size data can be contoured to produce an isopleth map. Figure 15.1 shows an example of pumice and lithic isopleth maps for the AD 180 Taupo ultra-Plinian deposit of New Zealand. Comparison of data sets collected by different workers on the same deposit suggest maximum clast size values can vary by a factor of 1.5 and even over 2 in some cases, a result of the difference in data collection techniques. For example, maximum clast sizes determined by Walker and Croasdale (1971) on the Fogo A Plinian

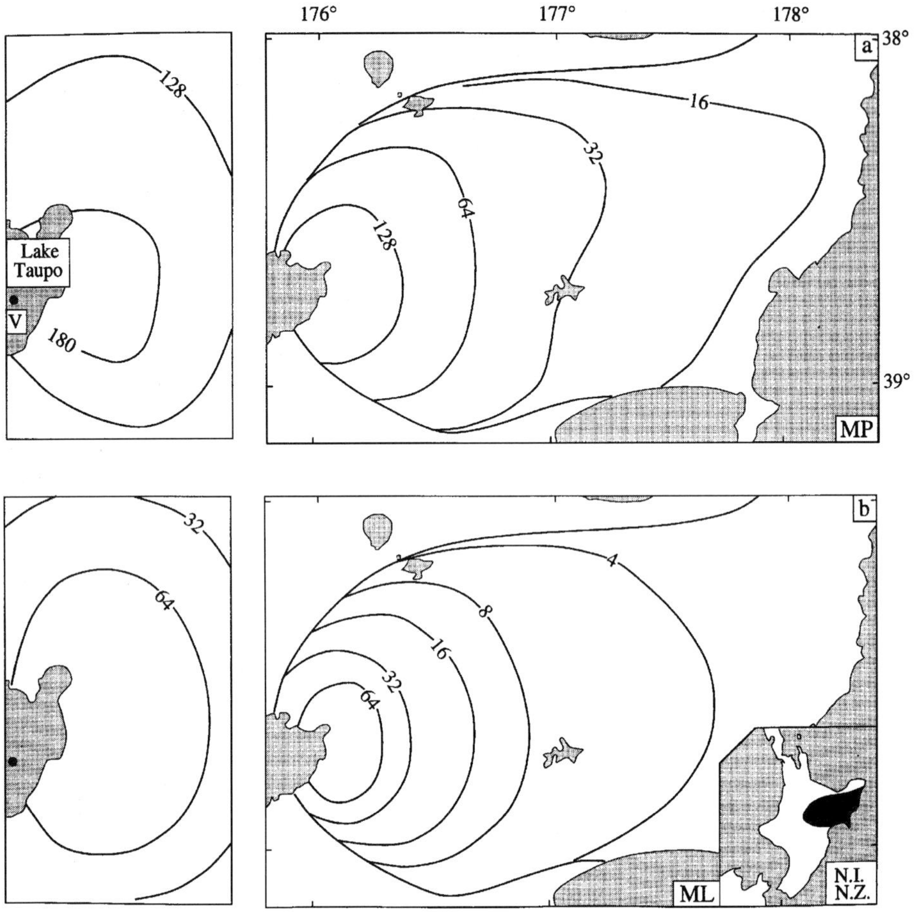

Figure 15.1 Maximum clast size isopleth maps of the AD 180 Taupo ultra-Plinian deposit, New Zealand: (a) maximum pumice diameter (MP) and (b) maximum lithic diameter (ML) in millimetres. V marks the position of the vent. Modified from Walker (1980)

deposit were an average of 1.6 times larger than determined by Bursik *et al.* (1992b). The discrepancy is explained because Walker and Croasdale (1971) used the average of the three maximum lengths, whereas Bursik *et al.* (1992b) used the average of the five largest clasts using the mean of the three orthogonal lengths of each clast.

Isopleth maps can be used to extract quantitative information about source plume characteristics and semi-quantitative information about wind strength. Carey and Sparks (1986) and Wilson and Walker (1987) have developed models to predict the distribution of maximum clasts from sustained Plinian-style eruptions. These models can be used to calculate peak eruption column height, the temporal evolution of column height and the magnitude of the cross-wind through which the column grew. In applying this method it is best to use lithic clasts because they display much less variation in density than pumice clasts.

15.2.1 Theoretical Considerations

A convective eruption column is highly turbulent and thus local velocities at any position will fluctuate irregularly and substantially with time. Studies of thermal plumes show, however, that the time-averaged vertical velocity, U_v, can be described as a symmetrical Gaussian function of distance from the plume axis

$$U_v = U_c \exp\left(\frac{-x^2}{b^2}\right) \tag{15.1}$$

where U_c is the centreline velocity at height h, x the radial distance and b the distance at which the velocity has decreased to $1/e$ times U_c (Figure 15.2).

The support of particles within the column and umbrella region can be understood with reference to the time-averaged internal velocity structure. At any position in the column there will be clasts which have a terminal fall velocity exactly equal to the ascent velocity. The terminal fall velocity for a clast of diameter d at high Reynolds number is given by

$$U_t = C_d\sqrt{\frac{dg\sigma}{\beta}} \tag{15.2}$$

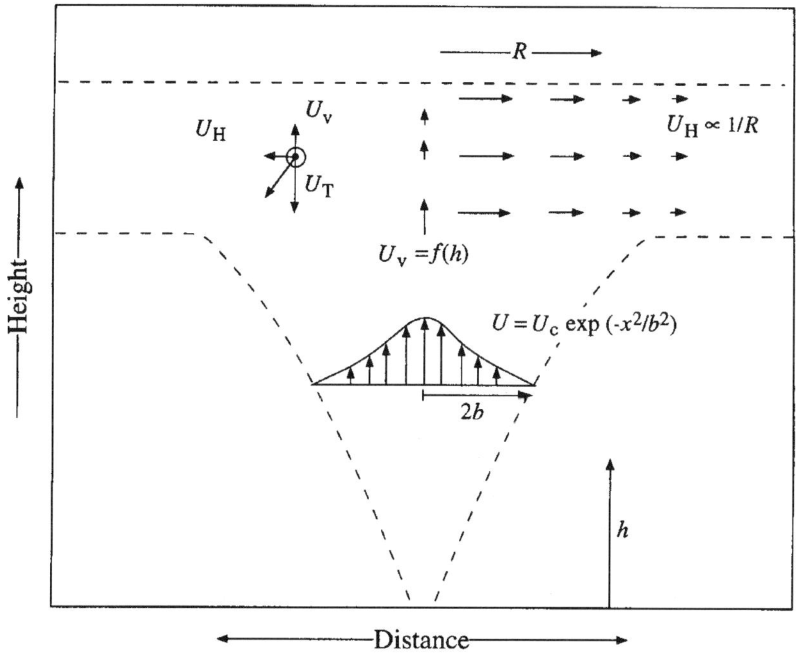

Figure 15.2 Cartoon showing the structure and velocity distribution of an eruption column and its effect on particle settling. The time-averaged velocity across the column of radius $2b$ is Gaussian. In the umbrella region flow is diverted horizontally. Three velocity components can be recognized acting on a clast: the vertical velocity of the plume, U_v, the terminal settling velocity of the clast, U_T, and the outward horizontal velocity in the umbrella region, U_H

where C_d is the drag coefficient, σ the clast density, β the density of the surrounding gas and g the acceleration due to gravity. Experiments indicate that a value of approximately unity is appropriate for C_d (Wilson and Huang 1979; see Chapter 14 for fuller discussion). A velocity surface within the plume can be defined where the mean vertical velocity is sufficient to just balance the settling velocity of a specific particle size and density. This surface cross-cuts the surfaces of constant velocity because the particle settling velocity is also related to the fluid density, which changes as a function of the height and radial position within the plume. In an axisymmetric plume, a two-dimensional cross-section shows this surface as a plume-shaped contour, which is referred to as a particle support envelope.

In order to calculate such surfaces a model is used that predicts the internal variations of velocity and density within an eruption plume (Chapter 4). The following discussion is based on the work of Carey and Sparks (1986) in which the column model of Sparks (1986) is used. The results of this model for maximum clast dispersal are compared with an alternative model of Wilson and Walker (1987). Most models suggest that vertical convective velocities can attain values up to 200 m s^{-1} and thus are strong enough to transport lithic clasts several centimetres in diameter to the top of most eruption columns.

Figure 15.3 shows an example of several particle support envelopes within a column of 21 km height for lithics (density of 2500 kg m^{-3}). Inside a specific support envelope the velocity is higher than the particle's terminal settling velocity and particles will migrate upwards. Outside of the envelope the average vertical velocity is less than the particle settling velocity and the particle can fall out. The actual transport mechanism of large clasts in the turbulent environment is likely to be complex. Clasts are probably shifted from the interior to outside their support region by large velocity fluctuations which are typically of the order of one-third of the average vertical velocity at any height.

The maximum lateral distance to which a certain size particle is supported within a plume is simply the maximum width of the particle support envelope from the plume

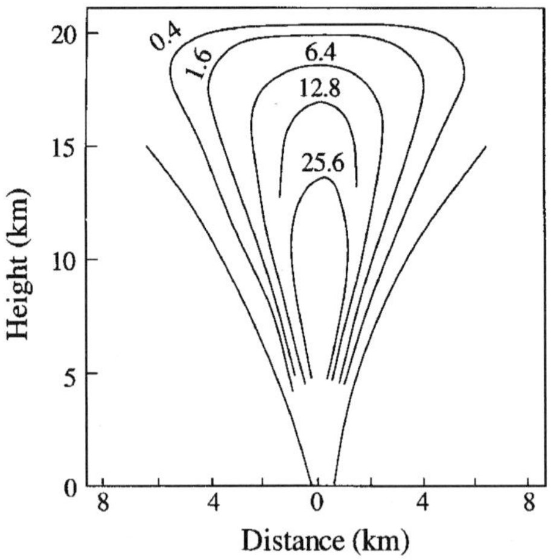

Figure 15.3 Clast support envelopes in a 21 km high column (from Carey and Sparks 1986). The contours represent the position in the column where a lithic clast (density 2500 kg m^{-3}) is just supported by the time-averaged vertical velocity. Contours are for the diameter in centimetres

centreline (Figure 15.3). A particle which leaves from this point will enter a flow regime where the vertical velocity is less than the terminal fall velocity and it will eventually be lost from the plume. If the plume is relatively low or the particles are very large they will not experience significant lateral transport. However, if the plume is very large once a particle leaves the edge of the support envelope it can be transported laterally before leaving the plume within the umbrella region. Under steady conditions there is a radial flow within the umbrella region as the plume spreads out between its level of neutral buoyancy, H_b, and the final plume height, H_t. The velocity, V_r, in this region is modelled by

$$V_r = \frac{M}{2\pi R\alpha(H_t - H_b)} \tag{15.3}$$

where M is the mass flow rate of air and ejecta at height H_b, and α the mean density of air between H_t and H_b. This radial flow transports particles laterally after they leave the maximum width of the support envelope to the time they reach the base of the umbrella region. The total maximum distance from the centreline of the plume for a certain particle size is therefore the maximum width of the support envelope plus the lateral transport within the spreading umbrella region. An improvement in the modelling for fallout from the umbrella region would be to use models for the slumping of a constant flux gravity current (Chapter 11).

Figure 15.4 shows particle support envelopes for eruption columns of 28, 35 and 43 km height. There is a dramatic increase in the support envelope width as the column height is increased, explaining why the dispersal of clasts is greater for higher eruption column heights. Figure 15.5 illustrates the great sensitivity of maximum grain size to column height. In a 20 km high column, for example, fist-sized lithic clasts (6.4 cm diameter) are transported to about 3 km, whereas in a 35 km high column the same sized clast is carried to nearly 10 km. The reason for this sensitivity is that the increase in column height by a factor of 1.75 corresponds to an increase in plume power of a factor of 9.4 (see discussions in Chapter 4) and an increase in the range of the clast of a factor of 3.3. This sensitivity means that, despite the crude nature of field data collection, column heights estimated from maximum clast size data are surprisingly accurate. For example, in Figure 15.5 fragments of 0.4 cm diameter reach 8 km distance for the 20 km high column, whereas fragments of 6.4 cm diameter reach slightly more than this distance for the 35 km high column. The diameter difference of a factor of 16 is very much larger than the discrepancies in field estimates of clast size. Thus the differences in the clast size measurements in the two studies of the Fogo A Plinian deposit (Walker and Croasdale 1971; Bursik *et al.* 1992b) led to height estimates of 30 and 27 km. The heights estimated from this method are thus considered to be generally much better than 5 km in accuracy.

Wilson and Walker (1987) approached the modelling of maximum clast dispersal by assuming that the vertical velocity of an eruption plume is constant across the plume at any particular height. This is the "top-hat" velocity profile. At the column margin there is an abrupt transition from the top-hat velocity to zero. Pyroclasts are assumed to be lost from the column margin at a height where the plume velocity is equal to the particle terminal settling velocity. No allowance is made for the horizontal transport of clasts that takes place within the expanding umbrella region of the plume. They used a numerical model which accounts for the progressive loss of clasts, and thus thermal energy, with increasing height in the column.

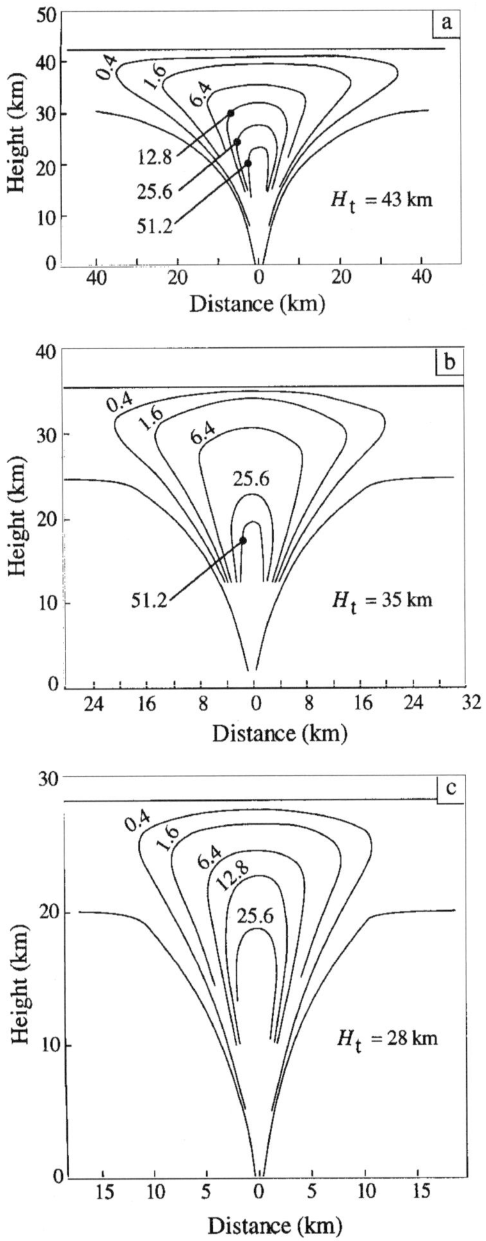

Figure 15.4 Clast support envelopes for columns with heights of 43 km (a), 35 km (b) and 28 km (c) after Carey and Sparks (1986). Contours are for lithic clast diameters in centimetres (density 2500 kg m^{-3})

Most plumes are affected by atmospheric winds in two ways (Chapter 11). First, wind causes a downwind deflection in the convective portion of the column which is a function

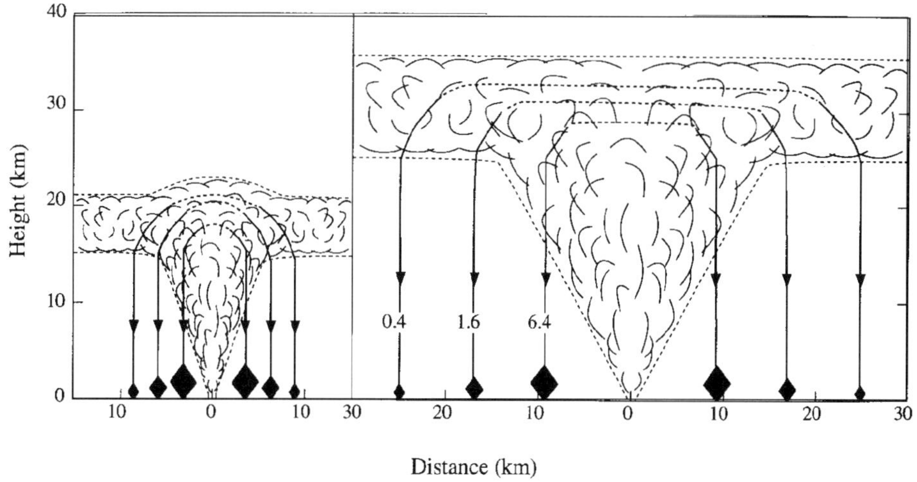

Figure 15.5 Cartoon showing trajectories of 6.4, 1.6 and 0.8 cm diameter lithic clasts (density 2500 kg m^{-3}) from columns of 20 and 35 km height, after Carey and Sparks (1986)

of the plume strength and wind speed. Second, the wind modifies motion in the umbrella region (Figure 15.6). In the upwind direction the umbrella region will spread until the radial velocity of the intrusion is equal to the wind speed. At this fixed position, called the stagnation point, the flow in the umbrella region is diverted laterally and swept around in the downwind direction.

The mass flux into the umbrella cloud must pass through a downwind plane defined by a width equal to $2R_w$ and the heights H_t and H_b (Figure 15.6). Consider the trajectories of

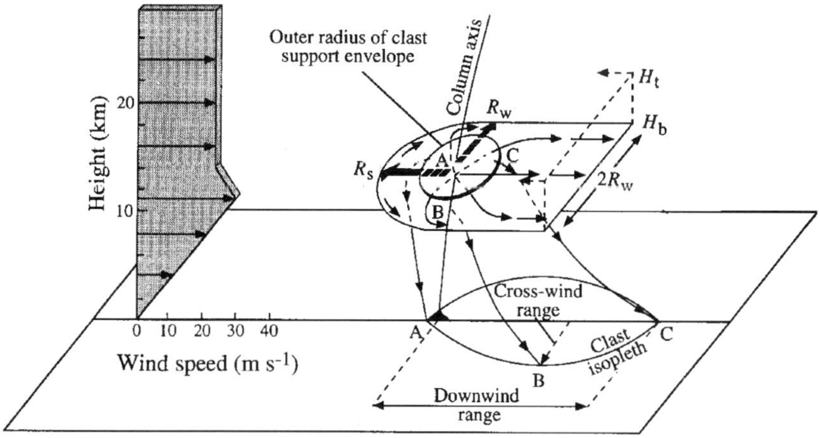

Figure 15.6 The principal features of clast fallout from a column interacting with a cross-wind. The figure represents a simple model for the maximum clast size range in the cross-wind and downwind directions. Details are discussed in the text. The wind profile on the left was used by Carey and Sparks (1986) to calculate downwind ranges of clasts

three particles that leave the edge of a particle support envelope first at the upwind stagnation point (A), second at a position 90° to the direction of the wind (B), and third at a position directly downwind (C) (Figure 15.6). Particle A leaves the umbrella region and is swept towards the column axis. Depending on the strength of the wind it may be carried past the source vent. Particle B leaves the support envelope and is transported outwards and downwind within the umbrella region. When it leaves the umbrella region it is then advected by the wind to some distance downwind of the source vent (Figure 15.6). Particle C leaves the support envelope and is transported downwind within the umbrella region. Below the umbrella region it is advected by the wind downwind of the source. The distance of particle B from the plume axis is a function of the plume height only and is not affected by the wind. This distance is called the cross-wind range and it is controlled by column height not wind strength. In weak plumes in a strong cross-wind the validity of this assumption becomes suspect. In such cases the overall width of the plume will be limited by the wind which could then have an influence on lateral clast range. Particle C is controlled by both column height and wind speed. Its distance from the source vent is the summation of transport within the umbrella region (column height effect) and transport in the atmosphere (wind effect).

The final depositional locations of the clasts constrain the shape and position of a clast isopleth. Carey and Sparks (1986) used this simple model to predict maximum size isopleths for pumice and lithic clasts as a function of column height and wind speed. Examples of model isopleths are shown in Figure 15.7 for column heights of 43 and 14 km. A substantial change in isopleth pattern occurs as a function of column height. Large column heights generate isopleth patterns that are highly circular even with strong winds. This is principally the result of most lateral transport occurring by intrusion of material within the umbrella region. The Taupo ultra-Plinian deposit (Figure 15.1) is an example of a very powerful column with an estimated column height of 51 km which produced symmetrical contours for lithic isopleths down to 0.8 cm diameter. For smaller clast sizes the dispersal becomes increasingly distorted downwind. As the column height decreases the ratio of transport within the umbrella region to the transport within the atmospheric wind field decreases. When the amount of transport in the atmosphere becomes large relative to the expansion in the umbrella region, as is the case for low eruption columns with strong winds, then the isopleths become substantially elongate in the downwind direction and take on a characteristically ellipsoidal shape (Figure 15.7). Figure 15.8 shows the maximum pumice isopleth map of an early phase of the 1875 eruption of Askja, Iceland, which was formed by a weak column (estimated height 14 km) in a strong wind (about 30 m s^{-1}).

15.2.2 Maximum Clast Method

The predicted isopleths of Carey and Sparks (1986) allow the column height and the wind strength to be estimated for ancient eruptions based on the geometry of clast isopleth patterns. Figure 15.9 shows plots of maximum cross-wind and downwind range for particles of 0.8 cm diameter and density of 2500 kg m^{-3}. A single point defines the column height (y-axis) and wind speed by interpolation between curves labelled from 0 to 30 m s^{-1}. Column heights and wind speeds are estimated for several particle sizes which are averaged to derive the source plume characteristics. Carey and Sparks (1986) used a

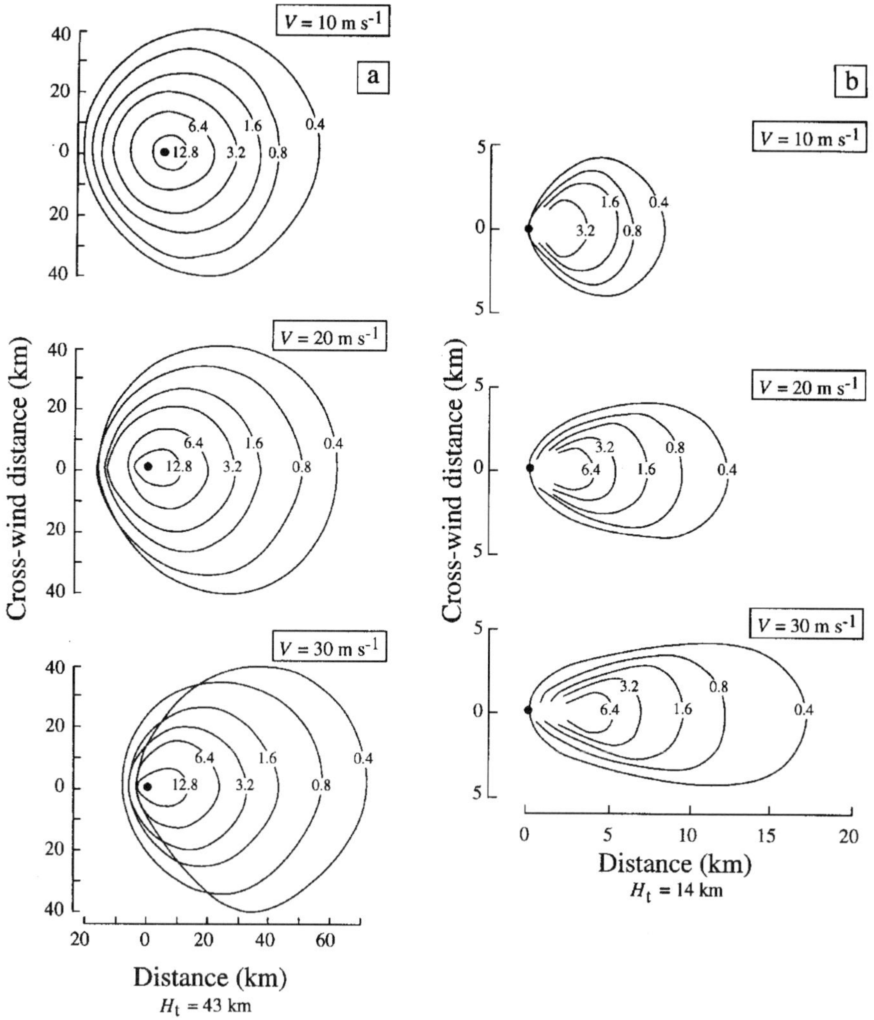

Figure 15.7 Model maximum lithic clast isopleth maps for (a) 43 and (b) 14 km high columns in winds of different strength (after Carey and Sparks 1986). The wind velocities are the maximum in the profile depicted in Figure 15.6. Contours are for clast diameters in centimetres with density 2500 kg m^{-3}

single wind profile in all model runs. There is, in fact, no unique solution to the wind speed because the vertical profile in the atmosphere is a function of geographic position and the particular day on which the eruption occurred. That is why wind speed must remain a semi-quantitative or qualitative interpretation based on the modelling results.

Wilson and Walker's (1987) method is portrayed in a different manner (Figure 15.10). They plot their results in terms of the product of $d\sigma$, known as the hydraulic size, where d is particle diameter and σ is particle density, and the cross-wind range. The solid curves on

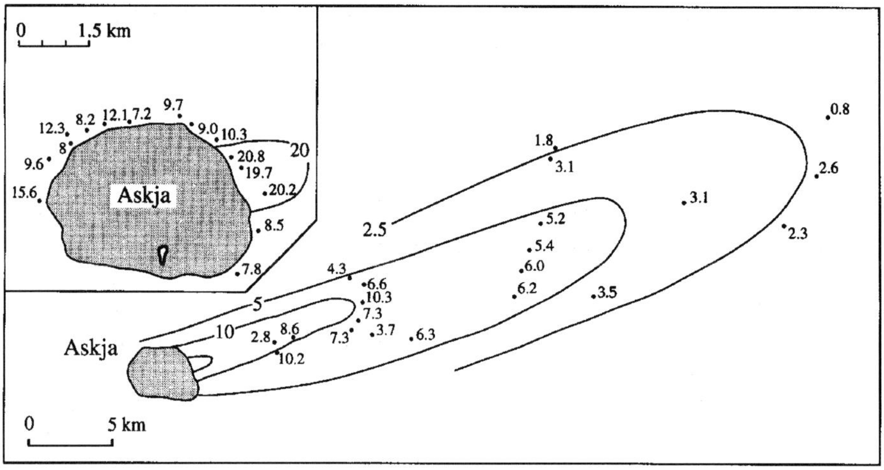

Figure 15.8 Maximum pumice clast isopleth map of layer C of the 1875 Askja eruption in Iceland (after Sparks *et al.* 1981) showing strong elongation of contours in downwind direction. Contours are in centimetres

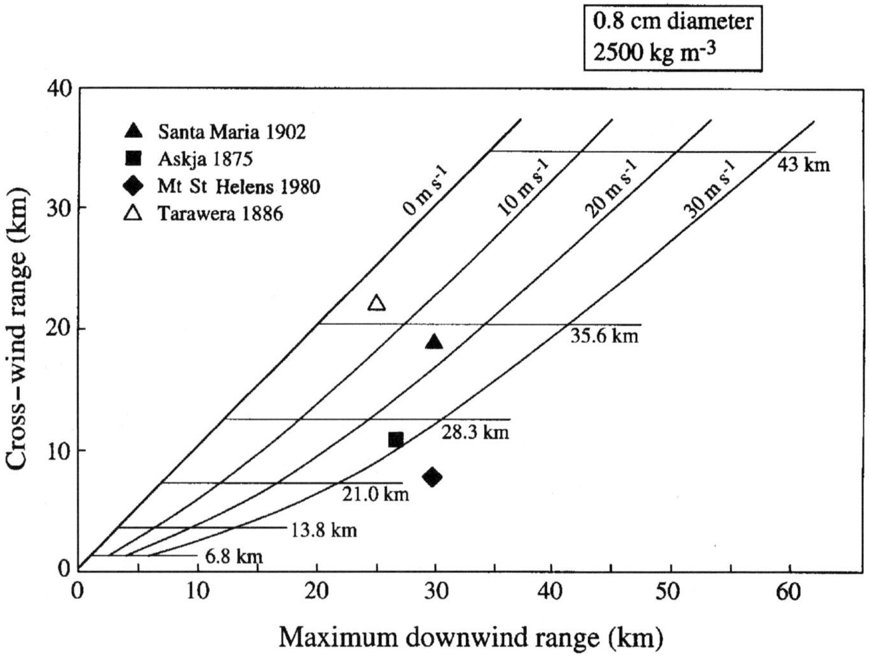

Figure 15.9 Relationship of maximum cross-wind and downwind ranges of 0.8 cm diameter, with density of 2500 kg m^{-3}, lithic clasts as a function of column height (indicated by horizontal lines) and wind velocity. Four examples of data from historic eruptions are shown for comparison. Observations on these eruptions give good agreement with the model. Similar diagrams for different clast diameters and densities can be found in Carey and Sparks (1986)

Figure 15.10 are the modelling results for different mass eruption rates. Column height is then related to mass eruption rate using the results from Chapter 4:

$$H_t = 0.236(M_o)^{1/4} \tag{15.4}$$

where H_t is column height in kilometres and M_o is the mass eruption rate in kilograms per second. Application of this method is limited to clasts whose hydraulic size lies above the inclined dashed line in Figure 15.10. The region above this line corresponds to particles that have fallen out of the lower and middle portion of an eruption plume. Smaller particles (below dashed line) are inferred to have travelled up into the umbrella region and Wilson and Walker's model does not treat the transport in this area.

Wilson and Walker (1987) treat wind speed estimates on an individual basis using atmospheric profiles that are appropriate for specific volcanoes. Downwind ranges are calculated using column spreading and atmospheric transport by the specified wind profile. These calculated distances are then compared with measured downwind ranges for several particle sizes in order to arrive at a wind factor. This factor is the average amount that the specified wind profile must be multiplied by to best fit the observed data. The wind factor is found by taking the ratio of the difference between the downwind and cross-wind range for the calculated and observed isopleths.

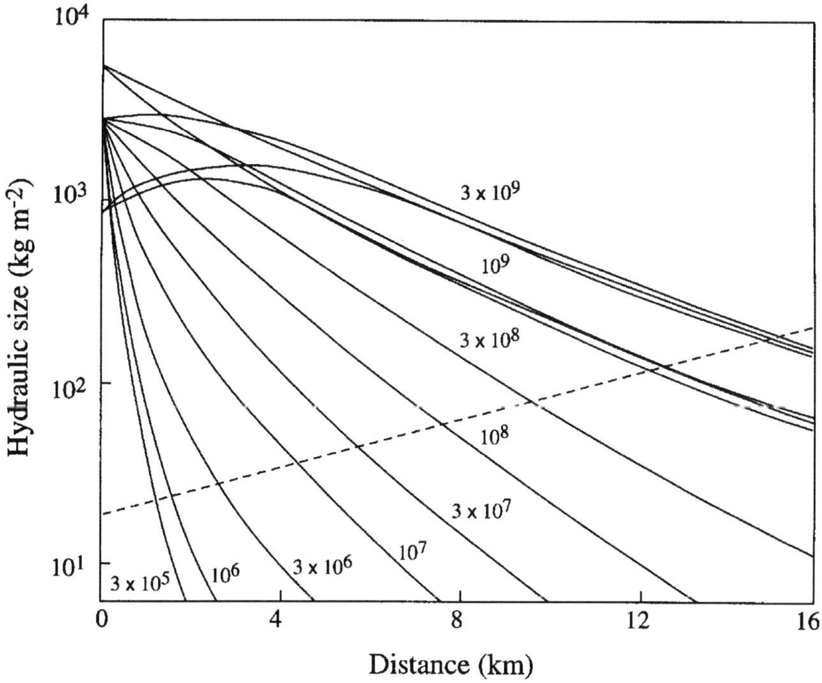

Figure 15.10 Hydraulic size in kilograms per square metre is plotted against cross-wind distance for different mass discharge rates in kilograms per second (after Wilson and Walker 1987). See text for details of significance of dashed line

15.2.3 Evaluation of Maximum Clast Method

The performance of models of clast dispersal can be evaluated from historical eruptions for which observations of column height and/or mass eruption rate are available. The number of eruptions that can be used is limited. There is in particular a lack of data from very large eruptions as a consequence of their infrequent occurrence. Eruptions in excess of VEI 7 occur about once every 250 years. Table 15.1 lists eight eruptions that have maximum clast data, observations of eruption column height and estimates of mass eruption rate. The list is limited to sustained Plinian-style events for which the models have been developed.

The May 18, 1980 eruption of Mount St Helens, Washington, USA, provides the best data set because column height was monitored by radar. Carey *et al.* (1990) determined the clast distribution for four levels within the May 18, 1980 fall deposit and correlated these levels to specific intervals during the eruption based on column height observations, grain size variations and lithological parameters. Column heights inferred from clast dispersal data show very good agreement with the radar-determined heights (Figure 15.11, Table 15.1). The Wilson and Walker (1987) model could only be applied to the B2 and B4 phases of the eruption. In both cases the inferred column height is less than the observed (Table 15.1). A systematic difference between the two models is that, for a given column height, the model of Carey and Sparks (1986) yields a lower mass eruption rate than that of Wilson and Walker (1987). This is related to an assumption of a higher efficiency of thermal energy utilization in the Carey and Sparks model.

Column height estimates for the March 2, 1982 eruption of El Chichon in Mexico and the November 13, 1985 eruption of Nevado del Ruiz in Colombia are based on the detection of a stratospheric aerosol layer by lidar measurements. Both models show very good agreement between the inferred column height from clast dispersal and the measured column height based on the aerosol layer (Table 15.1). The Ruiz eruption was relatively short-lived and the volume of erupted products and the duration can be used to independently check the estimates of mass eruption rate. Naranjo *et al.* (1986) calculated an average eruption rate of 3.0×10^7 kg s^{-1} based on the volume and duration. This com-

Table 15.1 Comparison of observed and inferred column heights for selected historic eruptions (column heights in kilometres)

Eruption	Observed column height	Carey and Sparks (1986)		Wilson and Walker (1987)	
		Column height	M_o	Column height	M_o
Mt St Helens (1980)					
B1	15.4	16.5	6.3×10^6		
B2	17.5	18.0	1.3×10^7	16.0	2.0×10^7
B4	19.0	16.0	3.6×10^7	16.0	2.0×10^7
El Chichon 1982 (B)	30.0	31.6	1.9×10^8	31.0	3.0×10^8
Nevado del Ruiz	29.0	31.0	5.0×10^7	28.0	2.0×10^8
Santa Maria 1902	28–48	34.0	2.2×10^8	32.0	3.2×10^8
Quizapo 1932	30.0	32.0	2.0×10^8	28.0	2.5×10^8

[a] M_o = mass eruption rate (kg s^{-1}).

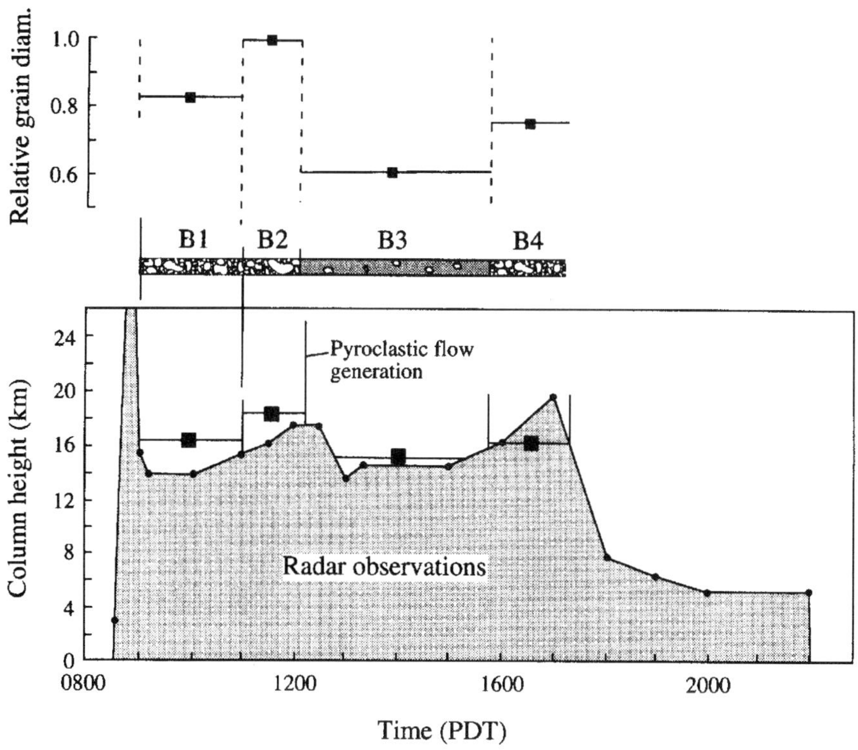

Figure 15.11 Column height variations with time for the Plinian phase of the May 18, 1980 Mount St Helens eruption, comparing observations from radar with height estimates from maximum clast size data for each of the four fall units (B1–B4) of the deposit (after Carey *et al.* 1990)

pares well with 5.0×10^7 kg s^{-1} peak mass eruption rate derived from clast dispersal data (Table 15.1).

The eruptions of Quizapu, Chile, in 1932 and Santa Maria, Guatemala, in 1902 provide two additional tests of the dispersal model, although the observational data on column height is of lower quality than the more recent eruptions. Both events were large Plinian eruptions that produced thick pumice fall deposits in proximal regions of the source vents. Hildreth and Drake (1992) provide a thorough analysis of the Quizapu Plinian eruption. A column height of 30 km was measured from a photograph. Both models provide good agreement with the observed column height for this event (Table 15.1). During the Santa Maria eruption, column height was estimated by two sets of observations. One placed the column between 27 and 29 km, whereas the other estimated a height of 48 km. Because of this discrepancy it can only be said that the inferred column heights based on clast dispersal fall within the very broad range of column height observations for this event.

In summary, the agreement between the models and observations are good considering all of the assumptions that go into the models and uncertainties in some of the column height observations. The methods of Carey and Sparks (1986) and Wilson and Walker (1987) produce roughly similar results except for some systematic differences explicable by initial assumptions of the respective models. Inferred column heights have, however, considerable uncertainties, perhaps plus or minus several kilometres, because of the subjective nature of isopleth contouring and the simplifying assumptions made in the models. The accuracy of the models as applied to eruptions with columns greater than 35 km remains untested because of the lack of well-documented historic eruptions of this scale.

The method also allows estimates of wind velocity. Carey and Sparks (1986) adopted a vertical wind profile similar to that of Shaw *et al.* (1974) in which the velocity increases linearly to a maximum velocity at 11 km altitude, then decreases to 0.75 of this maximum at 13 km and remains constant at greater heights (Figure 15.6). Comparison of the maximum wind speed with observation is quite good although the number of examples is still small. The method estimates a wind speed of 36 m s^{-1} for the 1980 Mount St Helens Plinian eruption which compares with radiosonde measurements near the tropopause of 27–33 m s^{-1} (Danielson 1981). These estimates can only be semi-quantitative because actual wind velocity profiles may be significantly different to those adopted in any model.

15.2.4 Application to Plinian Eruptions

Eruption duration

Column height estimates from maximum clast data allow estimates to be made of discharge rates (equation 15.4 and see Chapters 4 and 5). If ejecta volumes are known then the duration of an eruption can be calculated (Carey and Sparks 1986). There are, however, a number of pitfalls. Most isopleth maps do not discriminate between different levels in the deposit. Many Plinian fall deposits are reversely graded and thus the maximum grain size data provide information on peak discharge rates rather than average rates. Use of the peak discharge rate rather than the average results in an underestimate of the duration. Some eruptions are discontinuous so that the periods of weak or no activity are not taken into account. For example, Carey and Sparks (1986) estimated 13 hours for the Fogo 1563 eruption in the Azores, whereas the eruption lasted two days (Walker 1981c). This particular deposit is strongly stratified implying large fluctuations in discharge rate with maximum clast data only indicating peak rates. Finally, as discussed in Chapter 13, volumes of ejecta calculated from isopach maps can be significant underestimates.

Intensity variations of Plinian eruptions

Carey and Sigurdsson (1989) applied the maximum clast method to 45 Plinian eruptions of late Quaternary age to evaluate the range of eruption column heights based on the geological record. Column heights vary from 13 to 51 km. The largest estimated height was attained by the AD 180 Taupo eruption in New Zealand, which is close to the theoretical maximum height of 55 km predicted for eruption columns by Wilson *et al.* (1978).

There is a good correlation between the column height (or mass discharge rate) and the total amount of magma discharged (Figure 15.12). The correlation may reflect the important influence of magma chamber size. Models of volcanic eruptions envisage an initial overpressured chamber which increases in pressure with time either due to slow

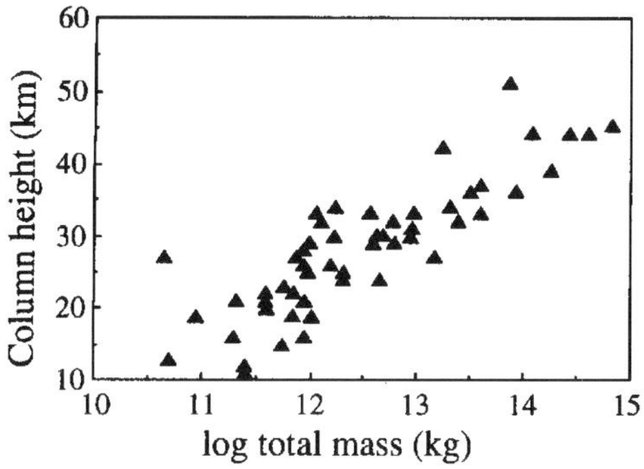

Figure 15.12 Variation of column height of the Plinian phase and total erupted mass in explosive eruptions. The column heights were estimated from the maximum clast size method. Data sources can be found in Carey and Sigurdsson (1989)

crystallization (Tait *et al.* 1989) or to replenishment (e.g. Sparks *et al.* 1984). When the overpressure exceeds the wall rock strength an eruption initiates and the volume of erupted material depends on both conduit evolution and chamber size. As magma is erupted, chamber pressure must decrease (Druitt and Sparks 1984; Stasiuk *et al.* 1993), provided that there is no further replenishment of the chamber from depth. For a given pressure decrease a large chamber can erupt larger volumes of magma and can sustain a more intense eruption. In addition, if the magma chamber becomes gas saturated then the decay of pressure with time is much slower because the exsolution of gas counteracts the loss of volume by eruption. The reasons why explosive eruptions stop are not well understood. However, plausible explanations are that when the chamber pressure drops too low, caldera collapse occurs closing off the conduit, or that the chamber is completely exhausted.

Temporal variations of column height

Plinian fall deposits commonly exhibit variations in particle size as a function of stratigraphic height. Many are reversely graded, whereas normal grading is surprisingly uncommon. Reverse grading has been interpreted as due to an increasing eruption intensity with time (Lirer *et al.* 1973; Wilson *et al.* 1980; Pescatore *et al.* 1987). Vertical variations of particle size at a particular depositional site are controlled by column height and the wind field (Wilson *et al.* 1980). If it can be demonstrated that wind intensity and direction did not fluctuate significantly during an eruption, then the maximum clast methods can be used to reconstruct the variations in eruption column height. Maximum clast isopleth maps are constructed for isochronous horizons within a deposit. Carey and Sigurdsson (1987) investigated column height variation of the AD 79 Plinian eruption of Vesuvius. Figure 15.13 shows the variations of pumice and lithics clasts in the fall deposit at Pompeii. Both type of clasts show strong reverse grading in passing from the lower white pumice fall deposit up into the overlying grey pumice fall deposit. Clast size then decreases abruptly as pyroclastic surge deposits enter the stratigraphy. Figure 15.13 shows column height evolution during the eruption as inferred from the dispersal of lithic clasts at different levels in the AD 79 fall deposit. Column height increased steadily from about

15 km at the beginning of the event to a maximum of 32 km during eruption of grey pumice. Towards the end of the event the column height diminished to a value similar to that at the start. Also shown on Figure 15.13 is the inferred magma discharge rate variation calculated from the inferred column heights. The results suggest that there was at least an order of magnitude change in the discharge rate during the Plinian phase of the eruption. Pescatore *et al.* (1987) carried out a similar analysis on the 3500 a Avellino pumice fall deposit of Vesuvius. This deposit also exhibits strong reverse grading and a compositional shift from white pumice at the base to grey pumice at the top. Application of the maximum clast method indicates that during eruption of the basal white pumice the column height increased from 18 to 27 km. A peak column height of 31 km was attained during eruption of the overlying grey pumice. The evolution of column height and magma discharge rate is therefore very similar to the AD 79 eruption.

The common increase in discharge rate with time in Plinian eruptions contrasts with the decrease of discharge rate with time in many lava eruptions (Stasiuk *et al.* 1993). Reverse grading has been attributed to conduit erosion which results in an increase of discharge rate with time (Wilson *et al.* 1980; Pescatore *et al.* 1987). However, the lithic clasts in Plinian deposits predominantly come from shallow levels (i.e. typically less than 1–2 km) and there is a paucity of lithics derived from the deeper parts of the conduit connecting the magma chamber at several kilometres depth with the surface (e.g. Varekamp 1993). While these observations support vent widening they do not support erosion at deeper levels. Therefore conduit widening is unlikely to be a major control on increase of discharge rate.

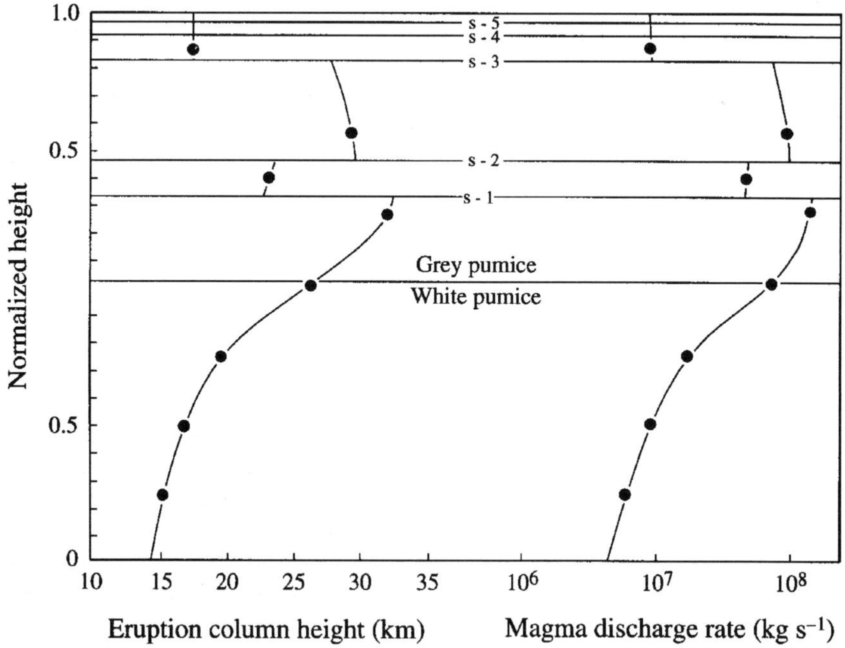

Figure 15.13 Variation of column height and magma discharge rate with stratigraphic height in the AD 79 Plinian deposit of Vesuvius inferred by applying the maximum clast size method to particle size data (after Carey and Sigurdsson 1987). The horizontal lines indicate the positions of surge deposits, s-1 to s-5, in the fall stratigraphy

A more plausible explanation is that during the eruption the region of fragmentation increases in depth and gas exsolution progressively decreases the density of the magma column. Both these effects decrease the hydrostatic head of the magma column with time. The viscous friction is also reduced as the fragmentation level falls. If the decrease of pressure head is faster than the decrease of chamber pressure then an increase in driving pressure gradient can be expected. Together with the reduced friction an increase in discharge rate is predicted. A fuller discussion of the underlying concepts is provided in Chapter 3.

Dynamics of magma withdrawal

Magma discharge rates derived from column height estimates can be used to investigate the dynamics of magma withdrawal from a chamber. Fluid dynamical modelling of magma withdrawal has demonstrated that different compositional zones of a magma chamber can be tapped simultaneously and mixed during explosive eruptions (Blake 1981; Spera *et al.* 1986; Trial *et al.* 1992). Apart from magma density and viscosity contrasts, eruption intensity is the most important parameter which determines the evacuation volume and hence the potential for mixing during magma withdrawal. Intensity variations during Plinian events can therefore be linked with compositional gradients exhibited in fall deposits to constrain magma withdrawal processes. A quantitative model of magma withdrawal during the AD 79 Plinian eruption of Vesuvius was developed by Sigurdsson *et al.* (1990). Figure 15.14 shows the relationship between eruption intensity, pumice density and Ba and Sr contents of pumice. There is strong geochemical evidence that the trends in Ba and Sr reflect mixing between two magma types. The abrupt shift to grey pumice about half-way through the deposit is interpreted as draw-up of denser mafic magma (grey pumice) through an overlying layer of more evolved magma (white pumice) as a result of the increasing magma discharge rate (Figure 15.14). Fluid dynamic models predict that withdrawal of magma from a two-layer stratified reservoir can result in drawing up of magma from the lower layer through the upper layer when a critical combination of magma discharge rate and contrasts in density and kinematic viscosity are attained.

Column stability

Magma discharge rate plays an important role in the behaviour of volcanic plumes. Models of volcanic eruption plumes indicate that the transition from a convecting plume to a collapsing fountain is controlled by a combination of magma discharge rate, volatile content and vent radius (Chapters 4 and 6). Many ignimbrite-forming eruptions show a characteristic stratigraphy which begins with a Plinian fall deposit and is followed by pyroclastic flow deposits. This is the geological manifestation of the transition from the convective to collapsing regime. In order to evaluate the theoretical models for column behaviour the critical parameters at the transition need to be estimated. Analysis of the 1980 Mount St Helens eruption (Carey *et al.* 1990) and the AD 79 eruption of Vesuvius (Sigurdsson *et al.* 1985; Carey and Sigurdsson 1987) has shown that the inferred magma discharge rates are consistent with the evolution of volcanic processes for each event when compared to model predictions of eruption column stability (e.g. Wilson *et al.* 1980). The

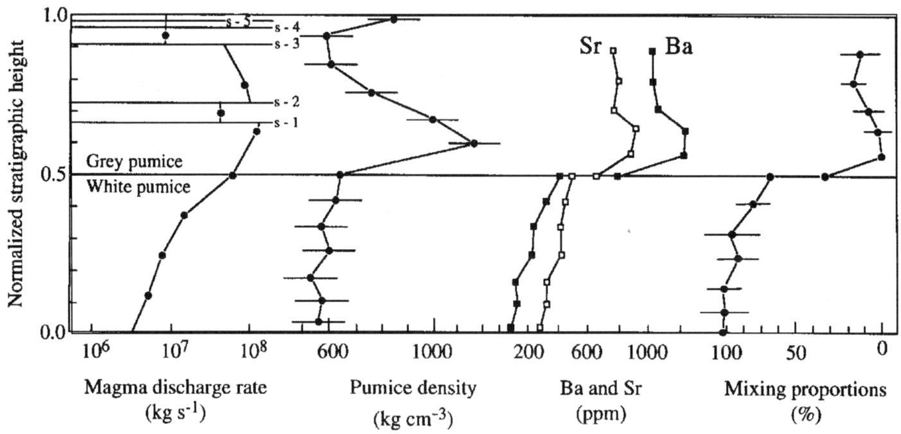

Figure 15.14 Variation of magma discharge rate, pumice density, and magma composition (represented by Ba and Sr) with height in the AD 79 Plinian deposit of Vesuvius after Carey and Sigurdsson (1987). The inferred proportions of the white pumice in the ejecta are shown and a fuller discussion of these estimates is given in the text

overall contrasting behaviour of the two eruptions can be attributed to differences in the initial volatile content and physical properties of the magmas.

Many more eruptions will be needed before any clear picture emerges about the stability conditions of eruption plumes. There are many simplifications which go into the theoretical models and there are uncertainties in determining the volatile contents of magmas. However, it is encouraging that some first-order observations of eruption plume behaviour are consistent with theoretical models.

15.3 APPLICATION OF PLUME SEDIMENTATION MODELS

Widespread tephra fall deposits are formed primarily by deposition of particles from the umbrella region and downwind plumes. The application of sedimentation models from Chapter 14 are now applied to the quantitative interpretation of the thickness variations and grain size distributions of fall deposits. These models are still in an early stage of development because of the complex processes of plume sedimentation (Chapter 14). Models have so far been largely concerned with symmetrical plumes in the absence of wind and with particles of uniform density. Little work has been done on the effect of wind with the exception of Bursik *et al.* (1992a). As yet these models cannot be applied too literally to the quantitative analysis of fall deposits. The new models, however, provide a conceptual framework for understanding the fundamental physical controls on thickness and grain size distributions, but direct comparison of model calculations with real data to estimate eruption conditions is perhaps premature.

15.3.1 Thickness Variations

Two approaches have been developed to interpret the thickness variations in pyroclastic fall deposits quantitatively. In advection–diffusion models (described in Chapter 14) a source function is specified and the deposition is modelled by numerical solutions to equations describing atmospheric diffusion and the wind velocity field. Good agreement between thickness observations and model predictions has been achieved (e.g. Hopkins and Bridgman 1985; Armienti *et al.* 1988; Macedonio *et al.* 1988, 1990). However, these models are highly empirical because several coefficients have to be specified for the source function, for atmospheric diffusion in horizontal and vertical directions, for the time dependence of atmospheric diffusion and for wind velocity variations with height in the atmosphere. Some of these empirical coefficients are not well constrained and have little relationship to dynamic conditions in eruption columns. Good agreement of predictions with observations can be achieved by best-fit solutions to the coefficients. Atmospheric diffusion may not be an appropriate way to model plume dispersal (Bursik *et al.* 1992b; see Chapter 14). Advection–diffusion models, however, have the advantage that large numbers of simulations can be run for many different atmospheric conditions and thus provide a powerful method for making volcanic risk assessments (e.g. Macedonio *et al.* 1990; Barberi *et al.* 1990). Figure 15.15 shows a simulation of the May 18, 1980 Mount St Helens Plinian fall deposit by Armienti *et al.* (1988) which shows very good agreement with the observed isopach map.

The second approach has been to develop simple sedimentation models for plume dispersal (Chapter 14). These models are less empirically based and relate deposit characteristics to important parameters of the eruption such as column height, column shape and wind velocity. They are, however, less developed than the advection–diffusion models. For example, they do not yet incorporate entrainment and they assume that the plume remains coherent.

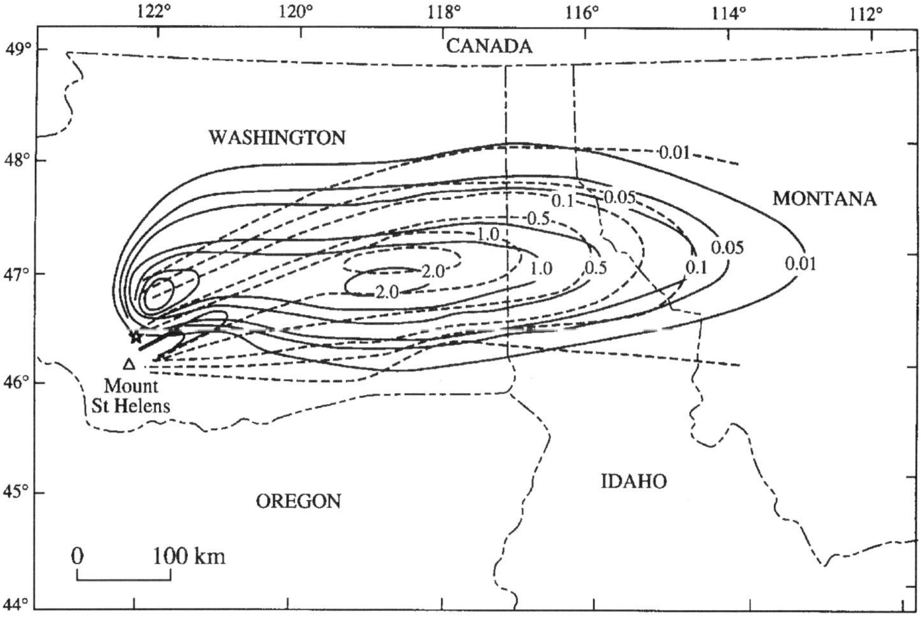

Figure 15.15 Simulation of the isomass map of the May 18, 1980 Plinian deposit of Mount St Helens by an advection–diffusion model (solid contours, after Armienti *et al.* 1988) in comparison with the observed deposit (dashed contours). Contours are in centimetres

Sparks *et al.* (1992) used the sedimentation models of the column margin and umbrella region to predict the accumulation of pyroclasts away from source for comparison with the characteristics of natural fall deposits. Numerical calculations were carried out using four different grain size distributions that approximate the range of natural eruptions. These calculations were done for an axisymmetric plume spreading in an environment with no wind. Figure 15.16 shows results for model calculations with column heights ranging from 10 to 35 km. The distinctive feature of these results is the marked break in slope that occurs in the proximal region for column heights greater than 15 km. The break-in-slope corresponds to the corner where the dominantly vertical motion of the eruption column is converted to largely horizontal motion in the umbrella region. The inner region, where fallout is from the column margin, has a steeper slope in accumulation than the area underlying the umbrella region (Figure 15.16).

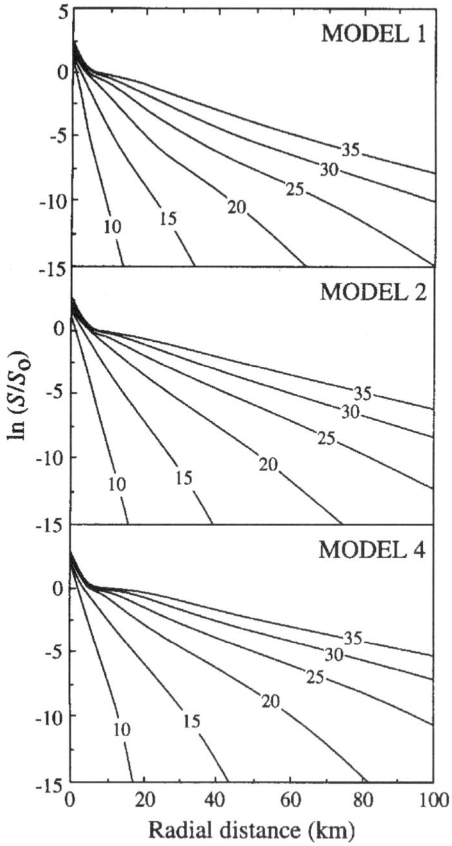

Figure 15.16 Model calculations of the variation of mass accumulation rate of ejecta, which is proportional to deposit thickness, as a function of distance from the vent for a range of column heights (labelled on the curves) from 10 to 35 km. The calculations are for wind absent conditions and use a total grain size distribution of ejecta which are typical of Plinian deposits. Note the break in slope at the distance where the model changes from sedimentation from the margin of the rising plume to sedimentation from the umbrella region. Further discussion of model assumptions and examples of calculations for different initial particle size distributions can be found in Sparks *et al.* (1992)

The results indicate that in the distance range 10–60 km the thickness decreases approximately exponentially with distance in accord with the behaviour of numerous fall deposits (Pyle 1989; Chapters 13 and 14). However, the results are not exactly exponential and the finest grain size distributions modelled by Sparks *et al.* (1992) can show significant departures. Koyaguchi (1994) has developed analytical solutions to the sedimentation model of Sparks *et al.* (1991) and has shown that true exponential behaviour can only occur for total particle size distributions with an explicit form, which is close to log normal with a variance (σ) of about 2.5. Distributions that differ from his analytical results cannot produce exponential thickness decay. The model distributions in Sparks *et al.* (1992) are based on empirical determinations on the total particle size distribution of ejecta. They are, in fact, quite similar to Koyaguchi's analytical results, but demonstrate that exponential thinning is not required by these models. The models of Sparks *et al.* (1992) are also not appropriate for particles falling at low Reynolds number. The models assume that the particles fall at high Reynolds number which is not correct for particles less than approximately 100 μm in diameter falling at sea-level and 500 μm in diameter at stratospheric altitudes. The sedimentation of fine-grained low Reynolds number particles is likely to yield markedly different thinning patterns, but models of this have not been published yet.

Exponential decay behaviour of tephra fall deposits still holds if data are plotted as thickness versus the square root of area, even if the fallout pattern has been highly distorted by wind (Pyle 1989). This suggests that the depositional process is not strongly affected by the wind and that only the geometry of the fallout area is modified relative to the no wind situation. Thus it is possible to compare model results computed for the no wind condition with actual field data. Figure 15.17 shows data for the Askja 1875 fall deposit in Iceland. Also plotted on this figure (small open circles) are model results for a 25 km high eruption column. The height was chosen based on independent estimates using the maximum clast technique (Carey and Sparks 1986). The agreement between the observed and the model is excellent considering that the wind effect is assumed to be negated in this type of plot. The distinctive break-in-slope of the Askja data can thus be attributed to transition from the inner column-dominated depositional regime and the outer umbrella cloud fallout. These results provide the first physical basis for an explanation of approximately exponential thickness decay in fall deposits.

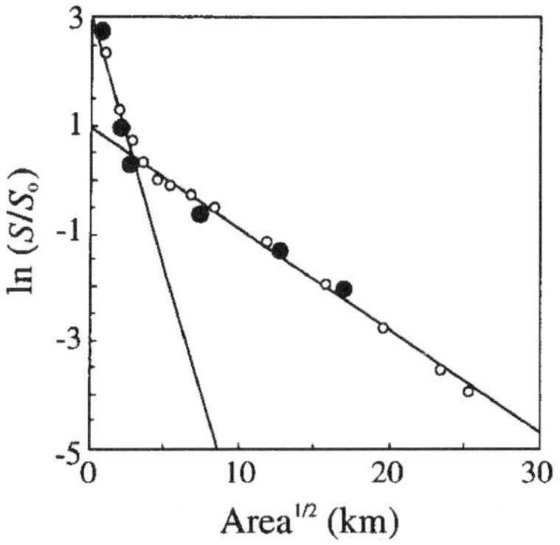

Figure 15.17 Model calculation of the variation of thickness against distance for a 25 km high column (open circles) is compared with data for the variation of thickness with the square root of the isopach area for the main Plinian layer D of the 1875 eruption of Askja, Iceland (closed circles)

Field measurements on tephra fall deposits are always compromises and approximations. Thicknesses fluctuate locally and inaccuracies in measurement can be caused by soil formation and downslope creep. The assumption that mass accumulation per unit area is equivalent to thickness is clearly an approximation since deposit porosity and the density and proportions of the components vary laterally. In the future, thickness may prove to be too crude a measurement for comparison with theoretical models. *In situ* bulk densities, individual component densities and proportions are needed to estimate masses per unit area.

15.3.2 Particle Size Variations

Tephra fall deposits show systematic relationships in their particle size characteristics (Walker 1971). These characteristics are commonly expressed in terms of maximum clast diameter, Md_ϕ (median diameter) and σ_ϕ (sorting coefficient). The particle size characteristics will be a function of eruption column height, wind strength, initial particle size and density distributions and the component proportions of pumice, crystals and lithics. Some insights can be gained into the grain size features of fall deposits by modelling the process with some simplifying assumptions. Sparks *et al.* (1992) modelled the grain size distribution of fall deposits using particles of uniform density and the condition of no wind. The models do not take account of low Reynolds number fall of fine particles and aggregation processes that have a large influence on the transport of fine ash (see Chapter 16).

Figure 15.18 shows modelled variations in median diameter and sorting coefficient for different column heights as a function of distance from source. Median diameter decreases exponentially away from source, as is observed (Pyle 1989; see Chapter 13). The slope of the decay is primarily a function of eruption column height, but total particle size distribution also plays a role. The decrease in median diameter is accompanied by a decrease in the sorting coefficient, indicating better sorting with distance from the vent. The sorting behaviour predicted by the model is not typical of many fall deposits. Most have sorting coefficients between 1.5 and 0.6, whereas the model predicts values between 2.0 and 1.4. This discrepancy is attributed to the effect of wind on the falling ash. In the proximal area the wind will act to winnow out fine ash, resulting in an improved degree of sorting compared to the model. The model results compare well, however, with eruptions which are inferred to have taken place in the presence of very weak winds. Figure 15.19 shows Md_ϕ and σ_ϕ data for the Fogo A and the Palulahua Plinian deposits. Both eruptions exhibit roughly circular patterns of clast dispersal and were thus not strongly influenced by wind. The sorting coefficients are similar in magnitude to the modelling results of Sparks *et al.* (1992) and display a decreasing trend with increasing distance from source.

Bursik *et al.* (1992b) have recently applied sedimentation theory (Chapter 14) to an analysis of the Fogo A Plinian deposit, San Miguel, Azores. The deposition in this powerful eruption in a weak wind should be controlled mainly by the strength of the source plume and the total particle size distribution of the ejecta. Mass accumulation data were collected for specific particle size intervals at different localities downwind from the source. The accumulation curves show maxima at distances between 6 and 8 km with only a slight dependence on particle size. Position of the maxima has been related to the area where the motion in the source plume changes from dominantly vertical (convective

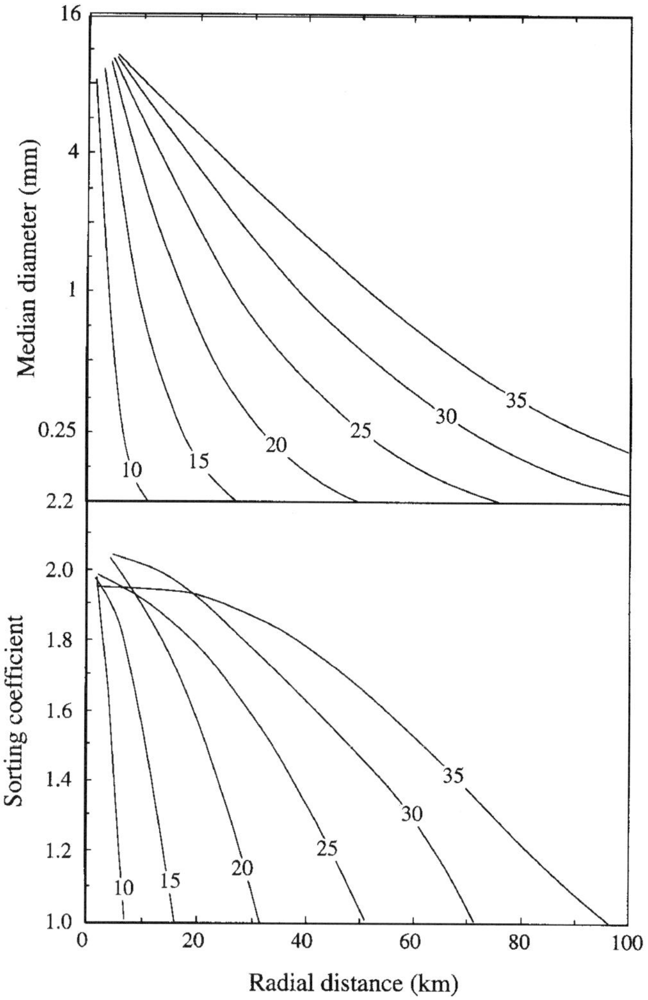

Figure 15.18 Variations of median diameter and sorting coefficient with distance from source for different column heights (10–35 km) using an initial particle size distribution typical of Plinian events. Details and assumptions of the models can be found in Sparks *et al.* (1992)

region) to horizontal (umbrella region). Beyond the maxima, each of the accumulation curves decays in accord with the general sedimentation theory (Figure 15.20). The slopes of the decay curves can be used to calculate the flow rate at the level of neutral buoyancy and the column height. For the upper part of the Fogo A deposit a column height of 21 km can be deduced from the accumulation data. This value is somewhat less than the column height of 27 km obtained independently for the Fogo A deposit based on maximum clast dispersal (Carey and Sparks 1986; Bursik *et al.* 1992b). The discrepancy could be due to the model involving all particles giving an average height, whereas the maximum clast size method gives a maximum height.

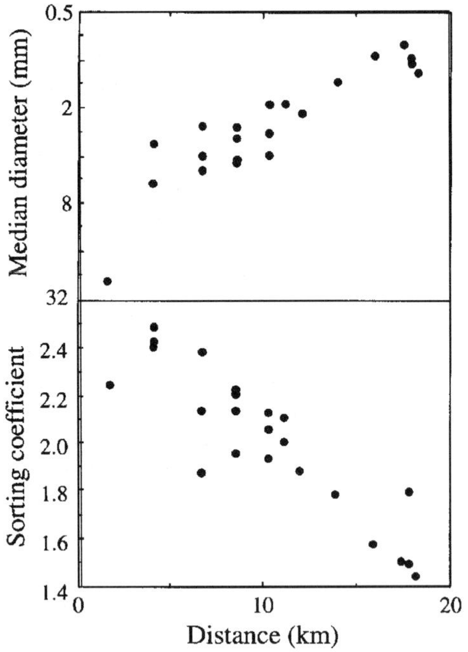

Figure 15.19 Data on the variation of median diameter and sorting coefficient for the Fogo A and Palulahua Plinian deposits (Papale and Rosi 1993), showing decrease in size and improvement in sorting with distance from the vent

15.3.3 Emplacement Temperature and Welding

Pyroclasts ejected into the atmosphere during explosive eruptions leave the vent at high temperatures (700–1200 °C). As they travel up in a convective plume and then fall back to the surface, they lose energy quickly to the surrounding atmosphere by convective and conductive heat transfer. Far from the vent, particles reach the ground close to ambient

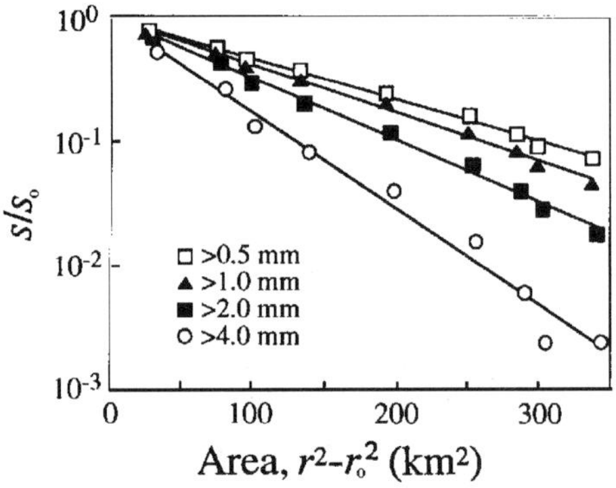

Figure 15.20 Total mass accumulated, S, beyond r_o divided by the total injected mass, S_o, versus the area beyond r_o, $r^2 - r_o^2$, for the Fogo A fall deposit

temperature. In areas close to the vent, however, some clasts reach the ground still hot. These hot clasts can pose a significant hazard due to the potential for injury to humans and animals and their ability to start fires. If conditions are suitable, then welding of the deposit can occur. Welding is a process by which hot, individual clasts are sintered together and vesicular glassy particles densify to form a dense rock. Upon close examination, however, individual clasts can still be recognized, although their shapes are typically deformed and flattened parallel to bedding by compaction. This deformation of juvenile clasts produces the distinctive eutaxitic rock texture. Welding is a relatively common feature of pyroclastic flow deposits which are emplaced at high temperature and can accumulate to several tens or hundreds of metres in thickness. Welded tephra fall deposits are restricted to proximal locations. Examples of welded fall include deposits of Santorini and Askja volcanoes (Sparks and Wright 1979) and on the island of Pantelleria (Wright 1980).

Thomas and Sparks (1992) have examined pyroclast cooling during fallout from convective eruption columns and the formation of welded tephra fall deposits using theoretical modelling. They assume clasts cool predominantly after they leave an eruption plume because the time spent in the rising plume is so much less than the fall time to the ground. Cooling during fallout occurs by a combination of forced convection by the atmosphere around the clast, radiation from the hot clast surface and conductive cooling within the clast. The temperature of a clast when it arrives at the ground is a function of its initial temperature, size, density, and height from which it fell. Figure 15.21 shows the results of heat loss calculations for various size pyroclasts released from elevations varying from 1 to

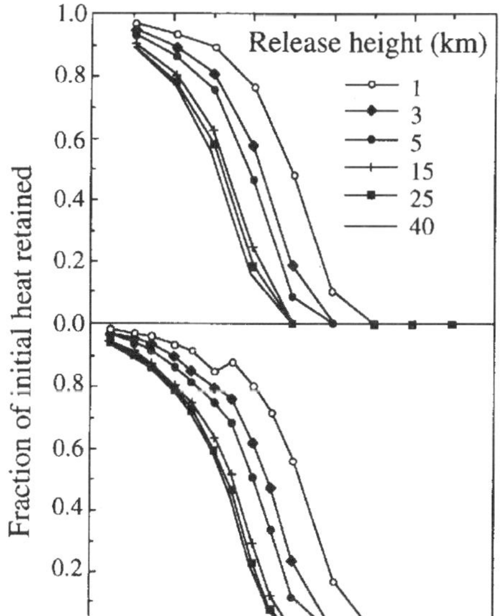

Figure 15.21 Heat loss represented as the fraction of the original temperature for clasts of different size as a function of the release height from the eruption column. Details of the calculation method can be found in Thomas and Sparks (1992)

40 km. Large clasts (> 25 cm in diameter) retain a large fraction of their initial heat even if they fall from altitudes as high as 40 km. Clasts between 25 and 1.6 cm in diameter lose significant amounts of heat depending on their size and release height. Small fragments (< 1.6 cm in diameter) lose virtually all of their heat no matter what the release height and consequently arrive at the ground cold. The particle size distribution of many Plinian eruptions contain a significant amount of material within the size range 25–1.6 cm diameter and thus the temperature of deposition will be a strong function of the release height of ejecta and total column height.

Thomas and Sparks (1992) utilized the plume sedimentation model of Bursik *et al.* (1992b) in conjunction with a cooling model for deposited particles in order to investigate the conditions that yield welding. The criterion they used for welding rhyolitic to dacitic clasts was to have the deposit temperature equal to or greater than 585 °C. Figure 15.22 shows calculated deposit temperatures for column heights of 5–20 km and particle size distributions with mean sizes from 0.2 to 6.4 cm. Welding is predicted only for size distributions with a mean size equal to or coarser than 3.2 cm. The area over which welding occurs, i.e. the distance from vent for symmetrical dispersal, increases with increasing column height. A 5 km high column has a predicted welding zone that extends to 500 m from the vent whereas a 20 km high column has a zone that extends to 2000 m. Higher columns disperse larger clasts which retain their heat further.

The modelling results constrain the conditions for welding and interpretation of welded tephra fall deposits. In contrast to the suggestion of Sparks and Wright (1979) it would appear that accumulation rate is not a particularly important factor in determining welding. Welding is most likely to occur for relatively high temperature dacitic and rhyolitic magma (> 850 °C). Many such magma types are erupted at lower temperatures and this may explain the limited number of documented welded fall deposits from Plinian-style eruptions. The zone of welding predicted by the modelling is quite restricted and even for large eruption columns (> 20 km high), welded deposits are not expected to form much further than 2 km from the vent. Deposits that exhibit welding to much greater distances, such as the Green Tuff of Pantelleria (7 km from the vent, Wolff and Wright 1981), probably have

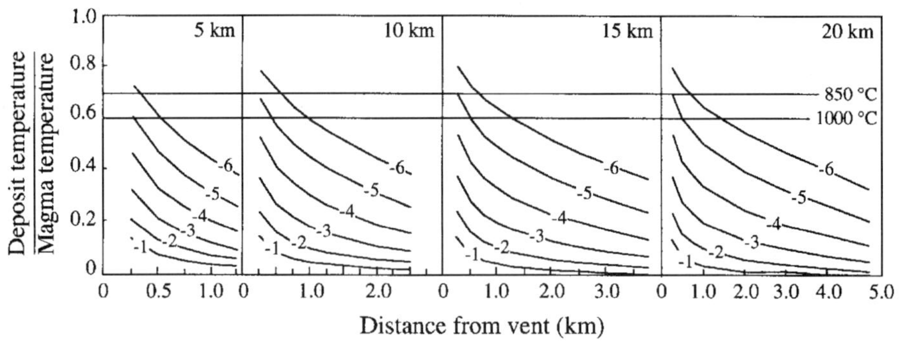

Figure 15.22 Deposit temperature as a function of distance from the vent for grain sizes of −1 to −6 ϕ. Calculations are shown for column heights ranging from 5 to 20 km. Horizontal lines mark the temperature necessary for welding at 850 and 1000 °C. Modified from Thomas and Sparks (1992)

a flow origin instead of fallout as suggested by Orsi and Sheridan (1984) and Mahood and Hildreth (1986).

15.4 SUMMARY

The grain size and thickness variations in tephra fall deposits contain information about the physical conditions in past volcanic eruptions. Methods have been developed to estimate eruption column heights and wind speeds from contoured isopleth maps of maximum grain size (Carey and Sparks 1986; Wilson and Walker 1987). These methods are based on models of the transport of large clasts in eruption columns assuming averaged velocities in the turbulent plumes. In the model of Carey and Sparks (1986) a support envelope is defined for each clast size by the positions where the mean upward plume velocity equals the clast terminal fall velocity. Maximum ranges are calculated as a function of column height and wind velocity. Clast range is a sensitive function of column height and therefore eruption rate. Column heights can be determined to an estimated accuracy of less than 5 km despite uncertainties in the methods of field data collection. Comparison between observations and models shows reasonable agreement.

Column height estimates derived from models of maximum clast dispersal can be applied to understanding the dynamics of Plinian eruptions and to reconstruct prehistoric volcanism for hazards assessments. Estimates of discharge rates and erupted volume allow eruption durations to be calculated, although the method is likely to give peak discharge rates and therefore to underestimate durations. Intensity variations within and between eruptions can be investigated. There is a positive correlation in Plinian eruptions between eruption rate and eruption magnitude which can be linked to the size of the magma chamber. Many Plinian deposits are reversely graded, implying an increase in eruption rate with time. Model-derived eruption rates from Plinian deposits just prior to generation of ignimbrite can be compared with models for the transition between stable convective columns and formation of collapsing fountains and pyroclastic flows (Chapters 4 and 6). The transition conditions derived from the application of the maximum clast dispersal models to deposits show reasonable agreement with the predictions of fluid dynamical models.

Variations of thickness and grain size distributions in tephra fall deposits can be understood in terms of the sedimentation models presented in Chapter 14. So far these models have only been developed for eruption columns in the absence of wind and for large and medium grain sizes that fall at high Reynolds numbers. Thus their application to estimating quantitative eruption parameters from deposits is restricted. Nevertheless, some insights can be obtained on the physical controls on exponential thinning, decrease of grain size with distance, the sorting of tephra fall deposits and emplacement temperatures. Eventually further development of these models should yield quantitative information on average column heights, in contrast to the maximum grain size methods which tend to give peak column heights.

16 Particle Aggregation in Plumes

NOTATION

A	average aggregation coefficient for range of particle sizes (dimensionless)
$c(p)$	concentration of particles of radius p (m^{-4})
$\Delta u(p)$	mean speed of aggregate relative to particles of radius p ($m\ s^{-1}$)
E	bulk aggregation coefficient for particles of radius p (dimensionless)
e	sticking coefficient (dimensionless)
ϕ	aggregate porosity (dimensionless)
f	collision coefficient (dimensionless)
g	acceleration due to gravity ($m\ s^{-2}$)
M	aggregate mass (kg)
$N(p)$	number of particles of radius p
p	radius of particle (m)
R	radius of aggregate (m)
ρ_a	air density ($kg\ m^{-3}$)
ρ_ϕ	pore density ($kg\ m^{-3}$)
ρ_p	particle density ($kg\ m^{-3}$)
ρ_s	aggregate density ($kg\ m^{-3}$)
t	time (s)
u	velocity of aggregate ($m\ s^{-1}$)
w	plume mass loading ($kg\ m^{-3}$)
x	distance (m)
z	height in plume (m)

16.1 INTRODUCTION

Direct observations have shown that many fine ash particles, typically less than 100 μm, fall from volcanic plumes as aggregates (Kittl 1933; Sorem 1982; Gilbert *et al.* 1991). Aggregates fall with higher velocities than their component particles (Lane *et al.* 1993) and this leads to relatively rapid fallout of fine-grained material from plumes. Aggregation can lead to anomalous deposit thicknesses (e.g. Carey and Sigurdsson 1982) and account for the polymodal grain size distribution often observed in these deposits (e.g. Brazier *et al.* 1982).

The mechanics of formation of aggregates are controlled by a complex interplay of physical and chemical processes within the plume involving solid, liquid and gas phases. Aggregates are classified according to the liquid mass fraction present during their formation. *Dry aggregates* have relatively high porosities; the component particles are held together by weak bonds. They collapse on impact with the ground and do not tend to be preserved in the geological record (Sorem 1982; Gilbert *et al.* 1991). *Accretionary lapilli* are of intermediate liquid content. They have relatively low porosities, the constituent particles are held together initially by strong liquid surface tension bonds and later by secondary mineral precipitates and they are often preserved intact in the geological record. They are thought to be particularly common in facies of phreatomagmatic eruptions, which have been associated with magma–(meteoric) water interaction and the generation of abundant fines. *Mud rain*, which represents the liquid-rich end member does not result in preservation of aggregate morphologies in the stratigraphic record.

In this chapter the evidence for particle aggregation is described by combining eye-witness accounts of eruptions with geological field observations of ancient deposits. The dominant mechanisms of aggregate formation are outlined. The way in which experiments and theory have been used to investigate aggregation processes is described. Research into aggregation of volcanic particles is at a youthful stage. As understanding of aggregation processes increases, improved models of the dispersion and sedimentation of fine-grained tephra from volcanic plumes will emerge. Coupled with field studies of deposits, these advances will allow increasingly accurate interpretations of plume and ambient atmosphere conditions during past volcanic eruptions.

16.2 GEOLOGICAL OBSERVATIONS

16.2.1 Anomalous Deposit Thicknesses

Conventionally, thicknesses of fall deposits are thought to decrease systematically with distance from the volcano. However, some deposits preserve one or more discrete regions of *anomalous thickening* away from the vent (Figures 16.1 and 16.2). For deposits with only one such region the term *secondary thickening* has been widely used (e.g. Sarna-Wojcicki *et al.* 1981). Observations of anomalous thickening are few, but this phenomenon is related to aggregation of fine particles in eruption plumes and is therefore likely to develop during most large eruptions. The reasons for the scarcity of examples may be due to: (1) regions of anomalous thickening have only slightly thicker deposits than surrounding regions so that, in the absence of a detailed study of the fresh deposit, they remain unidentified; (2) rapid post-depositional compaction and erosion of deposits obscure the anomalous thickening; and/or (3) part or all of the ash is deposited at sea.

The best example of anomalous thickening was displayed by the Mount St Helens, Washington, USA, fall deposit of May 18, 1980, which generated an extensive deposit in Washington, Idaho and Montana (Figure 13.12). For the first 140 km from the volcano the deposit thinned gradually from over 200 mm to $\sim$10 mm near Yakima. It then gradually thickened up to a distance of 325 km, with a second thickness maximum (Sarna-Wojcicki *et al.* 1981) of $\sim$35 mm near Ritzville (Figure 13.12). Further east the deposit thinned until it fell below measurable levels in central Montana. Carey and Sigurdsson (1982) carried out a computer simulation of ashfall from a volcanic plume to investigate the origin of the second thickness maximum. The model was constrained by observations of the eruption column height, elevation of the main region of particle transport, lateral spreading

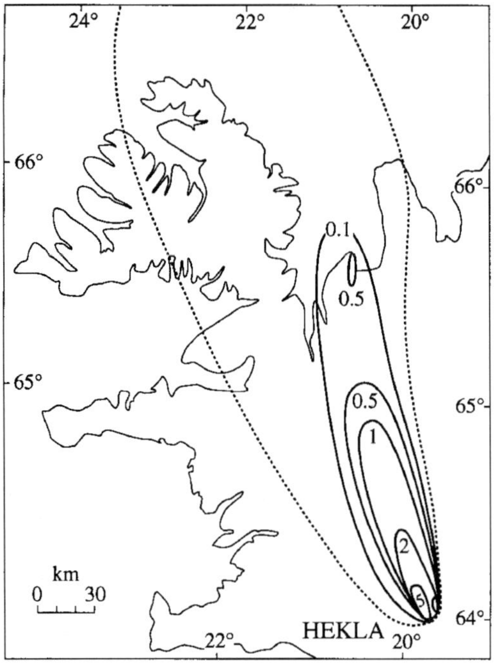

Figure 16.1 Isopach map for the 1970 Hekla eruption in Iceland. Isopachs are in centimetres. (After Thorarinsson and Sigvaldasson 1972, reproduced by permission of *Bulletin of Volcanology*.)

rate of the eruption plume and the atmospheric wind structure. These authors showed that the second thickness maximum could not be attributed to decreased wind velocities over central Washington, or injection of fine ash above the horizontal wind velocity maximum near the tropopause or the polymodal grain size distribution of the atmospherically dispersed particle-laden plume. They proposed that particles <63 μm wide formed aggregates several hundred micrometres in diameter which fell with a settling velocity of $0.35\ \mathrm{m\ s^{-1}}$. Rose and Hoffman (1982) made a similar suggestion, based on a comparison of settling velocities and ash arrival times, and Armienti *et al.* (1988) suggested that aggregation of <90 μm diameter particles had resulted in the anomalous deposit thickness.

Other examples of anomalous deposit thicknesses of fall deposits are rare. Kittl (1933) documented a second thickness maximum associated with the 1932 deposit of Quizapu volcano, in Chile, 550–700 km east of the vent. Larsson (1937) explained this by fluctuations in the atmospheric turbulence affecting rates of particle fallout. However, the downwind thickening could be explained by premature fallout of particles as dry aggregates. Hildreth and Drake (1992) could not identify the second thickness maximum and suggested that this was due to post-depositional compaction and erosion of the deposits. These authors found no evidence for aggregation, but established that 56–65 wt% of the fall deposit sampled in the originally documented area of secondary thickening comprised particles <63 μm in diameter.

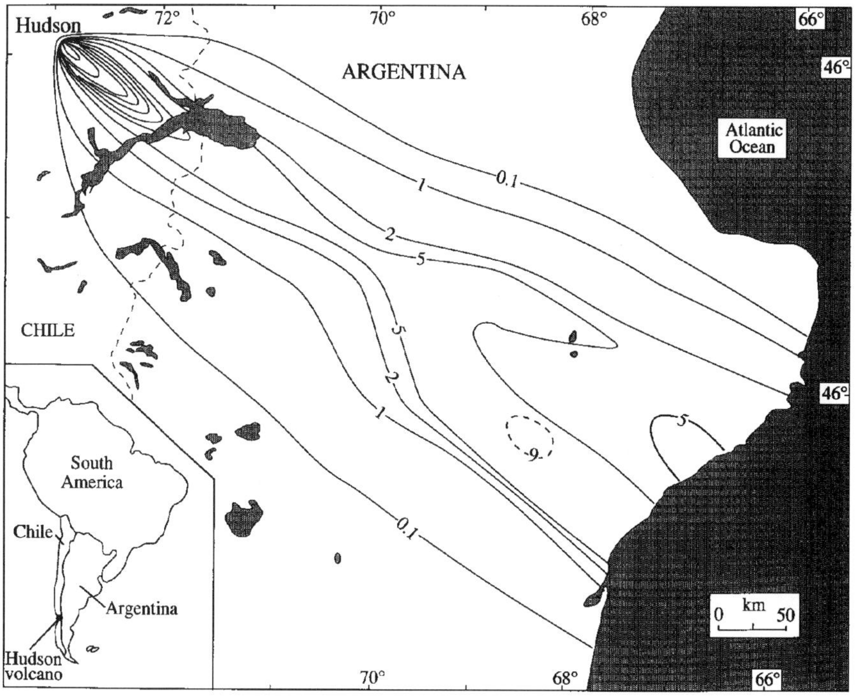

Figure 16.2 Isopach map for the August 12–15, 1991 eruption of Hudson volcano in Chile. Isopachs are in centimetres. (After Scasso *et al.* 1994, reproduced by permission of *Bulletin of Volcanology.*)

Thorarinsson and Sigvaldasson (1972) showed that the tephra layer formed by the 1970 Hekla eruption exhibited an anomalous deposit thickness with a secondary thickness maximum occurring 180 km north of the volcano (Figure 16.1).

Tephra from the August 12–15, 1991 eruption of Hudson volcano, in Chile, was deposited in a narrow east–south-east-trending sector of Patagonia (Scasso *et al.* 1994). The fall deposit exhibits unusually irregular thickness variations (Figure 16.2). It decreases up to 250 km from the volcano where it begins to thicken again. At approximately 360 km it attains a second thickness maximum and decreases in thickness before reaching a third maximum at the Atlantic coast at approximately 600 km. Scasso *et al.* (1994) suggested that particle aggregation was at least partly responsible for creating the unusual distribution pattern.

Some anomalous deposit thicknesses were found in an extensive marine fall deposit (the Roseau ash) produced by a major 30 000 a explosive eruption on Dominica (Carey and Sigurdsson 1980). Although Carey and Sigurdsson (1980) hypothesized that this was the result of sedimentary processes such as bioturbation, current reworking and slumping, particle aggregation in the atmosphere (and preserved in the water column) may have generated some of the complexity in the deposit thickness.

16.2.2 Particle Size Distributions

Particles falling as aggregates can lead to complex grain size distributions of fall deposits. For example, a number of workers (e.g. Fruchter *et al.* 1980; Rose and Hoffman 1980, 1982) sampled the May 18, 1980 Mount St Helens fall deposit and showed that the mean particle size decreased with distance from the volcano and that the deposit displayed a polymodal grain size distribution. Carey and Sigurdsson (1982) showed that it was polymodal (Figure 16.3) and poorly sorted, particularly in the distal sections. The total grain size distribution had modes at 10 and 180 μm with a weak one at 350 μm. A proximally dominant coarse mode migrated towards, and eventually merged with, a fine mode of fixed value (10 μm) with increasing distance from the volcano (Figure 16.3). Within 200 km of the vent the coarse mode was found to constitute the majority of the deposit. Beyond this point, in the vicinity of Ritzville (Figure 13.12), the bulk of the deposit was composed of the fine mode (10 μm) (Figure 16.4). An important characteristic in the region of secondary thickening (section 16.2.1) was the dominance of the fine ash component which exhibited only gradual changes in grain size with distance from the vent. By means of a computer simulation Carey and Sigurdsson (1982) showed that the polymodality could not be related to two types of source material, but resulted from deposition controlled by aggregation.

The 1991 fall deposit at Hudson volcano also shows marked bimodality in individual layers up to 270 km from the volcano (Scasso *et al.* 1994). The coarse population is

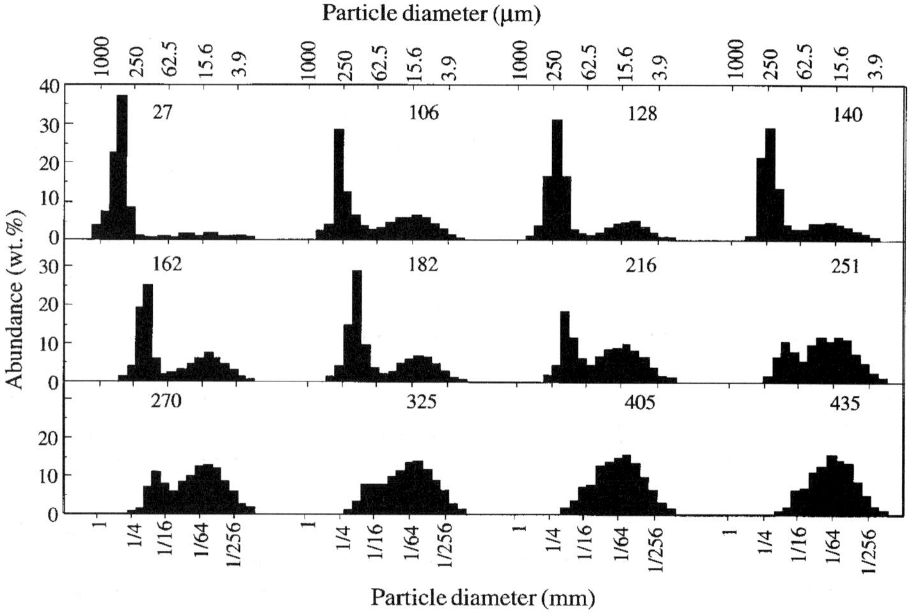

Figure 16.3 Particle size distribution for the bulk deposit of the May 18, 1980 Mount St Helens fall deposit along the dispersal axis. Numbers above histograms are distances, in kilometres, along the dispersal axis. (After Carey and Sigurdsson 1982, reproduced by permission of the American Geophysical Union.)

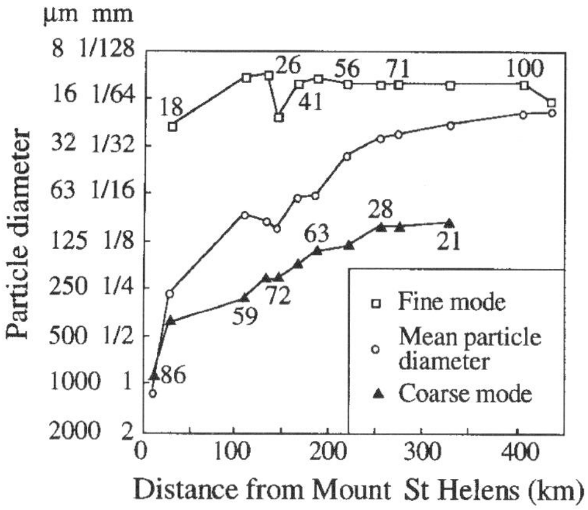

Figure 16.4 Particle diameter versus distance for the bulk deposit of the May 18, 1980 Mount St Helens fall deposit. Numbers on data points indicate the weight % of the total sample made up of that particular mode. (After Carey and Sigurdsson 1982, reproduced by permission of the American Geophysical Union.)

represented by a mode which shifts to finer sizes with distance from the vent. The fine population has a "fixed" mode in the region 15.6–32 μm. In distal areas the bulk deposit is almost unimodal and sorting shows systematic improvement with distance. The bimodality is interpreted to have resulted from particle aggregation which enabled fine particles to fall with coarse particles (Scasso *et al.* 1994).

Brazier *et al.* (1983) showed that several of the ash layers resulting from the Soufrière, St Vincent, explosions, between April 13 and 26, 1979 had bimodal grain size populations. The April 26 deposit showed a systematic decrease in particle size with distance from source (Figure 16.5) in the coarse population, but no systematic variation in the fine population. The same authors found that for the 2500 a Bridge River fall deposit of the north-west USA and the 7000 a Mount Mazama, Oregon, USA, fall deposit (Figure 16.5) the size of the fine mode remained uniform while that of the coarse mode decreased with distance from source. In both cases the two populations eventually merged to form a unimodal deposit. Brazier *et al.* (1983) showed that for the 3500 a Mount St Helens Yn ash between 100 and 172 km from source the coarse population generally fined (Figure 16.5) while that of the fine population varied randomly between 49 and 27 μm. Between 172 and 388 km the weight percentage of the sample contained within the fine population increased from 13.5 to 48.5%. The authors interpreted these widespread deposit characteristics as evidence for particle aggregation in eruption plumes.

The Minoan deep-sea ash layer of Santorini is another example of a bimodal deposit. It has a coarse population which fines systematically away from source and a fine population which remains uniform irrespective of distance (Sparks and Huang 1980). Similarly, the deep-sea ash layer of the Campanian Y5 (thought to be of the same age as the 38 000 a. Campanian ignimbrite (Barberi *et al.* 1978) has bimodal grain size characteristics (Sparks and Huang 1980). Cornell *et al.* (1983) investigated the grain size of the Campanian Y5 layer from deep-sea cores within 1600 km of the vent and showed that the deposit has a

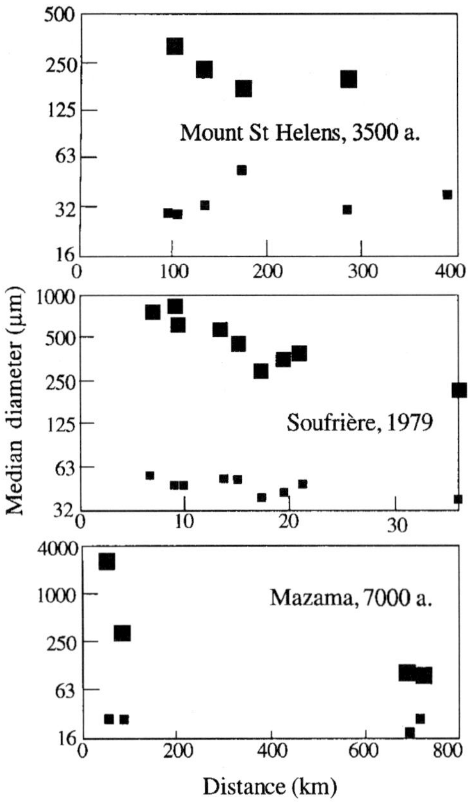

Figure 16.5 Particle size data for (a) Mount St Helens, (b) Soufrière and (c) Mazama fall deposits. Large symbols represent coarse fraction and small symbols represent fine fraction. (After Brazier *et al.* 1983, reproduced by permission of *Nature*.)

coarse mode which decreases with distance from source and progressively approaches a near-constant fine mode of approximately 13 μm. Distal samples were found to be unimodal. Grain size data were combined with a set of vertical wind profiles and used in a computer model to simulate fallout of tephra. In this study the downwind variation of grain size of the coarse mode could be accurately reproduced with transport of single particles but the fine mode could not. Such transport would have deposited virtually all of the fine ash beyond the studied area. In summary, it is likely that deposition of all of these fall deposits has been governed by particle aggregation.

16.2.3 Aggregates

A spectrum of morphological types of aggregates fall from plumes. Formation of a particular type depends on the amount of liquid available during aggregation. Aggregates can be divided into three main categories (Table 16.1): *dry aggregates* (Figure 16.6), *accre-*

tionary lapilli (Figure 16.7) and *mud rain* (Figure 16.8). The particular type of aggregate which forms, and efficiency of the aggregation process in a given eruption, depends upon a range of conditions including the relative humidity and temperature of the gas surrounding the particles, volume of entrained air or liquid into the plume, particle size distribution and charge on the ash particles and liquid drops. Observations of aggregates are now summarized from a number of eruptions.

Dry aggregates

Dry aggregates have been observed falling from plumes on several occasions. The most porous of these structures collapse into cone-shaped piles <5 mm in diameter on landing. Their geometries when falling are unknown. These aggregates typically comprise particles <200 μm in diameter and have porosities in the range 0.4–0.9 (Table 16.1). During the 1932 eruption of Quizapu volcano aggregates of particles up to 1 mm in diameter were reported to fall from the eruption plume at La Plata, 1160 km east of the volcano (Dartayet 1932; Lunkenheimer 1932; Kittl 1933). Similarly, during the May 18, 1980 eruption of Mount St Helens, ash was observed to fall from the plume in clusters 0.25–0.5 mm in diameter 390 km east of the volcano (Sorem 1982). These clusters consisted of angular fragments of freshly broken silicate minerals and glassy pumice ranging from 1 to 40 μm in diameter, and a small number of larger (40–60 μm diameter) crystals. The clusters appeared to be random collections of particles of different sizes and had no obvious growth pattern or central point of growth. During the same eruption, Hobbs *et al.* (1981) reported that 37 km downwind from the vent their aircraft was struck by fist-sized clumps of loosely aggregated ash. Ash falling like "snow clusters" was described by residents downwind of the August 12–15, 1991 eruption of Hudson volcano (Scasso *et al.* 1994).

Table 16.1 Dominant aggregate collision and binding mechanisms, and physical properties of aggregates

	Dry aggregate	Accretionary lapillus	Mud raindrop
Collision mechanisms			
Ambient plume motions	√	√	√
Fall velocity differences	√	√	√
Electrostatic forces	√		
Binding mechanisms			
Surface tension forces		√	√
Secondary mineral/ice crystal growth		√	
Electrostatic forces	√		
Van der Waals' forces			
Mechanical interlocking			
Physical properties			
Component particle diameter (μm)	<200	<90	<200
Aggregate diameter (mm)	<5	1–50 (but generally 1–10)	<5
Density (kg m^{-3})	220–1320	1200–1600	1000–1500
Porosity	0.4–0.9	0.3–0.5 (when dry)	0

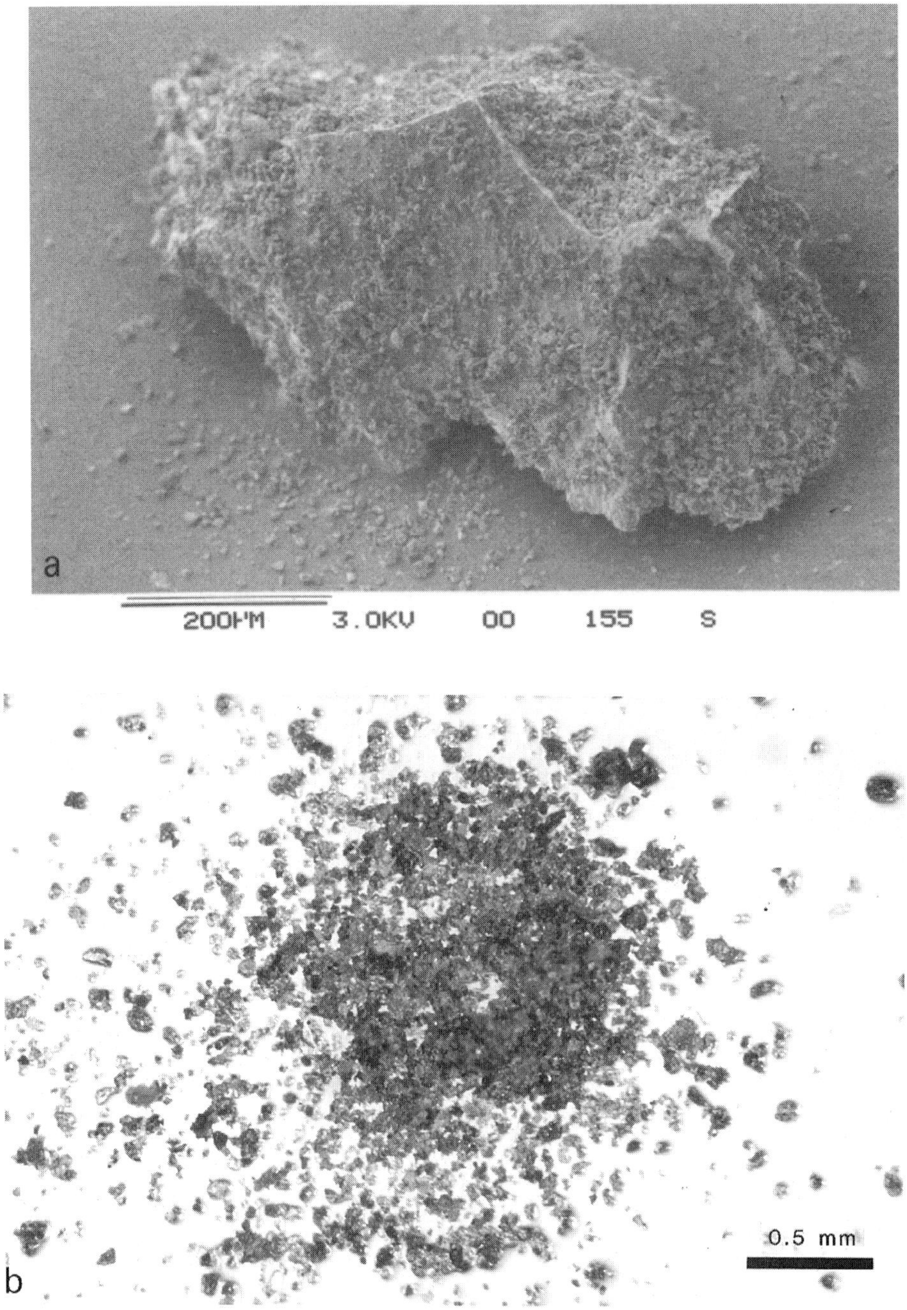

Figure 16.6 Microscope images of dry aggregates. (a) A > 200 µm diameter silicate particle with a coating of smaller particles which fell from the plume at Sakurajima volcano in Japan on to a varnish-covered glass slide. (b) A loosely bound dry aggregate collected on a glass slide as it fell from the plume at Sakurajima volcano. Note that the aggregate collapsed as it impacted the slide and therefore its true geometry is not preserved. (Photographs courtesy of J. S. Gilbert.)

Figure 16.7 SEM images of accretionary lapilli. (a) A side view of an accretionary lapillus which was collected on a varnish-covered glass slide while under the plume at Sakurajima volcano, Japan. (b) A broken accretionary lapillus from Sakurajima volcano showing its internal structure. (After Gilbert and Lane 1994a, reproduced by permission of *Bulletin of Volcanology*.)

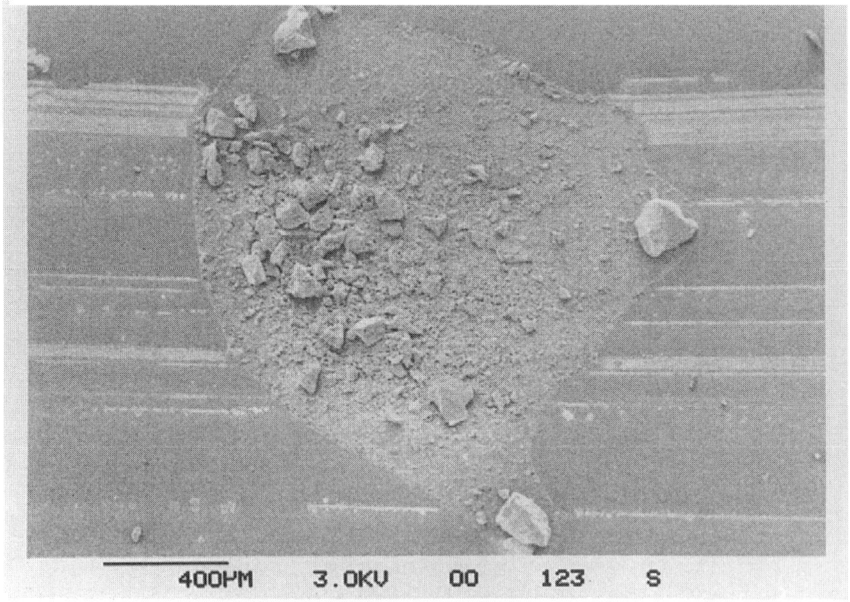

Figure 16.8 SEM photograph of evaporated drop of mud rain which was collected on a varnish-covered glass slide while under the plume at Sakurajima volcano in Japan. (Photograph courtesy of J. S. Gilbert.)

During the period 1990–94, observations of daily eruptions at Sakurajima volcano, in Japan, established that dry aggregates fall from the Sakurajima plumes when the relative humidity at ground level is $<80\%$ (Gilbert and Lane 1994a). These aggregates are of two types. First, particles >200 μm in diameter fall with a coating of small particles generally <20 μm wide (Figure 16.6a). Second, sub-micrometre to 200 μm wide particles fall as loosely bound aggregates <3 mm wide (Figure 16.6b). This latter type of aggregate is similar to those observed at Mount St Helens (Sorem 1982), is fragile and, on landing, collapses into piles <3 mm wide (Figure 16.6b).

Accretionary lapilli

Many authors have described accretionary lapilli in deposits but eyewitness accounts of them falling are rare and those that do exist provide only meagre data on the atmospheric conditions during their formation. Accretionary lapilli have been reported in fall deposits from many parts of the world and have been recognized in rocks as old as Archean (Heinrichs 1984). There are numerous accounts of accretionary lapilli in fall deposits and, more rarely, in pyroclastic flow and surge deposits (e.g. Fisher and Waters 1970; Schmincke *et al.* 1973; Hayakawa 1990; Gilbert 1991).

Accretionary lapilli typically consist of a core, which is composed of either relatively homogeneous aggregated ash or a single pumice fragment, surrounded by thin concentric layers of ash whose grain sizes may decrease towards the outer surface (Schumacher and Schmincke 1991). Other accretionary lapilli are homogeneous structures. Some accretionary lapilli contain irregular-shaped voids between particles, while others have spherical vesicles. Accretionary lapilli typically have porosities and densities of the order of 0.3–0.5 and 1200–1600 kg m^{-3} respectively (Schumacher and Schmincke 1991), and, with the exception of any nucleus fragments, are usually composed of particles ranging from sub-micrometre to 90 μm wide (Table 16.1).

Brazier *et al.* (1982) witnessed fallout of accretionary lapilli, on April 26, 1979, during the eruption of Soufrière, St Vincent. Most of the accretionary lapilli disintegrated on impact and there was "no direct evidence of water condensation in their formation. The night was cloudless and no rain was induced by the eruption." In contrast, Brazier *et al.* (1982) also reported fallout, on April 22, 1979, of wet accretionary lapilli which flattened or splashed on impact.

Tomita *et al.* (1985) reported fallout of accretionary lapilli on May 22, 1983, at Sakurajima volcano when the eruption column rose to approximately 4 km above sea-level. At this altitude in Kagoshima city, approximately 9 km west of the vent, the mean temperature and relative humidity were 4.3 °C and 18% respectively (Tomita *et al.* 1985). Because Sakurajima is for the most part surrounded by water, ground level relative humidities at this volcano are usually >40%. The accretionary lapilli collected by Tomita *et al.* (1985) were 1–5 mm in diameter, spherical, lacked concentric zones and contained angular fragments of volcanic glass, plagioclase, hypersthene, augite and andesite fragments with secondary calcium sulphate. Spherical vesicles were common.

Gilbert and Lane (1994a) witnessed fallout at distances between 2 and 5 km from the vent of <5 mm diameter accretionary lapilli from plumes at Sakurajima between 1990 and 1994. The plumes reached altitudes of up to 5 km. Accretionary lapilli were found to develop exclusively during periods of high (>80%) relative humidity (measured on the ground); they fell on overcast, cloudy days usually after a period of regional rainfall and were damp, with a coating of acid of pH less than one, when falling. On less humid days dry aggregates fell. The accretionary lapilli usually remained intact on impact and were composed of angular particles <200 μm in diameter, between which secondary calcium sulphate crystals (Figure 16.9) had grown.

Mud rain

Particles can be transported from the plume to the ground within liquid drops. Such drops are usually <5 mm in diameter and scavenge a wide range of particle sizes (Table 16.1). On particularly humid days during Vulcanian-style eruptions at Sakurajima volcano, particle-laden liquid drops fall from eruption plumes as brown mud rain (Lane *et al.* 1995). At this volcano, sometimes rainfall is restricted to the region downwind of the plume, suggesting that it is induced by condensation of entrained water vapour and magmatic gases. At other times, rainfall is more widespread, implying that the plume provides raindrop nuclei within a larger meteorologic system. Particles in these mud raindrops are angular and have a wide range of sizes although they are generally <200 μm in diameter (Figure 16.8). Mud raindrops were collected during a number of eruptions at Sakurajima during which there was regional rainfall. Directly downwind of the plume the mud rain was found to have a pH of <1 and high chlorine and sulphate contents in comparison to that off the dispersal axis (Figure 16.10), suggesting that there was efficient scavenging of

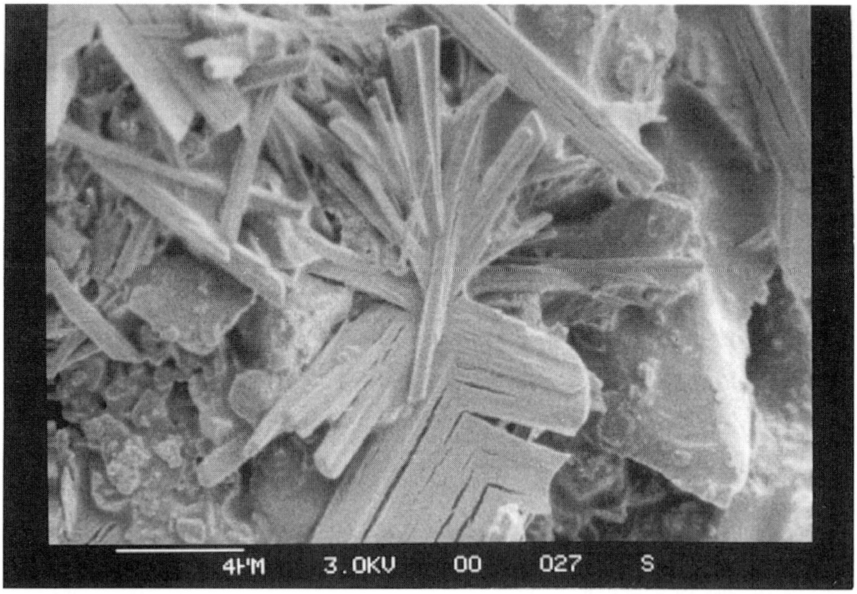

Figure 16.9 SEM image of calcium sulphate crystals on the surface and between particles of an accretionary lapillus collected at Sakurajima volcano, in Japan. (Photograph courtesy of J. S. Gilbert.)

volcanic gases by the liquid drops. After the mud raindrops had landed on the ground, calcium sulphate and sodium chloride crystals precipitated from solution.

16.3 AGGREGATION MECHANISMS

Aggregates within volcanic plumes grow via processes of *collision, binding* and *breakup* of solid and liquid phases. These processes are governed by a variety of variables such as

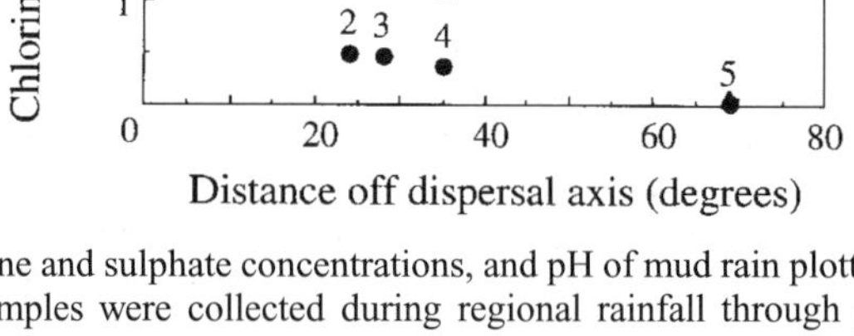

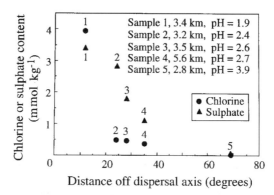

Figure 16.10 Chlorine and sulphate concentrations, and pH of mud rain plotted against distance off the dispersal axis. Samples were collected during regional rainfall through plumes at Sakurajima volcano, in Japan. The key shows the distances from the vent at which each sample was collected. (After Lane *et al.* 1995, reproduced with permission of Sakurajima Volcano Research Center.)

solid concentrations, shapes and dimensions, *liquid* compositions and volumes, *gas* compositions, velocities, relative humidities and temperatures. Aggregate growth rate is controlled by the rate of collision of particles and the efficiency of binding of particles subsequent to their collision. For growth to take place, the particle binding mechanism must be sufficient to dissipate any relative motion between colliding particles, and to withstand subsequent collisions with other particles and aggregates.

Collisions between solid and liquid bodies within plumes result from ambient plume motions, differences in terminal fall velocities and electrostatic forces (Table 16.1). Mechanisms which aid particle binding include surface tension forces resulting from interaction of particles and liquids, electrostatic forces, mineral growth and ice crystal growth (Table 16.1). The relative humidity of the air in the plume appears to play a pivotal role in determining the dominant binding mechanism of colliding bodies. In humid plumes, liquid films on particle surfaces bond particles by surface tension forces. This binding force is subsequently strengthened if secondary minerals precipitate from solution.

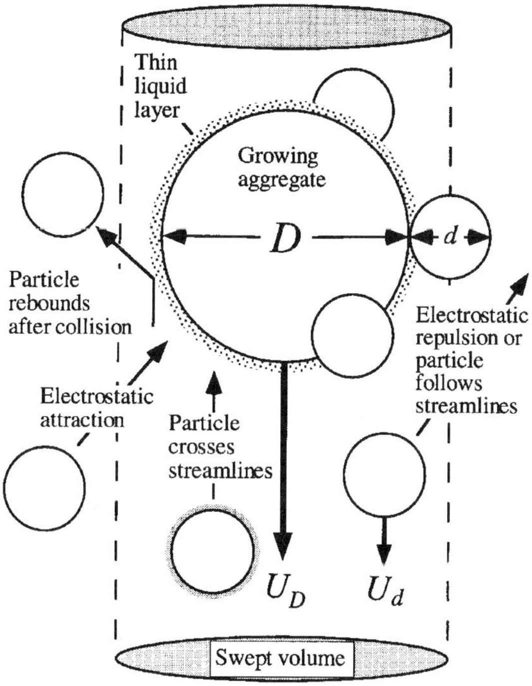

Figure 16.11 Schematic representation of aggregate growth processes. The high fall velocity (U_D) particle or aggregate (with diameter D) collides with low-velocity (U_d) particles (with diameter d) within the swept volume. Deviations from this assumption, accounted for by the aggregation coefficient, are caused by electrostatic repulsive or attractive forces and small particles following fluid streamlines around the growing aggregate. Particles which collide with the aggregate may rebound and not contribute to aggregate growth. The presence of liquids provides strong short-range bonding between particles. Secondary minerals may precipitate from liquids and generate more permanent bonds. (After Gilbert and Lane 1994a, reproduced by permission of *Bulletin of Volcanology*.)

In very humid environments, liquid drops scavenge particles from the plume as they rain out.

The important processes controlling aggregate formation and settling are considered below. Section 16.4 describes how these processes have been explored experimentally and combined into a simple model of aggregate formation.

16.3.1 Collision Mechanisms

Ambient plume motions

As particles and liquid drops are swept around by the turbulence in the plume they can collide. The rate at which bodies are brought sufficiently close to bind depends upon the concentration, shape and size distribution of the solid particles and liquid drops as well as the intensity of turbulent fluctuations. On collision, if the relative momentum between the bodies is sufficiently small, then this will enable aggregate growth to occur. If not, collision can cause aggregates to break up or particles to rebound. The relative momentum of the bodies is controlled by their aerodynamic properties which depend upon particle size, shape and density (Walker *et al.* 1971).

Differences in terminal fall velocities

Eruption plumes contain a spectrum of single particle, aggregate and liquid drop sizes. For example, sub-micrometre to metre-sized individual particles, sub-micrometre to centimetre wide aggregates and sub-micrometre to millimetre wide liquid or, at low temperatures, frozen drops, occur. The range of sizes and densities results in a very wide range of fall velocities (Walker *et al.* 1971). These differences produce collisions (Figure 16.12) which result in aggregate growth if binding takes place. Contact by collision is a complex process. Very small particles may follow streamlines, and flow around larger particles or aggregates. Large particles will not follow fluid streamlines, but may rebound after collision and therefore not adhere to growing aggregates.

Electrostatic forces

In relatively dry environments, it is thought that the electrostatic charge between particles plays a key role in particle aggregation. Field measurements of the charge carried by particles 1–250 μm in diameter, falling from plumes of Sakurajima volcano, have shown that dry particles can have surface charge densities as large as $\pm 10^{-5}$ C m^{-2} (Gilbert *et al.* 1991). This is close to the theoretical maximum surface charge density on a flat surface in the atmosphere of $\pm 2.6 \times 10^{-5}$ C m^{-2}, above which air breakdown occurs. Electric potential gradient monitoring (Hatakeyama and Uchikawa 1952; Hatakeyama 1958; Lane and Gilbert 1992; Lane *et al.* 1995) during explosive volcanic activity has shown that eruptions which generate considerable quantities of ash produce charged solid particles and volcanic gases (ions). Charge transfer and the generation of ions takes place during deformation (Enomoto and Hashimoto 1990) and fracture (Donaldson *et al.* 1988; Mathison *et al.* 1989) of materials and this phenomenon is known as fracto-emission or fracto-charging. Because particle-free steam eruptions produce much lower levels of

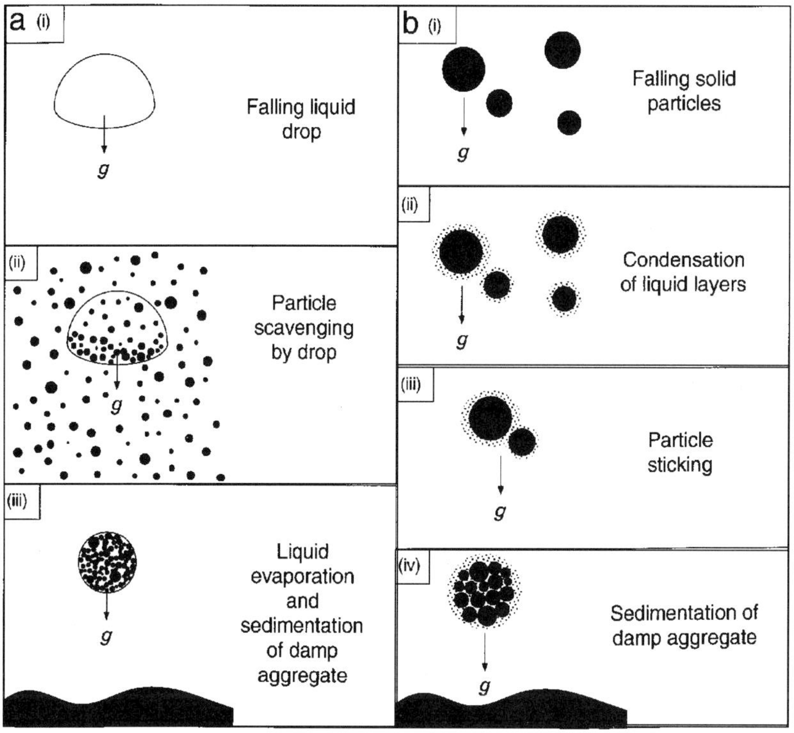

Figure 16.12 Hypothesis for the evolution of an accretionary lapillus. (a) A liquid drop falling through a plume scavenges particles and partial evaporation of the drop produces a damp aggregate. (b) Particles in a plume, coated with thin liquid layers, collide and coalesce. (After Gilbert and Lane 1994a, reproduced by permission of *Bulletin of Volcanology*.)

charge, Gilbert *et al.* (1991) proposed that the charge originates from fracto-emission during fragmentation. Explosions which generate very fine-grained particles, which have relatively large surface areas (such as during phreatomagmatic eruptions), are likely to create more charge per unit mass than the more coarse-grained Plinian-style counterparts.

The presence of electrostatic charge on particles in plumes may cause attraction and collision, or repulsion, of particles. The ratio of electrostatic to gravitational forces exceeds unity at separation distances between particle and aggregate of less than three particle diameters for a 100 μm particle, 10 diameters for a 10 μm particle and 30 diameters for a 1 μm particle (Gilbert and Lane 1994a). Aggregates can form where fall velocity differences are small when particles are charged. These aggregates can then form the nucleus of an aggregate which grows mainly due to fall velocity differences (Lane *et al.* 1993).

The electrostatic interaction of ice crystals and volcanic particles is an unexplored field. However, the charge on ice pellets is known to become polarized by the Earth's electric field and collisions with smaller ice crystals result in charge transfer (Iribarne and Cho 1980). At high altitudes and in cold climates, collisions must take place between volcanic particles and ice crystals as a result of electrostatic attraction.

16.3.2 Binding Mechanisms

Surface tension (capillary) forces

Particles in plumes act as condensation nuclei and may spend a substantial part of their airborne lifetime coated with liquid or solid condensate. When liquid-coated particles collide they have a high probability of binding due to the effect of strong short-range surface tension forces (Reimer 1983; Schubert 1984). In contrast, particles coated with solid condensed layers are less likely to adhere on collision than liquid-coated particles (Bacon and Sarma 1991).

The rate of condensation of liquid in a volcanic plume can be particularly high. Upward transport of evaporated groundwater, water vapour entrained from lower in the atmosphere and exsolved volatiles from the magma provide an ample source of vapour (Woods 1993b; Chapter 8). Furthermore, many hygroscopic compounds, which promote condensation, are found in volcanic plumes (Rose 1977; Rose *et al.* 1973, 1982; Gilbert and Lane 1994a). As a result, liquid drops and liquid films on the surface of particles can develop at low relative humidities. For example, sodium chloride solution forms at a relative humidity of 75–76% (Twomey 1977) and sulphuric acid solution develops at extremely low humidities (Ammann and Burtscher 1993). This condensate may be charged and typically has very low pH values (Lane and Gilbert 1992). Particularly large volumes of such condensate in humid conditions may produce liquid drops up to several millimetres in diameter which scavenge particles and lead to mud rain (Walker and Croasdale 1971).

Growth of secondary minerals and ice crystals

Acids which condense on particle surfaces within plumes can partially dissolve silicate minerals and glass. During the within-plume lifetime of particles, changes in ambient atmospheric conditions such as temperature and relative humidity can result in precipitation of secondary minerals from solution. Secondary minerals such as calcium sulphate (Figure 16.9) and sodium chloride have been observed within accretionary lapilli (Tomita *et al.* 1985; Gilbert and Lane 1994a) and on particles brought down by mud rain (Gilbert and Lane, unpublished data). This observation, combined with the fact that accretionary lapilli often survive impact with the ground, suggests that crystallization of secondary minerals can produce very strong bonds between particles during transport. However, the temperature conditions under which aqueous acid solutions in contact with silicate glass and mineral surfaces solidify are not known. In the case of mud rain, precipitation occurs after deposition when the liquid evaporates from the drop. Aggregates from explosive tests (Bacon and Sarma 1991), formed by collision and coalescence of particles coated with water condensed from entrained air, disintegrated on impact with the ground. Highly acidic vapours were presumably not available to cause secondary mineral growth and bind the aggregates sufficiently to allow survival on impact.

If aggregates are convected upwards through the atmosphere to particularly low-temperature regions, precipitation, not only of secondary minerals but also of ice, may occur. The atmospheric temperature typically falls below the freezing point of water within 2–3 km above the ground, depending upon the season and latitude and ash particles can provide nucleation sites for ice crystals. Solidification of liquid bridges between particles has the effect of greatly enhancing the strength of interparticle bonds. Therefore, lowering of plume temperatures will increase the probability of aggregates being preserved during transport. However, the collision of ice-coated particles does not typically lead to bonding (Bacon and Sarma 1991).

Electrostatic forces

If dry charged particles collide they may bind as a result of charge being unable to migrate, and thus neutralize, on the high resistivity surfaces of silicate materials. The bond strength is weak, but long range, compared to capillary and van der Waals' forces. Also, the electrostatic binding force exceeds that of gravity for particles smaller than approximately 600 μm in diameter (Reimer 1983), thereby preventing aggregate breakup due to the differential sedimentation rates of the constituent particles. Therefore, low-porosity, fragile dry aggregates can be mainly held together by electrostatic forces between particles during transport through the atmosphere (Gilbert *et al.* 1991; Schumacher 1994). However, aggregates landing on the ground tend to fragment either due to the impact or following the eventual decay to Earth of the charge. Note that if charged particles are coated with liquid layers, then on collision the charge will migrate across the liquid film and neutralize. In such cases surface tension can bind the particles.

Van der Waals' forces

These forces provide short-range attraction between particles (Reimer 1983). Van der Waals' forces are an order of magnitude smaller than capillary forces and become less than electrostatic forces at separations of > 1 μm.

Mechanical interlocking

Sorem (1982) postulated that mechanical interlocking of particles enhanced the strength of the dry aggregates observed during the May 18, 1980 eruption of Mount St Helens. Indeed, aggregation of snow crystals (where diffusion and sublimation produces hexagonal spikes) is enhanced by interlocking of adjacent spikes or dendrites, and this produces very large snowflakes in calm conditions (McIlveen 1992). The dry aggregates observed at Sakurajima (Gilbert *et al.* 1991) show no evidence for interlocking.

16.4 EXPERIMENTS AND THEORY

Analogue laboratory experiments and theoretical models have been developed to distinguish the relative importance of the mechanisms described in the previous section. These have led to several new insights into the formation of aggregates, particularly mud rain and accretionary lapilli. A review of this work is presented.

16.4.1 Laboratory Simulation

Data on the physical processes occurring during the growth of volcanic aggregates are sparse, and so a series of experiments were conducted by Gilbert and Lane (1994a) to investigate the controls upon growth of mud raindrops and accretionary lapilli. Gilbert and Lane (1994a) postulated two working hypotheses for the growth of accretionary lapilli. In the first, liquid drops and particles falling through plumes collide (due to differences in fall velocities) and bind due to surface tension forces. On partial evaporation of the drops, accretionary lapilli are formed (Figure 16.12a). In the second, particles in plumes coated with *thin* liquid layers collide (due to fall velocity differences) and bind due to surface tension leading to the formation of accretionary lapilli (Figure 16.12b). Due to the initially large mass of liquid per unit mass of solids envisaged in the first hypothesis, particles can migrate within the liquid drop during evaporation and this might have implications for the generation of the concentric zones of different grain sizes common in accretionary lapilli. In the second hypothesis, particles are substantially less mobile and generation of concentric zones requires transport of the evolving aggregate through regions of the plume containing different particle size populations. In both cases evaporation of liquids could potentially result in secondary mineral growth and electrostatic attraction could cause small particles, which would otherwise flow around the aggregate, to collide and accrete.

Apparatus and methods

A recirculating laboratory wind tunnel (Figure 16.13) in which the relative humidity, temperature and particle concentration could be controlled was used to investigate the hypotheses shown in Figure 16.12. In a series of experiments a shaped wire gauze created an air velocity minimum within the wind tunnel. Single liquid drops were injected into the velocity minimum, so that each drop (approximately 3 mm in diameter) was suspended in a roughly stationary position within the flow stream. Particles were introduced into the flow stream and the uptake of particles by the liquid drop and particle–liquid interactions during evaporation were observed as shown in Figure 16.12a. Evaporation of the liquid occurred by controlling the relative humidity at 10–15%. A variety of particle types <300 μm in diameter (andesitic volcanic ash from Sakurajima volcano, spherical glass ballotini, silicon carbide and fused alumina) were used in the experiments. The liquids were either distilled water or sodium chloride solutions.

In a second series of experiments a tube with an internal diameter of 7 mm was fitted centrally in the wind tunnel venturi (Figure 16.14). In the air stream at the top of the tube a polystyrene sphere 7.8 mm in diameter was suspended. Water vapour could be passed into the tube and condensed as a liquid layer on the sphere. Particles were fed into the flow stream and, on colliding with the liquid layer, many stuck to the sphere. In this way growth of an accretionary lapillus by the method shown in Figure 16.12b was accomplished.

Results

In the first experiments it was observed that, after loading a liquid drop (Figure 16.15a) with a small number of particles of any composition or size, rapid internal oscillations kept the particles in suspension. At higher particle concentrations sedimentation within the drop

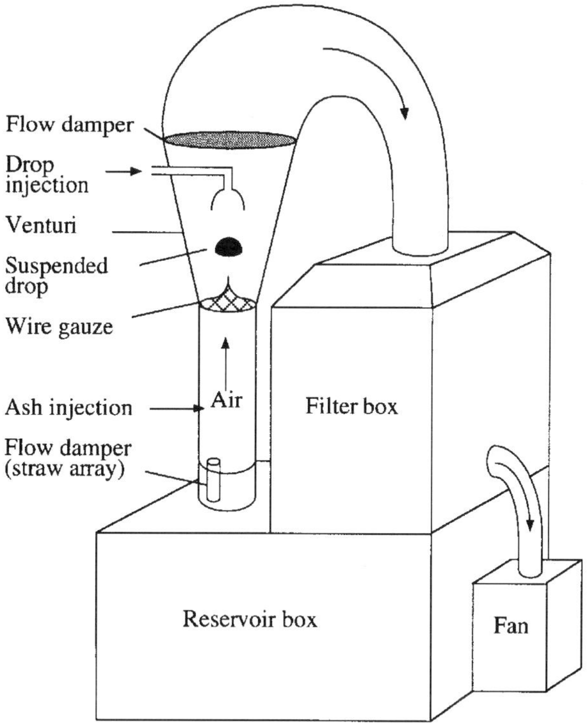

Figure 16.13 Schematic diagram of the wind tunnel used for experiments to explore the mechanisms of formation of mud raindrops and accretionary lapilli by particle scavenging of a liquid drop followed by evaporation of liquid, as shown in Figure 16.12a. (After Gilbert and Lane 1994a, reproduced by permission of *Bulletin of Volcanology*.)

commenced (Figure 16.15b). When particles of more than one composition were present the highest density particles sedimented to the bottom of the drop first, forming a horizontally layered particle-laden drop. At high particle loadings, the drops flattened, became unstable and fragmented. If a particle-laden drop evaporated in the air stream the resulting aggregate was hemispherical with a rounded lower surface and a flat (Figure 16.15c, d) upper surface. When distilled water was used the resultant dry aggregate, while stable in the air stream, was fragile and rarely survived collection within the apparatus. The aggregates were probably held together by liquid bridges or by precipitated impurities from the evaporating drop. Mechanical interlocking is particularly unlikely for aggregates made from spherical ballotini. When sodium chloride solutions were used, sodium chloride crystals precipitated on evaporation and acted as a binding agent for the particles. The resultant aggregates were substantially less fragile than those made with distilled water and were able to survive collision with the gauze and collection. When volcanic ash was used, which had a relatively wide range of grain sizes (sub-micrometre to 300 μm), the aggregates had irregular upper surfaces (Figure 16.16) and the most spherical aggregates generated were when spherical ballotini particles were used (Figure 16.17). This is because spherical particles slide over each other relatively easily and are reorganized by the liquid

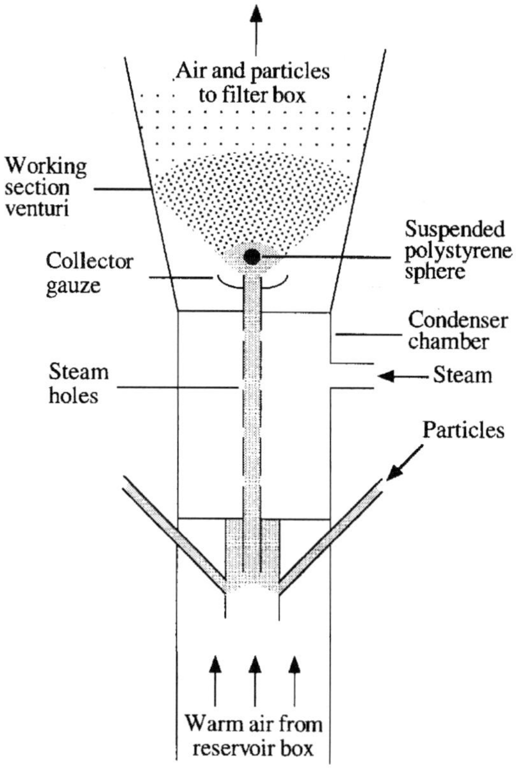

Figure 16.14 Schematic diagram of the working section of the wind tunnel used for experiments to study the growth of an accretionary lapillus by sticking of particles coated with thin liquid layers, as shown in Figure 16.12b. (After Gilbert and Lane 1994a, reproduced by permission of *Bulletin of Volcanology*.)

surface tension as the drop evaporates. Nevertheless, even the ballotini aggregates had rounded bases and flat upper surfaces. Only when very small liquid drops (<1 mm in diameter) were loaded with particles could perfectly spherical aggregates, similar to accretionary lapilli, be produced.

In the second series of experiments difficulties were encountered in sustaining a layer of water on the polystyrene sphere for extended periods. In practice, the sphere rapidly attained the air/water vapour stream temperature. This prevented condensation taking place and only a light coating of particles could be deposited on the sphere's surface. In plumes, particles may be coated with liquid layers well below 100% relative humidity due to the presence of hygroscopic compounds (section 16.3.2). To mimic this, the spheres were dipped in saturated sodium chloride solution, shaken to remove drops and evaporated dry to leave a coating of sodium chloride crystals on the surface. Figure 16.18 shows a sequence of photographs taken during a single experiment in which a sphere with a layer of sodium chloride solution is gradually coated with volcanic ash. Figure 16.18a shows the sphere suspended in the air stream at 15% relative humidity under which conditions sodium chloride is present as a solid. On raising the relative humidity to 85%, i.e. above

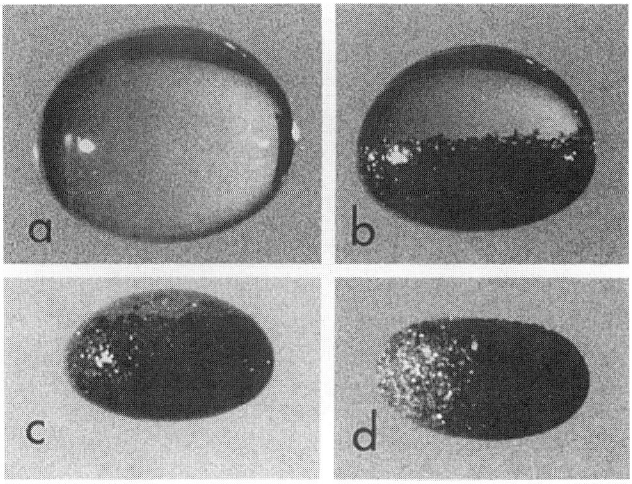

Figure 16.15 Sequence of events resulting from the gradual loading of a distilled water drop, suspended in the wind tunnel, with angular particles of silicon carbide (approximately 50 μm in diameter and with densities of 3200 kg m^{-3}) followed by liquid evaporation. Airflow is from bottom to top. (a) Particle-free liquid drop (3.25 mm in diameter). (b) Liquid drop (3.25 mm in diameter) in which particles have sedimented. (c) Particle-laden liquid drop (2.75 mm in diameter) undergoing evaporation with some water remaining. (d) Half-sphere aggregate (3.00 mm in diameter) with flat upper surface produced by evaporation of liquid. (After Gilbert and Lane 1994a, reproduced by permission of *Bulletin of Volcanology*.)

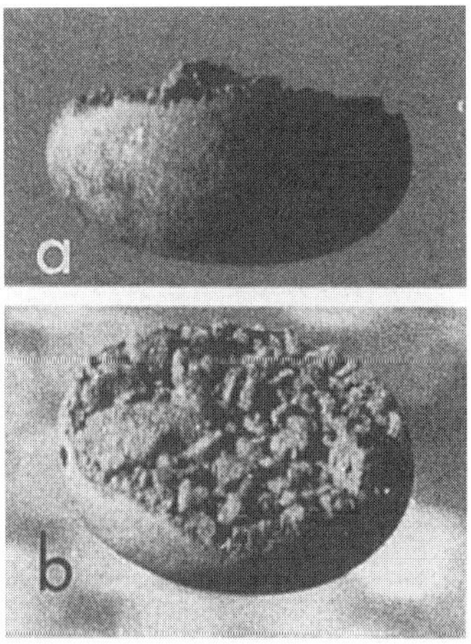

Figure 16.16 Aggregate (3.5 mm in diameter) generated by the gradual loading of a distilled water drop, suspended in the wind tunnel, with angular particles of volcanic ash (<300 μm in diameter with densities of approximately 2200 kg m^{-3}) followed by liquid evaporation. Airflow is from bottom to top. (a) Side view showing the hemispherical geometry. (b) View looking down on to the aggregate showing the irregular upper surface. (After Gilbert and Lane 1994a, reproduced by permission of *Bulletin of Volcanology*.)

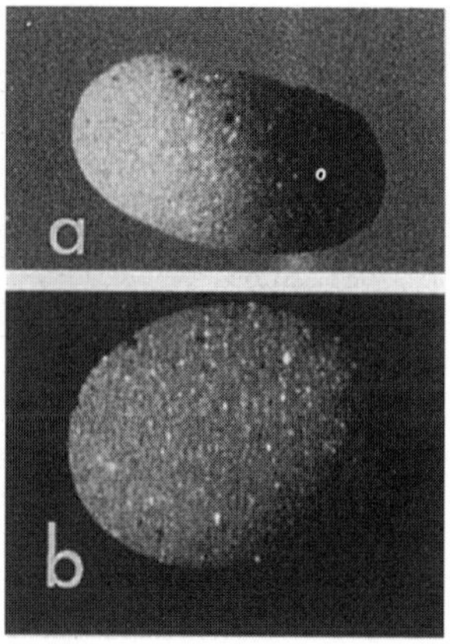

Figure 16.17 Aggregate (2.75 mm in diameter) generated by the gradual loading of a distilled water drop, suspended in the wind tunnel, with spherical ballotini particles (80 μm in diameter with densities in the range 2450–2550 kg m^{-3}) followed by liquid evaporation. Airflow is from bottom to top. (a) Side view showing the rounded lower and flat upper surface. (b) View looking down on to the aggregate showing the smooth upper surface. (After Gilbert and Lane 1994a, reproduced by permission of *Bulletin of Volcanology*.)

the hygroscopic phase transition for sodium chloride (Twomey 1977), a liquid layer was generated on the surface of the sphere (Figure 16.18b). At this point ash was injected into the air stream which, on colliding with the liquid layer, stuck to the sphere (Figure 16.18c, d). After the particles had been injected, the relative humidity was lowered to 15% in order to precipitate sodium chloride crystals. The crystals acted to bind the particles together. The end product was a sphere coated with a cemented aggregate layer (Figure 16.18e). Aggregate layers up to 0.5 mm thick were fabricated in this way, the thickness being related to the amount of liquid available on the surface of the sphere. Figure 16.19a is an SEM image of an aggregate made in the laboratory by sticking of volcanic particles to a sphere coated with sodium chloride solution. For comparison, Figure 16.19b shows part of the surface of an accretionary lapillus collected during fallout at Sakurajima. The porosities of the two aggregates appear to be similar.

Particle size analysis was carried out on the volcanic ash introduced into the air stream and on the ash which coated the suspended polystyrene spheres. Figure 16.20 shows a plot of particle number against equivalent spherical particle diameter for each sample. The ash adhering to the sphere was enriched in fines compared to the ash in the air stream.

Discussion

These experiments imply that formation of accretionary lapilli simply by particle scavenging of liquid drops is unlikely. This process results in hemispherical aggregates with flat upper surfaces, which have not yet been observed in nature. Furthermore, it takes approximately 30 minutes for a 5 mm diameter particle-laden drop falling at 10 m s^{-1} to evaporate to dryness at a relative humidity of 15% in the wind tunnel. This implies that a particle-laden drop would have to fall approximately 18 km at a relative humidity of

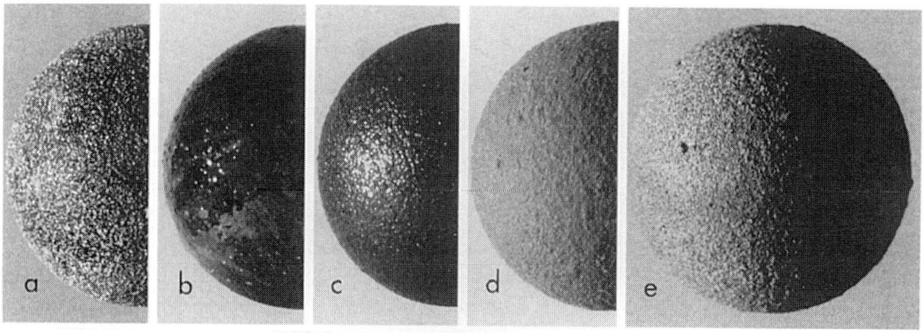

Figure 16.18 Sequential photographs of an experimental run showing the polystyrene sphere (7.8 mm in diameter) suspended in the air stream of the wind tunnel accumulating volcanic ash (< 300 μm in diameter). (a) Sphere suspended in a dry air stream (15% relative humidity) with sodium chloride crystals which have precipitated on the surface. (b) The same sphere in an air stream of 85% relative humidity. A layer of sodium chloride solution has now developed. (c) Sphere coated with volcanic ash at 85% relative humidity. (d) A thicker coating of ash than in (c), at 85% relative humidity. (e) Sphere coated with ash and sodium chloride crystals, the latter having cemented the particles together, in a relatively dry air stream (15% relative humidity) in the final stages of the experiment. (After Gilbert and Lane 1994a, reproduced by permission of *Bulletin of Volcanology*.)

< 15% if an accretionary lapillus was to be deposited on the ground. This aggregation process is unlikely during small eruptions, such as at Sakurajima, where the plumes rarely reach altitudes in excess of 6 km. Finally, raindrops with diameters exceeding 4 mm occur in thunderstorms where drops form from melting ice particles (Rauber *et al.* 1991). However, raindrops exceeding 3 mm in diameter are known from experiments to disintegrate upon collision with neighbouring drops provided that the diameter of the neighbour exceeds 1 mm. This implies that if accretionary lapilli form by the mechanism proposed in Figure 16.12a they would not have diameters in excess of 3 mm and accretionary lapilli size distribution should mirror that of raindrops (Willis 1984; Jones 1992). In the experiments spherical aggregates could only be formed from liquid drops < 1 mm in diameter. Accretionary lapilli occur with substantially larger diameters than this (e.g. Schumacher and Schmincke 1991). Therefore, the first hypothesis (Figure 16.12a) precludes the generation of concentrically zoned spherical aggregates and, as a result, is unlikely to have widespread implications for the growth of accretionary lapilli but it will result in mud rain.

The hypothesis of Figure 16.12b best explains accretionary lapilli growth provided that liquid layers on particle surfaces are solutions of hygroscopic phases as opposed to pure water. This mechanism allows build-up of concentrically zoned (with respect to grain size – depending on the supply of particles to the growing aggregate), cemented, spherical aggregates at relative humidities < 100%. Figure 16.20 shows the particle number distribution for ash injected into the wind tunnel air stream and ash adhered to the liquid-coated spheres for particles between 15 and 130 μm in diameter. The curves show that particles of different diameters have different aggregation coefficients. In the experiments the large particles must collide with the aggregate but they do not adhere. This is because large particles may have sufficient momentum or aerodynamic drag to overcome the

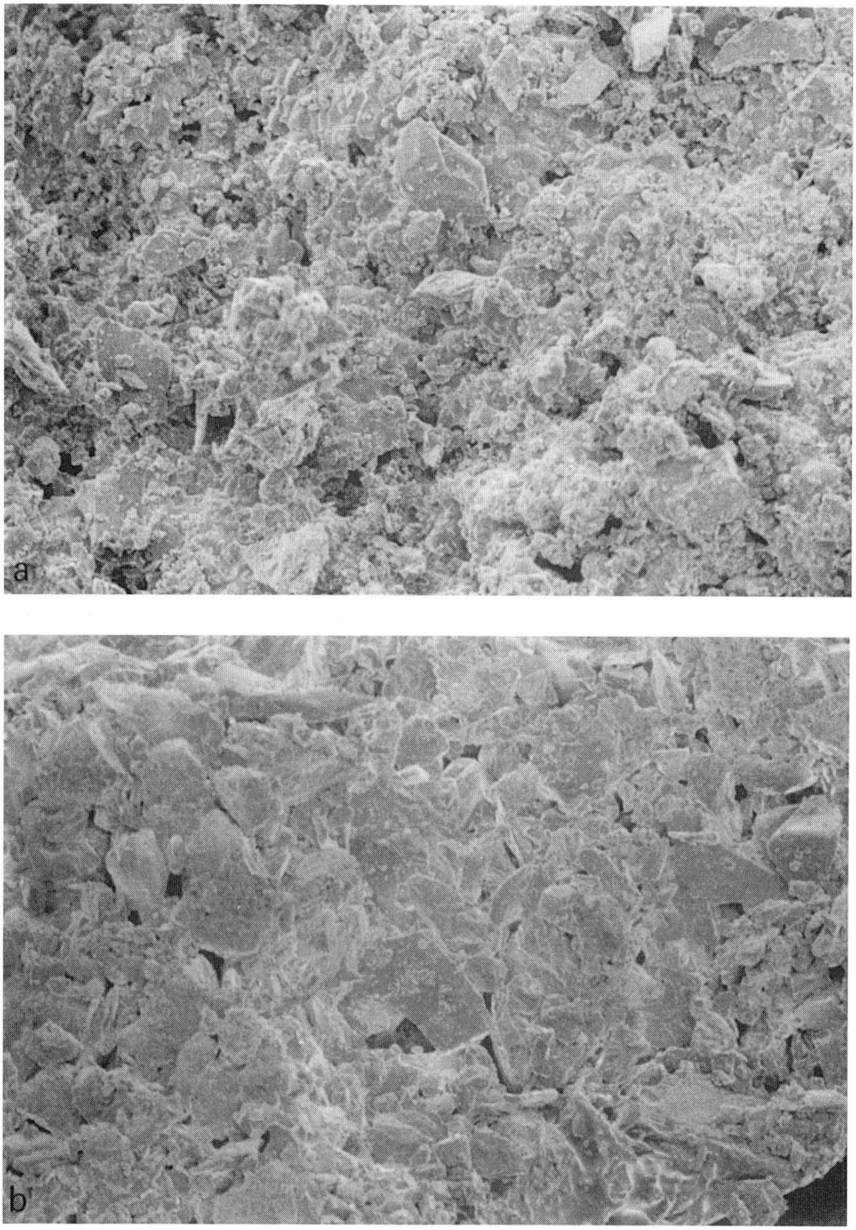

Figure 16.19 SEM images of: (a) part of an aggregate made in the wind tunnel by sticking of volcanic particles to a polystyrene sphere coated with sodium chloride solution (field of view is 80 μm), and (b) part of an accretionary lapillus collected during fallout from a Sakurajima plume in May 1990 (field of view is 300 μm). (After Gilbert and Lane 1994a, reproduced by permission of *Bulletin of Volcanology*.)

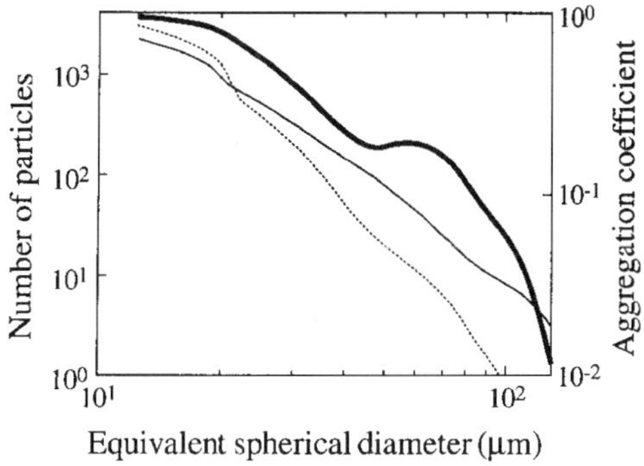

Figure 16.20 Number of particles against equivalent spherical particle diameter. The thin solid line represents volcanic ash introduced into the air stream during the wind tunnel experiment. The dotted line represents volcanic ash removed from the polystyrene sphere after the experiment. The thick solid line represents the aggregation coefficient calculated from the normalized ratio of number on sphere to number injected for each diameter channel. All lines are 33% weighted least squares fits to the data points. (After Gilbert and Lane 1994a, reproduced by permission of *Bulletin of Volcanology*.)

capillary binding force. The aggregation coefficient does not decrease at smaller particle sizes as may be expected if particles followed streamlines around a growing aggregate. This is consistent with data of May and Clifford (1967) which show that this effect is not important until particle size is <10 μm.

Schumacher and Schmincke (1991) found that at least 85 wt% of accretionary lapilli from the Laacher See tephra comprised particles <90 μm in diameter and reviewed other workers' grain size data to give a median grain size of approximately 36 ± 12 μm for accretionary lapilli. Tomita *et al.* (1985) found that accretionary lapilli from Sakurajima volcano were mainly composed of particles in the 6–22 μm range. Sheridan and Wohletz (1983) found that accretionary lapilli from the Pompeii and Avellino deposits of Vesuvius had median grain sizes of approximately 50 μm and that clasts larger than 125 μm were extremely rare. Brazier *et al.* (1982) showed that approximately 75 wt% of accretionary lapilli from the April 26, 1979 eruption of Soufrière, St Vincent, comprised particles <88 μm in diameter. The experiments suggest that these data are simply explained by the fact that accretionary lapilli aggregation processes are strongly particle size dependent and that only particles <80 μm in diameter have a significant probability (i.e. an aggregation coefficient >0.1) of being incorporated into a growing accretionary lapillus (Figure 16.20).

In summary, the formation of accretionary lapilli is controlled by ambient plume motions, fall velocity differences, surface tension forces and secondary mineral growth (Table 16.1). Electrostatic attraction may be important during nucleus formation and for the incorporation of very small particles which would otherwise follow streamlines around the falling aggregate. Mud raindrops form by scavenging of particles by liquid drops which have high fall velocities compared to most of the scavenged particles. The experiments and field data imply that particles of different diameters have different binding efficiencies. Large particles do not always adhere, but may rebound as they strike the aggregate if they have sufficient momentum to overcome the capillary binding force.

16.4.2 Theoretical Models of Aggregation

The laboratory experiments have established that some of the collision and binding mechanisms outlined in section 16.3 and listed in Table 16.1 are highly efficient. Therefore, a theoretical model for particle aggregation is now developed by combining a parameterization of the rate of particle–aggregate–liquid drop collisions with a model of the binding efficiency of the collisions.

Collision dynamics

The rate of collision is specified by considering the number of particles in the volume swept by an aggregate as it moves through space (Figure 16.11). However, this picture requires some refinement. First, very small particles act as passive tracers in the flow, and move with the gas around the aggregate. The tendency for particles to follow streamlines around objects can be quantified as the ratio of mass of particles of any one size impacting with the object to the mass that would have passed through its projected area if it were removed. From data of May and Clifford (1967) this ratio is approximately 0.95 for 50 μm, 0.9 for 20 μm, 0.75 for 10 μm and 0.5 for 7 μm diameter particles interacting with a 5 mm diameter sphere of fall velocity 10 m s^{-1}. Therefore, particles larger than 20 μm in diameter in general do not follow streamlines around objects but collide with them. Growth of an aggregate by this mechanism alone should cause a natural cut-off in the minimum size of aggregate particles. Second, if the collision force is electrostatic, then the effective volume of fluid from which particle collisions may originate is significantly larger than would take place solely by fall velocity differences, because particles may be drawn in and collide even if they are initially several particle radii outside the swept volume (Figure 16.11).

The rate of collision of an aggregate with particles of radius p per unit time t, $N(p)$, is given by

$$\frac{\mathrm{d}N(p)}{\mathrm{d}t} = f\pi \Delta u(p) R^2 c(p) \tag{16.1}$$

where $c(p)$ is the concentration of particles of radius p, $\Delta u(p)$ the mean aggregate speed relative to the particles of radius p, R the aggregate radius and the function f denotes the fraction of particles in the swept volume with which the aggregate collides and is known as the collision coefficient. This coefficient depends upon both the particle and aggregate sizes as well as the magnitude and sign of electrostatic charge. In the case of an electrostatic charge of attraction, f may exceed unity.

For particles falling freely under gravity, the relative velocity is given by a combination of the turbulent motion and settling speeds of the particles. Here, for simplicity, the collisions produced by the relative settling speeds of the particles are considered. This is the dominant process for larger clasts with settling speeds in excess of the mean turbulent velocities. A more complete analysis, particularly for small particles, would include the effects of ambient turbulence.

By defining the bulk density of the aggregate in terms of the mass contained in the enclosing sphere, the settling speed of the aggregate u was shown to be given by the approximate relationship

$$u = \sqrt{\frac{Rg(\rho_s - \rho_a)}{0.15\rho_a}} \tag{16.2}$$

where g is the acceleration due to gravity, ρ_s the aggregate density and ρ_a the air density.

Therefore the relative mean aggregate fall speed $\Delta u(p)$ is given by

$$\Delta u(p) = \sqrt{\frac{Rg(\rho_s - \rho_a)}{0.15\rho_a}} - \sqrt{\frac{pg(\rho_s - \rho_a)}{0.15\rho_a}} \tag{16.3}$$

If the porosity of the aggregate is ϕ, then the density of the aggregate ρ_s may be found from the expression for its mass M as a function of R, thus

$$M = \frac{4\pi}{3}R^3(\rho_p - \phi(\rho_p - \rho_\phi)) \tag{16.4}$$

where ρ_p is the particle density and ρ_ϕ is the pore density of the aggregate. Typically, aggregates are considerably larger than their component particles, and therefore have much larger settling speeds than their component particles. Therefore, for the purposes of modelling aggregate growth, the relative velocity (equation 16.3) may be approximated with the settling speed of the aggregate, u (equation 16.2).

Binding efficiency

The number of colliding particles sticking to the growing aggregate depends upon the binding mechanism and the relative momentum of the particles. Particles with high relative speed and momentum may be able to overcome the binding force and rebound. The binding mechanism is a function of temperature and hence position in the plume as well as particle size. For regions of the plume in which the temperature is above the freezing point of the liquid, liquid films will be present on particle surfaces. However, in other regions of the plume where temperature is sufficiently low that solidification of the condensate occurs, then the binding efficiency is massively reduced.

If the binding efficiency (or sticking coefficient) is denoted as $e(p, z)$, for particles of radius p, at a height in the plume of z, then the rate of increase of the mass of the aggregate is

$$\frac{dM}{dt} = \pi R^2 \cdot \int_0^\infty u(p)e(p, z)f(p, R)c(p)\left(\frac{4\pi}{3}p^3\right)\rho_p dp \tag{16.5}$$

In order to complete the model, values for the collision coefficient f and sticking coefficient e must be specified.

In regions of the plume where liquid films develop and surface tension forces play a role in holding particles together, particle binding will be particularly efficient. Based on their experimental data for surface tension driven binding, Gilbert and Lane (1994a) found that the bulk aggregation coefficient, defined as E, where $E = ef$, is close to unity for particles in the 15–20 μm diameter range but that it decreases to <0.1 for particles larger than 80 μm in diameter (Figure 16.20). Data of May and Clifford (1967) imply that for particles <7 μm in diameter, E again falls close to zero because the majority of particles in this size range are able to avoid collision. By integrating these empirical laws over all particle sizes, the average mass of particles aggregated per unit swept volume may be written as a fraction A of the total mass of particles in that volume, w. Gilbert and Lane (1994a) suggested that A has a value in the range 0.1–1.0. In particularly cold regions of the plume where the condensate is solid, A can be much smaller, and may be modelled as being approximately zero.

Using this approximation, equation (16.5) is then written in the simple form

$$\frac{\mathrm{d}M}{\mathrm{d}t} = A\pi u R^2 w \tag{16.6}$$

As the mass of the aggregate increases, so does its radius. By assuming that the porosity of the aggregate remains nearly constant, the radius of the aggregate may be related to the mass (equation 16.4). Therefore the mean radius of the aggregate may be found as a function of the time spent by the aggregate at temperatures in the range suitable to allow liquids to be present on particle surfaces. The model implicitly assumes that the porosity of an aggregate is constant, and this gives a leading order representation of the process.

If the plume mass loading is time independent then, from equations (16.4) and (16.6), we can derive an expression relating R to the distance travelled, thus

$$R = \frac{wAx}{4(\rho_\mathrm{p} - \phi(\rho_\mathrm{p} - \rho_\phi))} \tag{16.7}$$

The initial radius of the aggregate is typically very much smaller than its final radius, and is therefore negligible in this equation.

Discussion

Values of plume mass loadings for volcanic eruption columns vary from a few tens of kilometres per cubic metre at the vent to $< 10^{-2}$ kg m^{-3} for neutrally buoyant plumes (Harris *et al.* 1981; Lane and Gilbert 1992; Bursik *et al.* 1994). In a typical plume with mass loading of 6×10^{-3} kg m^{-3}, the size to which an aggregate, such as an accretionary lapillus, grows depends primarily upon the density of the solid phase, the bulk porosity of the aggregate, the average aggregation coefficient and the distance travelled. For typical porosities of accretionary lapilli (0.3–0.5), the accretionary lapillus radius is found (from equation 16.7) to be nearly independent of the air density and hence altitude. In Figure 16.21, the predicted size of the aggregate is shown as a function of distance, for aggregation coefficients of 0.1, 0.75 and 1.0 and a porosity of 0.4. As the aggregation coefficient decreases, the growth rate of the aggregate also decreases.

Figure 16.21 shows that for growth of accretionary lapilli during a single pass through a plume depth of 0.5–10 km, the size of the aggregate is predicted to be 0.7–20 mm. The maximum size of 20 mm corresponds well with field observations of Moore and Peck (1962). Spreading umbrella cloud thicknesses are several kilometres (Woods 1988). Brazier *et al.* (1982) reported the upper and lower altitudes of the April 26, 1979 eruption of the Soufrière, St Vincent to be 8.6 km and 3–4 km respectively. This gives a maximum within-plume fall distance of approximately 5 km. Accretionary lapilli fell from this plume with diameters ranging from 1 to 7 mm, with most in the 1–3 mm range. The maximum size predicted from Figure 16.21 for a 5 km plume depth, with an aggregation coefficient

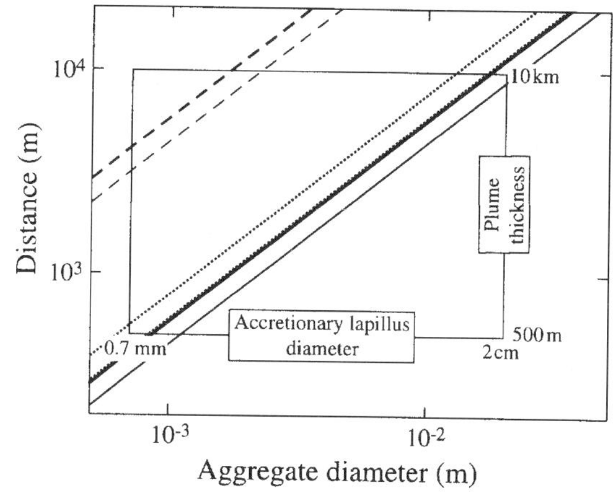

Figure 16.21 Distance travelled by aggregate versus aggregate diameter. Data are for an aggregate of porosity 0.4 falling through a volcanic plume of mass loading 6×10^{-3} kg m^{-3}. All bold lines represent liquid-filled pores and the other lines air-filled pores. The solid lines represent an aggregation coefficient of unity, the dotted lines of 0.75 and the dashed lines of 0.1. The box represents typical plume thicknesses of 0.5–10 km, which give a range for maximum accretionary lapillus diameter of 0.7–20 mm. (After Gilbert and Lane 1994a, reproduced by permission of *Bulletin of Volcanology*.)

of 0.75 (Figure 16.20), is approximately 7 mm. During eruptions at Sakurajima volcano the plane plume is normally approximately 1 km thick and plume heights are 3–4 km. From Figure 16.21 fallout through a 1 km thick plume would generate a maximum accretionary lapilli diameter of 2 mm. Tomita *et al.* (1985) reported accretionary lapilli in the size range 1–5 mm. The larger than predicted accretionary lapilli falling at Sakurajima may be explained by an increase in collision rate due to ambient turbulence within the plume. Accretionary lapilli have been observed falling within five minutes of eruption initiation at Sakurajima, indicating that they are probably growing in the eruption column where particle mass loadings are higher than at neutral buoyancy levels.

Accretionary lapilli are reported to reach sizes in excess of 5 cm (e.g. Schumacher and Schmincke 1991). These are much larger than predicted by a single pass even through a thick umbrella cloud. The formation of such large aggregates may be due to very high mass loadings or to significantly increased fall distances. Fall distances can be increased as a result of re-entrainment into the column after falling from the umbrella cloud or vigorous turbulent eddies within the umbrella cloud. Re-entrainment can explain the concentric layering reported for accretionary lapilli and multiple accretionary lapilli where a new accretionary lapillus nucleates on a pre-existing one.

16.5 SUMMARY

Aggregation of fine ash plays a critical role in controlling the dispersal of particles. All aggregates fall with higher fall velocities than their components. Aggregation produces complex grain size distributions, and can lead to enhanced thickening of fall deposits. Table 16.1 summarizes the dominant contact and binding processes inherent in the formation of volcanic aggregates. Aggregates may be classified according to the relative volumes of liquid they contain when falling, and range from dry aggregates through damp accretionary lapilli to mud raindrops.

The humidity of the host plume appears to dictate the aggregate growth mechanism. Rate of aggregate growth depends upon the rate of collision of aggregate, particles and liquid drops, and the efficiency of binding. Aggregation processes are strongly particle size dependent. Particles less than approximately 7 μm in diameter tend to avoid collisions, unless there are strong electrostatic forces acting. As particle size rises, it becomes increasingly difficult for the binding mechanisms to overcome the relative momentum of the colliding particles, and the particles separate after collision. The dominant grain diameters of both natural and laboratory simulated aggregates are <200 μm.

Particles falling from plumes in low humidity atmospheres carry high levels of electrostatic charge (section 16.3.1). The interplay of electrostatic forces and differences in fall velocities result in formation of dry aggregates. These types of aggregates do not remain intact when they are deposited. Nevertheless, their fingerprints remain in fall deposits in the form of enhanced thickening and polymodal grain size distributions (sections 16.2.1 and 16.2.2 respectively).

Accretionary lapilli form in relatively humid plumes, and the particles appear to bind on collision as a result of surface tension forces associated with thin films of condensate on particle surfaces. The formation of this condensate is greatly enhanced by the presence of hygroscopic compounds. While such aggregates move through the atmosphere, some of the liquid condensate may evaporate, producing minerals which cement the particles together. The formation of accretionary lapilli can be inhibited if particles are injected more than a few kilometres into the atmosphere because, as temperatures drop, the condensate tends to freeze with the result that the binding efficiency of colliding particles plummets. Electrostatic attraction may be important during nucleus formation and for the incorporation of very small particles which would otherwise follow streamlines and sweep around the falling aggregate. These types of aggregates are usually well cemented by the time they hit the ground and therefore become visibly locked into the geological record. Mud raindrops form in extremely humid plumes. Liquid drops develop as a result of condensation of the upward transport of atmospheric vapour, groundwater and/or magmatic gases. These drops then scavenge particles as they migrate through the atmosphere.

The classification of aggregates shown in Table 16.1 is not rigid; the complete spectrum of aggregate types exist in nature and one type of aggregate may evolve into another due to unsteady conditions within the plume. For example, a dry aggregate might evolve into an accretionary lapillus if the ambient relative humidity was raised. The resulting aggregate would be a high porosity structureless core surrounded by concentric zones of particles organized into a lower porosity spherical accretionary lapillus. These types of complex aggregates are commonly observed in volcanic sequences. Aggregates respond to subtle variations in the ambient atmosphere and, for this reason when examined in the geological record, provide "windows" of information into past plume conditions.

17 Environmental Hazards

17.1 INTRODUCTION

Volcanic eruptions can cause havoc to life on Earth. For land-based life the most catastrophic of volcanic hazards are lahars and pyroclastic flows. Volcanic plumes, in contrast to lahars and pyroclastic flows which cause rapid and absolute devastation within a confined area, are not generally fatal but their effects should not be underestimated. Volcanic plumes (Figure 17.1) comprise a mixture of solids, liquids and gases. The solid particles range in size from blocks tens of metres in diameter, which fall close to the volcanic vent, to millimetre and micrometre-sized particles which may be transported through the atmosphere for many hundreds or thousands of kilometres before being deposited on the ground. Rock fragments <2 mm in diameter are known as ash particles and these are hard, dense, angular, abrasive and pervasive. Volcanic gases such as water vapour, sulphur dioxide, hydrogen fluoride and hydrogen chloride, condense within plumes to form acidic solutions of corrosive liquids.

Fallout from plumes can cause respiratory problems for humans and animals, contamination of food and potable water, and result in collapse of buildings. It can damage machinery and engines. Sedimentation from plumes can also cause rapid infilling of reservoirs and problems for hydroelectric schemes. Thick, extensive ash fall deposits provide a ready sediment source for remobilization into potentially dangerous lahars. Volcanic plumes may be responsible for long-term hardship over vast regions. Fallout from the largest eruptions may cover millions of square kilometres (Chapter 13) and emplacement of volcanic pollutants into the atmosphere may drastically perturb the climate (Chapter 18). Eruption plumes have the potential to inject enormous quantities of rock particles and gases high into the atmosphere. Therefore, for airborne creatures such as birds and insects and for aircraft in flight, volcanic plumes constitute the most hazardous of eruption phenomena. With the ever increasing world population accommodated close to active volcanoes and the rising pressure on aviation travel, the hazards associated with plumes are becoming more frequently encountered. For this reason there is a need to be familiar with and able to mitigate plume-related environmental hazards.

The term *hazard* describes a phenomenon which poses a potential threat to persons, animals, vegetation or property. The magnitude of the hazard depends on many factors including: the type of plume (i.e. whether or not it is particle-laden and the composition of its gases), the degree of dispersal and dilution of the plume and the duration of the eruption. In this chapter the effects of volcanic plumes on human health, animals, vegetation, property and community infrastructure are considered. Due to the potentially large number of fatalities that may result from the encounter of a passenger aircraft with a

Figure 17.1 A typical Sakurajima plume, depositing particles on the surrounding countryside. The photograph was taken in May 1991, 5.5 km west of the active vent when the prevailing wind direction was north. (Photograph courtesy of S. J. Lane.)

volcanic plume, as well as the enormous monetary costs involved, a substantial part of this chapter is devoted to the impact of volcanic plumes on the aviation community.

17.2 HEALTH HAZARDS TO HUMANS

The products of volcanic eruptions are frequently directly perilous to humans. The hazards of volcanic plumes such as fallout of particles, mud rain and acid rain, the presence of noxious gases, lightning strikes and shock waves associated with eruptions are given in Table 17.1. Many of these hazards occur simultaneously during a single eruption and are complexly linked. The lack of autopsies on people killed in eruptions means that it is often difficult to isolate one hazard as the sole cause of death or injury. In this section incidents where fallout from eruption plumes has posed direct hazards to the health of humans are described. For detailed information on specific case studies the reader is referred to Blong (1984).

A major hazard to humans living close to volcanoes is particle fallout from eruption columns and dispersing plumes. The largest rock fragments, from several tens of metres to a few tens of centimetres in diameter, are thrown from the vent ballistically. Depending on their sizes and surface temperatures, impact by these may be fatal or cause lacerations and burns. Smaller particles of pumice, scoria and ash are transported within plumes to distances dependent on the densities of the particles and the prevailing wind velocities (see Chapters 4 and 13). The solid material then sediments from the plume and is deposited on

Table 17.1 Health hazards of plumes to humans and animals

Hazards	Effects
Particle fallout, including ballistics	Lacerations, burns, abrasion, burial, asphyxiation, respiratory ailments, eye irritation, collapse of buildings and crushing, contamination of food and potable water
Mud and acid rain fallout	Collapse of buildings and crushing, contamination of food and potable water, generation of lahars
Noxious gases	Asphyxiation, respiratory ailments, eye irritation
Lightning strikes	Electrocution and burns
Shock waves	Lacerations due to window glass breakage, deafness

the ground. Such particles comprise angular fragments of vesiculated glass, crystals and rock (lithic) fragments. Skin and eyes exposed to this type of fallout are extremely susceptible to lacerations, burns and abrasion, although exposure to fallout of particles <2 mm in diameter (i.e. volcanic ash) would be unlikely to result in lacerations and burns.

A heavy fallout of pumice or scoria may lead to burial, crushing or asphyxiation. Fallout of ash, on the other hand, would be unlikely to be of a sufficient rate to result in catastrophic burial, but respiratory problems could be serious. During the AD 79 eruption of Vesuvius, Italy, which buried Pompeii (Figure 17.2), some people survived as much as 2.8 m of pumice fall only to be killed by a pyroclastic surge (Francis 1979; Sigurdsson *et al.* 1982; Blong 1984; Baxter 1990). Therefore the total number of people killed by pumice fall at Pompeii is unknown but probably only a few hundred (Blong 1984).

Short-term medical problems experienced by those in regions affected by fallout include eye irritation, sore throat, nasal infections and aggravated asthma and bronchial conditions (Table 17.2). While none of these are considered fatal, they may cause considerable discomfort and exacerbate pre-existing ailments (Blong 1984). Much data were collected on the effects of volcanic plumes on human health as a result of the May 18, 1980 Mount St Helens eruption, in Washington, USA (refer to Figure 13.12 for a map of the Mount St

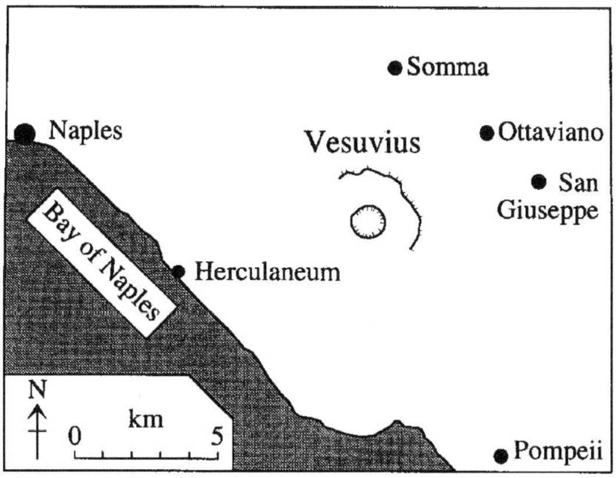

Figure 17.2 Map of Vesuvius, in Italy, showing locations of towns

Helens fallout region). Statistical data showed a sharp rise in admissions and emergency room visits for people with respiratory ailments in hospitals in the heaviest ashfall areas of eastern Washington (Anderson 1982). Many of those admitted, however, had long histories of respiratory difficulties that were aggravated by the fallout. More serious *long-term* conditions may be brought on by prolonged exposure to particles less than 15 μm in diameter (Blong 1984). These arise when silicate particles are inhaled and become lodged in the lungs, or other parts of the respiratory system. The particles may be subdivided into two groups, those with diameters 2.5–15 μm, and those with diameters <2.5 μm. The coarser of these may be deposited in the conducting airways of the human respiratory tract, and the finer in the terminal bronchial and alveolar portions of the lung. Measurement of the <15 μm suspended particles in Moscow, Idaho (Figure 13.12), indicated that dangerously high particle concentrations lasted for approximately one week after the May 18 eruption (Gage *et al.* 1982). Buist (1988) evaluated both the short- and long-term effects of exposure to volcanic ash from Mount St Helens and concluded that:

1. Ash acts as an irritant to the eyes;
2. There is no known effect of ash on the skin;
3. Fresh ash acts as an irritant to the upper and lower respiratory tract leading to mucous hypersecretion and bronchoconstriction that are likely to be reversible when exposure ceases;
4. The exposures experienced during the Mount St Helens eruptions were not sufficiently high to cause pulmonary fibrosis.

A major problem from particle fallout, and particularly mud rain, is that it may induce collapse of buildings (Figure 17.3), due to the weight of accumulated ash. Crushing as a result of building collapse is likely to be the largest cause of death and injury as a direct result of fallout from volcanic plumes (Table 17.3). For example, some of the skeletons

Table 17.2 Examples of short-term medical ailments triggered by specific eruptions

Volcanic eruption (year)	Reported ailments
Katmai, Alaska (1912)	Babies on Kodiak affected by severe coughs and eyes full of mucus (Erskine 1962)
Irazú, Costa Rica (1963–65)	Acute conjunctivitis, throat irritation, nasal irritation and discharge, and prolonged stress were experienced by people living 30 km west of the volcano. Severe bronchitic symptoms experienced by people with pre-existing chest complaints (Horton and McCaldin 1964)
Usu, Japan (1977)	Inflammation and discharging of eyes, as well as nose and throat irritations; particularly severe for those working out of doors (Saito *et al.* 1978)
Mount St Helens, USA (1980)	Increase in emergency room admissions by people with respiratory ailments (Anderson 1982)
Pinatubo, Philippines (1991)	Respiratory and eye problems forced ~5000 residents on west flank of volcano to leave their homes (SIBGVN 1991, vol. 16, no. 5)
Poás, Costa Rica (1994)	Nausea and coughing, irritated throat, eyes and skin (SIBGVN 1994, vol. 19, no. 5)

Figure 17.3 Patio area in the Officer's Club at Clark Air Base, in the Philippines on July 31, 1991 after the eruption of Pinatubo. Tables are covered with ash and some roofs have collapsed due to ash loading. (US Geological Survey photograph courtesy of T. J. Casadevall.)

retrieved from the deposits of the AD 79 eruption of Vesuvius, in Italy, had fractured skulls (Maiuri 1961), suggesting that roof collapse caused death. During the 1906 eruption, roof collapse produced approximately 200 fatalities in the town of Ottaviano (Figure 17.1) and in San Giuseppe 105 were killed as a result of the collapsed church roof (Lacroix 1906; Rittman 1962).

Mud rain and rainfall through "steam" plumes is acidic (section 16.2.3) with water, sulphuric acid and hydrochloric acid typically being dominant components. Medical problems related to acid rain include raw or burning throats and blisters on the lips (Erskine 1962), flesh burns (Griggs 1919) and loss of hair (Jagger 1925). Acid rain may

Table 17.3 Examples of deaths due to building collapse associated with specific eruptions

Volcanic eruption (year)	Number of deaths due to building collapse
Tarawera, New Zealand (1886)	Most of the 150 people killed in the towns of Te Ariki, Moura and Te Wairoa (Blong 1984)
Santa Maria, Guatemala (1902)	2000, out of a total of 5000 deaths (Anderson 1908)
Vesuvius, Italy (1906)	200 in Ottaviano (Lacroix 1906), 105 in San Giuseppe (Rittman 1962)
Vesuvius, Italy (1944)	17 (Blong 1984)
El Chichon, Mexico (1982)	Most of the 153 deaths reported following the eruption, coupled with fires started by incandescent fallout (Blong 1984)

react with zinc (sometimes present in galvanizing on roofs) releasing heavy metals which drain into drinking water supplies (Baxter *et al.* 1982). Acid rain at Poás, in Costa Rica, has resulted in enormous economic losses (of the order of several million US dollars) since 1988 through loss of timber, crops, machinery, grazing land, livestock, homes and human health (SIBGVN 1994, vol. 19, no. 5). Provided that direct contact with mud rain is avoided, rainfall during ashfall should be considered advantageous because it removes pollutants from the atmosphere at a relatively high rate (Lane *et al.* 1993).

Not only may noxious gases and particles induce asphyxiation and respiratory problems but they may also result in starvation and disease (Blong 1984). Starvation is not caused directly by ashfall, but rather by ashfall killing crops which feed humans and grazing livestock which are used for human consumption. Poor diets, which can occur as a result of water supplies stagnating and crops and livestock having been destroyed, leave people receptive to disease. The United Nations Development Programme reported that the death toll for the 1991 Pinatubo eruption had reached 722, of which 358 deaths were attributed to disease (SIBGVN 1991, vol. 16, no. 9).

Lightning is common in volcanic plumes (section 12.6.1). This results from high potential gradients generated by charge separation due to sedimentation of charged particles. Explosions which generate large quantities of small particles, such as during phreatomagmatic eruptions (Chapter 8), create relatively large amounts of charge per unit mass (Lane and Gilbert 1992). Although some deaths have occurred due to lightning strikes from plume to ground, such data are obscured by that of burns from pyroclastic flows. Hamilton (1772) reported that during the 1631 Vesuvius eruption, in Italy, men and beasts were struck dead by lightning, and Blong (1984) noted that some of the 1200 fatalities resulting from the 1814 eruption of Mayon, in the Philippines, were caused by electrical discharges. Also a lightning strike resulting from the 1883 Krakatoa eruption, in Indonesia, hit the lightning rod at First Point lighthouse in Sunda Strait, broke it and burned four convicts wearing iron rings (Simkin and Fiske 1983). Similarly, during the 1994 eruption of Rabaul, in Papua New Guinea, a man was struck by lightning (SIBGVN 1994, vol. 19, no. 8). Many explosive eruptions, such as that which preceded the lateral blast at Mount St Helens in May 1980 and those which occur regularly at Sakurajima, in Japan, generate atmospheric shock waves (e.g. Ishihara 1985). These cause windows to shatter and produce skin lacerations due to flying glass, as well as earaches and deafness (Blong 1984).

17.3 HAZARDS TO ANIMALS

The hazards of volcanic plumes to animals are similar to those of humans (Table 17.1) but are compounded when animals attempt to feed on vegetation coated with ash bearing toxic halogen precipitates. Also, teeth and "feet" are susceptible to serious abrasion, and starvation is likely if food supplies become buried with ash.

Sedimentation of particles from plumes is well known to generate respiratory and eye problems in animals as well as in humans. During the 1912 Katmai-Novarupta eruption, in Alaska (Figure 17.4), cattle suffered from severe nasal and eye infections while some bears were blinded by the ash and animals such as ermines, marmots, mice, rabbits and squirrels died (Martin 1913). Remobilization by wind of abrasive ash following the August 1991

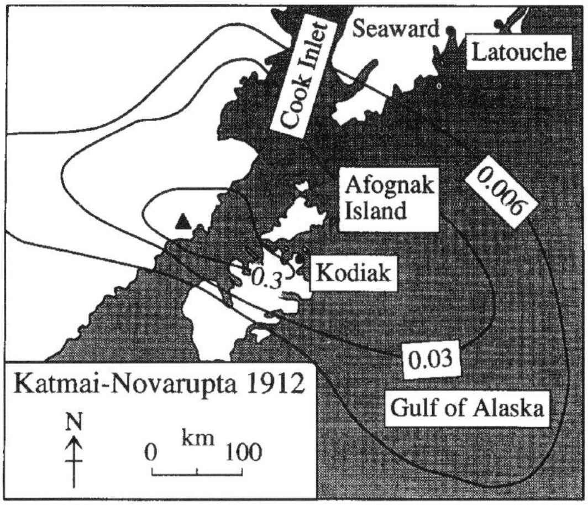

Figure 17.4 Isopach map for the 1912 eruption of Katmai-Novarupta, in Alaska. Solid triangle indicates location of volcano. Contour intervals are in metres. (After Wilcox 1959, reproduced by permission of the US Geological Survey.)

eruption of Hudson volcano, in Chile, caused extensive death and blinding of animals (SIBGVN 1991, vol. 16, no. 9; Scasso *et al.* 1994).

While impact with ballistic blocks and collapse of shelters due to loading with tephra can result in animal fatalities, burial by tephra is more common. The generation of mud rain as a result of rainfall during ashfall can result in movement of liquefied ash on steep slopes and generation of lahars. Burial by lahars may have particularly disastrous consequences.

Perhaps the most serious hazard for animals is contamination of the food supply. Depending on the chemistry of the eruption products, soluble fluorine may precipitate on surfaces of particles (Oskarsson 1980). Fluorine contents of grass in excess of 250 ppm are particularly toxic to grazing animals (Blong 1984). The high halogen contents of eruption products of many of the Icelandic volcanoes has had far-reaching consequences. Poisoning of approximately 1500 ewes and 6000 lambs followed the 1970 eruption of Hekla (Thorarinsson and Sigvaldasson 1972) when fluorine contents of up to 4000 ppm were recorded on ash-contaminated vegetation. These extreme levels affected only the sheep even though cattle, horses, dogs, cats and poultry lived in the area. Fluorine poisoning has been a common cause of deaths and injuries to livestock during the eruptions of Hekla 1693, 1776 and 1845 (Thorarinsson 1970), Grimsvotn 1934 (Nielsen 1937), Askja 1961, Surtsey 1963–64, and Heimaey 1973 (Blong 1984). The greatest of the Icelandic death tolls came after the Laki eruption in 1783, when approximately 50% of the horses and 79% of the sheep in the whole of Iceland died as a result of fluorinosis, other sicknesses and starvation (Stephensen 1785; Blong 1984).

Animals foraging for food can grind away their teeth as a result of eating ash-laden vegetation (Bolt *et al.* 1975). This was the fate of reindeer following the eruption of Katmai-Novarupta in 1912 (Martin 1913) and Aniakchak Crater, also in Alaska, in 1931 (Trowbridge 1976). In addition to tooth abrasion, animals may suffer from abrasion of paws and hooves if in contact with ash for long periods.

Most animal losses after ashfall occur as a result of starvation due to vegetation being covered with ash (Ross 1816; Huggins 1902). In these situations carnivorous animals tend to suffer less than herbivores. The 1991 eruption of Hudson volcano caused problems for livestock (principally sheep and cattle) and wildlife in Chile and Argentina where ashfall polluted water sources and covered pasture land. This resulted in starvation of many animals (SIBGVN 1991, vol. 16, no. 9). In some volcanic emergencies animals are evacuated, and this in itself leads to problems such as overcrowding and relatively rapid transfer of disease. Illnesses associated with moving animals are mainly increased abortion and of a nutritional nature, which can occur when animals acclimatized to high altitudes are brought down to lower altitudes. Problems of this type arose as a consequence of the 1963–65 Irazú eruption when 6000 cattle grazing on the slopes of the volcano had to be moved down to the plains (Horton and McCaldin 1964).

Animals are also susceptible to asphyxiation, lightning and shock waves. For example, livestock are known to have been killed by lightning during eruptions of Vesuvius 1631 in Italy (Hamilton 1772) and Soufrière, St Vincent, 1902 in the Caribbean (Anderson and Flett 1903) and will be affected in the same manner as humans by shock waves.

Several plume-related hazards are unique to flying animals such as birds and insects. Lacerations, abrasion, blinding and asphyxiation are the main problems for birds. In addition, during ashfall the clogging of feathers can severely restrict the mobility of birds. The blanketing of food sources by ash, pumice and scoria can also result in deaths and hardship. Examples of plume-related bird fatalities have been reported during the 1906 eruption of Vesuvius, in Italy, where birds with open beaks lay dead, as though asphyxiated (de Lorenzo 1906). The Tarawera 1886 eruption, in New Zealand, killed pigeons, ducks and sparrows. Surviving sparrows were temporarily blinded, with their eyelids gummed together by mud rain. Pheasants had so much wet ash in their feathers that they were unable to fly, and pheasants and quails came almost to house doors searching for food (Mair 1886). Tephra fall from the 1912 Katmai-Novarupta eruption, in Alaska, killed gulls, snipe, ptarmigan and ducks at Kodiak (Figure 17.4). Sea birds that survived the fallout could not find open water to land on, because of the presence of large sheets of floating pumice and many subsequently died (Erskine 1962). Dead swallows were observed in areas of heaviest ashfall at Mount St Helens, USA, in 1980 (Blong 1984). Cook *et al.* (1981) provided a stimulating summary of the effects of ashfall on insects around Mount St Helens as a result of the 1980 eruption and showed that they suffered from abrasion to the epicuticular wax layer which results in rapid desiccation and death. The problem for insects is compounded due to their large surface area to volume ratios. In similarity with birds, ash on the wings also affects the ability of insects to fly.

The effect of volcanic plumes on sea-dwelling animals is often overlooked yet there are many examples of eruptions where fallout from plumes has killed or damaged aquatic life by increasing water pH and turbidity. The extent of devastation depends on the magnitude and chemistry of fallout and the tolerance of the species present. For example, during the Hekla 1963 eruption, in Iceland, trout and salmon died (Blong 1984) and during the eruption of Tarawera, New Zealand, in 1886, many small fish were washed to the lakes' shores dead or dying (Mair 1886). The tephra fall at Kodiak, Alaska (Figure 17.4) in 1912, proved lethal to spawning salmon and it took several years for the spawning grounds to become fertile again (Erskine 1962). In Katmai Bay where more than 60 cm of tephra fell, barnacles and mussels down to the low tide mark died as did kelp as far as the eastern end of Afognak Island (Figure 17.4). Cod and halibut died in great numbers in the shallow waters of Cook Inlet (Martin 1913). The January 1979 eruption of Karkar, in Papua New Guinea, killed two species of fish, freshwater prawns and eels in rivers draining south-east from the caldera. While some fish, eels and prawns survived, all had a strong sulphurous smell but the local subsistence farmers claimed that they were edible (Blong 1984)! Around the active volcano Sakurajima, in Japan, in order to alleviate damage to fish and to improve offshore fish culture facilities, the removal of pumice from the sea is regularly carried out.

17.4 EFFECT ON VEGETATION

Many volcanic regions are well known for the long-term beneficial effects of rich soils and weathered volcanic deposits, particularly in tropical areas. However, sudden deposition of large amounts of ash, pumice and scoria can seriously damage or destroy both natural and cultivated vegetation. If tephra fall is very light then the effects may be insignificant. Damage to vegetation depends on many factors including the particle size of the tephra, the thickness of the deposit, the stage of plant growth and the tolerance of the species. Direct results of fallout are shown in Table 17.4.

At volcanoes with sustained activity over several tens of years, such as Sakurajima, in Japan (Figure 17.1), vegetation growth may approach a balance with ash deposition. Since 1955, Sakurajima has been in a state of near continuous activity, erupting most days and scattering ash on the vegetated flanks of the volcano (Figure 17.5). Beyond approximately 1 km of the vent the vegetation appears to have kept pace with sedimentation. Even so, in 1988 ashfall was estimated to have caused an annual 5 billion yen worth of damage to agricultural products such as vegetables, fruits and tea (Kobayashi *et al.* 1988). Research into ash-resistant plant breeding is under way at Sakurajima and protective measures, such as shielding crops from ashfall, are routinely undertaken.

During the 1943–52 eruption of Parícutin, in Mexico, large mature trees suffered from branch breakage due to loading with ash. Pine trees with basal diameters 10–30 cm survived best because their stems were sufficiently strong to resist excessive bending yet flexible enough for ash to fall off. Where more than 50 cm of tephra fell large branches were broken and pine seedlings and small trees died as a result of excessive bending and burial (Rees 1970). The reader is referred to Blong (1984; Table 7.3) for a summary of reported effects of ashfall on crops.

Table 17.4 Hazards of plumes to vegetation

Hazards	Effects
Particle fallout, including ballistics	Breakage, abrasion, perforation and defoliation, burns, burial and restriction of roots to oxygen, "caking" of leaves and flowers preventing photosynthesis and pollination
Mud and acid rain fallout	"Caking", burns, contamination of water supply
Noxious gases	Burns
Electrical strikes	Burns

Figure 17.5 Ash accumulation on vegetation at Sakurajima volcano, in Japan. The photograph was taken on May 12, 1990, 4 km northeast of the active vent. (Photograph courtesy of J. S. Gilbert.)

A variety of biological functions may be disturbed by tephra fall. Burial restricts access of roots to oxygen, and "caking" of leaves (Figure 17.5) and flowers with ash prevents photosynthesis and pollination. During the 1943–52 Parícutin eruption, fine ash entered avocado flowers thereby preventing pollination. Following the 1980 Mount St Helens eruption, increased apple drop was probably the result of a reduced rate of photosynthesis where ash clung to the leaves (Cook *et al.* 1981). Yamaguchi (1985) showed that the trees in the region around Mount St Helens, USA, suffered reduced annual growth due to stress associated with ash cover. Such interrupted growth results in tree ring anomalies which provide a useful technique for the precise dating of prehistoric explosive eruptions.

Mud rain, acid rain and gases can also adversely affect vegetation. Acid rain and noxious gases produced by the 1912 Katmai-Novarupta, Alaska, eruption killed vegetation several hundred kilometres to the north-east at Seaward (Figure 17.4). At Latouche, 480 km to the north-east, garden plants were destroyed and the leaves of many native perennials were burnt (Griggs 1922). Observations during eruptions of Mayon, in the

Philippines, in 1928 provide a graphic illustration of the kind of damage acid rain can do. Within 15 km of the volcano the leaves of abacá plants were wilted and browned. Acacia, santal and pilli trees lost most of their leaves while cogon and other grasses, including bamboo, were burnt brown. In some cases, however, only the portions of the plant facing into the wind were damaged (Faustino 1929). Abacá and papaya were particularly sensitive whereas coconut, coffee, banana, gabi and rice appeared unaffected (Galvez 1938; Galvez *et al.* 1939). Similarly, during the 1963–65 eruptions of Irazú, in Costa Rica, bean crops, corn, squash, tomato, onion, oat, rye, wheat and barley were all affected (Miller 1966). During 1986–90 acid rain from Poás, in Costa Rica, disrupted 20 000 m^2 of strawberry, dairy and coffee farms, affecting 681 farmers (SIBGVN 1994, vol. 19, no. 9). For details of the effect of different types of gases on plants the reader is referred to Blong (1984).

An indirect result of explosive volcanic activity are fires caused by either hot ballistic ejecta or lightning. For example, in 1980 following the eruption of Mount St Helens, in the USA, lightning strikes started an estimated 50–100 forest fires, most of which were subsequently doused by falling ash. Lightning strikes were also reported to have damaged tree crops at Soufrière, St Vincent, in the Caribbean in 1902 (Anderson and Flett 1903) and again in 1979 (Rowley 1979).

17.5 PROPERTY DAMAGE

Wherever heavy fallout from plumes takes place in populated areas it is inevitable that damage to property will occur (Figures 17.3 and 17.6). In some cases the monetary damage may be enormous and in all cases substantial personal cost is borne. Common types of property damage are listed in Table 17.5.

The most costly and dangerous problem is collapse of buildings resulting from extreme loading of ash (Figure 17.3). The amount of ash necessary to induce roof failure is controlled by several factors, such as ash/mud rain accumulation rate, wind speed, roof design and construction material (Blong 1984). Where mud rain falls, the problem becomes particularly acute due to the increased weight over that of dry ash. In addition, ground shaking may exacerbate the effects of ashfall and result in collapse of buildings where ash alone would not have caused collapse. Although the critical thickness of tephra above which collapse or damage will ensue has not been determined for all building types, Blong (1984) suggested that accumulation of a thickness of 10 cm of dry ash would be unlikely to cause roof collapse, but a thickness in excess of 10 cm would probably result in failure. As well as collapse of buildings resulting directly from ashfall, collapse can also be induced by the drifting of ash against walls, in which case walls can give way before the roof. In order to prevent collapse, ash must be removed at regular intervals from roofs and around walls. There are many examples of entire towns experiencing building collapse due to ashfall. For example, during the 1914 eruption of Sakurajima, in Japan, nearly 3000 buildings were affected (Omori 1922) and during the 1973 eruption of Heimaey, Iceland, even relatively strong buildings with flat roofs collapsed under loads of less than 1 m of tephra (Booth 1979). Those which survived the total tephra fall of more than 4 m had had their roofs cleared periodically. A major problem for the Heimaey emergency services and local population arose when the island's power plant building was destroyed through a combination of tephra fall and lava flow. During the August 1991 eruption of Hudson

Figure 17.6 Shrine gate at Kurokami, buried during the 1914 Sakurajima eruption. The shrine gate was originally 3 m high but now only the top metre is exposed. (Photograph courtesy of J. S. Gilbert.)

volcano, in Chile, ashfall caused collapse of roofs, and villagers were advised to sweep roofs after every heavy fall (SIBGVN 1991, vol. 16, no. 9). Tephra fall during the 1991 eruption of Pinatubo, in the Philippines, caused extensive structural damage to private residences, factories, storage facilities and aircraft hangars located within the area of major ashfall. Two factors exacerbated the effects of the ashfall. First, monsoon rains fell during the time of the eruption on June 14–15. The ash quickly became saturated with water and attained the consistency and density of wet cement. Second, there was strong ground shaking during the eruption which magnified the effects of ash loading alone, further undermining the structural supports of many buildings, especially large aircraft hangars and storage warehouses (T. Casadevall, personal communication).

In addition to the major structural affects described above, fallout may cause a whole range of types of damage, such as breakage, abrasion and perforation. For example, in the

Table 17.5 Hazards of plumes to property

Hazards	Effects
Particle fallout, including ballistics	Burial and collapse, breakage, abrasion, perforation, clogging of filters, electrical short-circuiting, fires
Mud and acid rain fallout	Burial and collapse, metal corrosion, clogging of filters, electrical short-circuiting
Noxious gases	Metal corrosion
Electrical strikes	Fires
Shock waves	Fracture of windows

1902 eruption of Soufrière, St Vincent in the Caribbean, 6–8 cm diameter bombs penetrated iron roofs and at Manam, Papua New Guinea, in 1957 bombs 5 × 7.5 cm wide perforated thatched roofs (Palfreyman and Cook 1976). At Sakurajima, in Japan, broken car windscreens are almost daily events (SIBGVN 1991, vol. 16, no. 6 and 7; SIBGVN 1992, vol. 17, no. 2).

Some volcanic bombs are sufficiently hot to start fires. At Heimaey, Iceland, in 1973, incandescent ballistics 0.1–2 m in diameter caused fires when they fell through buildings (Booth 1979). The Glacier Hut, 1200 m from Ruapehu's Crater Lake, in New Zealand, was severely damaged when hit by ballistics during the 1975 eruption. These landed hot and several burnt through the floor of the hut (Blong 1984). During the 1783 eruption of Asama, in Japan, house roofs at Karaizawa, 11 km from the vent, were damaged by red-hot pumice up to 0.5 m in diameter which fell from the plume and 52 of the houses were completely burnt (Aramaki 1956). During the 1914 eruption of Sakurajima, many buildings were ignited by bombs and, for this reason, school buildings and a civic hall at Sakurajima have recently been made non-flammable and their structures reinforced.

Volcanic ash is also capable of causing relatively superficial property damage. Small, angular ash particles are extremely abrasive and capable of scratching car paint and other exposed surfaces. Dillman and Roberts (1982) reported that almost half of the residents of eastern Washington, USA, experienced abrasive damage to interior floors and carpets from the 1980 Mount St Helens ashfall. Fine ash penetrates small cracks in protective structures and when contacting electrical devices can cause short-circuiting which, in some cases, results in fire (Blong 1984). Suspended ash can also play havoc with automobile or other engines that intake air.

Mud rain and acid rain, and noxious gases can lead to corrosion, premature ageing and weakening of building materials (such as galvanized roofs) and automobiles (Blong 1984). Acid rain generated by the 1912 eruption of Katmai-Novarupta, Alaska, corroded metalwork of buildings at Seaward (Figure 17.4) and at Cape Spencer, 1110 km away, and fumes tarnished freshly polished brass (Griggs 1922; Wilcox 1959). In houses in Kodiak, 160 km south-east of the volcano (Figure 17.4), bronze, brass and copper tarnished black, and laundry left on lines overnight turned yellow-red in colour (Erskine 1962). In Vancouver, Canada, a month after the 1912 eruption, acid rain was still capable of damaging clothes hung out to dry (Wilcox 1959).

Lightning strikes often result in fires that completely destroy property. Lightning appears to have played a role in the destruction of many buildings during the Caribbean eruptions of Mont Pelée in 1902 and Soufrière, St Vincent in the same year (Blong 1984). During the 1912 Katmai-Novarupta eruption the Woody Island naval station on Kodiak Island (Figure 17.4) was struck by lightning and burnt to the ground (Erskine 1962).

Explosion shock waves are also a hazard to property. There are many reports of fractured windows associated with this phenomenon. For example, shock waves during the Vesuvius, Italy, 1767 eruption burst open doors and windows in Naples (Hamilton 1772) (Figure 17.1). At Asama, in Japan, in 1783 shock waves caused windows to rattle in Tokyo over 300 km away and shook houses near to the volcano so violently that stone weights on roofs fell off (Aramaki 1956). Shock waves also caused notable damage to property during the 1883 eruption of Krakatoa, in Indonesia (Judd 1888), the 1888–90 eruption of Vulcano, in Italy (Bullard 1976), the 1907 eruption of Stromboli, in Italy (Perret 1912) and the 1958 eruption of Asama (SICSLP 1973).

17.6 DISRUPTION OF COMMUNITY INFRASTRUCTURE

Community activities have often been severely affected as a direct result of volcanic plumes. Ash, pumice and scoria pollutes water supplies, clogs sewers and water treatment plants and abrades and damages water purification machinery. Electric power cables can be damaged by the weight of ash and ignite either due to short-circuiting or lightning strikes. The demand on power can be drastically increased because of the need for emergency lighting and telecommunications. Road transport can be retarded by reduced visibility, slippery driving conditions and vehicle breakdowns resulting from clogged filters. Rail transport can be restricted due to burial of railway lines, electrical shorts, reduced visibility and clogged air filters. Hazards unique to ships include coatings of tephra to their decks, lightning strikes to their masts and floating pumice forming impassable barriers. Several volcanic eruptions and associated community infrastructure disruptions are presented in Table 17.6.

One of the most serious hazards with potentially far-reaching effects is contamination of the water supply in a populated region. Even minor amounts of ashfall may have adverse effects on reservoir turbidity and acidity of water supply systems. Chemical procedures such as buffering, dilution, filtration and flocculation can be used to return water to its normal state, but are time consuming and expensive. During the 1953 eruption of Mount Spurr, in Alaska, a 3–6 mm thick fall deposit in Anchorage caused the pH of the public water supply to fall to 4.5 and the turbidity to rise dramatically. Fortunately, the pH returned to normal after a few hours, but the turbidity took six days to revert to pre-contamination levels (Wilcox 1959). Similarly, during eruption of Irazú, in Costa Rica, which began in 1963, the water supply system in San José was affected. In June 1964, the main water pumping station went out of operation so that water had to be brought into the city in tankers. Particles clogged filters in the city's waterworks and residents' basins had to be unblocked by means of pumps (Clark and Lee 1965). These types of water problems during fallout are usually further exacerbated by the increased demand on water supply as a result of cleanup operations. For example, after the 1980 Mount St Helens, USA, eruption the use of water for clean-up tasks resulted in the demand increasing 2.5 times, with water consumption rates remaining above average for four days (Warrick *et al.* 1981). The compound effects of all these problems usually result in rationing.

Fallout can be particularly detrimental where sewage and storm water are collected into one pipe network (Blong 1984). This is because the many possible entry points allow tephra access to sewers and sewage treatment plants. After the Mount St Helens eruption of May 1980 fallout reached the waste treatment facility in Yakima (Figure 13.12) where, on May 19, approximately 15 times the normal amount of "grit" was removed in the pretreatment process (White *et al.* 1980). Even after pretreatment, particles were found in the raw sludge indicating that some tephra was passing into the primary clarifiers. On May 21, equipment failures and shut-downs occurred. Cleanup and repair of the plant took place sufficiently rapidly that it was possible to recommence treatment one week after the eruption. Damage to the waste-disposal system, however, was extensive (Blong 1984) and estimated to be of the order of US$4.06 million (White *et al.* 1980). At Sakurajima volcano, in Japan, where ashfall takes place daily, and in its closest city, Kagoshima, substantial investment has been made to "manage" ash in the community. Here, specially designed road sweepers keep the highways free of ash and pool cleaners have been

Table 17.6 Examples of effects of specific eruptions on community infrastructure

Volcanic eruption (year)	Reported effects
Sakurajima, Japan (1779)	Seto Straits filled with sheet of floating pumice 1.5 m thick which ships were unable to navigate (Koto 1916)
Tambora, Indonesia (1815)	Pumice rafts continued to be a nuisance to ship navigation four years after the eruption (Neumann van Padang 1971)
Krakatoa, Indonesia (1883)	Difficulties for ships negotiating pumice "fields" in Sunda Straits (Furneaux 1964). Ship's mast struck by lightning 80 km from volcano (Symons 1888)
Soufrière, St Vincent (1902)	Tram run off rails due to 5 mm of ash on Barbados (Blong 1984)
Vesuvius, Italy (1906)	Railway line from Naples to Salerno blocked for a few hours and Circumvesuvium Railway blocked for seven days (Anderson and Bonney 1917)
Katmai-Novarupta, Alaska (1912)	Boats unable to move through pumice rafts (Erskine 1962)
Kilauea, Hawaii (1924)	21 power poles destroyed by lightning associated with the eruption (Stearns 1925)
Lamington, Papua New Guinea (1951)	Electrical disturbance made radio transmission impossible (Taylor 1958)
Mount Spurr, Alaska (1953)	Water supply contaminated. Overloading of electrical network due to malfunction of automatic switching system (Wilcox 1959)
Irazú, Costa Rica (1963–65)	Water supply contaminated (Clark and Lee 1965)
Shiveluch, Kamchatka (1964)	Volcanic plume, which passed over village of Ust-Kamchatsk, was so charged that telephone and radio communication with other villages was not possible (Gorshkov and Dubik 1970)
Augustine, Alaska (1976)	Natural gas-fired turbines scoured clean (Miller 1976)
Mount St Helens, Washington (1980)	Ash reached sewage treatment plant, abrading pumps and pumping components (Blong 1984). Water rationing due to increased use for clean-up. Telephone exchanges closed down (Warrick *et al.* 1981). Traffic accidents due to reduced visibility (Blong 1984). Railway in Ritzville shut down due to 40 mm of ash (Warrick *et al.* 1981)

installed to remove ash from school swimming-pools. Ash removal from roads, sewers, drains and public parks is subsidized by the Japanese government.

Because volcanic ash is pervasive, abrasive, corrosive when damp (Lane *et al.* 1995) and carries a high static charge (Gilbert *et al.* 1991), electric power generation, transmission systems and radio communications may all be affected by ashfall. Ash deposited on electronic components can cause arcing, short-circuits and intermittent failures because it is conducting (Labadie 1994). Damp ash is relatively heavy and sticky, and is particularly prone to causing arcing and pole-fires on electrical-distribution systems. Blong (1984) dealt comprehensively with such problems and examples are given in Table 17.6. During the 1980 Mount St Helens, USA, eruptions, telephone exchanges closed down their air-cooling systems, windows were sealed and entry ways and exit ways limited. Even so, the particles, partly magnetic and highly conductive when wet, proved costly to remove from the electromechanical systems. Heavy rain was found to be beneficial because it washed ash from equipment, but light rain exacerbated the problem by causing the ash to form a conductive coating which led to numerous shorts and widespread power cuts (Warrick *et al.* 1981).

Sedimentation of tephra in reservoirs associated with hydroelectric power schemes and irrigation projects can have drastic consequences. The Brantas river basin development in Java, Indonesia, was conceived to produce a steady supply of irrigation water as well as water for municipal and industrial use, to protect the river course from the effects of flooding, for hydroelectric power generation and to control the high sedimentation rates comparable with other parts of Java (Suryono 1987). This project is unusual in that it has the additional task of managing ash from the eruptions of Kelud volcano which erupts approximately every 15 years. Eruptions in 1901, 1919, 1951 and 1966 caused serious damage to life and property, and disturbed the development of the basin. A number of sediment control structures over an area of 2000 km^2 have been constructed, including sand pockets and check dams, in an attempt to deal with ashfall from this volcano.

Adding to the problems of an already difficult situation during fallout from plumes, many of the necessary repairs to telephone, electrical and power services are often delayed due to disruption of transportation (Blong 1984). Driving both during and after ashfall is particularly hazardous due to the fact that ash becomes extremely slippery when wet. This problem is compounded by reduced visibility and because particles are not easily removed from windscreens without causing abrasion (Figure 17.7).

A particular hazard to road transport is lahars which may form when rain falls on unconsolidated ash. These fast-moving currents of mud may flow downhill from the volcano and across roads. Sakurajima volcano, in Japan, provides a fine example of how road transportation has evolved to cope with the consequences of daily ashfall. A busy

Figure 17.7 Accumulation of a few millimetres of ash on a car during an eruption at Sakurajima volcano, in Japan. The photograph was taken in May 1991, 5 km north-east of the active vent. Note the usefulness of umbrellas for shelter from ashfall! (Photograph courtesy of S. J. Lane.)

Figure 17.8 (a) Engineered lahar channel containing recent lahar debris at Sakurajima volcano, in Japan. The photograph was taken on May 16, 1990, 2.5 km south-south-west of the active vent. Note the white "steam" plume issuing from the vent in the background. (b) Notice showing engineered channel with Sakurajima in background. (Photographs courtesy of J. S. Gilbert.)

road circumnavigates the volcano at approximately 4 km from the vent. In order to prevent emergence of lahars on roads, specifically designed restrictive lahar channels have been constructed on the flanks of the volcano (Figure 17.8). These are routed under the main road and out to sea. Along the road a special warning system has been established with crossing gates which automatically stop road traffic during heavy fall or passage of lahars to ensure safety. To remove ash from roads the Sakurajima authorities employ road sprinklers and road-sweeping vehicles after every eruption.

In contrast to the frequent but small Vulcanian eruptions which occur at Sakurajima daily, less common Plinian eruptions, which generate large plumes over a period of a few days, have much more widespread but shorter-lived effects on road transport. For example, the unfamiliarity of the population with ashfall coupled with reduced visibility on highways and the slippery nature of the ash during the 1980 Mount St Helens, USA, eruption resulted in many accidents (Blong 1984). It also altered traffic volumes and speed restrictions, stranded travellers, affected food stocks and overloaded the automotive repair industry. Removal of the large volume of tephra deposited by the Mount St Helens eruption created difficulties for civil authorities in eastern Washington. Municipal-type street sweepers were used to sweep up the tephra after the bulk of material had been moved by road graders and snowploughs. These operations required large volumes of water to make the deposit workable and minimize dust (Markesino 1981). In Yakima (Figure 13.12), ash removed from the streets was stockpiled on sites throughout the city. Sprinkling systems were installed and the ash covered with topsoil and seeded to grass. In other places ash was bladed into ditches at the side of the road or into fallow fields ready to be ploughed under at the next cultivation. Fall deposits have played their part in disrupting rail as well as road traffic and examples (Blong 1984) include those from the eruptions of Soufrière (1902) in the Caribbean, Vesuvius (1906) in Italy as well as Mount St Helens (1980).

In addition to the difficulties met on land, transportation by sea has encountered hazards from large particle-laden plumes. During fallout a major problem for ships is keeping decks free of tephra, and floating pumice constitutes a serious problem for sea transportation. Vast fields of floating pumice generated by the 1883 Krakatoa eruption, in Indonesia, affected the route of the *Gouveneur General Loudon*, a steamship running down the Sumatran coast several days after the first explosions (Simkin and Fiske 1983) which found its way completely blocked by pumice 3 m thick (Furneaux 1964). For several weeks after the eruption, ships reported considerable difficulties negotiating pumice "fields" in the Sunda Straits. In January 1884 pumice rafts in the Java Sea off Jakarta reached several square kilometres in extent and were sufficiently thick to offer considerable resistance to steamships. An additional hazard for ships is lightning. While the *Gouveneur General Loudon* was in Lampong Bay, 80 km from Krakatoa, lightning struck the mainmast several times (Symons 1888).

17.7 AVIATION HAZARDS

Eruption plumes transport large amounts of particles, gases and aerosol droplets to and above the cruise altitudes of commercial airliners. All aircraft are designed to operate in particle- and acid aerosol-free environments and for this reason aircraft which encounter plumes are vulnerable to catastrophic failure of engines and deterioration of bodywork. Volcanic plume-related hazards for aircraft may be life-threatening and damage incurred can be extremely costly.

Evidence of an eruption during the past 500 years exists for 564 volcanoes (Casadevall and Thompson 1995). These volcanoes, and several hundred others which have not been active recently, are capable of generating eruption plumes. Many principal air routes are over volcanic chains. Because plumes drift in the atmosphere (Chapter 12), simple avoidance of airspace above volcanoes does not necessarily preclude an encounter between an aircraft and a volcanic plume. The problem is compounded by the fact that air travel is becoming increasingly popular and there is enormous pressure on air carriers to expand their services and offer efficient travel. Recent encounters of aircraft with plumes have triggered international efforts to address the problem. This section describes the effects of plumes on aircraft in flight and their ground-based support systems, and the hazard mitigation measures that are currently in use.

17.7.1 Disruptions of Airport Operations

In the past 50 years approximately 20 volcanic eruptions have produced fall deposits which have interrupted airport operations and caused about 40 airports to close temporarily for periods of hours to weeks. The majority of these have been since 1980 (Casadevall 1993) and selected cases are given in Table 17.7 (see Casadevall 1993, Table 1 for further details). Indeed, only a few millimetres of fall deposit on an airfield is sufficient to cause problems for aircraft. Manœuvring of aircraft during and soon after ashfall can be treacherous due to the reduced visibility and slippery nature of damp ash. Unconsolidated ash on the ground is readily redistributed by wind and by drawing of air through engines. Remobilization of ash causes both visibility as well as mechanical problems. The presence of wet ash on runways affects braking, turning and landing performance. Ashfall on stationary aircraft may require painstaking cleanup operations (Labadie 1994) and the weight of ash has been known to cause aircraft to settle back on their tail sections (Figure 17.9). Tephra fall can also affect airport facilities such as runways and communication systems, and induce collapse of airport buildings (section 17.5).

Cleanup of tephra at airports is expensive and difficult, particularly when fine ash is repeatedly reworked by wind. Ash on runways has proven to be most easily dealt with by wetting the ash, sweeping it up and moving it to a final dump site (Casadevall 1993). After removal of the bulk of the deposit, hydroblasting of surfaces with high-pressure water helps to remove final traces. Cleanup is particularly difficult on vegetated sites and this problem has sometimes been dealt with, not by removing the ash, but by watering it in, particularly along the critical areas each side of the runway.

17.7.2 Plume Encounters in Flight

While fallout from plumes can result in expensive cleanup operations the most serious problems occur when aircraft fly through plumes. Aircraft have been known to suffer major mechanical failure and external damage during the briefest of flights through volcanic plumes. Encounters have become increasingly common in the last decade, reflecting the opening up of new air routes and the increased volume of air traffic. The major incidents are summarized in Table 17.8. A discussion of the way in which volcanic ash affects various aircraft components is presented in section 17.7.3.

17.7.3 Effect of Ash and Aerosols on Aircraft

As early as 1953, following the eruption of Mount Spurr, in Alaska, the United States Civil Aviation Administration Office of Aviation Safety issued warnings to pilots working in the vicinity of Mount Spurr suggesting that they should take precautions to prevent damage by abrasive ash to aircraft engines and instruments and noting that filters, screens and pumps needed to be checked daily (Blong 1984).

Following the encounter of the DC-9-30 aircraft with the Mount St Helens, USA, plume on May 18, 1980 (Table 17.8) the United States Federal Aviation Administration (FAA) cited this incident in its general warning of hazards of contamination with volcanic ash, which it said could lead to "multiple engine failures", as well as contamination of pressurization/air condition systems and damage of pitot static systems (which provide an indication of air speed). In its warning, the FAA said, "the gritty dust will peel finish off leading edges... clog air filters and convert engine oil into a destructive abrasive compound and ruin engines" (*Aviation Week and Space Technology*, May 26, 1980). Three days after the main Mount St Helens event, the FAA issued an alert to air carriers and the

Figure 17.9 DC-10 sitting back on its tail section due to loading with volcanic ash at Cubi Point Naval Air Station, Philippines. (US Navy photograph courtesy of R. L. Rieger.)

general aviation community warning of the volcanic ash hazard. This alert is reproduced in full in Blong (1984). Other advisory information was issued by airframe and engine manufacturers (see *Aviation Week and Space Technology*, June 9, 1980 and *FAA General Aviation News*, September–October, 1980).

There have been several engine failures triggered by the intake of volcanic ash by jet engines (Table 17.8). Ingestion of ash may cause serious deterioration of engine performance due to both erosion of moving engine parts (Figure 17.10) and accumulation of partially melted ash as glass on turbine blades (Casadevall *et al.* 1991; Przedpelski and Casadevall 1994). A study which looked at both the fall deposits from the December 15, 1989 eruption of Redoubt, in Alaska, and particles in the engines of a Boeing 747-400 that flew into the plume, demonstrated that, while in-flight, ash melted in the hot combustor and turbine sections of the engine (Casadevall *et al.* 1991). Fluidal textures indicated that flowage of glass occurred during or after deposition. Analysis of the glass showed that, in addition to the plagioclase, pyroxene and Fe–Ti oxides found in the fall deposits, the glass also included metal particles derived from abrasion of engine parts. Ash ingestion by aircraft engines can lead to accumulation of glass which, in turn, results in compressor stall (Campbell 1994). Because the main components of volcanic ash are glass and minerals, they are unlikely to melt at temperatures less than 1000 °C. Therefore, during an in-flight plume encounter, the likelihood of engine failure can be limited by reducing aircraft engine power to the minimal setting, thus holding the engine temperature below that at which the ash is likely to melt (Casadevall *et al.* 1991; Przedpelski and Casadevall 1994).

Ash particles can also abrade *outer* parts of aircraft, particularly the leading edges of wings, nose cones and lights (Figure 17.11). Windscreens become "sandblasted" so that

Table 17.7 Examples of airport disruptions due to volcanic plume-generating eruptions

Volcanic eruption (year)	Reported problem
Kilauea, Hawaii (1924)	Kilauea Military Camp seriously affected by volcanic bombs (Blong 1984)
Vesuvius, Italy (1944)	~80 aeroplanes damaged by ashfall while on airfield near crater (Blong 1984)
Lamington, Papua New Guinea (1951)	Port Moresby airfield closed due to 1–2 mm of ash (Taylor 1958)
Mount Spurr, Alaska (1953)	Eruptions resulted in United States Air Force aircraft in Anchorage being evacuated to bases near Fairbanks (Blong 1984). Closure of Anchorage Airport and Elmendorf Air Base for several days (Juhle and Coulter 1955)
Agung, Indonesia (1963)	Surabaja Airport closed due to 10 mm of ash (Jennings 1969)
Sakurajima (1971)	Kagoshima Airport closed for less than one day due to 3 mm of ash (Casadevall 1993)
Etna, Italy (1979)	Catania Airport closed (Guest *et al.* 1980)
Mount St Helens, Washington (1980)	Closure of several airports as far east as Misoula in Montana where 1–2 cm ash fell. Grant County Airport 192 km from the volcano received 8–10 cm of ash which took 30 days to clean up and resulted in closure for 15 days (Warrick *et al.* 1981; Blong 1984; Bailey 1991)
Galunggung, Indonesia (1982)	Airport closure due to reduced visibility (Casadevall 1993)
Augustine, Alaska (1986)	United States Air Force aircraft evacuated from Anchorage for three days. Anchorage International Airport remained open but several major airlines cancelled incoming and outgoing flights for three days (Kienle 1994)
Redoubt, Alaska (1989–90)	Temporary suspensions of some services in mid-December at Merrill Field, Anchorage. At Elmendorf Air Force Base, Anchorage, jet fighter operations were limited on December 15–16 and some sorties cancelled. Military cargo flights diverted to either Fairbanks or Air National Guard Base at Anchorage International Airport. ~45 flights by military turbo-prop aircraft cancelled in December. At Anchorage International Airport in December–January both passenger carriers and air cargo operators curtailed operations, which resulted in reduced revenues at Anchorage International Airport by ~US$2.6 million (Brantley 1990; Casadevall 1994b)
Sakurajima, Japan (1990)	Closure of Kagoshima Airport for less than one day (Casadevall 1993)
Colima, Mexico (1991)	Closure of Colima Airport for several days (Casadevall 1993)
Pinatubo, Philippines (1991)	Seven airports in Philippines closed (Casadevall *et al.* 1996), including Manila International Airport for four days. Collapse of numerous aircraft hangars and maintenance facilities
Crater Peak, Mount Spurr, Alaska (1992)	Disruption of air traffic over Alaska, Canada, and northern and central United States. Temporary suspension of operations at Anchorage airports for several days for removal of ash and cleaning of airport facilities (Casadevall and Krohn 1995)
Sakurajima, Japan (1992)	Closure of Kagoshima Airport for one day due to < 1 mm of ash (Casadevall 1993)

Table 17.8 Examples of in-flight plume encounters

Volcanic eruption (year)	Reported problems
Spurr, Alaska (1953)	At least three F-94 military jets damaged after flying into plume (Casadevall and Krohn 1995). Sandblasting of paint from leading edges of wings and frosting of plexiglass canopy (Juhle and Coulter 1955)
Irazú, Costa Rica 1963–65	In 1963 Pan Am DC-6 aeroplane flew through plume and was forced to land in Panama. Aircraft suffered abrasion of its windows and engine problems (Barquero 1994)
Augustine volcano, Alaska 1976	On January 22, 1976, two F-4E Phantoms, on route from Galena to King Salmon, encountered a plume. Canopies of the aircraft became scoured, wing tips were sandblasted and particles entered the cockpit (Kienle and Shaw 1979). On January 25 a Japan Airlines cargo DC-8 *en route* to Tokyo suffered a scoured windscreen and abrasion to external radio parts, landing gears and air-conditioning system (Kienle 1994). A Boeing 747 and a DC-8 also bound for Tokyo reported ash adhering to the aircraft at cruising altitude (approximately 10 km), which caused minor damage (Kienle 1994)
Mount St Helens volcano, USA 1980	Hughes Airwest McDonnell-Douglas DC-9-30 encountered a plume on May 18, 1980. On May 25, 1980 a Lockheed L-100-30, belonging to Transamerica Airlines, incurred exposure to heavy fallout shortly after takeoff from McChord Air Force Base near Tacoma (Figure 13.12) while climbing through 3400 m. Number 2 and number 4 engines began to surge badly and had to be shut down. Two turbines destroyed and two badly damaged. Three windscreens replaced. Damage was ~US$0.5 million (*Aviation Week and Space Technology*, June 9, 1980). On the same day several light aircraft made forced landings as a result of engine failures due to ash in induction systems (*FAA General Aviation News*, September–October, 1980)
Galunggung volcano, Indonesia 1982	Domestic Indonesian Douglas DC-9 on Jakarta–Yogjakarta route flew through ash from Galunggung on April 5 1982. Aircraft required maintenance (Johnson and Casadevall 1994). On June 24, 1982, Australia-bound British Airways Boeing 747 aircraft carrying 240 passengers encountered a plume 150 km west-south-west of Galunggung volcano at ~11 300 m altitude at night. Four engines stalled and cockpit windscreen plus wing surfaces became abraded. St Elmo's fire produced spectacular display around the aircraft (E. H. J. Moody, personal communication). Aircraft lost 7500 m altitude in 16 minutes before engines could be restarted. Successful emergency landing on three engines at Jakarta. Crew of Singapore Airlines Boeing 747 on same flight path as BA aircraft, but 40 minutes later, on June 24 reported "smoke contamination" in the main cabin. Aircraft reached Perth without further incident where rock fragments found in all engines (Johnson and Casadevall 1994). On July 13, 1982 Singapore Airline Boeing 747 carrying 230 passengers encountered a plume at 9000 m altitude at night. Three engines stalled and aircraft lost 2400 m altitude before one engine restarted. Safe landing in Jakarta on two engines (Blong 1984). BA 747 aircraft on a diversionary route around Galunggung volcano ran into plume from Colo volcano, Indonesia, in July 1983!

Soputan volcano, Indonesia 1985	In May 1985 a Qantas Airways 747 aircraft on night-time flight between Hong Kong and Melbourne encountered a plume (Johnson and Casadevall 1994). No engine failures but aircraft was grounded for repairs for several days
Redoubt volcano, Alaska 1989–90	Between December 1989 and February 1990, five commercial jetliners suffered damage due to encounters with volcanic ash from Redoubt volcano in the Anchorage area. On December 15, an eruption at Redoubt at 1015 hours (AST) produced a plume which rose to ~12 000 m altitude and dispersed towards the north-east. A new Boeing 747-400 KLM passenger airliner encountered the plume 240 km north-east of volcano at 8000 m altitude when descending for landing in Anchorage. Four engines cut out, but restarted and safe landing made (Campbell 1994; Przedpelski and Casadevall 1994). Extensive damage included: compressor erosion and hot section damage to engines; contamination of fuel, oil, hydraulic and potable water systems; sandblasting of pilots' windscreens, cabin windows, navigation and landing light covers; plugging of all pitot and static probes and contamination of electronic equipment and furnishings. Cost of repair was >US$80 million (Steenblik 1990). Other encounters of aircraft with Redoubt plumes occurred on December 15 and 16, 1989 and February 21, 1990. On December 17, 1989, two jet airliners encountered the Redoubt plume 55 hours old, 5300 km from its source over west Texas (Casadevall 1994b)
Pinatubo volcano, Philippines 1991	16 damaging encounters between jet aircraft and plumes from June 12 and 15 eruptions. In-flight loss of power to one engine on each of two different aircraft. 10 engines damaged and replaced, including all four engines on a single jumbo jet (see Casadevall *et al.* 1996)
Hudson volcano, Chile 1991	Australian Airlines flight from Melbourne to Sydney reported encounter with "hazy cloud" 260 km north-east of Melbourne. Associated with the cloud was a strong smell of sulphurous gas which entered the aircraft and was noticed by the crew and passengers. Return flight from Sydney encountered "haze" at approximately the same place. Plume also reported by pilots from Qantas and Ansett on August 20 (SIBGVN 1991, vol. 16, no. 7)
Sakurajima volcano, Japan 1955–present	In Japan, 21 cases of volcanic ash encounters by aircraft were reported between 1973 and 1991 (Onodera and Kamo 1994). 12 involved ash plumes of Sakurajima volcano; other encounters were with plumes of Asama in 1973, Usu in 1977, Izu-Oshima in 1986 and Unzen in 1991. In November 1982 an Air Nauru jet carrying 39 passengers flew into Sakurajima's plume at ~3000 m altitude, six minutes after leaving Kagoshima Airport. Hairline cracks in three cockpit windows resulted from impact with particles. Aircraft returned to Kagoshima and landed safely. On August 5, 1991, when dense weather clouds prevented the pilot from seeing the plume, ejecta from an explosion struck the windscreen of a Boeing 737 at an altitude of 1200 m, 10 km north of the volcano. This was the first incident of in-flight damage near the volcano since June 24, 1986 (SIBGVN 1991, vol. 16, no. 7)

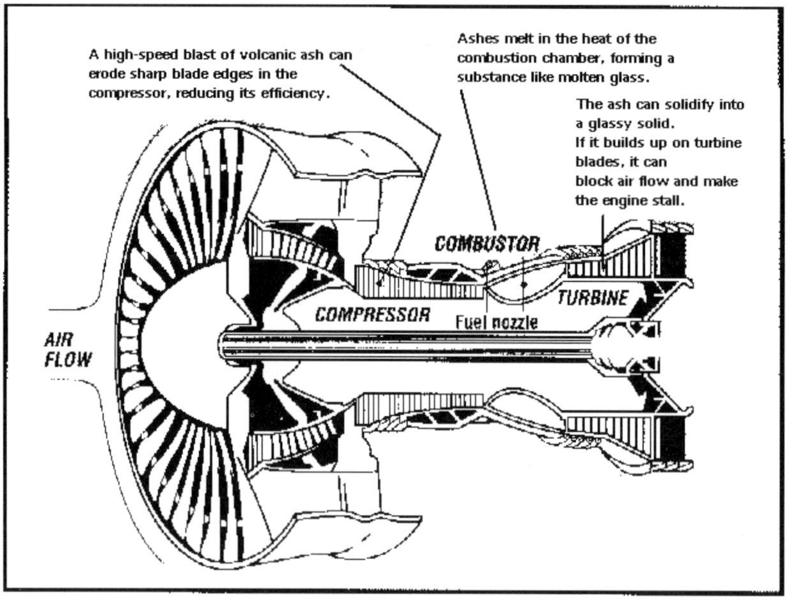

Figure 17.10 Jet engine cutaway showing areas of damage resulting from an encounter with a volcanic plume. (Reproduced with permission from T. J. Casadevall.)

visibility is severely reduced and ash has been found to block pitot tubes so that air-speed monitoring is hampered. Oil filtering systems may become clogged and oil contaminated. Particles can also be ingested into the cooling systems of aircraft electronics (Campbell 1994) and find their way into air-conditioning systems.

Damage to aircraft does not result exclusively from interaction with particulates. SO_2 released into the stratosphere by large eruption columns reacts to form H_2SO_4 aerosols (Chapter 18) which may spread globally and persist for months or years. These acid aerosols are highly corrosive. Aircraft which fly polar routes, where the tropopause is relatively low, are particularly susceptible to chemical attack from these aerosols. The major deterioration processes include: corrosion of metal parts and surface coatings in the hot sections of jet engines; contamination of plastic insulation and neoprene hosing in air distribution systems; corrosion of electrical contacts in avionics, and pitting, crazing and embrittlement of windows due to H_2SO_4 attack (Berner 1993). For example, following eruptions of El Chichon volcano, Mexico, in 1982, several airlines flying the northern polar route reported crazing of acrylic windows of aircraft due to increased concentrations of acid aerosols (Bernard and Rose 1990). Since the June 1991 eruption of Pinatubo in the Philippines, aircraft operating at altitudes of 10–12.5 km in the Northern Hemisphere have also shown signs of attack from H_2SO_4 aerosols, such as accumulation of sulphate minerals and trace amounts of quartz and feldspar in the turbines, premature fading of polyurethane paint on jetliners and crazing of acrylic aircraft windows (Casadevall and Rye 1994). Volcanic aerosols therefore pose a *long-term* threat to individual aircraft and those which fly polar routes require particularly rigorous monitoring.

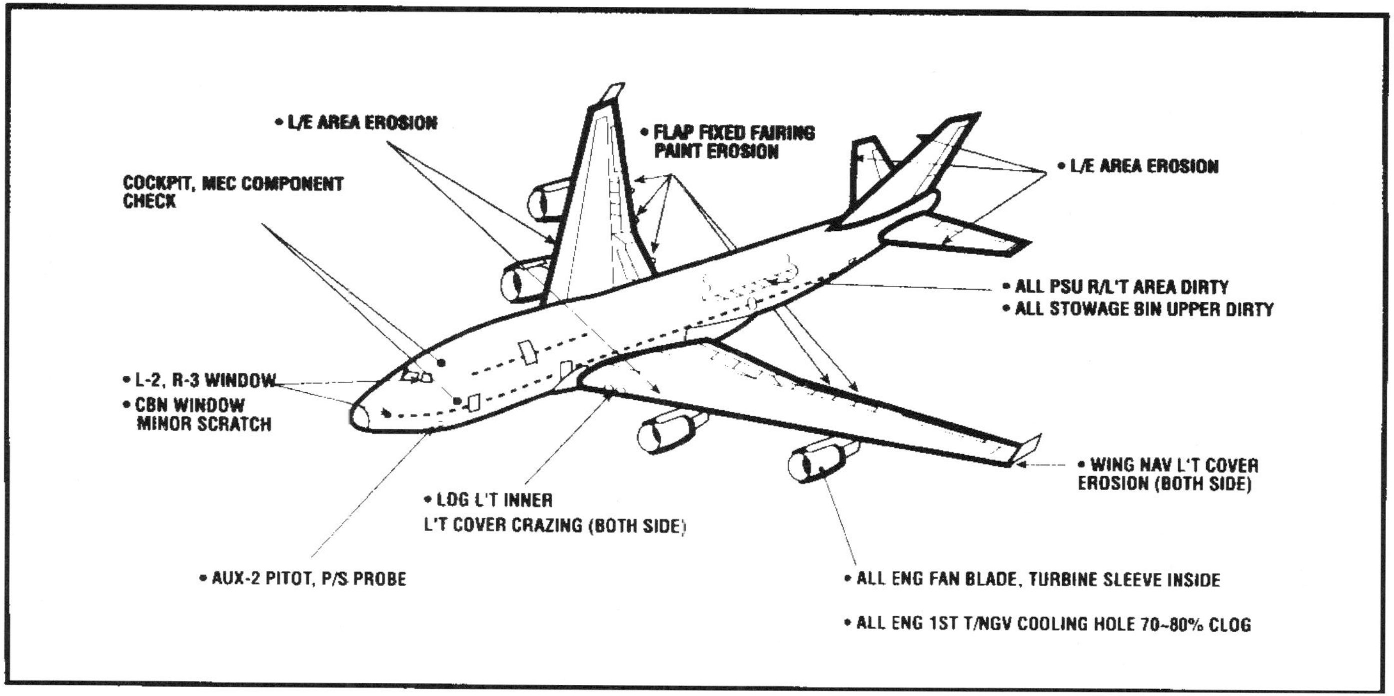

Figure 17.11 Diagram of B747-400 aeroplane showing exterior damage as a result of flying through a volcanic plume. (Reproduced with permission from T. J. Casadevall.)

17.7.4 Mitigation

The crux of the volcanic ash–aviation safety problem is that, while flying at night or in weather cloud, pilots are unable to detect an ash plume ahead. On-board radar systems cannot distinguish silicate particles from rainfall or aerosols. In addition, air-traffic control centres are not always able to provide pilots with sufficient warning of plumes in flight paths. This may either be due to poor communication between volcano observers and aviation authorities or due to the fact that many volcanoes in remote parts of the world are not monitored. Only about 170 active volcanoes are under continuous surveillance by volcano scientists (Casadevall and Thompson 1995). The communication problem is exacerbated by the fact that volcanic plumes can persist for hours to days after an eruption and are highly mobile. This results in distribution of plumes by winds hundreds of kilometres from their sources. Therefore, avoidance of airspace directly above an active volcano does not necessarily solve the problem, but instead, plumes require detailed tracking and airspaces need to be continuously monitored. Communication has to be in real time and, ideally, autonomous aircraft-borne sensor systems should be developed (Prata and Barton 1994; Barton and Prata 1994).

In recent years major advances have been made in the remote sensing and tracking of volcanic plumes (Chapter 12). A concerted effort has been made to improve communications between the aircraft industry and volcanologists. In 1982 the International Civil Aviation Organization (ICAO) in Montreal set up a Volcanic Ash Warnings (VAW) study group whose primary task has been to recommend revisions to the ICAO regulatory documents that deal with procedures associated with encounters. The VAW group has been active in drawing the attention of airlines to the plume problem, in developing a volcano reporting form for use by pilots, and in devising a voluntary international airways volcano watch.

Following the 1982 encounters of aircraft with the Galunggung plume (Table 17.8), the Australian Department of Aviation created the Airways Volcano Watch and the 1985 encounter with Soputan's plume (Table 17.8) prompted Australian and Indonesian aviation authorities to establish the Volcanological/Airspace Liaison Committee Australia–Indonesia (VULCAN). Both the VAW and VULCAN groups have given attention to the use of meteorological satellites in tracking plumes for the benefits of aviation safety and participated in international efforts to improve early warnings to pilots. VULCAN has prompted Australian scientists to devise the prototype of a small radiometer capable of being mounted on aircraft and looking ahead to discriminate volcanic plumes from normal weather clouds.

In 1988, the World Organization of Volcano Observatories (WOVO), in co-operation with ICAO and the International Association of Volcanology and Chemistry of the Earth's Interior (IAVCEI), requested the WOVO member institutions to improve contacts with civil aviation authorities in order to improve communications between ground-based observatories and air traffic. The outcome of the 1989–90 Redoubt incidents (Table 17.8) has been to increase interest in volcanic hazards and aviation safety among commercial and military air operators, aircraft manufacturers, aviation administrators, meteorologists and volcanologists. In order to focus this interest, the First International Symposium on Volcanic Ash and Aviation Safety was held in Seattle, USA, in July 1991. At this meeting it was emphasized that the key to mitigating in-flight disasters was to improve communication channels between volcanologists, meteorologists and airline personnel, and to effectively communicate experiences and practical solutions to the aircrews, airline carriers and aircraft manufacturers. The reader is directed to the *Proceedings of the First International Symposium on Volcanic Ash and Aviation Safety* (Casadevall 1994a) for further information.

In the same year as the symposium, a unique eruption monitoring warning communications system was installed linking the Sakurajima Volcanological Observatory and the Japan Airlines office at Kagoshima International Airport. The active volcano, Sakurajima (Figure 17.1), is located just 24 km south of the airport. The volcano is continually monitored by the Sakurajima Volcanological Observatory. A seismometer and infrasonic microphone installed on the flanks of the volcano record earthquakes and air shocks respectively. These signals are transmitted to the airport by commercial telephone lines where they are down-loaded on to a personal computer. The computer processes the signals and issues alarms in accordance with the seriousness of the seismic and eruptive activity. In addition, a video camera monitors the volcano and a real-time image is transmitted to the air traffic control centre. This system is used in order to alert aviation dispatchers and pilots, regardless of the time of day or the weather (Onodera and Kamo 1994), of an explosive eruption and generation of a particle-laden plume.

The eruptions of Mount Pinatubo, in the Philippines, in June 1991 injected enormous quantities of volcanic ash and acid aerosols into the stratosphere. The largest plume, from the June 15 eruption, was carried by upper-level winds to the west and circled the globe in 22 days. This plume spread to cover a band from 10°S to 20°N and therefore affected some of the world's busiest air traffic corridors. Sixteen damaging encounters between aircraft and the plumes were reported (Table 17.8) and ashfall damaged many aircraft on the ground (Figure 17.9) as well as forcing closure of several airports (Table 17.7). The majority of the Pinatubo encounters occurred at distances of up to 2000 km from the volcano with a plume that was at least 12 hours old. Despite information being available for plume locations, the large number of aircraft involved in the encounters implies that information either did not reach appropriate officials or that the pilots, air traffic controllers, and flight dispatchers who received this information were not sufficiently informed about the hazards of volcanic ash to react appropriately (Casadevall *et al.* 1996). Within days of the main eruption on June 15 the Philippine authorities established a plan to streamline data collection and flow of information between those on the ground and pilots. This involved co-operation between the Airline Operators' Council, the Board of Airline Representatives of Manila International Airport, the Philippine Institute of Volcanology and Seismology, the Philippine Atmospheric, Geophysical and Astronomical Services Administration and the Manila Airport Authority. This plan has been highlighted by the ICAO as an example of an operational model which could be used by other countries facing the volcanic threat to aviation safety (Casadevall *et al.* 1996).

Crater Peak vent on the south side of Mount Spurr in Alaska erupted on June 27, August 18 and September 16–17, 1992. Each eruption produced an ash column that rose to between 13 and 15 km above sea-level which was then dispersed by upper-atmosphere winds (see Figure 11.23 for a composite picture showing transport of the September plume). Although ash from the three eruptions disrupted air traffic over Alaska, western Canada and the United States, and considerable expenses were incurred as a result of cleanup operations and flight cancellations at Anchorage airports following the August 18 eruption, no aircraft suffered damage through encounters with ash-laden plumes (Casadevall and Krohn 1995). At the time of the Crater Peak eruptions, because encounters with volcanic plumes from the 1989–90 eruptions of Redoubt volcano, in Alaska, had caused serious damage to five passenger aircraft (Table 17.8), a high level of concern already existed in the aviation and volcanological communities about volcanic ash and its threat to aviation safety. The lack of in-flight encounters of the Crater Peak eruption plumes is a success story for the volcano monitoring and aviation authorities. This success hinged on the following developments which were implemented prior to the Crater Peak eruptions (Casadevall and Krohn 1995):

1. Improved warnings of eruptions provided by an increased level of monitoring of the Cook Inlet volcanoes by the Alaska Volcano Observatory;
2. Improved monitoring of the meteorological conditions in the Cook Inlet region by the United States National Weather Service;
3. Improvements in the detection and tracking of volcanic plumes, including increased awareness and reporting of ash plumes by pilots;
4. Improved education of pilots and flight dispatchers,
5. Streamlining of the ways in which information and warnings are communicated among agencies concerned with aviation safety.

Casadevall and Krohn (1995) provided an account of the lines of communications utilized during the Crater Peak eruptions and the response of the aviation community to the eruptions.

17.8 SUMMARY AND LESSONS LEARNED

Volcanic plumes are fluid dynamically fascinating, and produce deposits which reveal a minute-by-minute eruption history. However, their environmental hazards are serious and extraordinarily far-reaching. The fact that plumes migrate in the atmosphere and transport enormous quantities of solid particles, liquid drops and volcanic gases, results in them affecting not just the area proximal to the active volcano but many hundreds or thousands of square kilometres of land surface at distance from the volcano and very large volumes of the atmosphere. Many lessons have been learnt from the repeated interactions of volcanic plumes with man and animals and, for the most part, "common sense" approaches have been adopted during ashfall crises. There is, however, enormous scope for quantitative studies of the effect of ashfall on the health of humans and animals, the impact of ash on vegetation growth, and the damage ash causes to buildings and machinery. Until these studies have been carried out, guidelines for dealing with volcanic ash cannot be substantiated with scientific data. The environmental hazards of volcanic ash therefore warrant serious scientific study in the future. The basic lessons learned so far are summarized below.

In order to safeguard the health of humans it is essential that populations are kept away from the region of ballistic fallout, i.e. a few kilometres from the vent, during explosive volcanic eruptions. Inhalation of volcanic ash appears to exacerbate pre-existing respiratory problems. Therefore, people susceptible to such conditions should remove themselves from regions of fallout. All those exposed to ashfall should protect themselves by means of a dust mask and it is particularly important to cover exposed skin during fallout of mud rain, when liquids are likely to be acidic. Plume-to-ground lightning strikes are common and therefore it is wise to keep away from the crests of hills and tops of buildings when plumes are overhead.

Animals, like humans, should be kept out of the region affected by ballistic ejecta at volcanoes. Grazing animals are prone to poisoning by ash coated with halogen-rich compounds, and abrasion of hoofs, teeth and mouths. Movement of animals to alternative grazing may be hampered during heavy ashfall and can itself trigger additional health problems. However, in the ideal case, animals should either be moved out of the region of fallout or housed under cover, and fed uncontaminated food. In isolated cases it may be possible to cover areas of grazing land and keep them free from tephra. Animals should be kept away from hilltops when a plume is overhead due to the possibility of lightning strikes.

Most plants are not robust to heavy tephra fall. While it would not be sensible to attempt to protect vegetation from a major pumice and scoria fall, in areas repeatedly affected by light ashfall, such as around volcanoes which regularly generate small plumes, it makes economic sense to protect high-value crops. At Sakurajima, in Japan, individual satsuma oranges are routinely covered with plastic bags and many crops are grown in plastic tunnels.

Collapse of buildings, in particular roofs, is common during tephra fall and it cannot be overemphasized that roofs and walls should be kept free of ash. While sweeping a roof may be a dangerous occupation in itself, a roof laden with more than a few centimetres of ash, particularly wet ash, poses an even greater hazard. Because of the hard, abrasive nature of ash particles they can damage floors, carpets and furnishings, and therefore it is prudent to make every attempt to keep ash out of homes and offices. However, this is certainly not easy to achieve due to the pervasiveness of the small particles. Electrical and mechanical machinery are also at risk of damage from abrasion and short-circuiting so, in order to avoid expensive, time-consuming cleanup operations, valuable equipment should be kept free from ash.

Tephra fall is likely to result in contamination of water supplies. This means that water quality should be monitored and potable water may need to be brought into the affected area. To prevent abrasion of pipes and machinery, sewage should be kept separate from storm water. Automobile driving is hazardous during and soon after ashfall, both due to poor visibility and the slippery nature of ash, particularly when damp. Therefore, car journeys should not be made unless necessary. If driving during ashfall, beware of using windscreen wipers to remove ash because this can result in an opaque, scratched windscreen. It is useful to use bottled water, kept in the vehicle, to regularly pour over the windscreen. During heavy rain, soon after ashfall or during fallout of mud rain, it is important to be vigilant of lahars moving down the flanks of volcanoes, and therefore to keep out of topographic depressions. After ashfall the demand on power may be drastically increased because of the need for emergency lighting and telecommunications, and therefore all power should be used frugally. Power cables should be cleared of ash at the earliest opportunity.

In the past decade a great many lessons have been learnt about the interaction of plumes and aircraft, and their impacts on ground operations. The economic importance of keeping aircraft shielded from ashfall is clear. This can be done given sufficient warning of ashfall, by moving aircraft into hangars or alternative airfields outside the likely fallout zone. All pilots should be acutely aware of the slippery nature of ash on runways, the fact that dry ash may be drawn into engines and the damage such ash remobilization can do. The chore of cleaning up runways after ashfall can be made easier by damping down ash, sweeping it up, dumping it at a designated site and then hydroblasting surfaces to remove final traces of ash. Roofs of airport buildings should be kept free from tephra.

In-flight encounters of aircraft with plumes can have grave consequences. Aircraft should avoid plumes! If a particle-laden plume is encountered during a flight it is wise to reduce engine power so as to hold the engine temperature below that of the melting point of ash and to attempt to exit the plume. The key to combating dangerous plume encounters is for airline personnel, aircraft manufacturers, volcanologists and meteorologists to communicate both during and between crises. During crises, eruption information and anticipated plume trajectories need to be automatically given to air traffic controllers and pilots. Between crises, fundamental lines of communication need to be established, education about the physical and chemical properties of plumes should be provided and daily observations of plumes by pilots, volcanologists and others should be communicated to a central body. The effects of acidic aerosols on aircraft pose a long-term hazard and vigilance is required to routinely check for signs of chemical attack. In conclusion, real-time communication about the movement of volcanic plumes, underpinned by preparation and education, is the key to aircraft avoiding in-flight catastrophes and minimizing the detrimental effects of accumulation of tephra on the ground. Ideally, on-board plume-detection systems should be developed.

18 Atmospheric Effects

NOTATION

τ	optical depth (μm)
I/I_o	ratio of radiation emerging at bottom of aerosol layer to that incident at the top (dimensionless)
μ	pathlength of sunlight (m)
ΔT	temperature decrease (K)
S	sulphur mass (kg)
F_{solar}	solar radiation flux (kW m^{-2})
F_{IR}	infrared radiation flux (kW m^{-2})
ΔF_{net}	net radiation flux (kW m^{-2})
r_1	aerosol radius mode 1 (μm)
r_2	aerosol radius mode 2 (μm)
r_{eff}	area-weighted mean radius (μm)
N_1	aerosol concentration mode 1 (cm^{-3})
N_2	aerosol concentration mode 1 (cm^{-3})
V_{eff}	effective variance/width of particle size distribution (dimensionless)
τ_1	distribution width/variance mode 1 (dimensionless)
τ_2	distribution width/variance mode 2 (dimensionless)

18.1 INTRODUCTION

Volcanic plumes carry three major components: tephra, volcanic gases and entrained tropospheric air. The bulk of the tephra component settles out rapidly from the plume, aided by aggregation processes, and has a residence time of the order of only days or weeks in the Earth's atmosphere. Volcanic gases may either condense in the atmosphere to form liquids or solids, or undergo chemical reactions, such as in the case of SO_2 which forms volcanic sulphate aerosols, the most important contribution to volcanic forcing of climate. Volcanic forcing is the change in the Earth's radiation budget caused by volcanic eruptions. Volcanic aerosols diminish the solar energy available to the Earth's climate system, forcing the system to attempt to reach a new equilibrium state. Volcanic forcing is but one of many processes or forcing functions which can affect the climate system and cause it to shift from its present condition to a new state, either temporarily or long term. Volcanic aerosols from large explosive eruptions can have significant effects on climate. They are also known to accelerate destruction of the Earth's ozone layer. This chapter reviews current research on the effects of volcanic aerosols on the Earth's atmosphere, using the well-studied 1991 Pinatubo eruption in the Philippines as a primary example. This chapter also discusses large-scale atmospheric processes which disperse plumes on a global scale. *Volcanic aerosols* are defined as minute liquid and solid particles, typically in the size range 0.1–1 μm, that are suspended in the Earth's atmosphere. Liquid aerosols are primarily of sulphuric acid composition.

18.2 EARLY RESEARCH

The idea that volcanic eruptions can influence climate can be traced back at least two millennia, to the days of the Roman Empire. In 44 BC a large eruption took place at Etna on Sicily, and the haze from the eruption diminished sunlight in Rome. This phenomenon coincided with the assassination of Julius Caesar, and the death of the Emperor was thus directly related to the eruption in the minds of many Romans (Forsyth 1988). Plutarch wrote that the haze from the eruption had rendered the rays of the Sun so faint and cold that fruit could not ripen, but withered on the trees because of the frigid air. A number of sources indicate that climate was unusually severe in the Mediterranean region in 44 and 43 BC in the wake of the Etna eruption. A crop failure resulted in famine in Egypt, which in turn caused corn shortages in Rome. The crop failure in Italy was so severe, that the Senate appointed prominent leaders such as Cassius and Brutus to direct the harvesting and trade of corn. The dimensions of the volcanic aerosol produced during the 44 BC eruption are not known, but today Etna volcano is one of the largest emitters of sulphur dioxide gas to the atmosphere (Metrich and Clocchiatti 1989).

The history of research on the effects of eruptions on climate is divided into four main stages, and each stage was initiated or stimulated by a large eruption. The first scientific observations on the effects of volcanism were done in the wake of the Laki, Iceland, and Asama, Japan, eruptions in 1783, when volcanic haze from Iceland was carried south-east in the troposphere and lower stratosphere and spread over Europe (Sigurdsson 1982). That summer Benjamin Franklin was resident in France. Franklin noticed that a persistent haze lay over the land, which was dry and thus unlike ordinary fog. The dense haze so attenuated the strength of the Sun's rays that, when the rays were collected in the focal point of a magnifying glass, they were so faint that they could scarcely kindle brown paper. Consequently, he concluded, the ability of the Sun's rays to heat the Earth was greatly diminished (Franklin 1784). Franklin proposed that the source of the haze could be traced to volcanic eruptions in Iceland and thus laid the foundations for the scientific study of the effects of volcanism on climate. The year 1784 is reported as the coldest on record in France and England by Manley (1946).

Interest in the effects of volcanism on climate was rekindled by the 1815 Tambora eruption in Indonesia and the associated climate deterioration, which became known as the "year without summer" in North America and Europe. Thomas Mitchell (1817) described the haze over New York in the spring and summer of 1816 as a smoke-like haze, which was not moist and thus resembled a great deal the haze which spread over Europe 30 years earlier. In describing the mist over New York in 1816, Mitchell noted:

> "It had nothing of the nature of a humid fog. It was like that smoking vapour which overspread Europe about thirty years ago. The learned, who made experiments to ascertain its

nature, could only state its remarkable dryness, by which no polished surface or mirror could be obscured."

Here he is referring to the observations of Franklin on the haze from the great Laki eruption in Iceland in 1783 (Franklin 1784). Thus the haze of 1816 and also the associated climatic effects were already linked to volcanic activity, although these phenomena were not attributed to the eruption of Tambora volcano until research on this issue was stimulated again by the Krakatoa eruption.

When Krakatoa volcano erupted in Indonesia in 1883, the global optical effects in the atmosphere prompted the study of environmental effects of volcanic eruptions (Symons 1888). Many interesting speculations were put forward to explain the spectacular optical effects and other atmospheric phenomena. Virtually all scientists who studied these phenomena assumed, however, that the particles responsible were glassy volcanic ash particles in the stratosphere. This seemed logical at the time, because volcanic ash fell out in large quantities near the volcano. On the island of Ceylon (now Sri Lanka) Isis Pogson suggested, on the other hand, that the haze in the upper atmosphere from the eruption was derived from sulphur gases, and a similar suggestion was made by F. L. Clarke in Hawaii (Symons 1888). Later studies have shown that these speculations, contrary to general opinion of scientists at the time, were correct, but it was not until 1963 that scientists realized that Pogson was on the right track with respect to the important role of sulphur (Pogson 1883).

In 1901 the cousins Sarasin proposed that periods of cold climate and even ice ages could be caused by the effects of atmospheric haze in the stratosphere from volcanic eruptions, because the haze absorbs the Sun's radiation and cools the surface of the Earth (Sarasin and Sarasin 1901). The great explosive eruption of Katmai, Alaska, in 1912 dispersed volcanic haze widely in the stratosphere over the Northern Hemisphere. Precise measurements now showed for the first time that the haze greatly attenuated the thermal radiation which normally reached the Earth's surface (Abbot 1913), and therefore the Alaskan eruption stimulated research on the relationship between volcanism and climate. It was proposed that the haze can affect solar and terrestrial radiation in the atmosphere in three ways. Part of the solar radiation is reflected by the haze to space, part is absorbed in the haze and some of the radiation is scattered by particles in the haze. Thus the haze scatters radiation both down to the surface of the Earth and up into space.

W. J. Humphreys first attempted to establish quantitatively the relationship between volcanism and climate. He applied Rayleigh's equation to explore the complex interplay between the various components of the thermal radiation transmitted through the atmosphere, as a function of the wavelength (Humphreys 1913, 1920). He showed that first, the volcanic haze can absorb short-wavelength solar radiation, second, the short-wavelength solar radiation can be reflected by the haze, but this is highly dependent on aerosol size, and third, that the aerosol layer absorbs long-wave radiation or dark heat radiation which emanates from the warm Earth. Humphreys proposed that, on the basis of observations from the eruption of Krakatoa in 1883, volcanic aerosols in the stratosphere are formed by glassy volcanic particles, approximately $1–2 \times 10^{-6}$ m (1–2 μm) in size. Douglas Archibald calculated that the average diameter of aerosols was 1.5×10^{-6} m (1.5 μm) on the basis of scattering of sunlight in the "dust" layer in the stratosphere after the Krakatoa eruption (Symons 1888). Later work has shown that particles of this size should take about one year to settle out of the atmosphere, because the settling velocity of a 1.85×10^{-6} m (1.85 μm) glass particle is about 10^{-2} m s^{-1} (1 cm s^{-1}) at 40 km elevation and decreases to 4×10^{-4} m s^{-1} (0.4 mm s^{-1}) at 10 km due to the increasing viscosity of the atmosphere (Kasten 1968). Humphreys also concluded that backscattering of solar radiation to space from a volcanic "dust" layer was about 30 times greater than the absorption of thermal radiation in the "dust" layer, and consequently the volcanic layer led to cooling of the surface of the planet. Later studies have shown that several of Humphreys' fundamental assumptions are in error. He greatly exaggerated the aerosol cooling tendency, because he assumed that the aerosols scatter mainly in the backward direction (i.e. away from the Earth), and he also underestimated the greenhouse effect, because he assumed that the particles behaved as Rayleigh particles (particles which are small compared to the characteristic wavelength of the incident light radiation, and produce scattering of photons which is inversely proportional to the fourth power of the wavelength) in the thermal infrared. Later studies also showed that the volcanic aerosol layers in the stratosphere are not composed of glass particles or volcanic ash, but rather of a sulphuric acid, and that they approximate 10^{-7} m (0.1 μm) in diameter.

18.3 PHYSICAL PRINCIPLES

The early studies of the effects of volcanism were hampered by the fact that the volcanic layer in the atmosphere was inaccessible for sampling. In 1959 a stratospheric aerosol layer was first discovered at 16–24 km and sampled by balloon-borne instruments and research aircraft (Junge *et al.* 1961; Junge and Manson 1961; Junge 1963). In this pioneering study, which also formulated the principles of sedimentation rate of aerosol particles, the aerosol layer was shown to be composed of 10^{-7}–10^{-6} m (0.1–1 μm) diameter particles of sulphuric acid (75% $H_2SO_4 + 25\%$ H_2O). The formation of the layer was correctly attributed to oxidation reactions of SO_2 by Junge and colleagues, but at the time the gas was considered to have been derived from the troposphere and not from a volcanic source. This first volcanic aerosol layer was probably a remnant of the plume from the great Bezymianny eruption in Kamchatka in 1956. When Agung volcano on the island of Bali in Indonesia erupted in 1963, a major volcanic aerosol layer was observed to form for the first time and extensively sampled. The first and only isotopic measurements of the sulphate in the stratospheric aerosol attributed to the Agung eruption also gave proof that the sulphur was of volcanic origin (Castleman *et al.* 1974). Before quantitative models were developed, Lamb (1970) examined empirically the relationship between volcanic eruptions and climate change. On the basis of eruption magnitude and observed optical effects, he compiled a "dust-veil index" for calculating the relative importance of various historical volcanic eruptions on climate. Lamb argued that the volcanic "dust" had a "reverse greenhouse effect". The usefulness of the dust-veil index is, however, subject to severe limitations, in part due to the inadequate database of volcanic eruptions available to Lamb, but primarily due to the fact that the mass of the volcanic aerosol does not scale simply to the mass of erupted magma, as discussed below.

Pollack *et al.* (1976) carried out one of the first studies of the radiative effects of the volcanic aerosol on climate. They demonstrated the importance of aerosol size and showed that, although backscattering of solar radiation by the aerosol is important, the surface cooling is weak initially, due to the greenhouse effect of larger particles shortly after eruption. Of greatest importance, however, is the *optical depth*, sometimes referred to as the optical thickness, which is a measure of the depletion of the incoming solar radiation by the volcanic aerosols in the atmosphere. The absorption optical depth, τ, is defined as the ratio of the radiation emerging at the bottom of the aerosol-bearing atmosphere, I, to that incident at the top, I_o, and according to Beer's law

$$I/I_o = \exp(-\tau/\mu) \tag{18.1}$$

where μ is a measure of the length of the path of sunlight through the Earth's atmosphere as a function of the solar zenith angle. The mean value of μ is 0.58 (Sagan and Turco 1990). If $\tau = 0$ the layer is transparent, $I/I_o = 1$ and all of the sunlight incident on the aerosol layer is transmitted to the Earth's surface. If the volcanic aerosol layer has an optical depth of $\tau = 0.05$, $I/I_o = 0.92$ and only 92% of the sunlight incident on the atmosphere is transmitted through the layer to the Earth's surface. If $\tau = 1$, $I/I_o = 0.18$ and very little sunlight makes it through. For volcanic aerosols, however, both the absorption and scattering optical depths need to be considered.

Empirical studies have shown that the magnitude of sulphur output to the atmosphere from individual volcanic events shows some correlation with the observed climate response. Thus the sulphur yield to the atmosphere during a volcanic eruption correlates well with the observed decrease in Northern Hemisphere surface temperature (Devine *et al.* 1984; Sigurdsson 1990a; Rampino and Self 1982; Sigurdsson 1990b). As shown in Figure 18.1, the observed temperature decrease (ΔT) in the year following an eruption is related to the volcanic sulphur output by a power function, with the sulphur mass raised to a power close to a cube root law

$$\Delta T = 5.9 \times 10^5 (S^{0.31}) \tag{18.2}$$

where S is the volcanic sulphur mass yield to the atmosphere, in grams (Sigurdsson 1990a).

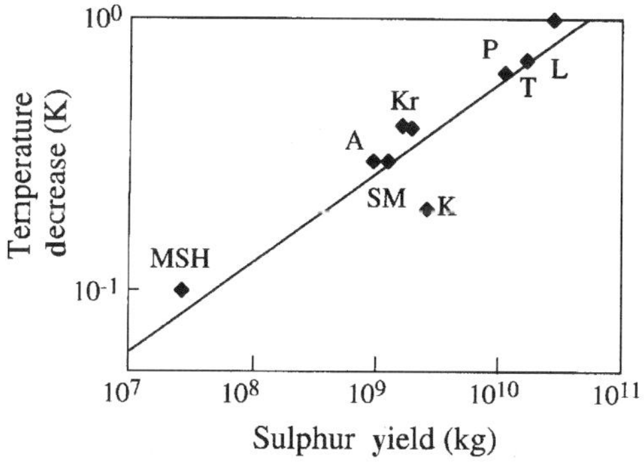

Figure 18.1 The observed relationship between sulphur yield to the atmosphere during large volcanic eruptions and the Northern Hemisphere temperature decrease following the event. Sulphur yield data are from Robock (1981), Sigurdsson (1982), Devine *et al.* (1984) and Palais and Sigurdsson (1989). Climatological data are from Rampino and Self (1982). MSH = Mount St Helens 1980; A = Agung 1963; SM = Santa Maria 1902; Kr = Krakatoa 1883; K = Katmai 1912; P = Pinatubo 1991; T = Tambora 1815; L = Laki 1783. Equation (18.2) in the text describes the best fit to the data, with a correlation coefficient of 0.92, but note that this correlation excludes the Pinatubo data. The Mount St Helens information defines the lower limit of the curve

Theoretical aspects of climate forcing by volcanic aerosols have most recently been addressed in detail by Lacis *et al.* (1992). The most important parameter for climate forcing appears to be the solar radiative flux through the Earth's atmosphere to the surface, or the heating rate of the Earth, which is expressed in terms of watts per square metre. It is the effect of volcanic aerosols on this flux which is of critical importance. The total radiative flux is divided into two components which act in opposing directions in terms of their effects on the surface temperature of the planet. One component is the solar flux of dominantly short-wave radiation, F_{solar}, and the other component is the long-wave thermal infrared radiation flux, F_{IR}. The infrared radiation is emitted from the Earth and its atmosphere out to space, but can be reflected back by the aerosol layer. It is the increase in greenhouse gases that has occurred since the Industrial Revolution that has decreased this infrared radiation loss and caused a heating by approximately 2 W m^{-2}. Lacis *et al.*

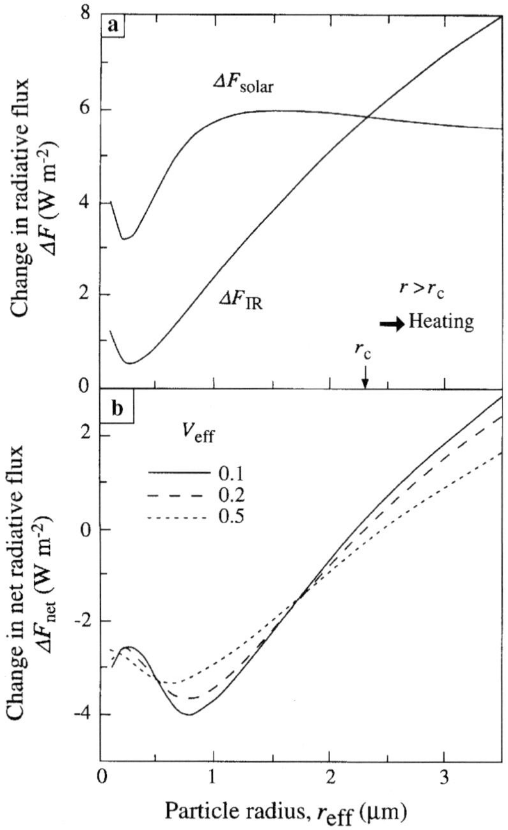

Figure 18.2 Theoretical models of climate forcing by a stratospheric aerosol layer (after Lacis *et al.* 1992). (a) Shows the change of solar flux, ΔF_{solar} (cooling), and infrared thermal flux, ΔF_{IR} (warming), at the tropopause, as a function of aerosol radius, caused by adding a stratospheric aerosol layer at 20–25 km altitude, with an optical depth $\tau = 0.1$. (b) Shows the change of the net radiative flux, ΔF_{net}, at the tropopause for an aerosol optical depth $\tau = 0.1$, as a function of aerosol radius, and for three values of effective variance or width of the particle size distribution, V_{eff}. Multiplying ΔF_{net} by 0.3 yields ΔT (°C), the equilibrium air temperature change at the Earth's surface

(1992) have shown that the size and even the sign of the climate forcing (i.e. Earth surface warming or cooling) is greatly influenced by the stratospheric volcanic aerosol size distribution, particularly the area-weighted mean radius, and to a lesser extent its chemical composition and the altitude of the aerosol layer.

Figure 18.2a shows the decrease in downward solar flux (cooling) and the increase in upward thermal flux (warming) as a function of aerosol radius for a particular stratospheric aerosol layer, with optical depth (τ_a) of 0.1. As shown in Figure 18.2, the change in both the solar flux and the downward infrared thermal flux at the tropopause is highly dependent on the size distribution of the volcanic aerosol. While the infrared forcing shows a marked increase (warming) with increasing aerosol radius, the change in the solar flux is minor, and consequently the size-dependent variation in the resultant net radiative flux, ΔF_{net}, is largely controlled by the infrared flux variation. Aerosols with critical radii $\geqslant 2 \times 10^{-6}$ m (2 μm) lead to tropospheric heating (Figure 18.2b), whereas aerosols of smaller size lead to cooling. As discussed below, the smaller aerosol particles are of greater climatological interest, because the lifetime of larger particles ($\geqslant 2 \times 10^{-6}$ m; 2 μm) in the stratosphere is in general too short to bring about such heating, except locally near the volcano.

Pollack *et al.* (1976) first demonstrated that the effects of aerosol silicate particles and sulphuric acid particles on the mean surface temperature are quite different. The dependence of the volcanic forcing (i.e. the change in the net radiative flux) on the aerosol composition and effective radius is shown in Figure 18.3, for sulphuric acid, basalt glass (analogous to volcanic ash particles) and desert dust (Lacis *et al.* 1992). In the submicrometre to micrometre-size range, the climate forcing is greater for basalt glass than for sulphuric acid, but the climate forcing is largest for desert dust due to its large refractive index. These authors also examined how the climate forcing and stratospheric warming depend on the altitude of the aerosol layer. Volcanic aerosol plumes emplaced above 30–35 km altitude cool both the stratosphere and the troposphere, whereas aerosol layers injected at lower altitudes bring about stratospheric heating, which increases as the layer subsides or drops in altitude. The most important factor in aerosol climate forcing is, of course, the optical depth (a measure of the aerosol mass) as shown in Figure 18.4. As discussed below, the optical depth varies as a function of the size of the eruption, or specifically its total output of sulphur dioxide gas.

18.4 SEDIMENTATION AND DISPERSAL OF VOLCANIC AEROSOLS

Volcanic aerosols which are composed predominantly of sulphuric acid are generally $< 10^{-5}$ m (10 μm) and typically in the range 10^{-7}–10^{-6} m (0.1–1 μm). Their settling behaviour from volcanic plumes in the atmosphere is significantly different from that of volcanic tephra. Sulphuric acid aerosols grow in the stratosphere by molecular condensation and coagulation processes until they attain a size and terminal fall velocity that brings about sedimentation through the tropopause. In the troposphere the aerosols are removed quickly by either washout, as they are swept up by falling drops and incorporated in them, or removed by rainout, where the aerosol particles act as the primary condensation nuclei in the precipitation formation process. Aerosols can also condense on ash particle surfaces and be rapidly removed because of the much larger settling velocity of most ash.

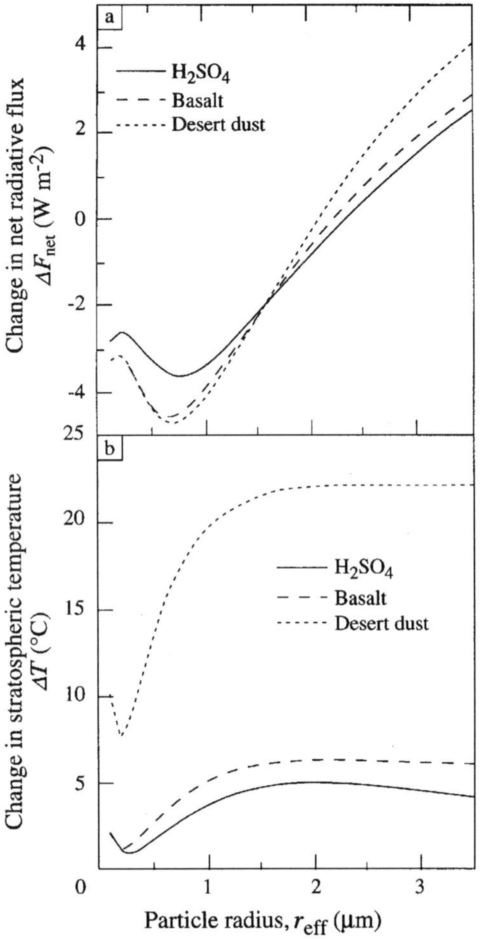

Figure 18.3 The effect of aerosol composition on (a) the net radiative flux, ΔF_{net}, in watts per square metre, and (b) the change in temperature in the stratospheric aerosol layer, as a function of effective aerosol radius, for optical depth $\tau = 0.1$ (after Lacis *et al.* 1992). Sulphuric acid is representative of a typical volcanic aerosol. A basalt dust layer is representative of the optical properties of glassy particles from most explosive volcanic eruptions. The large difference in the temperature change of an aerosol layer with desert dust in (b) is due to its strong absorption of solar radiation (a large real refractive index)

The first treatment of the settling of volcanic aerosols in the atmosphere was carried out by Humphreys (1920), who used Stokes' law to evaluate the fallout rate of particles from the 1883 eruption of Krakatoa volcano in Indonesia. He deduced, for example, that 1.85×10^{-6} m (1.85 μm) diameter glass particles took about one year to settle out of the stratosphere, and could thus conceivably have significant climate effects. As discussed in Chapter 16, however, small particles are likely to form relatively large aggregates with much higher settling velocities and therefore shorter atmospheric residence times.

When Junge *et al.* (1961) demonstrated that the stratospheric "dust" layer consists dominantly of sub-micrometre or micrometre-size sulphuric acid aerosol droplets, rather

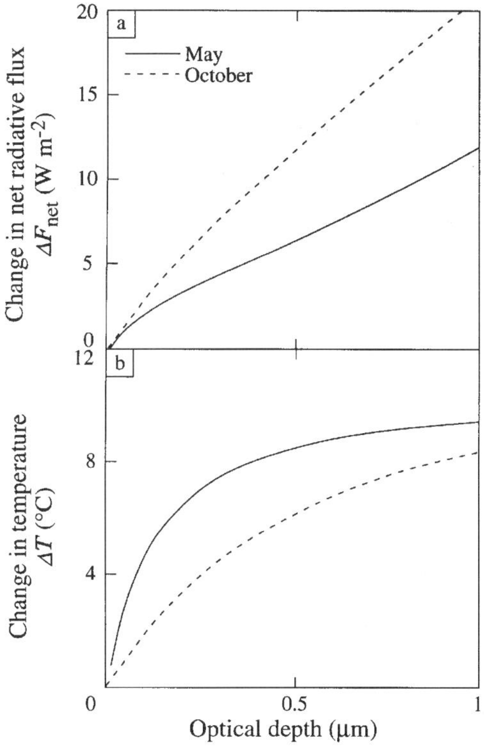

Figure 18.4 The effect of aerosol optical depth on climate forcing (after Lacis *et al.* 1992). (a) Change in the net radiative flux at the tropopause, ΔF_{net}, in watts per square metre, and (b) change in stratospheric temperature caused by addition of an aerosol layer at altitudes of 20–25 km for two aerosol size distributions typical of the El Chichon 1982 eruption. The "May" size distribution (one month after the eruption) contained a higher proportion of large particles (radius $\geqslant 1$ μm) than the "October" size distribution (six months after the eruption), which increased absorption of infrared thermal radiation, enhancing the stratospheric warming in the May scenario

than volcanic glass fragments, they calculated fall velocities of aerosol particles as a function of altitude using a similar formula to that applied by Humphreys. They obtained settling rates of the order of 10^{-3} m s^{-1} (1 mm s^{-1}) for a 2×10^{-6} m (2 μm) diameter particle of density 2000 kg m^{-3} in the stratosphere, and concluded that such particles typically settled out in half to one year (Junge 1963). A more recent evaluation of settling rates of aerosols was carried out by Kasten (1968), using Stokes' law modified by the Cunningham–Millikan correction (Millikan 1923). These settling rates and cumulative fallout time for sulphuric acid and glass aerosols in the 10^{-8}–10^{-5} m (0.01–10 μm) radius range are shown in Figures 18.5 and 18.6 respectively. These rates are likely to be appropriate for spherical sulphuric acid aerosol droplets, but less appropriate for irregularly shaped individual glass and mineral fragments $> 10^{-5}$ m (10 μm). In general the deviation from equant or spherical form decreases with decreasing grain size, and thus the trends shown in these figures should be a good indication of the settling process. On the other hand, calculated fallout times of individual glass particles in this size range may not be fully relevant, if their sedimentation in the atmosphere is influenced by aggregation.

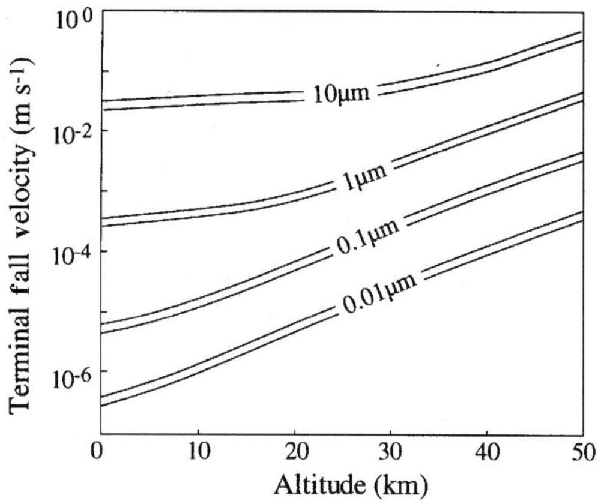

Figure 18.5 Settling velocity of aerosol particles as a function of altitude, for particles with radii in the range 0.01, 0.1, 1 and 10 μm. The upper of each pair of curves is for volcanic glass particles with a density of 2500 kg m^{-3}, and the lower is for sulphuric acid droplets with a density of 1788 kg m^{-3}. Calculated from the data of Kasten (1968) for a 1962 US standard atmosphere

leading to enhanced settling rates (Chapter 16). Furthermore, we should recall that the size population of the sulphate aerosol is dynamic, due to gas-to-liquid conversion and coagulation processes, which may lead to growth of particles during their stratospheric residence times.

A significant fraction of eruption plumes comprise relatively large particles, tens or hundreds of micrometres in diameter. The terminal velocity of volcanic particles in the diameter range 3×10^{-5}–5×10^{-4} m (30–500 μm) has been determined as a function of particle shape (Wilson and Huang 1979). Volcanic particles $\geqslant 10^{-4}$ m (100 μm) in dia-

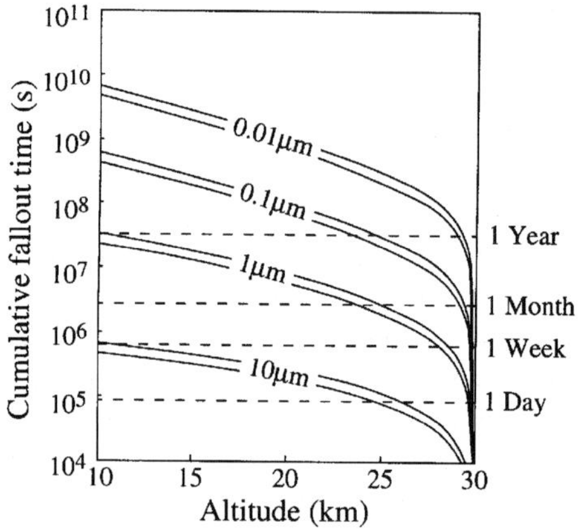

Figure 18.6 The cumulative fallout time (seconds) for sulphuric acid aerosols (the upper of each pair of curves) and volcanic glass particles, in the range 0.01, 0.1, 1 and 10 μm radius, as a function of altitude in the stratosphere, down to the tropopause (*c.* 10 km). These settling times are calculated for an initial aerosol injection at 30 km altitude, from the data of Kasten (1968) for a 1962 US standard atmosphere

meter have settling velocities $\geqslant 0.3$ m s^{-1} and thus fall out from a 25 km high eruption plume within one day. They cannot, therefore, be considered of interest in evaluation of climate forcing by volcanic eruptions. The Wilson–Huang grain size/terminal velocity relationship has been extrapolated to the 10^{-6}–10^{-5} m (1–10 mm) particle size range for fine ash from the 1982 El Chichon eruption, yielding settling rates a factor of two lower than the Stokes settling rates (Mackinnon *et al.* 1984). Such an extrapolation is questionable, however, because of the dependence of the shape factor on the particle diameter. Observed residence times of large particles in the El Chichon stratospheric aerosol plume are comparable to or somewhat larger than predicted by the Wilson–Huang model, with settling times for 5×10^{-6}–10^{-5} (5–10 μm) diameter particles of the order of 30–100 days from the 26 km level to the tropopause.

Dispersal of aerosols also depends on global circulation patterns in the atmosphere. Aerosols injected into low latitudes are distributed rapidly. For example, in the Pinatubo eruption of 1991 the aerosol layer formed a complete band around the Earth in the 15°S to 25°N region in only three weeks. However, the dispersal of aerosols in higher latitudes is a much slower process by atmospheric diffusion in the predominantly latitude-parallel global circulation system. The overall tendency of upper-level winds is to spiral towards the poles where downward movement of air masses aids sedimentation into the ice masses, which is why the polar ice caps contain an excellent record of global volcanism in the form of sulphate concentrations.

18.5 THE PINATUBO 1991 ERUPTION

The evolution of volcanic aerosols and their atmospheric effects are best demonstrated by the results of a variety of studies of the 1991 Pinatubo eruption. After a 400-year period of dormancy, Pinatubo volcano in the Philippines returned to activity on June 9, 1991 with a series of explosions, which reached a climax in the afternoon of June 15. Because of the exceptional size of the eruption, the explosive event was expected to lead to detectable climate forcing. It caused aerosol loading of the atmosphere about a factor of 1.7 greater than the El Chichon 1982 eruption, with an increase in the optical depth by about a factor of five over "normal" or pre-eruption conditions. The eruption and its stratospheric plume were monitored by satellite sensor and other instruments in greater detail than any other explosive eruption to date. NOAA infrared satellite images of the initial plume over the volcano on June 15 show a major temperature perturbation within a distance of 300–500 km from the vent (Figure 18.7). The perturbations shown in the figure have a wavelength of about 70 km near the source, corresponding to a period of about 230 s for 300 m s^{-1} sound velocity (Kanamori and Mori 1992). These oscillations observed in the eruption plume may have led to atmospheric pressure changes which in turn generated a seismic wave train of long-period Rayleigh waves recorded world-wide, due to the acoustic coupling of the atmospheric oscillation and the solid Earth (Kanamori and Mori 1992).

The growth and evolution of the eruption plume were documented by hourly satellite images (Figure 18.8), and plume height has been inferred from infrared images (Koyaguchi and Tokuno 1993). The main eruption plume appeared above the volcano at about 1420 hours Philippine Standard Time (PST) and expanded initially at rates up to 125 m s^{-1}, reaching a diameter of 400 km within one hour. The plume eventually

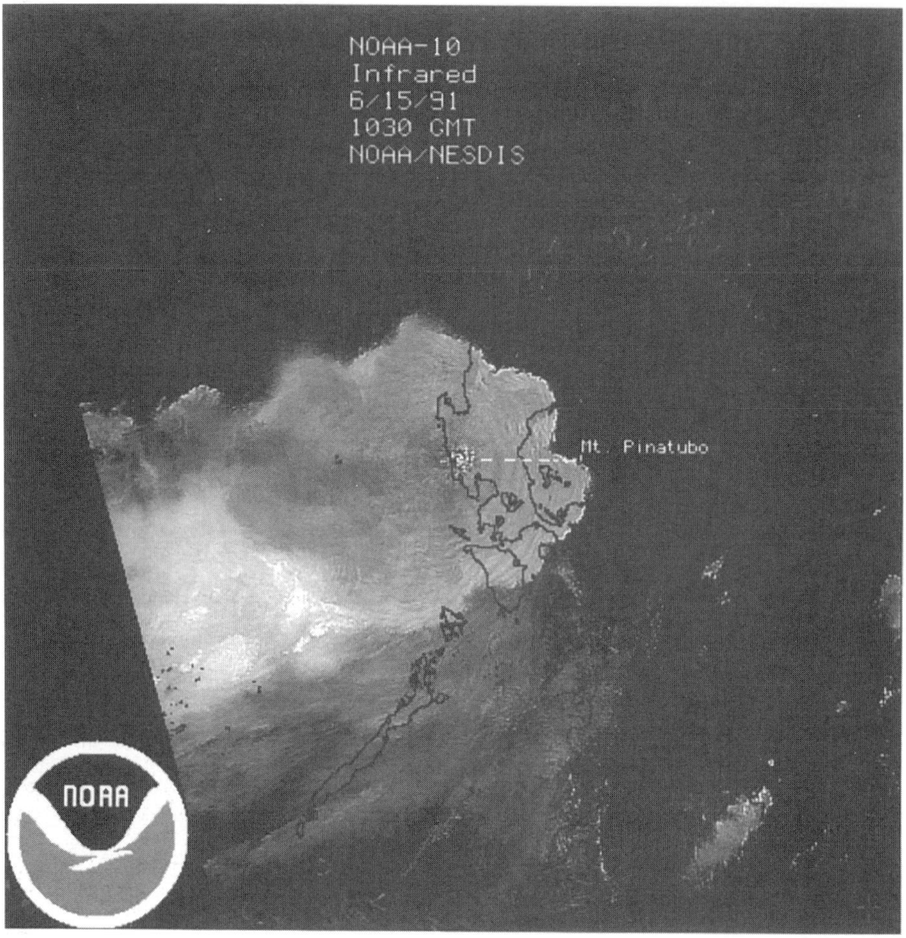

Figure 18.7 NOAA satellite infrared image of the Pinatubo eruption plume at 1030 hours GMT (1830 hours PST) on June 15, 1991. The image shows major wave-like temperature perturbations up to 500 km from the volcano, with wavelength of about 70 km. (Image courtesy of George Stephens, NOAA–NESDIS.)

expanded up to 250 km upwind after five hours, when it was about 3×10^{11} m^2 in area (Koyaguchi and Tokuno 1993). Satellite images show that during peak activity, the plume centre or axis of the eruption column was at an elevation of about 34 km and the eastern plume edge at 25 km, subject to a wind velocity of about 20 m s^{-1} south-westward. Both airborne and ground-based satellite observations, discussed later, indicate that the main aerosol cloud was largely confined to a relatively narrow band centred around 25 km altitude, and ranging from 20 to 29 km in altitude. This is in good agreement with the findings of the eruption column height above the volcano.

Given the magma temperature and eruption column dynamics (Koyaguchi and Tokuno 1993), the initial velocity of the column is likely to have been about 280 m s^{-1}, with a magma discharge rate of 1.6×10^9 kg s^{-1}. Thus the total erupted mass of 1.3–1.8×10^{13} kg is likely to have been discharged in two to three hours. The column

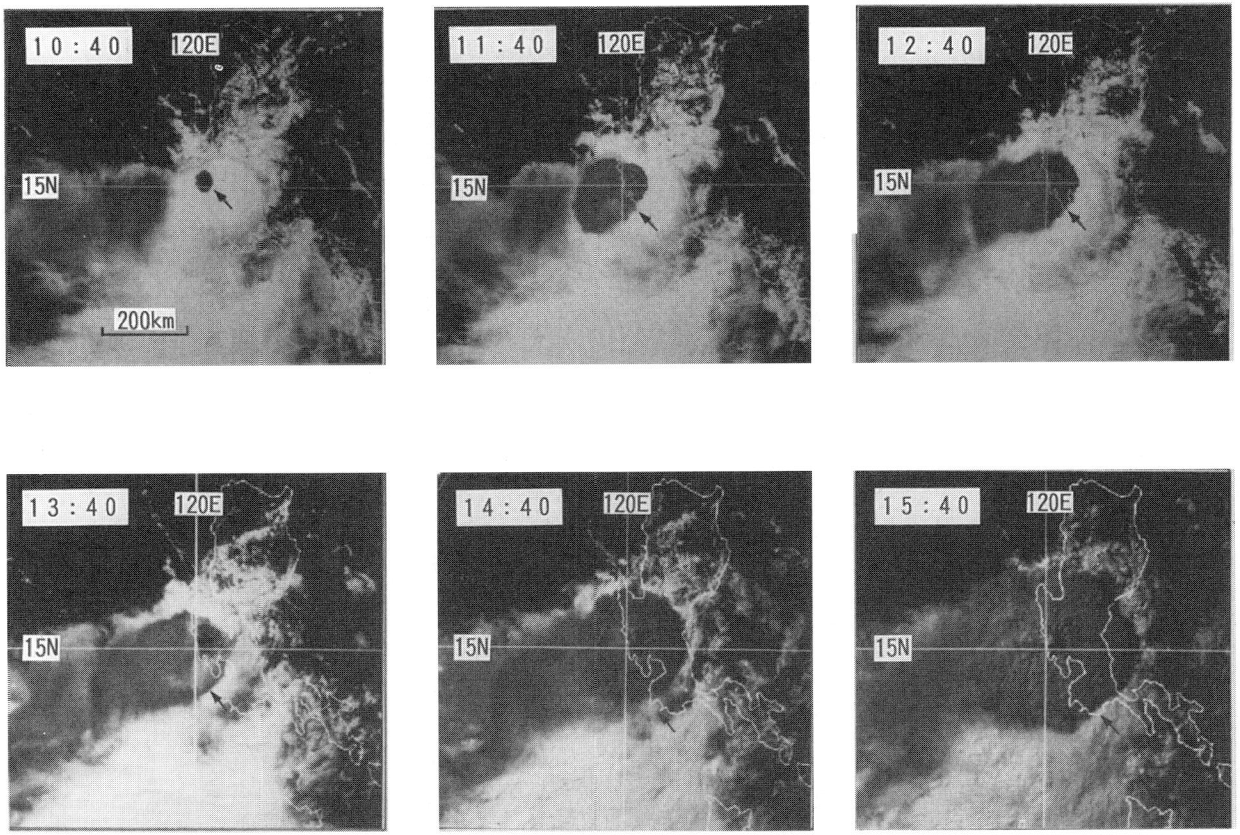

Figure 18.8 Expansion of the Pinatubo eruption plume and initial dispersal of the stratospheric plume on June 15, 1991, observed in hourly satellite images from 1040 hours to 1540 hours PST (Koyaguchi and Tokuno 1993)

dynamics calculations of Koyaguchi and Tokuno (1993) can be subject to considerable revision, however, because they utilize a lower temperature and a water content in the magma, than the observed values of 780 °C and 3.6 wt% (Rutherford and Devine 1993).

The origin of giant eruption plumes, such as the one which rose over Pinatubo in 1991, can be attributed either to Plinian activity, where the buoyant eruption column rises directly from the vent, or a co-ignimbrite origin, where the buoyant column is driven largely by the thermal energy derived from the pyroclastic flow. Large volumes of pyroclastic flows issued from the volcano during the eruption which, in combination with lahars, were mainly responsible for the death toll of 600 people. Koyaguchi and Tokuno (1993) have argued that the eruption started with the generation of pyroclastic flows, which produced part of the ash fallout, but that the main eruption plume and associated fallout was the result of a very large Plinian eruption. The Pinatubo eruption has been interpreted as a Plinian eruption (Koyaguchi and Tokuno 1993), but in terms of some features of the fall deposit, the eruption has the characteristics of a co-ignimbrite event. The deposit features include low thickness, bimodal grain size distribution and poor sorting of the proximal fall deposit (layer C; Figure 5 in Koyaguchi and Tokuno 1993), in contrast to thick, well-sorted and unimodal-sized deposits typically derived from Plinian eruptions at comparable distance from source.

Fallout of tephra began immediately after the major explosion and was largely completed in one hour (Koyaguchi and Tokuno 1993). Fallout was dominantly to the west and south-west of the volcano, dictated by the prevailing lower stratospheric wind. Thus the bulk of the tephra component of the eruption plume was decoupled within hours from the volatile or gas component, which included water vapour and sulphur dioxide. The latter component remained in the stratosphere near the 25 km level for several months, where it underwent a gas-to-liquid conversion to form the volcanic aerosol. The volatile-rich stratospheric plume is likely to have contained the bulk of the magmatic water erupted by Pinatubo. The magma contained 5.5 wt% H_2O dissolved in the melt (Rutherford and Devine 1993), but because the magma contained approximately 35% crystals, the bulk water content of the erupted magma was approximately 3.6 wt%, corresponding to a *potential* magmatic water mass of 4.5–6.5 $\times$ 10^{11} kg injected into the stratosphere. In addition, the plume would have contained entrained atmospheric water vapour and the volcanic sulphur dioxide component, as well as other magmatic gases (HCl, CO_2, etc.). The mass fraction of tropospheric air entrained in the eruption column is estimated to be 0.6 of the mixture of pyroclasts, magmatic volatiles and air (Koyaguchi and Tokuno 1993). Adopting 0.77 wt% H_2O as a reasonable value for tropospheric air at 0 °C at 5.5 km (Tabazadeh and Turco 1993), the potential mass of water entrained from the troposphere and into the eruption plume is in the range 1.5–2 $\times$ 10^{11} kg. Much of the water in the eruption plume was probably precipitated locally as rain or ice during plume rise in the cold upper troposphere. However, significant water may have entered the stratosphere, because infrared images (Figure 18.7) indicate that the main eruption plume was tens of degrees hotter than the ambient air, and therefore possibly above the water condensation temperature.

18.5.1 Sulphur Dioxide Emission

The Pinatubo magma is a crystal-rich dacite, erupted at a temperature of 780 °C, containing 3.5 wt% H_2O, and derived from a magma reservoir at a depth of about 7–8 km in the Earth's crust (Rutherford and Devine 1993). The magma also carried significant quantities of the sulphur-rich mineral anhydrite to the surface. In view of the large stratospheric sulphate aerosol generated by the eruption, the magma would be expected to carry a large quantity of sulphur, dissolved in the melt prior to eruption, as a free sulphur-rich volatile phase, or in sulphur-rich minerals (e.g. anhydrite) which could break down during the eruption to give off a sulphur-rich gas. The total volume of erupted products is estimated to be 5–7 km^3 dense-rock equivalent, or approximately $1.3–1.8 \times 10^{13}$ kg (Rutherford and Devine 1993; Koyaguchi and Tokuno 1993). In comparison, the mass of sulphur dioxide emitted to the stratosphere by the eruption is about 2×10^{10} kg. This figure was obtained from TOMS measurements and corresponds to 1–1.5 wt% of the erupted magma (Bluth *et al.* 1992). The amount of sulphur dissolved in the melt prior to eruption is on the other hand only approximately 70 ppm, and thus degassing of the magma could only account for a fraction (5×10^8 kg) of the observed sulphur emission. Clearly, the majority of the sulphur discharge to the atmosphere came from another component of the magma, other than dissolved sulphur. It has been demonstrated experimentally that anhydrite breakdown during magma ascent and eruption is a viable hypothesis for this observed excess sulphur emission (Baker and Rutherford 1992).

The levels of SO_2 emission to the atmosphere following the eruption were monitored by the TOMS instrument on the Nimbus-7 satellite (Bluth *et al.* 1992). The SO_2 emission from the cataclysmic event on June 15 was detected by TOMS the following day as a discrete mass about 1000 km west-south-west of the volcano (Figure 18.9 (Plate V)). During the following days the cloud stretched rapidly to the west into an elongate plume about 4 km in thickness, with an average speed of 35 m s^{-1} for the leading edge (Bluth *et al.* 1992). Two weeks later the leading edge had reached California, and the plume still contained about 60% of its original SO_2 content. The plume circled the globe in 22 days, and at a velocity very similar to that of the El Chichon volcanic aerosol plume in 1982. By this time the plume had also spread about 20° in latitude. The transport of the plume can be attributed to the stratospheric easterlies in the Northern Hemisphere (summer) and westerlies in the Southern Hemisphere (winter). In the summer there is a wind reversal near 20 km altitude, with easterlies above and westerlies below. The wind direction in the lower stratosphere undergoes a quasi-biennial oscillation, and at the time of the Pinatubo eruption, the change-over from westerly to easterly flow had already occurred, with mean zonal winds of 20–30 m s^{-1} at the 3×10^3 Pa level (30 mbar; ~24 km altitude; Labitzke and McCormick 1992).

The initial mass of the SO_2 emission is estimated as 2×10^{10} kg on the basis of the TOMS data, with a loss rate of $10^9–1.5 \times 10^9$ kg per day due to gas-to-liquid conversion and formation of a sulphate aerosol (Bluth *et al.* 1992). When converted to sulphuric acid (approximately 75% H_2SO_4 + 25% H_2O), the Pinatubo volcanic sulphur dioxide plume has a potential volcanic aerosol mass of about 4×10^{10} kg, but at no time was this entire aerosol mass present in the atmosphere. Three to four weeks after the eruption the column abundance of SO_2 in the atmosphere (i.e. the mass of SO_2 in the atmospheric column per square centimetre of the Earth's surface) was still of the order of $3–5 \times 10^{16}$ molecules cm^{-2} and no H_2S was found (Mankin *et al.* 1992; Goldman *et al.* 1992).

18.5.2 Sulphur Dioxide Decay and Sulphate Aerosol Evolution

The evolution of the SO_2 loading was monitored by the microwave limb sounder instrument on the UARS satellite, which measured SO_2 levels in the global atmosphere resulting from emission from the volcano (Figure 18.10 (Plate VI); Read *et al.* 1993). Three months after the eruption, the SO_2 gas plume formed an equatorial band, with peak SO_2 of about 15 parts per billion per unit volume of atmosphere near 26 km elevation, displaced slightly south of the equator. Although the plume was mainly confined to a relatively narrow altitude band, from about 20 to 30 km height, its global distribution was patchy, as shown in Figure 18.10. The removal rate of SO_2 gas after the eruption shows an exponential decay, with an *e*-folding decay time (decreasing to half its previous value) of 33–35 days (Figure 18.11). When extrapolated back to the day of the eruption, the SO_2 emission trend indicates an initial emission of the order of 1.7×10^{10} kg (MLS data) to 2×10^{10} kg (TOMS data).

The conversion of sulphur dioxide gas to a sulphate aerosol is a complex process, which includes a series of oxidation reactions of the type:

$$SO_{2(gas)} + OH \rightarrow HOSO_2$$
$$HOSO_2 + O_2 \rightarrow SO_3 + HO_2$$
$$SO_3 + H_2O \rightarrow H_2SO_{4(liq)}$$

There are a number of oxidants at work, but it is generally regarded that the most important of these is the hydroxyl (OH) radical. The observed rate of SO_2 decay for the Pinatubo gas plume is consistent with catalytic conversion of SO_2 to sulphuric acid by reaction with OH in the atmosphere, requiring OH levels of $1.2\text{–}3.3 \times 10^{12}\ m^{-3}$ at the altitude of the gas plume in order to oxidize the SO_2 (Read *et al.* 1993).

In the first week of the eruption, a stratospheric aerosol layer had already begun to form, as noted by the elevated optical depth in the Indonesia region (Stowe *et al.* 1992). Three weeks later the stratospheric aerosol had circled the globe, and spanned a band from 15°S to 25°N. The aerosol continued to grow in areal extent and depth as conversion of SO_2

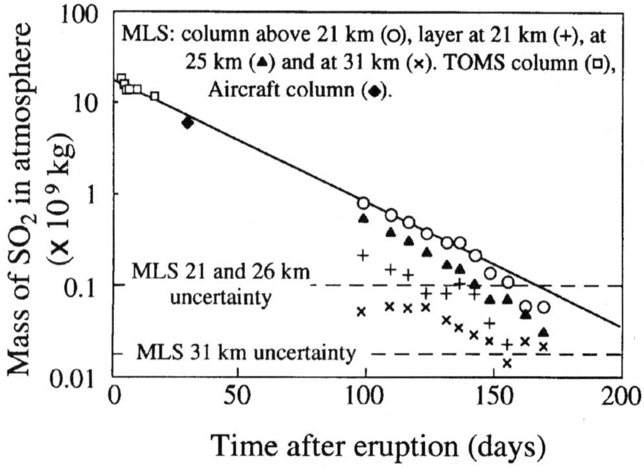

Figure 18.11 The decay of the total mass of sulphur dioxide in the global atmosphere following the 1991 Pinatubo eruption. When projected back to the time of eruption, the microwave limb sounder (MLS) measurements indicate a total initial SO_2 mass of 1.7×10^{10} kg, which is in good agreement with other satellite and aircraft observations of approximately 2×10^{10} kg (Read *et al.* 1993). Airborne observations reported by Mankin *et al.* (1992) are also shown

proceeded, covering 42% of the surface of the Earth only nine weeks after the eruption. The aerosol optical depth in the tropical zone peaked about three months after the eruption, but optical depths in mid-latitude zones remained high (Figure 18.12). The volcanic plume increased the atmospheric optical depth by about a factor of five over the tropical Pacific, to about 0.35 two months after the eruption. The increase in zonal mean optical depth in mid-latitude zones culminated about five months after the eruption (Minnis *et al.* 1993), and was beginning to decrease in October 1991 (Figure 18.13).

The most complete view of the evolution of the Pinatubo aerosol comes from stratospheric optical depth measurements with the SAGE II instrument on board the ERBS satellite, which has a near-global coverage from 50°S to 70°N (McCormick and Veiga 1992). In the two months following the eruption, the plume was mainly at 20–25 km altitude, with layers observed as high as 29 km altitude (Figure 18.14). Although it circled the globe equatorially (with an average easterly velocity of 30 m s^{-1}), the distribution of the aerosol plume was non-uniform, with an irregular pattern of advection of the plume to mid-latitudes. In one instance the South Atlantic high-pressure system at 20°S and a northward bulge in the polar jet stream advected the plume to mid-latitudes at 20–21 km altitude. Similarly, an Asian anticyclone advected the aerosol plume over eastern Asia at 16 km elevation, with a northward component of 2° per day (McCormick and Veiga 1992). The uneven transport of the aerosol plume from the tropics to high latitudes led to very different optical depth evolution in the two hemispheres of the Earth. Thus at 40°N an enhancement of the optical depth is noted at the end of June, but was not noted at 40°S until late July. The Pinatubo aerosol increased the optical depth of the atmosphere by two orders of magnitude between 40°N and 40°S in the first five months after the eruption.

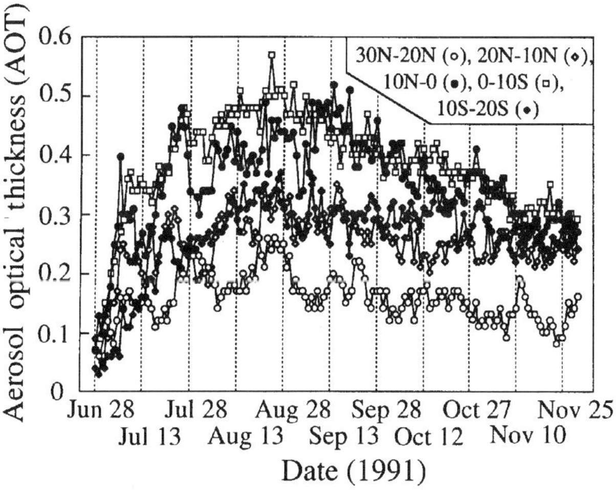

Figure 18.12 The change in aerosol optical depth following the Pinatubo 1991 eruption, for 10° latitude zones from 20°S to 30°N, covering the longitude range from 80°W to 100°E in the Pacific Ocean from June 28 to November 30, 1991 (Stowe *et al.* 1992). The rate of change in aerosol optical depth is most rapid in equatorial zones. Optical depth is a measure of the depletion of the incoming solar radiation due to aerosols in the atmosphere. It is an exponent function of the ratio of the radiation emerging at the bottom of the atmosphere to that incident at the top

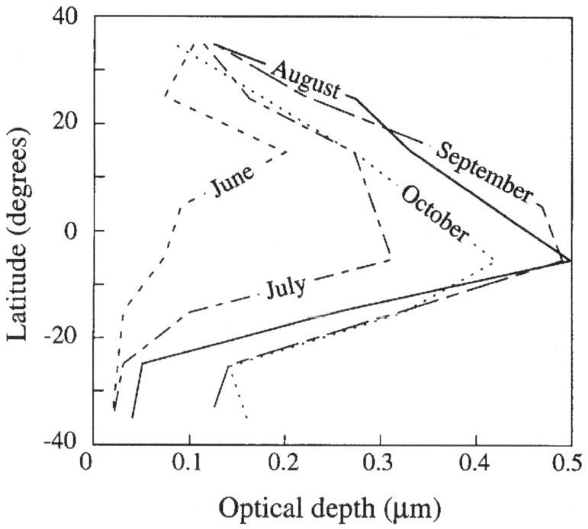

Figure 18.13 Zonal mean aerosol optical depths from NOAA satellite AVHRR for the five-month period following the Pinatubo 1991 eruption (after Minnis *et al.* 1993)

With mass loadings of 2×10^{10}–3×10^{10} kg of sulphate, the eruption injected into the atmosphere about three times the amount of the 1982 El Chichon eruption aerosol.

The volcanic aerosol size distribution has an important influence on a number of factors. It determines the total surface area of the stratospheric aerosol, which influences the rates of chemical reactions, for example, by providing surfaces for catalytic reactions. This parameter is typically defined as the total surface area of all aerosol particles within a unit volume of the aerosol-bearing air mass. When the total aerosol surface area concentration is high (50–100 $\mu m^2\ cm^{-3}$), substantial chemical perturbations, such as ozone (O_3)

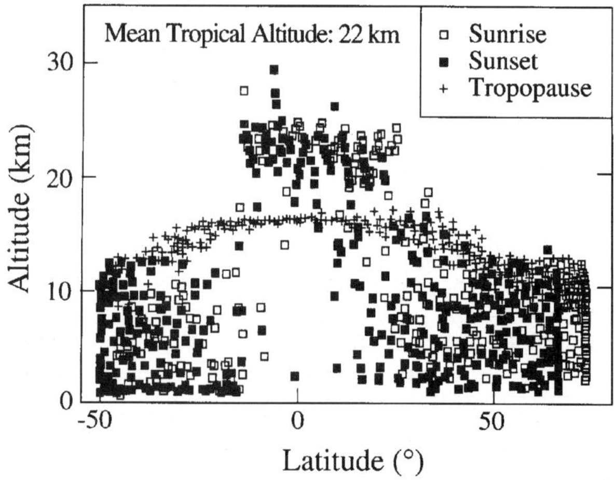

Figure 18.14 The Pinatubo aerosol plume top altitudes, as determined by SAGE II satellite measurements during June to August 1991, in the latitude range 50°S to 50°N (McCormick and Veiga 1992). The aerosol plume top is dominantly at approximately 25 km elevation, but ranges from 22 km to as high as 29 km altitude. During this period the aerosol is largely confined to equatorial regions, between 10°S and 30°N. Data points below 15 km altitude indicate the tropopause (crosses) and non-volcanic tropospheric aerosols

depletion (Hofmann and Solomon 1989), may occur in the stratosphere due to the catalysing effects of the aerosol. In the Pinatubo aerosol layer, peak surface area density was of the order of 40–90 $\mu m^2\ cm^{-3}$ (Grant *et al.* 1993; Deshler *et al.* 1992). The radiative or optical effects of aerosols also depend on shape and size of individual particles. Therefore the grain size and shape of the aerosol particles have an important effect on the volcanic forcing. Theory shows that when the effective radius of particles is less than 2.2×10^{-6} m (2.2 μm), the reflected short-wavelength forcing (SW) is dominant, as these small particles reflect the incoming solar radiation, which should result in surface cooling. For larger particles, on the other hand, the outgoing long-wavelength forcing (LW) is dominant, with backscatter of the terrestrial radiation by the larger particles in the aerosol layer, and this results in warming of the climate system (Lacis *et al.* 1992). Thus the sign of the volcanic forcing may change with time, during particle size evolution of the aerosol plume, and the warming effect of LW should be greatest in the initial stages. Similarly, the signal of the volcanic forcing may change spatially as a function of distance from source, if the aerosol size distribution so dictates.

The stratospheric aerosols from Pinatubo were initially very numerous and small, but also contained a larger-radius mode of particles that were either injected directly into the stratosphere from the eruption plume or were formed rapidly within the stratosphere (Larson and Michalsky 1992). Subsequent growth of particles <0.2 μm in diameter into particles ~ 1 μm in diameter occurred. There was a major increase from September 1991 to February 1992, and this enhanced the bi-modal size distribution to a peak in the aerosol loading. Larger particles of several micrometres in diameter were also present, but these were continually moving downward into the troposphere. Particle size distribution in the volcanic aerosol was measured when an important layer at 15–18 km and a subsidiary and relatively transient layer at 23 km altitude appeared over North America about one month after the eruption (Deshler *et al.* 1992). These layers consisted of about 95–98% sulphuric acid droplets. In the denser (lower) layer, the particles had a mode diameter of about 1.4×10^{-7} m (0.14 μm), while in the subsidiary (upper) layer the particles were significantly larger, with a mode diameter near 0.7×10^{-6} m (0.7 μm; Deshler *et al.* 1992). Evolution of the aerosol size and number density in the aerosol layer, as recorded by balloon-borne sampling over Wyoming (41°N), is shown in Figure 18.15. The aerosol was bimodal above 22 km, but generally unimodal at lower levels. There were two dominant modes, r_1 with radius ranging from 0.13 to 0.21×10^{-6} m (0.13 to 0.21 μm), and r_2 in the range $0.35–0.63 \times 10^{-6}$ m (0.35–0.63 μm). These particle size groups had characteristic concentrations (number of particles per unit volume of air). The r_1 group had a concentration N_1 in the region $> 10\ cm^{-3}$ and the r_2 group a concentration N_2 in the region $1\ cm^{-3}$.

During 1991 the size of the aerosols and their concentrations fluctuated substantially, but during 1992, these parameters remained relatively constant, with slight increases in r_1 and N_2 while N_1 decreased. The high variability of the aerosol size distribution in space and time is evident in the surface area versus particle diameter plots in Figure 18.16. These results show an initial aerosol that was highly variable in character (compare Figure 18.16b and 18.16c), with a transient very coarse mode which may have been the fallout of primary solid particles and secondary liquid (condensed) particles from higher layers. By approximately six months after the eruption the aerosol surface area had increased by an order of magnitude, and evolved to a unimodal-size distribution (Figures 18.16d and 18.16e).

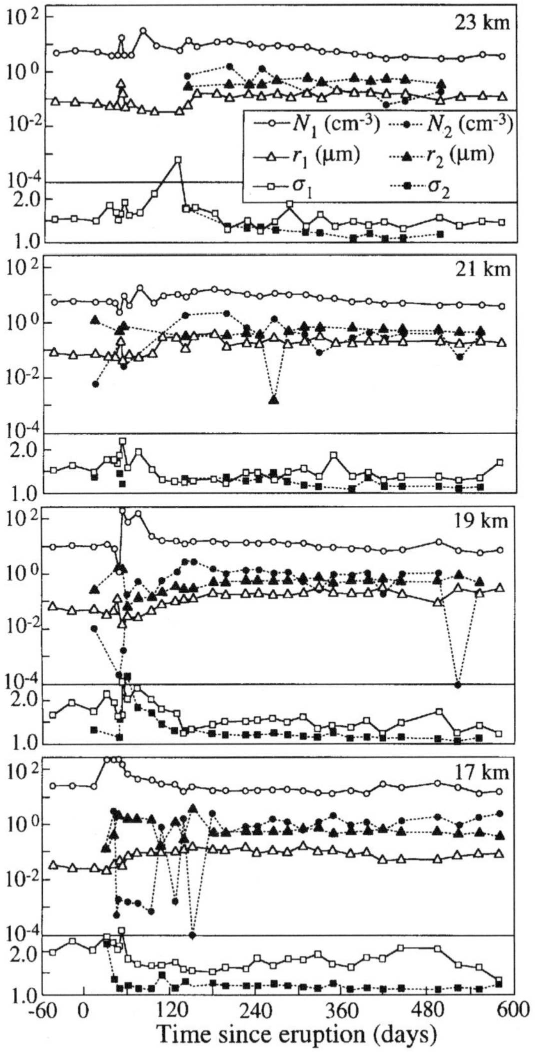

Figure 18.15 Evolution of the Pinatubo volcanic aerosol size parameters over Wyoming, as determined by balloon-borne sampling at altitudes of 17, 19, 21 and 23 km altitude (after Deshler *et al.* 1993). The aerosol was generally bimodal, with fine (r_1) and coarse (r_2) modes, whose number concentrations (particles per cubic centimetre) are shown by N_1 and N_2 respectively. The σ refers to distribution width or variance. The vertical axis shows the value of size modes ($r =$ radius) and number concentration of particles (N)

Sampling of the aerosol layer showed that the dominant particles were sulphuric acid droplets, devoid of any small solid particles that might have acted as condensation nuclei (Figure 18.17). The formation of the sulphate aerosol is thus most likely through homogeneous nucleation processes. Rare and large ($\leqslant 10$ μm diameter) "crustal" particles were present, particularly in the 23 km altitude layer, with aluminium-silicate composition, which were probably either volcanic glass or feldspar (Sheridan *et al.* 1992). The mass

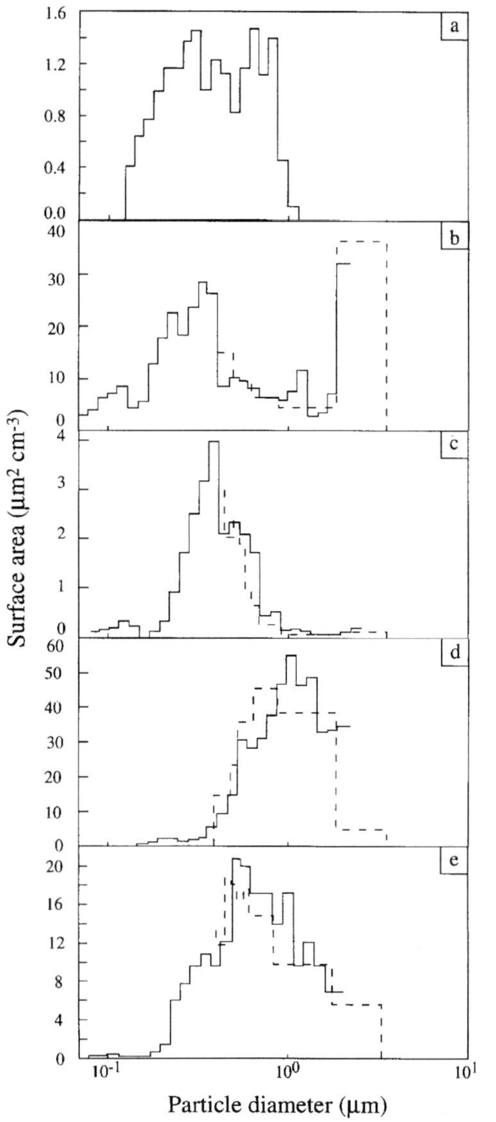

Figure 18.16 Aerosol particle surface area as a function of particle diameter, measured *in situ* with instruments on high-flying research aircraft (Brock *et al.* 1993). (a) Non-volcanic stratospheric unimodal aerosol prior to Pinatubo eruption at 19 km altitude, 37.5°N on December 29, 1988; (b) bimodal volcanic aerosol at 19.2 km, 30.2°N on August 22, 1991, with a coarse mode (2–4 μm) that may be fallout of primary and secondary (condensed) volcanic material from higher layers; (c) volcanic aerosol measured on the same day as (b) at 36.5°N at 16.4 km. Note the order of magnitude increase in surface area concentration compared to data given in (b), and the suppression or absence of the coarse mode present in (b); (d) volcanic aerosol at 43.4°N at elevation of 16.3 km on January 12, 1992 is dominantly unimodal; (e) unimodal volcanic aerosol at 65.8°N at elevation of 15.9 km on March 15, 1992

proportion of silicate particles in the aerosol increased with grain size, up to 40–70% for the >4 μm diameter range, but because this size range represents a very small fraction of the total distribution, the relative mass of the silicate fraction is very minor ($\leqslant 1\%$). Indication of silicate particles also came from the depolarization of lidar backscattered light in the aerosol plume. The 22 km altitude aerosol layer showed significant depolarization at its base (Winker and Osborn 1992b). Normally sulphuric acid aerosol droplets do not polarize light, unless frozen or crystalline, so the measurements may be consistent with silicate particles sedimenting at the base of the plume. Measurements of the optical depth over the Caribbean about one month after the eruption are consistent with a monomodal distribution of sulphate aerosol particles in the sub-micrometre range (0.36–0.70 μm in diameter; Valero and Pilewskie 1992). Some measurements indicate, however, a bimodal distribution, with a small sulphate aerosol mode (0.2 μm in diameter) and a large mode which may represent volcanic glass fragments (1.6 μm in diameter). The optical depth measurements indicate an atmospheric mass loading of 40–80 mg m^{-2}.

One of the optical effects of volcanic aerosols is a faint reddish-brown ring around the Sun. This was first noted after the 1883 Krakatoa eruption, and is known as Bishop's ring. This results from scattering of sunlight by aerosol particles and was noted in Japan about one month after the Pinatubo eruption. The occurrence and intensity of Bishop's ring has been shown to be highly sensitive to aerosol size distribution (Asano 1993), and indicates a relatively narrow aerosol diameter of 1.14–1.26 μm and small variance. With larger particles, the Bishop's ring becomes more reddish, and with smaller particles more yellowish. The unimodal size distribution observed over Japan contrasts strongly with the bimodal size distribution observed by lidar over Wyoming (Deshler *et al.* 1992). Thus the size distribution of the Pinatubo aerosol differed greatly with location.

The removal of volcanic aerosols from the stratosphere is generally attributed to sedimentation in response to aggregation into larger particles, which have settling velocities sufficiently high to induce removal from the stratosphere. Another important process of aerosol removal concerns tropopause folds. In certain conditions, the atmospheric circulation may lead to the folding of stratospheric air parcels into the underlying troposphere, or cause the generation of tropospheric folds in the region of the jet stream. The atmosphere overlying the mid-latitudes of the Earth's surface is characterized by a circulation pattern which is dominated by high-velocity westerlies. These are centred on a core or jet, with velocities up to 40 m s^{-1}, located just above the tropopause. The jet stream and the westerlies change position as a function of the season, at which times they take on a wave-like motion, the waves being known as Rossby or planetary scale waves. The meandering of these features in the atmosphere may lead to folds at the tropopause, and result in incorporation of stratospheric air masses into the underlying troposphere, often related to the interaction of large-scale or regional Rossby waves and the subtropical jet stream.

The importance of tropopause folds is indicated by the strong seasonality and coincidence in the integrated stratospheric backscatter level of both the El Chichon and Pinatubo aerosol plumes, with tropospheric backscatter increasing during tropopause fold events (Post *et al.* 1992). Thus atmospheric dynamics bring down stratospheric aerosol-laden material into the troposphere. This material is then purged very quickly from the troposphere by normal precipitation processes.

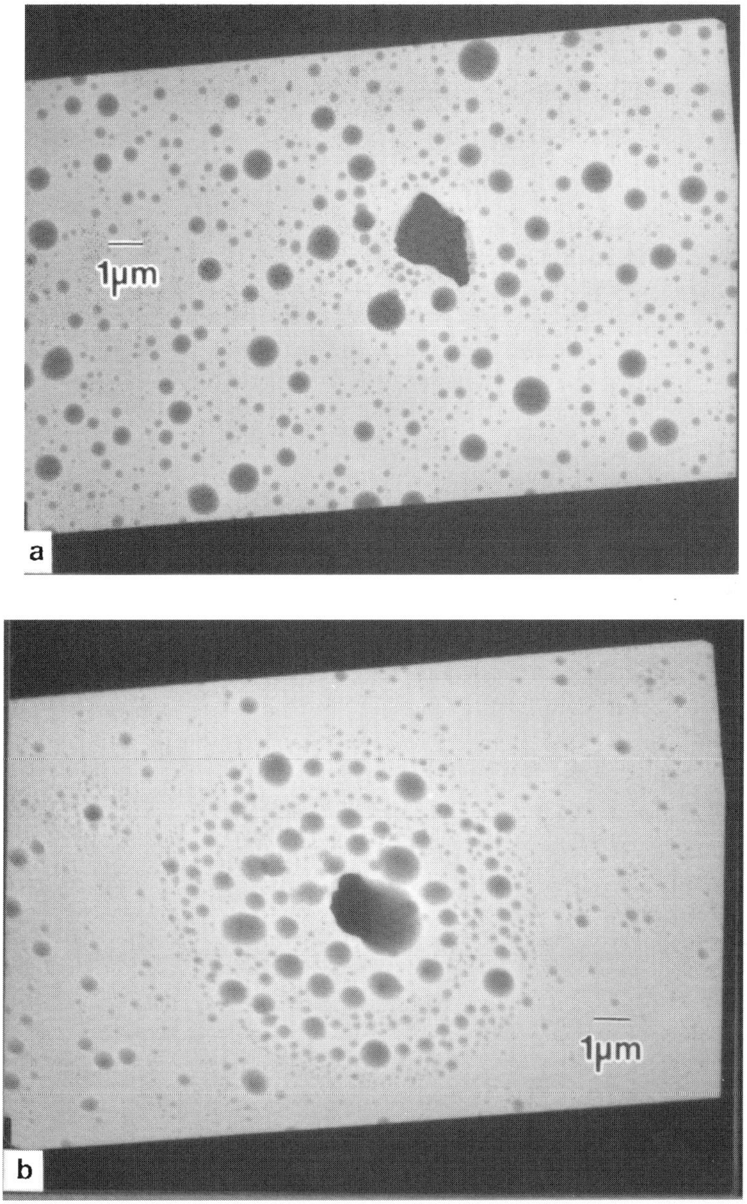

Figure 18.17 Sulphate particles from the Pinatubo volcanic aerosol layer collected during balloon flights over Wyoming on July 30, 1991 from 15.5 to 37.5 km elevation (Sheridan *et al.* 1992). These particles were originally collected as aqueous H_2SO_4, but were subsequently neutralized by ambient NH_3 to form $(NH_4)_2SO_4$ crystals

18.5.3 Stratospheric Warming

Volcanic sulphate aerosols scatter incoming solar radiation back to space, but are unable to absorb sufficient outgoing thermal radiation to offset the solar losses at the surface, and they therefore lead to net global near-surface cooling. Absorption of infrared solar radiation by the volcanic aerosol may affect stratospheric temperature. The stratosphere at the 21–24 km level exhibited a strong warming trend in the months after the Pinatubo eruption, with a temperature anomaly of the order of 4–4.5 °C. This was most notable above the subtropical and tropical regions, but was observed as far as 30°N (Labitzke and McCormick 1992). Balloon-borne sondes over the Congo (4°S) showed a marked temperature increase after the eruption (Grant *et al.* 1993). A maximum increase of about 8 °C at 18.5 km altitude occurred two months after the eruption, a 3–4 °C increase at 22.5 km altitude after three months, and increases of 3–4 °C at 26.5 and 28.5 km altitude were found. The stratospheric warming decreased to near pre-eruption levels about six months after the eruption (Figure 18.18).

The Pinatubo eruption occurred in the transition from a warm phase to a cool phase of the quasi-biennial oscillation (QBO) in stratospheric temperature, so that initially the volcanic warming of the stratosphere was diminished by cooling associated with the QBO (Angell 1993). Taking this into account, the nine-season temperature trends (Angell 1993) show a Northern Hemisphere and equatorial stratospheric warming following the Pinatubo eruption that is comparable to that of the eruptions of El Chichon (1982) and Agung (1963), but a significantly greater warming of the stratosphere above the Southern Hemisphere following the Pinatubo event, of the order of 2 °C (Figure 18.19). The skewed distribution of the warming to the southern stratosphere is unusual when compared to the other recent events, and may be related to an influence from the 1991 Hudson eruption in Chile.

18.5.4 Tropospheric and Surface Cooling

General models of volcanic forcing of climate change have assumed an alteration in the Earth's radiation balance due to the effects of the aerosol, and the response parameter that

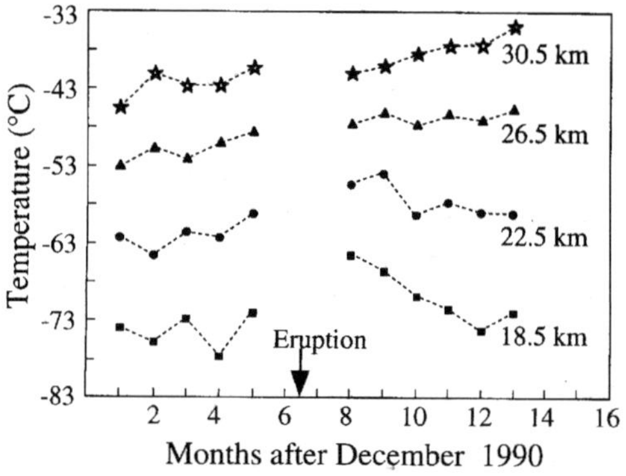

Figure 18.18 Monthly average stratospheric temperatures at four altitudes, measured using balloon sondes above the Congo (4°S) between January 1991 and January 1992 (Grant *et al.* 1993). A maximum warming of 8 °C occurred at 18.5 km altitude two months after the eruption

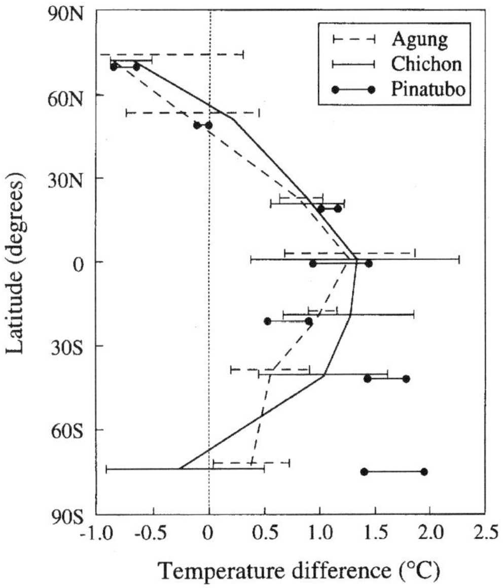

Figure 18.19 Lower stratosphere warming as a function of latitude, following the eruptions of Agung (dashed line), El Chichon (solid line) and Pinatubo (filled circles). Temperatures are estimated from the average temperature difference between the nine-season interval beginning with the season of eruption, and the two nine-season intervals both sides thereof (after Angell 1993). Horizontal lines are one standard deviation either side of the average temperature difference. For the Pinatubo event, the filled circles, connected by a horizontal line, show likely upper and lower bounds. Note the much greater Southern Hemisphere stratospheric warming following the Pinatubo eruption

has usually been measured as a result of volcanic forcing is the air or sea surface temperature. Uncertainties related to methods commonly used to estimate volcanic forcing, by calculation of the radiative transfer on the basis of measured aerosol properties, may now be reduced due to direct measurement of the Earth's radiation budget, which is now possible from satellites.

The increase in planetary albedo from sunlight scattered off the veil of volcanic aerosols in the stratosphere should temporarily cause a surface cooling effect, that may mask any greenhouse gas warming trend. To further complicate the issue, it is likely that the radiative properties of normal cirrus clouds are altered by the influx of stratospheric aerosols that are entering the upper troposphere, due to the decay and fallout of volcanic aerosols. Sedimenting volcanic aerosols may influence the formation and growth of cirrus ice particles, and thus a cooling from increased cirrus cloud solar scattering could reinforce the increased albedo effect of stratospheric aerosols.

Volcanic aerosols can thus alter the radiation balance both directly, by increasing the optical depth of the atmosphere, and indirectly by altering cloud microphysical characteristics. Since sulphate aerosols are efficient cloud condensation nuclei, they can act as ice nuclei and thus increase the number and reduce the diameter of water particles in clouds (Minnis *et al.* 1993). It was initially estimated that the observed global increase by about a factor of five in optical depth, to a value of up to 0.35, due to the Pinatubo aerosol would lead to a decrease in the global net radiation balance of 2.5 W m^{-2} which could in turn lead to about 0.5 °C global cooling over the following two to four years (Stowe *et al.* 1992). This estimate has been confirmed by direct global atmospheric radiance measurements made by microwave sounding units on NOAA satellites, which yield lower troposphere (*c.* 3 km altitude; 700 mbar) air temperatures (Dutton and Christy 1992). The microwave data show a pronounced cooling, beginning immediately after the June 1991 Pinatubo eruption, of about 0.5 °C for at least 14 months. This cooling began in the tropics and propagated out of the subtropics to mid-latitudes and polar regions. Most of the cooling (0.7 °C) was in the Northern Hemisphere, presumably due to a larger land mass.

Other direct satellite measurements of the radiation budget have been made with the Earth radiation budget experiment satellite radiometer instruments, which measure both reflected short-wave and total outgoing long-wave radiation (Minnis *et al.* 1993). In the tropics (5°S to 5°N) the Pinatubo aerosol in August 1991 had brought about an increase of 10 W m^{-2} in the monthly mean reflected short-wave radiation, leading to a corresponding decrease of almost 8 W m^{-2} in the net radiation reaching the Earth's surface, i.e. radiative cooling. By comparison, the Earth normally absorbs about 240 W m^{-2} of solar energy, and thus the volcanic aerosol brought about a decrease of 4%. When the latitude zone from 40°S to 40°N is considered, the net forcing or cooling that took place by August 1991 was −4.3 W m^2. In comparison, the net radiation flux anomaly in this latitude zone for the 1982 El Chichon eruption was only −3.5 W m^{-2}. A global radiative cooling that can be ascribed to the Pinatubo eruption up to October 1991 is a minimum of 2.7 ± 1 W m^{-2} (Minnis *et al.* 1993).

The distribution of the radiative anomaly closely followed the spread of the Pinatubo aerosol plume (Figure 18.20). The first signs of the forcing by the eruption occurred in July, when the short-wave radiative anomaly component increased dramatically in the equatorial zone, with a corresponding decrease in the long-wave component. By August the radiative anomalies had spread to 30°N and 20°S, with a significant negative net anomaly and cooling. The strongest effects were between 10°N and 30°S in September and October (Minnis *et al.* 1993). Within a given latitude zone, the effects of the aerosol are not uniform, however. Pinatubo's cooling effect tended to be greatest over areas that are generally cloud-free, such as the central equatorial Atlantic and Pacific oceans and central Australia. The results of measured radiative anomalies described above, with a minimum of 2.7 ± 1 W m^{-2}, are comparable to but somewhat lower than the predicted climate forcing due to the eruption estimated from global climate models. Hansen *et al.* (1992) estimated a global mean climate forcing up to 4 W m^{-2} by the aerosol, exceeding the total forcing due to anthropogenic greenhouse gases since the Industrial Revolution began (2–2.5 W m^{-2}). However, the results of climate models of the Pinatubo forcing match rather well with observed trends in global monthly surface, troposphere and stratosphere temperatures (Figure 18.21; Hansen *et al.* 1993).

18.5.5 Ozone Perturbation

Ozone is generated in the upper atmosphere by the catalysis of O_2. Short-term depletion of the Earth's ozone layer over several years may occur as a result of large explosive volcanic eruptions that inject material into the stratosphere and disrupt the chemical equilibrium

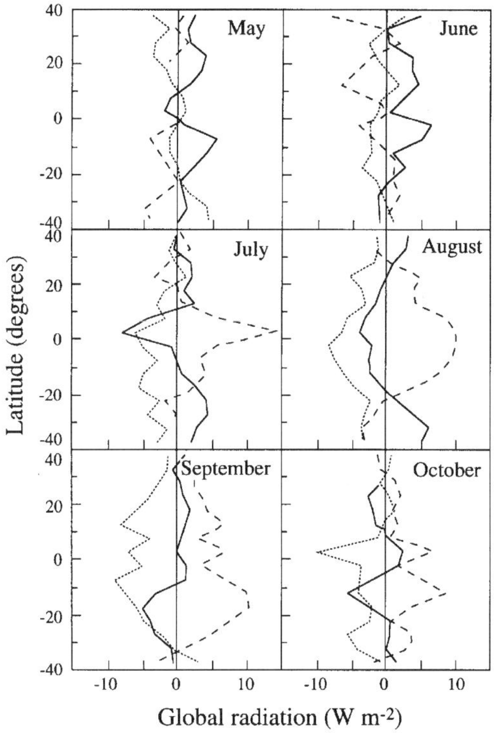

Figure 18.20 The distribution of global radiation anomalies before and after the June 1991 Pinatubo eruption, as a function of latitude (after Minnis *et al.* 1993). The zonal radiative anomalies are from May to October 1991, and show long-wave (LW; solid line), short-wave (SW; dashed line) and net flux anomaly (NET; heavy solid line) in watts per square metre at the Earth's surface

that maintains the ozone layer. The atmosphere's ozone layer absorbs ultraviolet radiation, including UV-B (290–320 nm wavelength), which is the most biologically harmful radiation reaching the Earth's surface. This radiation harms plants and animals by damaging cellular DNA.

Volcanic aerosols can bring about ozone destruction through heterogeneous reactions with chlorine species in the atmosphere (Hofmann and Solomon 1989), and also conceivably by direct injection of volcanic chlorine into the atmosphere. The level of volcanic chlorine emission to the atmosphere appears to be highly variable from one eruption to another. The abundance of HCl gas in the Pinatubo plume was measured over the Caribbean about one month after the eruption (Mankin *et al.* 1992). The observed column amount of HCl was close to 2×10^{15} molecules cm^{-2} or only barely above the background value, and thus no significant addition of this gas to the atmosphere occurred, in contrast to the case of the El Chichon 1982 eruption (Mankin and Coffey 1984). Any reduction in ozone concentration following the Pinatubo eruption cannot therefore be ascribed to volcanic chlorine.

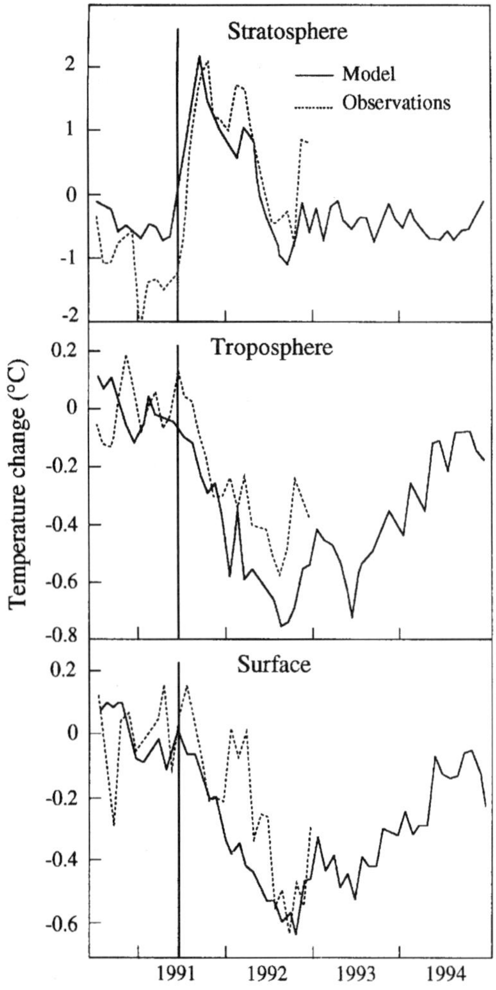

Figure 18.21 The observed versus modelled monthly temperature trends for stratosphere, troposphere and the Earth's surface before and after the 1991 Pinatubo eruption (Hansen *et al.* 1993)

Observations of ozone after the eruption of Mount Pinatubo in 1991 indicate that ozone formation and destruction can both be linked to photochemistry involving sulphur dioxide (Bekki *et al.* 1993). If there is enough SO_2, ozone can be catalytically produced as follows:

$$SO_2 + h\nu = SO + O$$
$$SO + O_2 = SO_2 + O$$
$$O + O_2 = O_3$$

However, SO_2 absorbs ultraviolet light strongly, thereby increasing the optical depth and inhibiting ozone formation by the catalysis of O_2. Sulphate aerosols may also serve as sites for heterogeneous chemical reactions which deplete ozone. Thus the sulphate aerosol furnishes sites for chemical reactions that release reactive chlorine radicals which destroy ozone.

The volcanic aerosol can also catalyse the conversion of reactive nitrogen oxides to nitric acid and liberate reactive chlorine compounds (Law and Pyle 1993). Chlorine is normally bound in nitrogen oxide compounds as $ClONO_2$. In the presence of sulphuric acid aerosols the nitrogen can be more readily converted to nitric acid, shifting the chlorine balance to production of ClO, which can destroy ozone

$$ClONO_2 + H_2O \rightarrow HOCl + HNO_3$$

This is consistent with the observed increase in nitric acid near volcanic aerosols, as discussed below. The hydrolysis of N_2O_5 may also play a role (Law and Pyle 1993). Ozone destruction could also be promoted by the creation of abundant nucleation sites for ice particles, particularly in the polar regions. The ice particles create reactive surfaces for ozone-destroying reactions and thus a greater number density of ice particles will enhance reaction rates. Finally, the role of water injected into the stratosphere may be important. Models of the atmosphere (Toumi *et al.* 1993) predict that about half of the ozone loss in the lower atmosphere is by reactions involving OH. Therefore injection of large amounts of water into the stratosphere by volcanic plumes may also contribute to ozone depletion.

A significant decrease in ozone concentration in the stratosphere was observed in the equatorial regions two months after the Pinatubo eruption. A maximum depletion occurred at 24–25 km elevation, coinciding with the peak aerosol distribution, but ranging from 19 to 28 km, and the ozone depletion at these altitudes was 18–20% (Grant *et al.* 1992). Brasseur (1992) also reported ozone depletion of the order of 5–8% above the tropics after the Pinatubo eruption, and the Antarctic ozone hole in 1991 was about 10–15% deeper than observed previously, implying volcano-induced ozone depletion. Globally, the period following the Pinatubo eruption shows a record low total ozone; thus for 1992 it was about 2–3% lower than any previous year and the depletion continued into 1993 (Gleason *et al.* 1993). The largest decreases were observed from 10°S to 20°S and from 10°N to 60°N, and for the first time a sustained ozone decrease has been observed simultaneously in both hemispheres.

Recently it has been proposed that the stratospheric ozone balance after the Pinatubo eruption was also influenced by production of ozone due to gas-phase sulphur dioxide reactions (Bekki *et al.* 1993). In the first month after the eruption, when high levels of SO_2 prevailed in the stratosphere, ozone production was largely balanced by depletion. When the sulphur dioxide reservoir was largely exhausted, the balance shifted to ozone depletion.

18.5.6 Depletion of Nitrogen Dioxide

The effects of widespread volcanic aerosols on stratospheric chemistry are numerous, including depletion of HNO_3 and H_2O_2 (Laj *et al.* 1990, 1993). The behaviour of nitrogen species in the stratosphere is greatly influenced by OH concentration and sulphate aerosols, which affect the rate of formation and removal of HNO_3. It has been demonstrated from studies of polar ice cores that increased concentrations of volcanic sulphate aerosols have favoured the condensation and removal of HNO_3 following the major eruptions of Tambora (1815), Laki (1783) and Katmai (1912; Laj *et al.* 1993). In the daytime, NO_2 is produced from N_2O_5 by photolysis reactions in the stratosphere, and the process is reversed at night. About two months after the Pinatubo eruption, a marked reduction in the daytime stratospheric NO_2 was observed over New Zealand, of about 40% lower than in previous years. The depletion reached a maximum in October and persisted to the end of 1991, exceeding the depletion following the 1982 eruption of El Chichon by more than a factor of two (Johnston *et al.* 1992). The decrease may be due to the effects of volcanic aerosols on either the photolysis rates of atmospheric chemical species, leading to reduction in NO_2 and a corresponding increase in NO, or due to an increased heterogeneous chemical conversion rate of reactive nitrogen to nitric acid on the surfaces of volcanic aerosol droplets. Heterogeneous reactions of the following type may occur on aerosol surfaces, leading to removal of active nitrogen:

$$N_2O_5 + H_2O \rightarrow 2HNO_3 \text{ (on sulphate aerosol)}$$

Perliski and Solomon (1992) concluded that the dominant factor in the observed nitrogen depletion is heterogeneous chemical conversion to nitric acid.

18.6 VOLCANOLOGIC PARAMETERS

The volcanologic and geochemical parameters which determine the magnitude and properties of volcanic aerosol plumes, and in turn their atmospheric effects, have been reviewed by several workers (e.g. Rampino *et al.* 1988; Palais and Sigurdsson 1989; Sigurdsson 1990a, 1990b). Explosive Plinian and ignimbrite-forming eruptions are by far the most efficient mechanism of volcanic plume generation and thus in the transfer of tephra and volcanic volatiles to the stratosphere. Volcanic plumes generated by major fissure eruptions of the flood-basalt type may be significant in rare cases, and in regions where the tropopause is at low altitude, such as during the Laki eruption in Iceland in 1783. Two important parameters which greatly influence the eruption dynamics are intensity (mass eruption rate), and magnitude (total erupted mass). The eruption column height, and thus the level of volcanic aerosol injection, in Plinian and co-ignimbrite eruptions is governed by the thermal energy flux of the volcano, which is directly proportional to the mass eruption rate of magma (Chapters 2–5). Maximum observed column heights in well-monitored historic eruptions are of the order 25–30 km, but heights up to 43 km have been inferred from the dispersal of ejecta from very energetic explosive eruptions, such as the 1815 Tambora event. A general positive correlation exists between intensity and magnitude for Plinian eruptions, but data are too sparse to determine if this relationship holds for ignimbrite-forming eruptions as well (Carey and Sigurdsson 1989). For a given mass eruption rate, the Plinian column will rise higher than a co-ignimbrite column, because the former includes most of the erupted material, and thus utilizes nearly all of the available thermal energy. A co-ignimbrite column, on the other hand, typically receives less than half of the total erupted material and proportionately less thermal energy (Woods and Wohletz 1991; Chapter 7). This difference in efficiency is, however, offset to a large extent by the much greater magnitude or total erupted mass of ignimbrite-forming eruptions, which form by far the largest deposits from explosive eruptions on Earth (10^{10}–10^{12} kg magma), and are therefore likely to have the greatest atmospheric impact.

As shown in Figure 18.3, the climate forcing is greater for volcanic glass particles than for sulphate aerosols, and furthermore the forcing due to glass particles changes sign with increasing size, to net heating at radii $\geqslant 2\ \mu m$. Unfortunately this potential effect is difficult to evaluate, because data are generally lacking on the dispersal of very fine and distal ash from major explosive eruptions. Grain size versus distance data for two prehistoric eruptions of exceptional magnitude (the Toba and Campanian eruptions) indicate transport of 10 μm radius glass particles of the order of 2000–3000 km (see Figure 2 in Sigurdsson 1990a). With probable mean zonal winds of 20–30 $m\ s^{-1}$, their deposition is likely to have occurred within two days, compared to the six days fallout time for individual particles, as inferred from the data in Figure 18.6. This difference between the calculated single-particle fallout rate and the observed rate is most likely due to particle aggregation (Chapter 16). Any climate forcing due to the coarse silicate glass component of the volcanic aerosol is therefore likely to be short-lived, and restricted to warming in the downwind region near source.

All lines of evidence indicate that the mass of sulphur released to the atmosphere is the dominant factor in volcanic forcing. This parameter has been subject to detailed scrutiny in recent eruptions, where sulphur dioxide or sulphate mass can be determined by remote sensing methods, but unfortunately the determination of sulphur output from eruptions predating the technological age has proven difficult and fraught with pitfalls. Such indirect estimates of volcanic sulphate aerosol mass for past eruptions have been made on the basis of the amplitude of acidity layers in polar ice cores and petrologic data from tephra deposits. Intuitively, one would expect that the mass of sulphur dioxide emission during eruption would correlate broadly with magnitude or total mass of erupted magma. While this holds for several eruptions, it is now evident that the sulphur mass fraction of magmas erupted from the Earth is highly variable. Thus the sulphur emitted, as a fraction of the total erupted magma, ranges by more than three orders of magnitude, from about 100 ppm for sulphur-poor magmas, to over 1 wt% in the case of anhydrite-bearing magmas, such as El Chichon 1982 and Pinatubo 1991. The sulphur output during eruption may be derived from one or more of at least four components or degassing processes, including the degassing of erupted magma (i.e. exsolution of gas which is dissolved in the magma at depth), degassing of the magma reservoir (non-erupted magma), liberation of a free sulphur-rich gas phase, and the breakdown of sulphur-rich minerals during magma ascent in the crust or at the time of eruption, such as anhydrite or pyrrhotite. In some events, such as the 1815 Tambora eruption, the mass of sulphur released, estimated from polar ice cores, agrees closely with petrologic estimates of the sulphur mass lost during degassing of the erupted magma. In certain other events the degassing of erupted magma falls far short of accounting for the sulphur mass released to the atmosphere. The resolution of this "sulphur excess" problem is one of the major challenges confronting the study of the volcanic forcing of climate. In the recent eruptions of Pinatubo 1991 and El Chichon 1982 it is likely that the excess can be accounted for by breakdown of the mineral anhydrite during eruption (e.g. Rutherford and Devine 1993).

The ultimate origin of sulphur in magmas emitted by volcanoes is another major topic for future research. Magmas which generate explosive eruptions are generally highly evolved end-products of a differentiation process, and one might expect that sulphur would be enriched in the melt by crystal fractionation. This is generally not the case, however, because sulphur solubility in most magmas correlates strongly with iron content, and is highest in relatively primitive melts, such as basalts, and very low in silicic magmas. However, in some evolved magmas characterized by high fugacities of oxygen, the sulphur solubility is enhanced as SO_4 species, and the sulphur-rich mineral anhydrite is stable (Carroll and Rutherford 1988). Anhydrite breakdown during magma ascent and eruption can potentially account for the "sulphur excess" observed in recent eruptions such as El Chichon 1982 and Pinatubo 1991, but the origin of the anhydrite is a subject of debate. The wide range in sulphur isotope ratios in Pinatubo anhydrites has been taken to indicate a xenocrystic origin, possibly from a remobilized hydrothermal system of the volcano (McKibben *et al.* 1993). In the case of El Chichon, the high sulphur content of the 1982 magma has been attributed to involvement of subducted volcanogenic massive sulphide deposits (Rye *et al.* 1984).

It is clear from recent studies and historic evidence that sulphate aerosol loading of Pinatubo-size ($\sim 2 \times 10^{10}$ g) causes a measurable climate forcing, including global surface cooling. A Tambora-size (3×10^{11} g) sulphate aerosol has a global impact, including severe regional effects on agricultural productivity due to low temperature or frosts during the growing season. A major concern for society is the depression in world grain production resulting from such an event. The global production of grain, which dominates human diets, has slowed significantly in the last decade to an annual growth rate of +0.7%, less than half the population growth rate (Brown 1993). As world carryover grain stocks only correspond to about 60 days of consumption, a 10% depression in global grain production due to a volcanic event could have extreme consequences. It is clearly desirable to evaluate the probability of occurrence of such events in the future. The forecasting of volcanic eruptions is in principle possible on the basis of statistics of volcanic eruption frequency for the past 1000 years (Simkin 1993). However, the existing database only applies to the magnitude of volcanic eruptions (the erupted mass of magma) and no comparable database has been developed for sulphur mass output.

18.7 SUMMARY

The ability of volcanic eruptions to influence climate was recognized as early as the days of the Roman Empire. Throughout historical times numerous descriptions of dry fogs, anomalous cool periods and unusual optical effects closely associated with large eruptions stimulated many workers to suggest a direct link. The analysis of modern explosive eruptions and their deposits indicates that volcanic plumes can inject large quantities of tephra, volcanic gases and entrained air to high levels (stratospheric) in the atmosphere. Most events are relatively short-lived and consequently the atmospheric response to these injections is evolutionary. Solid particles begin to fall back to the surface quickly while chemical reactions between volcanic gases and stratospheric components generate small liquid droplets that eventually coalesce and are removed by fallout or folding of stratospheric air parcels into the lower level troposphere.

Direct sampling of volcanic plumes at high levels in the atmosphere reveals that one of the most important components are small (micrometre to sub-micrometre diameter) droplets of liquid sulphuric acid, generated by conversion of sulphur dioxide gas on the time scale of several months. These droplets have long residence times in the atmosphere and are critical to climate perturbations from volcanic eruptions. The formation of a sulphuric acid aerosol veil following a large-scale explosive eruption can result in significant increases in the optical depth of the atmosphere, which is a measure of the depletion of incoming solar radiation by the aerosols. The energy balance is dependent on the amount of aerosol in the atmospheric layer and their size distribution. In many cases, however, the presence of a stratospheric acid layer results in a net cooling at the surface.

The eruption of Mount Pinatubo in 1991 provides an excellent example of the atmospheric response to a large explosive eruption. A globally dispersed sulphuric acid veil was formed several months after the event and led to a surface cooling of as much as 0.5 °C for 14 months. During the same time, strong warming of the order of 4–4.5 °C was detected in parts of the stratosphere as a result of the adsorption of infrared solar radiation by the volcanic aerosol layer. The important role of the sulphuric acid aerosol has been confirmed by studies of other large-scale historic eruptions where a correlation has been documented between the mass of sulphur dioxide injected into the atmosphere and the magnitude of surface cooling. Thus, an eruption's ability to trigger significant climate perturbation is linked to the amount of sulphur it is able to inject into the critical stratospheric zone. The total sulphur discharge from an eruption is related to magmatic chemistry, the presence or absence of a free gas phase, the breakdown of the sulphur-bearing phases such as anhydrite or pyrrhotite and the total volume of erupted magma.

References

Abbot, C.G. (1913) Do volcanic explosions affect our climate? *The National Geographic Magazine* **24**, 181–198.

Abraham, G. and Eysink, W.D. (1969) Jets issuing into fluid with a density gradient. *Journal of Hydraulic Research* **7**, 145–175.

Adilibirov, M. and Dingwell, D.B. (1996) Magma fragmentation by rapid decompression. *Nature* **380**, 146–148.

Allard, P., Carbonnelle, J., Metrich, N., Loyer, H. and Zettwoog, P. (1994) Sulphur output and magma degassing budget of Stromboli volcano. *Nature* **368**, 326–330.

Ammann, M. and Burtscher, H. (1993) Aerosol dynamics and light-scattering properties of a volcanic plume. *Journal of Geophysical Research* **98**, 19705–19711.

Anderson, A.T., Newman, S., Williams, S.N., Druitt, T.H., Skirius, C. and Stolper, E. (1989) H_2O, CO_2, Cl and gas in plinian and ash-flow Bishop rhyolite. *Geology* **17**, 221–225.

Anderson, J. (1982) Mount St Helens May 18 Ashfall: The human ecology of an unanticipated natural hazard. In: *Mount St Helens: One year later.* Keller, S.A.C. (ed.) Eastern Washington University Press, pp. 181–189.

Anderson, R., Bjornsson, S., Blanchard, D.C., Gathman, S., Hughes, J., Jonasson, S., Moore, C.B., Survilas, H.J. and Vonnegut, B. (1965) Electricity in volcanic clouds. *Science* **148**, 1179–1189.

Anderson, T. and Flett, J.S. (1903) Report on the eruptions of the Soufrière in St. Vincent, and on a visit to Montagne Pelée in Martinique. *Philosophical Transactions of the Royal Society of London* **A200**, 353–553.

Anderson, T. (1908) The volcanoes of Guatemala. *Royal Geographical Society*, 473–489.

Anderson, T. and Bonney, T.G. (1917) *Volcanic studies in many lands, vol. 2*. John Murray, London, 202 pp.

Andres, R.J., Kyle, P.R. and Chuan, R.L. (1993) Sulfur-dioxide, particle and elemental emissions from Mount Etna, Italy during 1987. *Geologisches Rundschau* **82**, 687–695.

Andres, R.J., Rose, W.I., Kyle, P.R., DeSilva, S., Francis, P.W., Gardeweg, M. and Moreno, H. (1991) Excessive sulfur-dioxide emissions from Chilean volcanoes. *Journal of Volcanology and Geothermal Research* **46**, 323–329.

Angell, J.K. (1993) Comparison of stratospheric warming following Agung, El Chichon and Pinatubo volcanic eruptions. *Geophysical Research Letters* **20**, 715–718.

Anilkumar, A.V., Sparks, R.S.J. and Sturtevant, B. (1993) Geological implications and applications of high-velocity two-phase flow experiments. *Journal of Volcanology and Geothermal Research* **56**, 145–160.

Apsimon, H.M., Gudiksen, P., Khitrov, L., Rodhe, H. and Yoshikura, T. (1988) Modeling of the dispersal and deposition of radionuclides; lessons from Chernobyl. *Environment* **30**, 17–20.

Aramaki, S. (1956) The 1783 activity of Asama volcano. Part I. *Japanese Journal of Geology and Geography* **27**, 189–229.

Aramaki, S., Hayakawa, Y., Fujii, T., Nakamura, K. and Fukuoka, T. (1986) The October 1983 eruption of Miyakejima volcano. *Journal of Volcanology and Geothermal Research* **29**, 203–229.

Arculus, R.J. and Wills, K.J.A. (1980) The petrology of plutonic blocks and inclusions from the Lesser Antilles Island arc. *Journal of Petrology* **21**, 743–799.

Armienti, P., Macedonio, G. and Pareschi, M.T. (1988) A numerical model for simulation of tephra transport and deposition: Applications to May 18 Mount St Helen's eruption. *Journal of Geophysical Research* **93**, 6463–6476.

Asaeda, T. and Imberger, J. (1993) Structure of bubble plumes in linearly stratified environments. *Journal of Fluid Mechanics* **249**, 35–57.

Asano, S. (1993) Estimation of the size distribution of Pinatubo volcanic dust from Bishop's ring simulations. *Geophysical Research Letters* **20**, 447–450.

Bacon, C. and Druitt, T. (1988) Compositional evolution of the zoned calcalkaline magma chamber of Mount Mazama, Crater Lake, Oregon. *Contributions to Mineralogy and Petrology* **98**, 224–256.

Bacon, D.P. and Sarma, R.A. (1991) Agglomeration of dust in convective clouds initialized by nuclear bursts. *Atmospheric Environment* **25**, 2627–2642.

Bagdassarov, N. and Dingwell, D.A. (1992) Rheological investigation of vesicular rhyolite. *Journal of Volcanology and Geothermal Research* **50**, 307–322.

Bailey, D.M. (1991) Cleanup of Grant County Airport after the May 18, 1980 eruption of Mount Saint Helens. *First International Symposium on Volcanic Ash and Aviation Safety.* Program and abstracts volume, p. 11.

Baille, M.G.L. and Munro, M.A.R. (1988) Irish tree rings, Santorini and volcanic dust veils. *Nature* **332**, 344–346.

Baines, W.D., Turner, J.S. and Campbell, I.H. (1990) Turbulent fountains in an open chamber. *Journal of Fluid Mechanics* **212**, 557–592.

Baker, E., Lavelle, J., Feely, R., Massoth, G. and Walker, S. (1989) Episodic venting of hydrothermal fluids from the Juan de Fuca Ridge. *Journal of Geophysical Research* **94**, 9237–9250.

Baker, E. and Massoth, G. (1987) Characteristics of hydrothermal plumes from two vent fields on the Juan de Fuca ridge, northeast Pacific Ocean. *Earth and Planetary Science Letters* **85**, 59–73.

Baker, E., Massoth, G. and Feely, R. (1987) Cataclysmic hydrothermal venting on the Juan de Fuca ridge, *Nature* **329**, 149–151.

Baker, E.T. (1994) A 6-year time series of hydrothermal plumes over the Cleft segment of the Juan de Fuca ridge. *Journal of Geophysical Research* **99**, 4889–4904.

Baker, L. and Rutherford, M.J. (1992) Anhydrite breakdown as a possible source of excess sulfur in the 1991 Mount Pinatubo eruption. *EOS, Transactions of the American Geophysical Union* **73**, 625.

Bandeen, W.R. and Fraser, R.S. (1982) Radiative effects of the El Chichon volcanic eruption: preliminary results concerning remote sensing. *NASA Technical Memorandum* **84959**, 102.

Barberi, F., Innocenti, F., Lirer, L., Munro, R., Pescatore, T. and Santacroce, R. (1978) The Campanian ignimbrite: a major prehistoric eruption in the Neapolitan area (Italy). *Bulletin of Volcanology* **41**, 10–32.

Barberi, F., Macedonio, G., Pareschi, M., and Santacroce, R. (1990) Mapping the tephra fallout risk: an example from Vesuvius, Italy. *Nature* **344**, 142–144.

Barclay, J., Riley, D.S. and Sparks, R.S.J. (1995) Analytical models for bubble growth during decompression of high viscosity magmas. *Bulletin of Volcanology* **57**, 422–431.

Barker, K.B. and Vonnegut, B. (1990) Network observations of CG lightning attending eruption of Volcano Redoubt. *EOS, Transactions of the American Geophysical Union* **71**, 1237.

Barquero, J. (1994) Volcanoes and aviation safety in Costa Rica. *Proceedings of the First International Symposium on Volcanic Ash and Aviation Safety, US Geological Survey Bulletin* **2047**, 9–12.

Barry, R.G. and Chorley, R.J. (1976) *Atmosphere, weather and climate.* Holt, Rinehart and Winston, New York, 320 pp.

Barton, I.J. and Prata, A.J. (1994) Detection and discrimination of volcanic ash clouds by infrared radiometry-II: Experimental. *Proceedings of the First International Symposium on Volcanic Ash and Aviation Safety. US Geological Survey Bulletin* **2047**, 313–318.

Barton, I.J., Prata, A.J., Watterson, I.G. and Young, S.A. (1992) Identification of the Mount Hudson volcanic cloud over SE Australia. *Geophysical Research Letters* **19**, 1211–1214.

Batchelor, G.K. (1967) *An introduction to fluid dynamics.* Cambridge University Press, London, 615 pp.

Batchelor, G.K. (1988) A new theory for the instability of a fluidised bed. *Journal of Fluid Mechanics* **193**, 75–110.

Batiza, R., Fornari, D., Vanko, D. and Lonsdale, P. (1984) Craters, calderas, and hyaloclastites on young Pacific seamounts. *Journal of Geophysical Research* **89**, 8371–8390.

Baxter, P.J. (1990) Medical effects of volcanic eruptions I. Main causes of death and injury. *Bulletin of Volcanology* **52**, 532–544.

Baxter, P.J., Flak, R.S., Falk, H., French, J. and Ing, R. (1982) Medical aspects of volcanic disasters: An outline of the hazards and emergency response measures. *Disasters* **6**, 268–276.

Baxter, P.J., Ing, R., Falk, H., French, J., Stein, G.F., Bernstein, R.S., Merchant, J.A. and Allard, J. (1981) Mount St. Helens eruptions, May 18 to June 12 1980: An overview of the acute health impact. *Journal of the American Medical Association* **246**, 2585–2589.

Begnin, P., Hopfinger, E.J. and Britter, R.E. (1981) Gravitational convection from instantaneous sources on inclined slopes. *Journal of Fluid Mechanics* **107**, 407–422.

Behringer, R.P., Meyers, S.D. and Swinney, H.L. (1991) Chaos and mixing in a geostrophic flow. *Physics of Fluids* **A3**, 1243–1249.

Bekki, S., Toumi, R. and Pyle, J.A. (1993) Role of sulphur photochemistry in tropical ozone changes after the eruption of Mount Pinatubo. *Nature* **362**, 331–333.

Bernard, A. and Rose, W.I. Jr (1990) The injection of sulfuric acid aerosols in the stratosphere by the El Chichón volcano and its related hazards to the international traffic. *Natural Hazards* **3**, 59–67.

Berner, P. (1993) Operators confront mounting window damage. *Aviation Equipment Maintenance*, November 1993, 34–37.

Bernstein, R.S., McCawley, M.A., Attfield, M.D., Green, F., Dollberg, D.D., Baxter, P.J. and Merchant, J.A. (1982) Epidemiologic assessment of the risk for adverse pulmonary effects from persistent occupational exposures to Mount St. Helens' volcanic ash (tephra). In: Keller, S.A.C. (ed.) *Mount St. Helens: One year later.* Eastern Washington University Press, pp. 181–189.

Bischoff, J.L. and Rosenbauer, R.J. (1985) An empirical equation of state for hydrothermal seawater, 3.2% NaCl. *American Journal of Science* **285**, 725–763.

Blackburn, E., Wilson, L. and Sparks, R.S.J. (1976) Mechanisms and dynamics of strombolian activity. *Journal of the Geological Society of London* **132**, 429–440.

Blake, S. (1981) Eruptions from zoned magma chambers. *Journal of the Geological Society of London* **138**, 281–287.

Blake, S. (1984) Volatile oversaturation during the evolution of silicic magma chambers as an eruption trigger. *Journal of Geophysical Research* **89**, 8237–8244.

Blong, R.J. (1981) Some effects of tephra falls on buildings. In: Self, S. and Sparks, R.S.J. (eds) *Tephra studies. Proceedings NATO Advanced Studies Institute, Laugarvatn and Reykjavik, June 18–29, 1980*, Reidel, Dordrecht, Series C **75**, 405–420.

Blong, R.J. (1984) Volcanic hazards. *A sourcebook on the effects of eruptions.* Academic Press, Sydney, 424 pp.

Bluth, G.J.S., Doiron, S.D., Schnetzler, C.C., Krueger, A.J. and Walter, L.S. (1992) Global tracking of the SO_2 clouds from the June, 1991 Mount Pinatubo eruptions. *Geophysical Research Letters* **19**, 151–154.

Blythe, A.R. and Reddish, W. (1979) Charges on powders and bulking effects. *Institute of Physics Conference Series* **48**, 107–124.

Bolt, B.A., Horn, W.L., Macdonald, G.A. and Scott, R.F. (1975) *Geological hazards.* Springer-Verlag, New York.

Bond, A. and Sparks, R.S.J. (1976) The Minoan eruption of Santorini, Greece. *Journal of the Geological Society of London* **132**, 1–16.

Bonnecaze, R.T., Huppert, H.E. and Lister, J.R. (1993) Sediment-driven gravity currents. *Journal of Fluid Mechanics* **250**, 339–369.

Booth, B. (1979) Assessing volcanic risk. *Journal of the Geological Society of London* **136**, 331–340.

Bottinga, Y. and Weill, D. (1972) The viscosity of magmatic silicate liquids: a model for calculation. *American Journal of Science* **272**, 438–475.

Bower, S.M. and Woods, A.W. (1996) On the dispersal of clasts from volcanic craters during small explosive eruptions. *Journal of Volcanology and Geothermal Research* **73**, 19–32.

Brantley, S.R. (1990) The eruption of Redoubt Volcano, Alaska December 14, 1989–August 31, 1990. *US Geological Survey Circular* **1061**, 33.

Brasseur, G. (1992) Volcanic aerosols implicated. *Nature* **359**, 275.

Brazier, S.A., Davis, A.N., Sigurdsson, H. and Sparks, R.S.J. (1982) Fall-out and deposition of volcanic ash during the 1979 explosive eruption of the Soufrière of St. Vincent. *Journal of Volcanology and Geothermal Research* **14**, 335–359.

Brazier, S.A., Sparks, R.S.J., Carey, S.N., Sigurdsson, H. and Westgate, J.A. (1983) Bimodal grain size distribution and secondary thickening in air-fall ash layers. *Nature* **301**, 115–119.

Bredow, J., Parco, R., Damson, M.S., Betty, C., Self, S. and Thordarson, T. (1995) A multifrequency laboratory investigation of attenuation and scattering from volcanic ash clouds. *IEEE Transactions on Geoscience and Remote Sensing* **33**, 1071–1082.

Briggs, G.A. (1969) *Plume Rise*. US Atomic Energy Commission, 80 pp.

Brock, C.A., Jonsson, H.H., Wilson, J.C., Dye, J.E., Baumgartner, D., Borrmann, S., Pitts, M.C., Osborn, M., DeCoursey, R.J. and Woods, D.C. (1993) Relationships between optical extinction, backscatter and aerosol surface and volume in the stratosphere following the eruption of Mt. Pinatubo. *Geophysical Research Letters* **20**, 2555–2558.

Brook, M., Moore, C.B. and Sigurgeirsson, T. (1974) Lightning in volcanic clouds. *Journal of Geophysical Research* **79**, 472–475.

Brown, L.R. (1993) *State of the world*. W.W. Norton, New York, 268 pp.

Brown, E.H. and Hall, F.F. (1978) Advances in atmospheric acoustics. *Review of Geophysical Space Physics* **16**, 47–110.

Brown, G.L. and Roshko, A. (1974) On density effects and large structure in turbulent mixing layers. *Journal of Fluid Mechanics* **64**, 775–816.

Bruce, P. and Huppert, H. (1989) Solidification and melting in dykes. In: Ryan, M.P. (ed.) *Magma transport and storage*. Wiley, New York, pp. 87–101.

Buddington, A.F. and Lindsley, D. (1964) Iron-titanium oxide minerals and synthetic equivalents. *Journal of Petrology* **5**, 310–357.

Buist, A.S. (1988) Evaluation of the short and long term effects of exposure to inhaled volcanic ash from Mt. St. Helens. *Kagoshima International Conference on Volcanoes*. Proceedings volume, pp. 709–712.

Bullard, F.M. (1976) *Volcanoes of the earth*. University of Texas, Austin, 441 pp.

Bunsen, R. (1847) Uber den inneren zusammenhang der pseudovulkanischen Erscheinungen Islands. *Liebigs Ann Chem u Pharm* **62**, 1–59.

Buresti, G. and Casarosa, C. (1989) One dimensional adiabatic flow of equilibrium gas-particle mixtures in long vertical ducts with friction. *Journal of Fluid Mechanics* **203**, 254–272.

Burnham, C.W. (1979) The importance of volatile constituents. In: Yoder, H. (ed.), *The evolution of igneous rocks*. Princeton University Press, Princeton, pp. 439–478.

Bursik, M.I. (1989) Effects of the drag force on the rise height of particles in the gas-thrust region of volcanic eruption columns. *Geophysical Research Letters* **16**, 441–444.

Bursik, M.I. and Woods, A.W. (1991) Buoyant, superbuoyant and collapsing eruption columns. *Journal of Volcanology and Geothermal Research* **45**, 347–350.

Bursik, M., Carey, S. and Sparks, R.S.J. (1992a) A gravity current model for the May 18, 1980 Mount St. Helens plume. *Geophysical Research Letters* **19**, 1663–1666.

Bursik, M.I., Sparks, R.S.J., Gilbert, J.S. and Carey, S.N. (1992b) Sedimentation of tephra by volcanic plumes: I Theory and its comparison with a study of the Fogo A plinian deposit, Sao Miguel (Azores). *Bulletin of Volcanology* **54**, 329–344.

Bursik, M.I., Sparks, R.S.J., Carey, S.N. and Gilbert, J.S. (1994) The concentration of ash in volcanic plumes inferred from dispersal data. *Proceedings of the First International Symposium on Volcanic Ash and Aviation Safety US Geological Survey Bulletin* **2047**, 19–29.

Bursik, M.I., Melekestsev, I.V. and Braitseva, O.A. (1993) Most recent fall deposits of Krudach volcano, Kamchatka, Russia. *Geophysical Research Letters* **20**, 1815–1818.

Calder, E.S., Sparks, R.S.J. and Woods, A.W. (1997) Dynamics of co-ignimbrite plumes generated from pyroclastic flows of Mount St. Helens (August 7, 1980). *Bulletin of Volcanology* (in press).

Caltabiano, T., Romano, R. and Budetta, G. (1994) SO_2 flux measurements at Mount Etna (Sicily). *Journal of Geophysical Research* **99**, 12809–12819.

Campbell, E.E. (1994) Recommended flight-crew procedures if volcanic ash is encountered. *Proceedings of the First International Symposium on Volcanic Ash and Aviation Safety. US Geological Survey Bulletin* **2047**, 151–155.
Cann, J. and Strens, M. (1989) Modeling periodic megaplume emission by black smoker systems. *Journal of Geophysical Research* **94**, 12227–12237.
Cardoso, S.S.S. and Woods, A.W. (1994) Mixing by turbulent plumes in a confined stratified region. *Journal of Fluid Mechanics* **250**, 277–306.
Carey, S.N. (1991) Transport and deposition of tephra by pyroclastic flows and surges. In: Sedimentation in volcanic settings, *SEPM Special Publication* no. **45**, 39–57.
Carey, S.N. and Sigurdsson, H. (1980) The Roseau ash: deep-sea tephra deposits from a major eruption on Dominica, Lesser Antilles Arc. *Journal of Volcanology and Geothermal Research* **7**, 67–86.
Carey, S.N. and Sigurdsson, H. (1982) Influence of particle aggregation on deposition of distal tephra from the May 18, 1980, eruption of Mount St. Helens volcano. *Journal of Geophysical Research* **87**, 7061–7072.
Carey, S.N. and Sigurdsson, H. (1985) The May 18, 1980 eruption of Mount St. Helens 2. Modeling of dynamics of the plinian phase. *Journal of Geophysical Research* **90**, 2948–2958.
Carey, S.N. and Sigurdsson, H. (1986) The 1982 eruptions of El Chichon volcano, Mexico (2): Observations and numerical modelling of tephra fall distribution. *Bulletin of Volcanology* **48**, 127–141.
Carey, S.N. and Sigurdsson, H. (1987) Temporal variations in column height and magma discharge rate during the 79 A.D. eruption of Vesuvius. *Geological Society of America Bulletin* **99**, 303–314.
Carey, S.N. and Sigurdsson, H. (1989) The intensity of plinian eruptions. *Bulletin of Volcanology* **51**, 28–40.
Carey, S.N. and Sparks, R.S.J. (1986) Quantitative models of the fallout and dispersal of tephra from volcanic eruption columns. *Bulletin of Volcanology* **48**, 109–125.
Carey, S.N., Sigurdsson, H. and Sparks, R.S.J. (1988) Experimental studies of particle-laden plumes. *Journal of Geophysical Research* **93**, 15314–15328.
Carey, S.N., Gardner, J., Sigurdsson, H. and Criswell, W. (1990) Variations in column height and magma discharge during the May 18, 1980 eruption of Mount St. Helens, *Journal of Volcanology and Geothermal Research* **43**, 99–112.
Carey, S., Sigurdsson, H., Mandeville, C. and Bronto, S. (1996) Pyroclastic flows and surges over water: an example from the 1883 Krakatau eruption. *Bulletin of Volcanology* **57**, 493–511.
Carroll, J.J. and Parco, S.A. (1996) *Social organization in a crisis situation: The Taal disaster.* Philippine Sociological Society, Manila.
Carroll, M.R. and Rutherford, M.J. (1987) The stability of igneous anhydrite: experimental results and implications for sulfur behaviour in the 1982 El Chichon trachyandesite and other evolved magmas. *Journal of Petrology* **28**, 781–801.
Carroll, M.R. and Rutherford, M.J. (1988) Sulfur speciation in hydrous experimental glasses of varying oxidation state: results from measured wavelength shifts of sulfur X-rays. *American Mineralogist* **73**, 845–849.
Cas, R.A. and Wright, J.V. (1987) *Volcanic successions modern and ancient.* Allen and Unwin, London, 528 pp.
Cas, R.A. and Wright, J.V. (1991) Subaqueous pyroclastic flows and ignimbrites: an assessment. *Bulletin of Volcanology* **53**, 357–380.
Casadevall, T.J. (1993) Volcanic hazards and aviation safety, lessons of the past decade. *FAA Aviation Safety Journal* **2**, 1–11.
Casadevall, T.J. (1994a) Volcanic ash and aviation safety. *Proceedings of the First International Symposium on Volcanic Ash and Aviation Safety. US Geological Survey Bulletin* **2047**.
Casadevall, T.J. (1994b) The 1989–1990 eruption of Redoubt Volcano, Alaska: Impacts on aircraft operations. *Journal of Volcanology and Geothermal Research* **62**, 301–316.
Casadevall, T.J. and Krohn, M.D. (1995) Effects of the 1992 Crater Peak eruptions on airports and aviation operations in the United States and Canada. In: Keith, T.E.C. (ed.) *The 1992 eruptions of Crater Peak vent, Mount Spurr volcano, Alaska. US Geological Survey Bulletin* **2139**, 205–220.

Casadevall, T.J. and Rye, R.O. (1994) Sulfur in the atmosphere: sources and effects on jet-powered aircraft. *EOS Transactions, American Geophysical Union* **75** (16), 75.

Casadevall, T.J. and Thompson, T.B. (1995) *World map of volcanoes and principal aeronautical features*. USGS Geophysical investigations map GP–1011.

Casadevall, T.J., Delos Reyes, P.J. and Schneider, D.J. (1996) The 1991 Pinatubo eruptions and their effects on aircraft operations. In: Newhall, C.G. and Punongbayan, R.S. (eds) *Fire and Mud: eruptions and lahars of Mount Pinatubo, Philippines*. Philippine Institute of Volcanology and Seismology, Quezon City and University of Washington Press, Seattle, pp. 1071–1088.

Casadevall, T.J., Meeker, G.P. and Przedpelski, Z.J. (1991) Volcanic ash ingested by jet engines. *First International Symposium on Volcanic Ash and Aviation Safety*. Program and abstracts volume, p. 15.

Casadevall, T.J., Rose, W.I. Jr, Fuller, W.H., Hunt, M.A., Moyers, J.L., Woods, D.L., Chuan, R.L. and Friend, J.P. (1984) Sulfur dioxide and particles in quiescent plumes from Poas, Arenal, and Colima volcanos, Costa Rica and Mexico. *Journal of Geophysical Research* **89**, 9633–9641.

Cashman, K.V. and Fiske, R.S. (1991) Fallout of pyroclastic debris from submarine volcanic eruptions. *Science* **253**, 275–280.

Castleman, A.W. Jr., Munkelwitz, H.R. and Manowitz, B. (1974) Isotopic studies of the sulfur component of the stratospheric aerosol layer. *Tellus* **26**, 222–234.

Caulfield, C.P. and Woods, A.W. (1995) Plumes with non-monotonic mixing behaviour. *Geophysical Astrophysical Fluid Dynamics* **79**, 173–199.

Cebeci, T. and Bradshaw, P. (1984) *Physical and computation aspects of convective heat transfer.* Springer-Verlag, pp. 1–487.

Chalmers, J.A. (1967) *Atmospheric Electricity.* Pergamon Press, London, 515 pp.

Chassaing, P., George, J. and Claria, A. (1974) Physical characteristics of a subsonic jet in a cross stream. *Journal of Fluid Mechanics* **62**, 41–64.

Chen, J.-C. (1980) Studies on gravitational spreading currents. *W.M. Keck Laboratory, California Institute of Technology, Report KH-R-40.*

Chouet, B.A., Hamiseviez, N.T. and McGetchin, T.R. (1974) Photoballistics of volcanic jet activity at Stromboli, Italy. *Journal of Geophysical Research* **79**, 4961–4976.

Christiansen, R.L. (1984) Yellowstone magmatic evolution: its bearing on understanding large-volume explosive volcanism. In: *Explosive volcanism inception, evolution and hazards*. National Academy Press, Washington, DC, pp. 84–95.

Christiansen, R.L. and Peterson, R.L. (1981) The 1980 Mount St. Helens eruption: Chronology of the 1980 eruptive activity. *US Geological Survey Professional Paper* **1250**, 17–30.

Chubb, J.N. (1990) Two new designs of "field mill" type fieldmeters not requiring earthing of rotating chopper. *IEEE Transactions on Industry Applications* **26**, 1178–1181.

Clark, D.E. and Lee, H. (1965) Ceniza-arena cleanup in San Jose, Costa Rica: Operational aspects as related to nuclear weapon fallout decontamination. *Stanford Research Institute, Project MU-5069.*

Clift, R., Grace, J.R. and Weber, M.E. (1978) *Bubbles, drops and particles*. Academic Press, New York, 380 pp.

Colwell, R.N. (1983) *Manual of Remote Sensing*. American Society of Photogrammetry, Falls Church, Virginia.

Converse, D.R., Holland, H.D. and Edmond, J.M. (1984) Flow rates in the axial hot springs of the East Pacific Rise 21°N: implications for the heat budget and the formation of massive sulphide deposits. *Earth and Planetary Science Letters* **69**, 159–175.

Cook, R.J., Barron, J.C., Papendick, R.I. and Williams, G.J. (1981) Impact on agriculture of the Mount St. Helens eruptions. *Science* **211**, 16–22.

Cornell, W., Carey, S.N. and Sigurdsson, H. (1983) Computer simulation of transport and deposition of the Campanian Y-5 ash. *Journal of Volcanology and Geothermal Research* **17**, 89–109.

Coulter, R.L., Martin, T.J. and Weckwerth, T.H. (1989) Minisodar measurements of rain. *JAOT* **6**, 3, 369–377.

Crabb, D., Durão, D.F.G. and Whitelaw, J.H. (1981) A round jet normal to a crossflow. *Transactions of the ASME, Journal of Fluids Engineering* **103**, 142–153.

Crank, J. (1956) *The mathematics of diffusion*. Oxford, Clarendon Press, 347 pp.

Criswell, W. (1987) Chronology and pyroclastic stratigraphy of the May 18, 1980 eruption of Mount St. Helens, Washington. *Journal of Geophysical Research* **92**, 10237–10266.

Crozier, W.D. (1964) The electric field of a New Mexico dust devil. *Journal of Geophysical Research* **69**, 5427–5429.

Csanady, D.T. (1980) *Turbulent diffusion in the environment*. Reidel, Dordrecht, 248 pp.

Curtis, A.R. (1994) *Space satellite handbook*, 3rd edn. Gulf Publishing Company, Houston, 346 pp.

Dade, W.B. and Huppert, H.E. (1996) Emplacement of the Taupo ignimbrite by a dilute turbulent flow. *Nature* **381**, 509–512.

Danielson, E.F. (1981) Trajectories of the Mount St. Helens eruption plume. *Science* **211**, 819–821.

Dartayet, M. (1932) Observacíon de la lluvia de cenizas del 11 de abril de 1932 en La Plata. *Rev. Astron. (Buenos Aires)* **4**, 183–187.

D'Asaro, E., Walker, S. and Baker, E. (1994) Structure of the two hydrothermal megaplumes. *Journal of Geophysical Research* **99**, 20361–20373.

Decker, R.W. (1990) How often does a Minoan eruption occur? In: Hardy, D. (ed.) *Thera and the Aegean World III*. **2**, 444–452. Thera Foundation, London.

Decker, R.W. and Christiansen, R.L. (1984) Explosive eruptions of Kilauea volcano, Hawaii. In: *Explosive volcanism, inception, evolution and hazards*. Natural Academy Press, Washington, 122–131.

De Lorenzo, G. (1906) The eruption of Vesuvius in April 1906. *Journal of the Geological Society of London* **62**, 476–483.

De Natale, G., Pingue, F., Allard, P. and Zollo, A. (1991) Geophysical and geochemical modelling of the 1982–1984 unrest phenomena at Campi Flegrei caldera (southern Italy). *Journal of Volcanology and Geothermal Research* **48**, 199–222.

Denlinger, R.P. (1987) A model for the generation of ash clouds by pyroclastic flows, with application to the 1980 eruptions of Mount St. Helens, Washington. *Bulletin of Volcanology* **92**, 10284–10298.

Deshler, T.A., Adriani, G.P., Gobbi, D.J., Hofmann, G., Di Donfrancesco and Johnson, B.J. (1992) Volcanic aerosol and ozone depletion within the Antarctic polar vortex during the austral spring of 1991. *Geophysical Research Letters* **19**, 1819–1822.

Deshler, T., Johnson, B.J. and Rozier, W.R. (1993) Balloonborne measurements of Pinatubo aerosol during 1991 and 1992 at 41°N, vertical profiles, size distribution and volatility. *Geophysical Research Letters* **14**, 1435–1438.

Devine, J.D., Sigurdsson, H., Davis, A.N. and Self, S. (1984) Estimates of the sulfur and chlorine yield to the atmosphere from volcanic eruptions and potential climate effects. *Journal of Geophysical Research* **89**, 6309–6325.

Dillman, J.J. and Roberts, M.L. (1982) The impact of the May 18 Mount St. Helens ashfall: Eastern Washington residents report on housing-related damage and cleanup. In: Keller, S.A.C. (ed.) *Mount St Helens: One year later*. Eastern Washington University Press, Cheney, pp. 191–198.

Dingwell, D.B. and Webb, S.L. (1989) Structural relaxation in silicate melts and non-Newtonian melt rheology in geologic processes. *Physics and Chemistry of Minerals* **16**, 508–516.

Dingwell, D.B. and Webb, S.L. (1990) Relaxation in silicate melts. *European Journal of Mineralogy* **2**, 427–449.

Dingwell, D.B., Bagdassarov, G., Bussod, G. and Webb, S. (1993) Magma rheology. In: *Mineralogical Association of Canada short course handbook on experiments at high pressure and applications to the Earth's mantle*, R.W. Luth (ed.), **21**, 131–196.

Dobran, F. (1992) Non-equilibrium flow in volcanic conduits and application of the eruption of Mt. St. Helens on May 18 1980 and Vesuvius in AD 79. *Journal of Volcanology and Geothermal Research* **49**, 285–311.

Dobran, F. and Papale, P. (1993) Magma–water interaction in closed systems and application to lava tunnels and volcanic conduits. *Journal of Geophysical Research* **98**, 14041–14058.

Dobran, F., Neri, A. and Macedonio, G. (1993) Numerical simulations of collapsing volcanic columns. *Journal of Geophysical Research* **98**, 4231–4259.

Docteurs van Leeuwen, W.M. (1936) Krakatau 1883–1933. *Annales du Jardin Botanique de Buitenzorg*, 46-37, E.J. Brill, Leiden.

Donaldson, I.G. (1968) The flow of steam–water mixtures through permeable beds: a simple simulation of a natural undisturbed hydrothermal region. *New Zealand Journal of Science* **11**, 3–23.

Donaldson, E.E., Dickinson, J.T. and Bhattacharya, S.K. (1988) Production and properties of ejecta released by fracture of materials. *Journal of Adhesion* **25**, 281–302.

Dowden, J., Kapadia, P., Brown, G. and Rymer, H. (1991) Dynamics of a geysir eruption. *Journal of Geophysical Research* **96**, 18059–18071.

Drexler, J.W., Rose, W.I. Jr, Sparks, R.S.J. and Ledbetter, M.T. (1980) The Los Chocoyos ash, Guatemala: a major stratigraphic marker in middle America and three ocean basins. *Quaternary Research* **13**, 327–345.

Druitt, T.H. (1992) Emplacement of the May 18, 1980, lateral blast deposit ENE of Mount St. Helens, Washington. *Bulletin of Volcanology* **54**, 554–572.

Druitt, T.H. and Bacon, C. (1986) Lithic breccia and ignimbrite erupted during the collapse of Crater Lake Caldera, Oregon. *Journal of Volcanology and Geothermal Research* **29**, 1–32.

Druitt, T. and Sparks, R.S.J. (1984) On the formation of calderas during ignimbrite eruptions. *Nature* **310**, 679–681.

Dunbar, N.W. and Kyle, P. (1993) Lack of volatile gradient in the Taupo plinian–ignimbrite transition: evidence from melt inclusion analysis. *American Mineralogist* **78**, 612–618.

Dunbar, N.W. and Hervig, R.L. (1992) Volatile and trace element composition of melt inclusions from the lower Bandelier tuff: implications for magma chamber processes and eruptive style. *Journal of Geophysical Research* **97**, 15151–15170.

Dutton, E.G. and Christy, J.R. (1992) Solar radiative forcing at selected location and evidence for global lower tropospheric cooling following the eruptions of El Chichon and Pinatubo. *Geophysical Research Letters* **19**, 2313–2316.

Edmonds, J., Measures, C., McDuff, R., Chan, E. and Collier, R. (1979) Ridge crest hydrothermal activity and the balances of the major and minor elements in the ocean: the Galapagos data. *Earth and Planetary Science Letters* **46**, 1–18.

Eichelberger, J.C., Carrigan, C.R., Westrich, H.R. and Price, R.H. (1986) Non-explosive silicic volcanism. *Nature* **323**, 598–602.

Elachi, C. (1987) *Introduction to the physics and techniques of remote sensing*. Wiley, New York, 413 pp.

Elder, J. (1981) *Geothermal systems*. Academic Press, London, 508 pp.

Endo, K., Fukuoka, T., Miyaji, N. and Sumita, M. (1986) Recent progress in tephra study. *Bulletin of Volcanology Society of Japan* **30**, 237–266.

Enomoto, Y. and Hashimoto, H. (1990) Emission of charged particles from indentation fracture of rocks. *Nature* **346**, 641–643.

EOSAT Landsat Data User Notes (1986) **1**, no. 3, 1–2.

Ernst, G., Davis, J. and Sparks, R.S.J. (1994) Bifurcation of volcanic plumes in a crosswind. *Bulletin of Volcanology* **56**, 159–169.

Ernst, G.G.J. (1996) Sedimentation from turbulent jets, plumes and gravity currents generated atop plumes in still environment and in crossflow. PhD thesis, Geology Dept., University of Bristol.

Ernst, G.G.J., Carey, S.N., Bursik, M.I. and Sparks, R.S.J. (1996a) Sedimentation from turbulent jets and plumes. *Journal of Geophysical Research* **101**, 5575–5589.

Ernst, G.G.J., Cave, R.R., German, C.R., Palmer, M.R., and Sparks, R.S.J. (1996b) Simultaneous vertical and lateral splitting of a hydrothermal plume: first ever evidence from water column sampling at Steinah. In: *Marine hydrothermal systems and the origin of life*. Kluwer, 242 pp.

Erskine, W.F. (1962) *Katmai, a true narrative*. Abelard-Schuman, London, New York, 223 pp.

Fagents, S.A. and Wilson, L. (1993) Explosive volcanic eruptions VII. The ranges of pyroclasts ejected in transient volcanic explosions. *Geophysical Journal International* **113**, 359–370.

Faivre-Pierret, R. and Le Guern, F. (1983) Health risks linked within inhalation of volcanic gases and aerosols. In: Tazieff, H. and Sabroux, J.-C. (eds) *Forecasting volcanic events*. Elsevier, New York, 635 pp.

Faustino, L.A. (1929) Mayon volcano and its eruptions. *Philippine Journal of Science* **40**, 1–43.

Feely, R., Levison, M., Massoth, G., Robert-Baldo, G., Lavelle, J., Byrne, R., Von Damm, K. and Curl, H. (1987) Composition and dissolution of black smoker particulates from active vents on the Juan de Fuca ridge. *Journal of Geophysical Research* **92**, 11347–11363.

Fierstein, J. and Nathensen, M. (1992) Another look at the calculation of fallout tephra volumes. *Bulletin of Volcanology* **54**, 156–167.

Fischer, H.B. (1979) Turbulent jets and plumes. In: Fischer, H.B., List, E.J., Koh, R.C.Y., Imberger, J. and Brooks, N.H. (eds) *Mixing in inland and coastal waters.* Academic Press (San Diego, California), pp. 315–389.

Fisher, R.V. (1979) Models for pyroclastic surges and pyroclastic flows. *Journal of Volcanology and Geothermal Research* **6**, 305–318.

Fisher, R.V. and Schmincke, H.U. (1984) *Pyroclastic rocks.* Springer-Verlag, Berlin, 472 pp.

Fisher, R.V. and Waters, A.C. (1970) Base surge bedforms in maar volcanoes. *American Journal of Science* **268**, 157–180.

Fogel, R. and Rutherford, M.J. (1990) The solubility of carbon dioxide in rhyolitic melts; a quantitative FTIR study. *American Mineralogist* **75**, 1311–1326.

Forsythe, P.Y. (1988) In the wake of Etna, 44 BC. *Classical Antiquity*, vol. 7, pp 49–57. Univ. Calif. Press.

Francis, P.W. (1979) Vesuvius inquest: August 1979. *Geographical Magazine* **51**, 750–755.

Franklin, B. (1784) Meteorological imaginations and conjectures. *The Literary and Philosophical Society of Manchester. Memoirs* **2**, 373–377.

Freier, G.D. (1960) The electric field of a large dust devil. *Journal of Geophysical Research* **65**, 3504.

Freundt, A. and Schmincke, H.U. (1985) Lithic-enriched segregation bodies in pyroclastic flow deposits of Laacher See Volcano (East Eifel, Germany). *Journal of Volcanology and Geothermal Research* **25**, 193–224.

Freundt, A. and Schmincke, H.U. (1992) Mixing of rhyolite, trachyte and basalt magma erupted from a vertically and laterally zoned reservoir, composite flow P1, Gran Canaria. *Contributions to Mineralogy and Petrology* **112**, 1–19.

Fric, T.F. and Roshko, A. (1994) Vortical structure in the wake of a transverse jet. *Journal of Fluid Mechanics* **279**, 1–47.

Fruchter, J.C., Robertson, D.E., Evans, J.C., Olsen, K.B., Lepel, E.A., Laul, J.C., Abel, K.H., Sanders, R.W., Jackson, P.O., Wogman, N.S., Perkins, R.W., van Tuyl, H.H., Beauchamp, R.H., Shade, J.W., Daniel, J.L., Erikson, R.W., Schmel, G.A., Lee, R.N., Robinson, A.V., Moss, O.R., Briant, J.K. and Cannon, W.C. (1980) Mount St. Helens ash from the 18 May 1980 eruption: chemical, physical, mineralogical and biological properties. *Science* **209**, 1116–1125.

Fudali, R.F. and Melson, W.G. (1972) Ejecta velocities, magma chamber pressure, and kinetic energy associated with the 1968 eruption of Arenal volcano. *Bulletin of Volcanology* **35**, 383–401.

Furneaux, R. (1964) *Krakatoa.* Prentice-Hall, New Jersey, 224 pp.

Gage, D.R., Jernegan, M.F. and Farwell, S.O. (1982) Characterization and quantification of inhalable particulate volcanic ash from Mount St. Helens. In: Keller, S.A.C. (ed.) *Mount St. Helens: One year later.* Eastern Washington University Press, pp. 181–189.

Galvez, N.L., Aquino, D.I. and Mamisao, J.P. (1939) Agricultural value of fine ejecta of Mayon volcano. *Philippine Agriculturist* **27**, 844–849.

Galvez, N.L. (1938) The chemical and physical composition of the fine ejecta of Mayon volcano. *Philippine Agriculturist* **27**, 765–794.

Gardner, J.E., Rutherford, M.J., Carey, S. and Sigurdsson, H. (1995) Experimental constraints on pre-eruptive water contents and changing magma storage prior to explosive eruptions of Mount St. Helens volcano. *Bulletin of Volcanology* **57**, 1–18.

Gass, I.G., Harris, P.G. and Holdgate, M.W. (1963) Pumice eruption in the area of the South Sandwich Islands. *Geological Magazine* **100**, 321–330.

Gerlach, T.M. and Graeber, E.J. (1985) Volatile budget of Kilauea volcano. *Nature* **313**, 273–278.

German, C.R. and Sparks, R.S.J. (1993) Particle recycling in the TAG hydrothermal plume. *Earth and Planetary Science Letters* **116**, 129–134.

German, C.R., Briem, J., Chin, C., Danielson, M., Holland, S., James, R., Jonsdottir, A., Ludford, E., Moser, C., Olafsson, J., Palmer, M.R., and Rudnicki, M.D. (1994) Hydrothermal activity on the

Reykjanes Ridge: the Steinholl vent-field at 63°06′N. *Earth and Planetary Science Letters* **121**, 647–654.

Giberti, G. and Wilson, L. (1990) The influences of geometry on the ascent of magma in open fissures. *Bulletin of Volcanology* **52**, 515–521.

Gifford, F.A., Barr, S., Malone, R.C. and Mroz, E.J. (1988) Tropospheric relative diffusion to hemispheric scales. *Atmospheric Environment* **22**, 1871–1879.

Gilbert, J.S. (1991) The stratigraphy of a proximal late-Hercynian pyroclastic sequence; the Vilancós region of the Pyrenees. *Geological Magazine* **128**, 111–128.

Gilbert, J.S. (1994) Experimental and theoretical modeling of volcanic processes. *Geoscientist* **4**, 12–14.

Gilbert, J.S. and Lane, S.J. (1994a) The origin of accretionary lapilli. *Bulletin of Volcanology* **56**, 398–411.

Gilbert, J.S. and Lane, S.J. (1994b) Electrical phenomena in volcanic plumes. *Proceedings of the First International Symposium on Volcanic Ash and Aviation Safety. US Geological Survey Bulletin* **2047**, 31–38.

Gilbert, J.S., Lane, S.J., Sparks, R.S.J. and Koyaguchi, T. (1991) Charge measurements on particle fallout from a volcanic plume. *Nature* **349**, 598–600.

Gill, A.E. (1982) *Atmosphere-Ocean Dynamics*. Academic Press, Orlando, 662 pp.

Glaze, L.S. and Baloga, S.M. (1996) The sensitivity of buoyant plume heights to ambient atmospheric conditions: Implications for volcanic eruption columns. *Journal of Geophysical Research* **101**, 1529–1540.

Glaze, L.S. and Self, S. (1991) Ashfall dispersal for the 16 September 1986, eruption of Lascar, Chile, calculated by a turbulent diffusion model. *Geophysical Research Letters* **18**, 1237–1240.

Glaze, L.S., Francis, P.W., Self, S. and Rothery, D.A. (1989) The 16 September 1986 eruption of Lascar volcano, north Chile: satellite investigations. *Bulletin of Volcanology* **51**, 149–160.

Gleason, J.F., Bhartia, P.K., Herman, J.R., McPeters, R., Newman, P., Stolarski, R.S., Flynn, L., Labow, G., Larko, D., Seftor, C., Wellemeyer, C., Kornhyr, W.D., Miller, A.J. and Planet, W. (1993) Record low global ozone in 1992. *Science* **260**, 523–526.

Goldman, A., Murcray, F.J., Rinsland, C.P., Blatherwick, R.D., David, S.J., Murcray, F.H. and Murcray, D.G. (1992) Mt Pinatubo SO_2 column measurements from Mauna Loa. *Geophysical Research Letters* **19**, 183–186.

Gorshkov, G.S. and Dubik, Y.M. (1970) Gigantic directed blast at Shiveluch volcano (Kamchatka). *Bulletin of Volcanology* **24**, 261–288.

Grace, J.R. and Mathur, K.B. (1978) Height and structure of the fountain region above spouted beds. *Canadian Journal of Chemical Engineering* **56**, 533–537.

Grant, W.B., Browell, E.V., Fishman, J., Brackett, V.G., Fenn, M.A., Butler, C.F., Nganga, D., Minga, A., Cros, B., Veiga, R.E., Stowe, L.L. and Long, C.S. (1993) Aerosol-associated changes in stratospheric ozone following the eruption of Mt. Pinatubo. *Journal of Geophysical Research* **99**, 8197–8211.

Grant, W.B., Fishman, J., Browell, E.V., Brackett, V.G., Nganga, D., Minga, A., Cros, B., Veiga, R.E., Butler, C.F., Fenn, M.A. and Nowicki, G.D. (1992) Observations of reduced ozone concentrations in the tropical stratosphere after the eruption of Mt. Pinatubo. *Geophysical Research Letters* **19**, 1109–1112.

Griggs, R.F. (1919) Scientific results of the Katmai expedition of the National Geographic Society: 4. The character of the eruption as indicated by its effects on nearby vegetation. *Ohio Journal of Science* **19**, 173–209.

Griggs, R.F. (1922) *The valley of ten thousand smokes*. National Geographic Society, Washington DC, 340 pp.

Guest, J., Murray, J., Kilburn, C. and Lopes, R. (1980) Eruptions of Mount Etna during 1979. *Earthquake Information Bulletin* **12**, 154–160.

Gunnarsson, F. (1973) *Volcano, ordeal by fire in Iceland's Westmann Islands*. Iceland Review Books, Reykjavik.

Hallworth, M.A., Phillips, J.C., Huppert, H.E. and Sparks, R.S.J. (1993) Entrainment in turbulent gravity currents. *Nature* **362**, 829–831.

Hallworth, M.A., Phillips, J.C., Huppert, H.E. and Sparks, R.S.J. (1996) Entrainment into two-dimensional and axisymmetric gravity currents. *Journal of Fluid Mechanics* **308**, 289–312.

Hamilton, W. (1772) *Observations on Mt Vesuvius, Mt Etna and other volcanoes*. Cadell, London.

Hammer, C.U., Clausen, H.B., Freidrich, W.L. and Tauber, H. (1987) The Minoan eruption of Santorini in Greece to date 1645 BC *Nature* **328**, 517–519.

Hansen, J., Lacis, A., Ruedy, R. and Sato, M. (1992) Potential climate impact of Mount Pinatubo eruption. *Geophysical Research Letters* **19**, 215–218.

Hansen, J., Lacis, A., Ruedy, R., Sato, M. and Wilson, H. (1993) How sensitive is the world's climate. *Research & Exploration* **9**, 142–158.

Harding, A.J., Orcutt, J., Kappus, M., Vera, E., Mutter, J., Buhl, P., Detrick, R. and Brocher, T. (1989) Structure of young oceanic crust at 13°N on East Pacific Rise from expanding spread profiles. *Journal of Geophysical Research* **94**, 12163–12196.

Harris, D.M. and Rose, W.I. Jr. (1983) Estimating particle sizes, concentrations, and total mass of ash in volcanic clouds using weather radar. *Journal of Geophysical Research* **88**, 10969–10983.

Harris, D.M. and Rose, W.I. Jr. (1987) Plume dynamics and instantaneous eruption rates during the July 22, 1980 eruption of Mount St. Helens. *EOS, Transactions of the American Geophysical Union* **68**, 1649.

Harris, D.M., Rose, W.I. Jr., Roe, R. and Thompson, M.R. (1981a) Radar observations of ash eruptions. In: Lipman, P.W. and Mullineaux, D.R. (eds) *The 1980 eruptions of Mount St. Helens, Washington. US Geological Survey Professional Paper* **1250**, 323–333.

Harris, D.M., Sato, M., Casadevall, T.J., Rose, W.I., Jr and Bornhorst, T.J. (1981b) Emission rates of CO_2 from the plume measurements. *US Geological Survey Professional Paper* **1250**, 201–207.

Hatakeyama, H. (1958) On the disturbance of the atmospheric electric field caused by the smoke-cloud of the volcano Asama-yama. *Pap. Met. Geophys.* **8**, 302–316.

Hatakeyama, H. and Uchikawa, K. (1952) On the disturbance of the atmospheric potential gradient caused by the eruption-smoke of the volcano. *Aso. Pap. Met. Geophys. Tokyo* **2**, 85–89.

Hay, R.L. (1959) Formation of the crystal-rich glowing avalanche deposits of St. Vincent. *Journal of Geology* **67**, 540–562.

Hayashi, T. (1972) Bifurcation of bent-over plumes in the ocean. *Coastal Engineering Japan* **15**, 153–165.

Hayakawa, Y. (1990) Mode of eruption and deposition of the Hachinohe phreatoplinian ash from the Towada volcano, Japan. *Geographical Reports Tokyo Metropolitan University* **25**, 167–182.

Haymon, R.M. (1983) Growth history of hydrothermal black smoker chimneys *Nature* **301**, 695–698.

Haymon, R. and Kastner, M. (1981) Hot spring deposits on the East Pacific Rise at 21°N: preliminary description of mineralogy and genesis. *Earth and Planetary Science Letters* **53**, 363–381.

Haymon, R., Fornari, D., Edwards, M., Carbotte, S., Wright, D. and MacDonald, K. (1991) Hydrothermal vent distribution along the East Pacific Rise crest (9°09′–54′N) and its relationship to magmatic and tectonic processes on fast-spreading mid-ocean ridges. *Earth and Planetary Science Letters* **104**, 513–534.

Haymon, R., Fornari, D., Von Damm, K., Lilley, M., Perfit, M., Edmond, J., Shanks, W., Lutz, R., Grebmeier, J., Carbotte, S., Wright, D., McLaughlin, E., Smith, M., Beedle, N. and Olson, E. (1993) Volcanic eruption of the mid-ocean ridge along the East Pacific Rise Crest at 9°45′–52′N: direct submersible observations of seafloor phenomena associated with an eruption event in April, 1991. *Earth and Planetary Science Letters* **119**, 85–101.

Head, J. and Wilson, L. (1987) Lava fountain heights at Pu'u' 'O'o Kiluaea, Hawaii: Indicators of amount and variations of exsolved magma volatiles. *Journal of Geophysical Research* **92**, 13715–13719.

Head, J.W. and Wilson, L. (1989) Basaltic pyroclastic eruptions: influence of gas-release patterns and volume fluxes on fountain structure, and the formation of cinder cones, spatter cones, rootless flows, lava ponds and lava flows. *Journal of Volcanology and Geothermal Research* **37**, 261–271.

Heffter, J.L. (1965) The variation of horizontal diffusion parameters with time for travel periods of one hour or longer. *Journal of Applied Meteorology* **4**, 153–156.

Heffter, J.L. (1983) *Branching atmospheric trajectory (BAT) model.* NOAA Technical Memorandum ERL ARL-121, 19 pp.

Heffter, J.L. and Stunder, B.J.B. (1993) Volcanic ash forecast transport and dispersion (VAFTAD) model. *Weather and Forecasting* **8**, 533–541.

Heiken, G. and Wohletz, K. (1985) *Volcanic ash.* University of California Press, Berkeley, 246 pp.

Heiken, G., Wohletz, K. and Eichelberger, J.C. (1988) Fracture fillings and intrusive pyroclastics, Inyo domes, California. *Journal of Geophysical Research* **93**, 4335–4350.

Heiken, G., Crowe, B., McGetchin, T., West, F., Eichelberger, J.C., Bartram, D., Peterson, R. and Wohletz, K. (1980) Phreatic eruption clouds: the activity of La Soufrière de Guadeloupe, French West Indies, August–October 1976. *Bulletin of Volcanology* **43**, 383–395.

Heilprin, A. (1903) *Mont Pelée and the tragedy of Martinique.* J.B. Lippincott, Philadelphia, 337 pp.

Heinrichs, T. (1984) The Umsoli chert, turbidite testament for a major phreatoplinian event at the Onverwacht/Fig Tree Formation (Swaziland Supergroup, Archaean, South Africa). *Precambrian Research* **24**, 237–283.

Helfrich, K.R. and Battisti, T.M. (1991) Experiments on baroclinic vortex shedding from hydrothermal plumes. *Journal of Geophysical Research* **96**, 12511–12518.

Henstock, T., Woods, A.W. and White, R.S. (1993) The accretion of oceanic crust by episodic sill intrusion. *Journal of Geophysical Research* **98**, 4143–4161.

Hildreth, W. (1979) The Bishop Tuff: evidence for the origin of compositional zonation in magma chambers. *Geological Society of America Special Paper* **180**, 43–75.

Hildreth, W. (1981) Gradients in silicic magma chambers: implications for lithospheric magmatism. *Journal of Geophysical Research* **86**, 10153–10192.

Hildreth, W. (1987) New perspectives on the eruption of 1912 in the Valley of Ten Thousand Smokes, Katmai National Park, Alaska. *Bulletin of Volcanology* **49**, 680–693.

Hildreth, W. and Drake, R.E. (1992) Volcano Quizapu, Chilean Andes. *Bulletin of Volcanology* **54**, 93–125.

Hill, L.G. and Sturtevant, B. (1989) An experimental study of evaporation waves in superheated liquid. In: Meier, G.E.A. and Thompson, P.A. (eds.) *Adiabatic waves in liquid-vapor systems.* Springer-Verlag, Berlin.

Hobbs, P.V., Radke, L.F., Eltgroth, M.W. and Hegg, D.A. (1981) Airborne studies of the emissions from the volcanic eruptions of Mount St Helens. *Science* **211**, 816–818.

Hobbs, P.V., Radke, L.F., Lyons, J.H., Ferek, R.J., Coffman, D.J. and Casadevall, T.J. (1991) Airborne measurements of particle and gas emissions from the 1990 volcanic eruptions of Mount Redoubt. *Journal of Geophysical Research* **96**, 18735–18752.

Hoblitt, R.P. (1986) Observations of the eruption of July 22 and August 7, 1980, at Mount St. Helens, Washington. *US Geological Survey Professional Paper* **1335**, 1–44.

Hoblitt, R.P. (1994) An experiment to detect and locate lightning associated with eruptions of Redoubt Volcano. *Journal of Volcanology and Geothermal Research* **62**, 499–517.

Hoblitt, R.P. and Murray, T.L. (1990) Lightning detection and location as a remote eruption monitor at Redoubt volcano, Alaska. *EOS, Transactions of the American Geophysical Union* **71**, 1701.

Hoblitt, R.P., Miller, C.D. and Vallance, J.W. (1981) Origin and stratigraphy of the deposit produced by the May 18 directed blast. *US Geological Survey Professional Paper* **1250**, 401–419.

Hoff, R.M. (1992) Differential SO_2 column measurements of the Mount Pinatubo volcanic plume. *Geophysical Research Letters* **19**, 175–178.

Hoff, R.M. and Gallant, A.J. (1980) Sulfur dioxide emissions from La Soufrière volcano, St. Vincent, West Indies. *Science* **209**, 923–924.

Hofmann, D.J. and Solomon, S. (1989) Ozone destruction through heterogenous chemistry following the eruption of El Chichon. *Journal of Geophysical Research* **94**, 5029–5041.

Holasek, R.E. and Rose, W.I. (1991) Anatomy of 1986 Augustine volcano eruptions as recorded by multispectral image processing of digital AVHRR weather satellite data. *Bulletin of Volcanology* **53**, 420–435.

Holasek, R.E. and Self, S. (1995) GOES weather-satellite observations and measurements of the May 18, 1980, Mount St. Helens eruption. *Journal of Geophysical Research* **100**, 8469–8487.

Holasek, R.E., Woods, A.W., and Self, S. (1996) Experiments on gas–ash separation processes in volcanic umbrella plumes. *Journal of Volcanology and Geothermal Research* **70**, 169–181.

Holland, T.J.B. and Blundy, J. (1994) Non-ideal interactions in calcic amphiboles and their bearing on amphibole-plagioclase geothermometry. *Contributions to Mineralogy and Petrology* **116**, 433–447.

Holm, N.G. (1992) *Marine hydrothermal systems and the origin of life*. Kluwer, 242 pp.

Holton, J.R. (1992) *An introduction to dynamics meteorology.* Academic Press, New York, 511 pp.

Hopkins, A. and Bridgman, C. (1985) A volcanic ash transport model and analysis of Mount St. Helens ashfall. *Journal of Geophysical Research* **90**, 10620–10630.

Horton, R.J.M. and McCaldin, R.O. (1964) Observations on air pollution aspects of Irazu Volcano, Costa Rica. *US Public Health Reports, Washington DC* **79**, 925–929.

Houghton, B.F. and Nairn, I.A. (1991) The 1976–1982 Strombolian and phreatomagmatic eruptions of White Island, New Zealand: eruptive and depositional mechanisms of a wet volcano. *Bulletin of Volcanology* **54**, 25–49.

Houghton, B.F. and Wilson, C.J.N. (1990) A vesicularity index for pyroclastic deposits. *Bulletin of Volcanology* **51**, 451–462.

Houze, R.A. (1993) *Cloud dynamics*. Academic Press, San Diego, 572 pp.

Huff, W., Bergstrom, S. and Kolata, D. (1992) Gigantic Ordovician volcanic ash fall in North America and Europe: biological, tectonomagmatic and event-stratigraphic significance. *Geology* **20**, 875–878.

Humphries, S., Zierenberg, R., Mullineaux, L. and Thomson, R.E. (1995) Seafloor hydrothermal systems: physical, chemical, biological and geological interactions. *American Geophysical Union, Geophysical Monograph* **91**, 1–466.

Humphreys, W.J. (1920) *Physics of the air*. Philadelphia, 665 pp.

Humphreys, W.J. (1913) Volcanic dust and other factors in the production of climatic changes and their possible relation to ice ages. *Bull. Mount Weather Observatory* **VI**, 1–34.

Hunt, J.C.R., Snyder, W.H. and Lawson, R.E. (1978) Flow structure and turbulent diffusion around a three-dimensional hill: fluid modeling study on effects of stratification, 1. Flow structure, EPA-600/4-78-041, *US Environmental Protection Agency*, Research Triangle Park, NC, 84 pp.

Huppert, H.E., Turner, J.S., Carey, S.N., Sparks, R.S.J. and Hallworth, M.A. (1986) A laboratory study of pyroclastic flows down slopes. *Journal of Volcanology and Geothermal Research* **30**, 179–199.

Hurwitz, S. and Navon, O. (1994) Bubble nucleation in rhyolitic melts: experiments at high pressure, temperature and water content. *Earth and Planetary Science Letters* **122**, 267–280.

Iribarne, J.V. and Cho, H.-R. (1980) *Atmospheric physics*. Reidel, Dordrecht, 212 pp.

Ishihara, K. (1985) Dynamical analysis of volcanic explosion. *Journal of Geodynamics* **3**, 327–349.

Ishihara, K.J. (1990) Pressure sources and induced ground deformation associated with explosive eruptions at an andesitic volcano: Sakurajima volcano, Japan. In: Ryan, M.P. (ed.) *Magma transport and storage*. Wiley, Chichester, pp. 335–356.

Italiano, F. and Nuccio, P.M. (1992) Volcanic steam output directly measured in fumaroles: the observed variations at Vulcano island, Italy between 1983 and 1987. *Bulletin of Volcanology* **54**, 623–630.

Izett, G.A. (1981) Volcanic ash beds: recorders of Upper Cenozoic silicic pyroclastic volcanism in the western United States. *Journal of Geophysical Research* **86**, 10200–10222.

Jagger, T.A. (1925) *Volcanoes declare war; Logistics and strategy of Pacific volcano science*. Paradise of the Pacific Ltd, Honolulu.

Jaupart, C. and Allegre, C. (1991) Gas content, eruption rate and instabilities of eruption in silicic volcanoes. *Earth and Planetary Science Letters* **102**, 413–429.

Jaupart, C. and Vergniolle, S. (1988) Laboratory models of Hawaiian and Strombolian eruptions. *Nature* **331**, 58–60.

Jennings, P. (1969) Disruptions of the environmental balance: The eruptions of Mt Agung and Mt Kelut, Indonesia. MA Geography thesis, University of Hawaii.

Johnson, R.W. (1991) Coping with the aircraft/ash-cloud problem in Australia. *First International Symposium on Volcanic Ash and Aviation Safety*. Program and abstracts volume, p. 26.

Johnson, R.W. and Casadevall, T.J. (1994) Aviation safety and volcanic ash clouds in the Indonesia–Australia region. *Proceedings of the First International Symposium on Volcanic Ash and Aviation Safety, US Geological Survey Bulletin* **2047**, 191–197.

Johnston, P.V., McKenzie, R.L., Keys, J.G. and Matthews, W.A. (1992) Observations of depleted stratospheric NO_2 following the Pinatubo volcanic eruption. *Geophysical Research Letters* **19**, 211–213.

Jones, D.M.A. (1992) Raindrop spectra at the ground. *Journal of Applied Meteorology* **31**, 1219–1225.

Judd, J.W. (1888) On the volcanic phenomena of the eruption, and on the nature and distribution of the ejected materials. In: Symons, G. (ed.) (1888) pp. 1–46.

Juhle, W. and Coulter, H.W. (1955) The Mt. Spurr eruption, July 9, 1953. *EOS, Transactions of the American Geophysical Union* **36**, 188–202.

Junge, C.E. (1963) Sulfur in the atmosphere. *Journal of Geophysical Research* **68**, 3975–3976.

Junge, C.E. and Manson, J.E. (1961) Stratospheric aerosol studies. *Journal of Geophysical Research* **66**, 2163–2182.

Junge, C.E., Chagnon, C.W. and Manson, J.E. (1961) Stratospheric aerosols. *Journal of Meteorology* **18**, 81–108.

Kahl, J.D., Schnell, R.C., Sheridan, P.J., Zak, B.D., Church, H.W., Mason, A.S., Heffter, J.L. and Harris, J.M. (1991) Predicting atmospheric debris transport in real-time using a trajectory forecast model. *Atmospheric Environment* **25A**, 1705–1713.

Kanamori, H. and Mori, J. (1992) Harmonic excitation of mantle Rayleigh waves by the 1991 eruption of Mount Pinatubo, Philippines. *Geophysical Research Letters* **19**, 721–724.

Kasten, F. (1968) Falling speed of aerosol particles. *Journal of Applied Meteorology* **7**, 944–947.

Kennett, J. (1981) Marine tephrochronology. In: Emiliani, C. (ed.) *The sea*. Wiley, New York, pp. 1373–1436.

Kennett, J., McBirney, A. and Thunnel, R. (1977) Episodes of volcanism in the circum-Pacific region. *Journal of Volcanology and Geothermal Research* **2**, 145–163.

Kerr, R.C. (1991) Erosion of a stable density gradient by sediment-driven convection. *Nature* **353**, 423–425.

Kieffer, S.W. (1982) Fluid dynamics and thermodynamics of Ionian volcanism. In: Morison D. (ed.) *Satellites of Jupiter*. University of Arizona Press, Tucson, pp. 647–723.

Kieffer, S.W. (1977) Sound speed in liquid–gas mixtures, water–air and water–steam. *Journal of Geophysical Research* **82**, 2895–2904.

Kieffer, S.W. (1981) Fluid dynamics of the May 18 1980 blast at Mt. St. Helens. *US Geological Survey Professional Paper* **1250**, 379–400.

Kieffer, S.W. (1984) Seismicity at Old Faithful Geyser: An isolated source of geothermal noise and possible analogue of volcanic seismicity. *Journal of Volcanology and Geothermal Research* **22**, 59–95.

Kieffer, S.W. and Sturtevant, B. (1984) Laboratory studies of volcanic jets. *Journal of Geophysical Research* **89**, 8253–8268.

Kienle, J. (1991) Volcanic ash-aircraft incidents in Alaska in the years prior to the December 15, 1989 747 Redoubt encounter. *First International Symposium on Volcanic Ash and Aviation Safety*. Program and abstracts volume, pp. 27–28.

Kienle, J. (1994) Volcanic ash–aircraft incidents in Alaska prior to the Redoubt eruption on 15 December 1989. *Proceedings of the First International Symposium on Volcanic Ash and Aviation Safety, US Geological Survey Bulletin* **2047**, 119–123.

Kienle, J. and Shaw, G.E. (1979) Plume dynamics, thermal energy and long-distance transport of vulcanian eruption clouds from Augustine volcano, Alaska. *Journal of Volcanology and Geothermal Research* **6**, 139–164.

Kienle, J., Davies, J.N., Miller, T.P. and Yount, M.E. (1986) 1986 eruption of Augustine volcano: public safety response by Alaskan volcanologists. *EOS, Transactions of the American Geophysical Union* **67**, 580–582.

Kikuchi, K. and Endoh, T. (1982) Atmospheric electrical properties of volcanic ash particles in the eruption of Mt. Usu volcano, 1977. *Journal of the Meteorological Society of Japan* **60**, 548–561.

Kim, S.L., Mullineaux, L.S. and Helfrich, K.R. (1994) Larval dispersal via entrainment into hydrothermal vent plumes. *Journal of Geophysical Research* **99**, 12655–12665.

Kittle, E. (1933) Estudio sobre los fenómenos volcánicos y material caído durante la erupcíon del grupo del "Descabezado" en el mes de abril de 1932. *Anal. Museo. Nac. Hist. Nat.* (Buenos Aires) **37**, 321–364.

Klinkhammer, G., Rona, P., Greaves, M. and Elderfield, H. (1985) Hydrothermal manganese plumes in the mid-Atlantic ridge rift valley. *Nature* **314**, 727–731.

Kneizys, F.X., Shettle, E.P., Abreu, L.W., Anderson, G.P., Chetwynd, J.H., Gallery, W.O., Selby, J.E.A. and Clough, S.A. (1988) Atmospheric transmittance/radiance: the LOWTRAN 7 model. *Air Force Geophysical Laboratory – Technical Report* **88-0177**.

Kobayashi, T., Ishihara, K. and Hirabayashi, J. (1988) *A guide book for Sakurajima Volcano*. Kagoshima International Conference on Volcanoes, Kagoshima, Japan, 88 pp.

Kokelaar, P. (1986) Magma–water interactions in subaqueous and emergent basaltic volcanism. *Bulletin of Volcanology* **48**, 275–289.

Kokelaar, P. and Busby, C. (1992) Subaqueous explosive eruption and welding of pyroclastic deposits. *Science* **257**, 196–200.

Korthapalli, A., Lourenco, L. and Buchlin, J.M. (1990) Separated flow generated by a vectored jet in a cross-flow. *AIAA Journal* **28**, 414–420.

Koto, B. (1916) The great eruption of Sakura-jima in 1914. *Journal of the College of Science Imperial University of Tokyo* **38**, 229.

Koyaguchi, T. (1994) Grain-size variations of the tephra derived from umbrella clouds. *Bulletin of Volcanology* **56**, 1–9.

Koyaguchi, T. and Tokuno, M. (1993) Origin of the giant eruption cloud of Pinatubo, June 15, 1991. *Journal of Volcanology and Geothermal Research* **5**, 85–96.

Koyaguchi, T. and Woods, A.W. (1996) On the explosive interaction of water and magma. *Journal of Geophysical Research* **101**, 5561–5574.

Krueger, A.F. (1982) Geostationary satellite observations of the April 1979 Soufrière eruptions. *Science* **216**, 1108–1109.

Krueger, A.J. (1983) Sighting of El Chichon sulfur dioxide clouds with the Nimbus 7 Total Ozone Mapping Spectrometer. *Science* **220**, 1377–1378.

Krueger, A.J. (1985) Detection of volcanic eruptions from space by their sulfur dioxide clouds. *AIAA 23rd Aerospace Sciences Meeting*.

Krueger, A.J., Walter, L.S., Bhartia, P.K., Schnetzler, C.C., Krotkov, N.A., Sprod, I. and Bluth, G.J.S. (1995) Volcanic sulfur-dioxide measurements from the Total Ozone Mapping Spectrometer instruments. *Journal of Geophysical Research* **100**, 14057–14066.

Krueger, A.J., Walter, L.S., Schnetzler, C.C. and Doiron, S.D. (1990) TOMS measurement of the sulfur dioxide emitted during the 1985 Nevado del Ruiz eruptions. *Journal of Volcanology and Geothermal Research* **41**, 7–15.

Kyle, P.R., Meeker, K. and Finnegan, D. (1990) Emission rates of sulfur dioxide, trace gases and metals from Mount Erebus, Antarctica. *Geophysical Research Letters* **17**, 2125–2128.

Labadie, J.R. (1994) Mitigation of volcanic ash effects on aircraft operating and support systems. *Proceedings of the First International Symposium on Volcanic Ash and Aviation Safety, US Geological Survey Bulletin* **2047**, 125–128.

Labitzke, K. and McCormick, M.P. (1992) Stratospheric temperature increase due to Pinatubo aerosols. *Geophysical Research Letters* **19**, 207–210.

Labitzke, K., Naujokat, B. and McCormick, M.P. (1983) Temperature effects on the stratosphere of the April 4, 1982 eruption of El Chichon, Mexico. *Geophysical Research Letters* **10**, 24–26.

Lacis, A., Hansen, J. and Sato, M. (1992) Climate forcing by stratospheric aerosols. *Geophysical Research Letters* **19**, 1607–1610.

Lacroix, A. (1906) The eruption of Vesuvius in April, 1906. *Smithsonian Institution Annual Report*, pp. 223–248.

Laj, P., Drummey, S., Spencer, M.J., Palais, J.M. and Sigurdsson, H. (1990) H_2O_2 depletion in a Greenland ice-core: implication for oxidation of volcanic SO_2. *Nature* **346**, 45–48.

Laj, P., Palais, J.M., Gardner, J.E. and Sigurdsson, H. (1993) Modified HNO_3 seasonality in volcanic layers of a polar ice core: snow-pack effect or photochemical perturbation? *Journal of Atmospheric Chemistry* **16**, 219–230.

Lamb, H.H. (1970) Volcanic dust in the atmosphere with a chronology and assessment of its meteorological significance. *Philosophical Transactions of the Royal Society of London* **266**, 425–533.
Lane, S.J. and Gilbert, J.S. (1992) Electric potential gradient changes during explosive activity at Sakurajima volcano, Japan. *Bulletin of Volcanology* **54**, 590–594.
Lane, S.J., Gilbert, J.S. and Hilton, M. (1993) The aerodynamic behaviour of volcanic aggregates. *Bulletin of Volcanology* **55**, 481–488.
Lane, S.J., Gilbert, J.S. and Kemp, A.J. (1995) Electrical and chemical properties of eruption plumes at Sakurajima volcano, Japan. In: *8th Report of Geophysical and Geochemical Observations at Sakurajima Volcano*, pp. 105–127.
Lane-Serff, G.F. (1995) Particle recycling in hydrothermal plumes: Comment on "Particle recycling in the TAG hydrothermal plume" by C.R. German and R.S.J. Sparks. *Earth Planetary Science Letters* **132**, 233–234.
Larson, N.R. and Michalsky, J.J. (1992) Evolution of Mount Pinatubo stratospheric aerosols above Washington State as measured by ground-based solar photometers. *EOS, Transactions of the American Geophysical Union*, abstr.
Larsson, W. (1937) Vulkanische asche vom Ausbruch des chilenischen vulkans Quizapu (1932) in Argentina gesammelt. *Geol. Inst. Upsala. Bull.* **26**, 27–52.
Law, K.S. and Pyle, J.A. (1993) Modeling of trace gas budgets in the troposphere. *Journal of Geophysical Research* **98**, 18377–18412.
Leus, X., Kintanar, C. and Bowman, V. (1981) Asthmatic bronchitis associated with a volcanic eruption in St. Vincent, West Indies. *Disasters* **5**, 67–69.
Levine, A.H. and Kieffer, S.W. (1991) Hydraulics of the August 7, 1980 pyroclastic flow at Mount St. Helens, Washington. *Geology* **19**, 1121–1124.
Liepmann, H. and Roshko, A. (1957) *Elements of gas dynamics*. Wiley, New York, 439 pp.
Lin, X.J. and Coakley, J.M. (1993) Retrieval of properties for semitransparent clouds from multispectral infrared imagery data. *Journal of Geophysical Research* **98**, 18501–18514.
Linden, P.F. and Simpson, J.E. (1990) Continuous two-dimensional releases from an elevated source. *Journal of Loss Prevention Processing Industry* **3**, 82–87.
Lipman, P.W. (1967) Mineral and chemical variations within an ash-flow sheet from Aso caldera, south-western Japan. *Contributions to Mineralogy and Petrology* **16**, 300–327.
Lipman, P. and Mullineaux, D. (1981) The 1980 eruptions of Mount St. Helens, Washington. *US Geological Survey Professional Paper* **1250**, 837 pp.
Lirer, L., Pescatore, T., Booth, B. and Walker, G.P.L. (1973) Two plinian pumice-fall deposits from Somma-Vesuvius, Italy. *Geological Society of America Bulletin* **84**, 759–772.
List, E.J. (1982) Turbulent jets and plumes. *Annual Review of Fluid Mechanics* **14**, 189–212.
Little, C.G. (1972) On the detectability of fog, cloud, rain and snow by acoustic echo sounding methods. *Journal of Atmospheric Science* **28**, 748–755.
Lockwood, J. and others (1984) Mauna Loa Volcano, Hawaii, *SEAN Bull. Smithsonian Institution* **9** (3), 2–9.
Long, R.R. (1953) A laboratory model resembling the "Bishop-wave" phenomenon. *Bulletin of the American Meteorological Society* **34**, 205–211.
Longmire, E.K. and Eaton, J.K. (1992) Structure of a particle-laden round jet. *Journal of Fluid Mechanics* **236**, 217–257.
Lonsdale, P. and Becker, K. (1985) Hydrothermal plumes, hot springs and conductive heat flow in the Southern Trough of Guaymas Basin. *Earth and Planetary Science Letters* **73**, 211–225.
Low, T.B. and List, R. (1982) Collision, coalescence and breakup of raindrops. Part 1–2: Experimentally established coalescence efficiencies and fragment size distributions in breakup-parameterization of fragment size distributions. *Journal of Atmospheric Science* **39**, 1591–1618.
Lowell, R.P. and Germanovich, L.N. (1995) Dike injection and the formation of megaplumes at Ocean Ridges. *Science* **267**, 1804–1807.
Lowenstein, J.B. and Mahood, G.A. (1991) New data on magmatic H_2O contents in pantellerites, with implications for petrogenesis and eruptive dynamics. *Bulletin of Volcanology* **54**, 78–83.
Lunel, T., Rudnicki, M., Elderfield, H. and Hydes, D. (1990) Aluminium as a depth-sensitive tracer of entrainment in submarine hydrothermal plumes. *Nature* **344** 137–139.

Lunkenheimer, F. (1932) La erupcíon del Quizapu en abril de 1932. *Revista Astrononica (Buenos Aires)* **4**, 173–182.

Lupton, J., Klinkhammer, G., Normark, W., Haymon, R., MacDonald, K., Wein, R. and Craig, H. (1980) Helium-3 and manganese at the 21°N East Pacific Rise hydrothermal site. *Earth and Planetary Science Letters* **50**, 115–127.

Lupton, J. and Craig, H. (1981) A major helium-3 source at 15°S on the East Pacific Rise. *Science* **214**, 13–18.

McBirney, A.R. and Murase, T. (1984) Rheological properties of magmas. *Annual Reviews of Earth and Planetary Science* **12**, 337–357.

McCormick, M.P. and Veiga, R.E. (1992) SAGE II measurements of early Pinatubo aerosols. *Geophysical Research Letters* **19**, 155–158.

McCormick, M.P., Swissler, T.J., Fuller, W.H., Hunt, W.H. and Osborn, M.T. (1984) Airborne and ground-based lidar measurements of the El Chichon stratospheric aerosol from 90°N to 56°S. *Geofisica Int.* **23**, 187–221.

McDonald, J.E. (1959) Rates of descent of fallout particles from thermonuclear explosions. *Journal of Meteorology* **17**, 380–382.

McDougall, T.J. (1990) Bulk properties of "hot smoker" plumes. *Earth and Planetary Science Letters* **99**, 185–194.

Macedonio, G., Dobran, F. and Neri, A. (1994) Erosion processes in volcanic conduits and application of the AD 79 eruption of Vesuvius. *Earth and Planetary Science Letters* **121**, 137–152.

Macedonio, G., Pareschi, M. and Santacroce, R.A. (1988) Numerical simulation of the Plinian fall phase of 79 A.D. eruption of Vesuvius. *Journal of Geophysical Research* **93**, 14817–14827.

Macedonio, G., Pareschi, M. and Santacroce, R. (1990) Renewal of explosive activity at Vesuvius: models for the expected tephra fallout. *Journal of Volcanology and Geothermal Research* **40**, 327–342.

McGetchin, T.R., Settle, M. and Chouet, B.H. (1974) Cinder cone growth modelled after Northeast crater, Mt Etna, Sicily. *Journal of Geophysical Research* **79**, 3257–3272.

McIlveen, J.F.R. (1992) *Fundamentals of weather and climate*. Chapman and Hall, London, 497 pp.

McIntyre, M.E. and Palmer, T.N. (1983) Breaking planetary waves in the stratosphere. *Nature* **305**, 593–600.

McKenzie, D. (1984) The generation and compaction of partially molten rock. *Journal of Petrology* **25**, 713–765.

McKibben, M.A., Eldridge, C.S. and Reyes, A.G. (1996) Sulfur isotopic systematics of the June 1991 Mount Pinatubo eruptions: A SHRIMP ion microprobe study (825–843). In: C.G. Newhall and R.S. Punongbayan (eds). *Fire and mud: eruptions and lahars of Mount Pinatubo, Philippines.* Philippine Institute of Volcanology and Seismology, Quezon City and University of Washington Press, Seattle, 1126 pp.

McMahon, H.F., Hester, D.D. and Palfery, J.G. (1971) Vortex shedding for a turbulent jet in a cross wind. *Journal of Fluid Mechanics* **48**, 73–80.

Mackinnon, I.D.R., Gooding, J.L., McKay, D.S. and Clanton, U.S. (1984) The El Chichon stratospheric cloud: solid particulates and settling rates. *Journal of Volcanology and Geothermal Research* **23**, 125–146.

Mader, H.M., Phillips, J.C., Sparks, R.S.J. and Sturtevant, B. (1996) Dynamics of explosive degassing of magma: I. Observations of fragmenting two-phase flows. *Journal of Geophysical Research* **101**, 5547–5560.

Mader, H., Zhang, Y., Phillips, J., Sparks, R.S.J., Sturtevant, B. and Stolper, E. (1994) Experimental simulations of explosive degassing of magma. *Nature* **372**, 85–88.

Maeda, S., Imayoshi, M. and Ohki, A. (1988) Tendencies of volcanic ashfall from Mt. Sakurajima and influences to living environment. *Kagoshima International Conference on Volcanoes 1988.* Proceedings volume, pp. 686–689.

Mahood, G. and Hildreth, W. (1986) Geology of the peralkaline volcano at Pantelleria, Strait of Sicily. *Bulletin of Volcanology* **48**, 143–172.

Mair, W.G. (1886) Notes on the eruption of Tarawera Mountain and Rotomahana, 10th June, 1886, as seen from Taheke, Lake Rotoite. *New Zealand Institute Transactions and Proceedings* **19**, 372–374.

Maiuri, A. (1961) Last moments of the Pompeians. *National Geographic* **120**, 651–669.

Malinconico, L.L., Jr (1979) Fluctuations in SO_2 emission during recent eruptions of Etna. *Nature* **278**, 43–45.

Malingreau, J.-P. and Kaswanda (1986) Monitoring volcanic eruptions in Indonesia using weather satellite data: The Colo eruption of July 28, 1983. *Journal of Volcanology and Geothermal Research* **27**, 179–194.

Mandeville, C.W., Carey, S.N., Sigurdsson, H. and King, J. (1994) Evidence for high temperature emplacement of the 1883 subaqueous pyroclastic flows from Krakatau volcano, Indonesia. *Journal of Geophysical Research* **99**, 9487–9504.

Mangan, M.T., Cashman, K.V. and Newman, S. (1993) Vesiculation of basaltic magma during eruption. *Geology* **21**, 157–160.

Mankin, W.G. and Coffy, M.T. (1984) Increased stratospheric hydrogen chloride in the El Chichon cloud. *Science* **226**, 170–172.

Mankin, W.G., Coffey, M.T. and Goldman, A. (1992) Airborne observations of SO_2, HCl, and O_3 in the stratospheric plume of the Pinatubo volcano in July 1991. *Geophysical Research Letters* **19**, 179–182.

Manley, G. (1946) Temperature trend in Lancashire 1753–1945. *Quaternary Journal of the Royal Meteorological Society* **72**, 1–21.

Markesino, J. (1981) Mount St. Helens ash cleanup. *Public Works*, January 1981, pp. 52–55.

Markham, J. and Kinoshita, W. (1980) Volcanic and seismic activity at Mount St. Helens. Monthly report of the Cascades Volcano Observatory (September and October 1980). *US Geological Survey.*

Martin, G.C. (1913) The recent eruption of Katmai volcano in Alaska. *National Geographic* **24**, 131–181.

Martin, D. and Nokes, R. (1988) Crystal settling in a vigorously convecting magma chamber. *Nature* **332**, 534–536.

Mathison, J.P., Langford, S.C. and Dickinson, J.T. (1989) The role of damage in post-emission of electrons from cleavage surfaces of single-crystal LiF. *Journal of Applied Physics* **65**, 1923–1928.

Matson, M. (1984) The 1982 El Chichon volcano eruptions – A satellite perspective. *Journal of Volcanology and Geothermal Research* **23**, 1–10.

Matson, M. (1985) Detection and tracking of volcanic ash clouds by meteorological satellite systems. *AIAA 23rd Aerospace Science Meeting, Reno, Nevada*, AIAA-85-00099.

Matthews, S.J., Gardeweg, M.C. and Sparks, R.S.J. (1997) The 1984–1996 activity of Lascar volcano, northern Chile: cycles of dome growth, dome subsidence, degassing and explosive volcanism. *Bulletin of Volcanology* (in press).

May, K.R. and Clifford, R. (1967) The impaction of aerosol particles on cylinders, spheres, ribbons and discs. *Ann. Occup. Hygiene* **10**, 83–95.

Mellors, R.A., Waitt, R.B. and Swanson, D.A. (1988) Generation of pyroclastic flows and surges by hot-rock avalanches from the dome of Mount St. Helens Volcano, USA. *Bulletin of Volcanology* **50**, 14–25.

Melson, W. and Saens, R. (1973) Volume, energy and cyclicity of eruptions of Arenal volcano, Costa Rica. *Bulletin of Volcanology* **37**, 416–437.

Melville, W.K. and Bray, K.N.C. (1979) A model of the two-phase turbulent jet. *International Journal of Heat and Mass Transfer* **22**, 647–656.

Metrich, N. and Clocchiatti, R. (1989) Melt inclusion investigation of the volatile behaviour in historic alkali basaltic magmas of Etna. *Bulletin of Volcanology* **51**, 185–198.

Middleton, J.M. and Thomson, R.E. (1986) Modelling the rise of hydrothermal plumes. *Canadian Technical Report of Hydrography and Ocean Sciences* **69**, 1–15.

Millan, M.M., Gallant, A.J., Chung, Y.-S. and Fanaki, F. (1985) COSPEC observation of Mt. St. Helens volcanic SO_2 eruption cloud of 18 May 1980 over southern Ontario. *Atmospheric Environment* **19**, 255–263.

Miller, C.D. and Hoblitt, R.P. (1981) Volcano monitoring by closed-circuit television. *US Geological Survey Professional Paper* **1250**, 335–341.

Miller, C.F. (1966) Operation ceniza-arena: the retention of fallout particles from Volcano Irazu (Costa Rica) by plants and people. Part 2. *Stanford Research Institute*, SRI project number MU-4890.

Miller, T.P. (1976) Augustine volcano, in Alaska's volcanoes northern link in the ring of fire. *Alaska Geographic* **4**, 16–29.

Millikan, R.A. (1923) The general law of fall of a small spherical body through gas, and its bearing upon the nature of molecular reflection from surfaces. *Physical Reviews* **22**, 1–23.

Minnis, P., Harrison, E.F., Stowe, L.L., Gibson, G.G., Denn, F.M., Doelling, D.R. and Smith, W.L. Jr. (1993) Radiative climate forcing by the Mount Pinatubo eruption. *Science* **259**, 1411–1415.

Mitchell, T.D. (1817) Atmospheric constitution of New York, from March to July 1816. *New York Medical Repository*, new series **3**, 304.

Modaress, D., Tan, H. and Elgobashi, S. (1984) Two-component LDA measurement in a two-phase turbulent jet. *AIAA Journal* **22**, 624–630.

Monfort, M. and Schultz, A. (1988) Time series measurements of hydrothermal vent temperature and diffuse percolation velocity: results from an Alvin submersible program, Endeavor segment, Juan de Fuca Ridge. *EOS* **69**, 1484.

Moore, J.G. and Melson, W.G. (1969) Nuees ardentes of the 1968 eruption of Mayon volcano, Philippines. *Bulletin of Volcanology* **33**, 600–620.

Moore, J.G. and Peck, D.L. (1962) Accretionary lapilli in volcanic rocks of the western continental United States. *Journal of Geology* **70**, 182–193.

Moore, J.G. and Rice, C.J. (1984) Chronology and character of the May 18, 1980, explosive eruptions of Mount St. Helens. *National Academy of Sciences, Special Volume on Explosive Volcanism*, 133–142.

Moore, J.C., Nakamura, K. and Alcaraz, A. (1966) The 1965 eruption of Taal volcano. *Science* **151**, 955–960.

Moreno, H. and Gardeweg, M.C. (1989) La erupcion riciente en el complejo volcanica Lonquimay (Dicembre 1988–). *Andes del Sur. Revista Geologica de Chile* **16**, 93–117.

Morton, B.R. (1957) Buoyant plumes in a moist atmosphere. *Journal of Fluid Mechanics* **2**, 127–144.

Morton, B.R. (1959) Forced plumes. *Journal of Fluid Mechanics* **5**, 151–163.

Morton, B.R. (1984) The generation and decay of vorticity. *Geophys. Astrophys. Fluid Dynamics* **28**, 277–308.

Morton, B.R., Taylor, G.I. and Turner, J.S. (1956) Gravitational turbulent convection from maintained and instantaneous sources. *Proceedings of the Royal Society of London Series* **A234**, 1–23.

Moussa, Z.M., Trischka, J.W. and Eskinazi, S. (1977) The near field in the mixing of a round jet with a cross-stream. *Journal of Fluid Mechanics* **80**, 49–80.

Naeser, C., Briggs, N., Obradovich, J. and Izett, G. (1981) Geochronology of Quaternary tephra deposits. In: Self, S. and Sparks, R.S.J. (eds) *Tephra studies*. Reidel, Dordrecht, pp. 13–47.

Nairn, I.A. (1976) Atmospheric shock waves and condensation clouds from Ngauruhoe explosive eruptions. *Nature* **259**, 190–192.

Nairn, I.A. and Self, S. (1978) Explosive eruptions and pyroclastic avalanches from Ngauruhoe in February 1975. *Journal of Volcanology and Geothermal Research* **3**, 39–60.

Nakada, S. and Fujii, T. (1993) Preliminary report on the activity at Unzen volcano (Japan). November 1990–November 1991 dacite lava domes and pyroclastic flows. *Journal of Volcanology and Geothermal Research* **54**, 319–333.

Naranjo, J., Moreno, H. and Banks, N. (1994) La erupcion del Hudson en 1991 (46°S), Region XI, Aisen, Chile. *Bolletino de Servicio Nacional de Geologia y Mineria – Chile* **44**, 1–50.

Naranjo, J.L., Sigurdsson, H., Carey, S. and Fritz, W. (1986) Eruption of the Nevado del Ruiz volcano, Colombia, on 13 November 1985: tephra fall and lahars. *Science* **233**, 961–963.

Nastrom, G.D. and Gage, K.S. (1985) A climatology of atmospheric wavenumber spectra of wind and temperature observed by commercial aircraft. *Journal of Atmospheric Sciences* **42**, 950–960.

Neiburger, M., Edinger, J.G. and Bonner, W.D. (1973) *Understanding our atmospheric environment.* W.H. Freeman, San Francisco, 293 pp.

Neri, A. and Dobran, F. (1994) Influence of eruption parameters on the thermofluid dynamics of collapsing volcanic columns. *Journal of Geophysical Research* **99**, 11833–11857.

Neumann and van Padang, M. (1971) Two catastrophic eruptions in Indonesia, comparable with the Plinian outburst of the volcano of Thera (Santorini) in Minoan time. Acta 1st International Scientific Congress on the Volcano of Thera, Greece, 15–23 September, 1969. *Archaeological Services of Greece, Athens*, pp. 51–63.

Newhall, C. and Self, S. (1982) The Volcanic Explosivity Index (VEI): An Estimate of Explosive Magnitude for Historical Volcanism. *Journal of Geophysical Research* **87**, 1231–1238.

Newman, S., Epstein, S. and Stolper, E. (1988) Water, carbon dioxide, and hydrogen isotopes in glasses from the ca. 1340 AD eruptions of the Mono craters, California: constraints on degassing phenomena and initial volatile content. *Journal of Volcanology and Geothermal Research* **35**, 75–96.

Nicholls, H.R. and Rinehart, J.S. (1967) Geophysical study of geyser action in Yellowstone National Park. *Journal of Geophysical Research* **72**, 4651–4663.

Nielsen, N. (1937) A volcano under an ice-cap, Vatnajokull, Iceland, 1934–36. *Geographical Journal* **90**, 6–23.

Ninkovich, D., Sparks, R.S.J. and Ledbetter, M.J. (1978) The exceptional magnitude and intensity of the Toba eruption, Sumatra: an example of the use of deep-sea tephra layers as a geological tool. *Bulletin of Volcanology* **41**, 286–298.

O'Lone, R.G. (1980) Volcano continues to snarl air traffic. *Aviation Week and Space Technology* **112**, 18–19.

Oberbeck, V.R., Danielsen, E.F., Snetsinger, K.B. and Ferry, G.V. (1983) Effect of the eruption of El Chichon on stratospheric aerosol size and composition. *Geophysical Research Letters* **11**, 1021–1024.

Omori, F. (1922) The Sakur-Jima eruptions and earthquakes, VI. *Bulletin of the Imperial Earthquake Investigation Committee* **8**, 467–525.

Onodera, S. and Kamo, K. (1994) Aviation safety measures for ash clouds in Japan and the system of Japan Air Lines for monitoring eruptions at Sakurajima volcano. *Proceedings of the First International Symposium on Volcanic Ash and Aviation Safety, US Geological Survey Bulletin* **2047**, 213–219.

Oppenheimer, C., Francis, P.W., Rothery, D.A., Carlton, R.W.T. and Glaze, L.S. (1993) Infrared image analysis of volcanic thermal features – Lascar volcano, Chile, 1984–1992. *Journal of Geophysical Research* **98**, 4269–4286.

Orsi, G. and Sheridan, M. (1984) The Green Tuff of Pantelleria: rheoignimbrite or rheomorphic fall? *Bulletin of Volcanology* **47**, 611–626.

Oskarsson, N. (1980) The interaction between volcanic glasses and tephra: Fluorine adhering to tephra of the 1970 Hekla eruption. *Journal of Volcanology and Geothermal Research* **8**, 251–266.

Palais, J.M. and Sigurdsson, H. (1989) Petrologic evidence of volatile emissions from major historic and pre-historic volcanic eruptions. In: Berger, A., Dickinson, R. and Kidson, J. (eds) *Understanding Climate Change.* AGU Geophysical Monograph **52**, 31–53.

Palfreyman, W.D. and Cooke, R.J.S. (1976) Eruptive history of Manam volcano, Papua New Guinea. In: Johnson, R.W. (ed.) *Volcanism in Australia.* Elsevier, Amsterdam, pp. 117–131.

Papale, P. and Rosi, M. (1993) A case of no-wind plinian fallout at Pululagua caldera (Ecuador): implications for models of clast dispersal. *Bulletin of Volcanology* **55**, 523–535.

Palmieri, L. (1873) *The eruption of Vesuvius in 1872.* Asher, London, 148 pp.

Papanicolaou, P.N. and List, E.J. (1988) Investigations of round vertical turbulent buoyant jets. *Journal of Fluid Mechanics* **195**, 341–391.

Papantoniou, D. and List, J.R. (1989) Large scale structure in the far-field of buoyant jets. *Journal of Fluid Mechanics* **209**, 151–190.

Parfitt, E.A. and Wilson, L. (1995) Explosive volcanic eruptions – IX. The transition between Hawaiian-style lava fountaining and Strombolian explosive activity. *Geophysical Journal International* **121**, 226–232.

Pasquill, F. and Smith, F.B. (1983) *Atmospheric diffusion*. Ellis Horwood, Chichester, 437 pp.

Peckover, R., Buchanan, D. and Ashby, D. (1973) Fuel-coolant interactions in submarine volcanism. *Nature* **245**, 307–308.

Perliski, L.M. and Solomon, S. (1992) Radiative influences of Pinatubo volcanic aerosols on twilight observations of NO_2 column abundances. *Geophysical Research Letters* **19**, 1923–1926.

Perret, F.A. (1912) Report of the recent eruption of the volcano "Stromboli". *Smithsonian Annual Report*, pp. 285–289.

Perret, F.A. (1913) Some Kilauean ejecta? *American Journal of Science* **35**, 611–618.

Perret, F.A. (1924) *The Vesuvius eruption of 1906*. Carnegie Institution of Washington, 151 pp.

Pescatore, T., Sparks, R.S.J. and Brazier, S. (1987) Reverse grading in the Avellino Plinian deposit of Vesuvius. *Bolletino Societa Geologica Italiano* **106**, 667–672.

Phillips, J.C., Lane, S., Lejeune, A.M. and Hilton, M. (1995) Gum rosin–acetone system as an analogue to the degassing of hydrated magmas. *Bulletin of Volcanology* **57**, 263–268.

Pierce, R.B. and Fairlie, T.D.A. (1993) Chaotic advection in the stratosphere: implications for the dispersal of chemically perturbed air from the polar vortex. *Journal of Geophysical Research* **98**, 18589–18595.

Pilot, M.J. and Ensor, D.S. (1970) Plume opacity and particulate mass concentration. *Atmospheric Environment* **4**, 163–173.

Pinkerton, H. and Stevenson, R. (1992) Methods of determining the rheological properties of magmas at sub-liquidus temperatures. *Journal of Volcanology and Geothermal Research* **53**, 47–66.

Pinto, J.P., Turco, R.P. and Toon, O.B. (1989) Self-limiting physical and chemical effects in volcanic eruption clouds. *Journal of Geophysical Research* **94**, 11165–11174.

Pitts, M.C. and Thomason, L.W. (1993) The impact of the eruptions of Mt. Pinatubo and Cerro Hudson on Antarctic aerosol levels during the Austral spring. *Geophysical Research Letters* **20**, 2451–2454.

Pogson, I. *Ceylon Observer*, September 18, 1883.

Pollack, J.B., Toon, O.B., Hofmann, D.J., Rosen, J.M. and Danielson, E.F. (1983) The El Chichon volcanic cloud: an introduction. *Geophysical Research Letters* **11**, 989–992.

Pollack, J.B., Toon, O.B., Sagan, C., Summers, A., Baldwin, B. and van Camp, W. (1976) Volcanic explosions and climatic change: a theoretical assessment. *Journal of Geophysical Research* **81**, 1071–1083.

Post, J.D. (1977) *The last great subsistence crisis in the western world.* Johns Hopkins Press, Baltimore, 240 pp.

Post, M.J., Grund, C.J., Langford, A.O., and Proffitt, M.H. (1992) Observations of Pinatubo ejecta over Boulder, Colorado, by lidars of three different wavelengths. *Geophysical Research Letters* **19**, 195–198.

Prandtl, L. (1954) *Essentials of fluid dynamics*. Blackie, Glasgow, 452 pp.

Prata, A.J. (1989) Observations of volcanic ash clouds in the 10–12 mm window using AVHRR/2 data. *International Journal of Remote Sensing* **10**, 751–761.

Prata, A.J. and Barton, I.J. (1994) Detection and discrimination of volcanic ash clouds by infrared radiometry – I: Theory. *Proceedings of the First International Symposium on Volcanic Ash and Aviation Safety. US Geological Survey Bulletin* **2047**, 305–312.

Pratt, W.E. (1911) The eruption of Taal volcano, January 30, 1911. *Philippines Journal of Science* **6**, 63–86.

Pratt, W.E. (1916) An unusual form of volcanic ejecta. *Journal of Geology* **24**, 450–455.

Press, F. and Siever, R. (1990) *The Earth*. W.H. Freeman, New York, 656 pp.

Proctor, F.H. (1987) The terminal area simulation system – Volume 1: theoretical formulation. *NASA Contractor Report*, CR-4046.

Proussevitch, A.A., Sahagian, D.L. and Anderson, A.T. (1993) Dynamics of diffusive bubble growth in magmas: isothermal case. *Journal of Geophysical Research* **98**, 22283–22308.

Przedpelski, Z.J. and Casadevall, T.J. (1994) Impact of volcanic ash from 15 December 1989 Redoubt Volcano eruption on GE CF6-80C2 turbofan engines. *Proceedings of the First International Symposium on Volcanic Ash and Aviation Safety. US Geological Survey Bulletin* **2047**, 129–136.

Pyle, D.M. (1989) The thickness, volume and grainsize of tephra fall deposits. *Bulletin of Volcanology* **51**, 1–15.

Pyle, D.M. (1995) Assessment of the minimum volume of tephra fall deposits. *Journal of Volcanology and Geothermal Research* **69**, 379–382.

Rampino, M. and Self, S. (1982) Historic eruptions of Tambora (1815), Krakatau (1883) and Agung (1963), their stratospheric aerosols and climatic impact. *Quaternary Research* **18**, 127–143.

Rampino, M.R. and Self, S. (1984) Sulfur-rich volcanic eruptions and stratospheric aerosols. *Nature* **310**, 677–679.

Rampino, M. and Self, S. (1992) Volcanic winter and accelerated glaciation follow the Toba super-eruption. *Nature* **359**, 51–53.

Rampino, M.R., Self, S. and Stothers, R.B. (1988) Volcanic winters. *Annual Reviews of Earth and Planetary Science* **16**, 73–99.

Rauber, R.M., Beard, K.V. and Andrews, B.M. (1991) A mechanism for giant raindrop formation in warm, shallow convective clouds. *Journal of Atmospheric Science* **48**, 1791–1797.

Read, W.G., Froidevaux, L. and Walters, J.W. (1993) Microwave Limb Sounder (MLS) measurement of SO_2 from Mt. Pinatubo. *Geophysical Research Letters* **20**, 1299–1302.

Realmuto, V.J., Abrams, M.J., Buongiorno, M.F. and Pieri, D.C. (1994) The use of multispectral thermal infrared image data to estimate the sulfur dioxide flux from volcanoes: A case study from Mount Etna, Sicily, July 29, 1986. *Journal of Geophysical Research* **99**, 481–488.

Rees, J.D. (1970) Parícutin revisited: A review of man's attempts to adapt to ecological changes resulting from volcanic catastrophe. *Geoforum* **4**, 7–25.

Rees, W.G. (1990) *Physical principles of remote sensing*. Cambridge University Press, New York, 247 pp.

Reid, J.R. (1982) Evidence of an effect of heat flux from the East Pacific Rise upon the characteristics of mid-depth waters. *Geophysical Research Letters* **9**, 381.

Reimer, T.O. (1983) Accretionary lapilli in volcanic ash falls: physical factors governing their formation. In: Peryt, T.M. (ed.) *Coated Grains*. Springer-Verlag, Berlin, pp. 56–68.

Richards, J.M. (1963) Experiments on the motions of isolated cylindrical thermals through unstratified surroundings. *International Journal of Air and Water Pollution* **7**, 17–34.

Richardson, L.F. (1922) *Weather prediction by numerical processes*. Cambridge University Press, London, 236 pp.

Richardson, L.F. and Proctor, D. (1926) Diffusion over distances ranging from 3 km to 86 km. *Memoirs of the Royal Meteorological Society* **1**, 1–8.

Richter, D.H., Eaton, J.P., Murata, K.J., Ault, W.U. and Krivoy, H.L. (1984) Chronological narrative of the 1959–60 eruption of Kilauea volcano, Hawaii. *US Geological Survey Professional Paper* **1350**.

Rittman, A. (1962) *Volcanoes and their activity*. Interscience, New York, 305 pp.

Robock, A. (1981) The Mount St. Helens volcanic eruption of May 18: minimal climatic effect. *Science* **212**, 1383–1384.

Robock, A. and Matson, M. (1983) Circumglobal transport of the El Chichon volcanic dust cloud. *Science* **221**, 195–196.

Rogers, R.R. and Yau, M.K. (1989) *Cloud physics*. Pergamon Press, Oxford, 123 pp.

Rona, P. and Speer, K. (1989) An Atlantic hydrothermal plume: Trans-Atlantic Geotraverse (TAG) area, mid-Atlantic ridge crest near 26°N. *Journal of Geophysical Research* **94**, 13879–13893.

Rona, P.A. and Trivett, D. (1992) Discrete and diffuse heat transfer at ASHES vent field, Axial volcano, Juan de Fuca Ridge. *Earth and Planetary Science Letters* **109**, 57–71.

Rose, W.I. (1977) Scavenging of volcanic aerosol by ash: atmospheric and volcanologic implications. *Geology* **5**, 621–624.

Rose, W.I. (1987) Interaction of aircraft and explosive eruption clouds: A volcanologist's perspective. *AIAA Journal* **25**, 52–58.

Rose, W.I. and Chesner, C. (1987) Dispersal of ash in the great Toba eruption, 75 Ka. *Geology* **16**, 913–917.

Rose, W.I. and Hoffman, M.F. (1980) Distal ashes of the May 8, 1980, eruption of Mt. St. Helens. *EOS, Transactions of the American Geophysical Union* **61**, 1137.

Rose, W.I. and Hoffman, M.F. (1982) The May 18, 1980 eruption of Mt. St. Helens: the nature of the eruption, with an atmospheric perspective. In: Deepak, A. (ed.) *NASA Symposium on Mount St. Helens eruption: its atmospheric effects and potential climatic impact.* Spectrum.

Rose, W.I., Bonis, S., Stoiber, R.E., Keller, M. and Bickford, T. (1973) Studies of volcanic ash from two recent Central American eruptions. *Bulletin of Volcanology* **37**, 338–364.

Rose, W.I., Chuan, R.L., Giggenbach, W.F., Kyle, P.R. and Symonds, R.B. (1986) Rates of sulfur dioxide and particle emissions from White Island volcano, New Zealand, and an estimate of the total flux of major gaseous species. *Bulletin of Volcanology* **48**, 181–188.

Rose, W.I., Chuan, R.L. and Woods, D.C. (1982) Small particles in plumes of Mount St. Helens. *Journal of Geophysical Research* **87**, 4956–4962.

Rose, W.I., Delene, D.J., Schneider, D.J., Bluth, G.J.S., Kreuger, A.J., Sprod, I., Mckee, C., Davies, H.L. and Ernst, G.G.J. (1995) Ice in the 1995 Rabaul eruption cloud – implications for volcano hazard and atmospheric effects. *Nature* **375**, 477–479.

Rose, W.I., Heiken, G., Wohletz, K., Eppler, D., Barr, S., Miller, T., Chuan, R.L. and Symonds, R.B. (1988) Direct rate measurements of eruption plumes at Augustine volcano: a problem of scaling and uncontrolled variables. *Journal of Geophysical Research* **93**, 4485–4499.

Rosenbaum, J.G. and Waitt, R.B. (1981) Summary of eyewitness accounts of the May 18 eruption. *US Geological Survey Professional Paper* **1250**, 53–67.

Ross, J.T. (1816) Narrative of the effects of the eruption from the Tomboro Mountain in the island of Sumbawa on the 11th and 12th of April 1815. *Lembaga Kebudajaan Indonesia Verhandelingen* **8**, 343–360.

Rothery, D.A. and Pieri, D.C. (1993) Remote sensing of active lava. In: Kilburn, C.R.J. and Luongo, G. (eds) *Active Lavas*. UCL Press, London, 374 pp.

Rothery, D.A., Francis, P.W. and Wood, C.A. (1988) Volcano monitoring using short wavelength infrared data from satellites. *Journal of Geophysical Research* **93**, 7993–8008.

Rouse, H., Yih, C.S. and Humphreys, H.W. (1952) Gravitational convection from a boundary source. *Tellus* **4**, 201–210.

Rowley, K. (1979) Soufrière: A volcano in the Caribbean environment. *Trinidad Naturalist* **2**, 19–27.

Rowley, P.D., Kuntz, M.A. and Macleod, N.S. (1981) The 1980 eruptions of Mount St Helens, Washington: pyroclastic flow deposits. *US Geological Survey Professional Paper* **1250**, 489–512.

Rudnicki, M.D., James, R.H. and Elderfield, H. (1994) Near-field variability of the TAG non-buoyant plume, 26°N, mid-Atlantic ridge. *Earth and Planetary Science Letters* **127**, 1–10.

Rutherford, M.J. and Devine, J. (1988) The May 18, 1980 eruption of Mount St. Helens, 3. Stability and chemistry of amphibole in the magma chamber. *Journal of Geophysical Research* **93**, 11949–11959.

Rutherford, M.J. and Devine, J.D. (1996) Preeruption pressure–temperature conditions and volatiles in the 1991 dacitic magma of Mount Pinatubo magma. In: C.G. Newhall and R.S. Punongbayan (eds), *Fire and mud: eruptions and lahars of Mount Pinatubo, Philippines.* Philippine Institute of Volcanology and Seismology, Quezon City and University of Washington Press, Seattle, 1126 pp.

Rutherford, M.J., Sigurdsson, H., Carey, S. and Davis, A. (1985) The May 18, 1980 eruption of Mount St. Helens 1, Melt composition and experimental phase equilibria. *Journal of Geophysical Research* **90**, 2929–2947.

Ryan, M.P. and Blevins, J.Y.K. (1987) The viscosity of synthetic and natural silicate melts and glasses at high temperature and 1 bar (10^5 Pascals) pressure and at higher pressure. *US Geological Survey Bulletin* **1764**, 1–563.

Rye, R.O., Luhr, J.F. and Wasserman, M.D. (1984) Sulfur and oxygen isotopic systematics of the 1982 eruptions of El Chichon volcano, Chiapas, Mexico. *Journal of Volcanology and Geothermal Research* **23**, 109–123.

Saffman, P. (1962) On the stability of laminar flow of a dusty gas. *Journal of Fluid Mechanics* **13**, 120–128.

Sagan, C. and Turco, R. (1990) *A path where no man thought.* Random House, New York, 499 pp.

Sahagian, D.L., Anderson, A.T. and Ward, B. (1989) Bubble coalescence in basalt flows: comparison of a numerical model with natural examples. *Bulletin of Volcanology* **52**, 49–56.

Saito, K., Taniguchi, N., Kumashiro, M., Sasaki, T., Sato, Y., Takakuwa, E., Murao, M., Osaki, J. and Abe, S. (1978) Effect of the 1977 eruption of Usu volcano on human living environment and health. In: *Usu eruption and its impact on environment.* Hokkaido University, pp. 169–206 (translation, US Geological Survey).

Sarasin, P. and Sarasin, F. (1901) *Verhandlungen der Naturforschenden Gesellschaftin Basel*, **XIII**, 603.

Sarna-Wojcicki, A.M., Morrison, S., Meyer, C. and Hillhouse, J. (1987) Correlation of upper Cenozoic tephra layers between sediments of the western United States and eastern Pacific Ocean and comparison with biostratigraphic and magnetostratigraphic age data. *Geological Society of America Bulletin* **98**, 207–223.

Sarna-Wojcicki, A.M., Shipley, S., Waitt, R., Dzurisin, D. and Wood, S. (1981) Areal distribution, thickness, mass, volume, and grain-size of airfall ash from the six major eruptions of 1980. *US Geological Survey Professional Paper* **1250**, 577–600.

Sato, H., Fujii, T. and Nakada, S. (1992) Crumbling dacite dome lava and generation of pyroclastic flows at Unzen Volcano. *Nature* **360**, 664–666.

Sawada, Y. (1983) Analysis of eruption clouds by the 1981 eruptions of Alaid and Pagan volcanoes with GMS images. *Papers in Meteorology and Geophysics* (Japanese) **34**, 307–324.

Sawada, Y. (1985) GMS observations of eruption clouds of the 1984 September–October Mayon eruption. *Philippine Journal of Volcanology* **2**, 143–155.

Scandone, R. and Malone, S. (1985) Magma supply, magma discharge and readjustment of the feeding system of Mount St. Helens during 1980. *Journal of Volcanology and Geothermal Research* **23**, 239–262.

Scasso, R., Corbella, H. and Tiberi, P. (1994) Sedimentological analysis of the tephra from the 12–15 August 1991 eruption of Hudson volcano. *Bulletin of Volcanology* **56**, 121–132.

Schatzmann, M. (1979) An integral model of plume rise. *Atmospheric Environments* **13**, 721–731.

Schlesinger, R.E. (1975) A 3-dimensional numerical model of an isolated deep convective cloud: Preliminary results. *Journal of Atmospheric Science* **32**, 934–957.

Schlesinger, R.E. (1978) A 3-dimensional model of an isolated thunderstorm. Part 1, Comparative experiments for variable ambient wind shear. *Journal of Atmospheric Science* **35**, 690–713.

Schlichting, H. (1979) *Boundary layer theory.* McGraw-Hill, New York, 817 pp.

Schmincke, H.-U. (1969) Ignimbrite sequence on Gran Canaria. *Bulletin of Volcanology* **33**, 1199–1219.

Schmincke, H.-U., Fisher, R.V. and Waters, A.C. (1973) Antidune and chute and pool structures in base surge deposits of the Laacher See area, Germany. *Sedimentology* **20**, 553–574.

Schneider, D.J. and Rose, W.I. (1992) Comparison of AVHRR and TOMS imagery of volcanic clouds from Pinatubo volcano. *EOS, Transactions of the American Geophysical Union* **73**, 624.

Schoeberl, M.R., Doiron, S.D., Lait, L.R., Newman, P.A. and Krueger, A.J. (1993) A simulation of the Cerro Hudson SO_2 cloud. *Journal of Geophysical Research* **98**, 2949–2955.

Schubert, H. (1984) Capillary forces – modeling and application in particulate technology. *Powder Technology* **37**, 105–116.

Schultz, A., Delaney, J. and McDuff, R. (1992) On the partitioning of heat flux between diffusion at point source sea floor venting. *Journal of Geophysical Research* **97**, 12299–12314.

Schumacher, R. (1988) Aschenaggregate in vulkaniklastischen transport systemen. PhD thesis, Ruhr Universität, Bochum, 139 pp.

Schumacher, R. (1994) A reappraisal of Mount St. Helens' ash clusters – depositional model from experimental observation. *Journal of Volcanology and Geothermal Research* **59**, 253–260.

Schumacher, R. and Schmincke, H.-U. (1991) Internal structure and occurrence of accretionary lapilli: a case study at Laacher See Volcano. *Bulletin of Volcanology* **53**, 612–634.

Sclater, J., Parsons, B. and Jaupart, C. (1981) Oceans and continents: similarities and differences in the mechanisms of heat loss. *Journal of Geophysical Research* **86**, 11535–11552.

Scorer, R.S. (1958) *Natural aerodynamics.* Pergamon Press, London, 312 pp.

Scorer, R.S. (1959) Theory of waves in the lee of mountains. *Quarterly Journal of the Royal Meteorological Society* **75**, 41–56.

SEAN Bulletin (1989) *Scientific Event Alert Network Bulletin* **14**, no. 11, 2–6.

Self, S. (1983) Large-scale phreatomagmatic silicic volcanism: a case study from New Zealand. *Journal of Volcanology and Geothermal Research* **17**, 433–469.

Self, S. and Rampino, M. (1981) The 1883 eruption of Krakatau. *Nature* **294**, 699–704.

Self, S. and Sparks, R.S.J. (1978) Characteristics of widespread pyroclastic deposits formed by the interaction of silicic magma and water. *Bulletin of Volcanology* **41**, 196–212.

Self, S. and Sparks, R.S.J. (1981) (eds) *Tephra studies*. Reidel, Dordrecht, 481 pp.

Self, S., Wilson, L. and Nairn, L. (1979) Vulcanian eruption mechanisms. *Nature* **277**, 440–443.

Self, S., Keinle, J. and Huot, J.P. (1980) Ukinrek Maars, Alaska; II, Deposits and formation of the 1977 craters. *Journal of Volcanology and Geothermal Research* **7**, 39–65.

Self, S., Sparks, R.S.J., Booth, B. and Walker, G.P.L. (1974) The 1973 Heimaey strombolian scoria deposit, Iceland. *Geology Magazine* **111**, 539–548.

Settle, M. (1978) Volcanic eruption clouds and the thermal power output of explosive eruptions. *Journal of Volcanology and Geothermal Research* **3**, 309–324.

Shapiro, A.E. (1954) *The dynamics and thermodynamics of compressible fluid flow.* Wiley, New York, 647 pp.

Shaw, D., Watkins, N., and Huang, T.C. (1974) Atmospherically transported volcanic glass in deep-sea sediments: theoretical considerations. *Journal of Geophysical Research* **79**, 3087–3094.

Shaw, H. (1972) Viscosities of magmatic silicate liquids: an empirical method of prediction. *American Journal of Science* **272**, 870–893.

Shepherd, J.B. and Sigurdsson, H. (1982) Mechanism of the 1979 explosive eruption of Soufrière volcano, St. Vincent. *Journal of Volcanology and Geothermal Research* **13**, 119–130.

Shepherd, J.B., Aspinal, W.P., Rowley, K.C., Pereira, J., Sigurdsson, H., Fiske, R.S. and Tomblin, J.F. (1979) The eruption of the Soufrière Volcano, St. Vincent, April–June 1979. *Nature* **282**, 24–28.

Sheridan, M.F. and Wohletz, K.H. (1983) Origin of accretionary lapilli from the Pompeii and Avellino deposits of Vesuvius. In: Gooley, R. (ed.) *Microbeam Analysis*. San Francisco Press, pp. 35–38.

Sheridan, M.F., Barberi, F., Rosi, M. and Santacroce, R. (1981) A model for plinian eruptions of Vesuvius. *Nature* **289**, 282–285.

Sheridan, P.J., Schnell, R.C., Hofmann, D.J. and Deshler, T. (1992) Electron microscope studies of Mt. Pinatubo aerosol layers over Laramie, Wyoming during summer 1991. *Geophysical Research Letters* **19**, 203–206.

Shy, S.S. and Breidenthal, R.E. (1990) Laboratory experiments on the cloud-top entrainment instability. *Journal of Fluid Mechanics* **214**, 1–15.

SIBGVN (1991) *Smithsonian Institution Bulletin of the Global Volcanism Network* **16**, no. 5.

SIBGVN (1991) *Smithsonian Institution Bulletin of the Global Volcanism Network* **16**, no. 6.

SIBGVN (1991) *Smithsonian Institution Bulletin of the Global Volcanism Network* **16**, no. 7.

SIBGVN (1991) *Smithsonian Institution Bulletin of the Global Volcanism Network* **16**, no. 9.

SIBGVN (1992) *Smithsonian Institution Bulletin of the Global Volcanism Network* **17**, no. 2.

SIBGVN (1994) *Smithsonian Institution Bulletin of the Global Volcanism Network* **19**, no. 5.

SIBGVN (1994) *Smithsonian Institution Bulletin of the Global Volcanism Network* **19**, no. 8.

SIBGAN (1994) *Smithsonian Institution Bulletin of the Global Volcanism Network* **19**, no. 9.

SICSLP (1973) *Smithsonian Institution Center for Short-lived Phenomena, Annual Report.*

Siegal, B.S. and Gillespie, A.R. (1980) *Remote Sensing in Geology.* Wiley, New York, 702 pp.

Sigurdsson, H. (1982) Volcanic pollution and climate: The 1783 Laki eruption. *EOS, Transactions of the American Geophysical Union* **10**, August, 601–602.

Sigurdsson, H. (1990a) Assessment of the atmospheric impact of volcanic eruptions. *Geological Society of America Special Paper* **247**, 99–110.

Sigurdsson, H. (1990b) Evidence of volcanic loading of the atmosphere and climate response. *Paleogeography, Paleoclimatology, Paleoecology*, Global and Planetary Change Section **89**, 2772–2789.

Sigurdsson, H. and Carey, S. (1981) Marine tephrochronology and Quaternary explosive volcanism in the Lesser Antilles arc. In: Self, S. and Sparks, R.S.J. (eds) *Tephra studies*. Reidel, Dordrecht, pp. 255–280.

Sigurdsson, H. and Carey, S. (1989) Plinian and co-ignimbrite tephra fall from the 1815 eruption of Tambora volcano. *Bulletin of Volcanology* **51**, 243–270.
Sigurdsson, H. and Carey, S. (1992) The eruption of Tambora in 1815: Environmental effects and eruption dynamics. In: Harington, C.R. (ed.) *The year without summer. World climate in 1816.* Canadian Museum of Nature, Ottawa, 16–45.
Sigurdsson, H., Carey, S., Cornell, W. and Pescatore, T. (1985) The eruption of Vesuvius in AD 79. *National Geographic Research* **1**(3), 332–387.
Sigurdsson, H., Cashdollar, S. and Sparks, R.S.J. (1982) The eruption of Vesuvius in AD 79: Reconstruction from historical and volcanological evidence. *American Journal of Archaeology* **86**, 39–51.
Sigurdsson, H., Cornell, W. and Carey, S. (1990) Influence of magma withdrawal on compositional gradients during the AD 79 Vesuvius eruption. *Nature* **345**, 519–521.
Sigurdsson, H., Carey, S. and Mandeville, C. (1991) Krakatau: submarine pyroclastic flows of the 1883 Krakatau eruption. *National Geographic Research and Exploration* **7**, 310–327.
Sigurdsson, H., Sparks, R.S.J., Carey, S. and Huang, T.C. (1980) Volcanogenic sedimentation in the Lesser Antilles arc. *Journal of Geology* **88**, 523–540.
Simkin, T. (1993) Terrestrial volcanism in space and time. *Annual Reviews of Earth and Planetary Science* **21**, 427–452.
Simkin, T. and Fiske, R.S. (1983) *Krakatau 1883. The volcanic eruption and its effects.* Smithsonian Institution Press, Washington 464 pp.
Simpson, J.E. (1987) *Gravity Currents.* Ellis Horwood, Chichester, 244 pp.
Sisson, T.W. (1995) Blast ashfall deposit of May 18, 1980 at Mount St. Helens, Washington. *Journal of Volcanology and Geothermal Research* **66**, 203–216.
Sisson, T.W. and Grove, T. (1993) Temperatures and H_2O content of low-MgO high alumina basalts. *Contributions to Mineralogy and Petrology* **113**, 167–184.
Slaughter, M. and Hamil, M. (1970) Model for deposition of volcanic ash and resulting bentonite. *Geological Society of America Bulletin* **81**, 961–968.
Slawson, P.R. and Csanady, G.T. (1971) The effect of atmospheric conditions on plume rise. *Journal of Fluid Mechanics* **28**, 33–49.
Smith, D. and Leeman, W. (1982) Mineralogy and phase chemistry of Mount St. Helens tephra sets W and Y as keys to their identification. *Quaternary Research* **17**, 211–227.
Smith, D. and Westgate, J. (1969) Electron probe technique for characterizing pyroclastic deposits. *Earth and Planetary Science Letters* **5**, 313–319.
Smith, R.L. (1979) Ash flow magmatism. *Geological Society of America Special Paper* **180**, 5–27.
Smith, R.L. (1960) Ash-flows. *Geological Society of America Bulletin* **71**, 95–842.
Smith, T.L. and Batiza, R. (1989) New field and laboratory evidence for the origin of hyaloclastite flow on seamount summits. *Bulletin of Volcanology* **51**, 96–114.
Smith, W.S. (1983) High-altitude conk out. *Natural History* **92**, 26–34.
Soo, S.L. (1967) *Fluid dynamics of multiphase systems.* Blaisdell, 524 pp.
Sorem, R.K. (1982) Volcanic ash clusters: tephra rafts and scavengers. *Journal of Volcanology and Geothermal Research* **13**, 63–71.
Sparks, R.S.J. (1976b) Grain size variations in ignimbrites and implications for the transport of pyroclastic flows. *Sedimentology* **3**, 147–188.
Sparks, R.S.J. (1978a) The dynamics of bubble formation and growth in magmas – a review and analysis. *Journal of Volcanology and Geothermal Research* **3**, 1–37.
Sparks, R.S.J. (1978b) Gas release rates from pyroclastic flows: an assessment of the role of fluidisation in their emplacement. *Bulletin Volcanologique* **41**, 1–9.
Sparks, R.S.J. (1986) The dimensions and dynamics of volcanic eruption columns. *Bulletin of Volcanology* **48**, 3–15.
Sparks, R.S.J. (1994) Magma generation in the earth. In: *Understanding the Earth*, Cambridge University Press, Cambridge, Ch. 5, pp. 91–114.
Sparks, R.S.J. and Huang, T.C. (1980) The volcanological significance of deep-sea ash layers associated with ignimbrites. *Geology Magazine* **117**, 425–436.

Sparks, R.S.J. and Walker, G.P.L. (1977) The significance of crystal-enriched air-fall ashes associated with crystal-enriched ignimbrites. *Journal of Volcanology and Geothermal Research* **2**, 329–341.

Sparks, R.S.J. and Wilson, C.J.N. (1989) The Minoan deposits: a review of their characteristics and interpretation. In: *Thera and the Aegean World III* **1**, 95–104.

Sparks, R.S.J. and Wilson, L. (1976) Model for the formation of ignimbrite by gravitational column collapse. *Journal of the Geological Society of London* **132**, 441–452.

Sparks, R.S.J. and Wilson, L. (1982) Explosive volcanic eruptions – V. Observations of plume dynamics during the 1979 Soufrière eruption, St. Vincent. *Geophysical Journal of the Royal Astronomy Society* **69**, 551–570.

Sparks, R.S.J. and Wright, J.V. (1979) Welded airfall tuffs. *Geological Society of America Special Paper* **180**, 155–165.

Sparks, R.S.J., Barclay, J., Jaupart, C., Mader, H., Phillips, J. and Sturtevant, B. (1994) Physical aspects of magma degassing I. experimental and theoretical constraints on vesiculation. *Mineralogical Society of America Review* **30**, 413–446.

Sparks, R.S.J., Bonnecaze, R.T., Huppert, H.E., Lister, J.R., Mader, H. and Phillips, J. (1993) Sediment-laden gravity currents with reversing buoyancy. *Earth and Planetary Science Letters* **114**, 243–257.

Sparks, R.S.J., Bursik, M.I., Ablay, G., Thomas, R.M.E. and Carey, S.N. (1992) Sedimentation of tephra by volcanic plumes. Part 2: controls on thickness and grain-size variations of tephra fall deposits. *Bulletin of Volcanology* **54**, 685–695.

Sparks, R.S.J., Carey, S. and Sigurdsson, H. (1991) Sedimentation from gravity currents generated by turbulent plumes. *Sedimentology* **38**, 839–856.

Sparks, R.S.J., Huppert, H. and Turner, J.S. (1984) The fluid dynamics of evolving magma chambers. *Philosophical Transactions of the Royal Society of London* **A310**, 511–534.

Sparks, R.S.J., Moore, J.G. and Rice, C.J. (1986) The initial giant umbrella cloud of the May 18, 1980 explosive eruption of Mount St. Helens. *Journal of Volcanology and Geothermal Research* **28**, 257–274.

Sparks, R.S.J., Self, S. and Walker, G.P.L. (1973) Products of ignimbrite eruptions. *Geology* **1**, 115–118.

Sparks, R.S.J., Wilson, L. and Hulme, G. (1978) Theoretical modeling of the generation, movement and emplacement of pyroclastic flows by column collapse. *Journal of Geophysical Research* **94**, 1867–1887.

Sparks, R.S.J., Wilson, L. and Sigurdsson, H. (1981) The pyroclastic deposits of the 1875 eruption of Askja, Iceland. *Philosophical Transactions of the Royal Society of London* **299**, 241–273.

Speer, K. and Helfrich, K. (1995) Hydrothermal plumes: a review of flow and fluxes. In: Parson, L.M., Walker, C.L., and Dixon, D.R. (eds) Hydrothermal vents and processes. *Geological Society of London Special Publication* **87**, 373–385.

Speer, K. and Rona, P.A. (1989) Model of an Atlantic and Pacific hydrothermal plume. *Journal of Geophysical Research* **94**, 6213–6220.

Spera, F., Borgia, A. and Strimple, J. (1988) Rheology of melts and magmatic suspensions 1. design and calibration of concentric cylinder viscometer with application to rhyolitic magma. *Journal of Geophysical Research* **93**, 10273–10294.

Spera, F., Yuen, D., Greer, J. and Granville, S. (1986) Dynamics of magma withdrawal from stratified magma chambers. *Geology* **14**, 723–736.

Spiess, F., MacDonald, K., Atwater, T., Ballard, R., Caronza, A., Cordoba, D., Cox, C., Diaz-Garcia, V., Francheteau, J., Guerrero, J., Hawkins, J., Hessler, R., Juteau, T., Kastner, M., Luzendyk, B., Macdougal, J., Miller, S., Normark, W., Oreutt, J. and Rangin, C. (1980) Hot springs and geophysical experiments. *Science* **207**, 1421–1433.

Squires, P. and Turner, J.S. (1962) An entraining jet model for cumulo-nimbus updrafts. *Tellus* **5**, 1–15.

Srivastava, R.C. (1987) A model of intense downdrafts driven by the melting and evaporation of precipitation. *Journal of Atmospheric Science* **44**, 1752–1773.

Stasiuk, M.V., Jaupart, C. and Sparks, R.S.J. (1993) On the variations of flow rate in non-explosive lava eruptions. *Earth and Planetary Science Letters* **114**, 505–516.

Stearns, H.T. (1925) The explosive phase of Kilauea Volcano, Hawaii, in 1924. *Bulletin of Volcanology* **5 & 6**, 1–16.

Steenblik, J.W. (1990) Volcanic ash: a rain of terra. *Air Line Pilot*, June/July 1990 **56**, 9–15.

Steinberg, G.S. and Lorenz, V. (1983) External ballistics of volcanic explosion. *Bulletin Volcanologique* **46**, 333–348.

Stephensen, M. (1785) Account of the volcanic eruption in Skaptefield's Syssel. In: Hooker, W.J. (1813) *Journal of a tour in Iceland in the summer of 1809*. Longman, Hurst, Rees, Orme and Brown, and John Murray, London, pp. 124–259.

Stevenson, D.S. (1993) Physical models of fumarolic flow. *Journal of Volcanology and Geothermal Research* **57**, 139–156.

Stoiber, R.E. and Bratton, G. (1978) Airborne correlation spectrometer measurements of SO_2 in eruption clouds from Guatemalan volcanoes. *EOS, Transactions of the American Geophysical Union* **59**, 1222.

Stoiber, R.E., Williams, S.N. and Huebert, B. (1987) Annual contribution of sulphur dioxide to the atmosphere by volcanoes. *Journal of Volcanology and Geothermal Research* **33**, 1–8.

Stoiber, R.E., Williams, S.N. and Huebert, B.J. (1986) Sulfur and halogen gases at Masaya Caldera Complex, Nicaragua: Total flux and variations with time. *Journal of Geophysical Research* **91**, 12215–12231.

Stolper, E. (1982) Water in silicate glasses: an infrared spectroscopic study. *Contributions to Mineralogy and Petrology* **81**, 1–17.

Stolper, E. and Holloway, J. (1988) Experimental determination of the solubility of carbon dioxide in molten basalt at low pressure. *Earth and Planetary Science Letters* **87**, 397–408.

Stolper, E. and Newman, S. (1994) The role of water in the petrogenesis of Mariana trough magmas. *Earth and Planetary Science Letters* **121**, 293–325.

Stommel, H. and Stommel, E. (1979) The year without a summer. *Scientific American* **240**, 176–186.

Stothers, R.B. (1989) Turbulent atmospheric plumes above line sources with an application to volcanic fissure eruptions on the terrestrial planets. *Journal of Atmospheric Science* **46** (17), 2662–2670.

Stothers, R., Wolff, J., Self, S. and Rampino, M. (1986) Basaltic fissure eruptions, plume heights, and atmospheric aerosols. *Geophysical Research Letters* **13**, 725–728.

Stowe, L.L. and Schwedfeger, A. (1982) Optical thickness measurements of El Chichon aerosol from NOAA 7 AVHRR. *EOS, Transactions of the American Geophysical Union* **63**, 898.

Stowe, L.L., Carey, R.M. and Pellegrino, P.P. (1992) Monitoring the Mt. Pinatubo aerosol layer with NOAA 11 AVHRR data. *Geophysical Research Letters* **19**, 159–162.

Stunder, B.J.B., Heffter, J.L. and Dayan, U. (1986) Trajectory analysis of wet deposition at Whiteface Mountain: a sensitivity study. *Atmospheric Environment* **20**, 1691–1695.

Sugioka, I. and Bursik, M.I. (1995) Explosive fragmentation of erupting magma. *Nature* **373**, 689–692.

Suryono, K. (1987) Controlling sediment in the Brantas river basin. *Water Power and Dam Construction* **6**, 25–29.

Sutherland, F.L. (1965) Dispersal of pumice supposedly from the March 1962 South Sandwich Islands on South Sandwich Island Shores. *Nature* **207**, 1332–1335.

Suzuki, T. (1983) A theoretical model for dispersion of tephra. In: Shimozuru, D. and Yokoyama, I. (eds) *Arc Volcanism, Physics and Tectonics*. Terra Scientific Publishing Company Terrapub, Tokyo, pp. 95–113.

Swanson, D.A. and Fabbi, B.P. (1973) Loss of volatiles during fountaining and flowage of basaltic lava at Kilauea volcano, Hawaii. *US Geological Survey Journal of Research* **1**, 649–658.

Swanson, D.A., Wright, T.L. and Helz, R.T. (1975) Linear vent systems and estimated rates of magma production and eruption for the Yakima Basalt on the Columbia Plateau. *American Journal of Science* **275**, 877–905.

Swissler, T.J., McCormick, M.P. and Spinhirne, J.D. (1983) El Chichon eruption cloud: Comparison of lidar and optical thickness measurements for October 1982. *Geophysical Research Letters* **10**, 885–888.

Symonds, R.B., Rose, W.I., Gerlach, T.M., Briggs, P.H. and Harmon, R.S. (1990) Evaluation of gases, condensates, and SO_2 emissions from Augustine volcano, Alaska: the degassing of a Cl-rich volcanic system. *Bulletin of Volcanology* **52**, 355–374.

Symons, G. (1888) *The eruption of Krakatoa and subsequent phenomena*. Royal Society of London, Krakatoa Committee Report, 494 pp.

Tabazadeh, A. and R.P. Turco 1993: Stratospheric chlorine injection of volcanic eruptions: HCl scavenging and implications for ozone. *Science* **260**, 1082–1086.

Tait, S., Jaupart, C. and Vergniolle, S. (1989) Pressure, gas content and eruption periodicity of a shallow crystallizing magma chamber. *Earth and Planetary Science Letters* **92**, 107–123.

Talbor, J.P., Self, S. and Wilson, C.J.N. (1994) Dilute gravity current and rain-flushed ash deposits in the A.D. 180 Hatape plinian pumice fall deposit, New Zealand. *Bulletin of Volcanology* **56**, 538–551.

Tanaka, S., Sugimura, T., Harada, T. and Tanaka, M. (1991) Satellite observation of the diffusion of Pinatubo volcanic dust to the stratosphere. *Journal of the Remote Sensing Society of Japan* **11**, 91–99.

Tatarski, S. (1961) *Wave propagation in a turbulent medium*. McGraw-Hill, 285 pp.

Taylor, G.A.M. (1958) The 1951 eruption of Mount Lamington, Papua. *Australian Bureau of Mineral Resources of Australia Geol. Geophys. Bulletin* **38**, 1–117.

Taylor, G.I. (1921) Diffusion by continuous movements. *Proceedings of the London Mathematical Society, Series 2* **20**, 196–212.

Taylor, G.I. (1950) The formation of a blast wave by an intense explosion. *Proceedings of the Royal Society* **201A**, 159–186.

Thomas, N., Jaupart, C. and Vergniolle, S. (1994) On the vesicularity of pumice. *Journal of Geophysical Research* **99**, 15633–15644.

Thomas, R.M.E. and Sparks, R.S.J. (1992) Cooling of tephra during fallout from eruption columns. *Bulletin of Volcanology* **54**, 542–553.

Thorarinsson, S. (1967) The eruption of Hekla 1947–1948, I. The eruptions of Hekla in historical times. A tephrochronological study. Visindafelag Islendinga, Reykjavik, 1–183.

Thorarinsson, S. (1969) The Laki eruption of 1783. *Bulletin of Volcanology* **33**, 910–929.

Thorarinsson, S. (1970) *Hekla, a notorious volcano*. Almenna Bókafélagid, Reykjavik, 62 pp.

Thorarinsson, S. (1971) Damage caused by tephra fall in some big Icelandic eruptions and its relation to the thickness of the tephra layers. *Acta 1st International Scientific Congress on the Volcano of Thera, Greece, 15–23 September 1969*. Archaeological Services of Greece, Athens 213–236.

Thorarinsson, S. and Sigvaldasson, G.E. (1962) The eruption of Askja 1962. A preliminary report. *American Journal of Science* **260**, 641–651.

Thorarinsson, S. and Sigvaldasson, G.E. (1972) The Hekla eruption of 1970. *Bulletin of Volcanology* **36**, 269–288.

Thorarinsson, S. and Vonnegut, B. (1964) Whirlwind produced by the eruption of Surtsey Volcano. *Bulletin of the American Meteorological Society* **45**, 440–444.

Thorarinsson, S., Einarsson, T., Sigvaldsonn, G. and Elisson, G. (1964) The submarine eruption off the Vestmann islands 1963–64. *Bulletin Volcanologique* **27**, 435–445.

Thordarson, T.H. and Self, S. (1993) The Laki (Skafter fires) and Grimsvotn eruptions in 1783–1785. *Bulletin of Volcanology* 233–263.

Tilling, R.I. (1989) *Volcanic Hazards: Short Course in Geology:* volume 1. American Geophysical Union, Washington, DC, 123 pp.

Tivey, M.K. (1992) Hydrothermal vent systems. *Oceanus* **34**(4), 68–74.

Tivey, M.K. and McDuff, R. (1989) Mineral precipitation in the walls of black smoker chimneys: a quantitative model of transport and chemical reaction. *Journal of Geophysical Research* **95**, 12617–12637.

Tomita, K., Kanai, T., Kobayashi, T. and Oba, N. (1985) Accretionary lapilli formed by the eruption of Sakurajima volcano. *Journal of the Japanese Association of Mineralogy, Petrology and Economic Geology* **80**, 49–54.

Toramoru, A. (1989) Vesiculation process and bubble size distribution in ascending magmas with constant velocities. *Journal of Geophysical Research* **94**, 17523–17542.

Toumi, R., Bekki, S. and Cox, R. (1993) A model study of ATMOS observations and the heterogeneous loss of N_2O_5 by the sulphate aerosol layer. *Journal of Atmospheric Chemistry* **16**, 135–144.
Trepte, C.R., Veiga, R.E. and McCormick, M.P. (1993) The poleward dispersal of Mount Pinatubo volcanic aerosol. *Journal of Geophysical Research* **98**, 18563–18573.
Trial, A., Spera, F., Greer, J. and Yuen, D. (1992) Simulations of magma withdrawal from compositionally zoned bodies. *Journal of Geophysical Research* **92**, 6713–6733.
Tritton, D.J. (1977) *Physical fluid dynamics*. Van Nostrand Reinhold, Wokingham, 362 pp.
Trivett, D.A. (1994) Effluent from diffuse hydrothermal venting: 1. A simple model of plumes from diffuse hydrothermal sources. *Journal of Geophysical Research* **99**, 18403–18415.
Trowbridge, T. (1976) Aniakchak Crater, in Alaska's volcanoes northern link in the ring of fire. *Alaska Geographic* **4**, 70–73.
Tunnicliffe, V. (1992) Hydrothermal-vent communities of the deep sea. *American Scientist* **80**, 336–349.
Turcotte, D., Ockendon, J., Ockendon, H. and Cowley, S. (1990) A mathematical model of Vulcanian eruptions. *Geophysical Journal International* **103**, 211–217.
Turner, J.S. (1960) A comparison between buoyant vortex rings and vortex pairs. *Journal of Fluid Mechanics* **7**, 419–432.
Turner, J.S. (1962) The starting plume in neutral surroundings. *Journal of Fluid Mechanics* **13**, 356–368.
Turner, J.S. (1966) Jets and plumes with negative or reversing buoyancy. *Journal of Fluid Mechanics* **26**, 779–792.
Turner, J.S. (1979) *Buoyancy effects in fluids*. Cambridge University Press, Cambridge, 368 pp.
Turner, J.S. and Campbell, I.H. (1987a) A laboratory and theoretical study of the growth of black smoker chimneys. *Earth and Planetary Science Letters* **82**, 36–48.
Turner, J.S. and Campbell, I.H. (1987b) Temperature, density and buoyancy fluxes in black smoker plumes and the criterion of buoyancy reversal. *Earth and Planetary Science Letters* **86**, 85–92.
Turner, J.S. and Yang, I.K. (1963) Turbulent mixing at the top of stratocumulus clouds. *Journal of Fluid Mechanics* **17**, 212–224.
Twomey, S. (1977) *Atmospheric Aerosols*. Elsevier, Amsterdam, 302 pp.
Valentine, G.A. (1987) Stratified flow in pyroclastic surges. *Bulletin of Volcanology* **49**, 616–630.
Valentine, G. and Wohletz, K.H. (1989) Numerical models of plinian eruption columns and pyroclastic flows. *Journal of Geophysical Research* **94**, 1867–1887.
Valentine, G.A., Wohletz, K.H. and Kieffer, S.W. (1992) Effects of topography on facies and compositional zonation in caldera-related ignimbrites. *Bulletin of the Geological Society of America* **104**, 154–165.
Valero, F.P.J. and Pilewskie, P. (1992) Latitudinal survey of spectral optical depths of the Pinatubo volcanic cloud-derived particle sizes, columnar mass loadings, and effects on planetary albedo. *Geophysical Research Letters* **19**, 163–166.
Varekamp, J.C. (1993) Some remarks on volcanic vent evolution during plinian eruptions. *Journal of Volcanology and Geothermal Research* **54**, 309–318.
Vergniolle, S. and Brandeis, G. (1994) Origin of the sound generated by strombolian explosions. *Geophysical Research Letters* **21**, 1959–1962.
Vergniolle, S. and Jaupart, C. (1986) Separated two phase flow in basaltic eruptions. *Journal of Geophysical Research* **91**, 12840–12860.
Vergniolle, S. and Jaupart, C. (1990) Dynamics of degassing at Kilauea volcano, Hawaii. *Journal of Geophysical Research* **95**, 2793–2809.
Vogel, J., Cornell, W., Nelson, D. and Southon, J. (1990) Vesuvius/Avellino, one possible source of seventeenth century BC climatic disturbances. *Nature* **344**, 534–537.
Walker, G.P.L. (1971) Grain-size characteristics of pyroclastic deposits. *Journal of Geology* **79**, 696–714.
Walker, G.P.L. (1972) Crystal concentration in ignimbrites. *Contributions to Mineralogy and Petrology* **36**, 135–146.
Walker, G.P.L. (1973) Explosive volcanic eruptions – a new classification scheme. *Geol. Rundsch.* **62**, 431–446.

Walker, G.P.L. (1979) A volcanic ash generated by explosions where ignimbrite entered the sea. *Nature* **281**, 642–646.

Walker, G.P.L. (1980) The Taupo Pumice: product of the most powerful known (Ultraplinian) eruption? *Journal of Volcanology and Geothermal Research* **8**, 69–94.

Walker, G.P.L. (1981a) Characteristics of two phreatoplinian ashes, and their water-flushed origin. *Journal of Volcanology and Geothermal Research* **9**, 395–407.

Walker, G.P.L. (1981b) Generation and dispersal of fine ash and dust by volcanic eruptions. *Journal of Volcanology and Geothermal Research* **11**, 81–92.

Walker, G.P.L. (1981c) Plinian eruptions and their products. *Bulletin of Volcanology* **44**-2, 223–240.

Walker, G.P.L. and Croasdale, R. (1971) Two Plinian-type eruptions in the Azores. *Journal of the Geological Society of London* **127**, 17–55.

Walker, G.P.L. and Croasdale, R. (1972) Characteristics of some basaltic pyroclastics. *Bulletin of Volcanology* **35**, 303–317.

Walker, G.P.L., Heming, R.F. and Wilson, C.J.N. (1980) Low-aspect ratio ignimbrites. *Nature* **283**, 286–287.

Walker, G.P.L., Self, S. and Wilson, L. (1984) Tarawera 1886, New Zealand, a basaltic plinian fissure eruption. *Journal of Volcanology and Geothermal Research* **21**, 61–78.

Walker, G.P.L., Wilson, L. and Bowell, E.G.L. (1971) Explosive volcanic eruptions – 1 the rate of fall of pyroclasts. *Geophysical Journal of the Royal Astronomy Society* **22**, 377–383.

Walker, G.P.L., Wilson, C.J.N. and Froggatt, P.C. (1980) Fines-depleted ignimbrite in New Zealand: the product of a turbulent pyroclastic flow. *Geology* **8**, 245–249.

Wallis, G.B. (1969) *One-dimensional, two phase flow.* McGraw-Hill, New York, 408 pp.

Warrick, R.A., Anderson, J., Downing, T., Lyons, J., Ressler, J., Warrick, M. and Warrick T. (1981) Four communities under ash – after the Mount St Helens. *Program on Technology, Environment and Man, Monograph* **34**, Institute of Behavioral Sciences, University of Colorado.

Waters, A.C. and Fisher, R.V. (1971) Base surges and their deposits: Capelinhos and Taal volcanoes. *Journal of Geophysical Research* **76**, 5596–5614.

Watkins, N.D., Sparks, R.S.J., Sigurdsson, H., Huang, T.C., Federman, A., Carey, S. and Ninkovich, D. (1978) Volume and extent of the Minoan tephra layer from Santorini volcano: new evidence from deep-sea sediment cores. *Nature* **271**, 122–126.

Webster, J.D., Taylor, R.P. and Bean, C. (1993) Pre-eruptive melt composition and constraints on degassing of a water-rich pantellerite magma, Fantale volcano, Ethiopia. *Contributions to Mineralogy and Petrology* **114**, 53–62.

Weill, A. (1991) Indirect measurements of fluxes using Doppler sonar. In: *Land surface evaporation.* Springer-Verlag, pp. 301–311.

Weill, A., Brandeis, G., Vergniolle, S., Baudin, F., Bilbille, J., Fèvre, J.-F., Pirron, B. and Hill, X. (1992) Acoustic sounder measurements of the vertical velocity of volcanic jets at Stromboli volcano. *Geophysical Research Letters* **19**, 2357–2360.

Weill, A., Klapisz, C. and Baudin, F. (1986) The CRPE minisodar: Applications in micrometeorology and in physics of precipitations. *Atmospheric Research* **20**, 317–335.

Wen, S. and Rose, W.I. (1994) Retrieval of sizes and total masses of particles in volcanic clouds using AVHRR bands 4 and 5. *Journal of Geophysical Research* **99**, 5421–5431.

Westrich, H.R. and Eichelberger, J.C. (1994) Gas transport and bubble collapse in rhyolitic magma: an experimental approach. *Bulletin of Volcanology* **56**, 447–458.

White, W.T., Day, T.G., Bertek, D. and Fisher, J. (1980) Mt. St. Helens volcanic ash vs Yakima wastewater treatment facility. *Institute for Water Resources Newsletter* **6**, 4–8.

Whitham, A.G. and Sparks, R.S.J. (1986) Pumice. *Bulletin of Volcanology* **48**, 209–223.

Whitlock, W.S. and Chalmers, J.A. (1956) Short period variations in the atmospheric electric potential gradient. *Quaternary Journal of the Royal Meteorological Society* **82**, 325–326.

Wilcox, R.E. (1959) Some effects of recent volcanic ash falls with especial reference to Alaska. *US Geological Survey Bulletin* **1028**-N, 409–476.

Williams, H. and Goles, G. (1968) Volume of the Mazama ash-fall and origin of the Crater Lake caldera. In: Dole, H.M. (ed.) *Andesite conference guide book, Oregon State Department Geological Mineral Industries Bulletin* **62**, 37–41.

Williams, H. and McBirney, A.R. (1979) *Volcanology*. Freeman, Cooper, San Francisco, CA, USA, 397 pp.
Williams, R.S. and Moore, J.G. (1973) Iceland chills a lava flow. *Geotimes* **18**, 14–17.
Williams, S.N. (1983) Plinian airfall deposits of basaltic composition. *Geology* **11**, 211–214.
Williams, S.N. and Self, S. (1983) The October 1902 plinian eruptions of Santa Maria volcano, Guatemala. *Journal of Volcanology and Geothermal Research* **16**, 33–56.
Willis, P.T. (1994) Functional fits to some observed drop size distributions and parameterization of rain. *Journal of Atmospheric Science* **41**, 1648–1661.
Wilson, C.J.N. (1980) The role of fluidisation in the emplacement of pyroclastic flows: an experimental approach. *Journal of Volcanology and Geothermal Research* **8**, 231–249.
Wilson, C.J.N. (1984) The role of fluidisation in the emplacement of pyroclastic flows, 2: experimental results and their interpretation. *Journal of Volcanology and Geothermal Research* **20**, 55–84.
Wilson, C.J.N. (1985) The Taupo eruption, New Zealand II. The Taupo ignimbrite. *Philosophical Transactions of the Royal Society of London, Series* **A314**, 229–310.
Wilson, C.J.N. (1986) Pyroclastic flows and ignimbrites. *Science Progress Oxford* **70**, 171–207.
Wilson, C.J.N. and Houghton, B.F. (1990) Eruptive mechanisms in the Minoan eruption: evidence from pumice vesicularity. *Thera Conference Volume II*, pp. 105–113.
Wilson, C.J.N. and Walker, G.P.L. (1980) Violence in pyroclastic flow eruptions. In: Self, S. and Sparks, R.S.J. (eds) *Tephra studies*. Reidel, Dordrecht, pp. 441–448.
Wilson, C.J.N. and Walker, G.P.L. (1982) Ignimbrite depositional facies: the anatomy of a pyroclastic flow. *Journal of the Geological Society of London* **139**, 581–592.
Wilson C.J.N. and Walker, G.P.L. (1985) The Taupo eruption, New Zealand. I. General aspects. *Phil. Trans. Roy. Soc. Lond.* **A314**, 199–228.
Wilson, L. (1972) Explosive volcanic eruptions – II The atmospheric trajectories of pyroclasts. *Geophysical Journal of the Royal Astronomy Society* **30**, 381–392.
Wilson, L. (1976) Explosive volcanic eruptions III. Plinian eruption columns. *Geophysical Journal of the Royal Astronomy Society* **45**, 543–556.
Wilson, L. (1980) Relationships between pressure, volatile content and ejecta velocity in three types of volcanic explosion. *Journal of Volcanology and Geothermal Research* **8**, 297–313.
Wilson, L. and Head, J. (1981) Ascent and eruption of magma on the Earth and Moon. *Journal of Geophysical Research* **86**, 2971–3001.
Wilson, L. and Huang, T.C. (1979) The influence of shape on the atmospheric settling velocity of volcanic ash particles. *Earth and Planetary Science Letters* **44**, 311–324.
Wilson, L. and Self, S. (1980) Volcanic explosion clouds: density temperature and particle content estimates from cloud motion. *Journal of Geophysical Research* **85**, 2567–2572.
Wilson, L. and Walker, G.P.L. (1987) Explosive volcanic eruptions – VI. Ejecta dispersal in Plinian eruptions: the control of eruption conditions and atmospheric properties. *Geophysical Journal of the Royal Astronomy Society* **89**, 657–679.
Wilson, L., Mouginis-Mark, P.J. and Glaze, L.S. (1996) Determination of eruption plume heights from photoclinometric analysis of weather satellite data. *Journal of Geophysical research* under revision.
Wilson, L., Parfitt, E.A. and Head, J.W. (1995) Explosive volcanic eruptions – VIII. The role of magma recycling in controlling the behaviour of Hawaiian-style lava fountains. *Geophysical Journal International* **121**, 215–225.
Wilson, L., Pinkerton, H. and Macdonald, R. (1987) Physical processes in Volcanic Eruptions. *Annual Reviews of Earth and Planetary Science* **15**, 73–95.
Wilson, L., Sparks, R.S.J. and Walker, G.P.L. (1980) Explosive volcanic eruptions – IV. The control of magma properties and conduit geometry on eruption column behaviour. *Geophysical Journal of the Royal Astronomy Society* **63**, 117–148.
Wilson, L., Sparks, R.S.J., Huang, T.C. and Watkins, N.D. (1978) The control of volcanic eruption column heights by eruption energetics and dynamics. *Journal of Geophysical Research* **83**, 1829–1836.
Wilson, M. (1989) *Igneous Petrogenesis: A Global Tectonic Approach*. Unwin Hyman, London, 466 pp.

Winker, D.M. and Osborn, M.T. (1992a) Airborne lidar observations of the Pinatubo volcanic plume. *Geophysical Research Letters* **19**, 167–170.

Winker, D.M. and Osborn, M.T. (1992b) Preliminary analysis of observations of the Pinatubo volcanic plume with a polarization-sensitive lidar. *Geophysical Research Letters* **19**, 171–174.

Winn, W.P. and Moore, C.P. (1971) Electric field measurements in thunderclouds using instrumented rockets. *Journal of Geophysical Research* **76**, 5003–5017.

Wohletz, K. (1983) Mechanisms of hydrovolcanic pyroclast formation: grain size, scanning electron microscopy and experimental studies. *Journal of Volcanology and Geothermal Research* **17**, 31–63.

Wohletz, K. (1986) Explosive magma–water interactions: thermodynamics, explosion mechanisms, and field studies. *Bulletin of Volcanology* **48**, 245–264.

Wohletz, K. and Heiken, G. (1992) *Volcanology and geothermal energy.* University of California Press, Berkeley, 432 pp.

Wohletz, K. and Sheridan, M.F. (1983) Hydrovolcanic explosions II. Evolution of basaltic tuff rings and tuff cones. *American Journal of Science* **283**, 385–413.

Wohletz, K., McGetchin, T.R., Sandford, M.T. and Jones, E.M. (1984) Hydrodynamical aspects of caldera forming eruptions, numerical experiments. *Journal of Geophysical Research* **89**, 8269–8286.

Wolfe, E. (1992) The 1991 eruptions of Mount Pinatubo, Philippines. *Earthquakes and Volcanoes* **23**, 5–37.

Wolfe, E.W., Garcia, O.M., Jackson, D.B., Koyanagi, R.Y., Neal, C.A. and Okamura, A.T. (1987) The Pu'u O'o eruption of Kiluaea Volcano, 1984, episodes 1–20, January 3, 1983–June 4, 1984. In: Decker, R.W., Wright, L. and Stauffer, P.H. (eds) *Volcanism in Hawaii. US Geological Survey Paper* **1350**, 471–508.

Wolff, J. and Wright, J.V. (1981) Formation of the Green Tuff, Pantelleria. *Bulletin of Volcanology* **44**, 681–690.

Wood, B.J. and Fraser, D.G. (1976) *Elementary thermodynamics for geologists*. Oxford University Press, Oxford, 303 pp.

Wood, B.J. and Banno, S. (1973) Garnet–orthopyroxene and orthopyroxene–clinopyroxene relationships in simple and complex systems. *Contributions to Mineralogy and Petrology* **42**, 109–124.

Woods, A.W. (1988) The dynamics and thermodynamics of eruption columns. *Bulletin of Volcanology* **50**, 169–191.

Woods, A.W. (1993a) A model of the plumes above basaltic fissure eruptions. *Geophysical Research Letters* **20**, 1115–1118.

Woods, A.W. (1993b) Moist convection and the injection of volcanic ash into the atmosphere. *Journal of Geophysical Research* **98**, 17627–17636.

Woods, A.W. (1995a) The dynamics of explosive volcanic eruptions. *Reviews of Geophysics* **33**, 495–530.

Woods, A.W. (1995b) A model of Vulcanian explosive eruptions. *Nuclear Engineering and Design* **155**, 345–357.

Woods, A.W. and Bower, S.M. (1995) The decompression of volcanic jets in a crater during explosive volcanic eruptions. *Earth and Planetary Science Letters* **131**, 189–205.

Woods, A.W. and Bursik, M.I. (1991) Particle fallout, thermal disequilibrium and volcanic plumes. *Bulletin of Volcanology* **53**, 559–570.

Woods, A.W. and Bursik, M.I. (1994) A laboratory study of pyroclastic flows. *Journal of Geophysical Research* **99**, 4375–4394.

Woods, A.W. and Cardoso, S.S.S. (1997) Triggering basaltic volcanic eruptions by bubble-melt separation. *Nature* **385**, 518–520.

Woods, A.W. and Caulfield, C.P. (1992) A laboratory study of explosive volcanic eruptions. *Journal of Geophysical Research* **97**, 6699–6712.

Woods, A.W. and Delaney, J. (1992) The heat and fluid transfer associated with flanges on hydrothermal venting structures. *Earth and Planetary Science Letters* **112**, 117–129.

Woods, A.W. and Kienle, J. (1994) The dynamics and thermodynamics of volcanic clouds: theory and observations from the April 15 and April 21, 1990 eruptions of Redoubt Volcano, Alaska. *Journal of Volcanology and Geothermal Research* **62**, 273–299.

Woods, A.W. and Koyaguchi, T. (1994) Transitions between explosive and effusive volcanic eruptions. *Nature* **370**, 641–644.

Woods, A.W. and Self, S. (1992) Thermal disequilibrium at the top of volcanic clouds and its effect on estimates of column height. *Nature* **355**, 628–630.

Woods, A.W. and K. Wohletz (1991) Dimensions and dynamics of co-ignimbrite eruption columns. *Nature* **350**, 225–227.

Woods, A.W., Holasek, R.E. and Self, S. (1996) Wind-driven dispersal of volcanic ash plumes and its control on the thermal structure of the plume-top. *Bulletin of Volcanology* **57**, 283–292.

Wright, J.V. (1980) Stratigraphy and geology of the welded air-fall tuffs of Pantelleria, Italy. *Geologisches Rundschau* **69**, 263–291.

Wright, J.V., Smith, A.L. and Self, S. (1980) A working terminology of pyroclastic deposits. *Journal of Volcanology and Geothermal Research* **8**, 315–336.

Wright, J.V., Roobal, M.J., Smith, A.L., Sparks, R.S.J., Brazier, S.A., Rose, W.I. Jr. and Sigurdsson, H. (1984) Late Quaternary explosive silicic volcanism on St. Lucia, West Indies. *Geological Magazine* **121**, 1–15.

Wright, S.J. (1977) Mean behavior of buoyant jets in a crossflow. *Proceedings of the ASCE, Journal of the Hydraulics Division*, **103** (HY5), 499–513.

Wright, S.J. (1984) Buoyant jets in density-stratified crossflow. *Journal of Hydraulic Engineering* **110**, 643–656.

Wu, J.M., Vakili, A.D. and Yu, F.M. (1988) Investigation of the interacting flow of nonsymmetric jets in crossflow. *AIAA Journal* **26**, 940–947.

Yamaguchi, D.K. (1985) Tree-ring evidence for a two-year interval between recent prehistoric explosive eruptions of Mt. St. Helens. *Geology* **13**, 554–557.

Yamanouchi, T., Suzuki, K. and Kawaguchi, S. (1987) Detection of clouds in Antarctica from infrared multispectral data of AVHRR. *Journal of the Meteorological Society, Japan* **65**, 949–962.

Zhang, Y. (1996) Dynamics of CO_2-driven lake eruptions. *Nature* **379**, 57–59.

Index